AF251532

EXCITATORY AMINO ACIDS AND NEURONAL PLASTICITY

ADVANCES IN EXPERIMENTAL MEDICINE AND BIOLOGY

Recent Volumes in this Series

Volume 263
RAPID METHODS IN CLINICAL MICROBIOLOGY: Present Status
and Future Trends
Edited by Bruce Kleger, Donald Jungkind, Eileen Hinks, and Linda A. Miller

Volume 264
ANTIOXIDANTS IN THERAPY AND PREVENTIVE MEDICINE
Edited by Ingrid Emerit, Lester Packer, and Christian Auclair

Volume 265
MOLECULAR ASPECTS OF DEVELOPMENT AND AGING OF
THE NERVOUS SYSTEM
Edited by Jean M. Lauder, Alain Privat, Ezio Giacobini,
Paola S. Timiras, and Antonia Vernadakis

Volume 266
LIPOFUSCIN AND CEROID PIGMENTS
Edited by Eduardo A. Porta

Volume 267
CONSENSUS ON HYPERTHERMIA FOR THE 1990s: Clinical Practice in
Cancer Treatment
Edited by Haim I. Bicher, John R. McLaren, and Giuseppe M. Pigliucci

Volume 268
EXCITATORY AMINO ACIDS AND NEURONAL PLASTICITY
Edited by Yehezkel Ben-Ari

Volume 269
CALCIUM BINDING PROTEINS IN NORMAL AND TRANSFORMED CELLS
Edited by Roland Pochet, D. Eric M. Lawson, and Claus W. Heizmann

Volume 270
NEW DEVELOPMENTS IN DIETARY FIBER: Physiological, Physicochemical,
and Analytical Aspects
Edited by Ivan Furda and Charles J. Brine

Volume 271
MOLECULAR BIOLOGY OF ERYTHROPOIESIS
Edited by Joao L. Ascensao, Esmail D. Zanjani, Medhi Tavassoli,
Alan S. Levine, and F. Roy MacKintosh

EXCITATORY AMINO ACIDS AND NEURONAL PLASTICITY

Edited by

Yehezkel Ben-Ari

Institut National de la Sante et de la Recherche Medicale
Paris, France

PLENUM PRESS • NEW YORK AND LONDON

Library of Congress Cataloging-in-Publication Data

European Neuroscience Association Satellite Symposium on Excitatory
   Amino Acids and Neuronal Plasticity (1989 : Fillerval, France)
     Excitatory amino acids and neuronal plasticity / edited by
   Yehezkel Ben-Ari.
          p.   cm. -- (Advances in experimental medicine and biology ; v.
   268)
     "Proceedings of the European Neuroscience Association Satellite
   Symposium on Excitatory Amino Acids and Neuronal Plasticity, held
   August 27-31, 1989, in Fillerval, France"--T.p. verso.
     Includes bibliographical references.
     ISBN 0-306-43534-9
     1. Excitatory amino acids--Congresses. 2. Neuroplasticity-
   -Congresses.   I. Ben-Ari, Yehezkel. II. Title. III. Series.
     [DNLM: 1. Amino Acids--physiology--congresses. 2. Neuronal
   Plasticity--congresses. 3. Synaptic Receptors--physiology-
   -congresses.   W1 AD559 v. 868 / QU 60 E795e 1989]
   QP364.7.E95   1989
   591'.188--dc20
   DNLM/DLC
   for Library of Congress   90-7177
                                                             CIP

Proceedings of the European Neuroscience Association
Satellite Symposium on Excitatory Amino Acids and
Neuronal Plasticity, held August 27-31, 1989,
in Fillerval, France

© 1990 Plenum Press, New York
A Division of Plenum Publishing Corporation
233 Spring Street, New York, N.Y. 10013

FOREWORD

Adult and immature nervous system are capable of considerable "plasticity" and unravelling the underlying mechanisms is one of the principal and most fascinating goals of Neurobiology. A major contribution to our understanding of neural plasticity has come from recent studies in excitatory amino acids - which are thought to mediate a large part of the excitatory synaptic transmission on the brain. Important steps in this explosive field are: 1) the synthesis of relatively specific antagonists of the N-methyl-D aspartate (NMDA) and non-NMDA receptors subtypes, 2) the characterization of the unique features of the NMDA receptor channel complex notably its voltage dependent Mg++ blockade, its permeability to calcium and its allosteric modulation by glycine, 3) the demonstration that by virtue of their Ca++ permeability NMDA receptors are involved in many - but not all - synapses in the initiation but not the maintennce of long term potentiation (LTP) an experimented model of learning and memory processes. More recent studies also indicate tha excitatory amino acids also play an important role in developmental plasticity in vivo; in cell cultures low levels of excitatory amino acids have trophic roles and can inhibit or promote neurite growth. Excitatory amino acids also play an important role also in other forms of neural plasticity such as the use dependent permanent changes in neural circuit produced by brief seizures (epileptogenesis) as well as the reactive sprouting and neosynapse formation which take place in epilepsy models and after deafferentiation or lesions. Finally, there has been considerable interest in the involvement of excitatory amino acids in anoxia.

A general message which emerges from this field is that perhaps similar cellular signals are involved in brain development, sprouting, and synapse formation and learning in adults and that these signals, as well as growth factors, may also play an important role in degenerative disorders.

This book contains the proceedings of the first symposium devoted to Excitatory Amino Acids (EAA) and neural plasticity. The international meeting held at the Château Fillerval (France) 27-31 August 1989, was attended by over 130 participants. Its organization and program were similar to the 2 previous meetings I have organized (or co-organized) in Fillerval (Ben-Ari 1981, Schwartz and Ben-Ari 1985). Like these previous meetings the present one was truly pluridisciplinary including sessions on pharmacology, physiology, morphology, neurochemistry, molecular biology and learning paradigms.

The meeting was divided in 4 principal topics:

1)   Introduction to EAA mechanisms.
2)   EAA and long term potentiation.
3)   EAA and developmental plasticity in vivo and in vitro.
4)   EAA and reactive plasticity and anoxia.

In many sessions the chairmen have prepared a brief commentary printed as an appendix after following the respective scientific papers in order to provide an insight into the lively discussions which took place in Fillerval.

This meeting was an official satellite symposium of the European Neuroscience Association meeting and was held under the auspices of INSERM

(French Medical Research Council) and IBRO (Mc Artur Foundation). This meeting could not have been held without the generous financial support of FIDIA Research Labs. (France), and I am very grateful to Dr. L. Bossi (FIDIA, France) for her support and enthusiasm.

The symposium also received some support from Sandoz, CIBA-CEIGY, UPJOHN and MERCK-SHARP and DOHME.

I am very grateful to my collaborators who have provided outstanding help for the organization, in particular to G. Charton (a real expert after his 3rd. symposium), E. Tremblay and S. Bahurlet. Last but not least, I am highly endebted to Tamara, Yasmina and Constance for their patience and understanding.

Paris, the 21th december 1989

Y. BEN-ARI

Y. Ben-Ari 1981 - "The amygdaloid complex". Elsevier North, Holland.
R. Schwarcz and Y. Ben-Ari 1985 -" Excitatory amino acids and epilepsy", Plenum press.

CONTENTS

SESSION I : MECHANISMS AND EXPRESSION OF RECEPTORS

GLUTAMATE RECEPTORS IN CULTURES OF MOUSE HIPPOCAMPUS STUDIED WITH FAST APPLICATIONS OF AGONISTS, MODULATORS AND DRUGS

Mark L. Mayer, Ladislav Vyklicky, Jr., and Doris K. Patneau

Unit of Neurophysiology and Biophysics, Bldg. 36 Room 2A21
LDN, NICHD, NIH, Bethesda, MD 20892 USA

INTRODUCTION

The study of excitatory synaptic transmission between nerve cells in mammalian brain slice preparations, and the use of binding assays to measure agonist and antagonist affinity, have suggested at least two subtypes of glutamate receptor channel complex. These have been named after the agonists with which they were first characterized, N-methylD-aspartic acid (NMDA) and quisqualic acid (for review see Watkins and Olverman, 1987; Mayer and Westbrook, 1987a). Although cultures of embryonic CNS have the disadvantage that neurons may express less mature forms of ion channel gene products, the high accessibility offered by dissociated cell preparations does allow experiments not possible with brain slices. In particular, whole cell recording and the use of fast perfusion techniques in CNS cultures greatly facilitates the study of receptor kinetics and ligand-gated ion channels; experiments of this type are not possible in intact preparations. Patch clamp techniques have now been applied to both brain slices (Edwards et al., 1989) and to isolated adult cells (Kiskin et al., 1986), but the success rate with these preparations is usually lower than with culture preparations, and the intact nature of slice preparations still hinders the use of rapid perfusion techniques.

Although both quisqualate and NMDA receptors contribute to the generation of excitatory synaptic responses in the brain and are colocalized in many areas of the CNS (Cotman et al., 1987), the contribution they make to synaptic transmission is not equal. Studies with antagonists indicate that quisqualate receptors mediate a fast-rising, fast-decaying response which contributes the largest part of the peak of the synaptic depolarization (Nelson et al., 1986; Blake et al., 1988; Kauer et al., 1988; Muller et al., 1988). In contrast, NMDA receptors mediate a slow-rising, long-duration response which contributes little to the peak of the synaptic depolarization during low frequency stimulation, but which sums with repetitive stimulation to produce a large depolarization (e.g., Dale and Roberts, 1985; Forsythe and Westbrook, 1988; Collingridge et al., 1988).

A more detailed understanding of how NMDA and quisqualate receptors work at central synapses depends on establishing both the concentration/response relationships for activation of quisqualate and NMDA receptors by transmitter candidates, and the anatomical distributions of these receptors in the subsynaptic membrane. Both of these are presumed to play major roles in shaping the relative contributions of quisqualate and NMDA receptors to excitatory synaptic responses. The concentration/response relationship for activation of receptor channel complexes can in principle be approached using fast perfusion techniques to study activation of excitatory amino acid receptors independent of desensitization, and we have begun a series of experiments to attempt this.

Rapid perfusion of agonists limits desensitization

Analysis of activation of receptor channel complexes by agonists is not easy to approach experimentally at any synapse, peripheral or central, due to several technical limitations; the most important are the onset of desensitization prior to obtaining an estimate of the equilibrium response to a given concentration of agonist, and the large size of the agonist-triggered ion current evoked by maximal doses of agonist. To reduce these problems we used fast applications of agonist to measure responses to excitatory amino acids before the onset of desensitization, coupled with brief periods of receptor activation to minimize the collapse of ion gradients during large agonist-evoked currents. In addition, the use of patch clamp electrodes of series resistance 5-10 $M\Omega$ together with a discontinuous voltage clamp for whole cell recording permitted a substantial reduction in the errors expected from measuring 15-20 nA currents with conventional continuous feedback amplifiers; at the gains used in our experiments this limited the maximum voltage escape to ≤ 3 mV.

Experiments were performed on neurons from the hippocampus of 17-18 day C57Bl/6J mice embryos plated on confluent astrocyte feeder layers (Mayer and Westbrook, 1987b). The perfusion system used for application of agonists consisted of an array of 9 glass tubes, each 400 μm diameter, mounted on a hydraulic manipulator driven by a stepping motor and positioned within 200 μM of neuronal cell bodies using a mechanical manipulator. A pump was used to drive solution through the tubes at a maximum rate of approximately 150 μm/ms, and latching solenoid valves used to direct the solution to a reservoir or onto nerve cells. Cells were always bathed in a rapidly flowing stream of control solution, except during application of amino acids. All experiments were performed at room temperature (25-27º C) using the whole-cell configuration of the patch clamp technique with a holding potential of -60 mV .

Pipettes for whole cell recording contained (mM) 125 CsMeSO3, 15 CsCl, 5 CsEGTA (or 5 Cs BAPTA), 0.5 CaCl2, 1 MgCl2, 5 Cs-HEPES, pH 7.2; osmolarity was adjusted to 310 mosm with sucrose. In some experiments 2 mM Mg-ATP was added. The extracellular solution contained (mM) 160 NaCl, 2.5 KCl, 2 CaCl2, 1 MgCl2, 10 Na-HEPES, 10 glucose, pH 7.3, osmolarity 325 mosm, with 400 nM TTX and 5 μM bicuculline methochloride. Solutions containing NMDA had glycine added as indicated (Johnson and Ascher, 1987), magnesium removed to prevent channel block (Mayer et al., 1984; Nowak et al., 1984), and the calcium concentration reduced to 0.2 mM to reduce Ca-mediated inactivation (Mayer and Westbrook, 1985). Amino acids were purchased from Sigma and Tocris Neuramin.

Because hippocampal neurons can develop an extensive dendritic tree during growth in dissociated culture, the response to application of agonist from a point source such as an ionto-phoretic or pressure pipette is complicated by diffusion of agonist to membrane areas adjacent to those activated during the initial period of agonist application. This makes it impossible to accurately define the spatial concentration profile and time course of agonist application. A perfusion apparatus capable of achieving a uniform agonist concentration over the entire dendritic tree would overcome this limitation, provided solution changes could be made quickly enough. To test our perfusion apparatus we exposed neurons to a solution containing 5 mM Na and 50 μM kainic acid, an amino acid which does not produce strong desensitization (Kiskin, Krishtal, and Tsyndrenko, 1986; Mayer and Vyklicky, 1989); this produced a small outward current. We then jumped the extracellular Na concentration to 165 mM, and measured the time course of the kainic acid-evoked inward current relaxation. Using small cells, and relatively high perfusion velocities, we were able to obtain solution exchange time constants of around 10 ms (Fig. 1). We interpret this as resulting from a rapid and uniform perfusion of the whole of the dendritic tree. With outside-out patches even faster perfusion is possible (Tang et al., 1989), most probably due to the smaller membrane area over which solution changes are made (e.g., Fig. 1).

Responses to fast applications of glutamate and its analogues

To characterize the action of transmitter candidates at quisqualate, NMDA and kainate receptors we initially studied responses to each of the "selective" agonists used

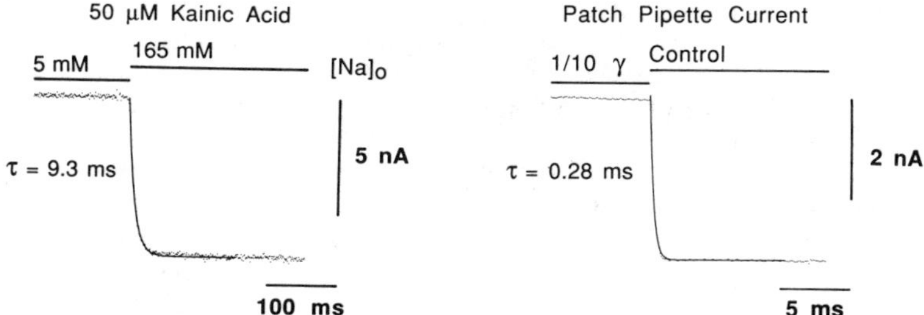

Fig. 1.   Fast perfusion of hippocampal neurons in dissociated culture. The membrane current response of a neuron voltage clamped at -60 mV and exposed to 50 µM kainic acid was recorded during a step increase in the extracellular sodium ion concentration from 5 to 165 mM. The line fitted to the data points is a single exponential function of time constant 9.3 ms. The response of an open patch pipette to a similar solution change was much faster, with a time constant of 0.28 ms.

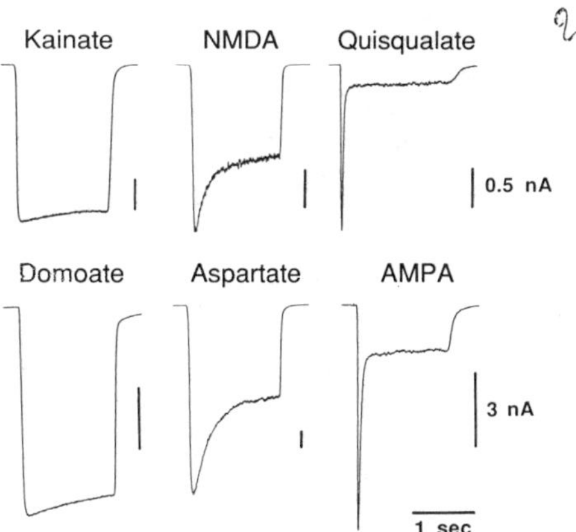

Fig. 2.   Responses to fast applications of 100 µM concentrations of kainate, domoate, NMDA, L-aspartate, quisqualate, and AMPA. All records were recorded in glycine-free solution containing 2 mM Ca and 1 mM Mg, except for the responses to NMDA and L-aspartate, which were recorded in a Mg-free solution containing 300 nM glycine and 0.2 mM Ca.

previously to define glutamate receptor subtypes, specifically focusing on the patterns of activation and desensitization at individual receptors (Fig 2). On the basis of these results, approaches were then developed to separate each component of the response to mixed agonists such as L-glutamate.

Responses to 100 µM kainate and domoate (a kainic acid analogue) showed fast activation and only a small amount of desensitization (Fig. 2). Responses to 100 µM NMDA and L-aspartate, studied with 300 nM glycine to allow activation of NMDA receptor channel activity, showed fast activation, with only moderate desensitization (Fig. 2). Responses to quisqualate and AMPA (DL-a-amino-3-hydroxy-5-methyl-4-isoxazoleproprionic acid), a quisqualate-like agonist currently used as a selective ligand in quisqualate receptor binding experiments, showed fast activation followed by very strong desensitization (Fig. 2).

The time constant of desensitization for quisqualate receptors was 30.4 ± 2.2 ms, and for NMDA receptors was 248 ± 22 ms; this difference was highly significant (unpaired t-test, p ≤ 0.001). When 1 mM NMDA was applied in the absence of glycine, at -60 mV holding potential with 1 mM extracellular Mg, essentially no response (≤ 100 pA) was recorded. When Mg was omitted, and 3 μM glycine added to the extracellular fluid, NMDA usually evoked currents ≥ 10 nA, with essentially no desensitization. Therefore responses mediated by activation of NMDA receptors can be either switched on by removing Mg and adding glycine, or completely blocked by adding Mg and removing glycine. These experimental manipulations, when applied with knowledge of the greater than 100-fold difference in potency of mixed agonists such as L-glutamate for these two receptor sub-types (see below), permits selective activation of either NMDA or quisqualate receptors without the use of antagonists of uncertain specificity. For example, it is now widely accepted that quinoxaline antagonists such as CNQX, which are potent antagonists of responses to kainate and quisqualate, also act as noncompetitive NMDA antagonists.

Glycine Modulates Desensitization at NMDA Receptors

Several groups have confirmed Johnson and Ascher's (1987) finding that glycine strongly potentiates responses to NMDA (Kleckner and Dingledine,1988; Kushner, Lerma, Zukin, and Bennett, 1988). Despite this the mechanism by which glycine acts remains largely unknown. In our experiments responses to NMDA showed strong desensitization at low concentrations of glycine (10 nM - 100 nM) but, as shown in Fig 3, essentially no desensitization at high concentrations of glycine (1 - 10 μM). At intermediate concentrations glycine produced a dosedependent reduction of NMDA receptor desensitization (Mayer et al., 1989). Glycine modulation of NMDA receptor desensitization was also observed with activation evoked by L-aspartate and L-glutamate. Our recent experiments suggest that glycine does not greatly alter the time constant of the onset of desensitization, but instead acts to speed up the rate of recovery from desensitized states. The block of NMDA receptor desensitization by μM concentrations of glycine allows the study of agonist potency at NMDA receptors to be pursued with relative ease. The rapid kinetics of desensitization at quisqualate receptors pose a greater challenge for similar measurements on this subtype of glutamate receptor.

Structure activity relationships for activation of NMDA receptors

After preliminary experiments to determine the foot of the dose/response curve for activation of NMDA and quisqualate receptors by a range of agonists, more detailed studies were made using an appropriate concentration range of selected agonists, and spanning a 3 log unit concentration range. For NMDA receptors experiments are in progress to determine the potency of a large number of endogenous excitatory amino acids. Although similar measurements have been attempted using displacement of radiolabelled antagonists to measure agonist potency, in experiments with other ligand gated ion channels the results obtained have proved to be poor predictors of agonist potency in physiological experiments. There are many possible reasons for this, but the binding of agonists to desensitized receptors, with much higher affinity than for activated receptors, has been well documented for the nicotinic acetylcholine receptor.

Our experiments, with NMDA receptor agonists, show a good correlation with results obtained in binding experiments with AP5 by Olverman et al., (1988), but with a systematic 3-fold lower affinity obtained from analysis of dose/response curves for hippocampal neurons under voltage clamp (Fig. 4). L-glutamate is the most potent ligand examined to date, and L-cysteate, the sulphonic acid analogue of L-aspartic acid, the least potent.

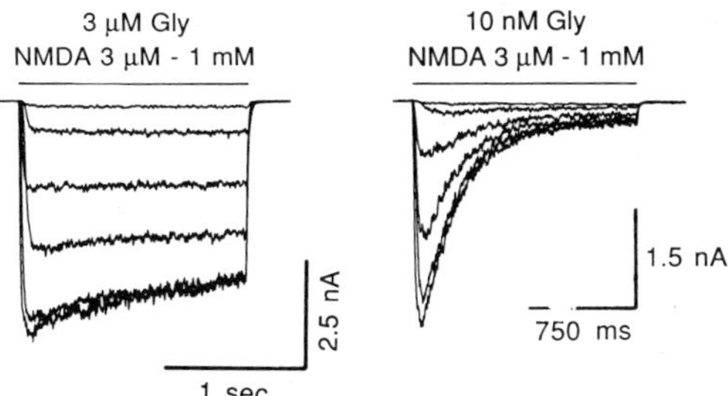

Fig. 3.   Responses to NMDA, 3 µM to 1 mM, recorded with either 3 µM or 10 nM glycine added to the extracellular solution. Desensitization is strong with 10 nM , but not 3 µM, glycine.

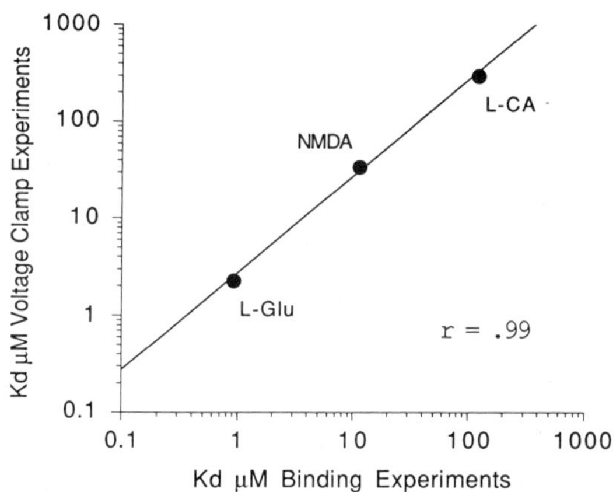

Fig. 4.   comparison of NMDA receptor agonist Kds. Correlation of NMDA receptor agonist equilibrium affinity constants estimated from voltage clamp analysis of responses to L-glutamate, NMDA and L-cysteic acid recorded from hippocampal neurons in dissociated culture (Y-axis), with affinity constants estimated from displacement of [3H]-D-AP5 from rat brain membranes (X-axis) and taken from Olverman et al. (1988).

Analysis of dose/response curves for activation of NMDA and Quisqualate receptors

Some endogenous ligands, such as L-glutamate, and L-homocysteic acid, are mixed agonists, and activate both NMDA and quisqualate receptors. Others, such as L-aspartate, appear to activate only NMDA receptors. Similar experiments are in progress with other endogenous amino acids such as L-homocysteinesulphinic, and quinolinic acids. In the case of mixed agonists such as L-glutamate we were able to distinguish between responses mediated by activation of quisqualate receptors, and responses mediated by activation of NMDA receptors because responses at NMDA receptors always occurred with much lower concentrations of agonist than responses at quisqualate receptors. In addition, responses mediated by activation of NMDA receptors can be distinguished from those due to quisqualate receptor activation by the following characteristic criteria: (1) Quisqualate receptor responses show fast desensitization, which is insensitive to glycine, whereas NMDA receptor responses show much slower desensitization, which is abolished at high concentrations of glycine. Thus by their temporal characteristics and different sensitivity of desensitization to glycine it is easy to distinguish responses mediated by activation of NMDA or quisqualate receptors. (2) The current evoked by NMDA has a high variance (due to gating of 50 pS channels with burst length distributions of 5-10 ms), while responses to quisqualate (and kainate) show very low variance (reflecting both a lower single channel conductance, and much briefer open times); thus spectral analysis of agonist evoked currents also helps to distinguish whether responses are mediated via activation of NMDA or quisqualate receptors. Such tests have shown that at concentrations of mixed agonist which produce saturation of NMDA receptor responses, the activation of quisqualate receptors is still on the foot of the dose/response curve. Examples of responses to L-glutamate and L-aspartate recorded with 3 µM glycine and no Mg (NMDA receptor responses) or zero glycine and 1 mM Mg (quisqualate receptor responses) are shown in Fig. 5.

Several interesting features emerged from analysis of these experiments: Hill slopes for activation of both quisqualate and NMDA receptors were ≥ 1, suggesting agonist cooperativity for receptor activation; L-glutamate was a very potent NMDA receptor agonist (Kd 2-3 µM), but a much weaker quisqualate receptor agonist (Kd around 300 µM). This difference in potency (Fig. 6) has important implications for the action of L-glutamate as an excitatory synaptic transmitter.

Implications for synaptic transmission

Based on work by Katz, Miledi, Stevens and their colleagues, a working model of synaptic transmission at the neuromuscular junction was developed in the early 1970s, and until recently it was appropriate to assume that the principles of this model also applied to the operation of central synapses. The basic features of synaptic transmission at the neuromuscular junction include the quantal release of packets of transmitter from synaptic vesicles, which then diffuses rapidly across the synaptic cleft to bind to and activate nicotinic acetylcholine receptors. The acetylcholine concentration in the synaptic cleft then drops rapidly, as a result of enzymatic hydrolysis, and on average individual molecules of transmitter are thought to activate only a single receptor channel complex before they are hydrolyzed. Because of this the decay time constant of the synaptic current is largely determined by the reciprocal of the closing rate constant for gating of the nicotinic acetylcholine receptor channel complex, rather than by the time course of decay of the concentration of acetylcholine in the synaptic cleft. More recent work with single channel recording suggests that individual receptor channel complexes open in bursts, and thus the decay time constant of the synaptic current is actually determined by the burst length distribution, although this does not substantially alter the operation of synaptic transmission at the level we are concerned with.

The action of central synapses utilizing excitatory amino acids is complicated by the presence of two distinct receptor channel complexes selectively activated by NMDA and quisqualate. Our observation that quisqualate receptors have a much lower affinity for mixed agonists than NMDA receptors (Fig. 6) is especially relevant for the activation of amino acid receptors at synaptic sites, if a mixed agonist such as L-glutamate is indeed the transmitter substance. If NMDA and quisqualate receptors are intimately colocalized in the subsynaptic membrane, our results imply that NMDA receptors would be saturated at concentrations of L-glutamate required to produce significant activation of quisqualate receptor responses. As a result it seems likely that the time course of activation of NMDA receptors during synaptic transmission will be prolonged compared to that for quisqualate receptors, because concentrations of transmitter high enough to activate NMDA receptors will persist for relatively long periods.

Of note for this model, we have found that L-aspartate is essentially devoid of activity at quisqualate receptors, and thus acts as a pure NMDA receptor agonist on mouse hippocampal neurons in culture. Whether this extends to other areas of the CNS, and is true *in vivo*, will have a profound impact on our thinking concerning the role of L-aspartate as a neurotransmitter candidate, since it implies that L-aspartate will be unable to produce conventional fast EPSPs. On the other hand, L-aspartate would be an ideal transmitter candidate where the nervous system requires pure slow, voltage-dependent EPSPs, associated with significant Ca influx through NMDA receptor channels.

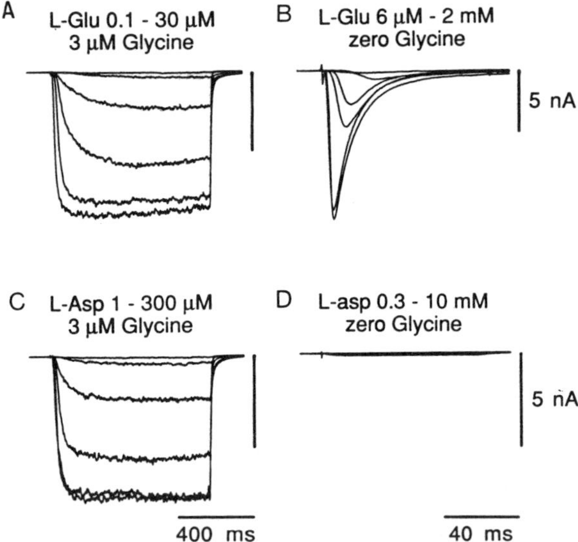

Fig. 5.   Responses to L-glutamate and L-aspartate acting at either NMDA receptors (top) or quisqualate receptors (bottom). The time scale for quisqualate receptor responses is 10 X faster than that for NMDA receptor responses.

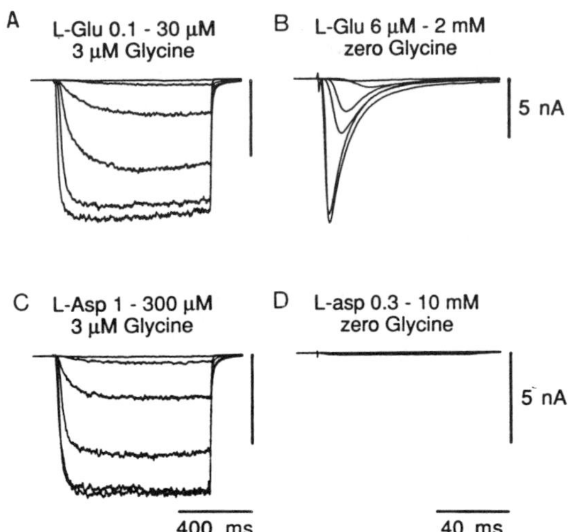

Fig. 6.   Concentration-response curves for activation of NMDA and quisqualate receptors by L-glutamate. The curves drawn through the data points are plotted according to the Hill equation, and were fitted using nonlinear regression analysis. The Kd estimated for activation of NMDA receptors was 2.8 µM, while the Kd estimated for activation of quisqualate receptors was 334 µM.

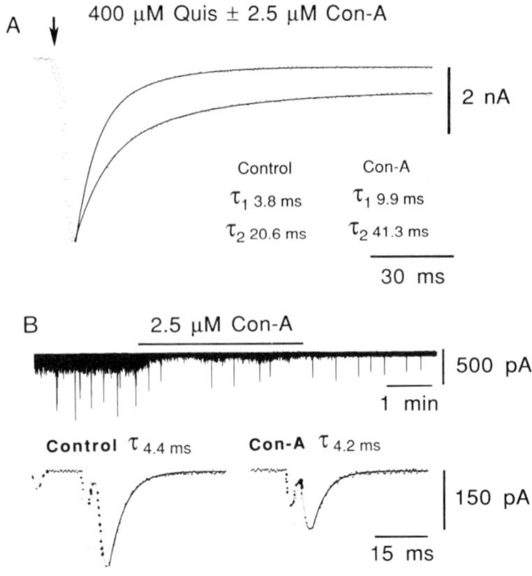

Fig. 7. Concanavalin-A slows desensitization of responses to quisqualate, but does not slow the decay of quisqualate mediated synaptic currents. A shows responses to 400 µM quisqualate recorded before and after a 10 minute application of 2.5 µM concanavalin-A; data points are fitted with double exponential curves, with time constants as indicated adjacent to the experimental records. B shows a chart record of excitatory monosynaptic currents evoked by stimulation of an adjacent hippocampal neuron before and following application of 2.5 µM concanavalin-A as indicated. The amplitude of the synaptic current is decreased by concanavalin-A, an effect also seen at the crayfish neuromuscular junction (Shinozaki and Ishida, 1979), but the decay time constant is unchanged by concanavalin-A. The expanded traces show exponential fits to averages of 20-40 synaptic currents.

Many questions remain concerning the behaviour of quisqualate and NMDA receptors at synaptic junctions. The decay time constant of the fast component of excitatory synaptic currents is around 1-3 ms. The rapid time course of desensitization at quisqualate receptors, which occurs with even low concentrations of agonist (Mayer and Vyklicky, 1989; Tang et al., 1989), is an obvious complication of the analysis of equilibrium responses, and suggests that a concentration jump analysis of ensemble responses recorded using single channel recording techniques may be required to characterize the gating kinetics of quisqualate receptors relevant to synaptic transmission (Tang et al., 1989). Whether desensitization per se contributes to the time course of excitatory synaptic currents is an open issue. Desensitization of quisqualate receptor responses is slowed by the lectin concanavalin-A, but the decay time constant of synaptic currents is unaltered by concanavalin-A (Fig. 7). On the other hand the time constant of desensitization of NMDA receptors, around 250 ms, is fast enough to impinge on long duration responses at NMDA receptors, although the degree to which this will be important will be strikingly sensitive to the concentration of glycine in the extracellular fluid .

REFERENCES

Blake, J.F., Brown, M.W. and Collingridge, G.L. (1988). CNQX blocks acidic amino acid induced depolarizations and synaptic components mediated by non-NMDA receptors in rat hippocampal slices. **Neurosci. Lett.** 89, 182-186.

Collingridge, G.L., Herron, C.E. and Lester, R.A.J. (1988). Frequency-dependent N-methyl-D-aspartate receptor-mediated synaptic transmission in rat hippocampus. **J. Physiol.** (Lond.) 399, 301-312.

Cotman, C.W., Monaghan, D.W., Ottersen, O.P. and Storm-Mathisen, J. (1987). Anatomical organization of excitatory amino acid receptors and their pathways. **Trends Neurosci.** 10, 273-280.

Dale, N. and Roberts, A. (1985). Dual-component amino acid-mediated synaptic potentials: excitatory drive for swimming in Xenopus embryos. **J. Physiol. (Lond.)** 363, 35-59.

Edwards, F.A., Konnerth, A., Sakmann, B. and Takahashi, T. (1989). A thin slice preparation for patch clamp recordings from synaptically connected neurones of the mammalian central nervous system. **Pflügers Arch.** (in press).

Forsythe, I.D. and Westbrook, G.L. (1988). Slow excitatory postsynaptic currents mediated by N-methyl-D-aspartate receptors on cultured mouse central neurones. **J. Physiol. (Lond.)** 396, 515-533.

Johnson, J.W. and Ascher, P. (1987). Glycine potentiates the NMDA response in cultured mouse brain neurones. **Nature** 325, 522-525.

Kauer, J.A., Malenka, R.C. and Nicoll, R.A. (1988). A persistent postsynaptic modification mediates long-term potentiation in the hippocampus. **Neuron** 1, 911-917.

Kiskin, N.I., Khrishtal, O.A. and Tsendrenko, A.Y. (1986). Excitatory amino acid receptors in hippocampal neurons: kainate fails to desensitize them. **Neurosci. Lett.** 63, 225-230.

Kleckner, N.W. and Dingledine, R. (1988). Requirements for glycine in activation of NMDA -receptors expressed in Xenopus oocytes. **Science** 241, 835-837.

Kushner, L., Lerma, J., Zukin, R.S. and Bennett, M.V.L. (1988). Coexpression of N-methyl-D -aspartate and phencyclidine receptors in Xenopus oocytes injected with rat brain mRNA. **Proc. Natl. Acad. Sci. USA** 85, 3250-3254.

Mayer, M.L. and Vyklicky, Jr. L. (1989). Concanavalin A selectively reduces desensitization of mammalian neuronal quisqualate receptors. **Proc. Natl. Acad. Sci. USA** 86, 1411-1415.

Mayer, M.L., Vyklicky, Jr. L. and Clements, J. (1989). Regulation of NMDA receptor desensitization in mouse hippocampal neurones by glycine. **Nature** 338, 425-427.

Mayer, M.L. and Westbrook, G.L. (1985). The action of N-methyl-D-aspartic acid on mouse spinal neurones under voltage clamp. **J. Physiol. (Lond.)** 361, 65-90.

Mayer, M.L. and Westbrook, G.L. (1987a). The physiology of excitatory amino acids in the vertebrate central nervous system. **Prog. Neurobiol.** 28, 197-276.

Mayer, M.L. and Westbrook, G.L. (1987b). Permeation and block of N-methyl-D-aspartic acid channels by divalent cations in mouse cultured central neurones. **J. Physiol. (Lond.)** 394, 501-527.

Mayer, M.L., Westbrook, G.L. and Guthrie, P.B. (1984). Voltage-dependent block by Mg2+ of NMDA responses in spinal cord neurones. **Nature** 309, 261-263.

Muller, D., Joly, M. and Lynch, G. (1988). Contributions of quisqualate and NMDA receptors to the induction and expression of LTP. **Science** 242, 1694-1697.

Nelson, P.G., Pun, R.Y.K. and Westbrook, G.L. (1986). Synaptic excitation in cultures of mouse spinal cord neurones: receptor pharmacology and behaviour of synaptic currents. **J. Physiol. (Lond.)** 372, 169-190.

Nowak, L., Bregestovski, P., Ascher, P., Herbet, A. and Prochiantz, A. (1984). Magnesium gates glutamate-activated channels in mouse central neurones. **Nature** 307, 462-465.

Olverman, H.J., Jones, A.W., Mewett, K.N. and Watkins, J.C. (1988). Structure/activity relations of N-methyl-D-aspartate receptor ligands as studied by their inhibition of [3H]D-2-amino-5-phosphonopentanoic acid binding in rat brain membranes. **Neuroscience** 26, 17-31.

Shinozaki, H. and Ishida, M. (1979). Pharmacological distinction between the excitatory junctional potential and the glutamate potential revealed by concanavalin A at the crayfish neuromuscular junction. **Brain Research** 161, 493-501.

Tang, C.M., Dichter, M. and Morad, M. (1989). Quisqualate activates a rapidly inactivating high conductance ionic channel in hippocampal neurons. **Science** 243, 1474-1477.

Watkins, J.C. and Olverman, H. J. (1987). Agonists and antagonists for excitatory amino acid receptors. **Trends Neurosci.** 10, 265-272.

11

MEASURING AND CONTROLLING THE EXTRACELLULAR GLYCINE

CONCENTRATION AT THE NMDA RECEPTOR LEVEL

Philippe Ascher

Laboratoire de Neurobiologie
Ecole Normale Supérieure
46, rue d'Ulm, 75231 Paris Cedex 05

INTRODUCTION

In 1987, Jon Johnson and I reported that glycine potentiated the NMDA responses of mouse embryonic neurones in primary cultures. This observation raised three questions :
- Is the link between glycine and NMDA a general feature of NMDA receptors ?
- What are the molecular mechanisms of the glycine-NMDA interaction ?
- What is the physiological significance of the glycine-NMDA interaction ?
Although important advances have been made towards answering these three questions, progress has been limited by the difficulty of measuring and of controlling the glycine concentration in the extracellular space.

GENERALITY OF THE GLYCINE-NMDA LINK

When we reported our initial observations, we were not aware of previous work indicating the presence in the brain of a strychnine-insensitive glycine binding site (de Feudis et al., 1978; Kishimoto et al., 1981; Bristow et al., 1986). A letter to Nature of N. Bowery (1987) brought to our attention the remarkable similarity between the anatomical distribution of the strychnine-insensitive glycine binding sites and the NMDA binding sites. This similarity has since been widely confirmed (Cotman et al., 1987), and has suggested a constitutive link between glycine and NMDA binding sites.

Physiological support for this link did not, however, follow immediately : while experiments performed on cultured cells rapidly revealed the presence of the glycine potentiation in a variety of cell types (see Thomson, 1989), experiments done on slices or with *in vivo* preparations seemed, at first, less convincing. For example, addition of exogenous glycine failed to produce a potentiation of NMDA responses in cortical slices (Fletcher and Lodge, 1988, Kemp et al., 1988), which could suggest a pharmacological difference between NMDA receptors in slices and in cultures. However, the discovery of glycine antagonists soon showed that the difference between the two preparations was due to the presence of endogenous glycine (or of a glycine-like compound) in the extracellular space of the slice.

Kynurenate (Kessler et al., 1987, 1989), 7-chlorokynurenate (Kemp et al., 1988) and HA-966 (Fletcher and Lodge, 1987; Fletcher et al., 1989; Foster and Kemp, 1989) were successively established as glycine antagonists. In various cases where exogenous glycine appeared unable to potentiate the NMDA responses, the three glycine antagonists were found to reduce the NMDA response, and the depressed response could then be reestablished by adding exogenous glycine (Birch et al., 1988; Kemp et al., 1989; Fletcher et al., 1989; Foster and Kemp, 1989). This strongly suggests that the glycine site is present in all the cases analyzed, and that the absence of potentiation by exogenous glycine is due to the saturation of the glycine site by endogenous glycine accumulated in the extracellular space.

This is not to say that in all cases the NMDA binding site has linkage with the glycine binding site. Monaghan et al. (1988) have provided evidence for some regional differences in the strength of the NMDA-glycine link, and more recently have shown that the regional differences in glutamate binding persist in the presence of saturating concentrations of glycine (Monaghan et al., 1989). There may thus be more than one type of NMDA receptor, but there is no evidence for the existence of a NMDA receptor insensitive to glycine.

MODELLING THE GLYCINE-NMDA INTERACTION

If one accepts that the NMDA receptor-channel complex possesses at least one glycine binding site and one NMDA binding site (probably an underestimate of its complexity), the reaction schemes linking the possible states of the NMDA complex have to incorporate at least four states to assume the possible existence of (1) an unbound receptor, (2) a glycine bound receptor, (3) a glutamate (NMDA) bound receptor and (4) a glycine and glutamate bound receptor. Introducing open, closed and possibly desensitized states will further complicate the picture.

Any attempt at testing a reaction model will have to rely on the use of known concentrations of glycine. It is particularly difficult to apply low concentrations of glycine in a controlled way. A most significant contribution in that direction is that of Kleckner and Dingledine (1988), who reported that, in oocytes expressing NMDA receptors, glycine was an absolute requirement for the development of NMDA responses. This suggested that only receptors having bound both the NMDA agonist and glycine were capable of a conductance change.

We have attempted to reproduce the observations of Kleckner and Dingledine (1988) on cultured neurones. Despite some success on isolated patches and, in a few cases, on isolated neurones, we found that in most cases of whole cell recording NMDA (100 mM) produced some response in the absence of added glycine, whatever the precautions taken to avoid glycine contamination in the perfusion apparatus. This could be taken to suggest that the NMDA receptors of cultured neurones do not have an absolute requirement for glycine; however, a more likely hypothesis is that some endogenous glycine remains trapped in the restricted extracellular space between the neurones and the underlying glial cells and between the dendrites and the glial cells which often cover them.

This restricted extracellular space may also explain the puzzling discrepancy between outside-out and whole-cell desensitization. Mayer et al., (1989) have reported that, in the whole-cell recording mode, NMDA induced currents show a marked densitization if the glycine concentration is low; if the glycine concentration is increased, desensitization decreases and eventually disappears at concentrations (1-10 µM) that saturate the glycine site (Johnson and Ascher, 1989a,b). In contrast, we have observed (Sather et al., 1989) that in outside-out patches desensitization is present whatever the concentration of glycine. A possible explanation for this discrepancy is that, in the whole-cell mode, slow diffusion of NMDA in the extracellular space delays the rise of the NMDA concentration, thus masking the persistent desensitization.

If this interpretation is adopted, it is clear that the evaluation of kinetic models of the glycine-NMDA interaction will depend on the availability of experimental methods allowing the application of known concentrations of glycine and of NMDA on the whole surface of the cell. In the absence of such methods, the use of outside-out patches appears to be the only means of obtaining precise enough control of agonist concentration.

EVALUATING THE EXTRACELLULAR GLYCINE CONCENTRATION

As discussed above, the fact that in many preparations the addition of exogenous glycine does not potentiate the NMDA response is now interpreted as indicating that the concentration of endogenous glycine is saturating the glycine site. It is in principle possible to calculate the endogenous glycine concentration by comparing the inhibition produced by a competitive antagonist in the presence of different glycine concentrations. If one denotes as K_i the apparent dissociation constant of the antagonist, and as I_{50} the concentration of the antagonist required to produce a 50 % inhibition in the presence of glycine at the concentration [Gly], the relation between K_i, I_{50} and [Gly] can be written

$$[Gly] = K_{Gly} \; ((\; I_{50} \; K_i)/ \; K_i)$$

where K_{Gly} is the apparent dissociation constant of the glycine potentiation. For example, in the case of cortical slices, Fletcher et al. found a value of $I_{50} = 200 \; \mu M$ for the action of kynurenate. Assuming that K_{Gly} and K_i had the values found in cultured neurones ($K_{Gly} = 340 \; nM$, Johnson and Ascher, 1987b; $K_i = 12 \; \mu M$, Ascher et al., 1988), one calculates [Gly] = 5.3 μM, which is a plausible value (despite the many approximations involved) and a clearly saturating concentration (Johnson and Ascher, 1987 a, b).

Is the glycine concentration always saturating, or are there cases in which a variation of the glycine concentration could act as a physiological or pathological signal ? Data suggesting that glycine is not always saturing have come both from experiments on slices and from experiments using *in vivo* preparations.

In some slices glycine produces a potentiation of the NMDA response (e.g. Thomson et al., 1989; Minota et al., 1989; Ben Ari et al., this volume). These observations are difficult to interpret : they clearly demonstrate that the glycine concentration is not saturating in the slice, but it is possible to argue that diffusion between the slice and the bath lowers the glycine concentration below that existing *in vivo*. Thus "thin slices", "cleaned slices", or slices possessing wider extracellular clefts could be expected to have lower glycine levels than thicker slices, without implying a difference in the glycine concentration *in vivo*, or a difference in the sensitivity to glycine of the NMDA receptor.

These objections do not apply to *in vivo* experiments showing potentiating effects of glycine. Salt (1989) observed a potentiating effect of glycine on the NMDA responses of thalamic neurones, and Larson and Beitz (1988) showed that intrathecally administered glycine potentiates convulsions induced by NMDA. These data suggest that the glycine concentration is not always saturating. They give some support to hypotheses implying interactions between glycinergic and glutamatergic synapses (Johnson and Ascher, 1987a) but also to the suggestion that the convulsions observed in non-ketotic hyperglycinemia could involve an increased stimulation of the NMDA receptors (Johnson and Ascher, 1988).

REFERENCES

Ascher, P., Henderson, G. and Johnson, J.W., 1988, Dual inhibitory actions of kynurenate on the N-methyl-D-aspartate (NMDA)-activated response of cultured mouse cortical neurones. **J. Physiol.** (London), 406:141P.

Birch, P.J., Grossman, C.J. and Hayes, A.G. 1988, Kynurenic acid antagonizes responses to NMDA via an action at the strychnine-insensitive glycine receptor. **Eur. J. Pharmacol.**, 154:85.

Bowery, N.G., 1987, Glycine binding sites and NMDA receptors in brain. **Nature**, 326:338.

Bristow, D.R., Bowery, N.G. and Woodruff, G.N., 1986, Light microscopic localization of [3H] glycine and [3H] strychnine binding sites in rat brain. **Eur. J. Pharmacol.**, 126:303.

Cotman, C.W., Monaghan, D.T., Ottersen, O.P., and Storm-Mathisen, J., 1987, Anatomical organization of excitatory amino acid receptors and their pathways. **Trends Neurosci.**, 10:273.

de Feudis, F.V., Orsensanz-Munoz, L.M. and Fando, J.L., 1978, High-affinity glycine binding sites in rat CNS : regional variation and strychnine sensitivity. **Gen. Pharmacol.**, 9:171.

Fletcher, E.J. and Lodge, D. 1988, Glycine reverses antagonism of N-methyl-D-aspartate by 1-hydroxy-3-aminopyrrolidone-2 (HA-966) but not by D-2-amino-5-phosphonovalerate (D-AP5) in rat cortical slices. **Eur. J. Pharmacol.**, 151:161.

Fletcher, E.J., Millar, J.D., Zeman, S. and Lodge, D. 1989, Non-competitive antagonism of N-methyl-D-aspartate by displacement of an endogenous glycine-like substance. **Eur. J. Neurosci.** 1: 196.

Foster, A.C. and Kemp, J.A., 1989, HA-966 antagonizes N-methyl-D-aspartate receptors through a selective interaction with the glycine modulatory site. **J. Neurosci.** 9:2191.

Johnson J.W. and Ascher, P., 1987b, Interaction of glycine with the N-methyl-D-aspartate receptor. **Soc. Neurosci. Abstr.**, 13:383.

Johnson, J.W. and **Ascher, P.** 1987a, Glycine potentiates the NMDA response in cultured mouse brain neurones. **Nature** , 325:529.

Johnson, J.W., and Ascher, P., 1988, The NMDA receptor and its channel. Modulation by magnesium and by glycine, in: "Excitatory amino acids in health and disease", D. Lodge, ed. John Wiley, Chichester.

Kemp, J.A., Foster, A.C., Leeson, P.D., Priestley, T., Tridgett, D., Iversen, L.L. and Woodruff, G.N. 1988, 7-Chlorokynurenic acid is a selective antagonist at the glycine modulatory site of the N-methyl-D-aspartate receptor complex. **Proc. Natl. Acad. Sci. USA**, 85:6547.

Kessler, M., Baudry, M., Terramani, T. and Lynch, G. 1987, Complex interactions between a glycine binding site and NMDA receptors. **Soc. Neurosci. Abstr.**, 13:760.

Kessler, M., Terramani, T., Lynch, G. and Baudry, M., 1989, A glycine site associated with N-methyl-D-aspartic acid receptors: characterization and identification of a new class of antagonists. **J. Neurochem.**, 52:1319.

Kishimoto, H., Simon, J.R., and Aprison, M.H., 1981, Determination of the equilibrium dissociation constants and number of glycine binding sites in several areas of the rat central nervous system using a sodium- independent system. **J. Neurochem.**, 37:1015.

Kleckner, N.W. and Dingledine, R. 1988, Requirement for glycine in activation of NMDA-receptors expressed in *Xenopus* oocytes. **Science**, 241:835.

Larson, A.A., and Beitz, A.J., 1988, Glycine potentiates strychnine-induced convulsions: role of NMDA receptors. **J. Neurosci.**, 8:3822.

Mayer, M.L., Vyklicky L. Jr. and Clements, J., 1989, Regulation of NMDA receptor desensitization in mouse hippocampal neurons by glycine. **Nature**, 338:425.

Minota, S., Miyazaki, I., Wang, M.Y., Read, H.L., and Dun, N.J., 1989, Glycine potentiates NMDA responses in rat hippocampal CA1 neurons. **Neurosci. Lett.**, 100:237.

Monaghan, D.T., Olverman, H.J., Nguyen, L., Watkins, J.C., and Cotman, C.W., 1988, Two classes of N-methyl-D)aspartate recognition sites : differential distribution and differential regulation by glycine. **Proc. Natl. Acad. Sci. USA**, 85:9836.

Monaghan, D.T., Lin, H.S., and Cotman, C.W., 1989, Two classes and two states of NMDA receptors, **Soc. Neurosci. Abstr.**, 15:199.

Salt, T.E., 1989, Modulation of NMDA receptor-mediated responses by glycine and D-serine in the rat thalamus *in vivo*. **Brain Res.**, 481:403.

Sather, W., Johnson, J.W., Henderson, G. and Ascher, P., 1989, Desensitization of NMDA responses in outside-out patches persists in low calcium, high glycine solution. **Soc. Neurosci. Abstr.** 15:325.

Thomson, A.M. 1989, Glycine modulation of the NMDA receptor-channel complex. **Trends Neurosci.** 12:34.

Thomson, A.M., Walker, V.E., and Flynn, D.M., 1989, Glycine enhances NMDA- receptor mediated synaptic potentials in neocortical slices. **Nature**, 338:422.

THE GLYCINE COAGONIST SITE OF THE NMDA RECEPTOR

Raymond Dingledine, Nancy W. Kleckner and Christopher J. McBain

Department of Pharmacology
University of North Carolina School of Medicine
Chapel Hill, NC 27599
USA

INTRODUCTION

The discovery of the glycine binding site on the NMDA receptor (Johnson and Ascher, 1987) is recognized as a highly significant event by those interested in receptor-channel mechanisms. It also offers a new target site in drug development for the treatment of neuropa-thologies associated with NMDA receptor activation. The list of neural dysfunctions mediated in part by synaptic activation of NMDA receptors very likely includes epileptiform seizures (Croucher et al., 1982) and brain damage induced by ischemia or hypoxia in some models (Simon et al., 1984). Growing evidence indicates however that NMDA receptor activation is important also for establishing certain neuronal connections during development (Tsumoto et al., 1987; Kleinschmidt et al., 1987; Lincoln et al., 1988) and has been implicated in some types of learning or memory (Morris et al., 1986). Thus both agonists and antagonists of the NMDA receptor may find clinical utility in the future. Currently available NMDA receptor blockers, acting either at the glutamate binding site or the open ion channel, either have difficulty penetrating the blood-brain barrier or have unacceptable side effects such as psychosis. The glycine site appears to present a pharmacology sufficiently different from that of the other binding sites on the NMDA receptor to offer a novel target for new drug discovery.

The preparation we have used is the **Xenopus** oocyte injected with rat brain mRNA. It is a large (1 mm diameter), hardy cell that can be cultured easily for a week or more after injection, during which time translation of the foreign mRNA, assembly of multisubunit recep-tors and right-side-out insertion into the plasma membrane occurs. Receptor properties are then studied under well-defined voltage clamp conditions. The reproducibility of this preparation offers a reliable bioassay for the quantitative study of receptor-coupled ion channel properties. We have shown previously that NMDA and kainate/quisqualate receptors with properties indistin-guishable from those in neurons are expressed by oocytes (Verdoorn et al., 1987, 1989; Verdoorn and Dingledine, 1988). Here we summarize some of our recent work on the pharmaco-logical characteri-zation of the glycine binding site on NMDA receptors (Kleckner and Dingledine, 1988, 1989; McBain et al., 1989).

Excitatory Amino Acids and Neuronal Plasticity
Edited by Y. Ben-Ari
Plenum Press, New York, 1990

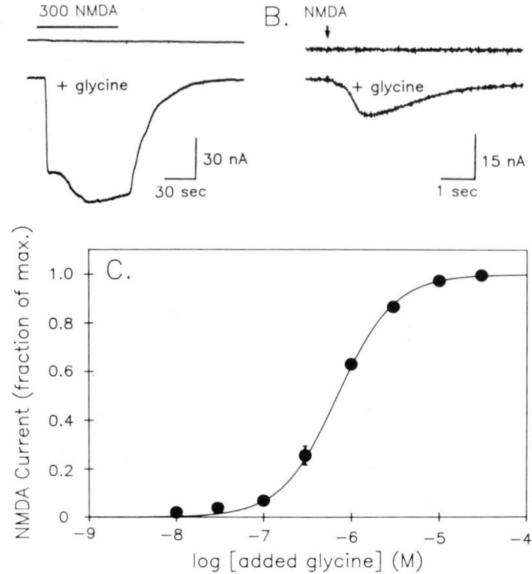

Fig. 1.   Glycine requirement for NMDA receptor activation. (A) Ionic currents produced in an mRNA-injected oocyte (voltage-clamped at -60 mV) by perfusion with 300 µM NMDA in the absence of added glycine (top) and in the presence of 3 µM glycine (bottom). (B) Ionic currents produced by brief pressure application of NMDA in the absence (top) and presence (bottom) of glycine. (C) Glycine concentration-response curve obtained as in (A) in the presence of 100 µM NMDA. Each point is the mean ± SEM current, as fraction of maximum, from five cells. From Kleckner and Dingledine, 1988.

AGONIST PHARMACOLOGY

Application of NMDA plus glycine elicited a biphasic current in oocytes, the peak response fading to a sustained plateau current. The relative contribution of either phase to the total current varied considerably among oocytes (cf. Fig 1A and Fig 5B). The initial peak decayed over 2-4 sec and therefore probably does not correspond to the much more rapid decay of NMDA-glycine current (time constant 250 msec) described by Mayer et al. (1989). Besides the difference in decay time constants, the rapid phase of decay described by Mayer et al. is minimized by high glycine concentrations whereas the slower initial phase of decay in our experiments is emphasized. The rapid phase of decay cannot at present be resolved by the relatively slow perfusion system used for oocyte experiments.

A key finding was that glycine seems *required* for the activation of this receptor by NMDA. A difficulty faced by all wishing to study the glycine site is that glycine is one of the most prevalent amino acids and as such is found in relatively high concentrations even in distilled water, especially water allowed to stand for some time. When precautions were taken to remove as much glycine as possible from the physiological saline, however, the application of even a high concentration of NMDA (e.g., an EC95) elicited no membrane current in oocytes (Fig. 1A,B). Conversely, glycine was equally ineffective in the absence of NMDA. Glycine concentration-response curves carried out in the presence of a fixed (EC80) concentration of NMDA extrapolated to zero current at sufficiently low (<10 nM) glycine concentrations (Fig. 1C). The action of glycine was not due to a shift in the affinity of NMDA for its binding site, but rather due to an increase in the maximum current NMDA can elicit (Fig. 2). These results indicate that glycine plays a role similar to that of the "conventional" agonists, NMDA and glutamate, rather than acting as a modulator of NMDA-induced current. We have carried out a structure-activity study of over 60 analogues of glycine to explore those structural features necessary to activate this site, with the following five conclusions. First, sterically unhindered and ionized carboxyl and amino-termini of the analogue appear essential for activation. Thus blockade of the carboxyl group (e.g., glycine methyl or ethyl ester), or its replacement by a phosphonic acid group, virtually abolished activity, as did substitutions on the amino terminus (e.g., N-methylglycine). Second, the distance separating the carboxyl and amino termini must be no greater than one carbon, since two carbons in the chain (ß-alanine for glycine, isoserine for serine, 3-aminobutyric acid or 3-aminoisobutyric acid for alanine) greatly reduced activity. Third, as previously recognized (Wong et al., 1987; Kleckner and Dingledine, 1988; Ransom and Deschenes, 1988; Snell et al., 1988), the glycine site is stereoselective; the D-isomers of serine and alanine were about 30 and 20 times as potent as the corresponding L-isomer (Fig. 3). Fourth, only small substitutions on the ß-carbon can be tolerated to retain activity. The potency of glycine is reduced about three fold when the ß-carbon proton is replaced with a methyl group (D-alanine) (Fig. 3), but larger hydrophobic substitutions (e.g., ethyl, isopropyl, butanyl, phenyl, hydroxyphenyl) result in drastically reduced activity. Fifth, the ß-carbon hydroxyl group of D-serine probably hydrogen bonds to a site in the receptor, since an isosteric substitution incapable of hydrogen bonding (2-aminobutyric acid) was inactive. It is likely that the receptor contains an hydrogen bond donor since halogenated D-serine analogues capable of

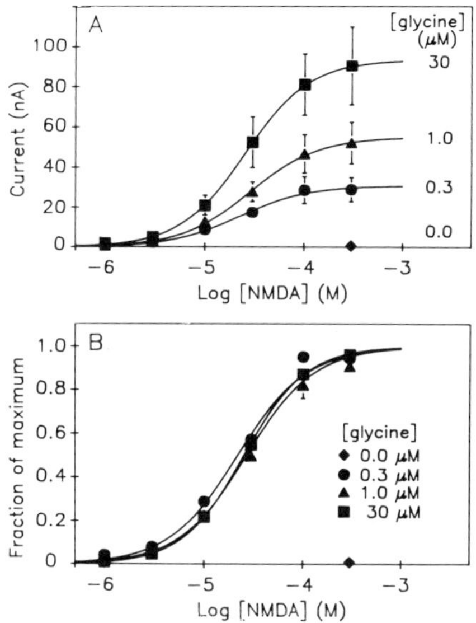

Fig. 2.   Glycine increases the maximal induced current but does not shift the EC_{50} of the NMDA concentration response curve. (A) Concentration response curves were generated in the presence of 0.0, 0.3, 1.0 and 30.0 µM glycine. Each data point represents the mean $\pm$ SEM current from 4-10 oocytes. B. The NMDA responses were expressed as the fraction of the maximal NMDA current for each concentration of glycine. From McBain et al., 1989.

only proton acceptance (ß-fluoroD-alanine, ß-chloro-D-alanine) were active whereas 1,2-diaminopropionic acid, which should only be able to donate protons at physiological pH, was inactive.

Our picture of the glycine recognition site, then, is that of a small, relatively polar pocket containing at least three points of possible attachment: positive and negative ionized sites capable of binding the carboxyl and amino termini of glycine, and an hydrogen bond donating site. This is quite similar to the view of the NMDA recognition site inferred from structure-activity studies (Watkins and Olverman, 1987), with the ionized omega-carboxyl group of glutamate having the same function as the omega-hydroxyl group of D-serine. The similarity between the two agonist binding sites raises the intriguing possibility that the glycine binding site may have been originally derived from the NMDA recognition site or vice versa. Since neither glycine nor glutamate acting alone can open NMDA ion channels, it seems reasonable to refer to them as "coagonists" of the NMDA receptor rather than agonist and modulator.

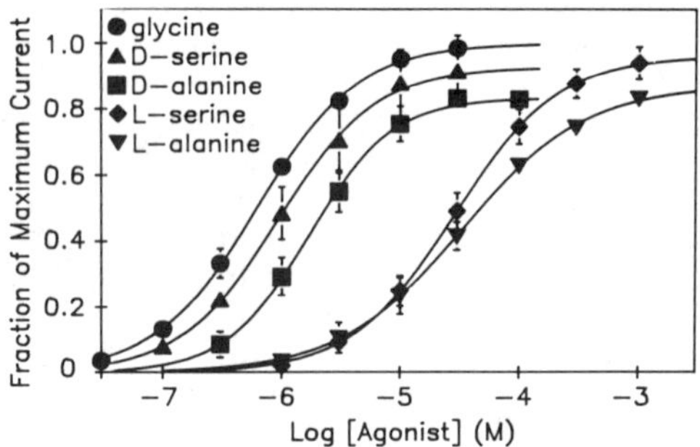

Fig 3.   Stereoselectivity of glycine site agonists. Dose response curves were constructed as illustrated in figure 1C. Each point is the mean (± SEM) response expressed as the fraction of maximum current induced by 30 µM glycine and NMDA (100 µM) determined in each oocyte. each curve represents data from 3-7 oocytes. From McBain et al., 1989.

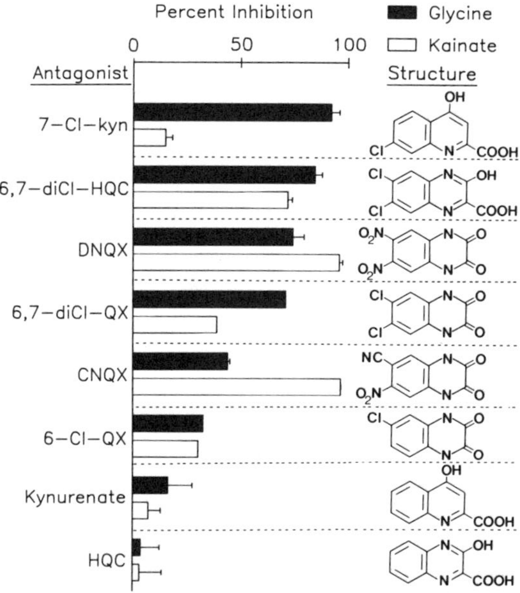

Fig. 4.   Effectiveness of quinoxaline and kynurenic acid derivatives at antagonizing inward currents evoked by NMDA/glycine and kainate (KA). Oocytes were perfused with 3 µM glycine and 15 µM antagonist for at least 5 min before application of 300 µM NMDA or KA in the presence of glycine and the antagonist. Bars represent the mean (± SEM) percent inhibition of the NMDA/glycine or KA induced current. Abbreviations: 7-Cl-kyn: 7-chlorokynurenate; 6,7-diCl-HQC: 6,7-dichloro-3-hydroxy2-quinoxaline carboxylic acid; DNQX: 6,7-dinitro-quinoxaline-2, 3-dione; 6,7-diCl-QX: 6,7-dichloroquinoxaline-2,3-dione; CNQX: 6-cyano-7-nitroquinoxaline-2,3-dione; 6-Cl-QX: 6-chloroquinoxaline-2,3-dione; HQC: 3-hydroxy-2-quinoxaline carboxylic acid. from Kleckner and Dingledine, 1989.

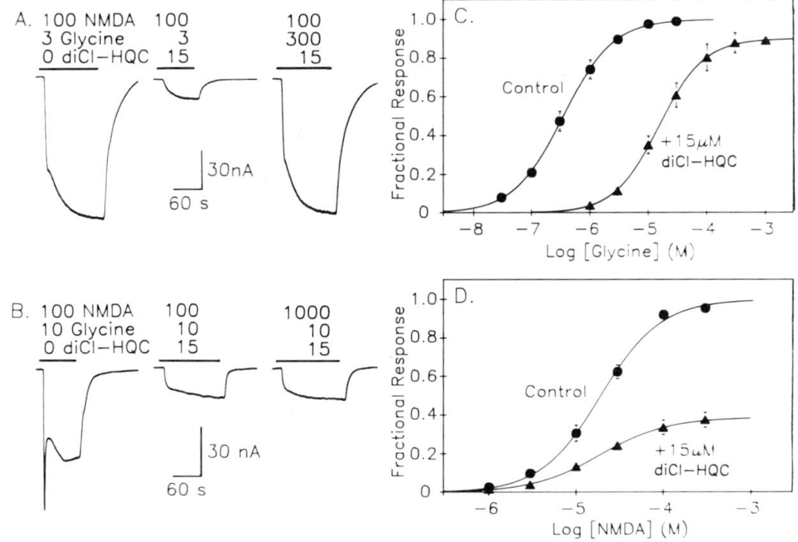

Fig. 5. 6,7-diCl-HQC antagonizes NMDA/glycine induced currents by an interaction at the glycine site. (A) and (B) The perfusion solution was changed to include NMDA and glycine, plus or minus the antagonist at the time indicated by the horizontal bar. The antagonist (15 µM) was perfused onto the oocyte for at least 5 minutes before drug application for the center and right traces. 6,7-diCl-HQC antagonism of the NMDA receptor was overcome by increasing the concentration of glycine (A), but not NMDA (B). Numbers represent the concentration of each drug in 15 µM. (C) Concentration-response curves for glycine generated in the absence and presence of 15 µM 6,7-diCl-HQC (n=3). The NMDA concentration was 100 µM. The glycine EC_{50} was shifted from 0.35 µM in the absence of 6,7-diCl-HQC to 15.8 µM in its presence. (D) Concentration - response curves for NMDA in the absence and presence of 15 µM 6,7-diCl-HQC (n=3). Glycine (10 µM) was present during all NMDA applications. The antagonist decreased the maximum response of the NMDA dose-response curve by 65% without influencing the NMDA EC_{50}. For both C and D the points are the mean ± SEM current (as a fraction of the maximum current in the absence of antagonist) for 3 cells. From Kleckner and Dingledine, 1989.

ANTAGONIST PHARMACOLOGY

In the past two years it has been recognized that many of the "non-selective" (i.e., able to block both kainate and NMDA responses) excitatory amino acid (EAA) antagonists act at the glycine site of NMDA receptors. This list includes kynurenic acid (Watson et al., 1988; Kessler et al., 1989), HA-966 (Fletcher and Lodge, 1988; Foster and Kemp, 1989), and has been expanded recently to encompass 7-chlorokynurenate (Kemp et al., 1988) as well as the quinoxaline derivatives CNQX and DNQX (Birch et al., 1988; Verdoorn et al., 1989). We assessed the potency, competitiveness and selectivity of eight potential glycine site antagonists in the oocyte preparation. Dose-response curves were constructed for glycine (at fixed NMDA concentration), NMDA (at fixed glycine concentration), and kainate in the presence and absence of these compounds. Where appropriate, Schild analysis was carried out.

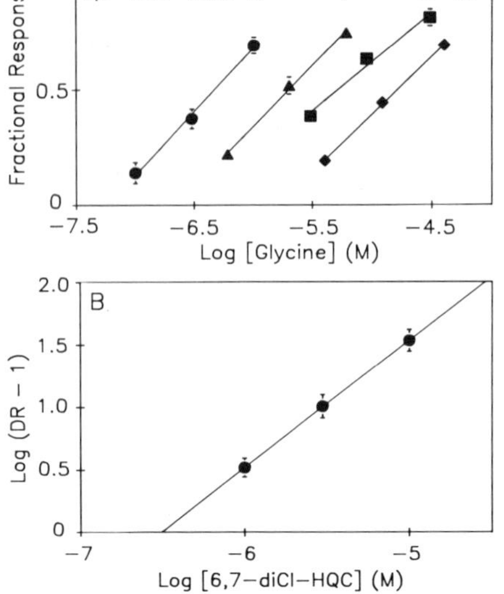

Fig. 6. Schild analysis indicating competitive antagonism of the glycine site on the NMDA receptor by 6,7-diCl-HQC. (A) Glycine dose-response curves were generated in the absence and in the presence of 1, 3, and 10 µM 6,7-diCl-HQC. The NMDA concentration was 100 µM throughout. The points represent the mean ± SEM current as a fraction of the maximal current in the absence of antagonist for 4 cells. (B) A linear (r=0.99) Schild plot of the above data, with slope equal to 1.02 ± 0.06, suggests that the antagonism is competitive. The pA2 was 6.52 ± 0.1. From Kleckner and Dingledine, 1989.

The most interesting finding was that every antagonist of the glycine site on NMDA receptors was shown also to be a kainate blocker (Fig. 4). As one detailed example, 15 µM 6,7-dichloro--3-hydroxy-2- quinoxaline carboxylic acid (6,7-diCl-HQC) produced a 45-fold parallel shift to the right of the glycine dose-response curve when the NMDA concentration was held constant (Fig. 5C). 6,7-DiCl-HQC, in contrast, simply reduced the maximum NMDA current that could be evoked in the presence of a fixed (10 µM) glycine concentration (Fig. 5D). A Schild plot (Fig. 6) of 6,7-diCl-HQC versus glycine had a slope of 1.02 ± 0.06 and a pA2 of 6.52 ± 0.10. These data indicate that the blocking action of 6,7-diCl-HQC at NMDA receptors (at least up to 15 µM antagonist concentration) can be explained completely by competitive block of the glycine site. 6,7-DiCl-HQC also proved to be a competitive antagonist of kainate (Fig. 7), with a pA2 of 5.52 ± 0.06. Thus this quinoxaline derivative exhibited a ten-fold selectivity for the glycine site over the receptor activated by kainate.

CNQX has been considered a potent and highly selective quisqualate receptor antagonist (Honore et al., 1988). CNQX, however, also blocks NMDA-induced currents in oocytes in a noncompetitive fashion (Fig. 8A,B). This block was found to result from a mixed competitive-noncompetitive block of the glycine site on NMDA receptors (Fig. 9). In similar experiments CNQX exerted a purely competitive block of the receptor activated by kainate, with a pA2 of 6.53. Since the block of the glycine site on the NMDA receptor by CNQX has a substantial noncompetitive component, the *selectivity* of CNQX as a kainate antagonist will depend on the prevailing concentrations of glycine and agonists acting at both NMDA and kainatesensitive receptors. As a guideline, if EAA receptors expressed on oocytes are identical to those in brain slice preparations, 2 µM CNQX would be expected to occupy up to 87% of the kainatesensitive receptors

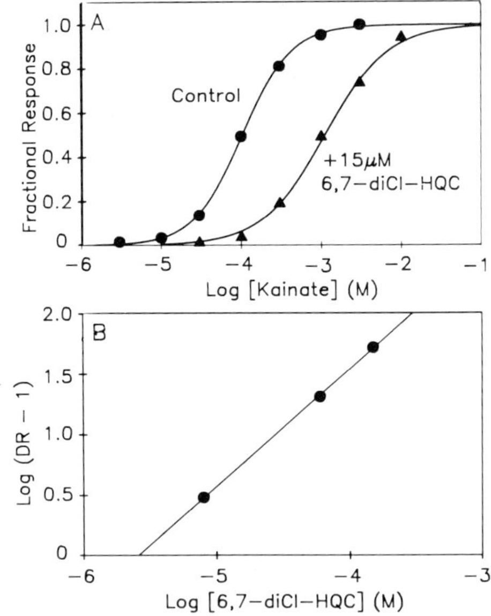

Fig. 7.   6,7-diCl-HQC is a competitive antagonist of the receptor activated by kainate in oocytes. (A) Dose-response curves for kainate were generated in the absence and presence of 15 µM 6,7-diCl-HQC. The dose-response curve was shifted 10.8-fold to the right in a parallel manner by 6,7-diCl-HQC. Points represent the mean ± SEM current for 5 cells. (B) Schild analysis was performed by constructing three-point dose-response curves in the absence of 6,7-diCl-HQC and in the presence of 8, 60 and 150 µM 6,7-diCl-HQC (n=4). The dose ratios (DR) were calculated from the EC_{50} of each curve and were plotted as the log (DR-1) against log[antagonist]. The slope of the regression line (0.98 ± 0.03) is consistent with competitive antagonism. The pA2 of 6,7-diCl-HQC for the kainate site was 5.52 ± 0.06. The points represent the mean ± SEM for 4 cells. From Kleckner and Dingledine, 1989.

in a brain slice (depending on local glutamate concentration), but block less than 20% of the NMDA receptors. Concentrations of CNQX of 10 µM or greater, however, would substantially block NMDA receptors.

The selectivity of compounds as competitive antagonists of the kainate and glycine sites can be compared by calculating the ratio of antagonist concentrations that produce a two-fold shift to the right in the respective agonist dose-response curves. By this measure, although 7-chlorokynurenate and 6,7-diCl-HQC have nearly the same potency against glycine (300 nM), 7-chlorokynurenate remains the most selective glycine site antagonist with a selectivity ratio of 40. The dehalogenated derivatives of these compounds were much less potent against both the glycine site and the quisqualate receptor. In addition, substitution of Br or I for Cl in 7-chlorokynurenate retained activity against glycine, but removal of the carboxyl group virtually abolished activity (W. Marshall, unpublished observations).

DISCUSSION

The molecular basis for the action of glycine on NMDA receptors remains unknown. Glycine seems to have several effects: it is required for activation of NMDA receptors and promotes recovery from a rapid phase of desensitization, yet it induces a slower phase of desensitization. Whether all these phenomena reflect a single primary action of glycine, multiple glycine binding sites on the NMDA receptor, or multiple NMDA receptors each influenced differently by glycine, are important topics for future study. A major unanswered question is whether the glycine site is saturated with agonist under physiological conditions, as the EC_{50} for glycine in oocytes, dialyzed neurons and isolated washed membranes preparations is 200-500 nM (Johnson and Ascher, 1987; Mayer et al. 1989; Bonhaus et al. 1987; Kleckner and Dingledine, 1988) whereas the glycine concentration in (human) cerebrospinal fluid is estimated to be 7 μM (Clarke et. al, 1989). Highly selective competitive antagonists are therefore necessary to investigate the physiological role of the glycine coagonist site in NMDA receptor mediated synaptic transmission. The most selective glycine site antagonist currently available, 7-chlorokynurenate, exhibits only 40-fold selectivity over the kainate-sensitive receptor in oocytes, suggesting that it will be only marginally useful for physiological studies in intact tissues. Nevertheless, over the concentration range of 5-30 μM 7-chlorokynurenate blocks both epileptiform burst-firing in hippocampal slices perfused with medium nominally free of Mg^{+2} (Kleckner and Dingledine, 1989), and NMDA-mediated cell death in cortical cultures (McNamara and Dingledine, 1989). Indeed,   NMDA-induced excitotoxicity was abolished by either 7-chlorokynurenate or D-APV, suggesting that antagonists of both the glycine and NMDA sites might be equi-effective for preventing ischemic brain damage. It remains to be seen whether the spectrum of side effects for glycine site antagonists is sufficiently different from those acting on the NMDA site to warrent their development.

Barnard et al. (1987) have speculated that excitatory and inhibitory amino acid receptors are possibly derived from a common ancestry that includes the nicotinic acetylcholine receptor. The pharmacological evidence reveals multiple similarities among the receptor channel family that supports this idea. Thus, some open channel blockers of the NMDA receptor are also nicotinic channel blockers (Lima-Landman and Albuquerque, 1988), glutamate is a good agonist at all EAA receptors, glycine site antagonists are also antagonists of the kainate sensitive receptor, and the structural features of agonist binding to the glycine site resemble those of the NMDA binding site. The glycine binding site on NMDA receptors may turn out to have molecular similarity to the kainate-sensitive receptor, the glutamate binding site on the NMDA receptor, as well as to the inhibitory glycine receptor. The molecular resolution of EAA receptors is therefore likely to prove quite interesting.

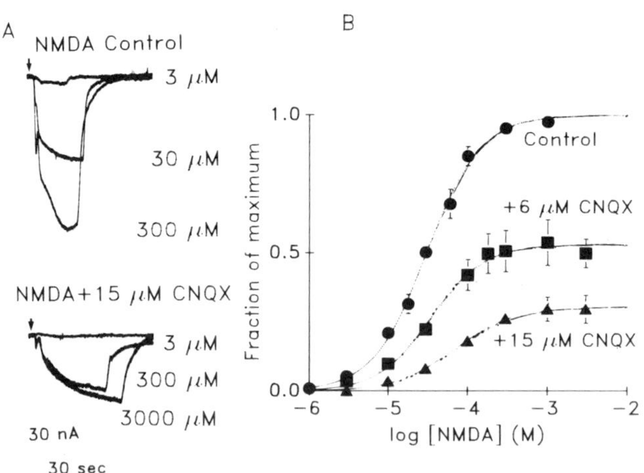

Fig. 8.   Non-competitive block of NMDA currents by CNQX. (A) Currents evoked by the indicated concentrations of NMDA in the absence (top) and presence (bottom) of 15 μM CNQX. (B) NMDA concentration-response curves in the absence and presence of 6 or 15 μM CNQX shows noncompetitive block of the NMDA site by CNQX. Each curve represents data from 3-6 cells. Glycine was 3 μM throughout. From Verdoorn et al., 1988.

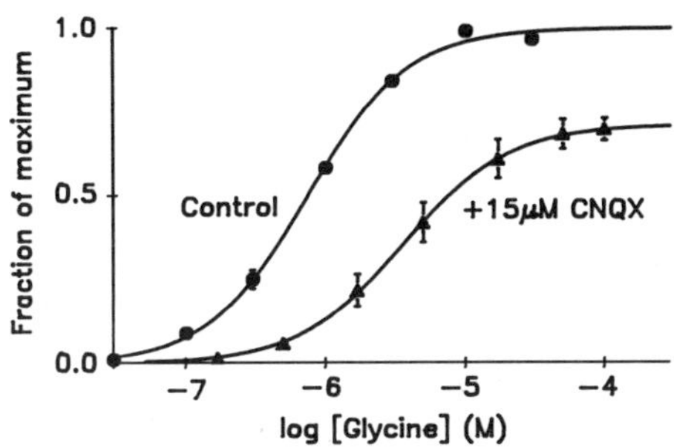

Fig. 9. Mixed competitive-noncompetitive block of the glycine site by CNQX. Glycine concentration-response curves were constructed in the absence and presence of 15 µM CNQX. Each curve represents data from 5 oocytes. NMDA was fixed at 100 µM throughout. From Verdoorn et al., 1988.

REFERENCES

Barnard, E.A., M.G. Darlison and P. Seeburg. Molecular biology of the GABA-A receptor: the receptor/channel superfamily. **Trends in Neurosci.**, 10, 502-509, (1987).

Birch, P. J., C. J. Grossman, and A. G. Hayes. 6,7-Dinitro-quinoxaline-2,3-dion and6-nitro,7-cyano-quinoxaline-2,3-dion antagonise responses to NMDA in the rat spinal cord via an action at the strychnine-insensitive glycine receptor. **Eur. J. Pharmacol.** 6, 177-180 (1988).

Bonhaus, D.W., B.C. Burge and J.O. McNamara. Biochemical evidence that glycine allosterically regulates an NMDA receptor-coupled ion channel. **Eur. J. Pharmacol.**, 142, 489-490, (1987).

Clarke, D.D., A.L. Lajthe and H.S. Maker. Intermediary metabolism. in G. Siegel, B. Agranoff, R.W. Albers and P. Molinoff (eds.). **Basic Neurochemistry**, 4th ed. Raven Press, New York, (1989).

Croucher, M.J., J.F. Collins, and B.S. Meldrum. Anticonvulsant action of excitatory amino acid antagonists. **Science** 216, 899-901, 1982.

Fletcher, E. J., and D. Lodge. Glycine reverses antagonism of N-methyl-D-aspartate (NMDA) by 1-hydroxy-aminopyrrolidone-2(HA-966) but not by D-2-amino-5-phosphono-valerate (D-APV) on rat cortical slices. **Eur. J. Pharmacol.** 1, 161-162 (1988).

Foster, A. C., and J. A. Kemp. HA-966 antagonizes N-methyl-D-aspartate receptors through a selective interaction with the glycine modulatory site. **J.Neurochem.** in press, 1989.

Honore, T. G., S. N. Davies, J. Drejer, E. J. Fletcher, P. Jacobsen, D. Lodge, and F. E. Nielsen. Quinoxalinediones: Potent competitive non-NMDA glutamate receptor antagonists. **Science** 241, 701-703 (1988).

Johnson, J. W. and P. Ascher. Glycine potentiates the NMDA response in cultured mouse brain neurons. Nature 325, 529-531 (1987).

Kemp, J.A., A.C. Foster, P.D. Leeson, T. Priestly, R. Tridgett, L.L. Iversen, and G.N. Woodruff. 7-Chlorokynurenic acid is a selective antagonist at the glycine modulatory site of the N-methyl-D-aspartate receptor complex. **Proc. Natl. Acad Sci.** USA 85, 6547-6550 (1988).

Kessler, M. T. Terramani, G. Lynch, and M. Baudry. A glycine site associated with N-methyl-D-aspartic acid receptors: Characterization and identification of a new class of antagonists. **J. Neurochem.** 52, 1319-1328 (1989).

Kleckner, N. W., and R. Dingledine. Requirement for glycine in activation of N-methyl-D-aspartate receptors expressed in **Xenopus** oocytes. Science 241, 835-837 (1988).

Kleckner, N. W., and R. Dingledine. Selectivity of quinoxalines and kynurenines as antagonists of the glycine site on N-methyl-D-aspartate receptors. **Mol. Pharmacol.** in press.

Kleinschmidt, A., M.F. Bear and W. Singer. Blockade of "NMDA" receptors disrupts experience -dependent plasticity of kitten striate cortex. **Science** 238, 355-358 (1987).

Lima-Landman, M.T. and E.X. Albuquerque. The novel neurotoxin H12-histrionicotoxin blocks the N-methyl-D-aspartate (NMDA) receptor of cultured hippocampus of the rat. **Soc. Neurosci. Abst.** 14, 96, 1988.

Lincoln, J., R. Coopersmith, E. W. Harris, C. W. Cotman, and M. Leon. NMDA receptor activation and early olfactory learning. **Dev. Brain Res.** 39, 309-312 (1988).

McBain, C. J., N. W. Kleckner, and R. Dingledine. Structural requirements for activation of the glycine coagonist site of NMDA receptors expressed in **Xenopus** oocytes. **Mol. Pharmacol.** in press.

McNamara, D. and R. Dingledine. Requirement for glycine in NMDA-induced excitotoxicity in rat cortical cultures. Soc. Neurosci. Abst. vol. , in press, 1989.

Mayer, M.L., L. Vyklicky and J. Clements. Regulation of NMDA receptor desensitization in mouse hippocampal neurons by glycine. **Nature.** 338, 425-427, (1989)

Morris, R.G.M., E. Anderson, G.S. Lynch and M. Baudry. Selective impariment of learning and blockade of long-term potentiation by an N-methyl-D-aspartate receptor antagonist, AP5. **Nature** 319, 774-776, 1986.

Ransom, R. W., and N. L. Deschenes. NMDA-induced hippocampal [3H]norepinephrine release is modulated by glycine. **Eur. J. Pharmacol.** 6, 149-5 (1988).

Simon, R.P., J.H. Swan, T. Griffiths and B.S. Meldrum. Blockade of N-methyl-D-aspartate receptors may protect against ischemic damage in the brain. **Science** 226, 850-852, 1984.

Snell, L.D., R.S. Morter, and K.M. Johnson. Structural requirements for activation of the glycine receptor that modulates the N-methyl-D-aspartate operated ion channel. **Eur. J. Pharmacol.** 6, 105-110 (1988).

Tsumoto, T., K. Hagihara, H. Sato, and Y. Hata. NMDA receptors in the visual cortex of young kittens are more effective than those of adult cats. **Nature** 327, 513-514 (1987).

Verdoorn, T.A. and R. Dingledine. Excitatory amino acid receptors expressed in Xenopus oocytes: agonist pharmacology. **Molec. Pharmacol.**, 34, 298-307, 1988.

Verdoorn, T.A., N.W. Kleckner, and R. Dingledine. Rat brain N-methyl-D-aspartate receptors expressed in *Xenopus* oocytes. **Science** 238, 1114-1116 (1987).

Verdoorn, T. A., N.W. Kleckner, and R. Dingledine. N-methyl-D-aspartate/glycine and quisqualate/kainate receptors expressed in *Xenopus* oocytes: Antagonist pharmacology. **Mol. Pharmacol.** 35, 360-368 (1989).

Watkins, J.C. and H.J. Olverman. Agonists and antagonists for excitatory amino acid receptors. **Trends in Neurosci.**, 10, 265-272, 1987.

Watson, G.B., W.F. Hood, J.B. Monahan, and T.H. Lanthorn. Kynurenate antagonizes actions of N-methyl-D-aspartate through a glycine-sensitive receptor. **Neurosci. Res. Comm.** 2, 169-174 (1988).

Wong, E.H.F., A.R. Knight, and R. Ransom. Glycine modulates [3H]MK-801 binding to the NMDA receptor in rat brain. **Eur. J. Pharmacol.** 142, 487-488 (1987).

THE PCP SITE OF THE NMDA RECEPTOR COMPLEX

J. F. MacDonald, M. C. Bartlett, I. Mody,
J. N. Reynolds, and M.W. Salter

Playfair Neuroscience Unit, The Toronto Hospital
Depart. Physiology and Pharmacology
University of Toronto, Toronto, Ontario

INTRODUCTION

Electropharmacological experiments by Lodge and his co-workers (Anis et al., 1983) first demonstrated that dissociative anaesthetics such as PCP and ketamine selectively block responses evoked by applications of NMDA agonists while sparing those mediated by other subtypes of excitatory amino acid receptors. Since that time others including ourselves demonstrated, using patch and voltage-clamp techniques, that this blockade is both dependent upon the membrane potential (voltage-dependent) and the presence of activated receptors (usedependent) (Honey et al., 1985; Huettner and Bean, 1987; Kemp et al., 1987; Kushner et al., 1988; MacDonald et al., 1987; Mayer et al., 1988; Rothman, 1988). These are characteristics which have most often been ascribed to an action of antagonists on voltage-dependent channels or on the channel domains of ligand-activated channels (see Hille, 1984). For the PCP site, a variety of binding studies have also shown that it is most likely located on or near the NMDA channel (Kemp et al., 1987; Monaghan et al., 1989). This site is most accessible when the receptor is in an activated state implying that the apparent affinity of PCP for its site is determined by channel gating. Nevertheless, much of this evidence is circumstantial and open to various interpretations.

Here we will review some aspects of the electrophysiological evidence suggesting that the PCP site is intimately associated with the NMDA channel pore and discuss some of the difficulties encountered in trying to interpret this evidence.

USE-DEPENDENT BLOCKADE AND TRAPPING

The block of NMDA currents by dissociative anaesthetics proceeds at an exceptionally slow rate unless the receptors are activated by exposure to an agonist. For example, the amplitude of the first response in the presence of a dissociative anaesthetic may be identical to that of the responses in the absence of drug. It is not until repeated applications of the agonist are made that the blockade develops and then proceeds to a steadystate level (Fig. 1A). The voltage dependence of the steady-state level of block can be demonstrated by changing the holding potential to a positive value and re-applying the agonist (Fig. 1B). Even though the antagonist is present, a use-dependent recovery is observed. This is not a full recovery to an unblocked state but it is merely a re-equilibration to a lesser level of blockade. Further evidence that the antagonists interact with the activated state of the receptor is manifest in the phenomenon of "trapping" (Lingle, 1973; Neely and Lingle, 1986). When the antagonist is washed out of the bathing chamber the responses

Excitatory Amino Acids and Neuronal Plasticity
Edited by Y. Ben-Ari
Plenum Press, New York, 1990

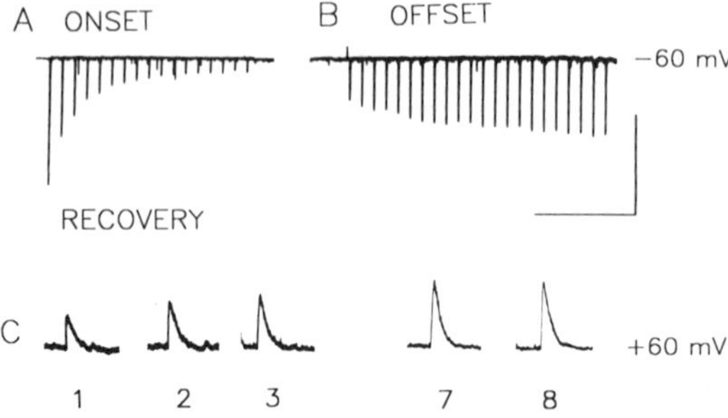

Fig. 1.   Blockade of NMDA currents by PCP shows use-dependency, trapping and voltagedependency. Whole cell currents recorded from two cultured hippocampal neurones using methods described in detail elsewhere (MacDonald et al., 1987). Brief applications of L-aspartic acid (250 µM) from a pressure pipette were repeated every 30 sec in order to evoke NMDA receptor-activated currents of consistent amplitude. A. The applications were halted while the bathing solution was replaced with one containing 5 µM PCP. The initial response in the presence of PCP was not depressed and the blockade only began to develop with repeated applications of the agonist (use-dependency). B. The applications were again stopped and the antagonist was removed from the bath. Nevertheless, the currents had not recovered (trapping) and further recovery was use-dependent. C. Individual NMDA currents from a different cell are shown. At a holding potential of +60 mV NMDA currents show use-dependent recovery in the continued presence of 1 µM PCP (voltage dependency). These currents had previously been blocked to a steady state level at a holding potential of -60 mV. The number below each response represents the pulse number in the series of responses at +60 mV.

to NMDA remain blocked until applications of agonist gradually cause a use-dependent recovery (Fig. 1B). Thus, the dissociative anaesthetics should be classified as uncompetitive antagonists (Pennefather and Quastel, 1982) rather than as either competitive or non-competitive antagonists which would show neither use-dependency nor trapping.

METHODOLOGICAL CONSIDERATIONS

Method of Agonist Application

The method of agonist application can significantly influence the interpretation of the actions of these blockers. One approach has been to employ continuous applications of relatively low concentrations of an NMDA receptor agonist with or without the inclusion of dissociative anaesthetic. The kinetics of the onset and offset of the blockade can then be used to estimate the microscopic rate constants (Huettner and Bean, 1988). It is tacitly assumed that in the absence of blocker that the NMDA currents would be maintained at a constant value. For example, if NMDA receptors were undergoing desensitization during the agonist application it would be necessary to include an additional kinetic component when calculating the rate constants. An advantage of applying the agonist at a low concentration is that desensitization of NMDA receptors is minimized, but in contrast the continuous exposure to the agonist will promote desensitization.

Another technical advantage of continuous application of agonist, considering the difficulties of making voltage- and patch-clamp recordings for long periods, is that the time required for making the measurements of the rates of blockade is on the order of seconds to minutes. However, in order to achieve this time scale of blockade, for an antagonist such as MK-801, it has also been necessary to use concentrations of the antagonist that are thousands of times greater than the K_i values determined from binding studies. These high concentrations may reveal non-specific effects of the antagonists.

An alternate approach which has been used by us (MacDonald et al., 1987) is to make very rapid and brief applications using high concentrations of agonist. This method of application takes advantage of the fact that the blockade by dissociative anaesthetics increases with higher concentrations of agonist. While desensitization during each individual response cannot be discounted, recovery from desensitization is rapid relative to the recovery from blockade by the antagonists and therefore it is easy to increase the time between applications to a point where there is no desensitization between responses (MacDonald, et al. 1987).

Alteration of NMDA Receptor State During Intracellular Dialysis

In many cases NMDA currents are recorded from neurones which are patch-clamped in the whole cell configuration. During such recordings the solution in the electrode will exchange with that in the interior of the cell. Previously it has been assumed that NMDA currents would be uninfluenced by this exchange but we have recently demonstrated that NMDA currents "wash-out" with an approximately exponential time course (Fig. 2). The time course of wash-out varies dramatically with the diameter of the patch electrode (from about 2 to 40 mins) and presumably depends upon many additional factors such as cell volume (Pusch and Neher, 1988). We have provided evidence that NMDA currents wash-out because of a loss of high energy phosphates from the cell interior (MacDonald et al., 1989). This hypothesis is supported by the demonstration that washout can be prevented (Fig. 2) or reversed by including a "phosphorylation support" solution in the patch electrode. Therefore, NMDA receptors are likely regulated by de/re-phosphorylation reactions.

Wash-out of NMDA currents presents a potential problem for the kinetic analysis of the blockade by dissociative anaesthetics. If a non-support solution is employed for recording there will be a high probability that the kinetics of wash out will be superimposed upon the kinetics of the blockade. This problem can be minimized by including the support solution in the recording pipette.

WHAT DOES THE VOLTAGE-DEPENDENCE OF THE BLOCKADE SIGNIFY ?

It appears as if the affinity of dissociative anaesthetics for their site of action depends directly upon the electric field across the membrane. Considering that these anaesthetics are positively charged (the amine group carries a net positive charge giving the molecules a dipole) at physiological values of pH it is tempting to speculate that each drug molecule must traverse a portion of the membrane field in order to reach its site of action. If this physical interpretation of the results is taken as correct then the distance (d) across the field that the blocker molecule must transit in order to reach this site can be calculated (Woodhull, 1973). Values of d for dissociative anaesthetics range from 0.5 to 1 (MacDonald et al., 1987; Mayer et al., 1988; Huettner and Bean, 1988) suggesting that the PCP site lies very deep within the channel, perhaps even near the intracellular mouth of the pore. Similar values of d (i.e. 1) have been calculated from the voltage-dependency of the block of NMDA channels by divalent cations (Ascher and Nowak, 1988). The calculation of d is based upon a number of assumptions including the "constant field assumption". Thus, the magnitude of the electric field across the membrane is assumed to decrease linearly with physical distance through the membrane (Hille, 1984). This is unlikely to be a valid assumption for many ion channels and it is usually the deviations from this assumption that are useful in delineating microscopic properties of these channels. For example, the NMDA channel may be occupied by more than a single permeant ion at a time, i.e. it may be a multi-ion pore (Hille, 1984). The presence of one cation in the channel would interact electrostatically with a second cation entering the channel (Ascher, 1988) thus causing a major deviation in linearity of the field. As a consequence the voltage-dependent block of NMDA channels either by divalent cations or by dissociative anaesthetics may tell us little about the actual physical location of their binding sites with the pore.

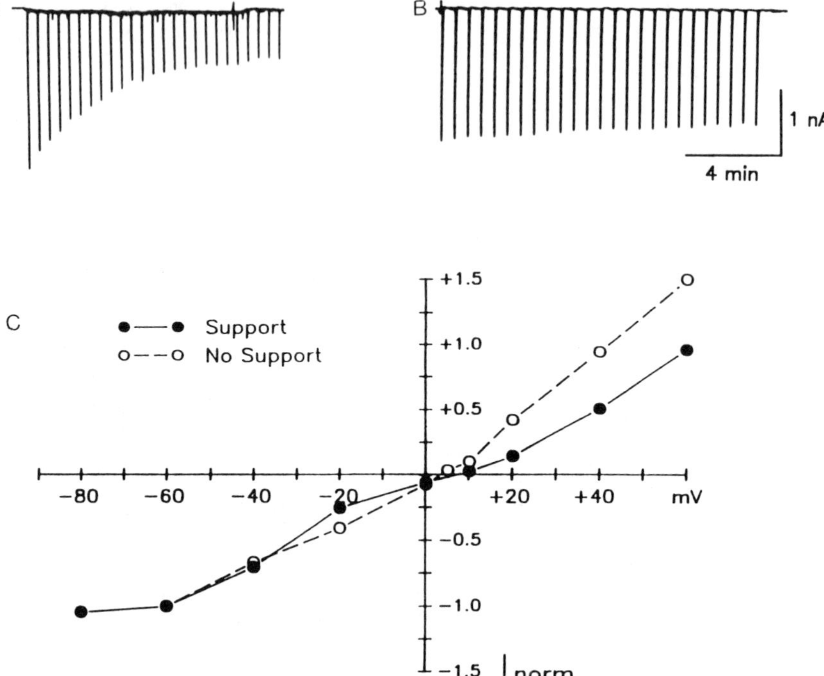

Fig. 2.   Wash-out of NMDA currents is prevented by an intracellular support
solution. Whole cell currents evoked by L-aspartate were recorded as described in Fig. 1. Care was taken to begin the applications of L-aspartate
within 30 sec after breaking the membrane patch. In A, the recording solution
did not contain a support solution and the currents demonstrated the
characteristic wash-out. However, in B the recording pipette contained an
intracellular support solution (ATP, phosphocreatine and creatine phosphok-
inase; see MacDonald et al., 1989) and wash-out was minimal. In C, the
current-voltage (I/V) relationship for the NMDA currents was determined
for two cells; the amplitudes are normalized with respect to that at a
holding potential of -60 mV. In one case the support solution was used. In the
other case the support system was not included and the I/V relationship was
examined after wash-out had occurred. The reversal potentials were similar
indicating that the wash-out cannot be attributed to a change in the driving
force on the permeating ions.

An alternative explanation is that the PCP site itself might demonstrate voltage-
dependence. In support of such a possibility is the demonstration that the block of NMDA currents
by pentobarbital is also voltage-dependent, even though this antagonist is neutrally charged (Fig.
3). Thus, voltage-dependency of blockade may be a feature of some aspect of receptor function and
not simply of blockade by a charged antagonist. This could arise in at least two general ways: a)
the configuration of the binding site might be altered by the membrane field directly or b) the
probability that a single channel remains open might be voltage-dependent.

One way to test whether the dissociative anaesthetic molecules are influenced by the
transmembrane field is to see if it is their charged form which is responsible for the blockade.
Dissociative anaesthetics are weak bases and therefore the concentration of charged antagonist
can be altered by changing the pH of the bathing solution. Using this approach we have recently
found that most of the blockade by ketamine can be attributed to the positively charged form of
the drug (MacDonald, et al. in preparation). Although there are a number of alternative

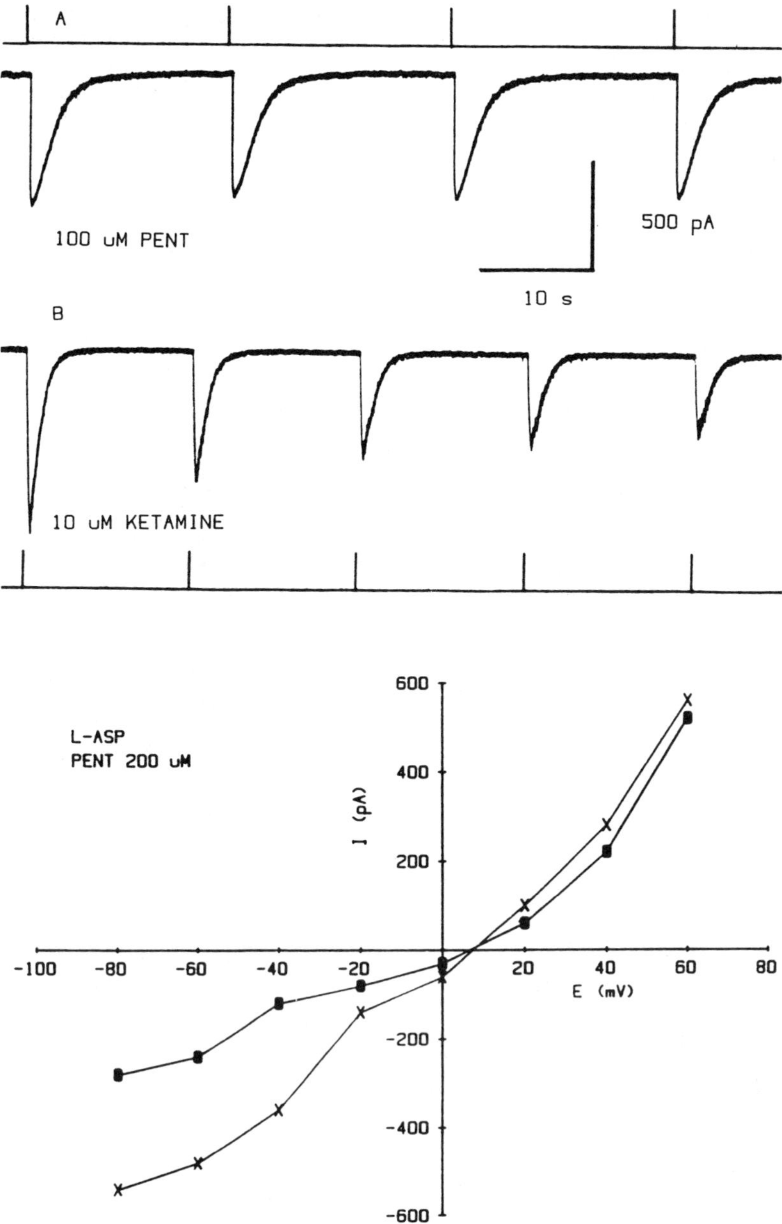

Fig. 3.　Blockade of NMDA currents by pentobarbital is voltage but not usedependent. A series of L-aspartate currents recorded in the presence of pentobarbital (A) is compared with that in the presence of ketamine (B). In each case the leftmost response is the first in the presence of the antagonist. Note that unlike ketamine no use-dependency was observed in the presence of a pentobarbital. The depression of these currents was however voltage-dependent (C).

interpretations of our results, including the possibility that the alteration in pH titrates some critical binding site of the NMDA receptor complex, the simplest explanation is that positively charged ketamine binds to negatively charged groups perhaps located in the channel mouth. The presence of negative sites in the mouth of the NMDA channel has already been proposed to account for the effects of altering extracellular calcium on the reversal potential of NMDA single channel currents (Ascher and Nowak, 1988).

However, does the effect of altering pH necessarily tell us anything about the possible origin of the voltage-dependency of the ketamine blockade? In fact it does not because altering the pH simply changes the concentration of the charged (i.e. blocking) form of the antagonist. To examine the voltage-dependency further it is necessary change the distribution of charge on the molecule. By choosing an appropriate structural analogue of PCP such as TCM, where a morpholino group has been added to reduce the dipole, the charge distribution on the molecule can be changed. TCM has been used by Aguayo et al. (1986) to examine whether or not the voltage-dependence of the block of nicotinic ACh channels is attributable to the charge dipole on PCP. They reported that TCM is much less potent in blocking these currents and that the block is *voltage-independent*. In preliminary experiments we have found that TCM is also considerably less potent than PCP in blocking NMDA currents but that the blockade by TCM is still highly *voltage-dependent* (MacDonald, et al. in preparation). This result is consistent with the site itself being voltage-dependent.

WHAT DO THE USE-DEPENDENCY AND TRAPPING SIGNIFY ?

The simplest, but by no means the only explanation for use-dependency and trapping, is that the antagonist molecule enters the channel pore when the channel is open (open channel blockade). If the occupied channel closes then the antagonist molecule will be physically trapped and will be unable to escape until the channel is re-activated. Considering the hydrostatic bulk of the dissociative anaesthetic molecules it seems unlikely that the closed channel could contain a cavity large enough to hold an entire molecule. Alternatively, it is possible that only a portion of the molecule (i.e. the amine group) is physically pinched by closure of the channel gate (Neely and Lingle, 1986).

The rates of association and dissociation with the open channel must be considerably greater than with closed channels in order for us to observe use-dependency and trapping. However, the dissociative anaesthetics may also have routes into or out of the closed channel via the lipid phase of the membrane as suggested by the work of Javitt and Zukin (1989).

DO ALL DISSOCIATIVE ANAESTHETICS CAUSE A VOLTAGE- AND USE-DEPENDENT BLOCK OF NMDA CURRENTS ?

There are studies which apparently contradict the observations that the blockade by dissociative anaesthetics is voltage- and use-dependent. Karschin, et al. (1988) have reported that the blockade of NMDA currents by dissociative anaesthetics is neither voltage- nor use-dependent in retinal ganglion neurones. Furthermore, Halliwell et al. (1988) have suggested that in cultured hippocampal neurones MK-801 fails to block NMDA currents in a voltage-dependent fashion. Nevertheless, they report that recovery from blockade is more rapid following membrane depolarization.

There are two reasons for the discrepancies between these results and our own (Honey, et al. 1985; MacDonald, et al. 1987) and those of Huettner and Bean (1988). First, the use of excessively high concentrations of antagonists can obscure both use- and voltage-dependency and second it is critical to achieve or at least approach equilibrium for the blockade.

The affinity of ketamine, PCP and MK-801 for their blocking site varies with membrane potential but it is also highly likely that the forward rate of association depends strongly upon antagonist concentration. As a consequence, if very high concentrations of antagonist are employed then the affinity of the antagonist will also become very high at all values of membrane potential (MacDonald, et al. 1987; Huettner and Bean, 1988). Furthermore, care must be taken to ensure that the reaction has actually reached, or at least has approached equilibrium, before the degree of blockade at a given potential is measured. In fact, as correctly surmised by Heuttner and Bean (1988), MK-801 causes a highly voltage-dependent blockade of NMDA currents in cultured hippocampal neurons when the concentration is close to the K_i (MacDonald, et al. in preparation). The combination of very high antagonist concentration and brief exposure of the agonist is the paradigm of drug application most likely to obscure both the voltage-and use-dependency of the blockade by dissociative anaesthetics.

Dissociative anaesthetics are highly lipophilic compounds and they must partition readily into the membrane lipid surrounding the NMDA receptor complex. It seems likely that lipophilic portions of the antagonist molecules would therefore interact with hydrophobic sites near or a part of the NMDA receptor. A similar mechanism has been proposed for the binding of dissociative anaesthetics to ACh channels (Aguayo et al. 1986). Thus, some dissociative anaesthetics may block closed NMDA channels and demonstrate no voltage-dependency what-so- ever. In this regard, the work of Manallack et al. (1988) has provided evidence for a three site interaction of PCP with the NMDA receptor complex. Two of these sites are hydrophobic sites (R1, R2) one of which binds the aromatic ring of the antagonist. It is the third site (R3) which binds to the protonated nitrogen atom found on these antagonists which might account for PCP's interaction with the open channel.

FAST OR SLOW CHANNEL BLOCKADE?

A surprising feature of the blockade of NMDA currents is the remarkably slow onset of the block in comparison, for example, to the block of voltage-dependent Na^+ channels by charged local anaesthetics (Starmer, et al. 1984). The slow kinetics of the blockade of macroscopic currents predicts that at a microscopic level an individual NMDA channel would have to open several thousand times before it became blocked. Once an antagonist molecule did bind to the channel the chance of it leaving the open channel would be even less. Therefore, we would anticipate no change in any single channel parameter such as conductance, mean open or closed times, etc. Instead, the activity of a single channel would simply appear to disappear from the patch recording and might not re-appear for a period of hours. However, a concentrationdependent reduction in the mean open time of the channel has been reported (Heuttner and Bean, 1988; Bertolino, et al. 1988) and Heuttner and Bean (1988) have suggested that the molecular rate of association of MK-801 with open NMDA channels may actually be comparable to that for a fast channel blocker. The apparent *macroscopic* rate of onset of the block only appears slow because it is the product of the probability of channel opening (which is apparently extremely low) and the *microscopic* rate of association.

SUMMARY

Evidence from electropharmacological experimentation favors the hypothesis that the PCP site is intimately associated with the channel domain of the NMDA receptor. But it is too early to state that this site lies deep within the NMDA channel pore. Determining the molecular details of the PCP site will require a complete and detailed kinetic analysis of NMDA single channel behavior. Furthermore, it is likely that hydrophobic receptor site(s) are responsible for some aspects of the blockade by at least some members of the dissociative anaesthetic family.

REFERENCES

Aguayo, L.G. and Albuquerque, E.X., 1986, Effects of phencyclidine and its analogs on the end-plate current of the neuromuscular junction, **J. Pharmacol. Exp. Ther.**, 239:15.

Anis, N.A., Berry, S.C., Burton, N.R. and Lodge, D., 1983, The dissociative anaesthetics, ketamine and phencyclidine, selectively reduce excitation of central mammalian neurones by N-methyl-Daspartate, **Br. J. Pharmacol.**, 79:565.

Ascher, P., 1988, Magnesium block of the NMDA receptor channel, in: "Frontiers in Excitatory Amino Acid Research," E.A. Cavalheiro, J. Lehmann and L. Turski, eds., Alan R. Liss, Inc., New York.

Ascher, P. & Nowak, L., 1988, The role of divalent cations in the NMDA responses of mouse central neurones in culture, . Physiol. (London), 399:247.

Bertolino, M., Vicini, S., Mazetta, J. and Costa, E., 1988, Phencyclidine and glycine modulate NMDA-activated high conductance cationic channels by acting at different sites, **Neurosci. Lett.,** 84:351.

Hille, B., 1984, "Ionic Channels of Excitable Membranes," Sinauer Associates Inc., Sunderland.

Halliwell, R.F., Peters, J.A. and Lambert, J.J., 1989, The mechanism of action and pharmacological specificity of the anticonvulsant NMDA antagonist MK-801: a voltage clamp study on neuronal cells in culture, **Br. J. Pharmacol.,** 96:480

Honey, C.R., Miljkovic, Z. and MacDonald, J.F., 1985, Ketamine and phencyclidine cause a voltage-dependent block of responses to L-aspartic acid, **Neurosci. Lett.,** 61:135.

Huettner, J.E. and Bean, B.P., 1988, Block of N-methyl-D- aspartate-activated current by the anticonvulsant MK-801: Selective binding to open channels, **Proc. Natl. Acad. Sci. USA,** 85: 1307.

Javitt, D.C. and Zukin, S.R., 1989, Biexponential kinetics of [3H]MK-801 binding: Evidence for access to closed and open N-methyl-D-aspartate receptor channels, **Mol. Pharmacol.,** 35:387.

Karschin, A., Aizenman, E. and Lipton, S.A., 1988, The interaction of agonists and non competitive antagonists at the excitatory amino acid receptors in rat retinal ganglion cells *in vitro*, **J. Neurosci.,** 8:2895.

Kemp J.A., Foster A.C. and Wong E.H.F., 1987, Non-competitive antagonists of excitatory amino acid receptors, TINS, 10:294.

Kushner, L., Lerma, J., Zukin, R.S. and Bennett, M.V.L., 1988. Coexpression of N-methyl-D-aspartate and phencyclidine receptors in Xenopus oocytes injected with rat brain mRNA, **Proc. Natl. Acad. Sci. USA,** 85:3250.

Lingle, C., 1983, Blockade of cholinergic channels by chlorisondamine on a crustacean muscle, **J. Physiol. (Lond.),** 339:395.

MacDonald, J. F., Miljkovic, Z. and Pennefather, P., 1987, Use-dependent block of excitatory amino-acid currents in cultured neurons by ketamine, **J. Neurophysiol.,** 58:251.

MacDonald, J.F., Mody, I. and Salter, M.W., 1989, Regulation of N-methyl-D-aspartate receptors revealed by intracellular dialysis of Murine neurones in culture, **J. Physiol. (Lond.),** 414:17.

Manallack, D.T., Wong, M.G., Costa, M., Andrews, P.R. and Beart, P.M., 1988, Receptor site topographies for phencyclidine-like and sigma drugs: predictions from quantitative conformational, electrostatic potential, and radioreceptor analyses, **Mol. Pharmacol.,** 34:863.

Mayer, M.L., Westbrook, G.L. and Vyklicky, L. Jr., 1988, Sites of antagonistic action on N-methyl-D-aspartic acid receptors studied using fluctuation analysis and a rapid perfusion technique, **J. Neurophysiol.,** 60:645.

Monaghan, D.T., Bridges, R.J. and Cotman, C.W., 1989, The excitatory amino acid receptors: Their classes, pharmacology, and distinct properties in the function of the central nervous system, **Ann. Rev. Pharmacol. Toxicol.,** 29:365.

Neely, A. and Lingle, C.J., 1986. Trapping of an open-channel blocker at the frog neuromuscular acetylcholine channel, **Biophy. J.,** 50:981.

Pennefather P. and Quastel D.M.J., 1982, Modification of dose response curves by effector blockade and uncompetitive antagonism, **Mol. Pharmacol.,** 22:369.

Pusch. M and Neher, E., 1988, Rates of diffusional exchange between small cells and a measuring patch pipette, **Pflugers Archiv.,** 411:204.

Rothman, S., 1988, Noncompetitive N-methyl-D-aspartate antagonists affect multiple ionic currents, **J. Pharmacol. Ther.,** 246:137.

Starmer, C.F., Grant, A.O. and Strauss, H.C., 1984, Mechanisms of use-dependent block of sodium channels in excitable membranes by local anaesthetics, **Biophy. J.,** 46:15.

Woodhull, A.M., 1973, Ionic blockage of sodium channels in nerve, **J. Gen. Physiol.,** 61:687.

ANTAGONISTS OF NMDA-ACTIVATED CURRENT IN CORTICAL NEURONS:[1]

COMPETITION WITH GLYCINE AND BLOCKADE OF OPEN CHANNELS

James E. Huettner

Department of Neurobiology
Harvard Medical School
220 Longwood Avenue
Boston, MA 02115
USA

There is increasing evidence that ion channels activated by L-glutamate underlie rapid excitatory synaptic transmission throughout the vertebrate central nervous system (Mayer and Westbrook, 1987) Nearly 30 years ago, glutamate was found to depolarize neurons in most areas of the brain and spinal cord (Hayashi, 1954 ; Curtis et al. 1960 ; Krnjevic and Phillis, 1963) but it was not until pharmacological experiments revealed the presence of several distinct receptors (Watkins and Olverman, 1987) for glutamate that significant progress was made toward understanding the mechanisms of glutamate excitation. Over the past ten years, the discovery of selective agonists and antagonists has played a major role in defining the subtypes of glutamate receptors (Dingledine et al. 1988) and recent studies (Jahr and Stevens, 1987 ; Ascher et al., 1988 ; Cull-Candy et al., 1988) of whole-cell and single channel currents using the patch clamp technique (Hamill et al., 1981) have begun to provide information about the channels gated by glutamate receptors.

N-methyl-D-aspartate (NMDA) serves as a selective agonist for the most thoroughly characterized subtype of glutamate receptor (Watkins and Olverman). Binding of glutamate or NMDA activates channels of 45-50 pS that are permeable to monovalent cations and calcium (Jahr and Stevens, 1987 ; Cull-Candy et al., 1988 ; Mayer et al., 1987 ; Ascher and Nowak, 1988). Currents gated by the NMDA receptor can be modulated in a number of ways. 1) $Mg2+$, $Co2+$ and several other divalent ions produce a direct, voltage-dependent blockade of the channel, (Ascher and Nowak, 1988 ; Nowak et al., 1984 ; Mayer and Westbrook, 1985) with block by external ions being less severe at positive potentials. 2) $Zn2+$ inhibits the current (Westbrook and Mayer, 1987 ; Peters et al., 1987) by binding to an external site that reduces the frequency of channel opening. 3) Low levels of extracellular glycine greatly enhance the current (Johnson and Ascher, 1987) by an increase in channel activation. All of these regulatory mechanisms are found in NMDA receptors expressed by *Xenopus* oocytes injected with rat brain mRNA, (Verdoorn and Dingledine, 1988 ; Kushner et al., 1988) which reinforces the proposal that the binding sites for glycine, $Mg2+$ and $Zn2+$ are integral components of the NMDA receptor and channel. There also may be further mechanisms for modulation, as suggested by a recent report (Aizenman et al., 1989) that the redox state of an external domain of the receptor can regulate current amplitude.

---

[1]*This work was supported by the Harvard University Society of Fellows, the Esther A. and Joseph Klingenstein Fund and by grants to Bruce P. Bean from the National Institutes of Health (HL-35034), the Culpeper Foundation and the Rita Allen Foundation. I am grateful to Bruce Bean for many helpful discussions and for support.

A number of organic compounds block the conductance increase elicited by NMDA. In addition to specific, competitive antagonists of the transmitter binding site (Watkins and Olverman, 1987) several blockers have been found that interact with the modulatory sites mentioned above. Following the discovery of potentiation by glycine, two compounds previously shown to antagonize responses to NMDA, Kynurenic acid (Mayer et al., 1988 ; Kemp et al., 1988) and HA-966, (Fletcher and Lodge, 1988 ; Foster and Kemp, 1989) were found to act predominantly at the glycine binding site. The dissociative anesthetics, ketamine and phencyclidine (PCP), (Anis et al., 1983 ; Honey et al., 1985 ; MacDonald et al., 1987) the anticonvulsant, MK-801, (Wong et al., 1986 ; Huettner, 1989) and the tricyclic antidepressants, promazine and desipramine (Sernagor et al., 1989), have been implicated as channel blocking agents.

This chapter summmarizes my recent work on derivatives of indole-2-carboxylic acid, (Huettner, 1989) which were found to compete for the glycine potentiation site, and work (Huettner and Bean, 1989) done in collaboration with Bruce Bean on blockade of NMDA gated channels by MK-801. The experiments were performed on rat cortical neurons maintained in primary dissociated cell culture (Huettner and Baughman, 1986).

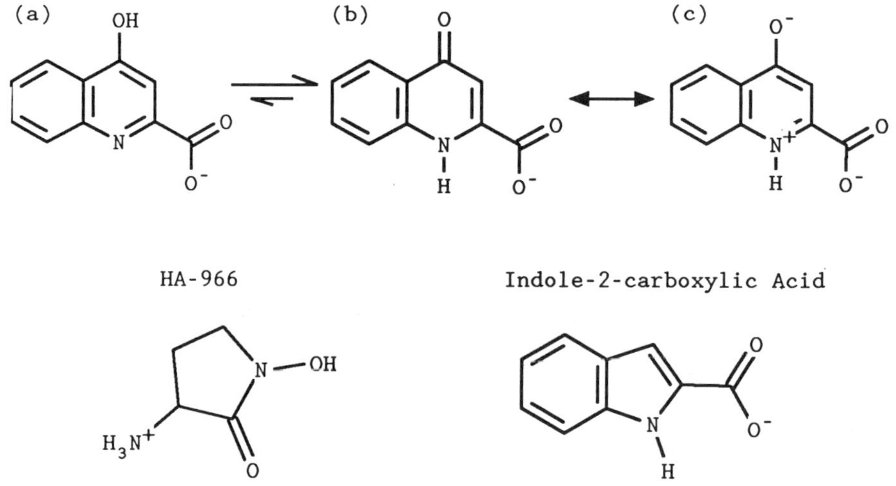

Fig. 1.   Chemical structures of kynurenic acid, HA-966 and indole-2-carboxylic acid.

Glycine Site Antagonists

The discovery that glycine enhances the current activated by NMDA in cultured central neurons and in oocytes injected with rat brain mRNA led to a reevaluation of antagonists that were known to be selective for NMDA. In particular, attention focused on kynurenic acid and HA-966, which produce noncompetitive block of responses to NMDA (Evan et al., 1978 ; Evan et al., 1987). Both antagonists were shown to inhibit ³H-glycine binding to rat brain membranes (Kemp et al., 1988 ; Foster and Kemp, 1989 ; Kessler et al., 1989) (Unpublished experiment.) and high concentrations of glycine were found to overcome the antagonism due to low doses of kynurenic acid (Mayer et al., 1988, Kemp et al., 1988) and HA-966 (Foster and Kemp, 1989). These results suggest that the principal action of these antagonists is to inhibit potentiation by glycine, although at concentrations above 100 µM kynurenic acid begins to block the NMDA binding site and also kainate and quisqualate receptors (Kemp et al., 1988).   Kemp and colleagues (Kemp et al., 1988) have found that substitution of chlorine for hydrogen at the 7 position enhances the selectivity of kynurenic acid for the glycine site.

Kynurenic acid is frequently shown as structure (a) in figure 1,5,22 however, the tautomeric form (b) and its contributing canonical form (c) are likely to predominate in physiological solutions (Elguero et al., 1976) and may therefore be the forms that actually interact with amino acid receptors. By screening various compounds with structural similarities to kynurenic acid, indole-2-carboxylic acid (I2CA) was found (Huettner, 1989) to antagonize NMDA when glycine levels are low. Dose response experiments (Huettner, 1989) indicate that I2CA competes directly for the glycine poten- tiation site and not for the transmitter binding site. Figure 2 shows that block by I2CA is equally potent at positive holding potentials as at negative potentials, and that the rates of onset of block and recovery from block are rapid.

In solutions containing low concentrations of glycine, both I2CA (Fig. 3 C) and 7-Cl-Kyn (Kemp et al., 1988) completely block responses to NMDA, but, with saturating levels of HA-966 a small response to NMDA remains (Fig. 4 A and Foster and Kemp, 1989). Kemp and colleagues (Kemp et al., 1988 ; Foster and Kemp, 1989) have considered two possible explanations for this difference. One sug-

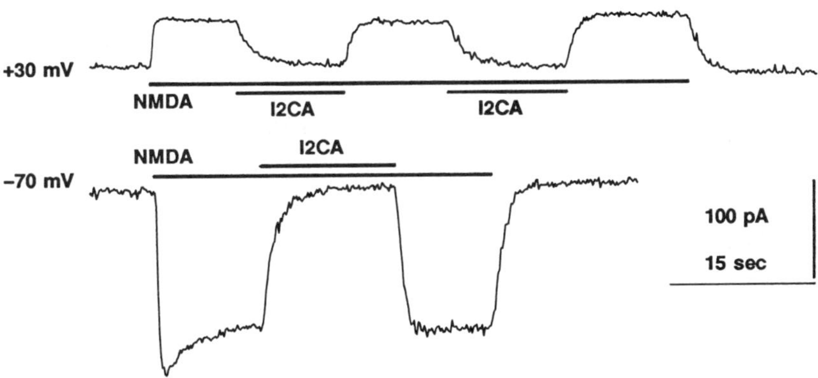

Fig. 2.   Whole-cell recordings of current activated by 500 µM NMDA plus 300 nM glycine applied by local perfusion from microcapillary tubes. Exposure changed from agonist alone to agonist plus 500 µM I2CA as indicated. Holding potential +30 mV, above; -70 mV, below. External solution: 160 mM NaCl, 10 mM HEPES (pH 7.4), 1 mM EDTA, 3 mM $CaCl_2$ and 300 nM tetrodotoxin. Internal solution: 120 mM $CsCH_3SO_3$, 5 mM CsCl, 10 mM EGTA, 10 mM HEPES, 5 mM MgATP, 1 mM Na_2GTP, pH 7.4 with CsOH.

gestion, based on an analogy to modulation of GABA responses by ligands of the benzodiazpine re- ceptor, (Haefely et al., 1985) is that 7-Cl-Kyn and, by extension, I2CA may serve as inverse agonists of the glycine potentiation site. In this scheme there should be a small, basal response to NMDA when the glycine site is unoccupied ; binding of 7-Cl-kyn or I2CA would act to suppress this basal level of channel activation while HA-966 would function as a simple competitive antagonist and therefore would not decrease the response below that obtained with NMDA alone.

The second possibility is that glycine is absolutely required for the channel to open and that HA-966 serves as a weak partial agonist while I2CA and 7-Cl-Kyn function as full antagonists. In this case, HA-966 should produce some degree of potentiation if glycine levels are low and there should be no response to NMDA in the complete absence of an agonist for the glycine site. Several lines of evidence appear to favor this second alternative; however, a complete resolution of this question remains pro- blematic. The high affinity of the glycine potentiation site has made it very difficult to study effects of NMDA in the complete absence of glycine. Cells in culture release glycine into the surrounding medium and unless extreme precautions are taken, most physiological solutions are contaminated by nannomolar levels of glycine. In a study of NMDA receptors expressed by *Xenopus* oocytes injected with mRNA from rat brain, Kleckner and Dingledine (1988) have shown that when rigorous measures are taken to avoid contamination by glycine there is virtually no response to NMDA alone. At early times in culture, cortical contamination by glycine there is virtually no

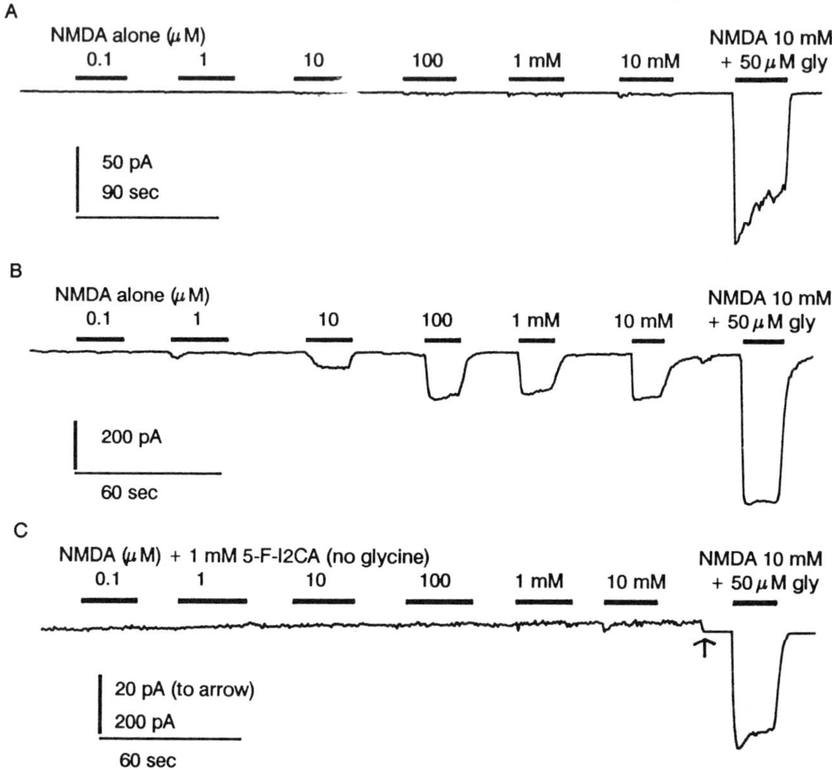

Fig. 3.   Whole-cell current gated by NMDA without added glycine increase time in culture. (**A**) At 4 d *in vitro*, NMDA alone evoked <2% of the c obtained with 50 μM glycine added. (**B**) After 12 d, NMDA alone gave nearly 30% of the maximum current recorded with saturating glycine. (**C**) The response to NMDA alone at 12 d *in vitro* was completely blocked by 1 mM 5-FluoroI2CA. (Unpublished experiment).

response to NMDA alone. At early times in culture, corticalneurons also show very little response to NMDA in solutions lacking added glycine (Huettner, 1989) (with contaminating levels of glycine less than 20 nM, as determined by HPLC; D. Sah, personal communication). Figure 3 A shows that even high doses of NMDA, up to 10 mM, produce only minimal current unless glycine is added. The increase observed with saturating glycine in these early cultures is up to 100 fold. This is the situation before significant proliferation of glial cells has occured. As cultures mature, the amplitude of whole cell current with saturating NMDA and glycine increases, suggesting an increase in receptor density. However, the response to NMDA without added glycine increases disproportionately, such that in a one to two week culture, which contains a confluent background layer of glial cells, there is only a 5 to 20 fold difference in the response to NMDA alone versus NMDA plus saturating glycine (Figure 3 B). This change is likely to be due to release of glycine by glial cells that cannot be overcome by rapid perfusion of solutions lacking added glycine.

Fig. 4 B shows that at very early times in culture, when contamination by glycine release is low, HA-966 can produce modest potentiation of current gated by NMDA, as would be expected for a partial agonist. Although it is difficult to rule out, there is no evidence to suggest that I2CA or 7-Cl-Kyn serve as inverse agonists rather than simple antagonists.

Channel Block by MK-801

Direct blockade of open channels by charged organic compounds has been described for both voltage- and transmitter-gated channels (Armstrong, 1971 ; Neher and Steinbach, 1978). A number of antagonists, including ketamine (Mayer et al., 1988 ; Honey et al., 1985 ; MacDonald et al., 1987). PCP26 and MK-801 (Wong et al., 1986 ; Huettner and Bean, 1988 ; Kloog et al., 1988) appear

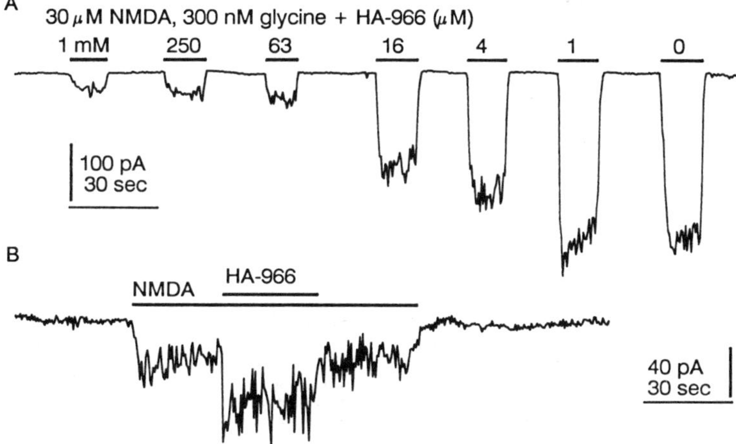

Fig. 4.   Inhibition (**A**) and potentiation (**B**) by HA-966. (**A**) Whole-cell recording of current activated by 30 µM NMDA plus 0.3 µM glycine and zero to 1 mM HA-966. (**B**) Current evoked by NMDA, without added glycine, plus 1 mM HA-966 as indicated. Three days *in vitro* , holding potential -70 mV. (Unpublished experiment.)

to act by this mechanism on channels gated by NMDA. Blockade by these agents is not competitive with NMDA or glycine; in fact, channel activation speeds the onset of block, a property which is sometimes refered to as "use-dependence" (Honey et al., 1985 ; MacDonald et al., 1987 ; Wong et al., 1986 ; Huettner and Bean, 1988). In the case of MK-801 (Huettner and Bean, 1988) simultaneous application of MK-801 and NMDA to cortical neurons gives rise, within a few seconds, to a long-lasting block of NMDA gated current, whereas even 4 to 5 minute applications of MK-801 alone do not lead to appreciable block.

Recovery from block is also greatly speeded by channel activation and is strongly voltage-dependent. On the average, the rate of recovery of whole-cell current during constant NMDA application is 50 fold more rapid at +30 mV than at -70 mV. In Figure 5 A, the response to NMDA was blocked by MK-801 and recovery during a continuous application of NMDA was monitored at -70 mV. Only partial recovery is apparent after 5 minutes in NMDA. Figure 5 B and C show two experiments in which, following block by MK-801, the cells were held at +30 mV and once each second stepped down to -70 mV for 80 milliseconds. In Fig. 5 B, recovery from block at +30 mV, with continuous application of NMDA, is essentially complete by 5 minutes. Fig. 5 C shows that recovery does not proceed at +30 mV without NMDA; although substantial recovery can be seen during the brief test pulses of NMDA, there is no change during the 5 minute period in control solution between the first and second application of NMDA.

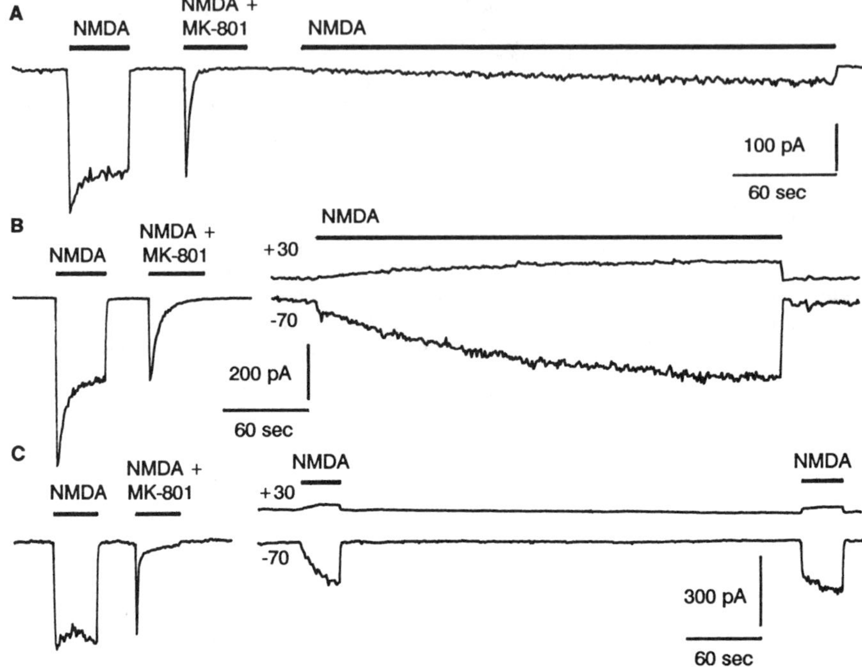

Fig. 5.   Recovery of whole-cell current evoked by 30 µM NMDA plus 1 µM glycine
from blockade by 10 µM MK-801. With continuous application of NMDA
recovery was much slower at a holding potential of -70 mV (**A**) than at +30
mV (**B**). (**C**) Recovery from block at +30 mV only occurred during application
of NMDA. (Unpublished experiment.)

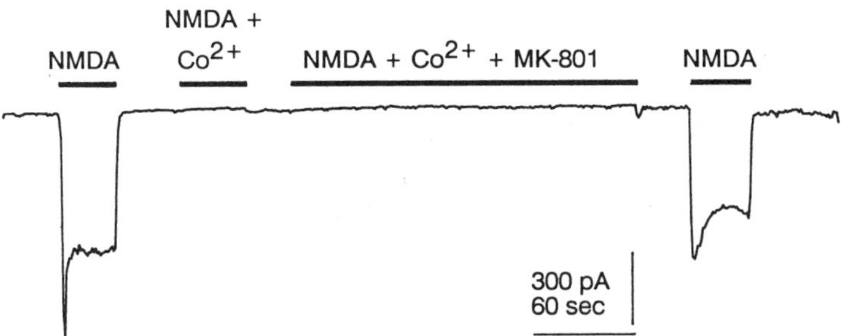

Fig. 6.   Cobalt, at 1 mM, completely blocks the current evoked by 30 µM NMDA plus
1 µM glycine and prevents long-lasting blockade by 2 µM MK-801.
(Unpublished experiment.)

Taken together, the results indicate that binding and unbinding of MK-801 are greatly accelerated when the channel has been opened by transmitter (see also Kloog et al., 1988); the rates of association and dissociation with closed channels, measured at room temperature in this physiological assay, appear to be at least 100 to 1000 fold slower than with open channels. Both the requirement for channel activation and the voltage dependence of unblock suggest that MK-801 may bind within the pore of the channel.

Further evidence for binding within the channel comes from experiments with the divalent cations, Mg^{2+} and Co^{2+}, which are known to block in the channel gated by NMDA (Ascher and Nowak, 1988 ; Nowak et al., 1984 ; Mayer and Westbrook, 1985). Both Mg^{2+} (Huettner and Bean, 1988) and Co^{2+} (Fig. 6), at concentrations sufficient to completely block whole-cell current due to NMDA, prevent MK-801 from binding to the channel. When the channel is blocked by Mg^{2+} or Co^{2+}, the binding site for MK-801 is inaccesible.

Although Mg^{2+} and MK-801 produce antagonism by essentially the same mechanism, the appearance of blockade by these two agents is strikingly dissimilar due to differences in their rates of binding and unbinding. Mg^{2+} binds and unbinds rapidly (Ascher and Nowak, 1988 ; Nowak et al., 1984) to produce a "flickery" block of NMDA-gated channels; these rapid kinetics, together with the fairly high concentrations of Mg^{2+} needed to achieve significant block, yield a rapid approach to equilibrium blockade, so that block of whole cell current develops nearly as fast as Mg^{2+} is applied. In contrast to Mg^{2+}, MK-801 binds rapidly to open channels but dissociates very slowly (Huettner and Bean, 1988). In outside-out patches exposed to NMDA plus MK-801, almost all of the channel openings occur within the first 30 to 60 seconds of application. Channels that become blocked by MK-801 remain blocked for many minutes and since NMDA-gated channels can close with MK-801 trapped inside, there is no increase in burst duration such as that seen with local anesthetic block of acetylcholine receptor channels (Neher and Steinbach, 1978). Eventually, all of the channels in a patch exposed to NMDA and MK-801 become blocked.

The binding of MK-801 to open channels gives rise to a dose-dependent reduction in the mean channel life time (Huettner and Bean, 1988) because openings can be terminated either by drug block or by normal closing. The reduction in mean open time provides a direct estimate of a, the forward rate constant for binding to open channels, of 3×10^7 M-1 sec-1 , a value which is close to the theoretical maximum allowed by diffusion (Neher and Steinbach, 1978). At the whole-cell level, blockade by MK-801 develops much more slowly, following a roughly exponential time course with a time constant of approximately 8 seconds for 10 µM MK-801 (see fig. 5). Since MK-801 binds preferentially to open channels, the approach to equilibrium blockade by a given concentration of MK-801 will largely depend upon the product of a and the fraction of open channels, p_o. The value for a of 3×10^7 M-1 sec-1 obtained from the reduction in mean single channel lifetime implies that the slow onset of macroscopic blockade is due to a low probability of opening, on the order of 0.002. This low value for p_o was confirmed (Huettner and Bean, 1988) by direct measurements in control patches with no overlapping channel events and also by counting the number of openings following application of NMDA plus MK-801. The rate constants for channel closing and for drug binding predict that about two-thirds of the openings in 10 µM MK-801 will be terminated by block rather than by normal closing. By counting the number of single channel events following MK-801 application, which varied from 15 to 335, an estimate can be made of the number of channels in the patch. This number, together with the channel activity obtained in control records, provides a value for p_o ; estimates made in this way ranged from 0.0002 to 0.001. It seems somewhat remarkable that p_o is so low even with saturating concentrations of NMDA and glycine; the possibility is raised that there may be mechanisms for biochemical or pharmacological modulation that greatly enhance NMDA activated current.

An interesting question is whether MK-801 produces a change in the open probability of blocked channels. With MK-801 lodged in the channel pore, either the open or closed blocked state could be stabilized or destabilized relative to unblocked channels (see Miller and al., 1987). In principal, it should be possible to measure the "stabilization factor" due to MK-801 binding (Miller and al., 1987). A shift in channel gating induced by MK-801 might result in a change in affinity for transmitter or for glycine, but this has not been reported. A large difference in open probability between blocked and unblocked channels could also produce a discrepancy between the dissociation constant for MK-801 as determined from binding of radiolabeled MK-801 with that determined from the time course of blockade and recovery of whole-cell current (Huettner and Bean, 1988). However, Kd values obtained from electrophysiological experiments (Huettner and Bean, 1988) and from the kinetics of radiolabeled MK-801 binding (Kloog et al., 1988) both agree well with the dissociation constant of 2 to 10 nM, determined from equilibrium binding (Wong et al., 1986 ; Kloog et al., 1988).

Conclusions

NMDA receptors are complicated entities, with multiple sites for regulation. Biophysical experiments with selective antagonists of modulatory sites on the NMDA receptor have provided insights into the mechanisms of control of channel activity. These antagonists should also prove useful for investigations of the role of modulation *in vivo* and they may possess clinical value for the treatment of stroke, epilepsy and other disorders, which are thought to involve excitotoxic damage via Ca2+ entry through NMDA-gated channels.

References

E. Aizenman, S. A. Lipton, and R. H. Loring, Selective modulation of NMDA responses by reduction an oxidation, **Neuron** 2:1257 (1989).

N. A. Anis, S. C. Berry, N. R. Burton, and D. Lodge, The dissociative anaesthetics, ketamine and phencyclidine, selectively reduce excitation of central mammalian neurones by N-methyl-aspartate, **Br. J. Pharmacol.** 79:565 (1983).

C. M. Armstrong, Interaction of tetraethylammonium ion derivatives with the potassium channels of giant axons, **J. Gen. Physiol.** 58:413 (1971).

P. Ascher, P. Bregestovski, and L. Nowak, N-methyl-D-aspartate activated channels of mouse central neurones in magnesium-free solutions. **J. Physiol.** 399:207 (1988).

P. Ascher and L. Nowak, The role of divalent cations in the N-methyl-D-aspartate responses of mouse central neurones in culture, **J. Physiol.** 399:247 (1988).

S. G. Cull-Candy, J. R. Howe, and D. C. Ogden, Noise and single channels activated by excitatory amino acids in rat cultured cerebellar granule neurons, **J. Physiol.** 400:189 (1988).

D. R. Curtis, J. W. Phillis, and J. C. Watkins, The chemical excitation of spinal neurones by certain acidic amino acids, **J. Physiol.** 150:656 (1960).

R. Dingledine, L. M. Boland, N. L. Chamberlin, K. Kawasaki, N. W. Kleckner, S. F. Traynelis, and T. A. Verdoorn, Amino acid receptors and uptake systems in the mammalian central nervous system, **CRC Crit. Rev. Neurobiol.** 4:1 (1988).

J. Elguero, C. Marzin, A. R. Katritzky, and P. Linda, The tautomerism of heterocycles, **Adv. Heterocyclic Chem.** Supp. 1 (1976).

R. H. Evans, A. A. Francis, and J. C. Watkins, Mg2+ -like selective antagonism of excitatory amino acid-induced responses by *a,e*-diaminopimelic acid, D-*a*-aminoadipate and HA-966 in isolated spinal cord of frog and immature rat, **Brain Res.** 148:536 (1978).

R. H. Evans, S. J. Evans, P. C. Pook, and D. C. Sunter, A comparison of excitatory amino acid antagonists acting at primary afferent C fibres and motoneurones of the isolated spinal cord of the rat, **Br. J. Pharmac.** 91:531 (1987).

E. J. Fletcher and D. Lodge, Glycine reverses antagonism of N-methyl-D-aspartate (NMDA) by 1-hydroxy-3-aminopyrrolidone (HA-966) but not by D-2-amino-5-phosphonovalerate (D-AP5) on rat cortical slices, **Eur. J. Pharmacol.** 151:161 (1988).

A. C. Foster and J. A. Kemp, HA-966 antagonizes N-methyl-D-aspartate receptors through a selective interaction with the glycine modulatory site, **J. Neurosci.** 9:2191 (1989).

W. Haefely, E. Kyburz, M. Gerecke, and M. Mohler, Recent advances in the molecular pharmacology of benzodiazepine receptors and in the structure activity relationships of their agonists and antagonists, **Adv. Drug Res.** 14:165 (1985).

O. Hamill, A. Marty, E. Neher, B. Sakmann, and F. J. Sigworth, Improved patch-clamp techniques for high-resolution current recording from cells and cell-free membrane patches, **Pfluegers Archiv.** 391:85 (1981).

T. Hayashi, Effects of sodium glutamate on the nervous system, **Keio J. Med.** 3:183 (1954).

C. R. Honey, Z. Miljkovic, and J.F. MacDonald, Ketamine and phencyclidine cause a voltage-dependent block of responses to L-aspartic acid, **Neurosci. Lett.** 61:135 (1985).

J. E. Huettner and B. P. Bean, Block of N-methyl-D-aspartate-activated current by the anticonvulsant MK-801: selective binding to open channels, **Proc. Natl. Acad. Sci. USA** 85:1307 (1988).

J. E. Huettner, Indole-2-carboxylic acid: a competitive antagonist of potentiation by glycine at the NMDA receptor, **Science** 243:1611 (1989).

J. E. Huettner and R. W. Baughman, Primary culture of identified neurons from the visual cortex of postnatal rats, **J. Neurosci.** 6:3044 (1986).

C. E. Jahr and C. F. Stevens, Glutamate activates multiple single channel conductances in hippocampal neurons, **Nature** 325:522 (1987).

J. W. Johnson and P. Ascher, Glycine potentiates the NMDA response in cultured mouse brain neurons, **Nature** 325:529 (1987).

J. A. Kemp, A. C. Foster, P. D. Leeson, T. Priestley, R. Tridgett, L. L. Iversen, and G. N. N. W. Kleckner and R. Dingledine, Requirement for glycine in activation of NMDA receptors expressed in *Xenopus* oocytes, **Science** 241:835 (1988).

M. Kessler, T. Terramani, G. Lynch, and M. Baudry, A glycine site associated with N-methyl-D-aspartic acid receptors: characterization and identification of a new class of antagonists, **J. Neurochem.** 52:1319 (1989).

Y. Kloog, V. Nadler, and M. Sokolovsky, Mode of binding of [3H]dibenzocycloalkenimine (MK-801) to the N-methyl-D-aspartate (NMDA) receptor and its therapeutic implication, **Febs. Lett.** 230:167 (1988).

K. Krnjevic and J. W. Phillis, Iontophoretic studies of neurones in the mammalian cerebral cortex, **J. Physiol.** 165:274 (1963).

L. Kushner, J. Lerma, R. S. Zukin, M. V. L. Bennett, Coexpression of N-methyl-D-aspartate and phencyclidine receptors in Xenopus oocytes injected with rat brain mRNA, **Proc. Natl. Acad. Sci. USA** 85:3250 (1988).

J. F. MacDonald, Z. Miljkovic, and P. Pennefather, Use-dependent block of excitatory amino acid currents in cultured neurons by ketamine, **J. Neurophysiol.** 58:251 (1987).

M. L. Mayer and G. L. Westbrook, The physiology of excitatory amino acids in the vertebrate central nervous system, **Prog. Neurobiol.** 28:197 (1987).

M. L. Mayer, A. B. MacDermott, G. L. Westbrook, S. J. Smith, and J. L. Barker, Agonist- and voltage-gated calcium entry in cultured mouse spinal cord neurons under voltage clamp measured with arsenazo III, **J. Neurosci.** 7:3230 (1987).

M. L. Mayer and G. L. Westbrook, The action of N-methyl-D-aspartic acid on mouse spinal neurones in culture, **J. Physiol.** 361:65 (1985).

M. L. Mayer, G. L. Westbrook, and L. Vyklicky, Jr., Sites of antagonist action on N-methyl-D-aspartic acid receptors studied using fluctuation analysis and a rapid perfusion technique. **J. Neurophysiol.** 60:65 (1988)

C. Miller, R. Latorre, and I. Reisin, Coupling of voltage-dependent gating and Ba++ block in the high-conductance, Ca++-activated K+ channel, **J. Gen. Physiol.** 90:427 (1987).

E. Neher and J. H. Steinbach, Local anaesthetics transiently block currents through single acetylcholine receptor channels, **J. Physiol.** 277:153 (1978).

L. Nowak, P. Bregestovski, P. Ascher, A. Herbet, and A. Prochiantz, Magnesium gates glutamate activated channels in mouse central neurons, **Nature** 307:462 (1984).

S. Peters, J. Koh, and D. W. Choi, Zinc selectively blocks the action of N-methyl-D-aspartate on cortical neurons, **Science** 236:589 (1987).

E. Sernagor, D. Kuhn, L. Vyklicky, Jr., and M. L. Mayer, Open channel block of NMDA receptor responses evoked by tricyclic antidepressants, **Neuron** 2:1211 (1989).

T. A. Verdoorn and R. Dingledine, Excitatory amino acid receptors expressed in *Xenopus* oocytes: agonist pharmacology, **Mol. Pharmacol.** 34:298 (1988).

J. C. Watkins and H. J. Olverman, Agonists and antagonists of excitatory amino acid receptors, **Trends Neurosci.** 10:265 (1987).

G. L. Westbrook and M. L. Mayer, Micromolar concentrations of Zn2+ antagonize NMDA and GABA responses of hippocampal neurons, **Nature** 328:640 (1987).

E. H. F. Wong, J. A. Kemp, T. Priestley, A. R. Knight, G. N. Woodruff, and L. L. Iversen, The anticonvulsant MK-801 is a potent N-methyl-D-aspartate antagonist, **Proc. Natl. Acad. Sci. USA** 83:7104 (1986).

Woodruff, 7-Chlorokynurenic acid is a selective antagonist at the glycine modulatory site of the N-methyl-D-aspartate receptor complex, **Proc. Natl. Acad. Sci. USA** 85:6547 (1988).

COMMENTARY - MECHANISMS AND EXPRESSION OF RECEPTORS

P. Ascher and M.L. Mayer

The meeting commenced with a session entitled "Mechanisms and Expression of Receptors", and initiated a trend of a nearly exclusive focus on NMDA receptors with continued for the rest of the meeting. This emphasis on the characterization of NMDA receptors appeared to arise from two sources of inspiration. On the one hand, the role of NMDA receptors in integrative behaviours such as LTP and the plasticity of neuronal circuits is clearly recognized by the majority of neurobiologists, and underlay many of the presentations in later sessions. On the other hand, cellular physiologists, one of whom stated no interest in obtaining results of "functional significance", clearly preferred to get their teeth into experiments amenable to more quantitative analysis. Whether this trend is to be encouraged might be questioned, given the lack of agreement concerning interpretation of nearly identical experiments presented by the 1st two speakers in the session.

Mayer and Ascher presented whole cell and single channel data concerning the regulation of NMDA receptor activity by glycine. Allowing for the warner climate in Washington versus Paris they agreed that NMDA receptor activity is very low at ambient concentrations of glycine (estimated at 20-50 nM), and reflects binding of glycine to a site of equilibrium dissociation constant 200-400 nM. The on rate for glycine is around 10^7 M^{-1} sec^{-1}, and the dissociation rate around 0.5 -1 (Paris) or 4 sec^{-1} (Washington). Mayer proposed that an NMDA evoked reduction in affinity for glycine underlay "desensitization" at NMDA receptors. He showed that the time constant for glycine at the same concentration, and that for ligands with low affinity for the glycine receptor desensitization was faster or absent, consistent with rapid dissociation of these analogues measured in a separate series of experiments. However, based on analysis of single channel data Ascher proposed that NMDA receptors show desensitization at all concentrations of glycine, and that in whole cell experiments buffered diffusion of NMDA in the subcellular region between the bottom of neurons and the underlying substrate generated an artificial glycine concentration dependence of desensitization. He suggested that the apparent off rate for NMDA should vary with glycine concentration, an effect not observed by Mayer. Further experiments by both groups are in progress to resolve this difference of interpretation.

Continuing the theme of analysing the interaction of glycine with NMDA receptors, Dingledine presented experiments with messenger RNA injected oocytes. These express NMDA receptors and allow quantitative pharmacological measurements of agonist and antagonist potency, experiments which are possible, but much more difficult to perform on single neurons. In agreement with results obtained by Mayer and Ascher, Dingledine reported a Kd for the glycine antagonist 7-chlorokynurenic acid of around 350 nM. Another competitive glycine antagonist, of 10-fold lower potency : 5-fluoro-indole-2-carboxylic acid, was described by Huettner, and data obtained with this and other ligands will help in defining the molecular features required for affinity at the glycine binding site. With this in mind, exploiting the satbility of oocyte recording, Dingledine then examined a large series of glycine analogues for activity in modulating NMDA receptors. Together with data obtained using antagonists these results suggest similarity in the binding sites for both NMDA and glycine, a proposal which eventually might be confirmed with structural data obtained from genetic analysis of glutamate receptor ion channel proteins.

Huettner and MacDonald both presented results obtained with cell culture preparations, and described experiments to examine the action of the NMDA antagonists MK801 (Huettner) and ketamine (MacDonald). Huettner showed convincingly that the rate of dissociation of MK801 from the NMDA receptor chanel was highly voltage dependent. He then showed that quaternary derivatives of MK801 is accessible only from the extracellular face of the membrane. However changing the nature of the intracellular cation from permeant (Cs) to impermeant (N-methyl-glucamine) slowed the rate of recovery of MK801 antagonism, suggesting that the voltage dependence of MK801 antagonism may result from an indirect effect : knock out of antagonist by permeant cations leaving the channel. In addition to studies with ketamine, MacDonald showed that ATP is required to maintain large amplitude responses to NMDA. Spectral analysis showed no change in open time or conductance, suggesting that channels become inactivated when dephosphorylated.

The results of this 1st session thus dealt nearly exclusively with receptor mechanisms and attempted quantitative analysis in terms of mechanism, with varying degrees of sophistication and success. Many further experiments of this type will ultimately be required to understand glutamate receptor function in detail, and this hardening of the neurobiological approach seems likely to continue, despite the difficulty in performing and interpreting such experiments.

<u>SESSION II</u>: TYPES OF EAA RECEPTORS

EXPERIMENTS WITH KAINATE AND QUISQUALATE AGONISTS AND ANTAGONISTS IN RELATION TO THE SUB-CLASSIFICATION OF 'NON-NMDA' RECEPTORS

J.C. Watkins , P.C.K. Pook , D.C. Sunter , J. Davies* and
T.Honore**

Department of Pharmacology, School of Medical Sciences
Bristol BS8 1TD, UK. *Department of Pharmacology, School
of Pharmacy, Brunswick Square, London, WC1N 1AX, UK
**Ferrosan Research Division, Sydmarken 5, DK-Soeborg
Denmark

The differentiation of "kainate" (K) and "quisqualate" (Q) sub-classes of non-NMDA excitatory amino acid (EAA) (Davies et al., 1979 ; Watkins, 1978) receptors was originally based on the selective depressant action of L-glutamic acid diethyl ester (GDEE) on quisqualate induced responses in ionophoretic experiments *in vivo* (McLennan and Lodge, 1979 ; Watkins and Evans, 1981). This concept was supported by the opposite selectivity shown by other antagonists subsequently developed, e.g., γ -D-glutamylglycine (DGG) (Davies and Watkins, 1985) and γ -D-glutamylaminomethyl sulphonate (GAMS)[7], and by the marked differences observed in the relative potencies of kainate and quisqualate in different tissues (Shinozaki, 1978 ; Agrawal and Evans, 1986). The results of binding studies with ^{3}H-kainate and ^{3}H-L-glutamate under various conditions also generally supported the notion of separate kainate and quisqualate receptors, as did those of later experiments with the quisqualate analogue, ^{3}H-AMPA (Foster and Fagg, 1984; Monaghan et al., 1989). In conformity with this concept, autoradiographic patterns obtained in brain sections with ^{3}H-kainate and ^{3}H-AMPA are quite distinct (Cotman et al., 1987).

Nevertheless, in electrophysiological experiments it is often difficult to distinguish pharmacologically responses mediated by kainate and quisqualate (Mayer and Westbrrok, 1987 ; Collingridge and Lester, 1989), and therefore to assign categorically a transmitter receptor involved in synaptic excitation to one or the other type. Hence the common reference to "K/Q" receptors. There are multiple reasons for this inability to differentiate K and Q receptors adequately, and include:

1) The relative inactivity of GDEE in experiments *in vitro* (Davies et al., 1979) and its low potency, lack of EAA-specificity (Davies et al, 1979 ; Davies and Watkins, 1979) and general unreliability even in experiments *in vivo* .

2) The failure of antagonists such as DGG and GAMS, which appear to be K receptor selective *in vivo* (Davies and Watkins, 1981 ; Davies and Watkins, 1985), to show much difference on kainate- and quisqualate-induced responses *in vitro* (Francis et al. 1981 ; Jones et al., 1984), and

3) The reported inability of the potent quinoxaline non-NMDA receptor antagonists, 6-cyano-7-nitro-quinoxaline-2,3-dione (CNQX) and 6,7-dinitroquinoxaline-2,3-dione (DNQX), which show differences in affinity for ^{3}H-kainate and ^{3}H-AMPA binding sites in rat brain membranes (Honore et al.), to discriminate adequately between

Excitatory Amino Acids and Neuronal Plasticity
Edited by Y. Ben-Ari
Plenum Press, New York, 1990

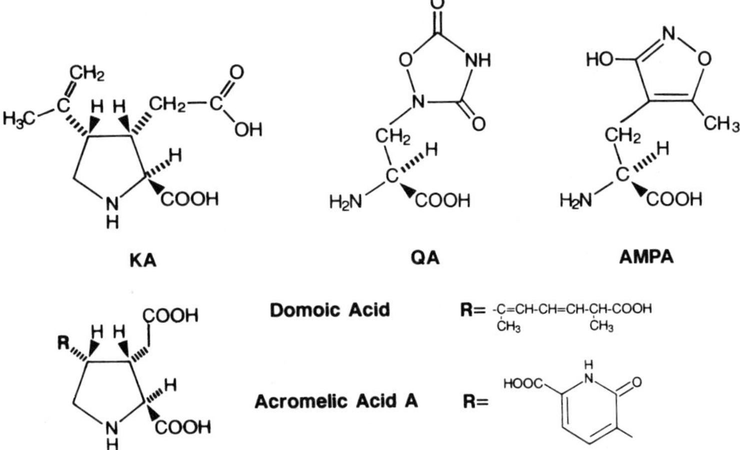

Fig. 1   Structures of some non-NMDA receptor agonists. Abbreviations: KA, kainic acid; QA, quisqualic acid; AMPA, γ-amino-3-hydroxy-5-methylisoxazole-4-propionic acid.

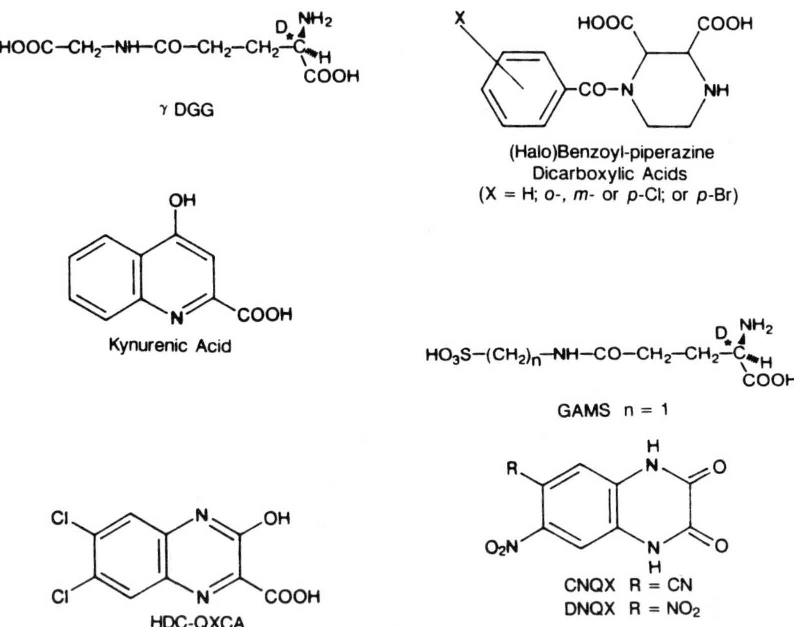

Fig. 2   Structures of some non-NMDA receptor antagonists. Abbreviations: DGG, γ-D-glutamylglycine; HDC-QXCA, 2-hydroxy-6,7 dichloroquinoxaline-2-carboxylic acid; GAMS, γ -D-glutamylaminomethylsulphonic acid; CNQX, 6-cyano-7-nitro-quinoxaline-2,3 dione; DNQX, 6,7-dinitroquinoxaline 2,3-dione.

kainate- and quisqualate-(or AMPA-) induced responses in electrophysiological experiments either *in vivo* (Honore et al.) or *in vitro* (spinal cord (Birch et al. 1988) and hippocampal slices (Blake et al., 1989).   However, in another study (Flechter et al. 1988), in rat cortical wedges and frog spinal cord, an approximately 4-fold differential in favour of quisqualate induced responses, relative to kainate, was observed for the two quinoxalines, with AMPA being somewhat more susceptible than quisqualate.

There are probably both individual and common causes for the above phenomena.   Thus, the mechanism of GDEE action, and possibly also that of DGG and GAMS, could well involve particular pH and ionic conditions which are different *in vivo* from those in the *in vitro* methods used. For example, GDEE is applied as a cation from acidic solutions in ionophoretic

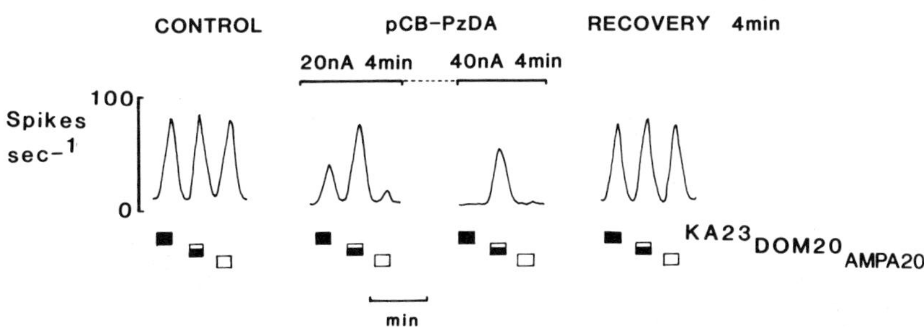

Fig. 3   Action of ionophoretically administered 1-(*p*-chlorobenzoyl) piperazine- 2,3-dicarboxylate (*p*CB-PzDA) (initially 20 nA, increased later to 40 nA) on excitatory responses of a cat spinal dorsal horn neurone induced by kainate (KA, 23 nA), domoate (DOM, 20 nA) and AMPA (20 nA) Responses induced by AMPA and kainate were more sensitive to the antagonist than responses induced by domoate.

Table 1. K_D values (μm) for antagonism of various agonist-induced responses in isolated spinal cord of neonatal rat.

	AMPA	ACRO	QUIS	KAIN	DOM	$\dfrac{QUIS}{AMPA}$	$\dfrac{KAIN}{AMPA}$	$\dfrac{DOM}{AMPA}$
γDGG	366		980	1034	2560	2.7	2.8	7.0
KYN			258	164				
pCB-PzDA	127		262	188		2.1	1.5	
pBB-PzDA	85		253	243	555	3.0	2.9	6.5
HDC-QXCA	5.9	8.1	13	25	24	2.3	4.3	4.1
DNQX	1.0	1.2	2.8	4.8	6.8	2.7	4.7	6.7
CNQX	1.4	3.4	5.0	5.3	7.9	3.5	3.7	5.5

(The $\dfrac{QUIS}{AMPA}$, $\dfrac{KAIN}{AMPA}$ and $\dfrac{DOM}{AMPA}$ columns are grouped under the heading K_D ratios.)

Abbreviations: KYN, kynurenate; pCB- and pBB-PzDA, 1-p-chloro (and p-bromo)-benzoyl piperazine-2,3-dicarboxylate; others, see Figs 1 and 2.

experiments, which may reduce local extracellular pH, and the selective action of GDEE on quisqualate-induced responses in such experiments could conceivably be pH dependent. This possibility has not yet been tested *in vitro*. Another possible difference between *in vivo* and *in vitro* unlikely as a contributing factor by the finding that GDEE (1-10 mM) had no depresssant effect on AMPA or kainate-induced responses on the neonatal rat spinal cord either in the presence (1 mM) or absence of Mg^{2+} (P.C.K. Pook and J.C. Watkins, unpublished observations).

Although differences in extracellular fluid composition *in vivo* and *in vitro* could likewise be a factor also in the case of DGG and GAMS, the relative failure of these substances to differentiate responses to kainate and quisqualate *in vitro* seems more likely to reflect a true non-selectivity of the antagonists for these agonist-induced responses, with the apparent selectivity (kainate>quisqualate) *in vivo* possibly reflecting marked differences in the dose response relations of the two agonists under ionophoretic conditions. Such agonist dose response plots are not usually constructed because of the difficulty of their interpretation, increased ejection currents leading to an increase in the number of more remote receptors being activated and nearer receptors being activated differentially by radially increasing concentrations of the agonist. The uncertainty of access of the antagonist to all the receptors activated by each of the two agonists would confuse the issue further, as would any uptake of one or the other agonist.

The failure of CNQX and DNQX adequately to differentiate between kainate-, quisqualate-and/or AMPA-induced responses in ionophoretic experiments despite the somewhat greater affinity of the quinoxalines for ^{3}H-AMPA than for ^{3}H-kainate binding sites could have a similar explanation. However, coupled to such technical problems as described above and to the probability of non-selective action of the antagonists at the different receptors is the additional likelihood that quisqualate and kainate themselves are non-specific in electrophysiological experiments, both of them activating more than a single type of receptor. This concept receives support from binding experiments, in which quisqualate was shown to have moderate affinity for ^{3}H-kainate binding sites (Foster and Fagg, 1984) and kainate for ^{3}H-AMPA binding sites (Honore et al., 1982).

In a search among known agents for the most selective agonists and antagonists for different types of non-NMDA receptors we have compared dose-ratios for antagonism by a range of antagonists against responses induced in motoneurones in the isolated hemisected spinal cord of the neonatal rat by a series of kainate and quisqualate- related agonists. The series of agonists used included domoic acid (Biscoe et al., 1976) and acromelic acid A, a new excitant isolated from the poisonous mushroom *Clitocybe acromelalga* (Biscoe et al., 1976), both these agonists being closely related structurally to kainic acid (Fig 1). The antagonists (Fig 2) included γ-D-glutamylglycine (DGG), kynurenate (Perkins and Stone, 1982 ; Ganong et al., 1983), p-chloro- and p-bromobenzoyl piperazine dicarboxylates (Davies et al., 1984), DNQX and CNQX, as well as 3-hydroxy-6,7-dichloro-quinoxaline-2-carboxylate (HDC-QXCA) which was reported to be kainate-selective (relative to quisqualate and AMPA) in sodium efflux and membrane binding studies (Frey et al., 1988).

It can be seen from Table 1 that responses induced by AMPA were the most sensitive, the order being similar for all the antagonists: AMPA > acromelic acid A > quisqualate > kainate > domoate. The ratio of the K_D values for domoate and AMPA was about 6. The simplest explanation of this uniformity of agonist rank order of sensitivity to such a variety of antagonists is that two receptors contribute to the responses, one receptor being more sensitive to each of the antagonists than the other, and activated to different extents by different agonists.

Since AMPA (as distinct from quisqualate) does not inhibit ^{3}H-kainate binding (Krogsgaard-Larsen et al., 1989), it seems likely that the responses induced by this isoxazole do not involve a component mediated by kainate receptors.   In this respect, AMPA is probably a selective agonist at a particular sub-type of "Q" receptor, and this sub-type should perhaps now be re-classified as the AMPA receptor in view of the several different types of receptor/ binding sites for which quisqualate, but not AMPA, has been shown to have moderate-high affinity (see ref.11).   The electrophysiological results described here suggest the possibility that domoate is likewise more selective for "K"-type receptors than is kainate itself.   However, it would appear that it is not totally specific since preliminary findings indicate that domoate inhibits ^{3}H-AMPA binding with a K_i of approximately 4.5 M (our unpublished observations), very similar to that of kainate[21]. Thus, part of the domoate response may also be mediated by AMPA receptors. However, the higher potency of domoate compared with kainate as an excitant, which is probably a reflection of their relative affinities for pure kainate (domoate) receptors (Agrawal and Evans, 1986 ; Foster and Fagg, 1984 ; Debonnel et al. 1989), would result in a lower proportion of AMPA receptors being activated by domoate, relative to kainate, for similar-amplitude responses at mixed AMPA/kainate receptors. This, in turn, would explain the greater resistance of domoate- than of kainate-induced responses to the action of the antagonists.

Table 1 also gives other useful information.   Thus the selectivity of DCH-QXCA for kainate observed in striatal slices (Frey et al., 1988) was not evident here in rat spinal cord; indeed, a two-fold selectivity in favour of quisqualate was observed.   Another finding was that, somewhat surprisingly, acromelic acid A, despite its kainate-like structure, is more AMPA- than kainate-like in this test system.   Thus it cannot be assumed that a kainate-like structure necessarily means a selective action at kainate (domoate) receptors.

Table 2.   Depression of AMPA, kainate and domoate-induced responses in cat spinal cord *in vivo* by CNQX

Antagonist	nA	Per cent Depression			
		AMPA	KAIN	QUIS	DOM
CNQX	0-10	74±6 (10)	70±7 (7)	35 (2)	32±5 (8)
pCB-PzDA	5-20	87±3 (6)	70±9 (4)	25 (2)	19±14 (5)

Micropipette solutions: CNQX 2 mM in 150 mM NaCl, pH 7

pCB-PzDA 200 mM (Na salt), pH 7

These results suggest that the use of AMPA and domoate as agonists for Q and K receptors, respectively, may be helpful in distinguishing the latter types of receptor from each other in electrophysiological experiments, especially *in vitro*.   In ionophoretic experiments *in vivo*, using extracellular recording, the use of quisqualate as a "Q" receptor agonist may be particularly misleading.   Thus, in cat spinal cord, the order of depression of agonist induced responses found in preliminary experiments using CNQX and *p*CB-PzDA as antagonists was AMPA > kainate > quisqualate > domoate (Table 2).   An example of the greater sensitivity to *p*CB-PzDA of responses to AMPA compared with those induced by domoate, with kainate responses having intermediate susceptibility, is shown in Fig 3.   These results support the view that the receptor involved in synaptic excitation that is depressed by such antagonists as used in the present work could well be of the Q(AMPA) rather than of the K(domoate) type even when kainate-induced and synaptic responses are clearly depressed in the absence of marked effects on quisqualate-induced responses (e.g. Davies and

Watkins, 1985).   The relative resistance of quisqualate-induced responses in this type of experiment perhaps reflects peculiarities in the dose-response relations of this agonist relative to AMPA, possibly due to uptake of quisqualate (Watkins and Evans, 1981 ; Davies et al., 1985)· Definitive classification of a receptor as a K-type will require the development of selective domoate antagonists, and/or the use of other criteria such as the very high domoate/kainate potency ratios seen in dorsal root C-fibre receptors (Agrawal and Evans, 1986) and in CA3 hippocampal neurones (Debonnel et al., 1989).

SUMMARY

1) The K_D values for a range of antagonists including DGG, *p*CB-PzDA, *p*BB-PzDA, HDC-QXCA, DNQX and CNQX have been determined using a series of non-NMDA receptor agonists in the isolated spinal cord of the neonatal rat.

2) CNQX and DNQX, and, to a lesser extent, DCH-QXCA, were by far the most potent antagonists, although the degree of selectivity did not vary much throughout the whole range of antagonists used.

3) AMPA and domoate were the most and least sensitive agonists, respectively, to the action of all the antagonists. Ionophoretic experiments in the cat spinal cord *in vivo* confirmed this order of susceptibility in the case of the antagonists CNQX and pCB-PzDA.

4) Acromelic acid A was a more AMPA-like than domoate-like agonist.

5) The results suggest that two receptors contribute to the responses induced by several of the agonists, and that quisqualate and kainate are less selective agonists at these receptors than are AMPA and domoate, respectively.

Acknowledgements

This work was supported by the Medical Research council and by US Public Health Service Grant NS 26540.

REFERENCES

Agrawal,S.G. and Evans,R.H, The primary afferent depolarizing action of kainate in the rat, **Br.J.Pharmac.** 87:345-355 (1986).

Birch,P.J.,Grossman,C.J. and Hayes,A.G, Kynurenate and FG 9041 have both competitive and non-competitive antagonist actions at excitatory amino acid receptors, **Eur.J.Pharmacol.**151:313-315 (1988).

Biscoe,T.J.,Evans,R.H.,Headley,P.M.,Martin,M. and Watkins,J.C, Structure-activity relations of excitatory amino acids on frog and rat spinal neurones, **Br.J.Pharmac.**58:373-382 (1976).

Blake,J.F.,Yates,R.G.,Brown,M.W. and Collingridge,G.L, 6-Cyano 7-nitroquinoxaline 2,3-dione as an excitatory amino acid antagonist in area CA1 of rat hippocampus, **Br.J.Pharmac.**97:71-76 (1989).

Collingridge,G.L. and Lester,R.A.J, Excitatory amino acid receptors in the vertebrate central nervous system, **Pharmacol. Revs.** (1989) in press.

Cotman,C.W.,Monaghan,D.T.,Ottersen,O.P. and Storm-Mathisen,J, Anatomical organization of excitatory amino acid receptors and their pathways, **TINS** 7:273-284 (1987).

Davies,J.,Evans,R.H.,Francis,A.A. and Watkins,J.C, Excitatory amino acid receptors and synaptic excitation in the mammalian central nervous system, **J.Physiol.**(Paris) 75:641-645 (1979).

Davies,J. and Watkins,J.C, Differentiation of kainate and quisqualate receptors in the cat spinal cord by selective antagonism with γ-D(and L)-glutamylglycine, **Brain Res.**206:172-177 (1981).

Davies,J. and Watkins,J.C, Depressant actions of γ-D-glutamyl-aminomethylsulfonate (GAMS) on amino acid-induced and synaptic excitation in the cat spinal cord, **Brain Res.** 327:113-120 (1985).

Davies,J. and Watkins,J.C, Selective antagonism of amino acid-induced and synaptic excitation in the cat spinal cord, **J.Physiol.(Lond.)** 297:621-636 (1979)

Davies,J.,Jones,A.W.,Sheardown,M.J.,Smith,D.A.S. and Watkins,J.C, Phosphonodipeptides and piperazine derivatives as antagonists of amino acid-induced and synaptic excitation in mammalian and amphibian spinal cord, **Neurosci.Lett.** 52:79-84 (1984).

Davies,J.,Francis,A.A.,Oakes,D.J.,Sheardown,M.J. and Watkins,J.C, Selective potentiating effects of β-*p*-chlorophenylglutamate on responses induced by certain sulphurcontaining excitatory amino acids and quisqualate, **Neuropharmacology** 24:177-180 (1985).

Debonnel,G.,Beauchesne,L. and De Montigny,C, Domoic acid, the alleged "mussel toxin," might produce its neurotoxic effect through kainate receptor activation: an electrophysiological study in the rat dorsal hippocampus, **Can.J.Physiol.Pharm.** 67:29-33 (1989).

Fletcher,E.J.,Martin,D.,Aram,J.A.,Lodge,D and Honoré,T, Quinoxalinediones selectively block quisqualate and kainate receptors and synaptic events in rat neocortex and hippocampus and frog spinal cord *in vitro*, **Br.J.Pharmac.**95:585-597 (1988)

Foster,A.C. and Fagg,G.E, Acidic amino acid binding sites in mammalian neuronal membranes: their characteristics and relationship to synaptic receptors. **Brain Res.Revs.** 7:103-164 (1984).

Francis,A.A.,Jones,A.W. and Watkins,J.C, Dipeptide antagonists of amino acid-induced and synaptic excitation in the frog spinal cord, **J.Neurochem.**35:1458-1460 (1980).

Frey,P.,Berney,,D.,Herrling,P.L.,Muelle,W. and Urwyler,S, 6-7-Dichloro-3-hydroxy-2 quinoxaline carboxylic acid is a relatively potent antagonist at NMDA and kainate receptors, **Neurosci.Lett.** 91:194-198 (1988).

Ganong,A.H.,Lanthorn,T.H. and Cotman,C.W, Kynurenic acid inhibits synaptic and acidic amino acid-induced responses in the rat hippocampus and spinal cord, **Brain Res.** 273:170-174 (1983).

Honore,T.,Davies,S.N.,Drejer,J.,Fletcher,E.J.,Jacobsen,P., Lodge,D. and Nielsen,F.E, Quinoxalinediones: potent competitive non-NMDA glutamate receptor antagonists, **Science** 241:701-703.

Honoré,T.,Lauridsen,J. and Krogsgaard-Larsen,P, The binding of [^{3}H]AMPA, a structural analogue of glutamic acid, to rat brain membranes, **J.Neurochem.** 38:173-178 (1982).

Ishida,M. and Shinozaki,H, Acromelic acid is a much more potent excitant than kainic acid or domoic acid in the isolated rat spinal cord, **Brain Res.**474:386-389 (1988).

Jones,A.W.,Smith,D.A.S. and Watkins,J.C, Structure-activity relations of dipeptide antagonists of excitatory amino acids, **Neuroscience** 13: 573-581 (1984).

Krogsgaard-Larsen,P.,Honoré,T.,Hansen,J.J.,Curtis,D.R. and Lodge,D, New class of glutamate agonists structurally related to ibotenic acid, **Nature** 284:64-66 (1980).

Mayer,M.L. and Westbrook,G.L, The physiology of excitatory amino acids in the vertebrate central nervous system, **Progr.Neurobiol.** 28:197-296 (1987).

McLennan,H. and Lodge,D, The antagonism of amino acid-induced excitation of spinal neurones in the cat. **Brain Res.** 169:83-90 (1979).

Monaghan,D.T.,Bridges,R.J. and Cotman,C.W, The excitatory amino acid receptors: their classes, pharmacology and distinct properties in the function of the central nervous system. **Ann.Rev.Pharmacol.Toxicol.** 29:365-402 (1989).

Perkins,M.N. and Stone,T.W, An ionophoretic investigation of the actions of convulsant kynurenines and their interaction with the endogenous excitant quinolinic acid, **Brain Res.** 247:184-187 (1982).

Shinozaki,H, Discovery of novel actions of kainic acid and related compounds, in "Kainic acid as a Tool in Neurobiology", E.G.McGeer,J.W.Olney and P.L.McGeer, eds.,Raven Press, New York (1978) pp 17-35.

Watkins,J.C. Excitatory amino acids, in "Kainic acid as a Tool in Neurobiology," E.G.McGeer, J.W. Olney and P.L.McGeer, eds., Raven Press, New York (1978) pp 37-69.

Watkins,J.C. and Evans,R.H, Excitatory amino acid transmitters, **Ann. Rev. Pharmacol. Toxicol.** 21:165-204 (1981).

HOMOCYSTEIC ACID AS TRANSMITTER CANDIDATE IN THE MAMMALIAN BRAIN

AND EXCITATORY AMINO ACIDS IN EPILEPSY

M. Cuénod[1], E. Audinat[1], K.Q. Do[1], B.H. Gähwiler[1],
P. Grandes[1], P. Herrling[3], T. Knöpfel[1], H. Perschak[2],
P. Streit[1], F. Vollenweider[1] and H.G. Wieser[2]

[1] Brain Research Institute, Zürich University
[2] Neurology Department, Zürich University and
[3] Sandoz Research Institute, Bern, Switzerland

In this paper, we will (I) review the available evidence suggesting that homocysteic acid
might be an excitatory transmitter in the mammalian central nervous system (CNS) and (II)
report preliminary results on extracellular changes of excitatory amino acid (EAA) concentration
in human epileptic foci.

I. Homocysteic acid as an excitatory transmitter

Homocysteic acid (HCA) has long been known as a strong excitant when exogenously applied
to neurons (Curtis and Watkins, 1963). It has a chemical structure analogous to glutamic acid, in
which the gamma-carboxyl group is substituted by a sulfonyl group. Recent evidence led to the
suggestion that HCA might be an excitatory neurotransmitter on the basis of the following findings: Endogenous HCA is released by depolarization of brain slices and it activates EAA receptors, preferentially the N- methyl-D-aspartate (NMDA) type wherever present.

(a) HCA release

Rat brain slices exposed to 50 mM K^+ released endogenous HCA as shown by Do et al (1986a),
who analyzed the tissue superfusate by HPLC following derivatization with 4-N,N-dimethyla-
mino-azobenzene-4'-isothiocyanate. This release was to a large extent Ca^{++}-dependent and was
present in all CNS regions investigated. It was most prominent in the cortex, where it amounted to
0.6 ± 0.1 pmol/mg protein/min, or 7.3 times the resting efflux (Do et al., 1986a). These observations
are compatible with the idea that HCA is present in and released from a neuronal compartment,
possibly a synaptic one.

(b) L-HCA as NMDA receptor agonist

L-HCA, before it was known to be an endogenous compound, was shown to act preferentially
on the NMDA receptor (Mewett et al., 1983; Meyer and Westbrook 1984). L-HCA , microiontophore-
tically applied to cat caudate neurons in vivo, induced a depolarization and firing pattern similar
to that elicited by NMDA. This effect could be blocked by the selective NMDA- antagonist D-2-
amino-7-phosphonoheptanoic acid (AP-7) (Do et al, 1986b). Moreover, L-HCA, as NMDA, induced

Excitatory Amino Acids and Neuronal Plasticity
Edited by Y. Ben-Ari
Plenum Press, New York, 1990

[³H]acetylcholine release from rat striatal slices, an effect which could be antagonized by AP-7 (Tsai et al, 1987). In neurons of rat neocortical slices, L-HCA had also been shown to act as NMDA receptor preferring agonist (Knöpfel et al, 1987, Thomson, 1989). In the rabbit retina, Neal and Cunningham (1989) presented some evidence for a role of L-HCA in the neurotransmission from the bipolar to the cholinergic amacrine cells: both light- and HCA-evoked release of acetylcholine were antagonized in a similar manner by substances interacting with NMDA receptors. Recently, Jones and Sillito (1989a and b) reported that, in the cat dorso-lateral geniculate nucleus, visual stimuli and iontophoretically applied L-HCA induced similar neuronal responses, both blocked by NMDA receptor antagonists.

(c) HCA induced neurotoxicity

High concentration of EAA induced specific patterns of cell damage in the CNS, a toxicity mediated by their receptors. L- HCA neurotoxicity in cortical neurons was blocked by another selective NMDA antagonist DL-2-amino-5-phosphonovalerate (AP-5) (Kim et al, 1987) and in chick retina by AP-7 (Olney et al, 1987). Moreover, the morphological pattern of cellular degeneration produced by L-HCA in chick retina was more similar to that obtained with NMDA than with glutamate or kainate (Olney et al, 1987; Pullan et al, 1987).

(d) Cerebellar HCA release and uptake after climbing fibers deprivation

In slices of the rat cerebellum, a small but significant K+- induced release of HCA was observed, amounting to 1.5-2.2 times the resting efflux. This release was completely abolished in cerebella deprived of climbing fibers by 3-acetyl-pyridine (3- AP) treatment, an observation suggesting that HCA might be released from climbing fibers (Vollenweider et al., 1990). This is the first evidence that HCA could be linked to a specific pathway.

Furthermore, there is circumstantial evidence for a specific HCA uptake system (Cox et al., 1977; Zeise et al, 1988; Tsai et al.,1989). In climbing fiber deprived cerebellum, the HCA uptake was reduced by about 50% (Lehmann, personal communication), an observation compatible with the hypothesis mentioned above.

(e) HCA immunohistochemical localization in cerebellar cortex

At the present time, immunohistochemistry is the method of choice to study the localization and distribution of endogenous neuroactive substances. The method pioneered by Steinbusch et al. (1978) for serotonin and applied to GABA and glutamate by Storm-Mathisen et al. (1983) has been developed for HCA. Highly specific polyclonal antibodies were obtained with glutaraldehyde-linked HCA-albumin conjugates and were tested by the technique introduced by Ottersen (1987). Immunohistochemistry was performed on semithin sections of perfusion-fixed rat cerebellar tissue by means of a silver-intensified peroxidase- antiperoxidase method (Liu et al., 1989).

The most conspicuous elements containing HCA-like immunoreactivity were, on the one hand, terminal-like dots and beaded fibrous profiles in close association with Purkinje cell perikarya and dendrites and, on the other, thin radial profiles extending throughout the molecular layer up to the pial surface. Moreover, the surroundings of capillaries in cerebellar cortex was occasionally immunoreactive (Grandes et al., 1989). This pattern was completely different from those observed with antibodies against glutamate or against GABA. Such a distribution is compatible with a localization of HCA in climbing fibers and/or in Bergmann glia, but the definitive definition of the immunoreactive elements has to be achieved at the electronmicroscopical level. The presence of HCA-like immunoreactivity in Bergmann glia could indicate that HCA, after having been released from a synaptic compartment, is then taken up by these glial cells, strategically located in close apposition to the Purkinje cells. Alternatively, the Bergmann glia might constitute the main compartment containing HCA and contribute to the HCA release observed.

(f) Purkinje cell response to HCA in cerebellar slice cultures

The finding that HCA might be present in climbing fibers of the rat cerebellum and released from their terminals raises the question of the receptor(s) involved in the mediation of the complex spikes. The participation of a classical NMDA receptor can be excluded, as NMDA does not directly depolarize mature Purkinje cells (Crepel et al., 1982; Kimura et al., 1985a, 1985b; Llano et al., 1988). However, the response of Purkinje cells to aspartate, another transmitter candidate in

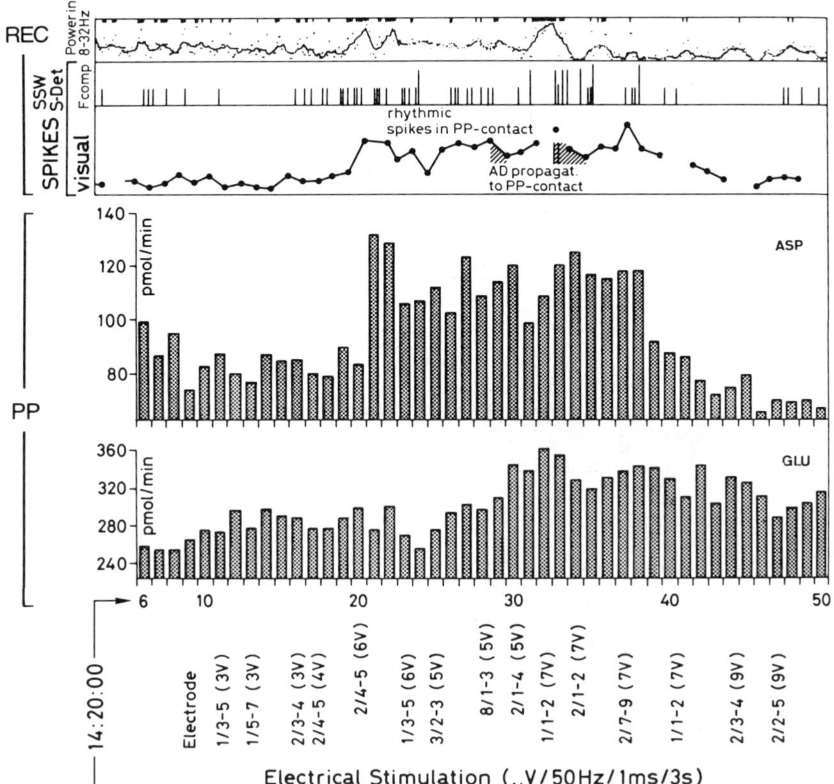

Fig. 1   Correlation of spike discharges and extracellular excitatory amino acid level in hippocampal focus of an epileptic patient over a 50 min period. The recording electrode (REC) and the push-pull cannula (PP) were located in the right hippocampus of patient BR, preoperatively. Power in 8-32 Hz: computer-assisted analysis of the depth EEG, showing the mean relative spectral power density in the frequency band 8-32 Hz. Spikes SSW, S-Det: frequency of spontaneous as well as electrically evoked epileptiform events detected either by a spike detection algorithm (Fcomp = Fricker's spike detection; the number of spikes is expressed in vertical bars each 8 sec; longest bars equal 5) spikes or by visual examination (visual); afterdischarge (AD) propagation was observed at times 29 and 33. The level of aspartate (ASP) and glutamate (GLU) in superfusate of consecutive one min fractions are given in pmol. Electrical stimulation (stimulus intensities are indicated in V) was applied with electrodes placed within the temporal lobe (electrode 2 is at the site of the push-pull cannula); frequency: 50 Hz; pulse duration: 1ms; train duration: 3s. Note the temporal correlation between the increased frequency of spikes and the elevation of aspartate levels in the superfusate and the glutamate level increase following the afterdischarge.

climbing fibers (Nadi et al.,1977; Toggenburger et al., 1983; Wiklund et al., 1984; Kimura et al., 1985c; Vollenweider et al., 1990), was blocked by APV and NMDA (Crepel et al., 1983; Dupont et al., 1987), while that of glutamate was little affected Furthermore, in cerebellar slices of the guinea pig, the climbing fiber response was blocked by both APV and NMDA (Kimura et al., 1985b).

In an attempt to gain more information on this question, the action and pharmocology of $_L$-HCA was compared with that of the climbing fibre response in co-cultured slices of cerbellum and inferior olive. In these cultures obtained from newborn rats and maintained in culture for up to five weeks (Gähwiler, 1981), Purkinje cells were identified by their morphology and location and by their immunoreactivity with the 28 kD calcium binding protein. Purkinje cells were single-electrode voltage- clamped at a potential of -55 to -65 mV and $_L$-HCA (500 μM), NMDA (50 μM to 1 mM) and 6-Cyano-7-nitroquinoxaline-2,3-diodine (CNQX, 2-10 μM) were superfused in Hanks, balanced salt solution containing 1 μM TTX. $_L$-HCA induced an inward current which was fully antagonized by the non-NMDA EAA receptor antagonist CNQX. As in adult slices, NMDA had no direct effect on Purkinje cells (Knöpfel et al., 1989). Climbing fibre responses, induced by electrical stimulation within the co- cultured inverior olive, were abolished by CNQX but were unaffected by the NMDA receptor antagonist D-APV (Audinat et al., 1989). These observations support the hypothesis that the excitation by $_L$-HCA as well as by the transmitter released by the climbing-fibre is mediated by non-NMDA receptors in Purkinje cells.

In conclusion, a Ca^{++}-dependent, K^+-induced release of HCA has been observed in various brain structures. In the cerebellar cortex, HCA is localized mostly in climbing fibers. Taken together, the results reviewed support the proposal that HCA is a transmitter acting preferentially on NMDA receptors in rat caudate and cortex and on non-NMDA receptors in Purkinje cell synapses.

II. Aspartate efflux from epileptic foci

The pathophysiology of epileptic discharges could involve either a deficit of inhibition or an excess of excitation, or both (Meldrum, 1987, Dingeldine et al., 1988). The information available documents changes in GABA and EAA transmission. Three lines of evidence suggest that EAA receptor activation contributes to seizure generation in experimental models of epilepsy. First, the extracellular accumulation of EAA receptor agonists in nervous tissue, particularly in the hippocampus, triggers focal seizures under some conditions (Fukuda et al., 1985; Anderson et al., 1987; Fischer and Alger, 1984; Westbrook and Lothman, 1983). Second, antagonists to EAA, particularly to NMDA receptors at least partly suppress epileptiform activity in experimental models in vivo and in vitro (Croucher et al., 1982; Meldrum et al., 1983; De Sarro et al., 1984; Chapman et al., 1986; Anderson et al., 1987; Sagratella et al., 1987). Third, during kindling of limbic or cortical seizures in rats, an increased release of glutamate and aspartate has been observed (Leach et al., 1985) as well as changes in excitatory receptors (Savage et al., 1984; Mc Namara et al., in press). Furthermore, an increase in CSF aspartate has been reported in adult patients with generalized seizures (Engelsen and Elsayed, 1984).

However, these results originate from measurements made in animal models, in plasma or cerebrospinal fluid of patients (usually collected outside the seizure period) or in biopsied or autopsied pathological brain tissue. To our knowledge, neurotransmitter changes in extracellular fluid of foci during epileptiform events have not been reported. The preliminary results of such an investigation, approved by the local ethics committee, will be briefly reported here. No damage was endured by the patients, who gave their informed written consent. In the course of stereo-EEG evaluation of candidates for epilepsy surgery (Wieser, 1987, 1988) a push-pull cannula was introduced into the lumen of a standard hollow-core multi-contact depth electrode placed in the hippocampus, which will be surgically resected at a later stage. This technique allows to perfuse the epileptic tissue at the tip of the electrode while recording simultaneously the local potentials. Sterilized Gey's solution was perfused through the push-pull cannula at a flow rate of 20μl/min and one minute fractions were collected over a period of one hour in patients under general anaesthesia. The perfusate was analyzed by o-phtalaldehyde precolumn derivatization HPLC to quantify putative amino acid transmitters. A computer-assisted analysis of the background depth-EEG activity and of spontaneous as well as electrically provoked epileptiform events were performed, and the results correlated with the biochemical measurements.

In three out of four patients, an increase in aspartate levels (approximately from 40 to 70 pmol/min) was observed in correlation with the epileptiform EEG events. Their correlation with changes in glutamate concentration was less clear, although there is possibly an elevation related to afterdischarges. Non-transmitter amino acid levels remained constant and unaffected by the epileptic discharges. These very preliminary observations show that extracellular changes in amino acid concentrations can be detected in chronic foci during epileptiform events as compared to the resting state. They suggest that the efflux of aspartate, probably reflecting the balance between release and uptake, is increased during epileptiform events. Of course, for ethical reasons it will be very difficult to obtain control values from non-epileptic brain tissue. Such studies, when completed, might throw light onto the pathophysiological events taking place in an epileptic focus and possibly contribute to the distinction of different types of disease processes. In analogy to EEG recordings, biochemical measurements might become part of the diagnostic and prognostic assessment of the patient.

AKNOWLEDGMENT

The authors wish to thank L. Heeb, Z.P. Jiang, L. Rietschin, B. Stierli and M. Zuber, for their expert technical assistance. This ivestigation was supported by grants of the Swiss National Foundation (3.389.86).

REFERENCES

Anderson W.W., Schwartzwelder H.S. and Wilson W.A., 1987, The NMDA receptor antagonist 2-amino-5-phosphonovalerate blocks stimulus train-induced epileptogenesis but not epileptiform bursting in the rat hippocampal slice, **J. Neurophysiol.**, 57:1- 21.

Audinat E., Knöpfel T. and Gähwiler B.H., 1989, Pharmacological characterization of excitatory postsynaptic potentials of Purkinje cells in organotypic co-cultures of cerebellum and inferior olive, Soc. Neurosci. Abstr. 15:612.

Chapman A.G., Faingold C.L., Hart G.P., Bowker H.M. and Meldrum B.S., 1986, Brain regional amino acid levels in seizure suspectible rats: changes related to sound-induced seizures, **Neurochem. Int.**, 8:273-279.

Cox D.W.G., Headley M.H. and Watkins J.C., 1977, Actions of L- and D-homocysteate in rat CNS: a correlation between low- affinity uptake and the time courses of excitation by microelectrophoretically applied L-glutamate analogues, **J. Neurochem.** 29:579-588.

Crepel F., Dhanjal S.S. and Sears T.A., 1982, Effect of glutamate, aspartate and related derivatives on cerebellar Purkinje cell dendrites in the rat: an in vitro study, **J. Physiol. (Lond.)** 329:297-317.

Crepel F., Dupont J.J. and Gardette R., 1983, Voltage clamp analysis of the effect of excitatory amino acids and derivatives on Purkinje cell dendrites in rat cerebellar slices maintained in vitro, **Brain Res.** 279:311-315.

Croucher M.J., Collins J.F. and Meldrum B.S., 1982, Anticonvulsant action of excitatory amino acid antagonists, **Science**, 216:899.

Curtis D.R., and Watkins J.C., 1963, Acidic amino acids with strong excitatory actions on mammalian neurones, **J. Physiol. (Lond.)** 166:1-14.

De Sarro G.B., Meldrum B.S. and Reavill C., 1884, Anticonvulsant action of 2-amino-7-phosphono- heptanoic acid in the substantia nigra, **Eur. J. Pharmacol.**, 106:175.

Dingledine R., Boland L.M., Chamberlin N.L., Kawasaki K., Kleckner N.W., Traynelis S.F. and Verdoorn T.A., 1988, Amino acid receptors and uptake systems in the mammalian central nervous system, **Critical Reviews in Neurobiol.** 4:1-96

Do K.Q., Mattenberger M., Streit P. and Cuénod M., 1986a, In vitro release of endogenous excitatory sulfur-containing amino acids from various rat brain regions, **J. Neurochem.**, 46:779- 786.

Do K.Q., Herrling P.L., Streit P., Turski W.A. and Cuénod M., 1986b, In vitro release and electrophysiological effects in situ of homocysteic acid, an endogenous N-Methyl-(D)-aspartic acid agonist, in the mammalian striatum, **J. Neurosci.**, 6:2226-2234.

Dupont J.L., Gardette R., Crepel F., 1987, Postnatal development of the chemosensitivity of rat cerebellar Purkinje cells to excitatory amino acids. An in vitro study., **Developmental Brain Res.** 34:59-68.

Engelsen B. and Elsayed S., 1984, Increased concentrations of aspartic acid in the cerebrospinal fluid of patients with epilepsy and trigeminal neuralgia: an effect of medication ?, **Acta Neurol. Scand.**, 69:70-76.

Fisher R.S. and Alger B.E., 1984, Electrophysiological mechanisms of kainic acid-induced epileptiform activity in the rat hippocampal slice, **J. Neurosci.**, 4:1312-1323.

Fukuda H., Tanaka T., Kaijima M., Nakai H. and Yonemasu Y., 1985, Quisqualic acid-induced hippocampal seizures in unanesthetized cats, **Neurosci. Lett.**, 59:53-59.

Gähwiler B.H., 1981, Organotypic monolayer cultures of nervous tissue, **J. Neurosci. Meth.**, 4:329-342.

Grandes P., Cuénod M. and Streit P., 1989, Localization of homocysteate-like immunoreactivity in cerebellar climbing fibers, **Eur. J. Neurosci.**, Suppl. 2:18.

Jones H.E. and Sillito A.M., 1989a, Effects of the putative neurotransmitters N-acetylaspartylglutamate (NAAG) and L-homocysteate (L-HCA) on cells in the anaesthetized feline dorsal lateral geniculate nucleus (dLGN), **J. Physiol.**, Oxford Physiol. Soc. Meeting, 30 June - 1 July:97P.

Jones H.E. and Sillito A.M., 1989b, The pharmacology and action of the putative retinal neurotransmitters NAAG and L- homocysteate on cat DLGN cells, **Eur. J. Neurosci.**, Suppl. 2:109.

Kim J.P., Koh J.-Y. and Choi D.W., 1987, L-Homocysteate is a potent neurotoxin on cultured cortical neurons, **Brain Res.**, 437:103-110.

Kimura H., Okamoto K. and Sakai Y., 1985a, Climbing and parallel fiber responses recoreded intracellularly from Purkinje cell dendrites in guinea pig cerebellar slices, **Brain Res.** 348:213-219.

Kimura H., Okamoto K. and Sakai Y., 1985b, Parmacological characterization of postsynaptic receptors for excitatory amino acids in Purkinje cell dendrites in the guinea pig cerebellum. **J. Pharmacobio.-Dyn.** 8:119-127.

Kimura H., Okamoto K. and Sakai Y., 1985c, Pharmacological evidence for L-aspartate as the neurotransmitter of cerebellar climbing fibers in the guinea pig, **J. Physiol. (Lond.)** 365:103- 119.

Knöpfel T., Zeise M.L., Cuénod M. and Zieglgänsberger W., 1987, L-Homocysteic acid but not L-glutamate is an endogenous N- methyl-D-aspartic acid receptor preferring agonist in rat neocortical neurons in vitro, **Neurosci. Lett.**, 81:188-192.

Knöpfel T, Audinat E., Staub C. and Gähwiler B.H., 1989, Excitatory amino acid receptors of purkinje cells and neurons of the deep nuclei in cerebellar slice cultures, **Europ. J. of Neurosci.**, Suppl. 2: 108.

Llano I, Mary A., Johnson J.W., Ascher P. and Gähwiler B.H., 1988, Patch-clamp recording of amino acid-activated responses in "organotypic" slice cultures, **Proc. Natl. Acad. Sci.**, USA, 85:3221-3225.

Leach M.J., Marden C.M., Miller A.A., O'Donnell R.A. and Weston S.B., 1985, Changes in cortical amino acids during electrical kindling in rats, **Neuropharmacol.**, 24:937-940.

Liu C.-j., Grandes P., Matute C., Cuénod M. and Streit P., 1989, Glutamate-like immunoreactivity revealed in rat olfactory bulb, hippocampus and cerebellum by monoclonal antibody and sensitive staining method. **Histochem.**, 90:427-445.

Mayer M.L., and Westbrook G.L., 1984, Mixed-agonist action of excitatory amino acids on mouse spinal cord neurones under voltage clamp, **J.Physiol. (Lond.)**, 354:29-53.

McNamara J.O., Yeh G., Bonhaus D.W., Okazaki M. and Nadler J.V., NMDA receptor plasticity in the kindling model, **In:** Excitatory Amino Acids and Neuronal Plasticity, Plenum Press, in press

Meldrum B.S., Croucher M.J., Badman G. and Collins J.F., 1983, Antieptileptic action of excitatory amino acid antagonists in the photosensitive baboon, Papio papio, **Neurosci. Lett.**, 39:101-104.

Meldrum B., 1987, Neurotransmitter amino acids in epilepsy, **In:** The London Symposia (EEG Suppl. 39) R.J. Ellingson, N.M.F. Murray and A.M. Halliday Eds, Elsevier Sci. Publ., 191-199.

Mewett K.N., Oakes D.J., Olverman H.J., Smith D.A.S. and Watkins J., 1983, Pharmacology of the excitatory actions of sulphonic and sulphinic amino acids. **In:** CNS Receptors - From Molecular Pharmacology to Behavior, P. Mandel and F.V. DeFeudis, eds, 163-174, Raven Press, New York

Nadi N.S., Kanter D., McBridge W.J. and Aprison M.H., 1977, Effects of 3-acetylpyridine on several putative neurotransmitter amino acids in the cerebellum and medulla of the rat, **J. Neurochem.**, 28:661-662

Neal M.J. and Cunningham J.R., 1989, L-homocysteic acid- a possible bipolar cell transmitter in the rabbit retina, **Neurosci. Lett.**, 102:114-119.

Olney J.W., Price M.T., Salles K.S., Labruyere J., Ryerson R., Mahan K., Frierdich G. and Samson L., 1987, L-Homocysteic acid: An endogenous excitotoxic ligand of the NMDA receptor, **Brain Res. Bull.**, 19:597-602.

Ottersen O.P., 1987, Postembedding light- and electron microscopic immunocytochemistry of amino acids: description of a new model system allowing identical conditions for specificity testing and tissue processing, **Exp. Brain Res.** 69:167-174.

Pullan L.M., Olney J.W., Price M.T., Compton R.P., Hood W.F., Michel J. and Monahan J.B., 1987, Excitatory amino acid receptor potency and subclass specificity of sulfur-containing amino acids, **J. Neurochem.**, 49:1301-1307.

Sagratella S., Frank C. and de Carolis A.S., 1987, Effects of ketamine and (+)cyclazocine on 4-aminopyridine and "magnesium free" epileptogenic activity in hippocampal slices of rats, **Neuropharmacol.**, 26:1181.

Savage D.D., Nadler J.V. and McNamara J.O., 1984, Reduced kainic acid binding in the rat hippocampal formation after limbic kindling, **Brain Res.**, 323:128-131.

Steinbusch H.W.M., Verhofstad A.A. and Joosten H.W.J., 1978, Localization of serotonin in the central nervous system by immunohistochemistry: description of a specific and sensitive technique and some applications. **Neuroscience,** 3:811-819.

Storm-Mathisen J., Leknes A.K., Bore A.T., Vaaland J.L., Edminson P., Haug F.M.S. and Ottersen O.P, 1983, First visualization of glutamate and GABA in neurons by immunocytochemistry, **Nature**, 301:517-520.

Thomson A.M., 1989, Glycine modulation of the NMDA receptor/channel complex, **TINS** 12:349-353.

Toggenburger G., Wiklund L., Henke H. and Cuénod M., 1983, Release of endogenous and accumulated exogenous amino acids from slices of normal and climbing fibre-deprived rat cerebellar slices, **J. Neurochem.** 41:1606-1613.

Tsai C., Wood P.L and Lehmann J., 1987, Homocysteate as a neurotransmitter candidate in the brain - presynaptic and postsynaptic characteristics, **Soc. Neurosci. Abstr.** 13:210.

Tsai C., Wood P.L. and Lehmann J., Homocysteic acid as a putative excitatory amino acid neurotransmitter. II. Presynaptic Characteristics of homocysteic acid uptake, **J. Neurochem.**, in press.

Vollenweider F.X., Do K.Q. and Cuénod M.: Effect of climbing fiber deprivation on release of endogenous aspartate, glutamate and homocysteate in slices of rat cerebellar hemispheres and vermis, **J. Neurochem.**, in press.

Westbrook G.L. and Lothman E.W., 1983, Cellular and synaptic basis of kainic acid-induced hippocampal epileptiform activity, **Brain Res.**, 273:97-109.

Wieser H.G., 1987, Stereo-Electroencephalography, **In:** Wieser HG and Elger CE (eds), Presurgical evaluation of epileptics, Springer-Verlag, Berlin, pp 192-204.

Wieser H.G., 1988, Selective amygdalo-hippocampectomy for temporal lobe epilepsy, **Epilepsia**, 29:100-113.

Wiklund L., Toggenburger G. and Cuénod M., 1984, Selective retrograde labelling of the rat olivocerebellar climbing fiber system with [^{3}H]-D-aspartate, **Neuroscience** 13:441-468.

Zeise M.L., Knöpfel T. and Ziegelgänsberger W., 1988, (±)- Parachlorophenylglutamate selectively enhances the depolarizing response to L-homocysteic acid in neocortical neurons of the rat: evicence for a specific uptake system, **Brain Res.**, 443:373-376.

SPECIFIC QUISQUALATE RECEPTOR LIGAND BLOCKS BOTH KAINATE AND QUISQUALATE RESPONSES

Tage Honore, Malcom Sheardown, Elsebet Ø. Nielsen, Jørgen Drejer,
Anker J. Hansen

Ferrosan CNS division, DK-2860 Søborg, DENMARK

INTRODUCTION

Excitatory amino acid (EAA) receptors were originally divided in NMDA receptors and non-NMDA receptors based on electrophysiological evidence [1]. The non-NMDA subtype of excitatory amino acid receptors can be further subdivided in two types, the kainate and the quisqualate (AMPA[*]) receptors. GDEE discriminate between the two agonists, being more active against quisqualate induced responses, when administered ionophoretically in vivo; however, this selectivity was not seen with bath application in vitro[1,6,7]. Another compound GAMS, although weak, show a reasonable degree of non-NMDA receptor antagonism[8]. Further development of selective kainate and/or AMPA receptor antagonists has been relatively slow.

Quinoxalinediones as non-NMDA antagonists

Recently a novel group of highly potent kainate and AMPA antagonists has been described, the quinoxalinediones[9]. DNQX and CNQX bind selectively to non-NMDA receptors as defined by inhibition of ^{3}H-AMPA and ^{3}H-kainate binding to rat cortical membranes. DNQX and CNQX are potent antagonists of electrophysiological responses mediated by non-MDA receptors[10-13]. Effects at NMDA receptors are much weaker. The latter are non-competitively with respect to NMDA receptors, acting instead via the modulatory glycine site of NMDA receptors[14,15]. Table 1 shows the selectivety of quinoxalinediones as inhibitors of binding to different EEA sites.

Complex structure of non-NMDA receptors

^{3}H-AMPA has been widely used for characterization of AMPA receptors. Molecular target size analysis using high energy irradiation technique has shown that high affinity AMPA receptors have a size of 51,600 daltons and are coupled to a membrane bound modulatory macromolecule of 128,400 daltons[16]. Scatchard plots of ^{3}H-AMPA binding (Fig. 1) revealed a high and a low affinity site which may be two different conformations of the same macromolecule[17]. Binding of ^{3}H-CNQX to a 51,800 dalton site revealed linear scatchard plots (Fig. 1), but biphasic inhibition

---

The quisqualate receptor is hereafter referred to as the AMPA receptor. AMPA is more specific in binding studies in that it has little affinity for ^{3}H-kainate binding site, andproduces responses in electrophysiological studies which, like those of quisqualate are GDEE-sensitive[2,3]. Quisqualate, besides interacting with the AMPA receptors in the mammalian CNS, also interact with excitatory amino acid receptors coupled to IP$_3$ turnover[4], and receptors in the invertebrate neuromuscular junction[5].

Excitatory Amino Acids and Neuronal Plasticity
Edited by Y. Ben-Ari
Plenum Press, New York, 1990

Table 1 Inhibition of binding to rat cortical membranes

IC_{50} μM ± SEM of three experiments[9,21]

Compound	(3H)AMPA	(3H)kainate	(3H)CPP	(3H)glycine
CNQX	0.30 0.15	1.5 ± 0.30	25	14
DNQX	0.5 ± 0.10	2.0 ± 0.10	40	9.5
CPP	450	-	0.50 ± 0.025	-
Glycine	-	-	-	0.16
AMPA	0.045 ± 0.005	45	>100	-
Kainate	20	0.0045 ± 0.001	>100	-
Quisqualate	0.025 ± 0.005	0.2 ± 0.01	40	-

curves using AMPA as inhibitor of the binding (table 2)[18,19]. These results substantiate the idea that the high and low affinity AMPA sites hare the same molecular entity, as both the high and low affinity AMPA sites have the same molecular target size. Furthermore scatchard plots of [3H]-AMPA binding and [3H]-CNQX binding revealed the same density of binding sites (Fig. 1).

The weak affinity for [3H]-glycine sites is for any practical reasons not important in functional experiments due to the presence of endogenous glycine or can be ruled out by addition of glycine to the bath media.

Similar to [3H]-AMPA binding, [3H]-kainate binding revealed biphasic scatchard plots[20]. However in this case the sites appear different; binding to the high affinity site is inhibited by

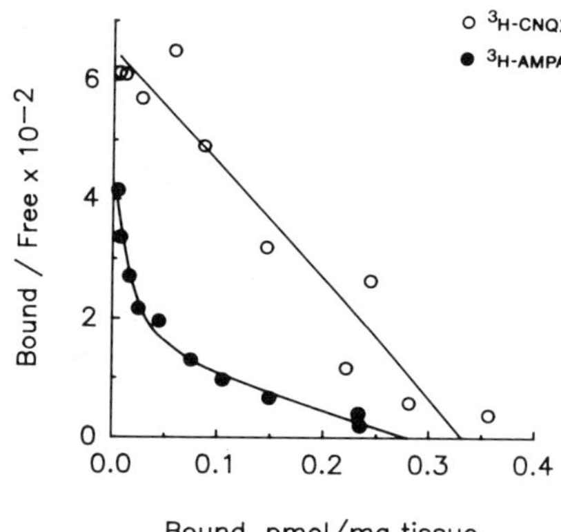

Fig. 1.   Scatchard plot of [3H]-AMPA and [3H]-CNQX binding to rat cortical membranes[21]

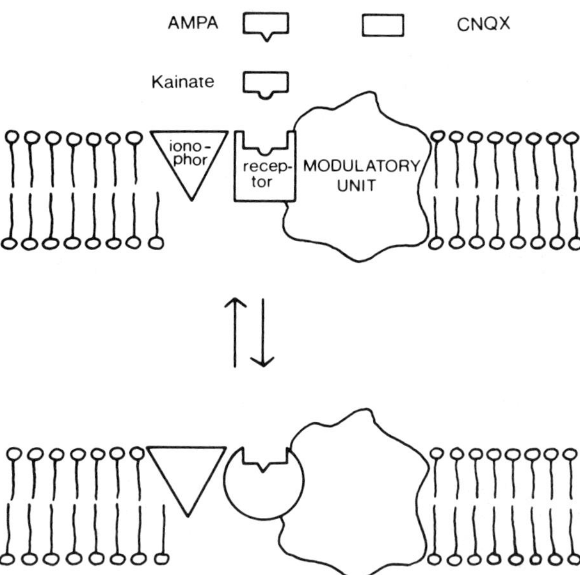

Fig. 2. Hypothetical model of the AMPA receptor.

calcium ions. Furthermore, the high and low affinity sites have different molecular target size, 76,600 and 52,400 daltons, respectively [20]. The similarity in the target site of the AMPA site and the low affinity kainate suggest that low affinity kainate sites are identical to AMPA sites.

When tested as an inhibitor of ^{3}H-CNQX binding, kainate, similar to AMPA, showed biphasic inhibition curves, however, kainate has the highest affinity (12 µM) to the conformation for which AMPA has the lowest affinity (22µM)[18] (table 2). The Ki's of quisqualate are 0.21 and 5.9 µM for the high and low affinity AMPA site, respectively [18]. Based on the above mentioned results, a model for the AMPA receptor is proposed (Fig. 2). The AMPA receptor subtype of EAA-receptors is coupled to an ionophor and a modulatory unit, which down regulates binding of agonist without effects on the binding of antagonists. The binding protein exists in two conformations, one with high affinity for AMPA and quisqualate and low affinity for kainate and one with higher affinity for kainate and low affinity for AMPA and quisqualate. Antagonists bind with equal affinity to the two conformations.

Table 2. Inhibition of 3H-CNQX binding to rat cortical membranes[18]

	Ki (µM ± SEM of n experiments)	
	Conformation A	Conformation B
CNQX	0.039† ± 0.002(3)	0.039† ± 0.002(3)
AMPA	0.3 ± 0.2(4)	22 ± 9(4)
Kainate	240 ± 70(3)	12 ± 4(3)

† KD from scatchard analysis

Table 3 Comparaison of the potency of NBQX and CNQX as inhibitors of binding to different EAA receptors in rat cortical membranes[21]

| | IC$_{50}$ (μM $\pm$ SEM of (n) experiments) | |
	NBQX	CNQX
^{3}H-AMPA	0.15 $\pm$ 0.01(3)	0.30 $\pm$ 0.15(3)
^{3}H-kainate	4.8 $\pm$ 0.5(3)	1.5 $\pm$ 0.3(3)
^{3}H-CNQX	0.02(2)	0.05 (2)
^{3}H-CPP	> 90(2)	25 (2)
^{3}H-glycine	> 100 (2)	14 (2)
^{3}H-TCP	>50 (2)	25 (2)

6-nitro-7-sulphamoyl-benzo(f)quinoxaline-2,3-dione (nbqx) a highly selective AMPA receptor ligand

NBQX (Fig. 3) is a novel analogue from quinoxalinedione type of non-NMDA receptor antagonist. NBQX is more selective for AMPA receptors as compared to CNQX (table 3)[21]. Thus, NBQX has 30-fold higher affinity for ^{3}H-AMPA binding than for high affinity ^{3}H-kainate binding, whereas CNQX has only 5-fold difference.

NBQX is the most potent inhibitor of ^{3}H-CNQX binding (IC$_{50}$ = 20 nM) and has very low affinity for ^{3}H-CPP, 3H-glycine and ^{3}H-TCP binding, indicating no interaction with the NMDA receptor complex.

Due to similarities in the responses in mammalian CNS tissue after stimulation with either quisqualate or kainate and similarities in the sensitivities to antagonists, it has always been an intriguing question whether non-NMDA receptors in a particular tissue are one or more subtypes. The effects of NBQX are studied in two functional in vitro models: Antagonism of EAA induced ^{3}H-GABA released in cultured cortical neurons and antagonism of EAA induced depolarizations in rat cortical wedge [10,12,21].

AMPA receceptor antagonism protects against delayed cell death after cerebral ischaemia

The selectivity and potency prompted investigations of in vivo activity of NBQX. Intravenous administration of the compound to NMRI mice antagonized clinic seizures induced by ICV administration of 20 μg quisqualate (ED 50 = 6 mg/kg). These results show that NBQX enters the brain after peripheral administration[21].

Antagonism of NMDA receptor neurotransmission has previously been shown to have protective effects on the neuronal cell death seen after cerebral ischaemia [22].

Recently, the effect of non-NMDA receptor blockade on ischaemic neuronal cell death has been investigated in Mongolian gerbils using NBQX[21].

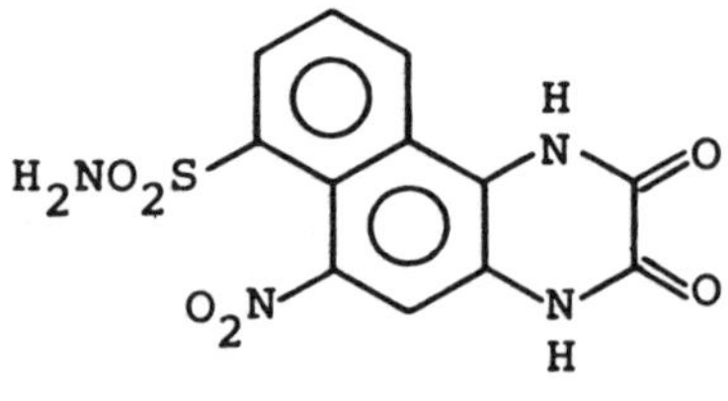

Fig. 3. Chemical structure of NBQX

Transient complete forebrain ischaemia was produced in female Mongolian Gerbils by occlusion of the carotic arteries for 5 min. Four days later the CA1 subfield of the hippocampus was assessed for neuronal cell death on a scale ranging from 0 for a normal hippocampus to 3 for an almost total loss of neuronal cells. Administration of NBQX (30 mg/kg i.p.) 15 and 5 min. before and 10 min. after the onset of occlusion gave a total protection against delayed neuronal cell death of CA1 neurons. Furthermore, dosing NBQX 60, 70 and 85 minutes after the onset of cerebral ischaemia gave total protection of neuronal cell death (Fig. 4).

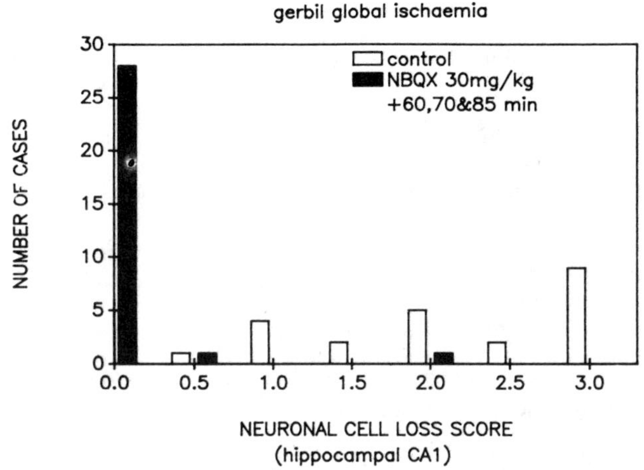

Fig. 4.   Protective effect of NBQX on delayed neuronal cell death following global ischaemia in Mongolian Gerbils.

Table 4 Potency of NBQX and XNQX as antagonist of EAA induced ^{3}H-GABA release in cultured cortical neurons

	EC50 (µM)	
Agonist	NBQX	CNQX
Quisqualate	0.03	1.0
Kainate	0.06	2.0

Table 4 summarizes the antagonistic effects of CNQX and NBQX on cultured cortical neurons. Despite the 5-fold and 30-fold selectivity of CNQX and NBQX for ^{3}H-AMPA and ^{3}H-kainate binding, the selectivity as antagonists of quisqualate and kainate induced ^{3}H-GABA release was much lower, 2-fold, for both compounds. Together with the above mentioned characteristics of ^{3}H-AMPA and ^{3}H kainate binding, the results suggest that kainate induced ^{3}H-GABA release in cortical cell cultures is mediated via low affinity kainate receptors = AMPA receptors.

Table 5 Potency of NBQX and CNQX as antagonist of depolarizations induced in rat cortical wedge

---

	EC_{50} (μM)	
Agonist	NBQX	CNQX
Quisqualate	0.02	0.63
Kainate	0.76	3.2

Table 5 summarizes the antagonistic effects of CNQX and NBQX on rat cortical wedge. Opposite to the results obtained in cortical cell cultures, the selectivity of CNQX and NBQX as antagonists of quesqualate and kainate induced depolarization is in accordance with the selectivities obtained in binding studies, 5-fold and 40-fold, respectively. These results suggest that kainate induced depolarization in the cortical wedge is mediated via high affinity kainate receptors. The explanation for the difference in antagonist sensitivities between cultured cortical cells and neocortical slices is at present not known, but it may be concluded that both high and low affinity kainate sites can mediate excitatory responses.

The findings suggest that delayed neuronal cell death following a period of global ischaemia is at least partially mediated by a mechanism involving AMPA and/or kainate receptors. Interestingly, it was recently shown[23] that quisqualate evoked dark cell degeneration in rat hippocampal slice could not be prevented by the simultaneously presence of the non-NMDA glutamate receptor antagonist CNQX. This compound was only effective when administered after the quisqualate challenge. The results suggested that activation of the AMPA receptor - which is blocked by CNQX - is responsible for neuronal degeneration. However, a receptor destinct from the AMPA receptor is initiating the chain of events leading to neuronal degeneration. A possible candidate is the recently described metabotropic quisqualate receptor which is insensitive to CNQX. Furthermore, influence neuronal Ca^{2+} homeostasis [24]. Quisqualate can evoke a biphasic increase in intracellular calcium levels, the second (plateau) phase being sensitive to CNQX and calcium antagonists. Based on these studies we hypothesize that glutamate released during ischaemia[25] triggers a post ischaemic release of an endogenous substance acting on ionotropic quisqualate receptors. This would evoke neuronal cell loss in a manner sensitive to NBQX administered up to 1 hour after the ischaemic insult.

REFERENCES

1. Davies, J., Evans, R.H., Francis, A.A., and Watkins, J.C., 1979, Exicatory amino acid receptors and synaptic excitation in the mammalian central nervous system, **J. Physiol.** (Paris), 75:641.
2. Krogsgaard-Larsen, P., Honore, P., Hansen, J.J.,Curtis, J.J., and Lodge, D., 1980, New class of glutamate agonist structurally related to ibotenic acid, **Nature**, 284:64.
3. Honore, T., Lauridsen, J., and Krogsgaard-Larsen, P., 1982, the binding of (^{3}H) AMPA, a structural analogue of glutamic acid, to rat brain membranes, **J. Neurochem.**, 38:173.
4. Sugiyama, H., Ito, I., and Hirono, C, 1987, A new type of glutamate receptor linked to inositol phospholipid metabolism, **Nature**, 325:531.
5. Shinozaki, H., and M. Ishida, 1988, Stizolobic acid, a competitive antagonist of the quisqualate-type receptor at the crayfish neuromuscular junction, **Brain Res.**, 451:353.
6. Watkins, J.C., and Evans, R.H., Excitatory amino acid transmitters, **Ann. Rev. Pharm. Tox.**, 21:165.
7. McLennan, H., 1983, Receptors for the excitatory amino acids in the mammalian central nervous system, **Prog. Neurobiol.**, 20:251
8. Drejer, J., Honore, and Schousboe, A., 1987, Excitatory amino acid-induced release of ^{3}H-GABA from cultured mouse cerebral cortex interneurons, **J. Neuroci.**, 7:2910.

9. Honore, T., Davies, S.N., Drejer, J., Fletcher, E.J., Jacobsen, P., Lodge, D., and Nielsen, F.E., 1988, Quinoxalinediones: Potent competitive non-NMDA glutamate receptor antagonist, **Science**, 241:701.

10. Fletcher, E.J., Martin, D., Aram, J.A., Lodge, D., and Honore, T., 1988, Quinoxalinediones selectively block quisqualate and kainate receptors and synaptic events in rat neocortex and hippocampus and frog spinal cord in vitro, **Br. J. Pharmacol.**, 95:585.

11. Andreasen, M., Lambert, J.D.C., and Jensen, M.S., 1988, Direct demonstration of an N-methyl-D-aspartate receptor mediated component of excitatory synaptic transmission in area CA1 of the rat hippocampus, **Neurosci. Lett.**, 93.61.

12. Drejer, J., and Honore, T., 1988, New quinoxalinediones show potent antagonism of quisqualate responses in cultured mouse cortical neurones, **Neurosci. Lett.**, 87/104.

13. Blake, J.F., Brown, M.W., and Collingridge, G.L., 1988, CNQX blocks acidic amino acid induced depolarizations and synaptic components mediated by non-NMDA receptors in rat hippocampal slices, **Neurosci. Lett.**, 89:182.

14. Birch, P.J., Grossman, C.J., and Hayes, A.G., 1988, 6,7-Dinitro-quinoxaline-2,3-dion and 6-nitro-7-cyano-quinoxaline-2,3-dion antagonise responses to NMDA in the rat spinal cord via an action at the strychnine-insensitive glycine receptor, **Eur. J. Pharmacol.**, 156:177.

15. Drejer, J., Sheardown, M., Nielsen, E.O., and Honore, T., 1989, Glycine reverses the effect of HA-966 on NMDA responses in cultured rat cortical neurons and in chick retina, **Neurosci. Lett.**, 98:333.

16. Honore, T., and Nielsen, M., 1985 Complex structure of quisqualate-sensitive glutamate receptors in rat cortex, **Neurosci. Lett.**, 54:27.

17. Honore, T., and Drejer, J., 1988, Chaotropic ions affect the conformation of quisqualate receptors in rat cortical membranes, **J. Neurochem.**, 51:457.

18. Honore, T., Drejer, J., Nielsen, E.O., and Nielsen, M., 1989, Non-NMDA glutamate receptor antagonist ^{3}H-CNQX binds with equal affinity to two agonist states of quisqualate receptors, **Biochem. Pharmacol**, in press.

19. Nielsen, E.O., Drejer, J., Cha, J.J., Young, A., and Honore, T., 1989, Autoradiographic characterization and localization of quisqualate binding sites in rat brain using the antagonist ^{3}H-CNQX: Comparison with ^{3}H-AMPA binding sites, **J. Neurochem**, in press.

20. Honore, T., Drejer, J., and Nielsen, M., 1986, Calcium discriminates two (^{3}H) kainate binding sites with different, **Neurosci. Lett.**, 65:47.

21. Unpublished.

22. Foster, A., Gill, R., and Woodruff, G.N., 1988, Neuroprotective effects of MK-801 in vivo : Selectivity and evidence for delayed degeneration mediated by NMDA receptor activation, **J. Neurosci.**, 8:4745.

23. Gartwaite, G., and Garthwaite, J., 1989, Quisqualate neurotoxicity: A delayed, CNQX-sensitive process triggered by a CNQX-insensitive mechanism in young rat hippocampal slices, **Neurosci. Lett.**, 99(1-2):113.

24. Murphy S.N., and Miller R.J., 1989, Two distinct quisqualate receptors regulate Ca^{2+} homeostasis in hippocampal neurons in vitro, **Mol. Pharmacol.**, 35:671.

25. Benveniste, H., Drejer, J., Schousboe, A., and Diemer, N.H. 1984, Elevation of the extracellular concentrations of glutamate and aspartate, **J. Neurochem.**, 43:1369.

MOLECULAR CHARACTERIZATION, ULTRASTRUCTURAL LOCALIZATION AND GENE CLONING OF THE CHICK CEREBELLAR KAINATE RECEPTOR

Vivian I. Teichberg, Nomi Eshhar, Ilana Maoz, Itzchak Mano, David Ornstein, Arturo Ortega and Paul Gregor

Department of Neurobiology, The Weizmann Institute of Science, Rehovot 76100 Israel

INTRODUCTION

The kainate receptors mediate some of the excitarory effects of glutamate by regulating the opening of voltage independent cation selective channels. Their presence on living brain cells is nowadays monitored electrophysiologically by measuring the changes of cell membrane conductance produced by kainic and domoic acids in the absence or presence of kainate receptor antagonists such as gammaglutamyglycine and CNQX.

On brain tissue sections, the kainate receptors are revealed by their ability to bind kainic acid (Monaghan and Cotman, 1982) and its analogues, domoic acid and kainyl-BSA (Eshhar et al., 1988).

Electrophysiological and histochemical studies have shown that the kainate receptors are widely distributed in the brain of vertebrates and are present both on neurons (Foster and Fagg, 1984) and glial cells (Backus et al., 1989; Gallo et al., 1989). Surprisingly, their brain tissue density vary to considerable extents from species to species. The brain of teleosts and birds contains as much as 100 times more kainate binding sites than the brain of mammals (London et al., 1980; Henke and Cuenod, 1980).

This very significant difference in site density and the availability of relatively large amounts of teleost and avian brain, tissues have been exploited and made possible the recent isolation and purification of kainate binding proteins from frog brain (Hampson and Wenthold, 1988), and cerebellum of chick (Gregor et al., 1988) and pigeon (Klein et al., 1988).

The isolated proteins display on SDS-PAGE similar molecular weights (around 49000) and most likely represent each species version of the same protein. Monoclonal antibodies directed against the isolated chick and frog kainate binding proteins interact, on Western blots of brain membranous proteins extracted from various species, with an epitope present both on a 48-49000 polypeptide and a 93000 polypeptide. These results suggest that the membrane bound kainate binding protein, in its native structure, may be composed of at least two different subunits relatively conserved throughout evolution.

Although the kainate binding protein displays some of the properties and excepted behavior of a neurotransmitter receptor, it remains to be determined whether it corresponds to the kainate receptor protein defined by the above mentioned electrophysiological and pharmacological criteria.

Excitatory Amino Acids and Neuronal Plasticity
Edited by Y. Ben-Ari
Plenum Press, New York, 1990

The present chapter deals with some of the attemps made in our laboratories to address this crucial question. We have studied the ultrastrutctural localization of the kainate binding protein to determine whether it is present, as expected, on the cell membrane. We have also clone the gene encoding the kainate binding protein to determine whether it codes for a protein displaying the same structural and topological properties as the already known ligand gated channels.

LOCALIZATION OF THE CHICK CEREBELLAR KAINATE BINDING PROTEIN ON BERGMANN GLIA.

We have used a monoclonal antibody (IX-50 mAb) recognizing on Western blots the 49000 and 93000 polypeptides of various species (Gregor et al. 1988) in order to localize, at the electron microscope level, its specific epitope in the cerebellar cortices of chick and rainbow trout.

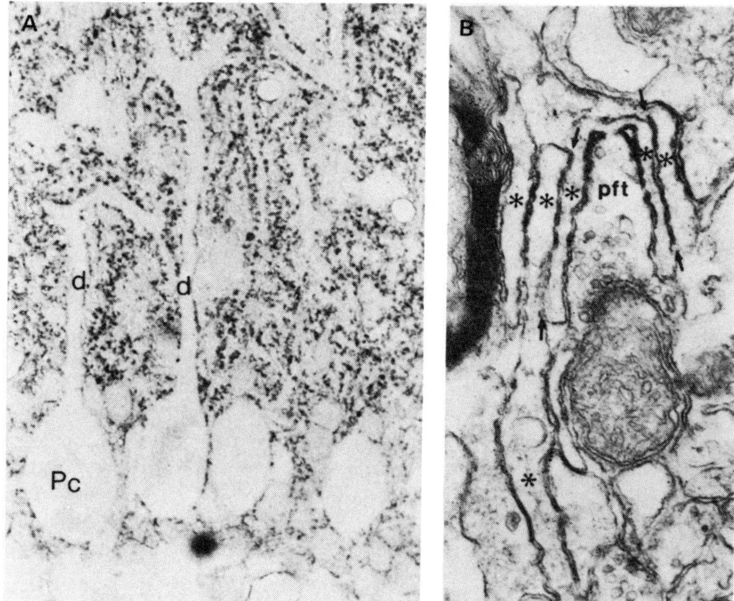

Figure 1. A: Chicken cerebellum, 1 day old, 1 um thick section, Postembedding immunocytochemistry with IX-50 mAb shows immunoreactivity outlining the Purkinje cells (Pc), their main ascending dendrite (d) as well as their spiny branchlets. B: Immunoperoxydase reaction with IX-50 mAb. This electron micrograph of trout cerebellum shows a presumed parallel fiber terminal (pft) in synaptic contact with a dendrite. Note that the glial membranes (asterisks) are immunoreactive along their contact with the terminal as well along faces where they contact other glial lamellae (arrows).

In an earlier immunocytochemical study (Eshhar et al., 1988; Gregor et al., 1988) at the light microscope, X-50 mAb binding sites were found exclusively in the cerebellar molecular layer but because of the almost homogeneous distribution of the labeling in this layer and the lack of resolution, it had not been possible to identify the nature of the cells carrying the interacting epitope i.e. the kainate binding protein. This question was answered in a light microscope study of the immunoreactivity of IX-50 mAbs, using, this time, semithin sections of the chick cerebellum. Although the immunoreactivity outlined the Purkinje cells and their dendrites, it was associated with the

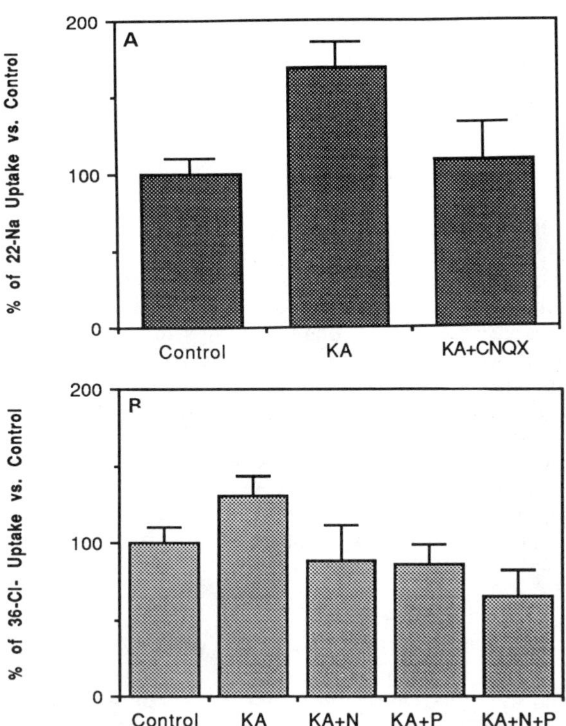

Fig. 2.   Ion fluxes evoked in chick cerebellar glial cells in culture by 0.1 mM Kainic acid. Controls set at 100% correspond to the basal ion influx in the absence of kainic acid. Both fluxes (A : 22Na$^+$ and B: 36Cl$^-$) were measured for three minutes. CNQX : 1 mM; picrotoxin (P) : 1mM; nipecotic acid (N) : 1mM. The 22Na$^+$ influx was measured in the presence of 0.5 mM ouabain.

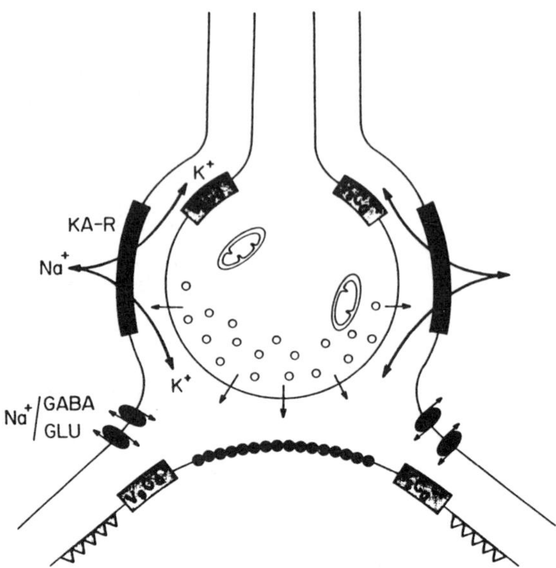

Fig. 3.   Model of the possible physiological role of the glial kainate receptor. The kainate receptor, activated by glutamate released from the nerve terminals of the parallel fibers (arrows), produces an efflux of potassium ions in the extracellular space and an influx of sodium ions in the glial cells. The extracellular K+ ions will further depolarize the nerve terminal and by the activation of voltage sensitive calcium channels thereby the synaptic efficacy. The intracellular Na+ ions will activate the sodium dependent reverse transport of GABA increasing the extracellular concentration of GABA and causing the activation of extrasynaptic GABA receptor (triangles). This model may explain why the stimulation of parallel fibers evokes in the Purkinje cells an EPSP (due to the release of glutamate) immediately followed by an IPSP (due to the release of GABA via the reverse action of the glial GABA transporter).

Bergmann glial cells which lie beneath, between and amongst the Purkinje cells. The same localization was found in the rainbow trout cerebellum. Furthermore an electron micrograph of a trout cerebellum showed the immunoreactivity along Bergmann glial lamellae (asterisks) surrounding the parallel fiber/spine synaptic complex. Similar observations were made with the chick cerebellum (Somogyi et al. 1990) confirming the Bergmann glia localization of the kainate binding protein.

THE FUNCTION OF THE GLIAL KAINATE RECEPTOR

Following the glial localization of the kainate binding protein, we investigated the presence of functional kainate receptors on chick cerebellar glial cells in culture. Fig. 1 shows that kainate is able to increase significantly the influx of radioactive sodium ions into glial cells, an effect blocked by CNQX. Interestingly, kainate induces also an influx of radioactive chloride ions which is blocked both by picrotoxin, a GABA receptor antagonist and by nipecotic acid, a GABA uptake blocker (Similar findings were reported by Gallo et al., 1989). Figure 2 outlines a plausible model.

This set of results indicate not only the presence of a functional kainate receptor on glial cells but suggest also what could be its function in the cerebellar circuitry.

CLONING THE GENE ENCODING THE KAINATE BINDING PROTEIN

The molecular cloning of the gene encoding the kainate binding protein was undertook to study its structure and determine whether the latter protein has any similarity with known receptors forming ligand gated channels. The latter form a family of proteins sharing common features (Stevens, 1987) which are therefore expected to be displayed as well by the kainate receptor.

The kainate binding protein was isolated by the method of Gregor et al. (1988) and submitted to a final step of purification consisting of ion exchange FPLC on mono Q. This purification afforded as single polypeptide of 49000 dalton with an estimated purity of 80-95%. This polypeptide was cleaved by cyanogen bromide and the resulting peptides separated by reverse phase-HPLC. A total of six peptides were microsequenced including the N-Terminus and the antisense oligonucleotides corresponding to two of these peptides were synthesized and used as probes in the screening of a chick cerebellar cDNA library. Several positive cDNA clones were identified in this screen and a 3.3 KB insert was chosen for further analysis. In Northern blots, both the synthetic oligonucleotides probes and the latter clone were found to strongly hybridize to cerebellar RNA and to a much lesser extent to RNA from cerebral hemispheres confirming the selective association of the kainate binding protein with the cerebellum.

DNA sequencing of the 3.3 KB insert revealed a strech of 1464 nucleotides in an open reading frame. The deduced amino acid sequence was analyzed both in terms of its hydropathy and resemblance to other channel forming receptors. Five hydrophobic putative transmembrane domains were identified, the first at the N terminal corresponding clearly to a signal peptide since it precedes the established N terminal sequence of the 49000 polypeptide. The other four putative transmembrane domains were found to be approximately at the same positions as the putative transmembrane domains of the nicotinic acetycholine receptor. Moreover, sequence homologies of 37% to 55% were found with the latter receptor in the putative transmembrane domains.

This set of data strongly suggest that the kainate binding protein gene encodes in fact the kainate receptor. Transfection experiments, now in progress, will provide the definite demonstration that the cerebellar kainate binding protein is a kainate receptor.

ACKNOWLEDGEMENTS

This work was supported by a grant from the Forscheimer Center for Molecular Genetics. VIT holds the Louis and Florence Katz-Cohen professionial chain in neuropharmacology.

REFERENCES

Backus, K.H., Ketternmann, H. and Schachner, M., 1989, Pharmacological characterization of the glutamate receptor in cultured astrocytes, **J. Neurosc. Res.**, 22:274.

Eshhar, N., Lederkremer, G., Beaujean, M., Goldberg, O., Gregor, P.,Ortega, A., Triller, A. and Teichberg, V.I., 1988, Kainyl-bovine serum albumin : a novel ligand of glutamate receptor with a very high binding affinity. **Brain Res**, 476 : 57.

Foster, A.C. and Fagg, G.E., 1984, Acidic amino acid binding sites in mammalian membranes : Their characteristics and relationship to synaptic receptors. **Brain Res . Rev .** 7 : 103.

Gallo, V., Giovamnini, C., Suergiu, R. and Levi, G., 1989, Expression of excitarory amino acid receptors bey cerebellar cells of the type-2 astrocyte cell leneage, **Neurochem.**, 52 : 1.

Gregor, P., Eshhar, N., Otega, A. and Teichberg, V.I., 1988, Isolation, immunochemical characterization and locallization of the kainate sub-class of glutamate receptor from chick cerebellum, **EMBO J.**, 7 : 2673.

Hampson, D.R. and Wenthold, R.J., 1988, A kainic acid receptor from frog brain purified using domoic acid affinity chromatography, **J. Biol. Chem.** 263 : 2500.

Henke, H. and Cuenod, M., 1980, Specific 3H-kainic acid binding in the vertebrate CNS. **In** Littauer, U.Z. et al editors, Neurotransmitters and their receptors, John Wiley & sons. pp 373-386.

Klein, A.U., Niederoest, B., Winterhalter, K.H., Cuenod, M. and Streit, P., 1988, A kainate binding protein in pigeon cerebellum : purification and localization by monoclonal antibody, **Neurosc. Lett.** 95 : 359.

London, E.D., Klemm, N. and Coyle, J. T., 1980, Phylogenetic distribution of ^{3}H-kainic acid receptor binding sites in neuronal tissue. **Brain Res** 192 : 463.

Monaghan, D.T. and Cotman, C.W., 1982, The distribution of 3H-kainic acid binding sites in rat CNS as determined by autoradiography, **Brain Res.** 252 : 91.

Stevens, C.F., 1987, Channel familiers in the brain, **Nature**, 328 : 198.

Somogyi, P., Eshhar, N., Teichberg, V.I., J.D.B. Roberts, Subcellular localization of a putative kainate receptor in Bergmann glial cells using a monoclonal antibody in the chick and fish cerebellar cortex". **Neuroscience**, 1990, in press.

INTRACELLULAR MESSENGERS ASSOCIATED WITH EXCITATORY

AMINO ACID (EAA) RECEPTORS

Joël Bockaert, Aline Dumuis, Olivier Manzoni, Joël Nargeot*, Kiyoshi
Oomagari, Jean-Philippe Pin, François Rassendren*, Michèle Sebben and
Fritz Sladeczek

Centre CNRS-INSERM de Pharmacologie-Endocrinologie,
Rue de la Cardonille, 34094 MONTPELLIER CEDEX 2, FRANCE. *CNRS -
CRBM, UPR 41, INSERM U. 249, Route de Mende, MONTPELLIER CEDEX,
FRANCE

INTRODUCTION

Glutamate (Glu) and L-aspartate (Asp) have now been recognized as the two main excitatory
neurotransmitters in vertebrate brain. In addition, to allow the fast excitatory neurotransmission,
these amino acids appear to play crucial roles in neuronal plasticity. These include changes in
neuronal development (Barnes, 1986 ; Balàzs et al., 1988 ; Didier et al., 1989), as well as changes in
synaptic efficacy, which probably underlie learning and memory (Nicoll et al., 1988). In addition,
they are likely to be implicated in neuronal death, associated with several neurological diseases
including strokes, hypoglycemia, epilepsy and Huntington's chorea.

Although some insights have been obtained into the mechanisms by which Glu triggers these
fundamental functions in neurons, our knowledge on this subject is still at its earliest stage. Four
years ago, it was classical to state that Glu acted on ionotropic receptors. Based on their
electrophysiological responses to specific agonists, these ionotropic receptors have been classified
as N-methyl-D-aspartate (NMDA), quisqualate (Q) and kainate (K) receptors (Watkins and
Olverman, 1987), although the distinction between Q and K is still in question (Pin et al., 1989).

However, at that time, two examples already showed that a single neurotransmitter, such as
acetylcholine (Ach) or γ-aminobutyric acid (GABA), was able to stimulate receptors having
different mechanisms of action and different primary structures. Indeed, Ach stimulates nicotonic
receptors which are ionotropic receptors having a pentameric structure and containing both
neurotransmitter binding sites and channel structures. Ach also stimulates metabotropic receptors
(muscarinic receptors) which are monomeric proteins, possessing seven transmembrane domains and
are positively or negatively coupled through G proteins to enzymes producing second messengers
(Fig. 1). The same situation is true for GABA which stimulates both ionotropic (GABA$_A$) and
metabotropic (GABA$_B$) receptors (Fig. 1). Similarly, 5-HT$_3$ receptors are ionotropic receptors,
whereas 5-HT$_1$ and 5-HT$_2$ receptors are metabotropic receptors coupled through G proteins to
adenylate cyclase and phospholipase C, respectively (Fig. 1).

With this idea in mind, we decided to investigate if Glu, in addition to activating ionotropic
receptors, also stimulates metabotropic receptors. The answer is yes, since, as indicated in Fig. 1,

Excitatory Amino Acids and Neuronal Plasticity
Edited by Y. Ben-Ari
Plenum Press, New York, 1990

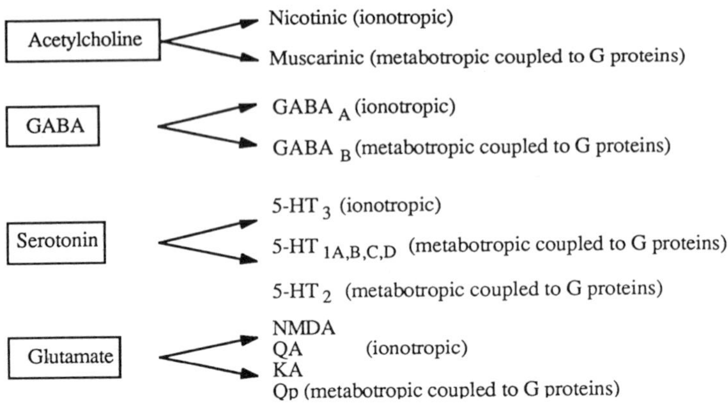

Fig. 1. Many classical neurotransmitters activate both ionotropic and metabotropic receptors

we (Sladeczek et al., 1985, 1988 ; Récasens et al., 1987), and others (Nicoletti et al., 1986 a,b ; Récasens et al., 1988 ; Sugiyama et al., 1987) clearly demonstrated that a quisqualate receptor, termed Qp, is directly coupled to phospholipase C.

We also recently demonstrated, as reported here, that this Qp receptor can also stimulate arachidonic acid release (Dumuis et al., in preparation). Considering ionotropic receptors, a new notion is emerging for at least NMDA and 5-HT$_3$ receptors (Reiser and Hamprecht, 1989) which indicates that in addition to modifying ionic fluxes, they can also indirectly stimulate the synthesis of intracellular signalling molecules. Therefore, these ionotropic receptors can trigger both rapid and slow responses. It is already known that NMDA receptors stimulate the production of cyclic GMP (Garthwaite and Garthwaite, 1987). More recently, it has been shown that the increase in cyclic GMP production is in fact the consequence of the production of nitric oxide (NO), a stimulator of soluble guanylate cyclase (Garthwaite et al., 1988). We have also found that NMDA stimulates arachidonic acid release (see Dumuis et al., 1988 and this report).

NMDA RECEPTORS STIMULATE ARACHIDONIC ACID RELEASE IN STRIATAL NEURONS

In cultures of striatal neurons after 6 days *in vitro* (DIV) when synapses have not yet been formed (Weiss et al., 1988), Glu (30 μM) rapidly stimulated basal arachidonic acid release. After a 10 min stimulation period, 1 to 2.5% of [3H]-arachidonic acid incorporated into the neurons were released (Dumuis et al., 1988). Glu and NMDA were the only excitatory amino acids (EAA) able to stimulate arachidonic acid release in these cultures in absence of Mg^{2+} (Fig. 2A). The effects of Glu and NMDA were blocked by APV (Fig. 2B), phencyclydine (PCP) and Mg^{2+} (data not shown). Mepacrine, an inhibitor of phospholipase A_2 (PLA$_2$) suppressed the NMDA induced arachidonic acid release (Fig. 2B). HPLC analysis of the radioactive [3H]- arachidonic acid and its putative metabolites released from striatal neurons, indicated that both NMDA and Glu stimulated almost exclusively [3H]-arachidonic acid release. No peak possibly corresponding to prostaglandins or to hydroxyeicosatetraenoic acid (HETE) being detectable after chloroform extraction of the [3H]-released metabolites (Fig. 3). The analysis has been made both by direct and reverse phase HPLC (Fig. 3).

The mechanism by which NMDA receptors stimulate arachidonic acid release is likely to be the consequence of their high Ca^{2+} conductance since it is known that Ca^{2+} activates phospholipase A_2. Indeed, omitting Ca^{2+} in the incubation medium suppresses the NMDA effect on arachidonic acid release and ionomycin, a Ca^{2+} ionophore, was able to trigger a massive release of arachidonic acid. Surprisingly, K^+ (56 mM) was unable to trigger arachidonic acid release. Recently, NMDA has also been found to trigger arachidonic acid release in primary cultures of cerebellar granule cells (Lazarewicz et al., 1988).

80

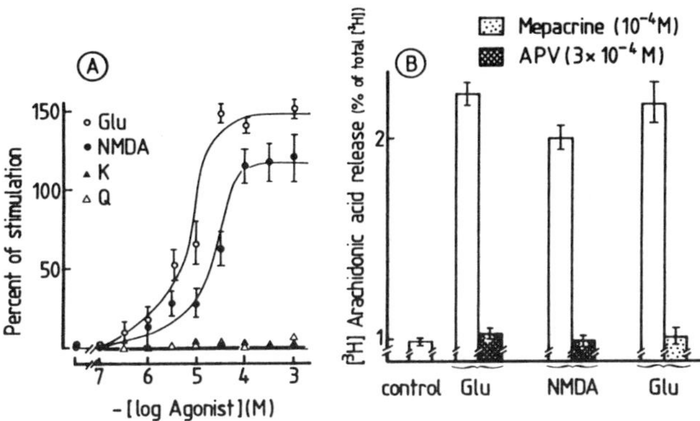

Fig. 2.   Stimulation of arachidonic acid release by EAA in 6 DIV striatal neurons. (A) Dose-dependent Glu, NMDA, K and Q effects. (B) Blockade of the Glu and NMDA effects by APV and mepacrine (modified from Dumuis et al., 1988).

THE STIMULATION OF PHOSPHOLIPASE C BY QUISQUALATE (Q) INVOLVES A NOVEL CLASS OF GLUTAMATE RECEPTORS, THE Qp METABOTROPIC RECEPTORS

Different models in which a Q-preferring receptor (Qp) stimulates Ins-P production

We first reported that in primary cultures of striatal neurons, EAA stimulate the production of total inositol phosphates (Ins-P) with the following order of potency : Q > Glu > NMDA ≥ K. The EC_{50}'s of these agonists were (µM) 0.16, 4, 15 and 10, respectively. The stimulation was a 3- to 4-fold increase over basal production. Whereas the small NMDA response was totally blocked by the NMDA receptor antagonist 2-amino-5-phosphono-valerate acid (APV), the Glu and Q responses were not, indicating that they did not result from activation of NMDA receptors (Sladeczek et al., 1985).

The Q-preferring stimulation of Ins-P production has now been reported in primary cultures of cerebellar granule cells (Nicoletti et al., 1986a), brain synaptoneurosomes (Récasens et al., 1987, 1988), astrocytes in culture (Pearce et al., 1986), brain slices from striatum (Doble and Perrier, 1989) and hippocampus (Schoepp and Johnson, 1988 ; Palmer et al., 1988).

In all these models, only total Ins-P responses were measured in response to Q and generally after long incubation periods in the presence of Li^+. However, no investigations have been made on the physiologically relevant Ins-P analogs which are formed in neurons in response to Q. It was of great importance to know if Ins(1,4,5)P3 and Ins(1,3,4,5)P4, the two main important analogs controlling the changes in Ca^{2+} movement in the cells, are really formed.
As shown in Fig. 4, after 30s a clear stimulation of Ins-P1, Ins-P2 as well as Ins(1,4,5)P3 and Ins(1,3,4,5)P4 productions were observed in striatal neurons in response to Q in the absence of Li^+.

This pattern of stimulation plus the observations that Ins-P production was not affected by the presence of TTX or removal of extracellular Ca^{2+} (Sladeczek et al., 1985), suggested a direct coupling between a new non-ionotropic Q receptor and phospholipase C. We proposed to designate these receptors as Qp (Sladeczek et al., 1988). In Xenopus oocytes injected with brain mRNA, a Q response resulting from the stimulation of the cascade : (1) activation of Q receptors, (2) production of Ins(1,4,5)P3, (3), release of intracellular Ca^{2+}, (4) activation of a Ca^{2+}-dependent Cl⁻ conductance can be demonstrated (Sugiyama et al., 1987 ; Verdoorn and Dingledine, 1989 ; Fong et al., 1988 ; Rassendren et al., 1989). Figure 5 shows that in these oocytes, Q induced two types of responses : (1) a smooth inward current which occurred rapidly after drug application (Fig. 5a) and which is mediated by ionotropic Q receptors (Qi) and (2) a delayed oscillatory response (Fig. 5a) due to the

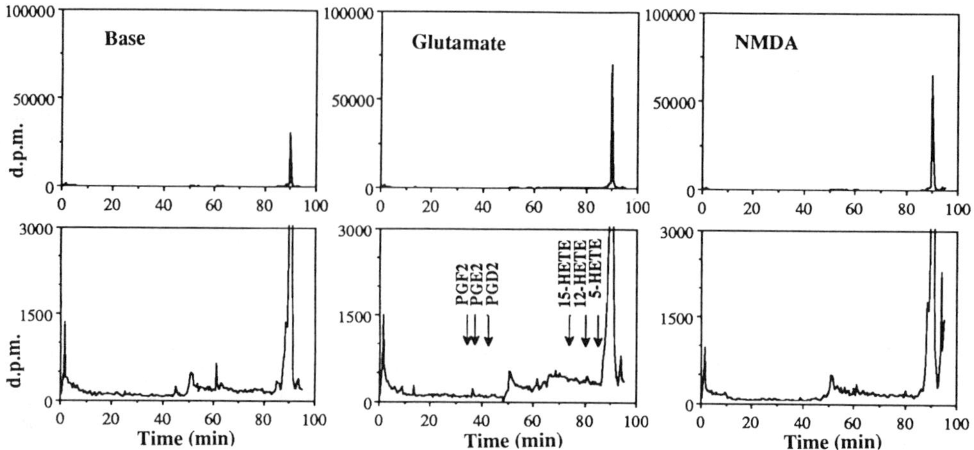

Fig. 3.   HPLC analysis of the [3H]-arachidonic acid metabolite release from 6 DIV striatal neurons under basal (Base), 100 μM glutamate-stimulated (Glutamate) or 100 μM NMDA-stimulated (NMDA) conditions. Striatal neurons grown for 6 DIV in 35 mm dish were labelled for 24h with 10 μCi [3H]-arachidonic acid. The release experiment was then conducted as described by Dumuis et al. (1988) by incubating the cells in 1 ml medium containing or not the stimulatory agent for 15 min after a 25 min washing period. [3H]-arachidonic acid and its [3H]-metabolites released from the cells were extracted twice with 5 vol. chloroform after acidification of the medium to pH 3.5 with formic acid. Chloroform fractions were reduced to dryness under vacuum, dissolved in 200 μl of water (pH 3.5 with acetic acid, 30%) / acetonitrile (70%) and then injected onto a reverse phase C18 column (Waters radial pack, resolve). Elution was performed at a flow rate of 3 ml / min with a mixture of buffer A (water equilibrated at pH 3.5 with acetic acid) and buffer B (acetonitrile). The gradient consists of 3 isocratic steps : from time 0 to 45 min, 28% B buffer ; from 50 to 85 min, 53% B buffer ; from 88 to 95 min 100% B buffer. Elution of standards are indicated by the arrows, and was followed by UV detection as previously described (Dumuis et al., 1988). Radioactivity eluted from the column was collected every 0.5 min. Chromatograms presented on the bottom are the same as those on the top except that the dpm values are cut off over 3000 dpm (see scales).

opening of a Ca^{2+}-dependent Cl^- conductance and typical in this model of activation of receptors coupled to phospholipase C. As expected, if Qp are indeed responsible for this response, intracellular injection of Ins(1,4,5)P3 mimicked the response (Sugiyama et al., 1987), whereas injection of EGTA suppressed it (Fig. 5b).

Pharmacological characterization of Qp receptors

Table 1 summarizes the main differences between the pharmacological characteristics of Qi and Qp receptors. One of the main features is that DL-α-amino-3hydroxy-5methyl4-isoxazole propionic acid (AMPA) an excellent agonist of Qi receptors, is a very weak agonist at Qp both in models measuring total Ins-P formation (Sladeczek et al., 1988 ; Récasens et al., 1988 ; Palmer et al., 1988), or in Xenopus oocytes, as illustrated in Fig. 5c (Rassendren et al.,1989).

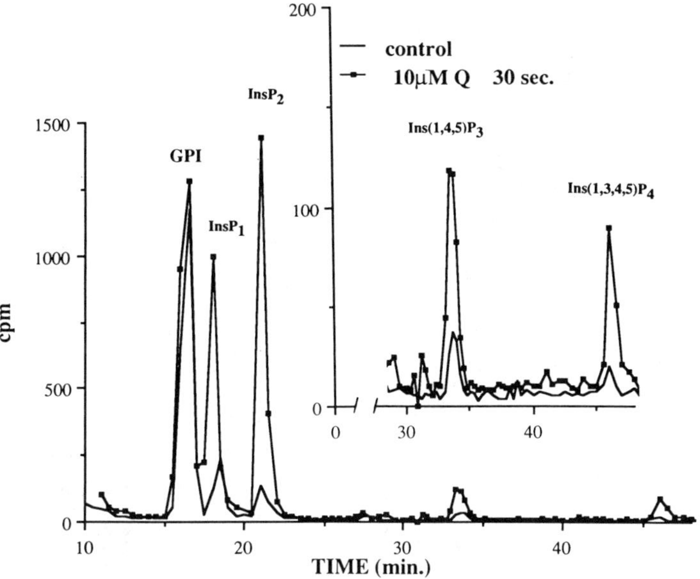

Fig. 4.   Separation of inositol phosphates by HPLC in the absence (control) and presence of Q in 14 DIV striatal neurons. The method used for HPLC separation has been described by Mouillac et al. (1989).

In addition, in each system described above, not one antagonist of Qi has been shown to block Qp responses (Sladeczek et al., 1988) (Table 1). This includes 6-cyano-7-nitro-2,3-dihydro-xyquinoxaline (CNQX) the newly discovered specific Qi antagonist (Drejer and Honoré, 1988).

Responses to receptors which are coupled to phospholipase C are often inhibited by phorbol esters, the potent activators of protein kinase C (PKC). We have shown that low concentrations of phorbol 12,13 dibutyrate (PdBu) did not inhibit the effect of Qi studied in two well-characterized models, the Qi induced GABA release (Pin et al., 1988, 1989) and Qi induced increases in intracellular calcium (Ca^{2+}_i) in striatal neurons (manuscripts in preparation). In contrast, PdBu potently inhibited the Qp response (Manzoni et al., 1989). Moreover, concanavalin A which potently increases the Qi effect on GABA release (Pin et al., 1989) did not affect the Qp response (data not shown). These differential modulations of Qp and Qi can be used to distinguish Qi-mediated from Qp-mediated responses (Table 1).

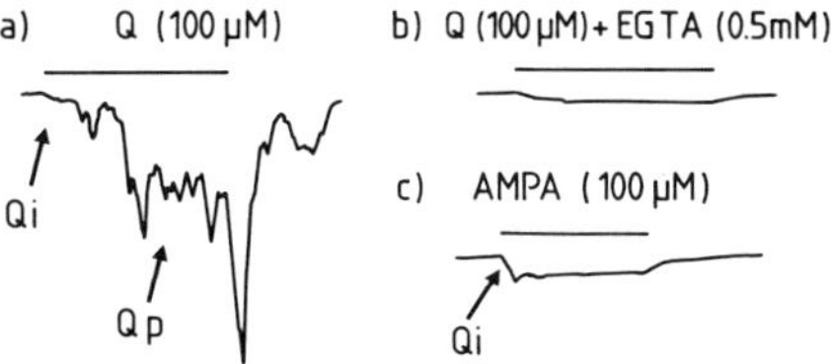

Fig. 5.   Expression of two distinct Q receptors (Qi and Qp) in Xenopus oocytes. The method used has been reported by Rassendren et al. (1989). The oocytes were voltage-clamped at -80 mV.

TABLE I

PHARMACOLOGICAL AND BIOCHEMICAL PROPERTIES OF QUISQUALATE IONOTROPIC (Qi) AND METABOTROPIC (Qp) RECEPTORS

	QUISQUALATE METABOTROPIC (Qp)	QUISQUALATE IONOTROPIC (Qi)
PHARMACOLOGY		
Agonists	• Quisqualate • Glutamate	• Quisqualate • AMPA • Kainate * • Glutamate
Competitive Antagonists	• None	• γ-DGG • GDEE • GAMS • CNQX
Non-competitive Antagonists	• None	• JSTX
Modulation :		
Inhibition by phorbol esters	+	−
Activation by concanavalin A	−	+
BIOCHEMICAL EVENTS DEMONSTRATED		
Increase in Ins-P1, Ins-P2, Ins-(1,4,5)P3 and Ins(1,3,4,5)P4 production	+	−
Activation of PKC	+	−
Increase in intracellular Ca^{2+} pools from intracellular pools	+	−
Increase in intracellular Ca^{2+} through VSCC	−	+
Increase in arachidonic acid release	+	−

*Induced a non-desensitizing response

Abbreviations : γ-DGG, γ-D-glutamylglycine ; GDEE, L-glutamic acid diethyl ester ; GAMS, α-D-glutamyl-aminomethyl-sulphonate ; JSTX, Joro spider toxin ; VSCC, voltage-sensitive Ca^{2+} channels.

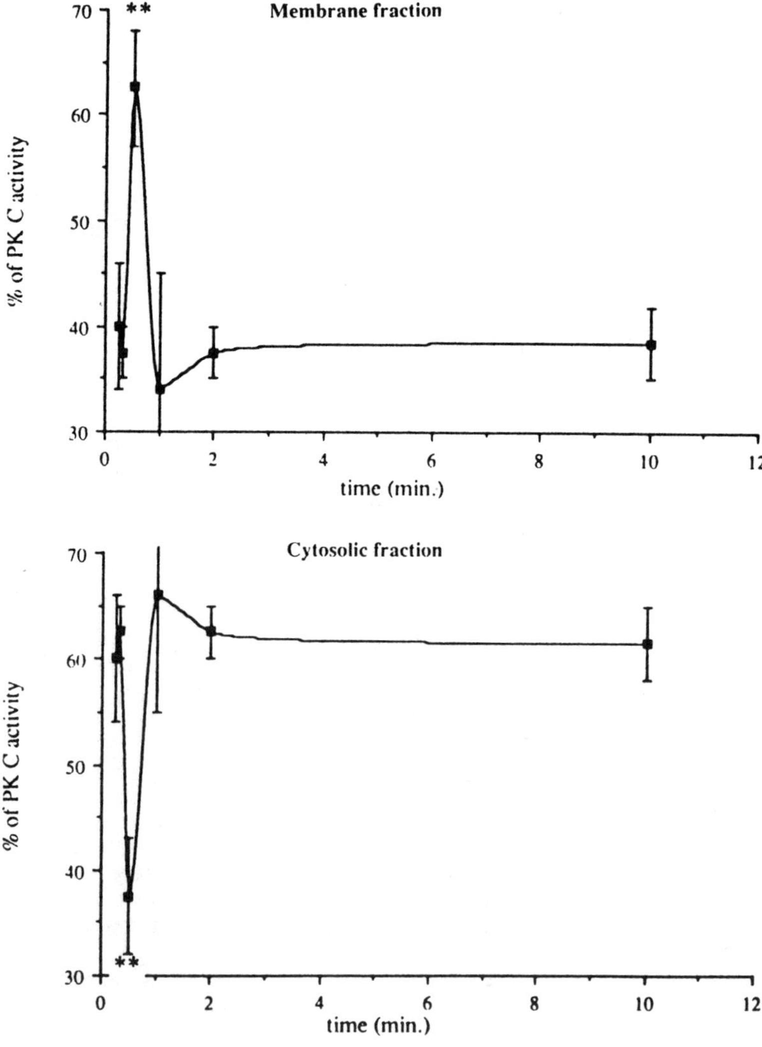

Fig. 6.   Translocation of PKC in 14 DIV striatal neurons after stimulation with Q (10 µM). Taken from Manzoni et al., 1989.

A further argument in favour of the non-identity between Qi and Qp receptors is that it is possible to obtain on agarose gel electrophoresis a high amount of purification of mRNA encoding for Qp, but not for Qi responses (Fong et al., 1988). This suggests that Qp is encoded by a single size class mRNA, whereas the Qi-gated channel is encoded by multiple species mRNA whose size distribution is heterogenous.

Qp receptors activate PKC and intracellular Ca^{2+} release in neurons

Activation of phospholipase C by Qp should produce both diacylglycerol (DAG) and Ins(1,4,5)P3 known to activate protein kinase C (PKC) and to release Ca^{2+} from

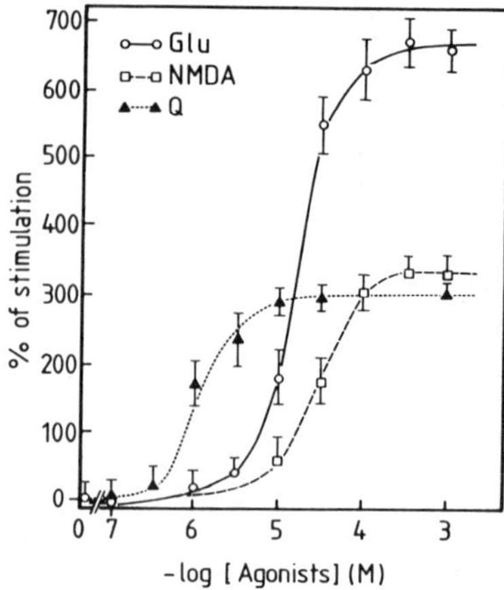

Fig. 7.   Stimulation of arachidonic acid release by EAA in 13 DIV striatal neurons. See Dumuis et al., 1989 for the method used to measure arachidonic acid release.

intracellular stores, respectively. In neurons from the central nervous system, both activation of PKC and release of Ca^{2+} from intracellular pools have been demonstrated as being the consequence of Qp receptor activation (Manzoni et al., 1989 ; Murphy and Miller, 1989).

Translocation of PKC from the cytosol to the membrane is generally considered as an indication of PKC activation. In striatal neurons, a 10 min stimulation with 100 nM PdBu increased the membrane bound activity from $37.2 \pm 2.9\%$ to $64.8 \pm 4.5\%$. Similarly, Q (10 µM) produced a similar translocation but which was transient, peaking at 30 s and disappearing after 40 s (Fig. 6). Such a transient activation has previously been observed for muscarinic receptors (Ter Bush et al., 1988). 10 µM AMPA which does not activate Qp receptors (Table 1) was unable to induce PKC translocation at 30 s (data not shown), indicating that this response probably results from Qp receptor activation.

Murphy and Miller (1989) recently showed that Q induced a rapid and transient $Ca^{2+}{}_i$ spike in hippocampal neurons due to a mobilization of $Ca^{2+}{}_i$ pools. These Ca^{2+} spikes are likely to be due to Qp since they were neither reproduced by AMPA, nor blocked by CNQX and observed in the absence of extracellular Ca^{2+} (Murphy and Miller, 1989).

Q also stimulates arachidonic acid release in striatal neurons

We have recently demonstrated (Dumuis et al., in preparation) that in striatal neurons maintained in culture for 13 DIV, another receptor, in addition to NMDA, is able to stimulate arachidonic acid release (Fig. 7). Indeed, in these neurons, Q was the more potent EAA able to stimulate arachidonic acid release ($EC_{50} = 1$ µM). NMDA was less potent, but had the same efficacy as Q (Fig. 7).

The effects of these two EAA were additive (data not shown). Glu, which is known to activate all EAA receptor types, had an efficacy which was twice that of NMDA and Q (Fig. 7). The Q receptor involved in this release is now being extensively studied

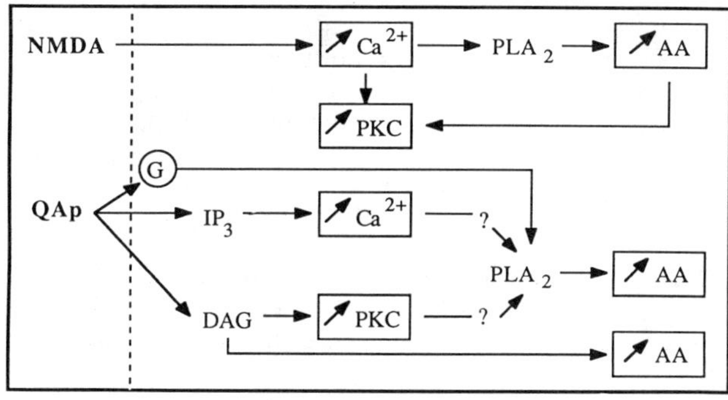

Fig. 8.   Signalling molecules associated with the action of Qp and NMDA receptors.

(Dumuis et al., in preparation). The characteristics of this Q receptor were identical to those for Qp, as described above (see Table 1). In particular, it was neither stimulated by AMPA, nor blocked by CNQX, nor stimulated by concanavalin A, but was inhibited by PdBu (data not shown). Therefore, it is likely that, in addition to stimulating Ins(1,4,5)P3 production, PKC activity and Ca^{2+}_i release, Qp receptors also stimulate arachidonic acid release.

The nature of arachidonic acid metabolites released were investigated using HPLC separation. They indicated that it is mainly arachidonic acid which is released (data not shown). The mechanisms by which Qp stimulates arachidonic acid release are still unknown. Many possibilities can be proposed (Fig. 8) : (1) activation of PLA_2 following increase in Ca^{2+}_i, or PKC activation, (2) deacylation of diacylglycerol (DAG), formed concomittantly to Ins(1,4,5)P3, by a DAG lipase, (3) activation of PLA_2 through a G protein. Experiments are now being conducted to know which one of these possible pathways is involved in Qp response.

DISCUSSION

It is now clear that Q stimulates a specific receptor coupled to phospholipase C that we propose to designate as a Qp receptor (Q receptors coupled to phospholipase C). We will refer to the ionotropic Q receptors as Qi. Only a few pharmacological tools are presently available to distinguish responses of these two receptors (see Table 1). Qp induces a pleiotropic set of signalling molecules (Fig. 8). Qp stimulates the formation of Ins-P including Ins(1,4,5)P3 and Ins(1,3,5)P4 (Fig. 4), the role of Ins(1,4,5)P3 is likely to release Ca^{2+} from an intracellular pool, as demonstrated in hippocampal neurons by Murphy and Miller (1989) and in Xenopus oocytes injected with brain mRNA, as demonstrated by several laboratories (Fig. 5). This increase in Ca^{2+}_i may activate or inhibit a number of Ca^{2+}-dependent ion channels that will alter the excitability of the neurons. Furthermore, several other Ca^{2+}-dependent processes can be triggered by this increase in Ca^{2+}_i (see below). By stimulating phospholipase C, Qp also produces DAG which is likely to result in translocation of PKC, as described in Fig. 6 (see Manzoni et al., 1989). An interesting observation is the rapid, but transient nature of PKC translocation. This could indicate that a feed-back mechanism occurs. Sphingosine is a possible candidate for such negative feed-back regulation (Hunnum and Bell, 1989). In addition to stimulating the classical Ins(1,4,5)P3 / DAG / Ca^{2+} pathway, Qp receptors also stimulate arachidonic acid release. The mechanisms by which Qp trigger this response could be multiple and further work is required to determine which one does operate (Fig. 8).

An interesting finding is that NMDA receptors also increase arachidonic acid release and that in both Qp and NMDA receptor systems, the essential product which is released is arachidonic acid. This is also true when arachidonic acid release is stimulated by ionomycin (data not shown). This suggests that in these neurons, lipoxygenase and cyclooxygenase activities are low. The observation that in 6 DIV, as well as in 13 DIV, K$^+$ (56 mM) did not produce any activation of arachidonic acid release, is intriguing (data not shown).

Preliminary experiments indicate that K$^+$ (56 mM) produces a sustained Ca^{2+} increase in cell bodies of striatal neurons. Although further experimental work will help solve this problem, one possibility could be that PLA$_2$ as well as Qp and NMDA receptors are co-localized in special compartments of neurons, for example, in spines of neurites in which VSCC's are not highly represented. In this context, it is interesting to note that strong depolarization does not induce long term potentiation (LTP) in the presence of APV (Nicoll et al., 1988). Therefore, the opening of VSCC by depolarization does not induce enough Ca^{2+} entry, at least in the right compartment, to trigger LTP.

The most pressing question remaining to be elucidated is : what is the precise function of Qp receptors ? We have already discussed (Sladeczek et al., 1988) the fact that in both hippocampal slices (Nicoletti et al., 1986b) and synaptoneurosomes (Récasens et al., 1987), Q- and Glu-induced Ins-P accumulations are maximal during rat brain synaptogenesis and then decline thereafter. One of the most exciting hypotheses is that these receptors could be implicated in neuronal development and plasticity. In this context, a role in LTP has to be considered with care. Indeed, as shown in Fig. 8, it is clear that both NMDA and Qp receptors are able to trigger a similar pattern of signalling molecules : (1) they both increase Ca^{2+}$_i$, (2) they both stimulate arachidonic acid release, and finally (3) both Qp (Manzoni et al., 1989) and NMDA receptors (Vaccarino et al., 1987) activate PKC.

Such a pattern of signalling molecules are likely to be involved in LTP and one has to consider the possibility that at some particular synapses, Qp may play a role in this phenomenon. It is particularly interesting that pertussis toxin (PTX), described to block Qp responses in some models (Sugiyama et al., 1987 ; Nicoletti et al., 1988), has been reported to block LTP in hippocampus, especially the induction of LTP in mossy fiber / CA3 synapses (Ito et al., 1988 ; Goth and Pennefather, 1989). Furthermore, since it is not yet possible to specifically block Qp in physiological experiments, a key role of Qp in LTP even in CA1, cannot be excluded.

It has been shown that 2-amino-4-phosphonobutyrate (APB) an antagonist of another EAA receptor (Ibop) coupled to Ins-P production (see Sladeczek et al., 1988), eliminates the late phases of LTP in rat hippocampus (Reymann and Metthies, 1989). In this context, the report that not only Ca^{2+} influxes, but also release of intraneuronal Ca^{2+} (a response classically mediated by Ins(1,4,5)P3) appear necessary to induce LTP is intriguing (Obenaus et al., 1989). LTP has also been shown to be associated with an increase in the labelling of phosphoinositides (Lynch et al., 1988).

Finally, one has to discuss the possible roles of arachidonic acid in NMDA (as well as in Qp) action. There have been several studies suggesting a role for arachidonic acid release in LTP. Mepacrine inhibits the time course of a persistent LTP response and this inhibition is reversed by cis-unsaturated fatty acids (Linden et al., 1987). William and Bliss (1988) reported that nordihydroguaiaretic acid (a lipoxygenase but also a

PLA$_2$ inhibitor) greatly reduces the induction of LTP in hippocampus. The interesting property of arachidonic acid is its ability to diffuse out of the cell and to have post- as well as pre-synaptic effects. In both neuronal compartments, arachidonic acid could stimulate PKC-γ (an enzyme which is almost exclusively expressed in brain and particularly in hippocampus and especially well stimulated by arachidonic acid) (Nishizuka, 1988). Arachidonic acid is a good candidate for a transynaptic role in LTP, being produced in post-synaptic elements and triggering some pre-synaptic modifications, such as PKC-induced phosphorylation of pre-synaptic proteins (F1, for example), known to be correlated with the persistence of LTP (Nelson et al., 1989), or increase in glutamate release found during LTP by Dolphin et al. (1982) but not by other investigators (Aniksztejn et al., 1989).

ACKNOWLEDGEMENTS

We would like to thank Angie Turner-Madeuf and Mireille Passama for their help in preparing this manuscript. Support was provided by grants from the CNRS, INSERM, DRET and from Bayer Troponwerke (France and FRG).

REFERENCES

Aniksztejn, L., Roisin, M.P., Amsellem, R., and Ben Ari, 1989, Long-term potentiation in the hippocampus of the anesthetized rat is not associated with a sustained enhanced release of endogenous excitatory amino acids, **Neuroscience**, 28:387-392.

Balàzs, R., Jørgensen, O.S., and Hack, N., 1988, N-methyl-D-aspartate promotes the survival of cerebellar granule cells in culture, **Neuroscience**, 27:437-451.

Barnes, D.M., 1986, Neurotransmitters regulate growth cones, **Science**, 234:1325-1326.

Didier, M., Roux, P., Piechaczyk, M., Verrier, B., Bockaert, J., and Pin, J.P., 1989, Cerebellar granule cell survival and maturation induced by K$^+$ and NMDA correlate with c-fos proto-oncogene expression, **Neurosci. Lett.**, in press.

Doble, A., and Perrier, M.L., 1989, Pharmacology of excitatory amino acid receptors cou- pled to inositol phosphate metabolism in neonatal rat striatum, **Neurochemistry**, 15:1-8.

Dolphin, A.C., Errington, M.L., and Bliss, T.V.P., 1982, Long-term potentiation of the perforant path *in vitro* is associated with increased glutamate release, **Nature**, 297:496-498.

Drejer, J., and Honoré, T., 1988, New quinoxalinediones show potent antagonism of quisqualate response in cultured mouse cortical neurons, **Neurosci. Lett.**, 87:104-108.

Dumuis, A., Sebben, M., Haynes, L., Pin, J.P., and Bockaert, J., 1988, NMDA receptors activate the arachidonic acid cascade system in striatal neurons, **Nature**, 336:68-70.

Fong, T.M., Davidson, N., and Lester, H.A., 1988, Properties of two classes of rat brain acidic amino acid receptors induced by distinct mRNA populations in Xenopus oocyte synapses, **Synapse**, 2:657-665.

Garthwaite, J., and Garthwaite, G., 1987, Cerebellar origins of cyclic GMP responses to excitatory amino acid receptor agonists in rat cerebellum *in vitro*, **J. Neurochem.**, 48:29-39.

Garthwaite, J., Charles, S.L., and Chess-Williams, R., 1988, Endothelium derived relaxing factor release on activation of NMDA receptors suggests a role as intracellular messenger in the brain, **Nature**, 336:385.

Goth, J.W., and Pennefather, P.S., 1989, A pertussis toxin sensitive G protein in hippocampal long-term potentiation, **Science**, 244:980-983.

Hunnun, Y.A., and Bell, R.M., 1989, Functions of sphingolipids and sphingolipid breakdown products in cellular regulation, **Science**, 243, 500-507.

Ito, I., Okada, D., and Sugiyama, H., 1988, Pertussis toxin suppresses long-term potentiation of hippocampal mossy fiber synapses, **Neurosci. Lett.,** 90:181-185.

Lazarewicz, J.W., Wroblewski, J.T., Palmer, M.E., and Costa, E., 1988, Activation of N-methyl-D-aspartate sensitive glutamate receptors stimulates arachidonic acid release in primary cultures of cerebellar granule cells, **Neuropharmacology,** 27:765-769.

Linden, D.J., Shen, F.S., Murakami, K., and Routtenberg, A., 1987, Enhancement of long-term potentiation by cis-unsaturated fatty acids : relation to protein kinase C and phospholipase A2, **J. Neurosci.,** 7:3783-3792.

Lynch, M.A., Clements, M.P., Errington, M.L., and Bliss, T.V.P., 1988, Increased hydrolysis of phosphatidyl inositol-4,5,-bisphosphate in long-term potentiation, **Neurosci. Lett.,** 84:291-296.

Manzoni, O.J.J., Finiels-Marlier, F., Sassetti, I., Bockaert, J., Le Peuch, C., and Sladeczek, F., 1989, The glutamate receptors of the Qp type activates protein kinase C and is regulated by protein kinase C, **Neurosci. Lett.,** in press.

Mouillac, B., Balestre, M.N., and Guillon, G., 1989, Transient inositol (1,4,5) trisphos- phate accumulation under vasopressin stimulation in WRK1 cells : correlation with intracellular calcium mobilization, **Biochem. Biophys. Res. Commun.,** 159:953-960.

Murphy, S.N., and Miller, R.J., 1989, Two distinct quisqualate receptors regulate Ca^{2+} homeostasis in hippocampal neurons, **Mol. Pharmacol.,** 35:671-680.

Nelson, R.B., Linden, D.J., Hyman, C., Pfenninger, K.H., and Routtenberg, A., 1989, The two major phosphoproteins in growth cones are probably identical to two protein kinase C substrates correlated with persistance of long-term potentiation, **J. Neurosci.,** 9:381-389.

Nicoletti, F., Wroblewski, J.T., Novelli, A., Alho, H., Guidotti, A., and Costa, E., 1986a, The activation of inositol phospholipid metabolism as a signal transducing system for excitatory amino acids in primary cultures of cerebellar granule cells, **J. Neurosci.,** 6:1905-1911.

Nicoletti, F., Iadarola, M., Wroblewski, J.T., and Costa, E., 1986b, Excitatory amino acid recognition sites coupled with inositol phospholipid metabolism : Developmental changes and interaction with $\alpha1$ -adrenoceptors, **Proc. Natl. Acad. Sci. USA,** 83:1931-1935.

Nishizuka, A., Nicoletti, F., Wroblewski, J.P., Fadda, E., and Costa, E., 1988, The molecular heterogeneity of protein kinase C and its implication for cellular regulation, **Nature,** 334:661-665.

Nicoll, R.A., Kauer, J.A., and Malenka, R.C., 1988, The current excitement in long-term potentiation, **Neuron,** 1:97-103.

Obenaus, A., Mody, I., and Baimbridge, A., 1989, Dantrolene-Na+ (dantrium) blocks induction of long-term potentiation in hippocampal slices, **Neurosci. Lett,** 98:172-148.

Palmer, E., Monaghan, D.T., and Cotman, C.W., 1988, Glutamate receptors and phosphoinositide metabolism : stimulation via quisqualate receptors is inhibited by N-methyl-D-aspartate receptor activation, **Mol. Brain Res.,** 4:161-165.

Pearce, B., Albrecht, J., Morrow, C., and Murphy, S., 1986, Astrocyte glutamate receptor activation promotes inositol phospholipid turnover and calcium flux, **Neurosci. Lett.,** 72:335-340.

Pin, J.P., Van Vliet, B.J., and Bockaert, J., 1988, NMDA and kainate-evoked GABA release from striatal neurons differentiated in primary culture : differential blocking by phencyclidine, **Neurosci. Lett.,** 87:87-92.

Pin, J.P., Van Vliet, B.J., and Bockaert, J., 1989, Complex interaction between quisqualate and kainate receptors, as revealed by measurement of GABA release from striatal neurons in primary culture, **Eur. J. Pharmacol.,** 172:81-91.

Rassendren, F.A., Lory, P., Pin, J.P., Bockaert, J., and Nargeot, J., 1989, A specific quisqualate agonist inhibits kainate responses in Xenopus oocytes injected with rat brain RNA, **Neurosci. Lett.**, 99:333-339.

Récasens, M., Sassetti, I., Nourigat, A., Sladeczek, F., and Bockaert, J., 1987, Characterization of subtypes of excitatory amino acid receptors involved in the stimulation of inositol phosphate synthesis in rat brain synaptoneurosomes, **Eur. J. Pharmacol.**, 141:87-93.

Récasens, M., Guiramand, J., Nourigat, A., Sassetti, I., and Devilliers, G., 1988, A new quisqualate receptor subtype (SAA2) responsible for the glutamate induced inositol phosphate formation in rat brain synaptoneurosomes, **Neurochem. Int.**, 13:463-467.

Reiser, G., and Hamprecht, B., 1989, Serotonin raises the cyclic GMP level in a neuronal cell line via 5-HT$_3$ receptors, **Eur. J. Pharmacol.**, 172:195-198.

Reymann, K.G., and Matthies, H., 1989, 2-amino-4-phosphonobutyrate selectively eliminates late phases of long term potentiation in rat hippocampus, **Neurosci. Lett.**, 98:166-171.

Schoepp, D.D., and Johnson, B.G., 1988, Excitatory amino acid agonist-antagonist interactions at 2-amino-4-phosphonobutyric acid-sensitive quisqualate receptors coupled to phosphoinositide hydrolysis in slices of rat hippocampus, **J. Neurochem.**, 50:1605-1613.

Sladeczek, F., Pin, J.P., Récasens, M., Bockaert, J., and Weiss, S., 1985, Glutamate stimulates inositol phosphate formation in striatal neurons, **Nature**, 317:717-719.

Sladeczek, F., Récasens, M., and Bockaert, J., 1988, A new mechanism for glutamate receptor action : phosphoinositide hydrolysis, **Trends Neurosci.**, 11:545-549.

Sugiyama, H., Ito, I., and Hirono, C., 1987, A new type of glutamate receptor linked to inositol phospholipid metabolism, **Nature**, 325:531-533.

TerBush, D.R., Bittner, M.A., and Holz, R.W., 1988, Ca^{2+} influx causes rapid translocation of protein kinase C to membranes, **J. Biol. Chem.**, 263:18873-18879.

Vaccarino, F., Guidotti, A., and Costa, E., 1987, Ganglioside inhibition of glutamate mediated protein kinase C translocation in primary cultures of cerebellar neurons, **Proc. Natl. Acad. Sci. USA**, 84:8707-8711.

Verdoorn, T.A., and Dingledine, R., 1989, Excitatory amino acid receptors expressed in Xenopus oocytes : agonist pharmacology, **Mol. Pharmacol.**, 34:298-307.

Watkins, J.C., and Olverman, H.J., 1987, Agonists and antagonists for excitatory amino acid receptors, **Trends Neurosci.**, 10:265-272.

Weiss, S., Pin, J.P., Sebben, M., Kemp, D.E., Sladeczek, F., Gabrion, J., and Bockaert, J., 1986, Synaptogenesis of cultured striatal neurons in serum-free medium : a morphological and biochemical study, **Proc. Natl. Acad. USA**, 83:2238-2242.

Williams, J.H., and Bliss, T.V.P., 1988, Induction but not maintenance of calcium-induced long term potentiation in dentate gyrus and area CA1 of the hippocampal slice is blocked by nordihydroguaiaretic acid, **Neurosci. Lett.**, 88:81-85.

THE GLYCINE SITE ON THE NMDA RECEPTOR: PHARMACOLOGY AND INVOLVEMENT IN

NMDA RECEPTOR-MEDIATED NEURODEGENERATION

A.C. Foster, A.E. Donald, C.L. Willis, R. Tridgett, J.A. Kemp and
T. Priestley

Merck, Sharp and Dohme Ltd., Neuroscience Research Centre, Terlings Park
Eastwick Rd., Harlow, Essex, CM20 2QR, U.K.

INTRODUCTION

It is now firmly established that the widespread actions of the excitatory amino acid neurotransmitters glutamate and aspartate in the mammalian central nervous system (CNS) are mediated by distinct receptor sub-types. Amongst these, the N-methyl-D-aspartate (NMDA) receptor
has been characterized most fully in terms of its pharmacological and biophysical properties
(Watkins and Evans, 1981; Mayer and Westbrook, 1987). The NMDA receptor can be divided into
three functional domains (Foster and Fagg, 1987): a transmitter recognition site, an ion channel and
an allosteric site activated by glycine. This chapter will review the glycine site on the NMDA
receptor, describe its pharmacology, including the identification of selective antagonists, and discuss experiments with these compounds which have attempted to determine the relevance of activation of the glycine site **in vivo** for neurodegenerative phenomena mediated by NMDA receptor
activation.

THE GLYCINE SITE ON THE NMDA RECEPTOR

Johnson and Ascher (1987) were the first to describe the ability of glycine to potentiate
NMDA receptor-mediated neuronal responses. Using patch clamp recording from neurons in culture
they identified an endogenous factor present in the culture medium which was responsible for the
full expression of NMDA responses. Further experiments identified this as glycine, which is
released into the medium by the cultured cells. The potentiating effect of glycine was specific for
NMDA receptor-mediated responses vis-a-vis those for other excitatory amino acid receptors, and
was insensitive to strychnine, ruling out a mechanism involving the classical inhibitory receptors
for glycine. Only low concentrations of glycine were required (threshold 10 nM, maximum effect 1 -
10 M), and serine and alanine also produced the effect but at higher concentrations than those
required for glycine. Glycine potentiation of NMDA receptor responses occured in isolated membrane patches, ruling out mediation by intracellular second messengers, and single channel recording indicated that the effect resulted from an increase in the frequency of channel opening. In another patch clamp study, Mayer et al (1989) have suggested that one effect of glycine is to increase
the rate of recovery of the NMDA receptor from desensitization.

Johnson and Ascher (1987) proposed that glycine may act at an allosteric site within the
NMDA receptor complex. This idea has rapidly gained acceptance from several lines of investigation: (1) high-affinity, strychnine-insensitive binding sites for [3H]glycine are present in the
mammalian CNS (see below), which have an anatomical distribution similar to that of the

Excitatory Amino Acids and Neuronal Plasticity
Edited by Y. Ben-Ari
Plenum Press, New York, 1990

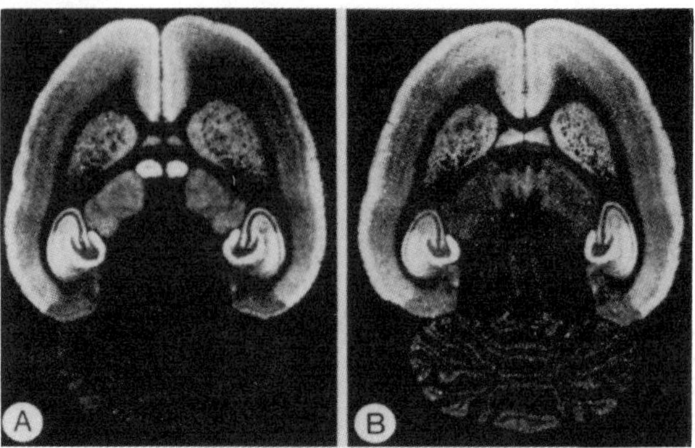

Fig. 1.   Autoradiograms of [3H]glycine (A) and NMDA-sensitive L-[3H]glutamate binding (B) to horizontal sections from rat brain. Note the similarity of binding site distribution (light areas).

NMDA receptor (Bristow et al, 1986; figure 1); (2) glycine facilitation occurs with NMDA receptors expressed in **Xenopus** oocytes (Verdoorn et al , 1987) and in soluble receptor preparations (McKernan et al, 1989), suggesting that the glycine site is an intergral part of the receptor protein; (3) allosteric interactions between the glycine site, and the ion channel and transmitter recognition sites have been observed (Wong et al, 1988; Bonhaus et al, 1988; Reynolds et al, 1988; Monaghan et al, 1988); (4) selective antagonists of the glycine site have been described (see below).

The magnitude of the enhancement of NMDA receptor responses by glycine is so great that when attempts are made to remove all glycine from the receptor vicinity little or no NMDA response can be elicited (Kleckner and Dingledine, 1988). This has prompted the suggestion that glycine may act as a co-agonist at the NMDA receptor (along with endogenous ligands at the transmitter recognition site, e.g. glutamate and aspartate). Thus, activation of the NMDA receptor may require the binding of two separate ligands at two distinct recognition sites (Kleckner and Dingledine, 1988). The magnitude of facilitation by glycine is, therefore, much greater than that for benzodiazepines at the $GABA_A$ receptor, the prototypical allosteric modulation identified for an amino acid neurotransmitter receptor. Currently, much attention is being focussed on the importance of the glycine site for NMDA receptor function under both physiological and pathophysiological conditions. The questions of whether the synaptic levels of glycine are sufficient to saturate the site on the NMDA receptor (cerebrospinal fluid levels of glycine are approx. 6 uM), or if mechanisms exist to control the degree of activation of the glycine site as a means of regulating NMDA receptor function, are being pursued in several laboratories (see Foster and Kemp, 1989a).

PHARMACOLOGY OF THE GLYCINE SITE

The pharmacological characteristics of this newly discovered recognition site for glycine on the NMDA receptor have been investigated with a variety of biochemical and physiological techniques, and several selective agonists and antagonists have been identified. High-affinity, strychnine-insensitive binding sites for [3H]glycine in rat forebrain regions were originally identified by Kishimoto et al (1981), which are now known to be those associated with the NMDA receptor. Our own studies with synaptic plasma membranes from rat cerebral cortex have revealed [3H] glycine binding sites with a dissociation constant (Kd) of 371 nM and a maximum number of sites (Bmax) of 7.23 pmol/mg protein (Donald et al, 1988). The structure-activity requirements for the [3H]glycine binding sites are rather strict. Amongst common amino acids, only glycine, serine and alanine have reasonable affinity, with a marked preference for the D-enantiomers of serine and alanine (Kishimoto et al, 1981; Snell et al, 1988; table 1). Derivatization of the amino group

Table 1.   Affinity and selectivity of glycine site ligands.

	[3H]Gly	L-[3H]Glu*	[3H]AMPA	[3H]KA	[3H]STR
Glycine	0.18				25
D-serine	0.57				>100
L-serine	30.3				>100
KYNA	41.0	184	101	>1000	>100
7-Cl KYNA	0.56	169	153	>100	>100
HA-966	17.0	>1000	>1000	>1000	>100

Values are IC50 in M. * NMDA-sensitive L-[3H]glutamate binding. KA = kainate; STR = strychnine. Values are from Kemp et al, 1988, Foster and Kemp, 1989 and unpublished observations.

of glycine, as in sarcosine, or the carboxyl group, as in glycine methyl ester, replacement of the carboxyl group with a phosphonate or sulphonate, as in 1-aminopropylphosponate or 1-aminopropylsulphonate, or increasing the carbon chain length, as in B-alanine, all lead to a large decrease of affinity (Donald, Tridgett and Foster - unpublished observations). 1-Amino-1-carboxycyclopropane (ACC) has been identified as a potent agonist at the glycine site (Nadler et al, 1988). The structure-activity profile of the [3H]glycine binding sites corresponds well with that for the facilitation of NMDA responses as measured in [3H]-(+)5-methyl-10,11-dihydro-5H dibenzo-[a,d]cyclohepten-5,10-imine maleate and [3H]thienyl phencyclidine (TCP) binding assays (Reynolds et al, 1987; Snell et al, 1988), further indicating that [3H]glycine binding sites represent the allosteric site associated with the NMDA receptor.

Several antagonists at the glycine site have now been identified (figure 2), some of which had been known as NMDA receptor antagonists for several years. The first was kynurenate (Watson et al, 1988; Kessler et al, 1989), a naturally-occuring substance, which is a broad-spectrum excitatory amino acid antagonist (Ganong et al, 1983). However, substitution of chlorine in the 7-position of this compound resulted in a selective 20-fold increase of affinity for the glycine site, making 7-chlorokynurenate (7-Cl KYNA) a moderately selective and potent glycine antagonist (Kemp et al, 1988; table 1). 3-Amino-1-hydroxypyrrolid-2-one (HA-966) was amongst the first substances reported as an excitatory amino acid antagonist (Davies and Watkins, 1972), and its relatively selective NMDA receptor antagonist effects are now known to be the result of an

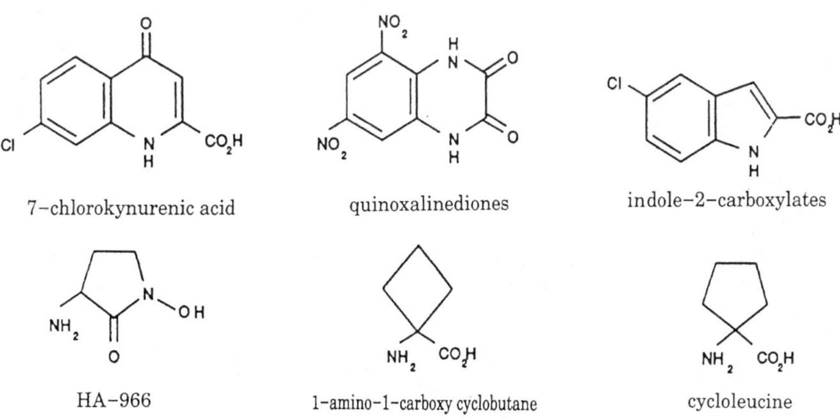

Fig. 2.   Structures of glycine antagonists.

interaction with the glycine site (Fletcher and Lodge, 1988; Foster and Kemp, 1989b; table 1).
A further series of active compounds include 1-amino-1-carboxycyclobutane (ACBC; Hood et al,
1989) and its weaker analogue cycloleucine (Snell and Johnson, 1988). Substituted indole-2-carbo-
xylates also possess glycine antagonist activity (Huettner, 1989). The quisqualate/kainate receptor
antagonists 6,7-dinitro-, 6-cyano, 7-nitro-, and 5,7-dinitro- derivatives of quinoxaline-2,3-dione
(DNQX, CNQX and MNQX, respectively) have, in addition, glycine antagonist effects (Birch et
al, 1988; Drejer et al, 1989), as does the related compound 6,7-dichloro-3-hydroxy-2-quinoxaline-
carboxylate (Birch et al, 1989).

In experiments using a functional measure of NMDA receptor activation, the following strategies have been used to determine whether a compound antagonises the receptor by competing at the
glycine site. With whole tissue slices, the extracellular levels of glycine may be sufficiently high
to maximally occupy the receptor (as indicated by the inability of glycine or D-serine to potentiate
NMDA receptor responses - see Foster and Kemp, 1989a), therefore antagonists which act via the
glycine site can block the receptor in the absence of any added glycine. Proof of a selective antagonism of the glycine site may be obtained by the demonstration of reversal by exogenous glycine or
D serine (Kemp et al, 1988; Fletcher and Lodge, 1988; Watson et al, 1988; Foster and Kemp, 1989b;
note that glycine or D-serine do not reverse the effects of competitive or un-competitive NMDA
receptor antagonists). This strategy may also be applicable **in vivo**. A more direct and quantitative
analysis can be made using patch clamp techniques with neurones in culture or **Xenopus** oocytes in
which receptors have been expressed by the injection of mRNA. In this experimental situation, the
levels of extracellular glycine can be more easily controlled, and competition between glycine and
the suspected antagonist demonstrated directly (see Kemp et al, 1988; Foster and Kemp, 1989b;
Huettner, 1989).

Functional assays of NMDA receptors have revealed some differences in the antagonist
effects of glycine site ligands. In rat cortical slice experiments, the antagonism of NMDA responses
by 7-Cl KYNA cannot be overcome by raising the NMDA concentration (Kemp et al, 1988), whereas
that produced by HA-966 is surmountable and results in a maximal shift in the NMDA concentration-response curve of 0.5 log units (Foster and Kemp, 1989b). In patch clamp experiments with cortical neurones in culture, 7-Cl KYNA blocks "basal" NMDA responses evoked in the absence of added
glycine, in addition to glycine facilitation of the responses (Kemp et al, 1988), whereas HA-966
blocks the glycine facilitation, but does not reduce basal NMDA responses (Foster and Kemp,
1989b). If one accepts that glycine is an absolute requirement for NMDA receptor activation
(Kleckner and Dingledine, 1988; see above), these results suggest that HA-966 may be a weak
partial agonist at the glycine site, and that 7-Cl KYNA is a full antagonist. There is also evidence
to indicate that D-cycloserine is a partial agonist at the glycine site (Hood et al, 1988), but this
compound has a greater degree of efficacy than HA-966 since it produces very little antagonism of
NMDA responses in rat cortical slices (Kemp, unpublished observations). Therefore, it appears
that, in common with many other neurotransmitter receptor sites, ligands at the glycine site on the
NMDA receptor may have differing efficacies, ranging from full agonists (glycine and D-serine) to
partial agonists (D-cycloserine and HA-966) to full antagonists (7-Cl KYNA). By analogy with
the GABA$_A$ receptor, the possibility exists that "inverse agonists" may also be found, although
given the dependence of NMDA receptor responses upon glycine site activation, these may prove
difficult to distinguish from full antagonists.

NEUROPROTECTIVE EFFECTS OF GLYCINE SITE ANTAGONISTS

The availability of selective antagonists for the glycine site on the NMDA receptor provides
the opportunity to investigate the role of activation of this site in neurodegenerative events. We
have begun such studies by assessing the ability of 7-Cl KYNA and HA-966 to prevent neuronal
degeneration in the rat brain caused by intracranial injection of excitotoxins. Quinolinate is an endogenous NMDA receptor agonist (Stone and Perkins, 1981) and causes typical excitotoxic "axon-
sparing" lesions when injected directly into the CNS, which are blocked by administration of
NMDA receptor antagonists (Schwarcz et al, 1983; Foster et al, 1988). Previous studies with the
uncompetitive NMDA receptor antagonist MK-801 demonstrated that the degeneration of hippocampal and striatal neurones caused by quinolinate (or NMDA) was of a delayed type, reminiscent
of that which occurs following a period of cerebral ischaemia (Foster et al, 1988).

Table 2.   Reversal of neuroprotective effects of 7-Cl KYNA and HA-966 by D-serine.

% Decrease of enzyme activity

2nd injection	CAT	GAD	N
Control (PBS)	69.0 + 6.1	73.4 + 3.4	12
D-ser 1 mol	48.5 + 6.4	67.3 + 7.6	5
L-ser 1 mol	56.6 + 8.4	77.6 + 5.8	5
7-CK 50nmol	4.4 + 3.3**	23.7 + 5.4**	5
7-CK 50nmol + D-ser 1 mol	50.1 + 15.6	61.2 + 12.6	5
7-CK 50nmol + L-ser 1 mol	19.3 + 5.0**	48.9 + 4.9*	5
HA 500nmol	21.9 + 3.9**	52.1 + 1.0*	4
HA 500nmol + D-ser 1 mol	75.0 + 7.4	74.7 + 4.4	5
HA 500nmol + L-ser 1 mol	27.7 + 4.7*	49.7 + 3.5*	5

Each animal received an intrastriatal injection of 200 nmol quinolinate followed 1 hour later by the compound(s) indicated. Values are mean + SEM of N animals. Significance of difference from respective control; * P<0.05, **P<0.01, Duncan's test. ser: serine; 7-CK: 7-chlorokynurenate; HA: HA-966; CAT: choline acetyltransferase; GAD: glutamate decarboxylase.

As an initial evaluation of its neuroprotective effects, 7-Cl KYNA was injected unilaterally into the rat striatum 1 hour after an intrastriatal injection of 200 nmol quinolinate (this time of antagonist administration was based on previous results with MK-801; Foster et al, 1988); control animals received an intrastriatal injection of phosphate-buffered saline (PBS; vehicle) instead of 7-Cl KYNA. Animals were killed 7 days later, and their striata dissected for measurements of choline acetyltransferase (CAT) and glutamate decarboxylase (GAD) activities, marker enzymes for striatal cholinergic and GABAergic neurones, respectively. Over a range of 10 - 50 nmol, 7-Cl KYNA gave a dose-dependent protection of striatal cholinergic and GABAergic neurones, with essentially complete protection at 50 nmol (table 2; Tridgett and Foster, 1988). HA-966 was tested in the same model and at a dose of 500 nmol gave significant, but incomplete protection of striatal neurones (table 2; Foster, Willis and Tridgett, submitted). Solubility constraints prevented the testing of higher doses of HA-966. Therefore, the delayed administration of these selective glycine antagonists attenuated neuronal degeneration in the striatum caused by NMDA receptor overactivation.

The compounds were also tested by direct injection into the rat hippocampus 1 hour after an intrahippocampal injection of 60 nmol quinolinate. Animals were perfused with fixative 7 days later and their brains processed for light microscopy. The extensive degeneration of pyramidal and granule neurones in the dorsal hippocampus caused by quinolinate was completely prevented by 50 nmol 7-Cl KYNA and partially protected by 500 nmol HA-966 (Foster, Willis and Tridgett, submitted).

The timecourse of neuroprotective effects was investigated following quinolinate (200 nmol) injection into the rat striatum. The temporal profile of protection by 7-Cl KYNA (50 nmol) and HA-966 (500 nmol) was similar to that previously observed with MK-801 (Foster et al, 1988), i.e. striatal cholinergic and GABAergic neurones were well protected when the antagonists were co-injected with quinolinate or administered 1 or 2 hours after quinolinate, but the effects diminished with administration at 5 hours and were absent at 8 hours following quinolinate (Foster, Willis and Tridgett, submitted).

The selectivity of 7-Cl KYNA and HA-966 for NMDA receptor-mediated neurodegenerative events was indicated by the fact that both compounds significantly protected against striatal neurodegeneration caused by intrastriatal injection of NMDA (200 nmol), but not that caused by the quisqualate receptor agonist, -amino-3-hydroxy-5-methylisoxazole-4-propionate (AMPA; 50 nmol) or kainate (5 nmol; Foster, Willis and Tridgett, submitted). To prove that the neuroprotective effects of 7-Cl KYNA and HA-966 are mediated by antagonism of the glycine site on the NMDA receptor, we attempted to reverse them by administration of D-serine. D-serine was chosen for these experiments (rather than glycine) because it has high affinity for the glycine site but is a poor ligand for strychnine-sensitive glycine receptors, and L-serine can be used as a control since its affinity for the glycine site is 60-fold lower than that of D-serine (see Table 1). As shown in Table 2, D-serine (1 µmol) completely reversed the antagonism of quinolinate-induced striatal neurodegeneration caused by 7-Cl KYNA and HA-966, but L-serine was either weaker or inactive.

CONCLUSIONS

Considering the short period of time since the effect of glycine on the NMDA receptor was first reported, rapid progress has been made to define the pharmacology of this interaction and develop specific agonists and antagonists. The extreme dependence of NMDA receptor function upon glycine site activation which is apparent from experiments using cultured neurones or **Xenopus** oocytes is also a feature of NMDA receptors in the living mammalian CNS, as inferred from previous studies with HA-966 *in vivo* (see refs. in Foster and Kemp, 1989). At present it is not clear whether glycine site activation constitutes a further means of regulating NMDA receptor function under physiological conditions, although this would seem likely unless this site is completely redundant. There is some evidence to indicate that the glycine site is not always maximally activated in vivo (see Foster and Kemp, 1989a). Similarly, further work is required to define the role of glycine in neuropathological events produced by excessive NMDA receptor activity. The results summarised above indicate that glycine site activation is required for the expression of the delayed neuronal degeneration caused by intracerebral injection of NMDA receptor agonists. Further insights can be anticipated from studies of the effects of glycine site antagonists in animal models of epilepsy, cerebral ischaemia and CNS trauma. Central to all of these issues is the need to understand the synaptic mechanisms which regulate the levels of glycine in the vicinity of NMDA receptors. The processes responsible for the uptake, release and metabolism of glycine in CNS areas where NMDA receptors are prominent must be evaluated, in addition to the possibility that endogenous glycine antagonists (kynurenic acid?) play a role in regulating the degree of glycine site activation. Such studies will shed further light on the fascinating interaction between glycine and the NMDA receptor.

REFERENCES

Birch, P. J., Grossman, C. J. , and Hayes, A. G., 1988, 6,7-Dinitro-quinoxaline-2,3-dione and 6-nitro-7-cyano-quinoxaline-2,3-dione antagonise responses to NMDA in the rat spinal cord via an action at the strychnine-insensitive glycine receptor, **Eur. J. Pharmacol.,** 156:177.

Birch, P. J., Grossman, C. J., and Hayes, A. G., 1989,Antagonist profile of 6,7-dichloro-3-hydroxy-2-quinoxaline carboxylate at excitatory amino acid receptors in the neonatal rat spinal cord, **Eur. J. Pharmacol.,** 163:127.

Bonhaus, D. W., Burge, B. C., and McNamara, J. O., 1987, Biochemical evidence that glycine allosterically regulates an NMDA receptor-coupled ion channel, **Eur. J. Pharmacol.,** 142:489.

Bristow, D. R., Bowery, N. G., and Woodruff, G. N., 1986, Light microscopic autoradiographic localization of [3H]glycine and [3H]strychnine binding sites in rat brain, **Eur. J. Pharmacol.,** 126:303.

Davies, J., and Watkins, J. C., 1972, Is 1-hydroxy-3-amino- pyrrolidone-2 (HA-966) a selective excitatory amino acid antagonist ? **Nature New Biol.,** 328:61.

Donald, A. E., Tridgett, R., and Foster, A. C., 1988, Characterization of [3H]glycine binding to a modulatory site within the N-methyl-D-aspartate receptor complex from rat brain, **Br. J. Pharmacol. Proc. Suppl.,** 95:892P.

Drejer, J., Jensen, L. H., Sheardown, M., and Honore, T., 1989, A new potent glycine antagonist FG9067 (MNQX) shows anticonvulsant activity, **J. Neurochem.** 52(Suppl.):S42.

Fletcher, E. J., and Lodge, D., 1988, Glycine reverses antagonism of N-methyl-D-aspartate (NMDA) by 1-hydroxy- 3-amino-pyrrolidone-2 (HA-966) but not by D-2-amino-5-phosphonovalerate (D-AP5) on rat cortical slices, **Eur. J. Pharmacol.,** 151:161.

Foster, A. C., and Fagg, G. E., 1987, Neurobiology: taking apart NMDA receptors, **Nature**, 329:395.

Foster, A. C., Gill, R., and Woodruff, G. N., 1988, Neuroprotective effects of MK-801 **in vivo**: selectivity and evidence for delayed degeneration mediated by NMDA receptor activation, **J. Neurosci.**, 8:4745.

Foster, A. C., and Kemp, J. A., 1989a, Neurobiology: glycine maintains excitement, **Nature**, 338:377.

Foster, A. C., and Kemp, J. A., 1989b, HA-966 antagonizes N- methyl-D-aspartate receptors through a selective interaction with the glycine modulatory site, **J. Neurosci.**, 9:2191.

Ganong, A. H., Lanthorn, T. H., and Cotman, C. W., 1983, Kynurenic acid inhibits synaptic and acidic amino acid-induced responses in the rat hippocampus and spinal cord, **Brain Res.**, 273:170.

Hood, W. F., Sun, E. T., Compton, R. P., and Monahan, J. B., 1989, 1-Aminocyclobutane-1-carboxylate (ACBC) : a specific antagonist of the N-methyl-D-aspartate receptor coupled glycine receptor, **Eur. J. Pharmacol.**, 161:281.

Hood, W. F., Compton, R. P., and Monahan, J. B., 1989, D-cycloserine: a ligand for the N-methyl-D-aspartate coupled glycine receptor has partial agonist characteristics, **Neurosci. Lett.**, 98:91.

Huettner, J. E., 1989, Indole-2-carboxylic acid: a competitive antagonist of potentiation by glycine at the NMDA receptor, **Science**, 243:1611.

Johnson, J. W., and Ascher, P., 1987, Glycine potentiates the NMDA response in cultured mouse brain neurones, **Nature**, 325:529.

Kemp, J. A., Foster, A. C., Leeson, P. D., Priestley, T., Tridgett, R., Iversen, L. L., and Woodruff, G. N., 1988, 7-Chlorokynurenic acid is a selective antagonist at the glycine modulatory site of the N-methyl-D-aspartate receptor complex, **Proc. Natl. Acad. Sci. (USA)**, 85:6547.

Kessler, M., Terramini, T., Lynch, G., and Baudry, M., 1989, A glycine site associated with the N-methyl-D-aspartic acid receptor: characterization and identification of a new class of antagonists, **J. Neurochem.**, 52:1319.

Kishimoto, H., Simon, J. R., and Aprison, M. H., 1981, Determination of equilibrium dissociation constants and number of glycine binding sites in several areas of the rat central nervous system, **J. Neurochem.**, 37:1015.

Kleckner, N. W., and Dingledine, R., 1988, Requirement for glycine in activation of NMDA receptors expressed in Xenopus oocytes, **Science**, 241:835.

Mayer, M. L., and Westbrook, G. L., 1987, The physiology of excitatory amino acids in the vertebrate central nervous system, **Prog. Neurobiol.**, 28:197.

Mayer, M. L., Vyklicky Jr., L., and Clements, J., 1989, Regulation of NMDA receptor desensitization in mouse hippocampal neurons by glycine, **Nature,** 338:425.

McKernan, R. M., Castro, S., Poat, J., and Wong, E. H. F., 1989, Solubilization of the N-methyl-D-aspartate receptor channel complex from rat and porcine brain, **J. Neurochem.**, 52:777.

Monaghan, D. T., Olverman, H. J., Nguyen, L., Watkins, J. C., and Cotman, C. W., 1988, Two classes of N-methyl-D-aspartate recognition sites: differential distribution and differential regulation by glycine, **Proc. Natl. Acad. Sci. (USA)**, 85:9836.

Nadler, V., Kloog, Y., and Sokolovsky, M., 1988, 1-Aminocyclopropane-1-carboxylic acid (ACC) mimics the effects of glycine on the NMDA receptor ion channel, **Eur. J. Pharmacol.**, 157:115.

Reynolds, I. J., Murphy, S. N., and Miller, R. J., 1987, [3H]Labelled MK-801 binding to excitatory amino acid receptor complex from rat brain is enhanced by glycine, **Proc. Natl. Acad. Sci. (USA)**, 84:7744.

Schwarcz, R., Whetsell Jr., W. O., and Mangano, R. M., 1983, Quinolinic acid: an endogenous metabolite that produces axon-sparing lesions in rat brain, **Science**, 219:316.

Snell, L. D., and Johnson, K. M., 1988, Cycloleucine competitively antagonizes the strychnine-insensitive glycine receptor, **Eur. J. Pharmacol.**, 151:165.

Snell, L. D., Morter, R. S., and Johnson, K. M., 1988, Structural requirements for activation of the glycine receptor that modulates the N-methyl-D-aspartate operated ion channel, **Eur. J. Pharmacol.**, 156:105.

Stone, T. W., and Perkins, M. N., 1981, Quinolinic acid: a potent endogenous excitant at amino acid receptors in CNS, **Eur. J. Pharmacol.**, 72:411.

Tridgett, R., and Foster, A. C., 1988, Prevention of NMDA receptor-mediated neurodegeneration in the rat striatum by an antagonist at the glycine modulatory site, **Br. J. Pharmacol. Proc. Suppl.**, 95:890P.

Verdoorn, T., Kleckner, N. K., and Dingledine, R., 1987, Rat brain NMDA receptors expressed in **Xenopus** oocytes, **Science**, 238:1114.

Watkins, J. C., and Evans, R. H., 1981, Excitatory amino acid transmitters, **Ann. Rev. Pharmacol. Toxicol.**, 21:165.

Watson, G. B., Hood, W. F., Monahan, J. B., and Lanthorn, T. H., 1988, Kynurenate antagonizes actions of N-methyl-D-aspartate through a glycine-sensitive receptor, **Neurosci. Res. Commun.**, 2:169.

Wong, E. H. F., Knight, A. R., and Ransom, R., 1987, Glycine modulates [3H]MK-801 binding to the NMDA receptor in rat brain, **Eur. J. Pharmacol.**, 142:487.

EVIDENCE FOR GLUTAMATE RECEPTOR SUBTYPES FROM IN VIVO ELECTROPHYSIOLOGY:

STUDIES WITH HA-966, QUINOXALINEDIONES AND PHILANTHOTOXIN

David Lodge and Martyn G. Jones

Department of Veterinary Basic Sciences
Royal Veterinary College
London NW1 0TU

INTRODUCTION

In this chapter we will briefly review some of the evidence from
electrophysiological studies of mammalian neurones in vivo which relates to
the subtypes of glutamate receptors. In particular we will concentrate on
recent experiments elucidating the mechanism of action of HA-966 as an
N-methyl-D-aspartate (NMDA) antagonist and of a new tricyclic
quinoxalinedione (NBQX) and a wasp venom, philanthotoxin, as quisqualate
antagonists. These studies were designed to test the hypothesis that the
NMDA subtype and one other non-NMDA subtype of glutamate receptor mediate
the excitation of rat spinal and brainstem neurones produced by acidic
amino acids.

Since the pioneering studies of Curtis and Watkins (1963), many
analogues of glutamate have been shown to excite neurones. Recent reviews
in the last ten years or so accept the proposition that three major
receptor subtypes mediate the postsynaptic depolarising actions of
glutamate. These are named the NMDA, quisqualate and kainate receptors
after the agonists which were described in the late 1970s to have
differential effects on neurones in various parts of the CNS, which were
later shown to have differential sensitivity to antagonists, and for which
distinct binding sites could be demonstrated in vitro (see Watkins and
Evans, 1981; McLennan, 1983; Foster and Fagg, 1984). It is now apparent
that quisqualate has been superceded by alpha-amino-3-hydroxy-5-methyl-4-
isoxazolepropionic acid (AMPA; Krogsgaard-Larsen et al., 1980) as a probe
for the ionotropic quisqualate receptor. During the 1980s there have been
many advances in excitatory amino acid pharmacology, particularly in
relation to the receptors and ion channels activated by NMDA (Watkins and
Collingridge, 1989).

Pharmacology of the NMDA Receptor-Ionophore Complex

Structure-activity studies of the NMDA recognition site have led to
the development of potent and selective antagonists. The most potent of
these are the piperazine and piperidine derivatives of D-alpha-aminoadipate
and D-alpha-aminosuberate in which the terminal carboxylic acid has been
replaced by a phophonate moiety (see Watkins, Krogsgaard-Larsen and Honore,
1990), although similar potency can be achieved by tetrazole replacement of
the phosphonate of CGS-19755 (Ornstein et al., 1989). None of these

compounds affect the actions of quisqualate or kainate to any great extent,
findings which are in keeping with a distinct NMDA-preferring subtype of
glutamate receptor.

Since the discovery that ketamine and phencyclidine were selective
NMDA antagonists (Lodge and Anis, 1972), several arylcyclohexylamines,
dioxalanes, benzomorphans, morphinans and propanolamines have been shown to
block excitatory actions of NMDA on central neurones in a non-competitive
manner (Lodge et al., 1988; Lodge and Johnson, 1990). The relative potency
of these compounds as NMDA antagonists correlates well with their potency
both at the phencyclidine (PCP) binding site and in behavioural studies
(see Aram et al. 1989; Martin and Lodge, 1989). A dibenzocycloheptenimine,
MK-801, is the most potent of this series of compounds and shows clear
use-dependency in its mode of action (Wong et al. 1986). This, together
with biophysical studies showing that the NMDA antagonism by PCP-like
compounds is voltage-dependent (see MacDonald this volume) and the agonist
dependency of PCP binding (Loo et al., 1986), suggests that these compounds
act at a site in the ion channel opened by activation of the NMDA receptor.

Although NMDA, quisqualate and kainate activate similar subconductance
levels of ion channels coupled to their receptors (Jahr and Stevens, 1987;
Cull-Candy and Usowicz, 1987), the selectivity of PCP and of magnesium
(Ault et al., 1980), which also produces a voltage-dependent channel block
of the NMDA activated channels (Nowak et al., 1984; Mayer et al., 1984),
suggests that these latter channels are pharmacologically distinct from
those activated by quisqualate and kainate.

GLYCINE MODULATION OF NMDA ACTIONS AND ITS BLOCK BY HA-966

In patch clamp studies of neurones in culture, Johnson and Ascher
(1987) showed that glycine greatly facilitated the action of NMDA, near
maximal effects being achieved with 1uM glycine. This action was mimicked
by D-, but not L-, serine and was not blocked by strychnine. Similar
results have been achieved in many isolated preparations and the presence
of a glycine modulatory site on the NMDA receptor channel complex is now
well established (see other chapters in this volume). However in cerebral
cortex slice preparations, we were unable to show that glycine or D-serine
would modulate the action of NMDA except in the presence of some
antagonists (see Fletcher and Lodge, 1988).

We have therefore investigated the interaction of glycine and NMDA on
trigeminal, cuneate, gracile and reticular neurones in the brainstem of
pentobarbitone-anaesthetised rats using microelectrophoretic application of
amino acids and drugs, while continuously recording extracellularly the
firing rate of single neurones from the centre barrel of the seven barrel
glass microelectrodes. The techniques have been more fully described
previously (Anis et al., 1983).

Neither glycine nor D-serine selectively increased NMDA actions. More
usually glycine caused some reduction of responses to all excitatory amino
acid agonists, an effect mediated via the strychnine-sensitive inhibitory
glycine receptor. Even during the ejection of strychnine to block its
inhibitory action, glycine did not potentiate responses to NMDA. On some
occasions, however, when the glycine was ejected from a 200mM solution at
pH less than 4, enhancement of NMDA-induced excitation was observed. This
facilitatory action was not confined to NMDA because responses to
quisqualate and kainate were also increased but not always to the same
extent as that of NMDA. Since this was not seen when glycine was ejected
from a solution at the more usual pH of 4-4.5, it is assumed that this
facilitation is due to the established excitatory action of hydrogen ions.

The parsimonious explanation for our failure to replicate Johnson and Ascher's observation is that in vivo the extracellular concentration of glycine or other glycine-like substances is sufficient to saturate the glycine modulatory site on the NMDA receptor channel complex. We therefore decided to see whether we could test this explanation by finding and testing glycine antagonists. During the course of these experiments, Kessler et al. (1987) reported that kynurenate displaced glycine binding. Independently, we found that, on brainstem neurones, kynurenate reduced preferentially responses to NMDA and this effect, but not the reduction of quisqualate and kainate, could be reversed by glycine. Because kynurenate also has actions at the NMDA and non-NMDA recognition sites, we also searched for compounds that might be more selective for the glycine site.

3-Amino-1-hydroxy-pyrrolidone-2 (HA-966) is a reasonably selective NMDA antagonist in vivo (Davies and Watkins, 1972), may be regarded as a glycine analogue and does not displace binding at the NMDA recognition site (Watkins and Olverman, 1988). In our experiments, HA-966 was a reasonably selective NMDA antagonist only reducing responses to quisqualate and to kainate at ejecting doses several times greater than those that antagonised NMDA. This selective antagonism by HA-966 could be partially reversed by co-ejection of glycine. Glycine did not enhance the actions of the other non-NMDA agonists but rather reduced them. Strychnine did not reverse the NMDA facilitatory effect of glycine but rather increased responses to all agonists. D-, but not L-, serine was also able to reverse the action of HA-966.

These results suggest that there is a glycine modulatory site on the NMDA receptor channel complex that can be demonstrated in vivo and that this site is normally fully saturated when testing responses to NMDA on brainstem neurones. These results correlate well with those from similar, but more quantitative, experiments performed on cortical wedges in vitro (Fletcher et al., 1989).

That glycine and its antagonists are selective for responses to NMDA may be taken as further evidence that the NMDA receptor channel complex is indeed a pharmacologically distinct subtype of glutamate receptor.

COMPETITIVE ANTAGONISTS FOR NON-NMDA RECEPTORS

Compared with progress on the NMDA receptor channel complex, the pharmacology of quisqualate (AMPA) and kainate receptors has developed more slowly. Glutamic acid diethylester (GDEE), gamma-amino-methane-sulphonic acid (GAMS) and other similar non-NMDA receptor antagonists (see Watkins, this volume) have been too weak and non-selective to be of any real value. The discovery that quinoxalinediones were reasonably selective and potent non-NMDA receptor antagonists (Honore et al., 1988; and see Honore this volume) was an important step forward. The original compounds had, however, only a threefold separation between kainate and AMPA binding and have recently been shown to be glycine antagonists on the NMDA receptor complex (Birch et al., 1988). Recently Honore's group have synthesised other novel quinoxalinediones including the tricyclic, NBQX, which shows a more than ten fold separation between kainate and AMPA in binding studies and very low affinity for the glycine and NMDA recognition sites (see Honore this volume).

In our in vivo studies on spinal neurones with the original quinoxalinediones, we found that they reduced both quisqualate and kainate responses in parallel and that this specificty was not improved by substituting AMPA for quisqualate (Honore et al., 1988). In that paper, we proposed that these three non-NMDA agonists increased the firing rate of

mammalian neurones by acting at a single receptor subtype. (In these same
experiments responses to NMDA were largely unchanged which may reflect the
relatively high levels of glycine in vivo compared with more isolated
neuronal preparations). The development of NBQX as a more selective AMPA
antagonist has allowed us to re-examine the above proposal.

Effect of NBQX on responses to AMPA, quisqualate, kainate and NMDA

NBQX was prepared as a 1mM or 2mM solution in 200mM NaCl at pH 8. In
more acid or more concentrated solutions and particularly at lower
temperatures, NBQX, like the earlier quinoxalinediones, would precipitate
out of solution and block the electrode. When ejected with currents
usually less than 20nA into the vicinity of brainstem neurones, NBQX
selectively reduced responses to AMPA, quisqualate and kainate and left
responses to NMDA relatively unaffected. Increasing the AMPA antagonist
current 5-10 fold did not reduce NMDA actions by more than 50%. With
respect to selectivity between the non-NMDA agonists, NBQX reduced
responses to AMPA to a somewhat greater extent than those to kainate but
these differences were small and occasionally the reverse selectivity was
seen.

These results are in contrast to the apparent selectivity of NBQX for
AMPA, compared with kainate, binding sites in biochemical experiments in
vitro. Several explanations are possible of which the two most likely are:-
i) that the in vitro conditions required for optimal binding change the
pharmacological profile of at least one of the distinct kainate and AMPA
binding sites, and ii) that the high affinity kainate receptor makes only
minor contribution to the excitatory effects of kainate on brainstem
neurones in vivo.

For a more thorough discussion of the properties of binding sites,
other chapters in this book and particularly that by Honore should be
consulted. It may, however, be briefly noted here that the binding
experiments demonstrate that there there are distinct high affinity kainate
and AMPA (quisqualate displaceable) binding sites. This [^{3}H]-kainate
binding is sensitive to low millimolar concentrations of calcium and
therefore might not be expected to be functional under in vivo conditions?
Furthermore the AMPA/quisqualate binding site exists in two affinity
states, conversion from low to high being aided by the presence of
thiocyanate ions. Pertinent to the present observations, kainate is,
however, able to displace AMPA binding from its low affinity site.

It therefore, seems quite plausible that the low affinity state of the
[^{3}H]-AMPA binding site may be the mediator of the depolarising actions of
kainate, as well as those of quisqualate and AMPA, on central neurones in
vivo. Our results with NBQX support this contention.

NON-COMPETITIVE ANTAGONISTS FOR NON-NMDA RECEPTORS

Pentobarbitone and other barbiturates are probably non-competitive
antagonists of quisqualate (Simmonds and Horne, 1988) but, because of their
weak and non-specific actions, barbiturates are not likely to be very
useful tools for exploring excitatory amino acid pharmacology. Possibly
with more potential are the arthropod toxins (Jackson and Usherwood, 1988).
Several arthropod toxins, mostly from various species of spider, have been
shown to reduce both the action of glutamate and the amplitude of the
synaptic potential on insect muscles. Although there may be some
differences, the receptors at these neuromuscular junctions are similar to
the mammalian quisqualate subtype (Jackson and Usherwood, 1988). Since the
block appears to be both voltage- and use-dependent (Kerry et al., 1988),

it is likely that these compounds act as channel blockers. There are also reports of these spider toxins blocking non-NMDA receptor mediated synaptic transmission in the mammalian CNS (eg. Saito et al., 1985) Recently the philanthotoxin from the Egyptian digger wasp has been shown to have similar actions on insect muscles (Eldefrawi et al., 1988). We therefore decide to examine its effects on brainstem neurones in an attempt to clarify further the pharmacology of responses to quisqualate and kainate.

<u>Philanthotoxin and responses to AMPA, quisqualate, kainate and NMDA</u>

Philanthotoxin was dissolved at a concentration of 10mM in 200mM NaCl, pH 4.3. Little difficulty was experienced passing current through the barrels containing this compound despite its large molecular weight and long polyamine tail. Ejection currents, usually less than 20nA, philanthotoxin produced a slow but selective antagonism of quisqualate and kainate but not of NMDA. In attempts to study the use-dependency of onset of antagonism by philanthotoxin, frequent and infrequent rates of agonist administration were employed. Although there was a hint that the rate of onset of antagonism was related to the frequency of agonist application, this was difficult to verify. The rate of recovery from the effects of philanthotoxin on termination of the ejecting current could, however, be demonstrated to be dependent on agonist application rate. Such a use-dependent recovery suggests trapping of the toxin in the closed channel.

Responses to quisqualate and kainate were reduced in parallel and in further experiments with AMPA there again appeared to be no selectivity between any of the non-NMDA agonists. Even when the rate of administration of quisqualate was different from that of kainate during a single test with philanthotoxin, the rates of onset and recovery of responses to both agonists were parallel. This result supports the proposition that kainate and quisqualate depolarise neurones by the same conductance channels.

CONCLUSIONS

The two simple conclusions from these experiments are:

i) NMDA depolarises brainstem and presumably most other neurones by activating a pharmacologically distinct subclass of glutamate receptor. Activation of this receptor can be modulated by chemicals that act at the glutamate recognition site, at the glycine modulatory site, at the divalent cation site in the channel and at the PCP binding site also in the channel. All these sites are unique to the NMDA subtype of glutamate receptor and do not influence the activity of non-NMDA subtypes. The NMDA selective action of zinc, tricyclics and alcohol, as proposed by other workers from in vitro studies, was not confirmed in our in vivo studies.

ii) Using more potent and selective compounds than have hitherto been available, it appears that quisqualate, AMPA and kainate increase the firing rate of these neurones by pharmacologically indistinguishable mechanisms. Indeed we postulate, from the above experiments with NBQX, that the responses to the above agonists are mediated via the low affinity state of the AMPA recognition site and, from the experiments with philanthotoxin, that they utilise the same conductance channels.

Acknowledgements: We would like to thank Dr N.A. Anis who helped with some of the experiments and supplied the philanthotoxin and to Dr T. Honore for helpful discussions and for supplies of NBQX.

REFERENCES

Anis, N.A, Berry, S.C., Burton, N.R. and Lodge, D. The dissociative anaesthetics, ketamine and phencyclidine, selectively reduce excitation of central mammalian neurones by N-methylaspartate. Br J Pharmacol 1983; 79: 565-575.

Aram. J.A., Martin, D., Tomcczyk, M., Zeman, S., Millar, J., Pohler, G. and Lodge, D. Neocortical epileptogenesis in vitro: studies with N-methyl-D-aspartate, phencyclidine, sigma and dextromethorphan receptor ligands. J Pharm Exp Ther 1989; 248: 320-328.

Ault, B, Evans, R.H., Francis, A.A., Oakes, D.J. and Watkins, J.C. Selective depression of excitatory amino acid induced depolarizations by magnesium ions in isolated spinal cord preparations. J Physiol 1980; 307: 413-428.

Birch, P.J., Grossman, C.J. and Hayes, A.G. Kynurenate and FG9041 have both competitive and non-competitive actions at excitatory amino acid receptors. Eur. J. Pharmacol 1988; 151: 313-315.

Cull-Candy, S.G. and Usowicz, M.M. Multiple conductance channels activated by excitatory amino acids in cerebellar neurones. Nature 1987; 325: 527-528.

Curtis, D.R. and Watkins, J.C. Acidic amino acids with strong excitatory actions on mammalian neurones. J Physiol 1963; 166: 1-14.

Davies, J. and Watkins, J.C. Is 1-hydroxy-3-amino-pyrrolidone-2 (HA966) a selective excitatory amino acid antagonist? Nature New Biol 1972; 328: 61-63.

Eldefrawi, A.T., Eldefrawi, M.E., Konno, K., Mansour, N.A., Nakanishi, K., Oltz, E. and Usherwood, P.N.R. Structure and synthesis of a potent glutamate receptor antagonist in wasp venom. Proc Natl Acad Sci USA 1988; 85: 4910-4913.

Fletcher, E.J. and Lodge, D. Glycine reverses antagonism of N-methyl-D-aspartate (NMDA) by 1-hydroxy-3-pyrrolidone-2 (HA-966) but not by D-2-amino-5-phosphonovalerate (D-AP5) on rat cortical slices. Eur J Pharm 1988; 151: 161-162.

Fletcher, E.J., Millar, J.D., Zeman, S. and Lodge, D. Non-competitive antagonism of N-methyl-D-aspartate by displacement of an endogenous glycine-like substance. Eur J Neurosci 1989; 1: 196-203.

Foster, A.C. and Fagg, G.E. Acidic amino acid binding sites in mammalian neuronal membranes: their characteristics and relationships to synaptic receptors. Brain Res Rev 1984; 7: 103-164.

Honore, T., Davies, S.N., Drejer, J., Fletcher, E.J., Jacobsen, Lodge, D. and Nielsen, F.E. Quinoxalinediones: Potent competitive non-NMDA glutamate receptor antagonists. Science 1988; 241: 701-703.

Jackson, H. and Usherwood, P.N.R. Spider toxins as tools for dissecting elements of excitatory amino acid transmission. Trends Neurosci 1988; 11: 278-283.

Jahr, C.E. and Stevens, C.F. Glutamate activates multiple single channel conductances in hippocampal neurones. Nature 1987; 325: 522-525.

Johnson J.W. and Ascher P. Glycine potentiates the NMDA response in cultured mouse brain neurones. Nature 1987; 325: 529-531

Kerry, C.J., Ramsey, R.L., Sansom, M.S.P. and Usherwood, P.N.R. Single channel studies on non-competetive antagonists of a quisqualate-sensitive glutamate receptor by argiotoxin$_{636}$ - a fraction isolated from the orb-weaver spider venom. Brain Res 1988; 459: 312-327.

Kessler, M., Baudry, M., Terramani, T. and Lynch, G. Complex interactions between a glycine binding site and NMDA receptors. Soc Neurosci Abstr 1987; 13: 760.

Krogsgaard-Larsen, P., Honore, T., Hansen, J.J., Curtis, D.R. and Lodge, D. New class of glutamate agonist structurally related to ibotenic acid. Nature; 1980: 284: 64-66.

Lodge, D. and Anis, N.A. Effects of phencyclidine on excitatory amino acid acid activation of spinal interneurones in the cat. Eur J Pharmacol 1982; 77: 203-204.

Lodge, D. Aram, J.A., Church, J., Davies, S.N., Martin, D., Millar, J. and Zeman, S. Sigma opiates and excitatory amino acids. In: Excitatory Amino Acids in Health and Disease. Ed. D. Lodge. John Wiley, London. 1988; pp. 237-259.

Lodge, D. and Johnson, K.M. Non-competitive excitatory amino acid antagonists. Trends Pharm Sci 1990; in press.

Loo, P., Braunwalder, A., Lehmann, J. and Williams, M. Radioligand binding to central phencyclidine recognition sites is dependent on excitatory amino acid receptor agonists. Eur J Pharmacol 1986; 123: 467-468.

Martin, D and Lodge, D. Phencyclidine receptors and N-methyl-D-aspartate antagonism: Electrophysiologic data correlates with known behaviours. Pharmacol Biochem Behav 1989; 31: 279-286.

Mayer, M., Westbrook, G.L. and Guthrie, P.B. Voltage-dependent block by Mg^{2+} of NMDA responses in spinal cord neurones. Nature 1984; 309: 261-263.

McLennan, H. Receptors for excitatory amino acids in the mammalian central nervous system. Prog Neurobiol 1983; 20: 251-271.

Nowak, L. Bregestovski, P., Ascher, P., Herbet, A. and Prochiantz, A. Magnesium gates glutamate-activated channels in mouse central neurones Nature 1984; 307: 462-465.

Ornstein, P.L., Schoeppe, D.D., Leander, J.D., Wong, D.T., Lodge, D. and Mason, N.R. In vitro and in vivo characterization of LY233053: a structurally novel competitive antagonist. Soc Neurosci Abstr 1989; in press.

Saito, M., Kawai, N., Miwa, A., Pan-Hou, H and Yoshioka, M. Spider toxin (JSTX) blocks glutamate synapse in hippocampal pyramidal neurons. Brain Res 1985; 346: 397-399.

Simmonds, M.A. and Horne, A.L. Antagonism of excitatory amino acids by barbiturates. In: Excitatory Amino Acids in Health and Disease, Ed. D. Lodge. John Wiley, London. 1988; pp. 219-237.

Watkins, J.C. and Collingridge, G.L. (Eds) The NMDA Receptor. Oxford Univ Press. 1989; in press.

Watkins J.C. and Evans R.H. Excitatory amino acid transmitters. Annu Rev Pharm Tox 1981; 21: 165-204.

Watkins, J.C., Krogsgaard-Larsen, P. and Honore, T. Structure-activity relations in the development of excitatory amino acid receptor antagonists. Trends Pharm Sci 1990; in press.

Watkins, J.C. and Olverman, H. Structural requirements for activation and blockade of EAA receptors. In: Excitatory Amino Acids in Health and Disease, Ed. D. Lodge. John Wiley, London. 1988; pp. 13-45.

Wong, E.H.F., Kemp, J.A., Priestley, T., Knight, A.R., Woodruff, G.N. and Iversen, L.L. The anticonvulsant MK-801 is a potent N-methyl-D-aspartate antagonist. Proc Natl Acad Sci USA 1986; 83: 7104-7108.

CHARACTERIZATION OF MEMBRANAL AND PURIFIED

NMDA RECEPTORS

A.F. Ikin, V. Nadler, Y. Kloog and M. Sokolovsky

Laboratory of Neurobiochemistry, Department of
Biochemistry, Faculty of Life Sciences
Tel Aviv University, Tel Aviv 69978, Israel

Over the past years considerable progress has been made towards
understanding the biochemical and physiological significance of the N-
methyl-D-aspartate (NMDA) type of excitatory amino acid receptor (for
review see Cotman & Iversen, 1987). Receptor binding and electrophysiological studies have indicated that the NMDA-receptor complex comprises
a number of distinct structural domains through which its function may
be regulated or pharmacologically modified. The domains identified so
far are: (i) the glutamate-binding site; (ii) an allosteric site, where
glycine and/or 1-aminocyclopropane-1-carboxylic acid (ACC) regulate agonist-induced channel opening; (iii) a site (or sites) through which Mg^{2+}
modulates the passage of ions; (iv) a site or sites that recognizes Zn^{2+}
and prevents channel activation, and (v) a site or sites within the receptor channel for binding of drugs such as phencyclidine (PCP)-like,
MK-801 and ketamine.

Despite these advances, little is known about the structure and
function of the NMDA receptor at a molecular level. Accordingly, we
undertook the purification of the solubilized receptor and its biochemical characterization. We also attempted to identify the structural requirements for the binding of PCP-like drugs to the membranal receptor.
The results of these studies are described in this report.

Purification of the NMDA/PCP receptor complex

In a recent report we described the solubilization of high-affinity
PCP-binding sites from rat forebrain membranes, employing the anionic
detergent sodium cholate (Ambar et al., 1988). In the present work we
proceeded to purify this solubilized NMDA/PCP-receptor complex by means
of affinity chromatography on amino-PCP agarose (Ikin et al., 1990).
The receptors were adsorbed onto the affinity column, which was then
well washed with buffer (800 ml) containing protease inhibitors and AP-5
(10^{-5} M). Receptors were eluted from the column upon addition of buffer
containing 10 μM PCP, 10 μM glutamate and 1 μM glycine.

Table I summarizes the data from a typical purification experiment.
Solubilization was followed by a 24% recovery of [^{3}H]TCP-binding sites,
which showed a decrease of 60% in specific activity. The amount of purified receptor obtained from 10 rat forebrains (530 pmol of binding

Table 1. Solubilization and purification of the
NMDA/PCP-receptor complex

Fraction	[^{3}H]TCP binding site (pmol)	Protein (mg)	Specific activity (pmol/ mg)	Degree of purifi- cation	Yield (%)
Crude membrane	530	336	1.6	1.0	100.0
Na-cholate extract	132	218	0.6	0.38	24.0
Affinity- purified receptor	38	0.0065	5850	3.700	7.0

sites, 336 mg protein) was 38 pmols of [^{3}H]TCP-binding sites, as measured by filtration assay with 30 nM [^{3}H]TCP). Purification of 3700-fold
was achieved.The total purified receptor yield was 7% and its specific
activity was 5800 pmol/mg protein. The instability of the purified receptor presented a technical difficulty: a considerable percentage of
the binding activity was lost after 20-24 hr at 4°C, and after freezing
the thawed receptors were completely inactive.

The purified receptor was subjected to SDS-PAGE followed by silver
staining, which revealed the presence of four polypeptides of Mr=67, 57,
46, and 33 kDa. The photoaffinity probe [^{3}H]-AZ-PCP, which acts as a
blocker of the NMDA channel, irreversibly labeled the first three of
these polypeptides upon their exposure to UV-illumination. The largest
amounts of radioactivity were taken up by a 52-kDa and a 42-kDa polypeptides. A labeled peak representing the 33-kDa polypeptide could sometimes be detected usually after short periods of incubation (5 min).
These results are in line with previously reported findings in crude
membrane preparations subjected to photoaffinity labeling of polypeptides by [^{3}H]-AZ-PCP (Haring et al., 1986; 1987), except that a 90-kDa
polypeptide labeled in the crude preparation was absent from the purified preparation. Since [^{3}H]-AZ-PCP, like other PCP-like ligands, binds
to a distinct site in the NMDA-receptor channel, one may reasonably
deduce that the channel is formed by the four polypeptides which were
labeled by this ligand and visualized by silver staining. It is possible that the 33-kDa polypeptide is only weakly associated with the
channel, and that its affinity labeling depends on the conformation of
the receptor.

The sum of the Mr values of the four polypeptides is 203 kDa; this
is close to 209±19, which is the Mr value estimated for the NMDA-receptor complex by the radiation inactivation method (Honore et al., 1987).
It should be noted that all ligand-gated receptor-channel complexes examined to date have been found to be oligomeric proteins with Mr values
of 100-300 kDa (Tallman & Gallager, 1985; Grenningloh et al, 1987; Guy &
Hucho, 1987).

The purified receptor showed similar binding characteristics and
pharmacological features to those found in crude membrane preparations
and in soluble extracts (Table 2).

Table 2. Pharmacological profile of membranal, solubilized
and purified NMDA/PCP receptor complex

Ligand	K_{app} (nM)		
	Membranal	Soluble	Purified
[^{3}H]TCP	23	31	60
PCP	67	150	220
MK-801	35	50	80
Dexoxadrol	220	230	110
Levoxadrol	32000	>10000	600
	ED_{50} (nM)		
Glutamate	250	250	150
Glutamate (+1 μM gly)			90
NMDA		800	500
NMDA (+1 μM gly)			200
Glycine	250	250	50
Glycine (+1 μM glu)		200	30

K_{app} of unlabeled ligands was derived from competition curves
Ed_{50} values were derived from dose response experiments where
the effect of glutamate, NMDA and glycine on [^{3}H]TCP binding
was measured.

     Solubilization and purification of the receptor had no significant
effect on either the binding affinities of the uncompetitive channel
blockers or the stereoselectivity of their binding sites (dexoxadrol vs.
levoxadrol). From the binding of [^{3}H]TCP it was evident that the receptor complex was as sensitive to NMDA-receptor ligands in the purified
preparation as in the crude membranes. Glycine enhanced the glutamate-
induced or the NMDA-induced binding of [^{3}H]TCP, while glutamate enhanced
its glycine-induced binding; this maintenance of interaction between the
different ligand sites (glutamate, glycine) and the channel indicated
that the entire NMDA/PCP-receptor complex was functionally purified.

Structural requirements for the binding of PCP derivatives to the NMDA
receptor

     Recently kinetic and equilibrium binding studies with [^{3}H]TCP (Kloog
et al., 1988a) and [^{3}H]MK-801 (Kloog et al., 1988b) showed that the en-
hancment in binding of these two labeled ligands to the NMDA receptor
results from an acceleration in the rates of both association and dissociation of the receptor-ligand complex. In addition, the nature of the
interactions between agonist-binding and PCP-binding sites on the NMDA-
receptor pointed to the operation of a mechanism involving steric blockage of the NMDA-receptor channel by PCP derivatives. In order to verify
the existence of such a mechanism we examined and compared the structure-activity relationships of the PCP-like drugs [^{3}H]MK-801, [^{3}H]TCP,
[^{3}H]PCP, [^{3}H]NH$_2$PCP, [^{3}H]AZ-PCP and [^{3}H]NO$_2$-PCP. This was done by performing kinetic- and equilibrium-binding experiments, at 25°C, using

homogenates of rat cortical membranes which had been washed repeatedly
in order to remove endogenous glutamate and glycine. With each of the
labeled ligands, kinetic parameters were measured in the absence (basal)
and in the presence (induced) of 1 µM glutamate and 1 µM glycine; this
combination of agonists was previously found to result in maximum enhancement of the rate of [^{3}H]TCP binding (Johnson et al., 1987).

Results of the experiments, expressed as association and dissociation half-times, are summarized in Table 3 for all the ligands except
[^{3}H]NO$_2$PCP (for which the kinetic analysis was not successful). Marked
differences were observed among ligands in their association half-times
of basal binding: their rank of order was [^{3}H]MK-801>[^{3}H]TCP>[^{3}H]NH$_2$PCP>
[^{3}H]PCP>[^{3}H]AZ-PCP. In contrast, the association half-time of induced
binding were similar for all the ligands (2-3.5 min), and were considerably lower than those of basal binding (Table 3). Obviously, then,
glutamate and glycine accelerated the binding of all these PCP-like
drugs. Similarly, the presence of the two agonists was associated with
a significant acceleration in the dissociation of the labeled ligands
from their binding sites. The induced dissociation rates of [^{3}H]AZ-PCP,
[^{3}H]PCP and [^{3}H]NH$_2$PCP from their receptors were higher than those obtained for [^{3}H]TCP and [^{3}H]MK-801 (see Table 3).

The behavior of each of the labeled ligands was then examined in
equilibrium-binding experiments (1-80 nM [^{3}H]-ligand) in order to find
out whether the differences observed in their kinetic-binding characteristics were also to be found in their equilibrium binding. In the
absence of added glutamate and glycine, equilibrium conditions are difficult to achieve because the binding of these uncompetitive blockers is
relatively slow. In the presence of 1 µM glutamate and 1 µM glycine,
equilibrium was reached within 2 hr even for the slowly associating
ligands [^{3}H]MK-801, [^{3}H]TCP and [^{3}H]NH$_2$PCP. Since the presence of these
two agonists does not affect the equilibrium binding of the [^{3}H]ligands
(Kloog et al., 1988a; 1988b), they were included in all the equilibrium-
binding assays.

It should be noted that the existence of low-affinity sites for
PCP—like drugs may interfere with binding assays using the ligand [^{3}H]-
AZ-PCP, because of its relatively low affinity for the NMDA receptor.
This might also explain why we failed to detect direct binding of
[^{3}H]NO$_2$PCP. We therefore evaluated the binding properties of each ligand

Table 3. Effects of glutamate and glycine on the association and
dissociation kinetics of the binding of uncompetitive
blockers to the NMDA-receptor channel

	Association half-time (min) with 9 nM [^{3}H]ligand		Dissociation half-time (min)	
	Basal	Induced (glu+gly)	Basal	Induced (glu+gly)
[^{3}H]MK-801	132.0±25.0	2.6±0.8	180±50	6.5±2.1
[^{3}H]TCP	91.0±16.0	3.0±1.0	123±43	5.1±2.3
[^{3}H]NH$_2$PCP	76.0±15.0	3.5±1.2	38± 6	4.2±0.6
[^{3}H]PCP	15.5± 3.0	2.0±0.9	20± 5	1.7±0.4
[^{3}H]AZ-PCP	11.0± 3.0	2.5±1.6	7± 2	1.0±0.5

Table 4. Parameters of binding of PCP-derivatives to
          the NMDA-receptor channel

	B_{max} (pmol/mg prot)	Direct binding K_D (nM)	Competition I_{50} (nM)
MK-801	2.1	3± 1	13± 2
TCP	2.3	10± 4	20± 3
NH$_2$PCP	2.1	24± 6	50± 3
PCP	2.5	35± 8	65± 12
AZ-PCP	2.5	100±20	300± 60
NO$_2$PCP	-	-	2500±400

in terms of its ability to compete with 10 nM [^{3}H]TCP for binding to the
receptor. The IC_{50} values obtained are in good agreement with the Kd
values, and confirm the relatively low affinity of AZ-PCP (IC_{50} = 300
nM) and the considerably lower affinity of NO$_2$PCP (IC_{50} = 2500 nM).

The data from this and from previous studies (Kloog et al, 1988a,
1988b) can be interpreted by assuming that PCP-like drugs are steric
blockers of the NMDA receptor and gain access to their sites in the receptor ion channel as a consequence of agonist-induced channel opening.
This would mean that when the channels are closed the binding of the
uncompetitive channel blockers would be limited since it would proceed
by a slow diffusion process, and the first-order rate constants for the
basal association and dissociation reactions (k_b) would thus be comparable. The half-times of basal association and basal dissociation of the
ligands tested in this study were indeed in close agreement, thus supporting the notion that in the absence of added agonists the diffusion
of ligand (L) out of the channel (characterized by k_b) is the rate-
limiting step of the overall binding process.

$$L_o \underset{k_b}{\overset{k_a}{\rightleftharpoons}} L_i + R \underset{k_{-1}}{\overset{k_1}{\rightleftharpoons}} RL$$

Also in line with the suggested mechanism is the accelerattion of
both association and dissociation rates of the tested ligands in the
presence of added agonists, since the binding process itself would then
become the rate-limiting step once the channel has opened and L_o has
been allowed to equilibrate with L_i.

The various PCP-like drugs differed from one another mostly in
their half-times of basal association and dissociation (k_b) and of induced dissociation (k_{-1}): values obtained for k_b (10^{-3} min^{-1}) were 2.3,
5.1, 12.4, 44 and 79 for MK-801, TCP, NH$_2$PCP, PCP and AZ-PCP, respectively. In addition, it would seem that the greater the affinity of the
ligand for the receptor, the higher its basal and induced dissociation
half-times (MK-801>TCP>NH$_2$PCP>PCP>AZ-PCP) - i.e. there are inverse relationships between K_D on the one hand, and k_b and k_{-1} on the other.
According to the two-step model, the equilibrium-binding constant of the
ligand is $K_D = (k_{-1}k_b)/(k_1k_a)$; product $k_{-1}k_b$ may thus be assumed to be
the major factor contributing to affinity differences between the blockers. A significant correlation was indeed found between k_bk_{-1} and K_D
(r=0.91, p<0.05)), thus suggesting that differences between the ligands
in the product of their association time constants are rather small.

We suggest that charge-transfer interactions may play an important role in the association of PCP-like ligands with their receptors. In the ligands tested here, the electrons of the thienyl ring in TCP would be the most readily available for participation in such interaction, while the least accessible would be the electrons of NO_2-PCP. The order of potency predicted on this basis would then be $TCP>NH_2PCP\simeq PCP>AZ$-$PCP>NO_2PCP$, which is similar to that obtained from the binding experiments.

The overall results would appear to indicate that the NMDA-receptor channel possesses a specific binding domain for PCP-like drugs, with a distinctive structural requirement for the binding of uncompetitive blockers. Charge transfer interactions might play an important role in the binding processes that lead to the NMDA-channel block induced by these ligands and to their slow rate of dissociation from their binding sites.

REFERENCES

Ambar, I., Kloog, Y. and Sokolovsky, M., 1988, Solubilization of rat brain phencyclidine receptors in an active binding form that is sensitive to NMDA receptor ligands. J. Neurochem., 51:133-140.

Cotman, C.W. and Iversen, L.L., (Eds.), 1987, Excitatory amino acids in the brain-focus on NMDA receptors. Trends in Neurosci., 10:263-265.

Foster, A.C. and Wong, E.H.F., 1987, The novel anticonvulsant MK-801 binds to the activated state of the N-methyl-D-aspartate receptor in rat brain. Br. J. Pharmacol., 91:403-409.

Grenningloh, G., Rienitz, A., Schmitt, B., Methfessel, C., Zensen, M., Beyrenther, K., Gundelfinger, E.D. and Betz, H., 1987, The strychnine-binding subunit of the glycine receptor shows homology with nicotinic acetylcholine receptors. Nature (London), 328:215-220.

Guy, H.R. and Hucho, F., 1987, The ion channel of the nicotinic acetylcholine receptor. Trends in Neurosci., 10:318-321.

Haring, R., Kloog, Y. and Sokolovsky, M., 1986, Identification of polypeptides of the phencyclidine receptor of rat hippocampus by photoaffinity labeling with [^{3}H]-azido phencyclidine. Biochemistry, 25:612-620.

Haring, R., Kloog, Y., Kalir, A. and Sokolovsky, M., 1987, Binding studies and photoaffinity labeling identify two classes of phencyclidine receptors in rat brain. Biochemistry, 26:5854-5861.

Honore, T., Drejer, J., Nielsen, M., Watkins, J.C. and Olverman, H.J., 1987, Molecular target size of NMDA antagonist binding sites. Eur. J. Pharmacol., 136:137-138.

Ikin, A.F., Kloog, Y. and Sokolovsky, M., 1990, N-Methyl-D-aspartate (NMDA)/phencyclidine receptor complex of rat forebrain: purification and biochemical characterization. Biochemistry, submitted.

Johnson, K.M., Snell, L.D. and Morter, R.S., 1987, N-methyl-D-aspartate enhanced [^{3}H]TCP binding to rat cortical membrane: effects of divalent cations and glycine. in: "Sigma opiod phencyclidine-like compounds as molecular probes in biology," E.F. Domino and J.M. Kamenka, eds., NPP Books, Ann Arbor, MI, pp. 259-268.

Kloog, Y., Haring, R. and Sokolovsky, M., 1988a, Kinetic characterization of the phencyclidine-N-methyl-D-aspartate receptor interaction: evidence for a steric blockade of the channel. Biochemistry, 27:843-88.

Kloog, Y., Nadler, V. and Sokolovsky, M., 1988b, Mode of binding of [^{3}H]-dibenzocycloalkenimine (MK-801) to N-methyl-D-aspartate (NMDA) receptor and its therapeutic implication. FEBS Lett., 230:167-170.

Tallman, J.F. and Gallager, D.W., 1985, The GABA-ergic system: a locus of benzodiazepine action. Ann. Rev. Neurosci., 8:21-44.

<u>SESSION III</u> : Ca^{++} AND EXCITATORY AMINO ACIDS

MECHANISMS UNDERLYING EXCITATORY AMINO ACID-EVOKED CALCIUM ENTRY IN

CULTURED NEURONS FROM THE EMBRYONIC RAT SPINAL CORD

Amy B. MacDermott, David B. Reichling, and Ottavio Arancio

Department of Physiology and Cellular Biophysics and the Center for
Neurobiology and Behavior, Columbia University, 630 W. 168th St., New
York, NY 10032

INTRODUCTION

Intracellular calcium is a ubiquitous and potent regulator of many cellular enzymes, channels, pumps, and structural elements. To perform such a variety of functions, the concentration of intracellular calcium ions ($[Ca^{2+}]_i$) must be carefully regulated. In vertebrate neurons, resting calcium levels are generally fixed between 5 and 10×10^{-8} M. However in an intact nervous system, neurons are seldom at rest and $[Ca^{2+}]_i$ can vary with both time and spatial distribution within a neuron. Fluctuations of $[Ca^{2+}]_i$ over orders of magnitude have been recorded from neurons in slices and in culture, in the absence of any external stimulus (Tank et al, 1988, Connor et al, 1987, Womack et al, 1988). These calcium transients can be driven by intrinsically-generated membrane oscillations or by synaptic activity. The studies described in this chapter address some of the mechanisms by which excitatory amino acid-mediated synaptic transmission might produce transient elevations of $[Ca^{2+}]_i$.

Many neurons in the central nervous system generate fast excitatory synaptic potentials that are mediated through one or more of the three operationally defined excitatory amino acid receptors: kainate, quisqualate and N-methyl-D-aspartate (NMDA) (Watkins and Evans, 1981, Mayer and Westbrook, 1987). Monosynaptic transmission involving both NMDA and non-NMDA excitatory amino acid receptors was first demonstrated by Dale and Roberts (1985) at synapses between interneurons and motoneurons in the spinal cord of *Xenopus* embryos. These investigators observed that the fast excitatory postsynaptic potential (fast EPSP) often had two components; the fast one was mediated through non-NMDA excitatory amino acid receptors while the slower component was mediated through NMDA receptors. Subsequently, similar dual component EPSPs have been observed in cultures of spinal cord and hippocampal neurons (Forsythe and Westbrook, 1988), as well as in neurons of cortical cultures (Huettner and Baughman, 1988), cortical slices (Jones and Baughman, 1988), and hippocampal slices (Collingridge et al, 1988). Many more examples of synaptic transmission mediated by NMDA receptors alone (Thomson, 1986) or non-NMDA receptors (Mayer and Westbrook, 1987) continue to be identified with increasing precision as the pharmacological and electrophysiological tools to study them become more refined.

Glutamate, an agonist at the kainate, quisqualate and NMDA receptors (Westbrook and Mayer, 1984), is thought to be an endogenous excitatory amino acid transmitter in the spinal cord, particularly in the dorsal horn (Davies and Watkins, 1979, Schneider and Perl, 1985, Jahr and Jessell, 1985). Consistent with this idea, the dorsal horn contains high levels of glutamate immunoreactivity (De Biasi and Rustioni, 1988), and high levels of glutamate binding sites (Greenamyre et al, 1984, Monaghan and Cotman, 1985). Many synapses in the dorsal horn of the mammalian spinal cord are axo-dendritic and axo-somatic (Brown, 1981), as are synapses between spinal cord cells and dorsal root ganglion/spinal cord cells grown in culture (Neale et al, 1983). Therefore the dendrites of spinal cord neurons should have relatively high densities of excitatory amino acid receptors.

Voltage-gated conductances on dendrites of central neurons may be activated following synaptic, depolarizing activation of neurons. Voltage-gated calcium channels have been localized to dendrites, a site where synaptic inputs are likely to be most dense (Llinas, 1988). Dendritic localization of these channels in spinal cord cells has not been established definitively but is suggested by analogy to other populations of neurons. It has been shown that spinal cord neurons in culture can generate calcium-dependent action potentials (Heyer et al, 1981). Further, high and low threshold calcium-dependent action potentials similar to those first identified by Llinas and colleagues and localized to cerebellar Purkinje cell dendrites (Llinas and Sugimori, 1980) have been identified in dorsal horn neurons in slices of immature rat spinal cord (Murase and Randic, 1983). Consistent with these observations of multiple forms of regenerative activity is the more recent demonstration of multiple types of calcium channels in identified dorsal horn neurons acutely dissociated from immature rat spinal cord (Huang, 1989). Therefore, it is likely that excitatory amino acid-mediated postsynaptic potentials stimulate calcium entry through dendrites of dorsal horn neurons via one or more of these voltage-gated calcium channels.

Excitatory amino acids can not only cause calcium entry into neurons via voltage-gated calcium channels, but also via NMDA-gated channels (MacDermott, 1986, Mayer et al, 1987). Although the NMDA channel is not activated by voltage in the absence of agonist, current flow through the channel is voltage-dependent (MacDonald and Wojtowicz, 1982). This property is due to voltage-dependent block of the NMDA channel by free magnesium ions (Mg^{2+}) (Nowak et al, 1984, Mayer et al, 1984). Thus calcium current through the NMDA channel is voltage dependent as has been illustrated using a combination of optical and voltage clamp recording techniques (MacDermott et al, 1986, Mayer et al, 1987). Consequently, during excitatory amino acid activation of neurons, both routes of calcium entry, voltage-gated calcium channels and NMDA channels, are voltage-dependent.

RESULTS

A simple illustration of how one or both routes of voltage-dependent calcium entry can be activated by excitatory amino acids is shown in Fig. 1. When both kainate and NMDA were applied to a neuron, $[Ca^{2+}]_i$ was elevated when no voltage control was imposed on the membrane potential. However, when the cell was voltage-clamped, only NMDA evoked a large increase in $[Ca^{2+}]_i$ (although not as large as in current clamp). These data suggest that activation of NMDA receptors, in low $[Mg^{2+}]$ solution, can result in depolarization of the membrane accompanied by calcium entry directly through NMDA-gated channels and indirectly through voltage gated calcium channels.

Normally, the amount of depolarization following activation of NMDA receptors will depend both on the number of receptors activated and the amount of Mg^{2+} present in the extracellular environment. The number of receptors activated will depend on the concentration of both the agonist and the modulator, glycine, near the receptors. Glycine increases the frequency of NMDA channel opening (Johnson and Ascher, 1987) and may be absolutely required for channel opening (Kleckner and Dingledine, 1988). With only a small number of receptors activated in the presence of normal $[Mg^{2+}]$, there will be little depolarization of the membrane potential due to channel block. Increased concentrations of NMDA will cause a greater net current leakage through blocked channels, leading to membrane depolarization. Since Mg^{2+} block of NMDA gated channels is voltage-dependent, this depolarization will partially unblock the channels, allowing greater flow of depolarizing current. Consequently, the relationship between agonist concentration and effective depolarization will be highly nonlinear. Furthermore, recruitment of voltage-gated channels will add still more complexity to the apparent dose-response relationship between [NMDA] and membrane potential. The amount of calcium entry, therefore, will also be nonlinearly related to agonist concentration, reflecting the multiple, interactive channel properties involved.

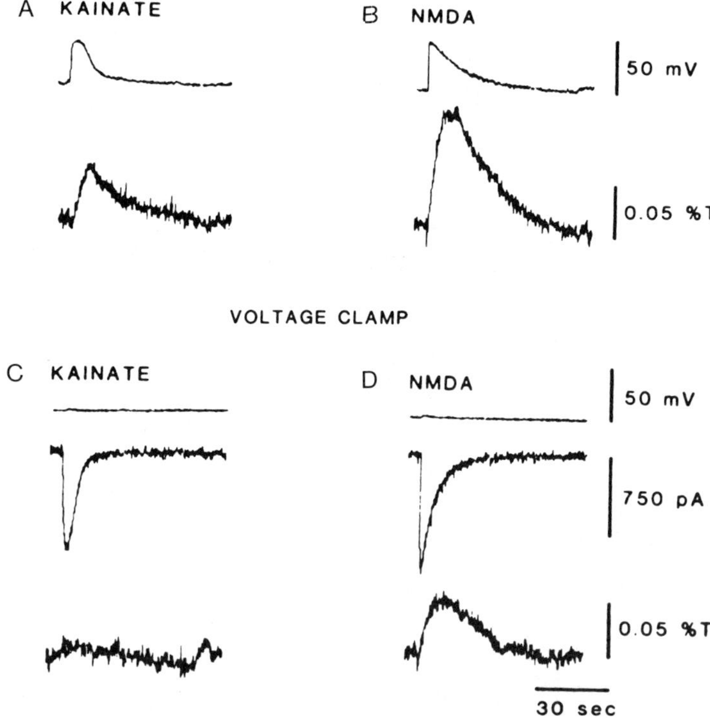

Fig. 1.   Comparison of kainate and NMDA elevation of $[Ca^{2+}]_i$ in a spinal cord neuron in culture under either current clamp or voltage clamp. All of the measurements were made on the same cell, using whole cell recording to monitor electrical activity and arsenazo III to measure changes in $[Ca^{2+}]_i$ (from Mayer et al., 1987).

We have attempted to quantitate $[Ca^{2+}]_i$ changes associated with NMDA receptor activation under conditions of variable agonist and Mg^{2+} concentrations. To do this, spinal cord neurons were grown in culture on glass coverslips for 1-3 weeks (Womack et al, 1988 for methods). Cells were loaded with the calcium indicator dye, indo-1 (Grynkiewicz et al, 1985), by exposure to 1×10^{-5}M of the acetoxymethyl ester form of indo-1 at room temperature for 30-45 min. Coverslips with loaded cells were then placed in a recording chamber on the stage of an inverted microscope. Dual photometers were used to measure changes in the ratio fluorescence at two wavelengths; 405 and 485 nm. Background measurements were made from a nearby field. The system was calibrated using indo-1 acid and standardized calcium/EGTA solutions *in vitro*. All recordings were made in a bath solution containing 5×10^{-7} M tetrodotoxin to suppress action potentials and 5×10^{-6}M glycine.

Figure 2A illustrates the change in $[Ca^{2+}]_i$ caused by applications of different concentrations of NMDA. In these experiments, a rapid perfusion system was used which simultaneously delivered agonist to the soma and neurites of the neuron under study. The bath solution had no added Mg^{2+}. The agonist-evoked change in $[Ca^{2+}]_i$ recorded in the soma varied with NMDA concentration between 5×10^{-6} and 5×10^{-5}M and became maximal between 5 and 10×10^{-5}M NMDA. The maximal value of $[Ca^{2+}]_i$ activated by NMDA was in the micromolar range, although the peak concentration varied from cell to cell. These data are similar to those observed by Murphy et al, 1987.

The presence of millimolar Mg^{2+} in the bath can block NMDA- and glutamate-evoked calcium entry into neurons (MacDermott et al, 1986, Kudo and Ogur, 1986, Mayer et al, 1987). More recently, we have tried to quantitate the interaction between $[Mg^{2+}]$ and agonist-evoked calcium entry. Adding increasing concentrations of Mg^{2+} to the bath solution depressed the elevation of $[Ca^{2+}]_i$

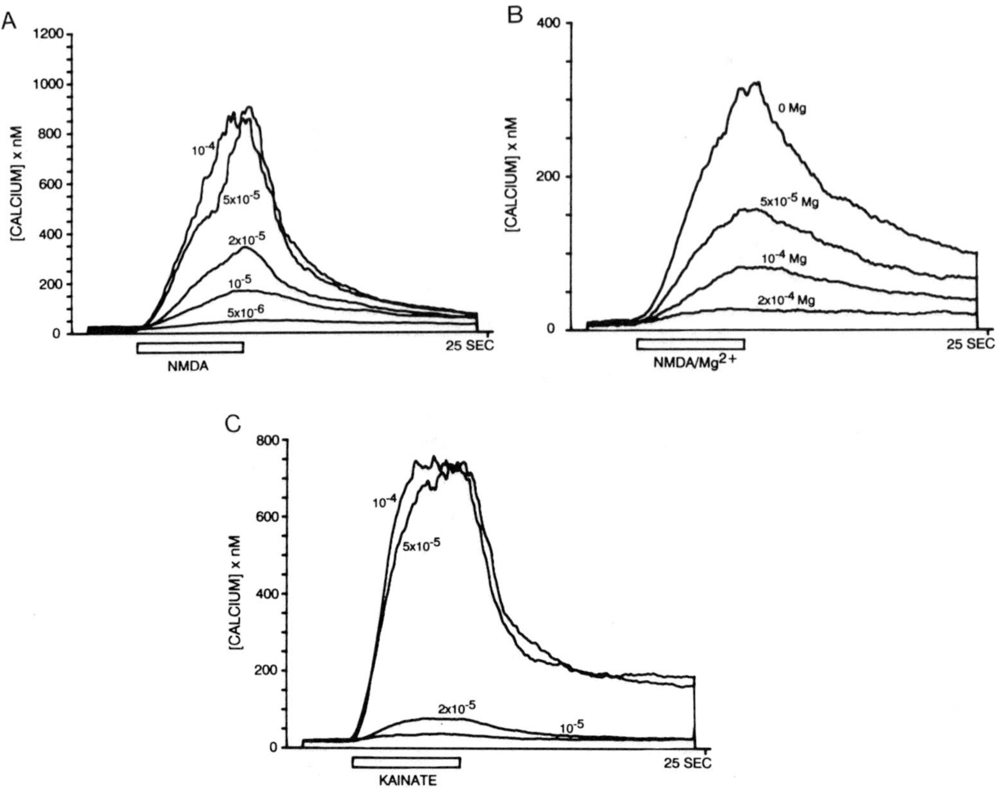

Fig. 2.   Agonist-induced changes in $[Ca^{2+}]_i$ measured in three separate neurons using indo-1. A shows the change in $[Ca^{2+}]_i$ induced by different concentrations of NMDA. B shows the depression of NMDA-induced calcium signals by different concentrations of Mg^{2+}. C shows the change in $[Ca^{2+}]_i$ induced by different concentrations of kainate.

evoked by non-saturating doses of NMDA. Fig. 2B shows that 5×10^{-5}M Mg^{2+} was sufficient to cause a decrease of more than 1×10^{-7}M in the NMDA-evoked change in $[Ca^{2+}]_i$. At 1×10^{-4}M Mg^{2+}, the NMDA-evoked elevation of $[Ca^{2+}]_i$ was further depressed while at 2×10^{-4}M Mg^{2+}, the calcium signal was almost completely blocked. The dose of NMDA used in this experiment was 2×10^{-5} M which, as can be seen in Fig. 2A, was about half maximal for elevation of $[Ca^{2+}]_i$. At concentrations of NMDA which saturate all available receptors, as may occur during synaptic activation (Jack et al, 1981), a much higher $[Mg^{2+}]_o$ may be required to completely block calcium entry.

The effect of different concentrations of kainate on $[Ca^{2+}]_i$ is very different from that of NMDA as illustrated in Fig. 2C. Kainate has a very sharp, dose-response effect on $[Ca^{2+}]_i$. While 2×10^{-5}M kainate doubled the resting level of calcium, 5×10^{-5}M kainate caused rapid, maximal, elevations of $[Ca^{2+}]_i$, ten-fold greater than the resting concentration. The rate of rise of $[Ca^{2+}]_i$ caused by maximal [kainate] (5×10^{-5}M) was always faster than the rate of rise evoked by maximal [NMDA]. The most simple interpretation of these data is that the calcium response to kainate is due to the activation of voltage-gated calcium channels.

The slowly-inactivating, voltage-gated calcium currents in acutely dissociated dorsal horn neurons exhibit peak current flow between -10 and 0 mV (Huang, 1989), comparable to the peak elevation of $[Ca^{2+}]_i$ elicited by voltage steps in cultured spinal cord neurons (Mayer et al, 1987). A non-desensitizing activation by high kainate concentrations is likely to be clamping membrane potential near the kainate reversal potential between -10 and 0 mV, producing optimal activation of calcium channels. These data do not exclude the possibility that some of the calcium is entering through kainate-gated channels. Murphy and Miller (1989) have reported that at higher concentrations of kainate ($>1 \times 10^{-4}$M), there is a component of calcium entry which cannot be accounted for by entry through voltage-gated calcium channels alone. However in our experiments, 1×10^{-4}M kainate was the maximum dose used. Furthermore, our application times were less than 10 sec while those of Murphy and Miller (1989) were for several minutes. Therefore, the two sets of experiments may be testing neurons in very different physiological states.

In all of our experiments to date, $[Ca^{2+}]_i$ measurements have been made at the soma, where both NMDA receptors and voltage-gated calcium channels appear to be present. However ultimately, it will be important to examine calcium regulation in the confined spaces of dendrites where synaptic excitatory amino acid receptors and voltage-gated calcium channels have the greatest opportunity to interact. The degree of interaction would be partly determined by the distribution and clustering of excitatory amino acid receptors. In this regard, hot spots of glutamate sensitivity have been demonstrated in cultured spinal cord cells (Ransom et al, 1977) and it has been suggested that for cultured chick motoneurons, synaptic contacts are required before such hot spots are evident (O'Brien and Fischbach, 1986b) or before any aspartate receptors are expressed (O'Brien and Fischbach, 1986a). In addition, a correspondence between glutamate, kainate and quisqualate sensitivity on neurites has been reported (Trussell et al, 1988). Observations of dual component excitatory amino acid-mediated synapses in the spinal cord and between spinal cord cells in culture also suggest that NMDA and non-NMDA receptors might be distributed in parallel over neurites. However, many other configurations of receptor distribution would be consistent such synaptic data. To address this issue, we have mapped the distributions of NMDA and non-NMDA receptors on neurites of spinal cord neurons in culture (Arancio and MacDermott, 1988) and acutely dissociated dorsal horn neurons from immature rat spinal cord (Arancio et al, 1989).

Figures 3 A & B illustrate that responses to NMDA and kainate can be heterogeneously distributed along the neurites of rat spinal cord neurons grown in culture. Whole cell recordings were made under voltage clamp using patch electrodes filled with (in mM) 140 K-gluconate, 5 HEPES, 1 EGTA/KOH, 0.1 $CaCl_2$, 2 $MgCl_2$ (pH 7.1, 300 mOsm). The cells were placed in a vigorously flowing bath and 1×10^{-3}M NMDA was applied through fine tip pressure pipettes placed within 2um of the neurite under study.

In Fig. 3 A & B, drug applications were made at 12 um intervals along two different neurites. The drug stream crossing the neurite was estimated to be 25-30 um in diameter (see Arancio andMacDermott, 1989, for methodological details). Figure 3A shows an example of the pattern of response amplitudes obtained by testing with NMDA. An area of high sensitivity to NMDA can be seen on a 75 µm region of the proximal neurite. Figure 3B shows a similar pattern of sensitivity to kainate, with a large proximal region of particularly high sensitivity. The most proximal region of many neurites (within the first 12 µm) was often, but not always, more sensitive than the soma to applications of NMDA and kainate. Figures 3 A & B also illustrate that applications of NMDA and kainate at distances from 75 to 150 µm evoked constant amplitude responses of -25 pA and -40 pA, respectively. Since current amplitude is expected to decrease with distance from the recording site at the soma, this maintained high response amplitude implies increasing densities of receptors distally along the neurites. Although these two examples show some similarity in the pattern of response to kainate and NMDA, the patterns of sensitivity vary widely among cells and can also vary between agonists on a single neurite (Aranio and MacDermott, 1989, Arancio et al, 1989).

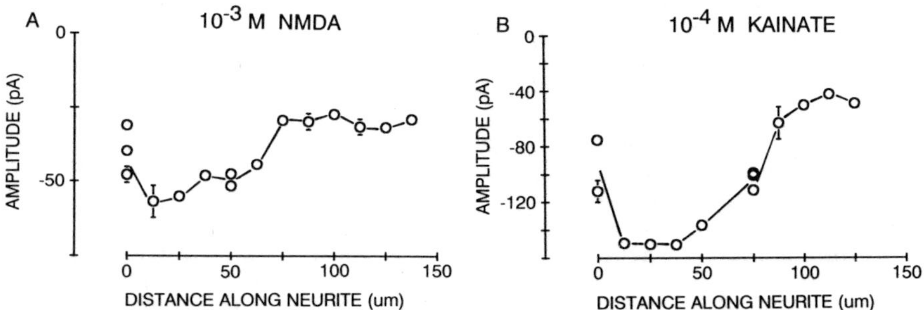

Fig. 3.   Mapping receptor density with NMDA (A) and kainate (B) reveal regions of high agonist sensitivity (see Arancio and MacDermott, 1989 for details).

DISCUSSION

These experiments demonstrate that there are high densities of NMDA and kainate receptors extending to the greatest distances from the soma that we were able to test. Furthermore, this distribution is punctuated by regions of even higher NMDA or kainate sensitivity. Although the existence of voltage-gated calcium channels on these same neurites can only be inferred at this time, it is possible that there is significant interaction between synaptic activation of the excitatory amino acid receptors and activation of voltage-gated calcium channels (MacDermott and Dale, 1987). In motoneurons of lamprey (Grillner and Wallen, 1985) and neocortical neurons (Flatman et al, 1986), for example, NMDA can cause an oscillatory membrane behavior which is dependent on calcium entry and the subsequent activation of calcium dependent potassium channels. The different types and numbers of excitatory amino acid receptors activated *in vivo* will vary with the number and frequency of synaptic inputs onto spinal cord neurons. Our data indicate that these factors, in turn, will strongly affect the kinetics and amplitude of changes in $[Ca^{2+}]_i$.

ACKNOWLEDGEMENT

We would like to thank Dr. K. Murase for introducing the fast drug perfusion technique to the laboratory, Isaac Alexis for assistance with data analysis, and Dr. S. Helm for providing some of the cultured neurons. This work has been supported by NIH/NINCDS grant NS 25215.

REFERENCES

Arancio, O. and MacDermott, A. B., 1989, Differential distribution of excitatory amino acid receptors on embryonic rat spinal cord neurons in culture. In preparation.

Arancio, O., Murase, K., Yoshimura, M, and MacDermott, A. B., 1989, Heterogeneous distribution of excitatory amino acid receptors on postnatal neurons acutely dissociated from rat dorsal horn, **Neuroscience Abstracts**, 15:943.

Brown, A. G., 1981, **Organization in the Spinal Cord**, Springer-Verlag, New York.

Collingridge, G. L., Herron, C. E. and Lester, R. A. J., 1988, Synaptic activation of N-methyl-D aspartate receptors in the schaffer collateral-commissural pathway of rat hippocampus, **J. Physiol.**, 399:283.

Connor, J. A., Tseng, H-Y., and Hockberger, P. E., 1987, Depolarization- and transmitter-induced changes in intracellular Ca^{2+} of rat cerebellar granule cells in explant cultures, **J. Neurosci.**, 7:1384.

Dale, N. and Roberts, A., 1985, Dual-component amino-acid-mediated synaptic potentials: Excitatory drive for swimming in **Xenopus** embryos, **J. Physiol.**, 363:35.

Davies, J. and Watkins, J. C., 1979, Selective antagonism of amino acid-induced and synaptic excitation in the cat spinal cord. **J. Physiol.**, 297:621.

De Biasi, S. and Rustioni, A., 1988, Glutamate and substance P coexist in primary afferent terminals in the superficial laminae of spinal cord, **Proc. Natl. Acad. Sci.**, 85:7820.

Flatman, J.A., Schwindt, P.C. and Crill, W.E. (1986) The induction and modification of voltage sensitive responses in cat neocortical neurons by N-methyl-D-aspartate. **Brain Res.** 363:62.

Forsythe, I. A. and Westbrook, G. L., 1988, Slow excitatory postsynaptic currents mediated by N-methyl-D-aspartate receptors on cultured mouse central neuronis, **J. Physiol.**, 396:515.

Greenamyre, J. T., Young, A. B., and Penney, J. B., 1984, Quantitative autoradiographic distribution of L-[^{3}H] glutamate binding sites in rat central nervous system. **J. Neurosci.**, 4:2133.

Grillner, S. and Wallen, P., 1985, The ionic mechanisms underlying N-methyl-D-aspartate receptor-induced, tetrodotoxin-resistant membrane potential oscillations in lamprey neurons active during locomotion, **Neurosci. Lett.**, 60:289.

Grynkiewicz, G., Poenie, M. and Tsien, R. Y., 1985, A new generation of Ca^{2+} indicators with greatly improved fluorescence properties. **J. Biol. Chem.** 260:3440.

Heyer, E. J., MacDonald, R. L., Bergey, G. K., and Nelson, P. G., 1981, Calcium-dependent action potentials in mouse spinal cord neurons in culture, **Brain Res.**, 220:408.

Huang, L-Y. M., 1989, Calcium channels in isolated rat dorsal horn neurones, including labelled spinothalamic and trigeminal cells, **J. Physiol.**, 411:161

Huettner, J. E., and Baughman, R. W., 1988, The pharmacology of synapses formed by identified corticocollicular neurons in primary cultures of rat visual cortex, **J. Neurosci.**, 8:160

Jack, J. J. B., Redman, S. J., and Wong, K., 1981, The components of synaptic potentials evoked in cat spinal notoneurones by impulses in single group Ia afferents, **J. Physiol.**, 321:65.

Jahr, C. E. and Jessell, T. M., 1985, Synaptic transmission between dorsl horn neurons in culture: antagonism of monosynaptic EPSPs and glutamate excitation by kynurenate, **J. Neurosci.**, 5:2281.

Johnson, J. W. and Ascher, P., 1987, Glycine potentiates the NMDA response in cultured mouse brain neurons, **Nature**, 325:529.

Jones, K. A. and Baughman, R. W., 1988, NMDA- and non-NMDA-receptor components of excitatory synaptic potentials recorded from cells in layer V of rat visual cortex, **J. Neurosci.**, 8:3522.

Kleckner, N. K. and Dingledine, R., 1989, Glycine is required for activation of NMDA receptors in Xenopus oocytes injected with rat brain mRNA, **Science**, 241:835.

Kudo, Y. and Ogura, A., 1986, Glutamtate-induced increase in intracellular Ca^{2+} concentration in isolated hippocampal neurones, **Br. J. Pharmac.**, 89:191.

Llinas, R. R., 1988, The intrinsic electrophysiological properties of mammalian neurons: insights into central nervous system function, **Science**, 242:1654.

Llinas, R. and Sugimori, M., 1980, Electrophysiological properites of in vitro Purkinje cell dendrites in mammalian cerebellar slices, **J. Physiol.**, 305:197.

MacDermott, A. B., Mayer, M. L., Westbrook, G. L., Smith, S. J., Barker, J. L., 1986, NMDA receptor activation elevates cytoplasmic calcium in cultured spinal cord neurones, **Nature**, 321:519.

MacDermott, A. B. and Dale, N., 1987, Receptors, ion channels and synaptic potentials underlying the integrative actions of excitatory amino acids, **TINS**, 10:280.

MacDonald, J. F. and Wojtowicz, J. M., 1982, The effects of L-glutamate and its analogues upon the membrane conductance of central murine neurones in culture, **Can. J. Physiol. Pharmacol.**, **60:282.**

Mayer, M. L., MacDermott, A. B., Westbrook, G. L., Smith, S. J., and Barker, J. L., 1987, Agonist- and voltage-gated calcium entry in cultured mouse spinal cord neurons under voltage clamp measured using arsenazo III, **J. Neurosci.**, 7:3230.

Mayer, M. L., Westbrook, G. L. and Guthrie, P. B., 1984, Voltage-dependent block by Mg^{2+} of NMDA responses in spinal cord neurones, **Nature** 309:261.

Mayer, M. L. and Westbrook, G. L., 1987, The physiology of excitatory amino acids in the vertebrate central nervous system, **Prog. Neurobiol.**, 28:276.

Monaghan, D. T. and Cotman, C. W., 1985, Distribution of N-mehtyl-D-aspartate-sensitive l-[3H]Glutamate-binding sites in rat brain, **J. Neurosci.**, 5:2909.

Murase, K. and Randic, M., 1983, Electrophysiological properties of rat spinal dorsal horn neurones in vitro: calcium-dependent action potentials, **J. Physiol.**, 334:141.

Murphy, S. N. and Miller, R. J., 1989, Regulation of Ca^{2+} influx into striatal neurons by kainic acid, **J. Pharmacol. Exp. Ther.**, 249:184.

Murphy, S. N., Thayer, S. A., and Miller, R. J., 1987, The effects of excitatory amino acids on intracellular calcium in single mouse striatal neurons in vitro, **J. Neurosci.**, 7:4145.

Neale, E. A., Nelson, P. G., MacDonald, R. L., Christian, C. N., and Bowers, L. M., 1983, Synaptic interactions between mammalian central neurons in cell culture. III Morphological correlates of quantal synaptic transmission, **J. Neurophysiol.**, 49:1459.

Nowak, L., Bregestovski, P., Ascher, P., Herbert, A.and Proshiantz, A., 1984, Magnesium gates glutamate-activated channels in mouse central neurones, **Nature** 307:462.

O'Brien, R. J. and Fischbach, G. D., 1986, Characterization of excitatory amino acid receptors expressed by embryonic chick motoneurons in vitro, **J. Neurosci.**, 6:3275.

O'Brien, R. J. and Fischbach, G. D., 1986, Modulation of embryonic chich motoneuron glutamate sensitivity by interneurons and agonists. **J. Neurosci.**, 6:3290.

Ransom, B. R., Bullock, P. N., and Nelson, P. G., 1977, Mouse spinal cord in cell culture. III. Neuronal chemosensitivity and its relationship to synaptic activity, **J. Neurophysiol.**, 40:1163.

Schneider, S. P. and Perl, E. R., 1985, Selective excitation of neurons in the mammalian spinal dorsal horn by aspartate and glutamate in vitro: correlation with location and excitatory input, **Brain Res.**, 360:339.

Tank, D. W., Sugimori, M., Connor, J. A., and Llinas, R. R., 1988, Spatially resolved calcium dynamics of mammalian Purkinje cells in cerebellar slice, **Science**, 242:773.

Thomson, A. M., 1986, A magnesium-sensitive post-synaptic potential in rat cerebral cortex resembles neuronal responses to N-methylaspartate, **J. Physiol.**, 370:531.

Trussell, L. O., Thio, L. L., Zorumski, C. F., and Fischbach, G. D., 1988, Rapid desensitization of glutamate receptors in vertebrate central neurons, **Proc. Natl. Acad. Sci.**, 85:4562.

Watkins, J.C. and Evans, R.H. (1981) Excitatory amino acid transmitters. **Ann. Rev. Pharmacol. Toxicol.** 21 165-204.

Westbrook, G. L. and Mayer, M. L., 1984, Glutamate currents in mammalian spinal neurons: resolution of a paradox, **Brain Res.**, 301:375.

Womack, M. D., MacDermott, A. B., and Jessell, T. M., 1988, Sensory transmitters regulate intracellular calcium in dorsal horn neurons, **Nature**, 334:351.

TOPOGRAPHICAL HETEROGENEITY OF GLUTAMATE AGONIST-INDUCED CALCIUM

INCREASE IN HIPPOCAMPUS

Yoshihisa Kudo[1], Etsuro Ito[2] and Akihiko Ogura[1]

[1]Department of Neuroscience, Mitsubishi Kasei Institute of Life Sciences
Minamiooya ,Machida, Tokyo 194 and [2]Advanced Research Center for
Human Sciences, Waseda University, Tokorozawa, Saitama 359, JAPAN

INTRODUCTION

Although the mechanisms underlying the establishment of long term potentiation (LTP) in
the hippocampus have not been understood yet, it is widely accepted that the increase in
intracellular Ca^{2+} concentration ($[Ca^{2+}]_i$) due to the activation of L-glutamate receptors plays a
crucial role in the early phase of LTP (Kudo et al., 1987 ; Malenka et al., 1988). It is therefore
important to specify the location of L-glutamate receptors responsible for the $[Ca^{2+}]_i$ elevation.

Activation of N-methyl-D-aspartate (NMDA)-preferred subspecies of L- glutamate receptor
has long been known to produce an influx of Ca^{2+} into the cell (Kudo and Ogura, 1986). Recently,
quisqualate (QUIS)-preferred subspecies of L- glutamate receptor is shown to elevate $[Ca^{2+}]_i$
through the mobilization of intracellularly stored Ca^{2+} coupled with phosphoinositide
metabolism (Sugiyama et al., 1987). To add this, kainate (KAI)-preferred subspecies is also
capable of elevating $[Ca^{2+}]_i$ via an unspecified pathway (Murphy and Miller, 1989).

The diffusion of Ca^{2+} in the neuronal cytoplasm is very slow and the state of elevated $[Ca^{2+}]_i$
remains in a limited volume after the activation of receptors for neurotransmitters or of vol-
tage-operated Ca^{2+} channels. Combining this fact with the above-mentioned knowledge on the
$[Ca^{2+}]_i$ elevations induced by L-glutamate receptor agonists, we can expect that the measurement
of $[Ca^{2+}]_i$ should give information concerning the distribution of each of subspecies of L-glutamate
receptor.

Digital image analysis of low-power pictures from fluorescence microscope which covers a
wide area of rat hippocampal slice should be especially helpful to know which receptor is
localized in which region. This type of study has conventionally been done by autoradiography
using a radioactive ligand for each receptor (Cotman et al., 1987). But the binding sites of ligands
do not necessarily mean the functional receptors. It is difficult in autoradiography to reveal the
distributions of multiple species of receptors in a single preparation. Development of silver grains
needs considerable time. The digital Ca^{2+} fluorography should be free of those problems, since a
single living preparation can be stimulated by different agonists in a repeated manner.

Moreover, comparison of the magnitudes of $[Ca^{2+}]_i$ elevation induced by agonists should
become possible by the Ca^{2+} fluorography, which is important in elucidating the role of Ca^{2+} in
the establishment of LTP. Suggestions might be obtained to the problems as follows.

Excitatory Amino Acids and Neuronal Plasticity
Edited by Y. Ben-Ari
Plenum Press, New York, 1990

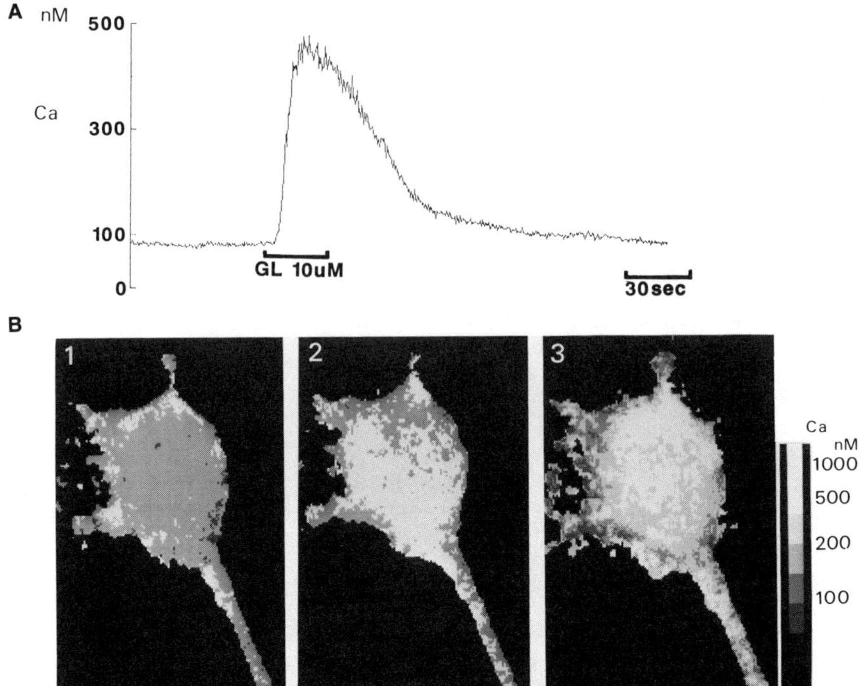

Fig. 1.   Elevation of $[Ca^{2+}]_i$ induced by L-glutamate in single hippocampal neurons under culture after treatment with tetrodotoxin. A, Time course of the $[Ca^{2+}]_i$ change. B, Two dimensional analysis of $[Ca^{2+}]_i$ distribution. 1, before stimulation, 2, 10 s after an application of L-glutamate (10 μM), 3, 30 s after the application.

Is the magnitude of $[Ca^{2+}]_i$ elevation sufficient for the activation of Ca^{2+}- dependent enzymes, including protein kinases or proteases? How long does the state of high $[Ca^{2+}]_i$ last? Can KAI- or QUIS-subspecies replace the role of NMDA-subspecies in mossy fiber/CA3 pyramidal synapses which are believed to be scarce in NMDA-receptor?

Present study is the first documentation of the technique of digital Ca^{2+} fluorography applied to hippocampal slice. The results obtained were compared with those obtained from single cultured neurons. Problems and limitations in the application of this technique are also discussed.

MATERIALS AND METHODS

Dissociation cell culture of rat hippocampal neurons Hippocampal neurons were isolated from rat embryos of gestational day 18 and maintained for >7 days on a thin glasscoverslip attached to a silicon rubber wall. The neurons were treated with 5 M fura-2 acetoxymethylester (fura-2/ AM) for **ca.** 60 min (37 C) and then placed on an inverted fluorescence microscope. The culture was continuously perfused (2 ml/min) with warmed (32° C) balanced salt solution (BSS) composed of (mM): NaCl 130, KCl 5.4, $CaCl_2$ 1.8, glucose 5.5, HEPES-NaOH 20 (pH 7.3). For detailed method see previous report (Ogura et al., 1988).

Preparation of rat hippocampal slice Hippocampal slices of ca. 300 m in thickness were obtained from rats of ca. 1 month of age. The slice was treated with fura-2/AM (7.5 - 10 M) for 45 min (37 C) and placed in a chamber made of a thin glass coverslip and a leucite wall. The chamber was continuously perfused with oxygenated artificial cerebrospinal fluid (32 C; 2 ml/min). Population spikes were recorded from the pyramidal layer of CA1 or CA3 region on stimulating Schaffer-collateral or mossy fiber, respectively. LTP in the population spike was induced by tetanization (50 Hz, 5 s) of each pathway.

Device for measurement of [Ca2+]i The device is composed of an inverted fluorescence microscope (Olympus IMT-2), a silicon-intensified- target video camera (Hamamatsu photonics ; C2400-8) and a 16-bit image processor (Hamamatsu photonics; ARGUS 100). This was applied to both of cultured neurons and slice preparations. For details see previous description (Ogura et al., 1988). Briefly, a fluorescence image of a fura-2-loaded specimen obtained under the illumination light of 340 nm in wavelength was divided by that obtained under the light of 360 nm. The specific software used for this pixel-to-pixel division includes correction for photobreaching of fura-2 and provides a series of processed (rationized) pictures at an arbitrary (> 1/30 s) sampling interval.

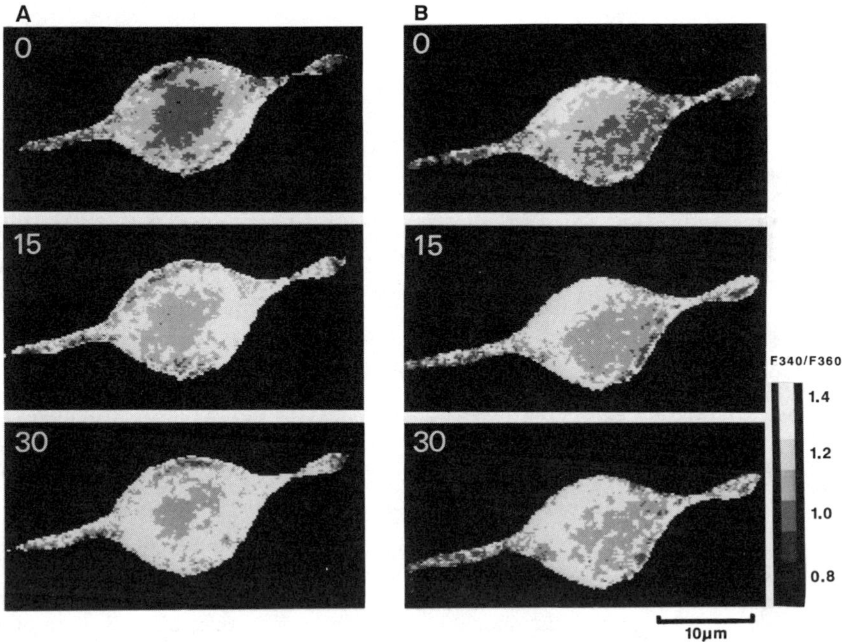

Fig. 2.　Changes in the distribution of $[Ca^{2+}]_i$ induced by NMDA and KAI in the same single neuron after treatment with tetrodotoxin and La^{3+} (50 µM). A, Effect of NMDA (10 µM); B, Effect of KAI (10 µM).

Objective lens used for the epifluorescence observation of cultured neurons was x40 or x100 (Olympus, Fluor), while that for viewing slices was x10. Conversion of the data of fluorescence ratio (forming a image) to the absolute levels of $[Ca^{2+}]_i$ (according to a calibration curve previously obtained by illuminating the standard solutions of known Ca^{2+} concentrations) was applied to the cultured neurons but not to the slices. This is because autofluorescence of the slice (due presumably to respiratory coenzymes) diluted the Ca^{2+}-dependent signal of the fluoroprobe in variable proportion. To overcome the autofluorescence, we had to load the slice with fura-2 sufficiently heavily until the fluorescence signals under excitation lights of 340 and 380 nm behaved counterwise (autofluorescence moves in the same direction). In some preparations, population spikes were recorded simultaneously.

127

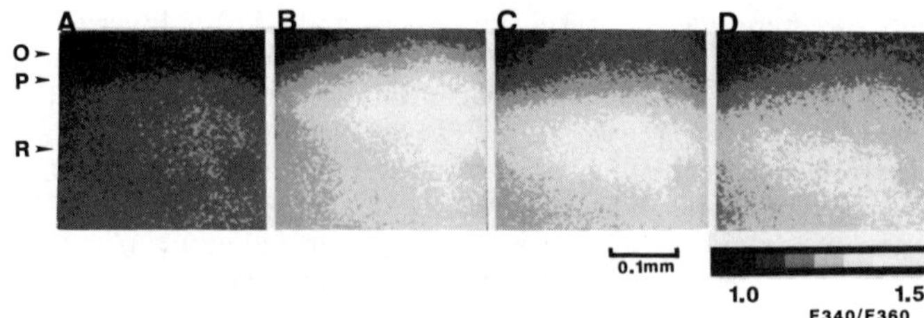

Fig. 3.   Topographically heterogeneous changes in $[Ca^{2+}]_i$ 30 s after exposures to L-glutamate receptor agonists (50 μM) in CA1 region of a rat hippocampal slice. Actually displayed is the ratio of fluorescence intensities under 2 illumination lights of 340 and 360 nm in wavelengths, which qualitatively represents $[Ca^{2+}]_i$ (see text). A, before exposure; B, NMDA; C, QUIS; D, KAI. O, P and R in this and following figures represent strata oriens, pyramidale and radiatum, respectively.

RESULTS

The $[Ca2+]i$ changes in single cultured neurons induced by agonists for L-glutamate receptor subspecies When a cultured hippocampal neuron was stimulated by L-glutamate (10 μM) under the presence of tetrodotoxin (1 μM), $[Ca^{2+}]_i$ increased markedly (Fig. 1A). A rationized image of the stimulated neuron showed that the $[Ca^{2+}]_i$ increase in the cytoplasm was uneven (Fig. 1B). This type of heterogeneity, though its pattern differed from cell to cell, was preserved after the blockade of voltage-sensitive Ca^{2+} channels by La^{3+} or by combination of nitrendipine and -conotoxin (Ogura et al., unpublished observation; cf. Fig. 2), suggesting that L-glutamate receptors are distributed heterogeneously in the neuronal surface membrane. Then we applied agonists for L-glutamate receptor subtypes, NMDA, QUIS and KAI, at the concentration of 10 μM to reveal the distribution of each receptor subspecies. There often occurred a differential distribution of the preferred sites of $[Ca^{2+}]_i$ elevation, according to the species of agonist. As seen in Fig. 2, the sites of $[Ca^{2+}]_i$ increase after exposures to NMDA and KAI were not identical, though some overlap was there (data for QUIS are not shown here).

Present results confirmed that our system of two-dimensional analysis of $[Ca^{2+}]_i$ can be used in detecting the topographical distribution of the L-glutamate receptor subtypes. The results additionally suggest that the receptor subtypes are expressed separately in neuronal membranes, casting a question to the theory of 'functional pair' which hypothesizes colocalization of NMDA-receptor and QUIS- or KAI-receptor (Collingridge and Bliss, 1987).

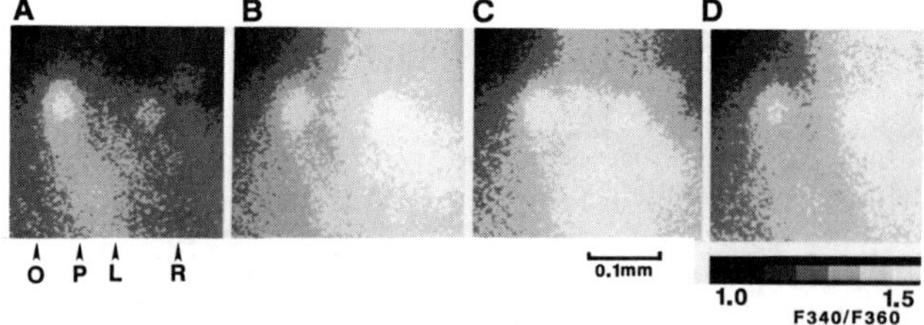

Fig. 4.   Topographically heterogeneous changes in $[Ca^{2+}]_i$ 30 s after exposures to L-glutamate receptor agonists (50 μM) in CA3 region of a rat hippocampal slice. A, before exposure; B, NMDA; C, QUIS; D, KAI. L indicates stratum lucidum.

There is possibility, however, that the receptors are expressed in the brain *in vivo* (where inherent synaptic connections were formed in the restricted regions of dendrites) differently from those in the neurons grown *in vitro* (where synaptic connections would be formed randomly. So we applied our video fluorographic technique to the slice preparation of rat hippocampus.

It should be repeated here that absolute $[Ca^{2+}]_i$ level of neurons within a slice cannot be obtained, due firstly to the specimen's autofluorescence and secondly to contamination of fluorescence signals from dead cells and non-neuronal cells. Nevertheless the changes in the ratio of fluorescence intensities displayed below should be regarded as a **qualitative** measure for $[Ca^{2+}]_i$ alterations in neurons, especially that in postsynaptic rather than presynaptic compartment as judged from anatomical volumes (see DISCUSSION).

Differential distribution of L-glutamate receptor subtypes as revealed by the $[Ca2+]i$ rises after applications of subtype-specific agonists in the slice preparation   As seen in Fig. 3A, the fluorescence ratio differed in layers. When the slice preparation was stimulated by NMDA, QUIS or KAI (50 M), the $[Ca^{2+}]_i$ level in the *st. radiatum* significantly increased. Among the agonists, NMDA was the most potent (Fig. 3B), though QUIS and KAI were also effective in elevating $[Ca^{2+}]_i$ in the CA1 region. In the cases of QUIS and KAI, the preferred sites of $[Ca^{2+}]_i$ rise tended to be more restricted (in the distal dendritic layer) than in the case of NMDA (Fig. 3C, D). 2-Amino-5-phosphonovalerate (APV; 10 M) reduced the NMDAinduced $[Ca^{2+}]_i$ increase in the CA1 region, but the effect of APV on QUIS or KAI was less obvious.

Each of the L-glutamate receptor agonist increased the $[Ca^{2+}]_i$ in the CA3 region as well. But there was an apparent difference in the preferred layers of $[Ca^{2+}]_i$ increase among the agonists (Fig. 4). QUIS (50 M) caused the increase in the somatic and proximal dendritic layers of CA3 (Fig. 4C), while NMDA induced a marked $[Ca^{2+}]_i$ increase in the distal dendritic layer (Fig. 4B). KAI also produced an increase in the same layers as in the case of NMDA (Fig. 4D).

The $[Ca2+]i$ changes by electrical stimulations related to LTP Figs.5 and 6 show the changes in $[Ca^{2+}]_i$ during tetanic stimulation given to the orthodromic synaptic pathways. When Schaffer-collateral was stimulated at 50 Hz for 5 s, the $[Ca^{2+}]_i$ level in the proximal layer of *st. radiatum* increased markedly, but the increase in *st. oriens* was less prominent (Fig. 5B). The increase lasted for about 2 min, during which time LTP in the population spikes developed markedly (Fig. 5A).

When mossy fiber was stimulated tetanically at 50 Hz for 5 s, the $[Ca^{2+}]_i$ increase in the distal layer of *st. radiatum* in CA3 region was evoked, which lasted for about 2 min (Fig. 6). The increase in the proximal layer of *st. radiatum* and in the *st. lucidum* was little observed.

The increase of $[Ca^{2+}]_i$ in CA1 region induced by stimulation of Schaffer-collateral was greatly reduced by the treatment with APV (1 M) which is known to block the establishment of LTP[9], but it was resistant to the treatment with 6-cyano-7-nitroquinoxaline-2,3-dione (CNQX; 10 M). On the other hand, the increase in $[Ca^{2+}]_i$ in the distal layer of *st. radiatum* induced by the stimulation of mossy fiber was little affected by APV (10 M), but was markedly blocked by CNQX (1 M)(Fig. 7).

DISCUSSION

The present study on cultured hippocampal neurons suggests that the subtypes of L-glutamate receptor are not necessarily expressed coincidentally in the neuronal membrane. The Ca^{2+} influx through voltage- dependent Na^+ channels, which can occur in some occasions, was blocked by tetrodotoxin. The Ca^{2+} influx via voltage-dependent Ca^{2+} channels are also blocked by the treatment with La^{3+} (50 M), which was ascertained by a total suppression of a $[Ca^{2+}]_i$ elevation triggered by high K^+ (50 mM)- depolarization (not shown here).

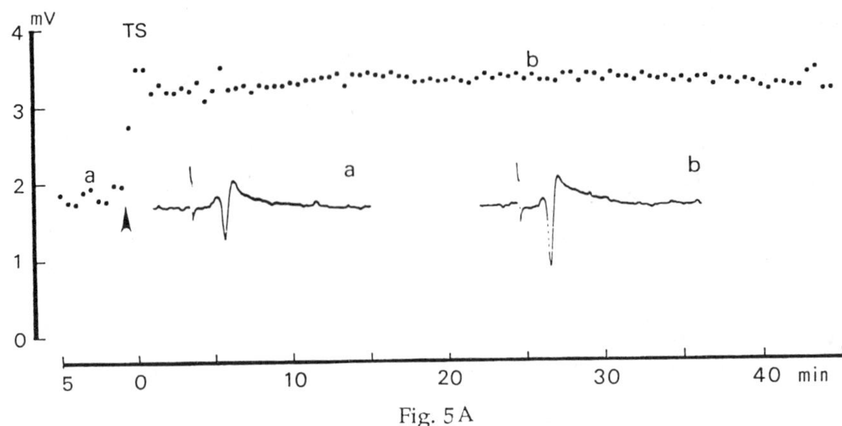

Fig. 5.   Simultaneous monitoring of fura-2 fluorescence signals in the CA1 region and electrical activity after a tetanic stimulation of Schaffer collateral. A, Time course of the change in the amplitude of population spikes recorded at the pyramidal cell layer; B, Ratio of fluorescence intensities representing $[Ca^{2+}]_i$ in CA1 region.

Fig. 6.   Time-lapse video fluorography representing the $[Ca^{2+}]_i$ change at CA3 region after a tetanic stimulation of mossy fiber (50 Hz, 5 sec).

Fig. 7.   The $[Ca^{2+}]_i$ rises after tetanic stimulation on mossy fiber in CA3 region and the effects of antagonists on them. A, Time lapse images showing the change in $[Ca^{2+}]_i$ after tetanic stimulation; B, effects of APV on the tetanus-induced change in $[Ca^{2+}]_i$; C, effects of CNQX on the tetanus-induced change in $[Ca^{2+}]_i$.

Thus the observed $[Ca^{2+}]_i$ increase should be ascribed to the activation of receptor-coupled cationic channels permeable to Ca^{2+} or to the liberation of Ca^{2+} from intracellular storage sites. NMDA and KAI evoked the $[Ca^{2+}]_i$ increase preferentially in the submembraneous compartments. QUIS tended to produce $[Ca^{2+}]_i$ increases in both submembraneous and deep cytoplasmic compartments. This difference may reflect distinct sources of Ca^{2+} according to the species of agonists. Since micrograph of a neuron is 2- dimensional projection of a sphere, the shellformed distribution of high $[Ca^{2+}]_i$ volume should result in an image of ring-formed distribution of high $[Ca^{2+}]_i$ area.

Knowing that activation of not only NMDA receptor but also non-NMDA receptors induces the $[Ca^{2+}]_i$ increase, we applied the Ca^{2+} fluorography to hippocampal slice preparation. This is to prove or disprove the theory of 'functional pairing' of L-glutamate receptor subtypes (Collingridge and Bliss, 1987). The theory claims that activation of nonNMDA receptor produces depolarization of postsynaptic spine membrane so as to make NMDA-receptor present in neighbor functionable. Extreme opinion insists that the NMDA- receptor and non-NMDA receptor share the same molecular entity activated differentially by different species of agonist (Mayer 1987).

130

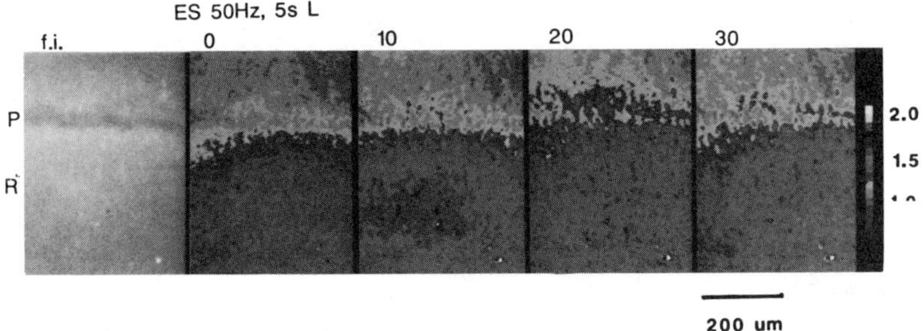

Fig. 5B

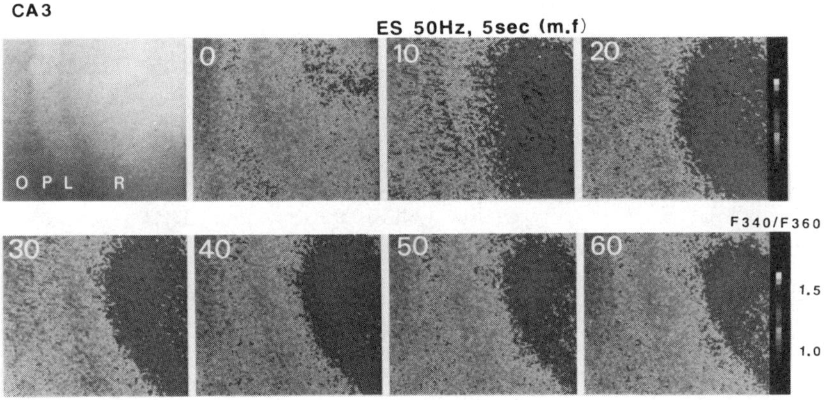

Fig. 6

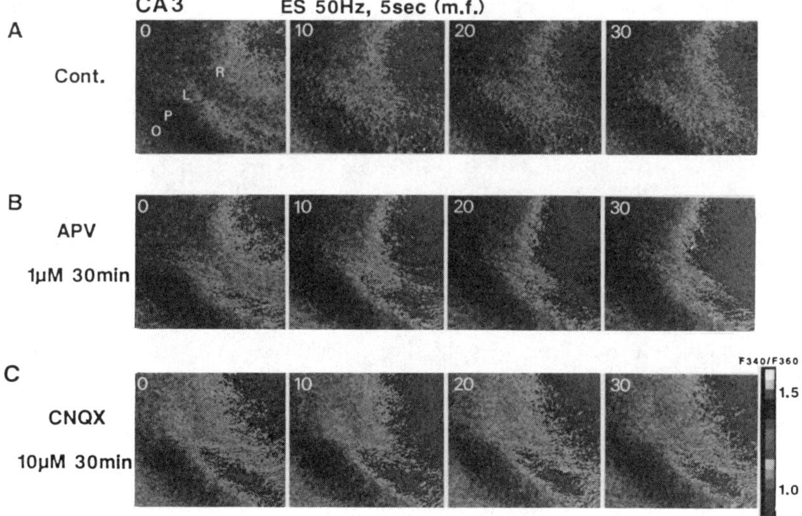

Fig. 7

We have shown above that the theory does not stand in the case of cultured neurons. Molecules of NMDA-, QUIS and KAI-receptors should be distinct. However, the receptors expressed in the cultured neuron are not necessarily arranged in the same way as those in the brain *in situ*, since synaptic activity should be crucial in determination of expressed receptor subtypes and of their distribution, as is the well- known case of neuromuscular synapse.

We were at first pessimistic that no difference in the topographical patterns of $[Ca^{2+}]_i$ rise would be observed after the application of different species of agonist to the slice preparation. This was since the La^{3+} treatment known to suppress the voltage-sensitive Ca^{2+} channels, which was effective in cultured neurons, was invalid in slices. This would result in a heterogeneous $[Ca^{2+}]_i$ rise due to possible heterogeneous distribution of this channel irrespective of the agonist species (remember that all agonists produce membrane depolarization).

The results obtained were, however, encouraging. The difference in layers of large $[Ca^{2+}]_i$ rise in response to distinct agonist species should represent the differential distribution of receptors, be that the participation of voltage-sensitive Ca^{2+} channels, if any, might have diluted the topographical characteristics of $[Ca^{2+}]_i$ rise. With some overlap, NMDA receptors are dense in the proximal dendrite, while the non-NMDA receptors distributed dominantly in the distal dendrite in CA1 region. These results are apparently not in favor of the theory of 'functional pair' in a strict sense.

It is interesting that the tetanic stimulation to the Schaffer collateral to cause LTP resulted in the increase in the fura-2 fluorescence in the same layer as in the case of NMDA administration. We do not insist that the non-NMDA receptors are totally absent in this layer. The 'functional pairing' of L-glutamate receptor subtypes during the establishment of LTP in a looser sense might be accepted.

Situation is different in CA3 region. The topological analysis of $[Ca^{2+}]_i$ rise showed that and KAI- receptors are distributed in the layer of distal dendrite in *st. radiatum*. QUIS-receptors are found to be distributed in the layer of proximal dendrite in *st. radiatum* and *st. lucidum*. Tetanic stimulation of mossy fiber induced the increase in fluorescence signal in the distal dendritic layer. Treatment with CNQX, a non-NMDA receptor antagonist, blocked the fluorescence increase due to the tetanic stimulation. But treatment with APV, an NMDA receptor antagonist, had little effect on the tetanus-induced $[Ca^{2+}]_i$ increase. These results suggest the receptor responsible for the LTP at mossy fiber/CA3 pyramidal cell synapse should be KAI-subtype, assuming that Ca^{2+} plays a key role in LTP in this synapse. However, NMDA-subtype receptor, which is blocked by APV, does exist in the same layer, and the the magnitude of NMDA-induced $[Ca^{2+}]_i$ increase seemed to be rather higher than the KAI-induced one. The role of NMDA-receptor in this layer remains unknown.

Possibility has been discussed that LTP at the mossy fiber/CA3 pyramidal cell synapse would be independent of $[Ca^{2+}]_i$ unlike LTP at the Schaffer collateral/CA1 pyramidal cell synapse (Nicoll et al., 1988). However, this discussion is based firstly on the autoradiographic observation that NMDA-receptor is scarcely distributed in the mossy fiber/CA3 pyramidal cell synapse and secondly on the assumption that NMDA-receptor is the sole species of L-glutamate receptor that allows the $[Ca^{2+}]_i$ increase. Both of these bases were not supported by the present study. We showed here that APV failed to block the $[Ca^{2+}]_i$ increase accompanied by mossy fiber stimulation. So we admit that NMDA-receptor does not play a dominant role in the LTP at mossy fiber/CA3 pyramidal cell synapse. But is even an auxiliary role denied?

We mentioned that the fura-2 signal in the slice experiment should have come not only from neurons but also from glial cells or even from debris. But dynamic changes due to electrical stimulation or agonist application should reflect the neuronal activity, let alone the static signal. Discrimination between the signals related to presynaptic activity and those of postsynaptic one is difficult. From the relative volume and from the agonists' and antagonists' effects we tentatively assumed that the majority of the dynamic fluorescence signal originates from postsynaptic compartment. But if presynaptic terminals possess L- glutamate receptors with their pharmacological nature common to the postsynaptic ones, and if the $[Ca^{2+}]_i$ rise in presynaptic cytoplasm much surpasses that in postsynaptic one to compensate its small volume, our assumption will collapse. In this context, comparison of the results obtained by the present method with the data from independent technique, such as a direct injection of Ca^{2+} fluoroprobe into postsynaptic cytoplasm (Tank et al., 1988), is required.

With the above reservation in mind, we conclude that the subtypes of L-glutamate are expressed and distributed separately in the neuronal membranes in the hippocampus both *in vitro* and *in vivo*. The activation of not only NMDA-subtype receptor, but also of nonNMDA receptors induces the $[Ca^{2+}]_i$ elevation. The $[Ca^{2+}]_i$ rise owing to non-NMDA receptor cannot be neglected in elucidating the mechanism underlying LTP. In other words, absence of the effect of APV to the establishment of LTP does not explicitly mean the absence of the role of Ca^{2+}. Two- dimensional Ca^{2+} fluorography demonstrated here will provide valuable information for the study of LTP in neural pathways other than hippocampus as well.

ACKNOWLEDGMENTS

This work was supported by grant-in-aid from Japanese Ministry of Education, Science and Culture. Technical assistance by staff of Hamamatsu Photonics Co. and by Ms. K. Akita is acknowledged.

REFERENCES

Collingridge, G.L. & Bliss, T.V.P. NMDA receptors - their role in long-term potentiation. **Tr. Neurosci.**, 10, 288-293 (1987).

Cotman, C.W., Monaghan, D.T., Ottersen, O.P., Storm-Mathisen, J. Anatomical organization of excitatory amino acid receptors and their pathways. **Tr. Neurosci.**, 10, 273-280 (1987).

Kudo, Y., Ito, K., Miyakawa, H., Izumi, Y., Ogura, A. & Kato, H. Cytoplasmic calcium elevation in hippocampal granule cell induced by perforant path stimulation and L-glutamate application. **Brain Res.**, 407, 168-172 (1987).

Kudo, Y. & Ogura, A. Glutamate-induced increase in intracellular Ca^{2+} concentrationin isolated hippocampal neurones. **Br. J. Pharmacol.**, 89, 191-198 (1986).

Malenka, R.C., Kauer, J.A., Zucker, R.S. & Nicoll, R.A. Postsynaptic calcium is sufficient for potentiation of hippocampal synaptic transmission. **Science (Wash.)**, 242, 81-84 (1988).

Mayer, M. Two channels reduced to one. **Nature (Lond.)**, 325, 480-481 (1987).

Murphy, S.N. & Miller, R.J. Regulation of Ca^{2+} influx into striatal neurons by kainic acid. **J. Pharmacol. Exp. Therap.**, 249, 184-193 (1989).

Nicoll, R.A., Kauer, J.A. & Malenka, R.C. The current excitement in long-term potentiation. **Neuron**, 1, 97-103 (1988).

Ogura, A., Miyamoto, M. & Kudo, Y. Neuronal death in vitro. **Exp. Brain Res.**, 73, 447-458 (1988).

Sugiyama, H., Ito, I, & Hirono, C. A new type of glutamate receptor linked to inositol phopholipid metabolism. **Nature (Lond.)**, 325, 531-533 (1987).

Tank, D.W., Sugimori, M., Connor, J.A. & Llinas, R.R. Spatially resolved calcium dynamics of mammalian Purkinje cells in cerebellar slice. **Science (Wash.)**, 242, 773-776 (1988).

PROTECTION BY NATURAL AND SEMISYNTHETIC GANGLIOSIDES FROM CA^{2+}-DEPENDENT

NEUROTOXICITY CAUSED BY EXCITATORY AMINO ACID (EAA) NEUROTRANSMITTERS

A. Guidotti, H. Manev, M. Favaron, G. Brooker, and
E. Costa

FGIN, Georgetown University School of Medicine
3900 Reservoir Road, N.W.
Washington, D.C.   20007 (USA)

Introduction

Neuronal death is a frequent occurrence during nervous system ontogenesis and continues, though at a slower rate, throughout life. EAA neurotransmitters, when added (in appropriate concentrations) to primary neuronal cultures, are the only neurotransmitters known to cause neuronal death by destabilizing homeostasis of free Ca^{2+} (Olney, 1969; Coyle, 1987; Vaccarino et al., 1987; Favaron et al., 1988; Connor et al., 1988).   In neurons, EAA increase Ca^{2+} influx through specific cationic channel activation, thereby stimulating Ca^{2+}-dependent enzymes, including protein kinase C (PKC) (Wroblewski and Danysz, 1989). Presumably, these actions of EAAs can be mediated at NMDA-, kainate- or quisqualate-sensitive synaptic receptors (Wroblewski and Danysz, 1989). Normally, intraneuronal Ca^{2+} homeostasis is maintained by the equilibration of free Ca^{2+}, with Ca^{2+} compartmentalized in endoplasmic reticulum and mitochondria and the operation of channels and pumps allowing Ca^{2+} influx (cationic channels) and efflux (Na^{+}-Ca^{2+} exchanger and ATPase-operated Ca^{2+} pumps) (Carafoli, 1987). It is, therefore, not surprising that many have attempted to evaluate whether EAA receptors are also operative in physiologically programmed neuronal death.   In central nervous system, an alteration of EAA transmission may occur in a variety of acute (stroke, ischemia, and CNS traumas) or chronic (autoimmune responses, Huntington's chorea, amyotrophic lateral sclerosis (ALS), olivopontocerebellar atrophy, lathyrism, and the Guamanian complex) neuropathological processes (Rothman, 1984; Kostic et al., 1989; Spencer et al., 1987; Plaitakis et al., 1988).

EAA transmission is the single most frequent type of synaptic signalling in CNS;   probably only an insignificant number of brain neurons fail to receive EAA synaptic input. This readily suggests that the various acute and chronic neurological diseases listed above cannot be treated by indiscriminately blocking or stimulating a specific subtype of EAA-sensitive synaptic receptor.   In fact, due to the ubiquity and functional importance of EAA synapses, the indiscriminate stimulation or blockade of a given glutamate receptor subtype throughout the brain is associated with untoward side effects, perhaps including psychotomimetic symptoms and/or neuronal degeneration (Olney et al., 1989). These considerations suggest that drugs used in the therapy of EAA dysfunction may not always be targeted to a specific receptor subtype but, particularly in the case of dysfunctions due to a local insult, may be better if directed at the mixed population of EAA receptors directly operative in causing the neuropathological process to be treated.

Targeting Drugs to EAA Receptors Responsible for Acute and Chronic Neuropathology

Structural diversity of EAA receptors and their associated specificity for various ligands (Costa et al., 1989) is neither located exclusively in any specific brain area nor related to a specific function (Young et al., 1989; Cotman et al., 1989). Thus, by using ligands for the primary transmitter recognition site of a specific structural receptor subtype, one cannot always achieve drug targetting for specific therapeutic purposes. When receptors for a class of neurotransmitters operative in at least 30% of brain synapses are the cause of neuronal death, the question that immediately arises is how can an exclusive drug targetting to the sites affected by the disease be achieved, leaving intact the function of identical receptors located in brain areas that must be kept functioning because they are still healthy and contributing to function, or even maintaining an important role compensatory to the function disrupted by the pathological process (Abrahams et al., 1988). For the reasons given above, in order to solve this problem, one cannot consider the structural diversity of the primary transmitter recognition sites and use ligands that, by virtue of their affinity characteristics, bind preferentially or even exclusively to receptors operative in causing specific neuropathology. EAA receptors, similarly to $GABA_A$ receptors, include a diversity in the characteristics of the effector systems (ionotropic, metabolotropic, and mixed type receptor) and of the allosteric modulatory centers for primary transmitter action (Wroblewski and Danysz, 1989; Costa, 1989). Since, as discussed above, drug targetting dependent on the characteristics of receptor recognition sites is not possible, we directed our attention to events downstream from these sites as prospective targets for specific drug actions. EAAs not only have the ability to open cationic channels which can carry ionized Ca^{2+} with degrees of efficacy related to channel conductance (Vicini et al., 1989; Cull-Candy et al., 1989), but they can also activate phospholipases (C and A_2) and increase cGMP content (Wroblewski and Danysz, 1989; Costa, 1989). Since neurotoxicity elicited by brain ischemia, stroke, and traumas is associated with paroxysmic stimulation of EAA receptors, we wondered whether it was possible to reduce the consequences of such stimulation by acting on a specific process in the chain of events triggered by paroxysmal activation of EAA receptors, a process not involved in their physiological functions. In order to select such a neuronal target for this appropriate therapy of pathological continuous stimulation of EAA receptors, we analyzed the chain of metabolic events associated with EAA-induced neurotoxicity. The ultimate goal of this analysis was the detection of an event of crucial significance in the pathological action -- but of marginal importance in the physiological action -- of EAA receptor stimulation. In other words, we were looking for drugs which could specifically affect an action dependent on the paroxysmal activation (abuse) of EAA receptors. To the best of our knowledge, pharmacologists have never described a class of antagonists whose action depends on abusive stimulation of EAA receptors. We, therefore, embarked on the search of this new class of compounds that we define as RADA (Receptor Abuse Dependent Antagonists) drugs of EAA.

<u>Duration of Free Cytosolic Ca^{2+} Increase and EAA Neurotoxicity</u>

Neuronal membranes are spanned by various cationic channels which, by allowing Ca^{2+} influx, transiently increase cytosolic concentrations of ionized Ca^{2+} (Connor et al., 1988). Some cationic channels are gated by voltage changes, others by ligand occupancy of specific recognition sites for transmitters or neuromodulators. Physiologically, the transient increase of ionized Ca^{2+} spells out a precise metabolic message for the neuron in which these channels are located. However, a protracted neuronal membrane depolarization and/or the paroxysmal activation of specific transmitter-gated ionotropic receptors can prolong the duration of free Ca^{2+} increase (Connor et al., 1988); this not only stimulates catalytic activity of Ca^{2+}-dependent enzymes but can cause the translocation of such enzymes from one to another neuronal compartment. Hence, the persistent stimulation of EAA receptor recognition sites may lead to irreversible neuronal structural changes due to changes in the compartmentalization of specific enzymes. Are these structural changes associated with neuronal death? Many transmitter receptors which include an increase of ionized cytoplasmic Ca^{2+} as part of their effector activity fail to cause long lasting increases in free Ca^{2+} and neuronal death, even when stimulated paroxysmally (Vaccarino et al., 1987). Hence, one must consider that paroxysmal activation of EAA receptors not only increases Ca^{2+} influx but also disrupts intraneuronal Ca^{2+} homeostasis through reduction of the buffering capacity of one or more of the neuronal mechanisms regulating ionized Ca^{2+} levels. The Na^{+}-Ca^{2+} exchanger and the Ca^{2+} ATPase pump located in neuronal membranes play a substantial role in the regulation of neuronal Ca^{2+} efflux which is pivotal in the homeostasis of ionized intracellular Ca^{2+} (Carafoli, 1987). Hence, such mechanisms may become of crucial importance when the steady state of ionized Ca^{2+} is abnormally elevated. The efficiency of the Ca^{2+} ATPase pump can be increased by its phosphorylation by cAMP-dependent, and probably other, protein kinase(s) PK (Carafoli, 1987). In addition, this pump is activated by a direct interaction with calmodulin (Carafoli, 1987). Though the sites for this interaction are not known, they could be modified by phosphorylation of appropriate PK consensus sites located in the intracellular domain of the pump protein. Hence, we studied whether EAA receptor activation could induce translocation of a cytosolic PK to the neuronal membrane (Vaccarino et al., 1987) which would change the phosphorylation profile of the Ca^{2+} ATPase pump and thereby interfere with the single molecular interaction between the pump and calmodulin. Theoretically, it is possible to consider that, if the regulation of the Ca^{2+} ATPase pump were to be curtailed, its buffering capacity to challenges of Ca^{2+} homeostasis would be reduced and untoward consequences to neuronal function and survival might result.

Using primary cultures of granular and cortical cells prepared from neonatal rat brain, receptor stimulation by carbachol or glutamate can activate PIP$_2$ turnover (Costa et al., 1988), but only glutamate can destabilize Ca^{2+} homeostasis and cause translocation of PKC in a dose dependent manner (Vaccarino et al., 1987). Moreover, the time course of this PKC translocation depends both on the extent and duration of EAA receptor stimulation (Manev et al., 1987) and can outlast the EAA pulse by longer than 60 minutes (Table 1). We have also shown that the translocated enzyme is fully functional (Vaccarino et al., 1987) and, therefore, can phoshorylate the new proteins it reaches in new intracellular surroundings. Since the Ca^{2+} ATPase pump which spans the membrane includes an important cytosolic regulatory domain, it is possible to postulate that this may serve as a substrate for the catalytic activity of the translocated PKC.

Table 1. Relationship between glutamate-induced neurotoxicity, translocation of PKC and $^{45}Ca^{2+}$ uptake in primary culture of cerebellar granule cells

	^{3}H-PDBu BINDING (f mol/mg prot.) minutes after GLU pulse			UPTAKE OF $^{45}Ca^{++}$ 30 min after GLU pulse	%VIABILITY 24 hours after GLU pulse
	0	30	90		
Control	330 ± 31	325 ± 18	301 ± 11	11 ± 0.56	95 ± 4
Glutamate 1 μM	655 ± 27*	291 ± 18	-	12 ± 1.6	91 ± 5
Glutamate 50 μM	799 ± 35*	625 ± 25*	495 ± 17*	22 ± 1.6*	19 ± 1*

Cerebellar granule neurons were exposed for a 15 min pulse to GLU (glutamate) in Locke's buffer without Mg^{++}. ^{3}H-PDBu binding to the monolayer of neurons was measured by 15 min incubation with 2nM ^{3}H-phorbol-12,13-dibutyrate (PDBu) (Vaccarino et al., 1987). Ca^{++} uptake (nmol/mg prot/1min) was measured at 37°C by adding 1 uCi of $^{45}CaCl_2$ (Manev et al., 1989). Viability (% cells alive) was estimated by fluorescein diacetate/propidium iodide staining (Favaron et al., 1988).

* P<0.01 when compared with control group

We are beginning to measure the duration of ionized cytosolic Ca^{2+} increases following stimulation of EAA receptors in the membrane of granule cells in primary culture. We have seen in this situation a persistent increase of high free Ca^{2+} levels after a glutamate application lasting 15 minutes. This enhanced content of ionized cytosolic Ca^{2+} rapidly returns to basal values if the transmitter pulse is followed for about 90 minutes by the addition of culture medium without Ca^{2+} and with EGTA (Manev et al., 1989). Presumably, under these conditions, the function of the processes attending to Ca^{2+} efflux is facilitated. Moreover, glutamate fails to cause neurotoxicity if Ca^{2+} is absent during the glutamate pulse. Under these conditions, the increase of ionized Ca^{2+} is greatly attenuated (Manev et al., 1989). Moreover, after the addition of medium containing Ca^{2+}, ionized Ca^{2+} fails to increase following the glutamate pulse. Hence, the increase of ionized Ca^{2+} observed after a glutamate pulse in the presence of Ca^{2+} may not be due to an activation of Ca^{2+} influx which follows activation of cationic channels by residual glutamate. Such a lack of residual activation of Ca^{2+} channels was also documented by direct measures of Ca^{2+} channel activation by voltage changes or by EAA transmitters. Probably a Ca^{2+} activated process is responsible for the increase of ionized Ca^{2+} after the glutamate pulse.

Thus, Ca^{2+} must be present during glutamate receptor activation and during the first 90 minutes following the glutamate pulse in order for EAA neurotoxicity to be observed (Manev et al., 1987). We have also found that if free Ca^{2+} levels fail to increase for a prolonged time period, PKC translocation is fleeting and variable and returns to prestimulation levels within minutes after the glutamate pulse. Hence, several lines of investigation suggest that EAA receptor neurotoxicity follows a paroxysmal activation if, after the transmitter pulse, the increase in ionized cytolosic Ca^{2+} content is sustained and PKC remains translocated for 1 hour or longer after the termination of the EAA transmitter pulse.

Inhbition of PKC Translocation by Natural Gangliosides

The report that sphingosine and sphingoglycolipids inhibit PKC activity (Hannun and Bell, 1987,1989) prompted us to investigate whether natural gangliosides inserted into membranes of cultured neurons could inhibit EAA-induced neurotoxicity (Favaron et al., 1988).

Table 2. Ganglioside pretreatment prevents glutamate-induced neuronal death in pirmary cultures of cerebellar granule cells

VIABILITY 24 HRS AFTER GLUTAMATE (% neurons alive)

Ganglioside pretreatment (μM/2 hrs)	Glutamate pulse (15 min) immediately after ganglioside pretreatment	Glutamate pulse (15 min) 12 hrs after ganglioside pretreatment
none	20 ± 5.2	18 ± 2.2
GM_1		
(10)	15 ± 7.0	-
(30)	32 ± 12	-
(60)	65 ± 9.1^x	30 ± 6.0
(100)	78 ± 4.0^x	-
GT_{1b}		
(10)	47 ± 9.4^x	-
(30)	67 ± 14.2^x	-
(60)	80 ± 5.0^x	60 ± 10^x
(100)	89 ± 4.1^x	-

Cultures were preincubated for 2 hours with gangliosides. After the removal of gangliosides and washing three times with Locke's solution, the cells were challenged with glutamate (50μM in Mg^+ free medium at 25° for 15 min). At the end of glutamate exposure, the cells were incubated for 24 hours in culture conditioned medium (Favaron et al., 1988). Viability was estimated with fluorescein diacetate/propidium iodide staining (Favaron et al., 1988). The results are the mean $\pm$ SE of at least three experiments. P<0.01 when compared with controls.

It became clear that GM_1 and GT_1b, if preincubated with granule cells for about 2 hours and then thoroughly removed from the media, can protect from EAA neurotoxicity (Table 2). This protection is dose and time dependent and lasts for about 18 hours after termination of a ganglioside pulse lasting 2 hours (Favaron et al., 1988). This inhibition of EAA neurotoxicity is probably due to ganglioside insertion in the membrane because it is linearly related to the increase in membrane ganglioside content (Fig. 1). This correlation can be shown to be statistically significant, not only for the protection of neuronal death elicited by EAA but also for the duration of the PKC translocation. However, incubation of primary neuronal cultures for 2 hours with ganglioside concentrations (GM_1 or GT_1b) which antagonize EAA-induced neurotoxicity fail to change glutamate thresholds for cationic channel opening, phosphatidyl inositol increase of metabolism, or phospholipase A_2 stimulation (Favaron et al., 1988). Hence, GM_1 and GT_1b target an important step in the chain of events triggered by the application of neurotoxic doses of EAA to primary neuronal cultures. This step, which is essential for the action of these excitotoxic acidic amino acids, appears to be an increase of cytosolic ionized Ca^{2+} and a persistent translocation of cytosolic PKC to the neuronal membranes (Manev et al., 1989).

<u>Pharmacological Profile of Receptor Abuse Dependent Antagonists (RADA)</u>

As discussed earlier, it is not possible to block indiscrimminately the activation of every EAA sensitive synaptic receptor in CNS without eliciting very important side effects, including psychotomimetic actions and an impairment in the regulation of some

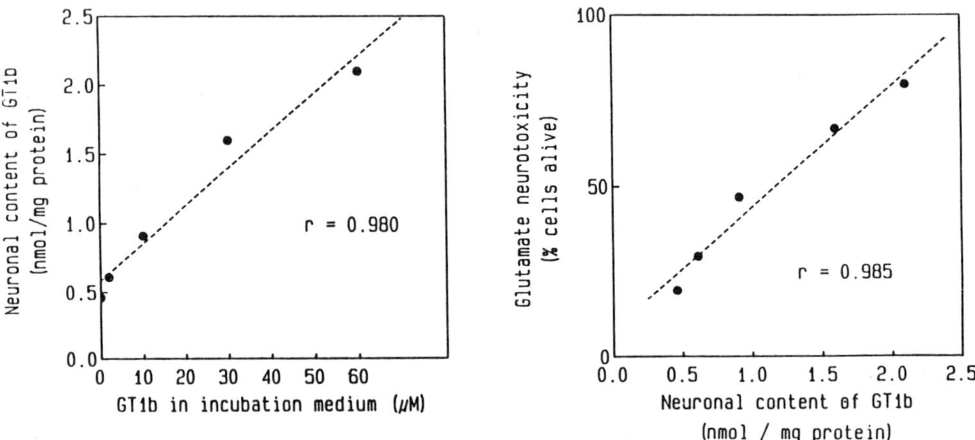

Figure 1.   Monolayers of cerebellar granule neurons in culture were preincubated for 2 hours at 37°C with GT_{1b} in Locke's buffer. At the end of 2 hours, the dishes were washed several times and gangliosides were extracted and their content was determined by high-performance thin-layer chromatography (Favaron et al., 1988), followed by densitometric analysis. Neuronal death induced by glutamate (50µM; 15 min) applied after ganglioside washout was quantified 24 hours later and expressed as % neurons alive.

aspects of respiration or blood pressure functions regulated by EAA transmission. Moreover, non competitive channel blockers of ionotropic EAA receptors appear to cause neurotoxicity "per se" (Olney et al., 1989).

Gangliosides fail to change EAA-mediated cationic channel opening or the EAA activation of phospholipase C and A_2 (Favaron et al., 1988). However, gangliosides block glutamate neurotoxicity, presumably by inhibiting the increase of free cytosolic Ca^{2+}, which outlasts the glutamate pulse by 90 minutes. This latter response and neuronal death are presumably related to the intervening translocation of cytosolic PKC (Manev et al., 1989). However, since both events appear to be strictly time related to each other, it is difficult to distinguish which event is causal and which is consequent. Nevertheless, GM_1 and GT_1b selectively inhibit the responses elicited by paroxysmal activation of EAA receptors but not responses due to the physiological action of EAA, including PKC activation and the induction of transcriptional activation of c-fos mRNA (Manev and Favaron, this laboratory, unpublished). Since gangliosides inhibit the responses which specifically depend on paroxysmal (abusive) stimulation of EAA receptors, they are RADA drugs for EAA receptor responses. Specifically, it is possible now to test whether GM_1 and GT_1b can relieve in vivo the consequences of EAA receptor paroxysmal and persistent stimulation, which presumably occur in those EAA receptors located in the area perifocal to brain ischemia. Preliminary data indicate that in animal models of brain ischemia, gangliosides injected before the injury can reduce the consequences of brain ischemia. This protective action is exerted without impairing any of the physiological actions that could be related to glutamatergic, GABAergic, or any other transmitter-related, function. This targetting of a drug to the site of a neuropathological process by using its ability to antagonize a rate limiting neuronal event represents a conceptual novelty in neuropharmacology. Thus, the RADA effect of gangliosides against EAA neurotoxicity represents the first case of a drug that can be targeted "in vivo" and "in vitro" to a specific pathological change causing degeneration and death of neurons.

Semisynthetic Gangliosides with Pharmacological Profiles More Suitable for the Development of RADA Drugs against EAA Neurotoxicity
====

Since natural gangliosides cannot be administered by mouth, and also since they reach their site of action slowly because they insert and concentrate in neuronal membranes very slowly, they are not ideally suited to be used in acute neuropathology due to paroxysmal stimulation of EAA receptors. We have searched for structural derivatives with an ideal pharmacological profile for relief of acute neurotoxicity associated with stroke, brain ischemias, and traumas. The synthesis of ganglioside derivatives which possess physicochemical properties facilitating absorption after oral administration, improve blood-brain barrier permeability and facilitate rapid insertion into neuronal membranes was initiated. The preliminary results so far achieved are reported here. Figure 2 shows the structure of the two best RADA compounds and gives important in vitro data defining their pharmacological profiles. We do not know yet whether these compounds are active in vivo in man in limiting the perifocal damage of brain ischemic areas. However, we know that they appear to be active in reducing severity of neuronal damage in experimental models of ischemias. Moreover, when administered in active doses, they are well tolerated by experimental animals. These results encourage us to proceed in the systematic study of the pharmacological and toxicological profiles of LIGA-4 and LIGA-20.

Conclusions
====

In conclusion, we believe that LIGA-4 and LIGA-20 belong to a new class (RADA drugs) of pharmacological antagonists of EAA-induced neuronal death. They are the first RADA drugs to mimic the effect of endogenous gangliosides which, when injected into animals and man, appear to adhere to the model of a RADA compound for EAA neurotoxicity. It is of interest that anti-GM_1 IgM antibodies are present in certain forms of ALS having a high content of tissue glutamate (Pestronk et al., 1989). We are now studying whether this class of antibodies can change susceptibility of primary neuronal cultures to EAA-induced neurotoxicity. In preliminary clinical experiments, immunosuppressants cause a relief of symptoms of ALS in patients with high titer of ganglioside antibodies (Pestronk, personal communication). It is of interest that endogenous constituents of neuronal membrane can limit the

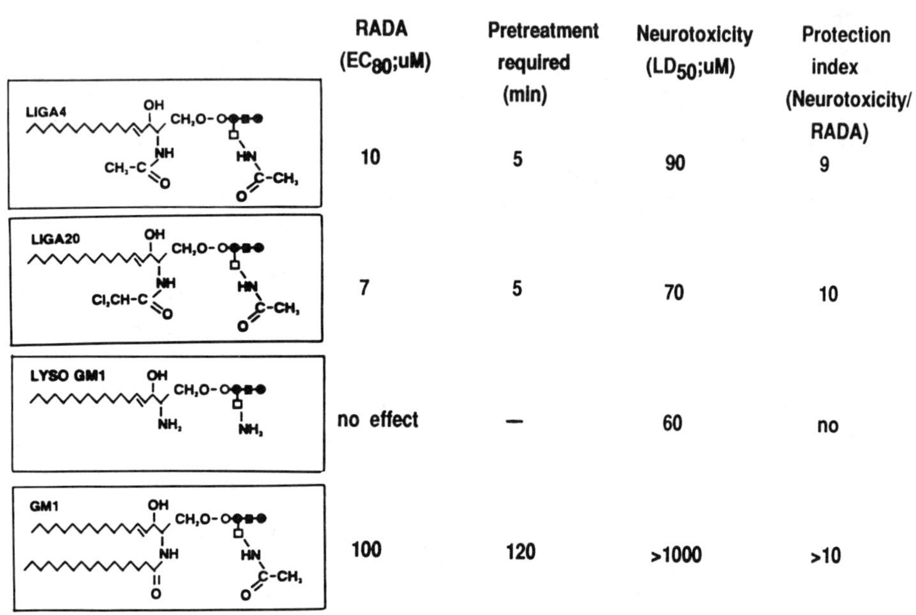

Fig. 2.   RADA (receptor abuse dependent antagonism) of naturally occurring sphingolipids and their synthetic derivaties. RADA is expressed as the concentration of sphingolipid (μM) required to protect 80% of granule neurons in culture from glutamate neurotoxicity. Cells were pretreated with corresponding compound for a time period required to obtain the maximal RADA effect. In absence of protective effect incubation was limited to 120 min. Neurotoxicity was measured at the corresponding incubation time; for Lyso GM$_1$ it was 10 minutes. After preincubation, sphingolipids were washed out and cultures were exposed to 50μM glutamate for 15 min. Following washout of glutamate, cultures were returned to the culture-conditioned medium and 24 hours later were stained with fluorescein-propidium iodide (Favaron et al., 1988).

process of cytosolic enzyme translocation to neuronal membrane directed by paroxysmal activation of EAA receptors. Perhaps there are physiological reasons for the structural diversity of the gangliosides in the membranes of various cell types and in neurons of different brain areas; one such reason could be the limitation of the danger ensuing excessive translocation of PKC. If this is the case, RADA action of drugs against EAA receptor stimulation, as discussed in this report, could be considered as drug action that follows where nature leads.

REFERENCES

Abrahams, T.P., Gillis, R.A., Hamosh, P., Taverie, D.A. and Silva, A.M., 1988, Systematic administration of the NMDA receptor antagonist MK$_{801}$ produces pronounced changes in cardiorespiratory function, Soc. Neurosci. Abstr., 14:937.

Carafoli, E., 1987, Intracellular Ca homeostasis, Ann. Rev. Biochem., 56:395-433.

Costa, E., Alho, H., Favaron, M. and Manev, H., 1988, Pharmacological implications of the allosteric modulation of neurotransmitter amino acid receptors, in: "Allosteric Modulation of Amino Adid Receptors: Therapeutic Implications," Barnard, E.A. and Costa, E., eds., Raven Press, New York.

Costa, E., Fadda, E., Kozikowski, A.P., Nicoletti, F. and Wroblewski, J.T., 1988, Classification and allosteric modulation of excitatory amino acid signal transduction in brain, slices and primary cultures of cerebellar neurons, in: "Neurobiology of Amino Adics, Peptides and Trophic Factors," J. Ferendelli, R. Coduis and E. Johnson, eds., Martinus Hijhoff Publ., Boston, Massachusetts.

Costa, E., 1989, Allosteric modulatory centers of transmitter amino acid receptors, Neuropsychopharm., 2:in press.

Cotman, C.W., Bridges, R.J., Geddes, J.w., Monagham, D.t. and Catron L.D., 1988, Anatomical organization of plasticity of excitatory amino acid receptors in the rodent and human brain: predictions on their functional state in neurodegenerative diseases, in: "Allosteric Modulation of Amino Acid Receptors: Therapeutic Implications," E.A. Barnard and E. Costa, eds., Raven Press, New York.

Coyle, J.T., 1987, Excitotoxins, in: "Psychopharmacology, The Third Generation of Progress," H.Y. Meltzer, ed., Raven Press, New York.

Cull-Candy, S.G. and Usowicz, M.M., 1988, Multiple conductances of glutamate receptor channels, in: "Allosteric Modulation of Amino Acid Receptors: Therapeutic Implications," E.A. Barnard and E. Costa, eds., Raven Press, New York.

Favaron, M., Manev, H., Alho, H., Bertolino, M., Ferret, B., Guidotti, A. and Costa, E., 1988, Gangliosides prevent glutamate and kainate neurotoxicity in primary neuronal cultures of neonatal rat cerebellum and cortex, Proc. Natl. Acad. Sci. USA, 85:7351-7355.

Hannun, Y.A. and Bell, R.M., 1987, Lysoshingolipids inhibit protein kinase C: implication for the sphingolipidases, Science, 235:670-674.

Hannun, Y.A. and Bell, R.M., 1989, Functions of sphingolipids and sphingolipid breakdown products in cellular regulation, Science, 243:500-507.

Kostic, V.S., Mojsilovic, L., and Stojanovic, M., 1989, Degenerative neurological disorders associated with a deficiency of glutamate dehydrogenase, J. Neurol., 236:111-114.

Manev, H., Favaron, M., Guidotti, A. and Costa, E., 1989, Delayed increase of Ca^{2+} elicited by glutamate: role in neuronal death, Molec. Pharmacol., 36:106-112.

Olney, J.W., 1969, Brain lesions, obesity and other disturbances in mice treated with monosodium glutamate, _Science_, 164:719-721.

Olney, J.W., Labruyere, J. and Price, M.T., 1989, Pathological changes induced in cerebrocortical neurons by phencyclidine and related drugs, _Science_, 244:1360-1362.

Pestronk, A., Adams, R.N., Cornblath, D. Kuncl, R.W., Drachman, D.B., Clawson, L., 1989, Patterns of serum IgM antibodies to GM_1 and GD_{1a} gangliosides in amyotrophic lateral sclerosis, _Ann. Neurol._, 25:98-102.

Plaitakis, A., Constantakakis, E. and Smith, J., 1988, The neuroexcitotoxic amino acid glutamate and aspartate are altered in the spinal cord and brain in amyotropic lateral sclerosis, _Ann. Neurol._, 24:446-449.

Rothman, S.M., 1984, Synaptic release of excitatory amino acid neurotransmitter mediated anoxic neuronal death, _J. Neurosci._, 4:1884-1891.

Spencer, P.S., Nunn, P.B., Hugon, J., Ludolph, A.C., Ross, S.M., Roy, D.N. and Robertson, R.C., 1987, Guam amyotropic lateral sclerosis linked to a plant excitant neurotoxin, _Science_, 237:517-522.

Vaccarino, F., Guidotti, A. and Costa, E., 1987, Ganglioside inhibition of glutamate mediated protein kinase C translocation in primary cultures of cerebellar neurons, _Proc. Natl. Acad. Sci. USA_, 84:8707-8711.

Vicini, S., Bertolino, M. and Costa, E., 1988, Modulation of glutamate activated cationic channels: a patch-clamp study, _in_: "Allosteric Modulation of Amino Acid Receptors: Therapeutic Implications," E.A. Barnard and E. Costa, eds., Raven Press, New York.

Wroblewski, J.T. and Danysz, W., 1989, Modulation of glutamate receptors: molecular mechanisms and functional implications, _Ann. Rev. Pharmac. Toxicol._, 29:441-474.

Young, A.B., Janh-Ho, J., Cha J., Greenamyre, T., Meragas, W.F. and Penney, J.B., 1989, Ditribution of PCP and glutamate receptors in normal and pathologic mammalian brain, _in_: "Allosteric Modulation of Amino Acid Receptors: Therapeutic Implications," E.A. Barnard and E. Costa, eds., Raven Press, New York.

SESSION IV : EAA AND IN VIVO DEVELOPMENT

PERIODIC INWARD CURRENTS TRIGGERED BY NMDA IN IMMATURE CA3 HIPPOCAMPAL NEURONES

E. Cherubini, Y. Ben Ari and K. Krnjevic*

INSERM, U. 029, 123 Bd. de Port Royal, 75014 Paris, France.*Present address: Anaesthesia Research Department, Mc Gill University, 3655 Drummond St., Montreal, Canada

INTRODUCTION

There is widespread interest in the function of N-methyl-D-aspartate (NMDA) receptors in the CNS, especially in the hippocampus where they are higly concentrated and may contribute to long term potentiation, a model of memory. More recent studies performed primarily in the visual system suggest a preferential involvement of NMDA receptors in developmental plasticity (Tsumoto et al., 1987 ; Kleinschmidt et al. 1987). Using a single electrode voltage clamp technique, we have found that NMDA generates slow inward currents in immature CA3 hippocampal neurones, already a few hours after birth, confirming that NMDA receptors are present and active very early in postnatal life. In addition, repeated applications of NMDA, initiated in immature (but not in adult) neurones, TTX-insensitive periodic inward currents (PICs), which persisted for hours as a prominent on-going phenomenon, whose frequency was independent of membrane potential. These currents were observed under conditions in which K+ conductances were much reduced by external Cs+, tetraethylammonium (TEA) or 4-aminopyridine (4-AP). Though TTX-insensitive, PICs appear to have a synaptic origin, since they were reduced by bicuculline and abolished by the excitatory amino acid antagonist kynurenic acid. It is suggested that NMDA may generate PICs by initiating in immature neurones periodic fluctuations of intracellular calcium, leading to pulsatile release of glutamate and GABA from nerve terminals.

METHODS

The present study is based on intracellular recordings from 13 adult and 54 immature CA3 hippocampal neurones. The experiments were done on 600 µm thick slices of hippocampus obtained from male Wistar either at an adult stage or during the immediate postnatal period (0-10 days). The slices were cut and transferred to a submerged type recording chamber as described previously (Ben-Ari et al., 1989). The slices were superfused (2.5-3 ml/min) at 34° C with oxygenated artificial cerebrospinal fluid (ACSF) of the following composition (mM): NaCl 126, KCl 3.5, CaCl$_2$ 2, MgCl$_2$ 1.3, NaH$_2$PO$_4$ 1.2, NaHCO$_3$ 25 and glucose 11 (pH 7.3). Microelectrodes filled with 3M KCl, 4M K-acetate or 2M CsCl (with tip resistance of 40-80 M) were used for intracellular recordings.

Neurones were voltage clamped using a single electrode voltage clamp amplifier (Axoclamp-2). The sampling frequency was 3-4 KHz, 30% duty cycle. To ensure correct operation of the clamp, the voltage at the head stage amplifier was monitored on a separate oscilloscope. Currents were recorded in the presence of tetrodotoxin (TTX, 1-3 µM) ant tetraethylammonium (TEA, 10 mM) to abolish network synaptic transmission, Na$^+$ currents, inward rectification (Q current) and the predominant outward currents. NMDA was bath applied for 2-3 min. with 10-15 min. intervals between consecutive applications. All the values in the text are expressed as mean ± S.E.M.

RESULTS

Bath application of NMDA (10 µM) produced in adult hippocampal neurones a TTX-insensitive slow inward current (0.56 ± 0.06 nA, n=5, at holding potential, V_H, between -30 and -50 mV) which was blocked by the selective NMDA receptor antagonist 2-amino-phosphono-valeric acid, D-APV. Similar D-APV-sensitive inward currtents were evoked in immature neurones (at V_H between -30 and -50 mV, 0.45 ± 0.04 nA, n=16). In addition to the slow inward current, repeated (2-5 times) applications of NMDA evoked in every immature neurone (n=33) TTX-insensitive periodic inward currents which persisted for several hours after NMDA was washed out. NMDA was essential for the generation of PICs since they were not observed in long lasting recordings (1-2 hours) made from neurones bathed with a medium containing TTX, TEA and Cs$^+$ (n=6) or TTX, Cs$^+$ and 4-AP (n=3) without NMDA. PICs were seen only if in addition to Cs$^+$ either TEA or 4-AP were present in the perfusion fluid, but not when Cs$^+$ was injected intracellularly through the recording micropipette (n=8).

Although application of D-APV prior to (and during) NMDA prevented the induction of PICs (n=3), once elicited, PICs were insensitive to D-APV, indicating that the activation of NMDA receptors is necessary for their induction but not for their maintenance. In contrast repeated applications of NMDA to adult neurones, in similar conditions, failed to evoke persistent inward currents (n=8). Being persistent and highly reproducible, PICs lent themselves to systematic investigation. When observed at faster speed, PICs proved to be discrete inward currents of rather stable amplitude and duration, generated at a frequency of 0.11 ± 0.01 Hz, n=33). Rythmic fluctuations in amplitude were occasionally observed. When switching from voltage to current clamp mode, recurrent spike like potentials were recorded at the same frequency. Although the frequency of PICs was voltage independent, their amplitude was a monotonic function of the holding potential over the range of -70 mV to +20 mV, with a reversal near 0 mV (-3.4 ± 5 mV, n=7).

Although it is known that NMDA can evoke TTX-insensitive periodic oscillations in membrane potentials of some central neurones (Wallen and Grillner, 1985), PICs sharply differ in that their frequency is independent of membrane potential. The closest parallel is perhaps that of the largely voltage independent oscillatory membrane currents that can be evoked in oocytes (Miledi et al. 1987) by several transmitters. Oscillatory transient inward currents observed in lacrymal gland (Evans and Marty, 1986) also bear some similarities. In all these examples, however, the oscillations usually decay very rapidly. The periodic inward currents evoked by NMDA in immature CA3 neurones are thus exceptional in being highly persistent (for several hours).

The following additional observations throw some light on the mechanisms of PICs. In contrast to the oscillatory currents observed in oocytes (Miledi et al., 1987), Cl$^-$ was not involved in PICs generation since they were readily observed in neurones recorded with K-acetate containing microelectrodes.

As postulated for the comparable currents in cardiac cells (Mechmann and Pott, 1986 ; Kimura et al., 1986) PICs could be produced either by an electrogenic Na-Ca exchange or a non specific cationic current activated by a rise in intracellular calcium (Yellen 1982). The Na-Ca exchange mechanism is unlikely to contribute to the generation of PICs since substitution of external Na$^+$ (126 mM) with TEA (n=3) did not change the frequency or the reversal potential of PICs (in two neurones the reversal was -15 and + 2 mV). In sympathetic neurones (Kuba et al., 1980), in skeletal muscle Weber and Herz, 1968) and in cardiac myocytes (Connors et al., 1983), large oscillations in intracellular calcium are produced by caffeine, a methylxanthine which releases calcium from the sarcoplasmatic reticulum (Weber and Herz, 1968). In two neurones caffeine (1 mM) was added to the bath in the presence of the phosphodiesterase inhibitor. 3-isobutyl-1-methylxantine (IBMX, 200 µM). In both cells caffeine induced an inward current and periodic oscillatory currents similar to those elicited by NMDA. These remained for prolonged periods after caffeine was washed out. To examine the involvement of intracellular calcium in PICs generation, 5 neurones were loaded with the calcium chelators ethyleneglycol-bis-(-aminoethyl ether) -N, N, N', N' -tetracetic acid (EGTA) or 1,2-bis (2-aminopehnoxy) ethane N,N,N'< N'- tetracetic acid (BAPTA). The calcium dependent afterhyperpolarization following action potentials was fully blocked a few minutes after impalement, indicating that EGTA and BAPTA were efficient in chelating intracellular calcium at the recording site. Unlikely current oscillations in oocytes (Miledi et al., 1987), PICs were still generated under these conditions by repeated applications of NMDA, suggesting that oscillations in intracellular calcium at the recording site (soma) do not play a major role. However it is possible that fluctuations in intracellular calcium occur at relatively distal, dendritic sites, not adequately buffered by EGTA or BAPTA. Another possibility however, is that PICs are generated indirectly, for example by transmitter release from presynaptic nerve terminals, initially triggered by NMDA. Ca^{2+} influx may then fluctuate periodically if the presynaptic calcium currents are not damped down by potassium currents (in the presence of TEA and 4-AP) thus causing corresponding variations in terminal $[Ca]_i$. Hence, in spite of the presence of TTX in concentrations which fully block fast action potentials and synaptic activity, it remains possible that oscillations in intracellular calcium induced by NMDA or caffeine generate a periodic synchronous transmitter release directly from presynaptic terminals of adjacent cells. The immature nervous system has a number of morphological and physiological features, which would tend to facilitate such an effect including frequent electrical coupling between adjacent neurones (Connors et al., 1983 ; Christie et al., 1989), which would facilitate synchronization of transmitter release. In keeping with the involvement of presynaptic transmitter in the generation of PICs are several other observations: PICs could not be elicited when only the postsynaptic potassium channels were blocked by intracellular Cs^+; they were abolished by a calcium free solution; reduced in frequency by the GABA A antagonist bicuculline (10-30 µM) and abolished by the broad spectrum excitatory amino acid antagonist kynurenic acid (0.5-1 mM).

CONCLUSIONS

Repeated applications of NMDA induce in immature CA3 neurones a pulsatile release of transmitter probably mediated by oscillations in intracellular calcium in presynaptic nerve terminals. The release of glutamate and GABA could thus be an important signal in the inhibition of dendritic outgrowth and the initiation of synaptogenesis (Mattson 1988). This constitutes a new model of long lasting effects produced by NMDA in immature neurones.

REFERENCES

Ben Ari, Y., Cherubini, E., Corradetti, R. and Gaiarsa, J.L., Giant synaptic potentials in immature rat CA3 hippocampal neurones. **J. Physiol.**, 416, 303-325, 1989.
Christie, M.J., Williams, J.T. and North, R.A., Electrical coupling synchronizes subthreshold activity in locus coeruleus neurons in vitro from neonatal rats. **J. Neurosci.** 9, 3584-3589, 1989.
Connors, B.W., Benardo, L.S. and Prince, D.A., Coupling between neurons of the developing rat neocortex. **J. Neurosci.**,3 , 773-782, 1983.
Evans, M.G. and Marty, A., Potentiation of muscarinic and α-adrenergic responses by an analogue of guanosine 5'-triphosphate. **Proc. Natl. Acad. Sci.**, 83, 4099-4103, 1986.
Kimura, J., Noma, A. and Irisawa, H., Na-Ca exchange current in mammalian heart cells. **Nature**, 319, 596-597, 1986.
Kleinschmidt, A., Bear, M.F. and Singer, W., Blockade of NMDA receptors disrupts experience dependent plasticity of kitten striate cortex. **Science**, 328, 355-358, 1987.
Kuba, K.J., Release of calcium ions linked to the activation of potassium conductance in a caffeine-treated sympathetic neurone. **J. Physiol.**, 298, 251-270, 1980.

Mattson, M.P., Neurotransmitters in the regulation of neuronal cytoarchitecture. **Brain Research Reviews,** 13, 179-212, 1988.

Mechmann, S. and Pott, L., Identification of Na-Ca exchange current in single cardiac myocytes. **Nature,** 319, 597-599, 1986.

Miledi, R., Parker, I., Sumikawa, K., Oscillatory chloride current evoked by temperature jumps during muscarinic and serotonergic activation in Xenopus oocyte. **J. Physiol.,** 383, 213-229, 1987.

Tsumoto, T., Hagihara, K., Sato, H. and Hata, Y., NMDA receptors in the visual cortex of young kittens are more effective than those of adult cats. **Nature,** 327, 513-514, 1987.

Wallen P. and Grillner, S., The effect of current passage on N-methyl-D-aspartate-induced, tetrodotoxin -resistant membrane potential oscillations in lamprey neurons active during locomotion. **Neurosci. Lett.,** 56, 87-93, 1985.

Weber, A. and Herz, R., The relationsheep between caffeine contracture of intact muscle and the effect of caffeine on reticulum. **J. Gen. Physiol.,** 52, 750-759, 1968.

Yellen, R., Single Ca^{2+} -activated non selective cation channels in neuroblastoma. **Nature,** 296, 357-359, 1982.

GABA MEDIATED SYNAPTIC EVENTS IN NEONATAL RAT CA3 PYRAMIDAL NEURONS

IN VITRO : MODULATION BY NMDA AND NON-NMDA RECEPTORS

J.L. Gaïarsa,   R. Corradetti,   Y. Ben-Ari and   E. Cherubini

INSERM U029, 123 BD. de Port-Royal, 75014 Paris, France

INTRODUCTION.

It is generally assumed that the synaptic inhibition in adult hippocampus is mediated by GABA, which acts on GABA-A (Ben-Ari and al 1981, Alger and Nicoll 1982, Kehl and Mc Lennan 1985), and GABA-B receptors, coupled to chloride and potassium channels respectively (Alger-and Nicoll 1982, Dutar and Nicoll 1988). When the inhibition is blocked by GABAergic antagonist, interictal discharges , mediated by excitatory amino acids appear (Wong and Traub 1983, Neuman and al 1988 b). Extensive theorical and experimental data indicate that the well known propensity of CA3 neurons to generate synchronized burst is due to the presence of recurrent excitatory collaterals between pyramidal neurons (Mc Vicar and Dudeck 1980, Wong and Traub 1983, Miles and Wong 1986).

If the synaptic transmission and the local network are relatively well documented in adult hippocampus, there is little information available in the immature hippocampus. A previous study performed *in vivo* on kitten hippocampus suggests that inhibition is the predominant form of synaptic activity (Purpura and Pappas 1968). The introduction of the *in vitro* slices preparations has made the hippocampus more accessible to electrophysiological investigations. Several studies performed in   hippocampal slices have shown a late development of the inhibition in the CA1 region (Dunwiddie 1981, Schwartzkroin 1982, Harris and Teyler 1983) , whereas inhibitory postsynaptic potential (ipsp) can be recorded at birth in rabbit CA3 neurons (Schwartzkroin 1982, Schwartzkroin and Kunkel 1982) and from postnatal day 6 in CA3 rat neurons (Swann and al 1989) . However, no data are available concerning the development of the CA3 area during the first postnatal week. This period is particularly important since most of the extrinsic pathway are not yet completely developed (i.e. the mossy fibers) and a synaptic rearrangement takes place at the end of the first postnatal week.

We have concentrated therefore our efforts on the synaptic activity during the first postnatal week. We have found that giant excitatory potentials constitute the predominant form of synaptic activity early in postnatal life. These events are mediated by GABA, which is depolarizing at this stage of development, and modulated presynaptically by NMDA and non NMDA receptors.

METHODS

Experiments were performed on CA3 hippocampal neurons in slices obtained from 0-18 days old Wistar rats (P0-P18, 0 taken as the day of birth). The methods for preparing and maintening

Excitatory Amino Acids and Neuronal Plasticity
Edited by Y. Ben-Ari
Plenum Press, New York, 1990

the slices have been extensively reported (Ben-Ari and al 1989). Briefly, slices (600 μM thick), were kept completely submerged in a recording chamber, and superfused (2.5-3 ml/min at 34°C) with artificial cerebrospinal fluid (ACSF) of the following composition (mM): NaCl 126, KCl 3.5, $CaCl_2$ 2, $MgCl_2$ 1.3, NaH_2PO_4 1.2, $NaHCO_3$ 25, glucose 11 (pH 7.3), gassed with 95% O_2 and 5% CO_2. Intracellular recording were made with glass microppettes filled with 3 M KCl (resistance of 50-80 MΩ) or 2M K-methylsulfate (resistance of 80-150 MΩ). Current was passed through the recording electrode by means of an Axoclamp 2 amplifier. Bridge balance was checked repeatedly during the experiments and capacitative transients with the electrode tip outside the neuron were reduced to a minimum by negative capacity compensation. Membrane potential was estimated from the potential observed upon withdrawal of the electrode from the cell. Voltage clamp of the neuronal soma was accomplished by means of an Axoclamp-2A pre-amplifier, switching between voltage recording and current injection at 3-4 KHz (30 % duty cycle). To ensure correct operation of the

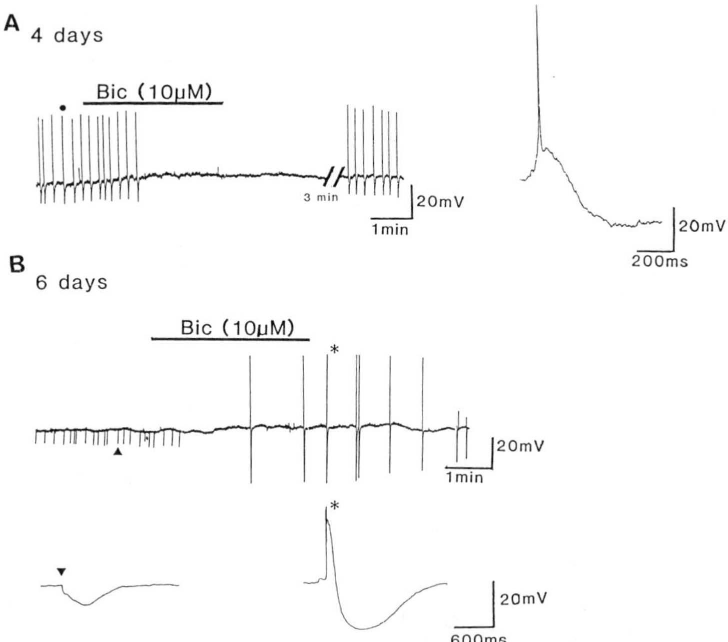

Fig. 1   GDPs and LHPs are GABA mediated.
A: GDPs recorded with K-methyl sulfate electrodes. Bath applied bicuculline (Bic) reversibly blocked the GDPs. The right trace show a GDP at an expanded time scale. Membrane potential: -70 mV.
B: LHPs recorded with K-methyl sulfate electrodes. The LHPs were blocked by bicuculline (Bic), which induced spontaneous synchronized interictal discharges. LHP (triangle) and an interictal discharge (star) are shown with an expanded time scale. Membrane potential: -60 mV.

clamp, the headstage amplifier was monitored on a separate oscilloscope. Signals were digitized and displayed on a digital oscilloscope and on a computer driven chart recorder.

Drugs.

Drugs were disolved in ACSF and superfused via a three-way tap system. Drugs used were : glutamate (Sigma); quisqualate (CRB); N-methyl-D-aspartate (NMDA, CRB); γ-aminobutyric acid (GABA, Sigma); glycine (Sigma); D-serine and L-serine (gifts of Dr. P. Ascher); D(-)2-amino-5-phosphonovalerate (AP-5, CRB or Tocris); DL-2-amino-7-phosphonoheptanoate (AP-7, gift of Dr. P.L. Herrling, Sandoz); (±)3-(2-carboxypiperazin-4-yl)-propyl-1-phosphonic acid (CCP, CRB); phencyclidine (gift of Dr. M. Lazdunski); Ketamine (Parke-Davis); 6-cyano-7-nitroquino xaline-2,3-dione (CNQX, gift of Dr. T. Honore, Ferrosan); strychnine (Sigma); tetrodotoxin (TTX, Sigma); bicuculline (Sigma); 7-Cl-kynurenate (gift of Dr. P.L. Herrling, Sandoz).

RESULTS

Long lasting intracellular recordings were made from over 300 CA3 hippocampal neurons

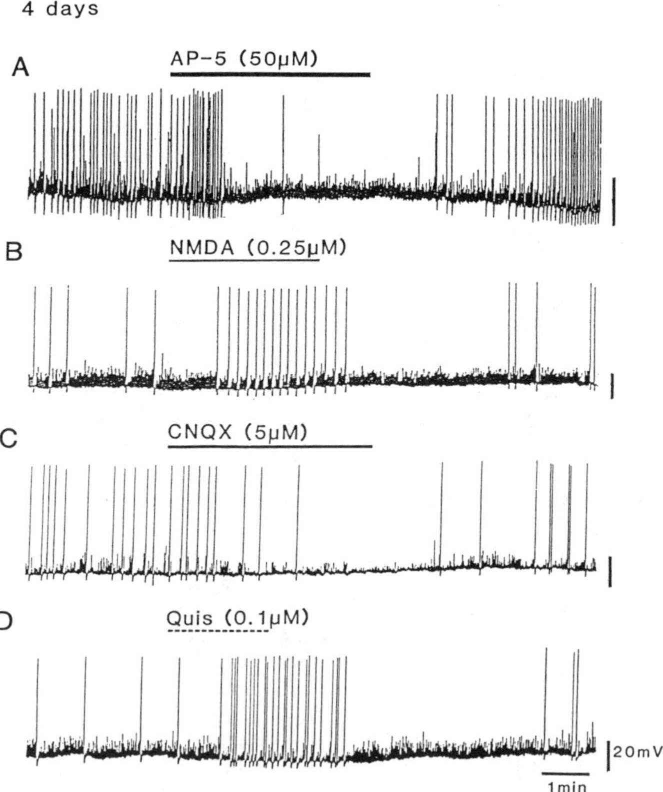

Fig. 2   GDPs are presynaptically modulated by NMDA and non NMDA receptors.
A: The GDPs were blocked by AP-5. Membrane potential: -80mV
B: Bath applied NMDA, even at low concentration, increased the frequency of GDPs. Membrane potential: -76 mV.
C: In another neuron, CNQX blocked GDPs. Membrane potential: -60 mV.
D: Quisqualate (Quis) enhanced the frequency of the GDPs. Membrane potential: -65 mV.
A,B,C and D recorded with K-Cl filled electrodes. A,B and D 4 days old rat. C 5 days old rat.

having a resting menbrane potential between -53 and -68 mV, action potentials greater than 55 mV and resting input resistance from 80 to 200 MΩ.

As already reported elsewhere (Ben-Ari and al 1989) during the first postnatal week (P0-P5) the majority of rat CA3 pyramidal neurons (more than 90 %) exhibited spontaneous giant depolarizing potentials (GDPs). These GDPs were constitued of a large depolarization (15-50 mV) with superimposed fast action potentials. GDPs had a duration ranging from 300 to 500 ms, and occur at a mean frequency of 0.14 Hz. They were network driven events since i> They were synchronous with

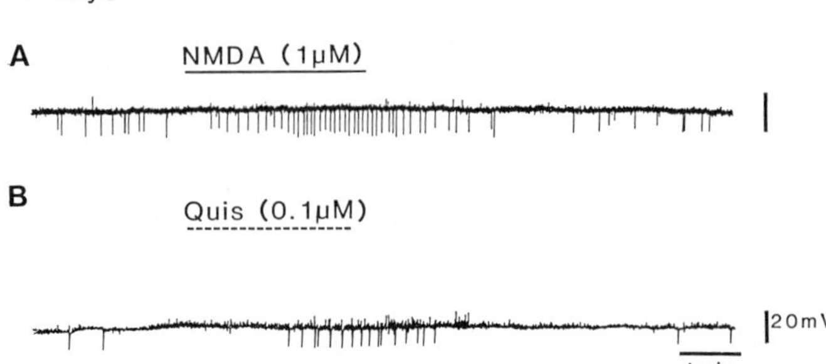

Fig. 3   LHPs are presynaptically modulated by NMDA and non NMDA receptors.
A: Bath application of NMDA increased the LHPs frequency without change in the membrane potential. Membrane potential: -60 mV.
B: Quisqualate (Quis) at low concentration increased the frequency of the LHPs. No variation in the membrane potential is observed. Membrane potential: -65 mV.
A and B recorded with K-methyl sulfate filled electrodes, 7 days old rat.

field potentials recorded with extracellular electrodes positioned in the stratum radiatum ii> They were synchronous in pairs of CA3 neurons recorded simultaneousely with intracellular electrodes iii> They were blocked by TTX (1µM) or by a high magnesium (6mM) low calcium (0.2 mM) solution known to preferentially block the polysynaptic activity (Berry and Pentreath 1976) iv> Their amplitude, but not their frequency , was dependent on the membrane potential.

The GDPs are GABA mediated synaptic events.

GDPs were mediated by GABA, acting on GABA-A receptors (Ben-Ari and al 1989) since i> they were blocked by bicuculline (10 µM) , a GABA-A receptor antagonist (Fig. 1A). ii> GDPs and the response to bath applied GABA reversed polarity at the same potential (-50 mV with K-methylsulfate filled electrodes and -20 mV with K-Cl filled electrodes). The shift of the reversal potential to more depolarized value with KCl electrodes suggest that a chloride conductance is involved in GDPs.

At the end of the postnatal week, during a transitional period (P6-P9), the percentage of pyramidal cells displaying GDPs decreased. They were replaced by large spontaneous hyperpolarizing potentials (LHPs). Like GDPs, LHPs were generated by a network since they were blocked by TTX (1 µM) and their frequency was independent from the membrane potential. As GDPs, LHPs were mediated by GABA acting on GABA-A receptors. They were blocked by bicuculline (10 µM) which, at this stage of development, induced for the first time interictal discharges (Fig. 1B). These interictal discharges were mediated by excitatory amino acids since i> they reversed around 0 mV, ii> they were reduced by the NMDA receptors antagonist AP-5 (50 µM), and iii> they were blocked by kynurenic acid (0.5-1 mM).

After postnatal days 10-12, neither GDPs nor LHPs were present and bicuculline induced interictal discharges similar to those seen in adult slices in the same conditions.

GDPs and LHPs are modulated by NMDA and non NMDA receptors.

GDPs were blocked or reduced in frequency by NMDA receptors antagonists AP-5, AP-7, CPP (30-50 µM,Fig. 2A) and NMDA channel blockers such as phencyclidine 1(µM) or ketamine (20µM)

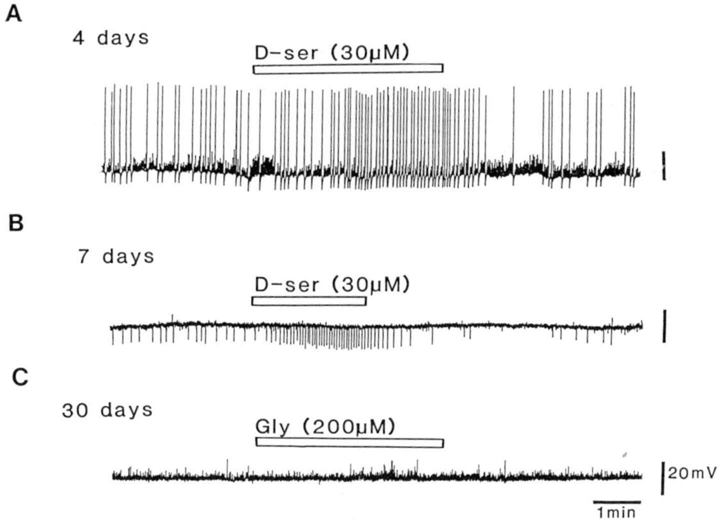

Fig. 4   D-serine enhances the frequency of GDPs and LHPs.
D-serine (D-ser) enhanced the frequency of GDPs (A) and HPs (B). Glycine
even at higher concentration had no effects on adult cells (C). A and C
recorded with KCl and B with K-methyl sulfate filled electrodes.
Membrane potential: -65 mV (A), -60 mV (B) and -70 mV (C).

(Corradetti and al 1988, Ben-Ari and al 1989). Furthermore, bath application of low concentrations of NMDA (0.25-1µM) increased the frequency of GDPs (Fig. 2B). This effect was prevented by bicuculline (10 µM) or TTX (1 µM), suggesting that it occured on presynaptic located GABAergic neurons. NMDA at these concentrations did not change membrane potential or input resistance of the cells. However in presence of TTX, bath application of NMDA at higher concentration (5-10 µM) induced a membrane depolarisation or in voltage clamped neurons a large inward current which reversed polarity around 0 mV (Ben-Ari and al 1988). The fact that GDPs and NMDA mediated currents reversed polarity at different potentials suggest that NMDA modulated presynaptically the release of GABA . In the same way, the LHPs were modulated by NMDA receptors since they were reduced in frequency or blocked by AP-5 or AP-7 (50 µM). Furthermore, bath application of NMDA (0.5-2 µM) induced an increase in frequency of LHPs without any effect on membrane potential or input resistance   (Fig 3A).

In some cases, CNQX (5 µM, n=4), a rather specific non NMDA receptor antagonist (Drejer and Honore 1988, Neuman and al 1988a, Harris and Miller 1989) blocked or reduced the frequency of the GDPs (Fig.2C), suggesting that also non-NMDA receptors participate to GDPs generation. In order to better characterize the presynaptic glutamate receptors controlling the GABA release, we have investigated the effects of bath applied glutamate and quisqualate. Bath application of quisqualate at low concentrations (50-100 nM), increased the frequency of the GDPs (Fig. 2D) and LHPs (Fig. 3B). These effects were not affected by AP-5 (50µM) but were blocked by CNQX (5 µM). Gutamate at very low concentrations (10-50 µM) enhanced the frequency of both GDPs and LHPs. The action of glutamate was partially antagonized by AP-5 (50 µM), and completely blocked by concomitant application of CNQX and AP-7 . As for NMDA also the effects of quisqualate and glutamate were prevented by TTX (1 µM) or bicuculline (10 µM), suggesting that both drugs act on GABAergic interneurons to modulated GABA release.

Glycine potentiate the frequency of GDPs and LHPs.

The NMDA receptor complex has a allosteric facilitatory site which is activated by submicromolar concentrations of glycine (Johnston and Ascher 1987). We have therefore investigated the

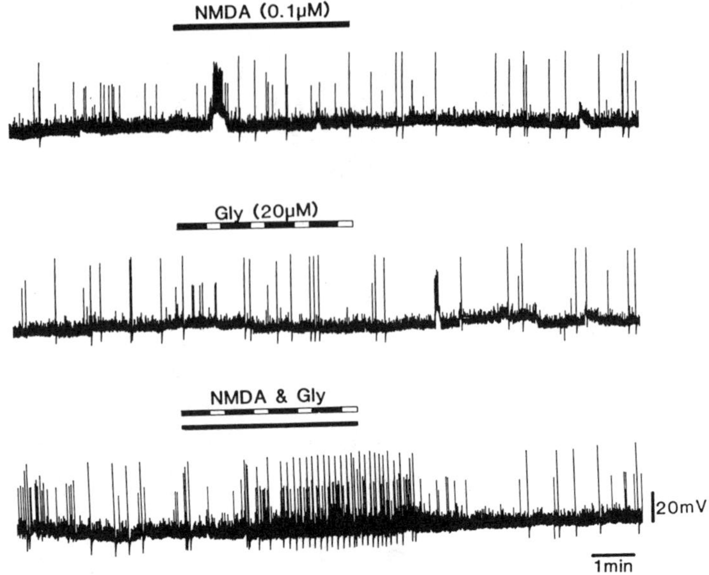

Fig. 5 Glycine enhanced the NMDA mediated events.
NMDA (Filled bar) or glycine (Gly, dotted bar), applied separately in control conditions, had no effect on GDPs or synaptic noise. Concomitant application of NMDA and glycine induced an increase in the frequency of GDPs and in synaptic noise. Membrane potential: -75mV. 5 day old rat. KCl filled electrode.

effects of glycine on GDPs and LHPs. Micromolar concentrations of glycine (10-30 μM) increased the frequency of the GDPs . The effect of glycine was mimicked by D-serine (10-30μM; Fig. 4A), but not by L-serine (10-30 μM). The potentiating effect of glycine was insensitive to strychnine (0.5-1 μM), but was blocked by NMDA receptors antagonists AP-5, AP-7 (50 μM) and by bicuculline (10 μM). Between P6-P9, glycine or D-serine (10-30 μM, Fig. 4B) enhanced the frequency of the LHPs in a strychnine resistant way and this effect was prevented by AP-5 (50 μM). Bicuculline (10 μM) blocked LHPs and induced interictal discharges, which were insensitive to glycine. After P10-P12 glycine, even at higher concentration (≤ 200 μM), had no effect on the synaptic noise, the input resistance or the membrane potential (Fig. 4C).

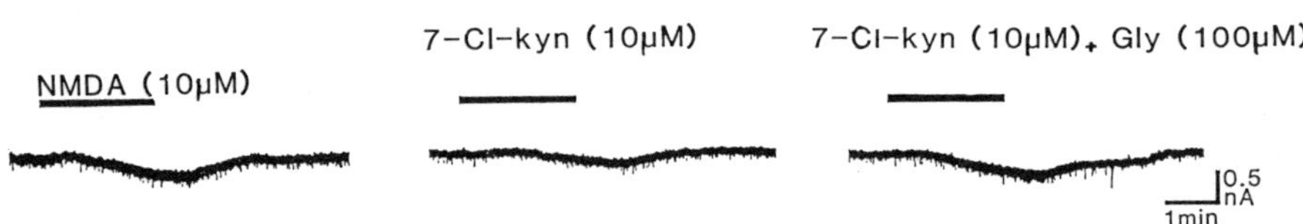

Fig. 6 Glycine reversed the effect of 7-Cl-kynurenate on NMDA mediated inward current. In presence of tetrodotoxine (1 μM) bath application of NMDA (filled bars) induced an inward current which was reduced by 7-Cl-kynurenate (7-Cl-kynu). Glycine (Gly) reversed the effect of 7-Cl-kynurenate. 4 day old rat. The cell was voltage clamped at a holding potential of -50 mV.

Since bicuculline blocked the effects of glycine it is likely that glycine act on NMDA receptors located on GABAergic interneurons. In order to test this hypothesis, we have studied the effects of glycine on bath applied NMDA.

Bath application of subthreshold concentration of NMDA (0.5-1 μM) , in presence of glycine (10-20 μM) which per se had no effect, induced a marked increase of the GDPs or HPs frequency marked and synaptic noise (Fig. 5). The potentiating effect of glycine was strychnine insensitive (1μM), but was blocked by AP-5 (50μM) and bicuculline (10 μM).

In presence of TTX, bath application of NMDA (10μM), induced a large inward current. Both glycine or D-serine ($\leq$ 50μM) failed to potentiate this current. However, glycine (50-100μM) reversed the reduction of the NMDA mediated current induced by 7-Cl-kynurenate (100-200μM, Fig 6) , a competitive glycine antagonist on the allosteric site (Kemp and al 1988). This suggests that the allosteric site is present on NMDA receptors located on pyramidal neurons, but is saturated by endogenous glycine levels.

DISCUSSION

From the present results, two main conclusions can be drawn:

-The GDPs and LHPs in neonatal CA3 pyramidal neurons are mediated by GABA, acting on GABA-A receptors.
-Both GDPs and LHPs are presynaptically controlled by NMDA and non NMDA receptors located on GABAergic interneurons.

In neonatal CA3 pyramidal cells, excitatory synaptic events are primarily mediated by GABA.

The GDPs are mediated by GABA acting on GABA-A receptors since they are blocked by bicuculline, a GABA-A antagonist, and they reversed polarity at the same potential as bath applied GABA.

GABA at this stage of development is depolarizing and excitatory (when recorded with methylsulfate containing electrodes GDPs reach the threshold for spike activation) and in fact constitutes the main excitatory drive to the pyramidal neurons since the spontaneous and evoked activity were blocked by bicuculline. We may wonder whether excitatory amino acid mediated collateral connections between pyramidal neurons responsible for interictal discharges (Mac Vicar and Dudeck 1980, Miles and Wong 1986) are already developed at birth. We know that these connections are at least in part already present, since they can be activated in presence of bicuculline with pharmacological manipulations (high potassium, low magnesium, NMDA). A recent study using intracellular injection of HRP confirms the presence of a recurrent collateral network (Gaiarsa and Ben-Ari in preparation). However in normal conditions this network is quiescent and the synaptic drive is primarily provided by GABAergic interneurons. . At the end of the first postnatal week at a time when extrinsic and intrinsic excitatory connections develop rapidily, GABA shifts from the depolarizing to the hyperpolarizing direction and at this moment bicuculline blocks the effect of GABA and induces interictal discharges. It is possible that the synaptic rearrangement and development of new connections constitute the signal of this shift.

If GABA is depolarizing what will repolarize the cells at the early developmental stage ? Cell excitability is regulated by a number of extrinsic and intrinsic factors such as neurotransmitters acting on specific receptors and voltage dependent conductances. It is likely that potassium conductances which are well developed at birth play an important role in the repolarization, notably the calcium dependent potassium conductances responsible for the afterhyperpolarization following the GDPs.

GDPs and LHPs are presynaptically modulated by NMDA and non NMDA receptors

The second major conclusion is that glutamate receptors control the GABA release responsible for the GDPs and LHPs and that these glutamate receptors located on GABAergic interneurons are of the NMDA (Corradetti and al 1988, Ben-Ari and al 1989) and non NMDA subtypes. The NMDA receptors are facilitated by bath application of glycine through is allosteric site (Gaiarsa and al 1990 *in press*). We have already proposed a schematic diagram (Ben-Ari and al 1989) as a working hypothesis : GABAergic interneuron oscillate following activation of NMDA receptors by glutamate released by axon collaterals of pyramidal cells. The increased in intracellular calcium (at least through NMDA activated channels) will activate a calcium dependent-potassium conductance which will repolarize the cell. The oscillatory GABAergic neuron will release GABA which in turn will depolarize the pyramidal cell and promote a positive feed back loop. We now add to the scheme also a non-NMDA receptor. We know that NMDA and non NMDA receptors are often co-localized on the same neuron and their activation often produce synergistic effects. Activation of non-NMDA receptors may depolarize the cells and facilitate the activation of NMDA receptors, a mechanism which has been proposed for LTP (Collingridge and Bliss 1987).

Presynaptic NMDA and non-NMDA receptors have been already described. They control GABA release in adult hippocampal (Taube and Schwartzkroin 1987) and in dorsolateral septal neurons (Gallagher and Hasuo 1989) or glycine release in neonatal sympathetic neurons (Dun and Mo 1989). In adult hippocampus however they are less conspicuous and they necessitate for their activation relatively high doses of glutamate. The very low concentrations of excitatory amino acids agonists used in our experiments are compatible with the hypothesis that interneurons at this stage of development are particularly enriched in glutamate receptors. The high density of glutamate receptors in immature nervous system (Baudry and al 1981) with a transient incease in NMDA (Tremblay and al 1898) and non-NMDA receptors (Erdo and Wolf) is in keeping with this hypothesis.

Althrough at present we do not know the mechanisms underlying GDPs generation and the effect of GABA depolarizing it is likely that these events constitute a signal in cell growth and differentiation.

REFERENCES

Alger, B.E. and Nicoll, R.A. (1982). Pharmacological evidence for two kinds of GABA receptor on rat hippocampal pyramidal cells studied in vitro. **J. Physiol. (Lond.)** 328,125-141.

Baudry, M., Arst, D., Oliver, M. and Lynch, G. (1981). Development of glutamate binding sites and their regulation by calcium in rat hippocampus. **Dev. Brain Res.** 1, 37-48.

Ben-Ari, Y., Cherubini, E., Corradetti, R. and Gaiarsa, J.L. (1989). Giant synaptic potentials in immature rat CA3 hippocampal neurones. **J. Physiol. (Lond.)** 416, 303-325.

Ben-Ari, Y., Krnjevic, K., Reiffenstein, R.J. and Reinhardt, W. (1981). inhibitory conductance changes and action of -aminobutyrate in rat hippocampus. **Neurosci.** 6,2445 -2463.

Berry, M.S. and Pentreath, V.W. (1976). Criteria for distinguishing between monosynaptic and polysynaptic transmission.**Brain Res.** 105, 1-20.

Collingridge, G.L. and Bliss, T.V.P. (1987). NMDA receptors-their role in long-term potentiation. **TINS** 10(7), 288-293.

Corradetti, R., Gaiarsa, J.L. and Ben-Ari, Y. (1988). D-amino-phosphonovaleric acid-sensitive spontaneous giant EPSPs in immature rat hippocampal neurones. **Eur. J. Pharmacol.** 154(2), 221-222.

Drejer, J. and Honore, T. (1988). New quinoxalinediones show potent antagonism of quisqualate responses in cultured mouse cortical neurons. **Neurosci. Lett.** 87, 104-108.

Dun, N.J. and Mo, N. (1989). Inhibitory postsynaptic potentials in neonatal rat sympathetic preganglionic neurones in vitro. **J. Physiol. (Lond.)** 410, 267-281.

Dunwiddie, T.V. (1981). Age related differences in the in vitro rat hippocampus : development of inhibition and effects of hypoxia. **Dev. Neurosci.** 4, 165-189.

Dutar, P. and Nicoll, R.A. (1988). A physilogical role for GABA-B receptors in the central nervous system. **Nature** 332, 156-158.

Forsythe, I.D., Westbrook, G.L. and Mayer, M.L. (1988). Modulation of excitatory synaptic transmission by glycine and zinc in cultures of mouse hippocampal neurons. **J. Neurosci.** 8(10), 3733-3741.

Gaiarsa, J.L., Corradetti, R., Cherubini, E. and Ben-Ari, Y. (1990) The allosteric glycine site of the N-methyl-D-aspartate modulates GABAergic-mediated synaptic events in neonatal rat CA3 hippocampal neurons. **Proc. Natl. Acad. Sci. USA** in press.

Gallagher, J.P. and Hasuo, H. (1989). Bicuculline- and phaclofen-sensitive components of N-methyl-D-aspartate-induced hyperpolarizations in rat dorsolateral septal nucleus neurones. **J. Physiol.** 418,367-377.

Harris, M.K and Miller, J.R. (1989). CNQX (6-cyano-7-nitroquinoxalone-2,3-dione) antagonizes NMDA-evoked (^{3}H)GABA release from cultured cortical neurons via an inhibitory action at the strychnine-insensitive glycine site. **Brain Res.** 489, 185-189.

Harris, M.K. and Teyler T.J. (1983). Evidence for a late development of inhibition in area CA1 of the rat hippocampus. **Brain Res.** 263, 339-343.

Johston, J.W. and Ascher, P. (1987). Glycine potentiates the NMDA response in cultured mouse brain neurons. **Nature** 325, 529-531.

Kehl, S.J. and Mc Lennan, H. (1985). A pharmacological characterization of chloride- and potassium- dependent inhibitions in the CA3 region of rat hippocampus in vitro. **Exp. Brain Res.** 60, 309-317.

Kemp, J.A., Foster, A.C., Leeson,P.D., Priestley, T., Tridgett, R., Iversen, L.L. and Woodruff, G.N. (1988). 7-Chlorokynurenic acid is a selective antagonist at the glycine modulatory site of the N-methyl-D-aspartate receptor complex. **Proc. Natl. Acad. Sci. USA** 85, 6547-6550.

Mac Vicar, B.A. and Dudeck, F.E. (1980). Local synaptic circuits in rat hippocampus : interactions between pyramidal cells. **Brain Res.** 184, 220-223.

Miles, R. and Wong, R.K.S. (1986). Excitatory synaptic interactions between CA3 neurones in the guinea pig hippocampus. **J. Physiol. (Lond.)** 373, 397-418.

Neuman, R.S., Ben-Ari, Y., Gho, M. and Cherubini, E. (1988a). Blockage of excitatory synaptic transmission by 6-cyano-7-nitroquinoxaline-2,3-dione (CNQX) in the hippocampus in vitro. **Neurosci. Lett.** 92, 64-68 .

Neuman, R., Cherubini, E. and Ben-Ari, Y. (1988b). Epileptiform burst elicited in CA3 hippocampal neurons by a variety of convulsants are not blocked by N-methyl-D-aspartate antagonists. **Brain Res.** 459, 267-274.

Purpura, D.P. and Pappas, G.D. (1968). Structural characteristics of neurones in the feline hippocampus during postnatal ontogenesis. **Exp. Neurol.** 22, 379-393.

Schwartzkroin, P.A. (1982). Development of rabbit hippocampus: physiology. **Dev. Brain Res.** 21, 469-486.

Schwartzkroin, P.A. and kunkel, D.D. (1982). Electrophysiology and morphology of the developing hippocampus of fetal rabbits. **J. Neurosci.** 2 (4), 448-462.

Sawnn, J.W., Brady, R.J. and Martin, D.L. (1989). Postnatal development of GABA-mediated synaptic inhibition in rat hippocampus. **Neurosci.** 28(3), 551-561.

Taube, J.S. and Schwartzkroin, P.A. (1987). Hyperpolarizing responses to application of glutamate in hippocampal CA1 pyramidal neurons. **Neurosci. Lett.** 78, 85-90.

Tremblay, E., Roisin, M.P., Represa, A., Charriaut-Marlangue, C., and Ben-Ari, Y. (1988). Transient increased density of NMDA binding sites in the developing rat hippocampus. **Brain Res.** 461, 393-396 .

Wong, R.K.S. and Traub, R.D. (1983). Synchronized burst discharges in disinhibited hippocampal slices. I. initiation in the CA2-CA3 region. **J. Neurophysiol.** 49, 442-458.

NEURAL NETWORKS AND SYNAPTIC TRANSMISSION IN IMMATURE

HIPPOCAMPUS

John W. Swann, Karen L. Smith and Robert J. Brady

Wadsworth Center for Laboratories & Research
New York State Department of Health
Empire State Plaza, Albany, NY 12201-0509

INTRODUCTION

During the last decade numerous reports have described a previously unappreciated but nonetheless common process of central nervous system development: the transient overproduction of axonal projections early in postnatal life. Several groups have shown that during maturation terminal axonal branches are pruned and long axonal collaterals degenerate (for review see Cowan et al., 1984; Easter et al., 1985; Stanfield, 1984). With axon elimination, associated synapses would also be lost. Consistent with this are results of ultrastructural studies that have shown that the density of synapses in cortex early in postnatal life is higher than in the adult (Huttenlocher et al., 1982; Huttenlocher, 1984; Rakic et al., 1986). Elimination of functional synapses has been shown to occur at both the neuromuscular junction and in autonomic ganglion during maturation (Purves and Lichtman, 1980). One example of functional synapse regression in the central nervous system is the transient multiple innervation of Purkinje cells by climbing fibers (Crepel, 1982).

One long-standing goal of developmental neurobiology has been to delineate those factors that determine whether a synapse will form and then be maintained. Evidence suggests that if a synapse is used early in life it is likely to persist. However, recent studies of the formation of ocular dominance columns in the visual system also suggest that activation of an NMDA (n-methyl-D-aspartate)-preferring excitatory amino-acid receptor can play an important role (Cline et al., 1987; Kleinschmidt et al., 1987). There is now ample evidence that multiple forms for many neurotransmitter receptors exist in the central nervous system (Grenningloh et al., 1987; Schofield et al, 1987). A neonatal form of the nicotinic acetylcholine receptor has been sequenced and characterized physiologically in muscle (Mishina et al., 1986). Similarly, unique forms of excitatory amino acid receptors may exist in the immature brain. If neonatal forms of the NMDA receptor do exist they might possess unique characteristics that permit them to play key roles in synaptogenesis in the developing nervous system.

With the advent of *in vitro* slice techniques, local circuits within the central nervous system and the neurotransmitter receptors they use became available for detailed electrophysiologic study. The questions of whether synapse elimination takes place and unique neurotransmitter receptors exist within networks early in life are entirely new avenues of investigation. A few laboratories have recently turned their attention to the study of the ontogeny of cortical local circuitry. Our laboratory has studied local circuits of the hippocampus (Swann et al., 1988). Results reviewed here are consistent with the hypothesis that during the second postnatal week in hippocampal development there is a transient overproduction of recurrent excitatory synapses between CA3 hippocampal neurons. The NMDA receptor found on these cells appears to have unique physiological characteristics.

Excitatory Amino Acids and Neuronal Plasticity
Edited by Y. Ben-Ari
Plenum Press, New York, 1990

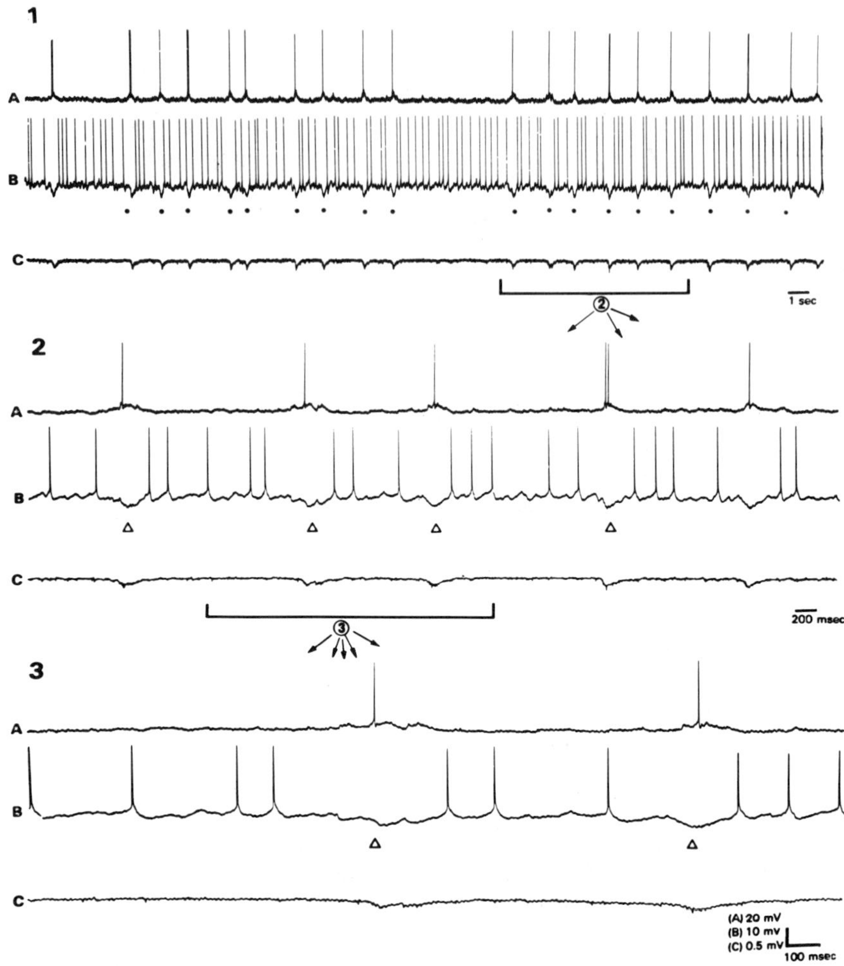

Fig. 1.   Spontaneous bursts of potential recorded simultaneously in a pair of CA₃
           neurons in immature hippocampus. Intracellular recordings are shown in
           Traces A and B. These were recorded simultaneously with an extracellular
           field recording (Traces C) for the proximal portions of the basilar dendritic
           layer. Segments of recordings in panel 1 and 2 that are marked by a line are
           shown in the succeeding panel at a faster time base. Recordings were made on
           postnatal day 11.

INITIAL INTRACELLULAR OBSERVATIONS.

Whenever CA₃ hippocampal neurons are impaled during the second week of life, large
spontaneous epsp- and ipsp-like potentials are commonly observed. If two cells are impaled
simultaneously synchronous bursts of such potentials are recorded in these cells. Figure 1 shows an
example of such recordings. Traces A and B are simultaneous intracellular recordings from two
neurons. Segments of Panel 1 are shown below in Panels 2 and 3 at faster time bases. What is clear
from these recordings is that, while the bursts of potentials are likely initiated by a single
common source, they are quite different in the two cells. In Traces A the potentials are primarily
depolarizing and produce action potentials. The potential bursts in Traces B hyperpolarize the

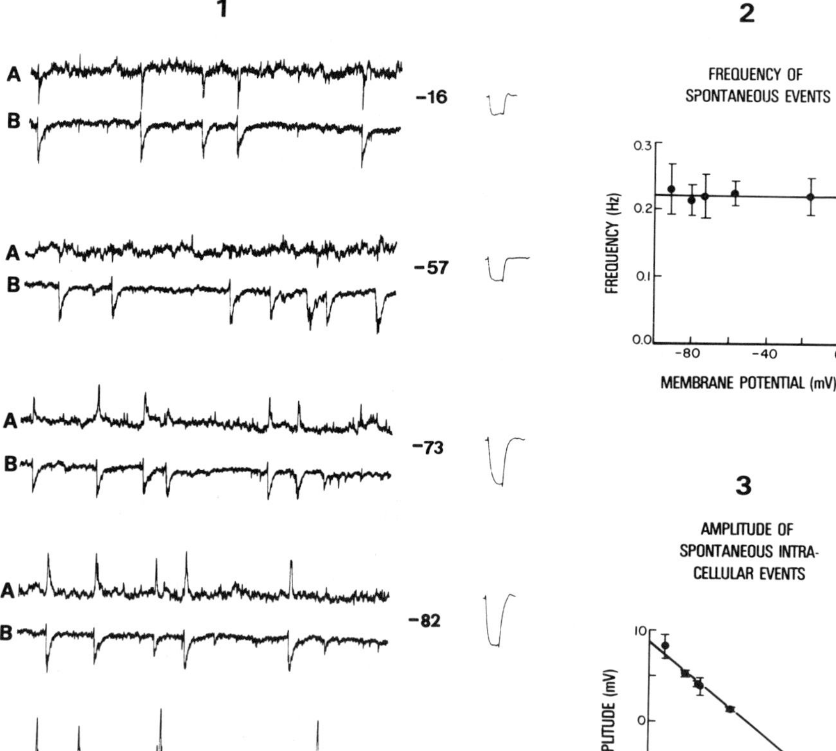

Fig. 2.   Effects of changing membrane potential on the frequency and amplitude of intracellularly recorded spontaneous bursts of potentials. Panel 1 shows representative intracellular (Traces A) recordings at 5 different membrane potentials (number to the right of traces). The simultaneously recorded extracellular field is shown in Traces B. To the right of these traces are changes in membrane potential produced by a hyperpolarizing step of injected intracellular current that was used to monitor membrane resistance at the different holding potentials. In Panel 2 the frequency of spontaneous bursts at the different potentials are plotted. In Panel 3 the amplitude of these events are plotted versus membrane potential.

membrane and temporarily suppress spontaneous spiking. In many ways the recordings from these two cells were fortuitous since in the majority of cells (e.g. see Fig. 2) the bursts of potentials are a complex mixture of inhibitory and excitatory events. Despite this complexity, the synchrony of these events is sufficient to produce a field potential (Traces C). These are most easily recorded in the proximal portion of the basilar dendritic layer. At this location they are negative although never more than 1 mV in amplitude.

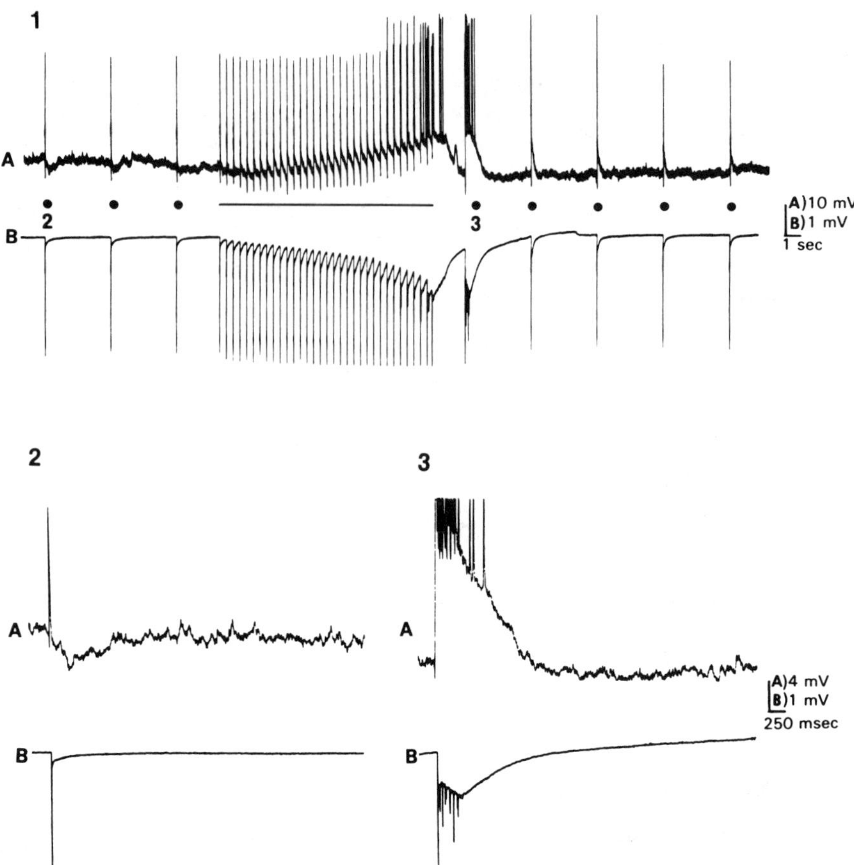

Fig. 3.   Effects of repetitive stimulation on orthodromically elicited responses. Simultaneously intracellular (Traces A) and extracellularly (Traces B) recordings are shown. Times of orthodromic stimulation are denoted by dots below traces in Panel 1. Lines below the trace shows duration of the train of stimuli. Events marked 2 and 3 are shown below at an increased gain and a faster time base.

Since these events have the appearance of bursts of synaptic potentials we studied changes in their amplitude and frequency with alterations in membrane potential. Synaptic potentials would not be expected to change in frequency but only in amplitude as membrane potential is altered. Figure 2 shows this to be the case. The frequency of the spontaneous events did not change at holding potentials from -90 to -15 mV. At the same time, their amplitudes decreased as a monotonic function of membrane potential. In this particular case the potential reversed polarity near -43 mV. However, the reversal potentials of these spontaneous synaptic events varied from cell to cell. This is probably a reflection of variations in the mixture of the inhibitory and excitatory synaptic potentials present in each cell.

Similar synchronous bursts of synaptic potentials have been reported by others recording from slices of mature hippocampus and cortex (Schwartzkroin and Haglund, 1986; Schneiderman, 1986; Miles and Wong, 1987a; Miles and Wong, 1987b). However, at least in the hippocampal studies, GABAergic synaptic transmission was partially suppressed either pharmacologically or by

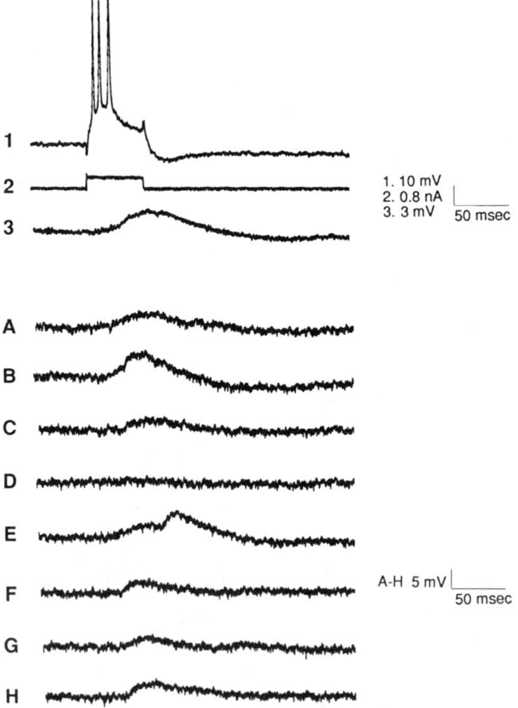

Fig. 4.   Polysynaptic epsps recorded in one cell in response to a burst of action potentials produced in another CA₃ hippocampal neuron. Traces 1 show the response of the presynaptic neuron to a depolarizing step of current monitored in Trace 2. Trace 3 is the averaged epsp produced postsynaptically. Traces A through H are the individual responses used to generate the average in 3.

tetanic stimulation. Thus the existence of these events under normal conditions in immature hippocampus might be expected to reflect a late development of GABAergic synaptic inhibition. However, intracellular studies from kitten, rabbit and rat do not support this proposition (Swann et al., 1989; Schwartzkroin, 1982; Purpura et al., 1968). In the CA₃ subfield of rats and rabbits ipsps are well developed early in postnatal life. For instance, in the rat measurements of the conductance of antidromically elicited ipsps on postnatal days 5-7 were not different from those on days 31 and 32. Indeed, on days 11 and 12 ipsp conductances were 3-4-fold larger. The reversal potentials of the ipsps were not different at the three ages (Swann et al., 1989). Thus it seems unlikely that a lack of synaptic inhibition during the second postnatal week can account for the presence of a synchronous burst of synaptic potentials. Recently Traub et al. (1988) have studied the origins of similar synchronized synaptic potentials with computer simulation. Their model of the CA₃ hippocampal neuronal network predicts that suppression of GABAergic synaptic transmission would lead to an increase in the amplitude of synchronized synaptic events. However, their model also shows that these synchronized events are highly dependent on the presence of recurrent excitation. If recurrent excitatory synapses are eliminated from the model, synchronized bursting of synaptic potentials do not occur. If the strength of recurrent excitation is increased the amplitude of these spontaneous synaptic events increases. Thus, we considered the possibility that enhanced recurrent excitation may exist during the second postnatal week compared to that present in mature animals.

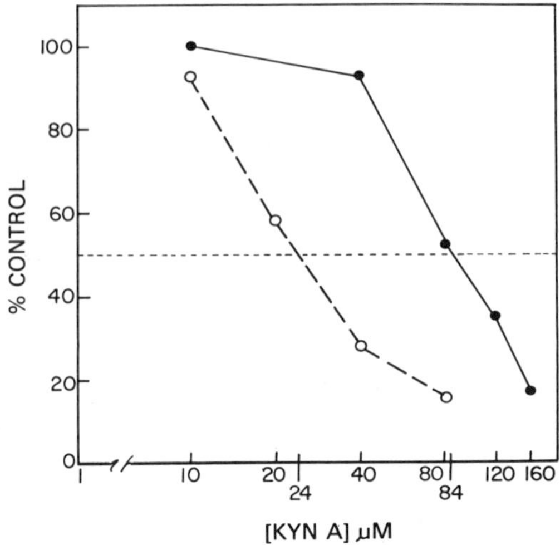

Fig. 5.   Comparison of the effects of kynurenic acid on iontophoretic responses to
NMDA recorded on postnatal day 12 (open circles) and day 31 (closed circles).
Local negative field potentials (Ganong et al., 1983) were produced in slices
bathed in Ca^{+2}-free media containing TTX (5 x 10^{-7}G/L) and Mn^{+2} (2 mM).
Different concentrations of kynurenic acid were bath-applied. Graphed are
amplitudes of signals expressed as percent of that recorded during the control
period.

EFFECTS OF REPETITIVE STIMULATION AND GABA ANTAGONISTS

Indirect evidence to support the notion of enhanced recurrent excitation in immature
hippocampus comes from several observations. First, is the response of the hippocampal CA_3
subfield to repetitive orthodromic stimulation. Figure 3 shows the response of a slice to such
stimuli. Prior to tetanization, a single stimulus results in an orthodromic epsp that initiates an
action potential. This is followed by a prolonged hyperpolarization which is actually a biphasic
inhibitory postsynaptic potential (Swann et al., 1989). Here the biphasic nature of the event is
obscured by background synaptic noise. During the train of stimuli the cell initially hyperpolari-
zes and then undergoes a gradual depolarization. One second after the train an identical stimulus
was again given. Now instead of an ipsp, the cell undergoes an intense 1-2 sec depolarization. The
amplitude of the depolarizing potential decreases with each succeeding stimulus. Within 5 min an
ipsp very similar to that obtained under control conditions, is recorded. In experiments performed
in mature slices similar trains of stimuli failed to produce such large prolonged depolarizations.

Repetitive stimulation is known to attenuate GABAergic synaptic transmission in hippocampus (Ben-Ari et al., 1979; McCarren and Alger, 1985). Consequently it would be of interest to
compare the effects of GABA antagonists in immature and mature tissue. Here again dramatic
differences are recorded (Swann and Brady, 1984; Brady and Swann, 1984). For instance, when
penicillin is applied to slices from rats 24 days of age or older, 100 msec burst discharges are
recorded. In contrast, when applied during the second postnatal week, penicillin produces
rhythmic repetitive discharges that can be up to 30 sec in duration. Recurrent excitation likely
plays an important role in the generation of these latter events since they are abolished by
excitatory amino-acid antagonists (Brady and Swann, 1986; Brady and Swann, 1988). Taken
together these observations support the view that excitatory synaptic transmission is very well
developed in immature hippocampus. The enhanced excitability produced either by repetitive
stimulation or GABA antagonists may be the product of an increase in recurrent excitatory
interactions at this critical stage in development.

RECURRENT EXCITATION - DUAL INTRACELLULAR RECORDING

Miles and Wong (1986) conducted a series of elegant neurophysiological experiments in which they examined the presence and salient features of recurrent epsps in mature guinea pig CA_3 hippocampal pyramidal cells. For the most part their conclusions are based on simultaneous intracellular recordings from 2 CA_3 hippocampal pyramidal cells. They examined synaptic potentials produced in one cell in response to an action potential occurring in the second impaled neuron. We recently completed a similar series of dual intracellular recordings in immature rat hippocampus (Smith et al., 1988). Because the conditions under which the studies were performed were not identical direct comparisons are difficult to make. However, differences in the results obtained in the two studies appear to support the notion that greater numbers of recurrent excitatory synaptic interactions take place in the developing hippocampus.

Miles and Wong (1987b) have shown that when synaptic inhibition is suppressed in mature hippocampus, pairs of CA_3 pyramidal cells which were previously not synaptically coupled became linked through polysynaptic excitatory pathways. They suggest that inhibitory interneurons normally prevent the spread of excitation through these local excitatory networks. Since we were interested in assessing the degree of recurrent excitation in the immature hippocampus all our recordings were done in the presence of the GABA antagonist, penicillin. We also chose to record from small segments (referred to as minislices) of the CA_3 subfield. This procedure allowed us to eliminate any effects other cell populations (granular cells and CA_1 pyramidal cells) may have had on intracellular recordings.

Similar to intact slices, the minislices undergo prolonged repetitive synchronized discharges as described above. Moreover, as in intact slices, these events are suppressed by excitatory amino acid antagonists. Using similar procedures, Miles and Wong (1983) have reported that 30% of individual pyramidal cells were able to initiate synchronized discharging of the entire CA_3 population. That is, when one pyramidal cell is depolarized by intracellular current injection and undergoes a burst of spikes, immediately thereafter the entire pyramidal cell population synchronously discharges. They showed that recurrent epsps produced by this cell in other pyramidal cells initiated the population event and suggested that the population discharge was in large part the product of the spread of this discharge through a network of pyramidal cells that make recurrent excitatory contacts with each other.

In immature rat hippocampus individual cells are also able to entrain the population. However, during the second postnatal week the majority of neurons, not 30%, demonstrate this capacity. In immature hippocampus nearly 60% can entrain the CA_3 population. If individual neurons initiate the discharges through recurrent excitation then this would suggest that epsps early in life are larger or more numerous than in adulthood. Dual intracellular recordings address this possibility directly. From data obtained in over 100 paired recordings we have concluded that recurrent excitation is unusually well developed early in postnatal life. Figure 4 shows an example of epsps produced in one cell in response to a burst of action potentials in another CA_3 pyramidal cell. This epsp is polysynaptically mediated, since its latency is long (> 10 msec) and variable, while its probability of occurrence is low (< 10%) when single action potentials were elicited presynaptically. Indeed we have recorded such epsps in 35% of paired recordings. In addition, 4% of pairs were monosynaptically coupled and 7% were disynaptically coupled. In 6% of pairs action potentials in either cell of the pair produced epsps in the other cell. All of these observations suggest that the frequency of excitatory synaptic interaction is very high in immature hippocampus, and higher than that in its mature counterpart (Miles and Wong, 1986; Miles and Wong, 1987). One factor that may explain the high degree of polysynaptic interaction in the immature hippocampus is that unitary epsps appear to be nearly twice as large as those reported in mature hippocampus. The larger an epsp the more likely it is to bring a postsynaptic cell in a network to spike threshold and permit the transmission of the signal along that path. The average amplitude of the mono- and disynaptic epsps at membrane potentials between -60 and -70 mV is 2.7 mV. In mature hippocampus the average amplitude has been reported to be 1.4 mV.

Numerous explanations may be proposed for the size of the epsps in developing hippocampus. One of the simplest explanations is that the input impedance of immature neurons exceeds that of mature cells. On average, the membrane resistance during the second week of life is 60 Mohms. This is somewhat higher than that recorded in mature hippocampus but does not appear sufficient to explain differences in the recurrent epsps. Another possibility is that more synaptic contacts exist between individual pyramidal cells. Estimates of the number of release sites or nerve terminals may be obtained from quantal analysis of monosynaptic epsps. This type of analysis also provides estimates of the amount of transmitter released (mean quantal content). Both are important parameters of the presynaptic properties of the recurrent synapses. However, estimates of these statistical parameters have yet to be completed.

EXCITATORY AMINO ACID RECEPTORS ON IMMATURE HIPPOCAMPAL NEURONS

The large recurrent epsps recorded in the immature hippocampus could also occur because the postsynaptic receptor is different. For instance, it might have a higher affinity for the endogenous neurotransmitter. Since CA_3 hippocampal pyramidal cells are thought to use an excitatory amino acid as their neurotransmitter, investigations of the properties of these types of receptor during postnatal development could be important. Ben Ari et al. (1988) recently reported that during the first postnatal week the response of some CA_3 neurons to NMDA was less voltage-dependent than is reported in mature hippocampus. However, we have found during the second postnatal week that NMDA responses are always voltage-dependent and appear very similar to those recorded in mature hippocampus. Pharmacologically, NMDA receptors are blocked by the same drugs that are known to antagonize receptors in mature hippocampus (Brady and Swann, 1986; Brady and Swann, 1988). However, we have also found that NMDA receptors in immature hippocampus have a higher affinity for the receptor antagonist, kynurenic acid (KYN). Figure 5 shows the results of one experiment in which different concentrations of KYN were bath-applied. The effects on field potentials produced by iontophoretically applied NMDA were monitored. The IC_{50} for NMDA responses recorded on day 12 was 24 µM. On day 32 the IC_{50} was 84 µM. Similar differences in the sensitivity of the NMDA receptor to KYN were recorded in two immature and three mature slices. Another technique used to assess receptor affinity, estimate of A', yielded a comparable difference in eight slices. Thus clearly measurable differences in the NMDA receptor exist in hippocampus between 1-week-old and mature animals. Whether similar differences in sensitivity to drugs can be detected for quisqualate and kainic acid receptors remains to be determined. Such differences could reflect developmental variations in postsynaptic receptors that could contribute to age-dependent changes in recurrent excitatory synaptic transmission in the CA_3 hippocampal neuronal networks.

Recent studies in our laboratory have uncovered a unique and potentially important property of NMDA receptors on immature CA_3 hippocampal neurons. These studies suggest that the voltage-dependency of the NMDA responses is generated in a novel way (Brady and Swann, 1988). Results show that, when Ca^{+2} is reduced in the bathing medium (but the Mg^{+2} concentration (2 mM) is unaltered), NMDA responses are increased. This increase is produced by a dramatic reduction in NMDA voltage dependency. Thus, in large measure Ca^{+2} appears to replace Mg^{+2} at the receptor channel for imparting voltage dependency. The impact of such a sensitivity of an NMDA receptor to extracellular Ca^{+2} in the developing hippocampus could be substantial since Ca^{+2} is commonly reduced during repetitive neuronal activation. Under these circumstances, NMDA receptor -mediated epsps could be enhanced not only because the cell depolarizes but also because the concentration of Ca^{+2} is reduced extracellularly. The contribution of such a synaptic mechanism to neuronal function could be significant since Ca^{+2} is thought to enter cells through the NMDA channel and have an important role intracellularly as a second messenger (for review see Cotman et al., 1989). Indeed a positive feedback system should exist at the NMDA receptor in immature hippocapus, where Ca^{+2} entry would lead to a reduction in Ca^{+2} extracellularly, enhance synaptic potentials, and further Ca^{+2} entry. Such a postsynaptic mechanism may be particularly important during critical periods in development when synaptogenesis is activity-dependent.

SUMMARY

The results reviewed in this chapter indicate that local circuit synaptic interactions are surprisingly well-developed in the rat hippocampal CA_3 subfield during the second postnatal

week. Intracellular recordings reveal large spontaneous epsps and ipsps. Synchronized bursts of
synaptic potentials are observed in most paired intracellular recordings. Antidromic and orthodromic electrical stimulation evokes synaptic responses that are reminiscent of recordings from
mature hippocampus. However, following brief trains of electrical stimuli and during bath application of a $GABA_A$ receptor antagonist, large prolonged depolarizations are recorded. These are
suppressed by excitatory amino acid antagonists. In comparison, slices from mature rats do not
produce these events under the same conditions. Thus, a hypothesis has been presented that the
degree of recurrent synaptic interaction between pyramidal cells may be enhanced during this
critical period in hippocampal development. An analysis of recurrent epsps using dual intracellular recordings is consistent with this contention. The degree of excitatory synaptic interaction
between CA_3 neurons appears to be at least equal to and likely in excess of that reported in mature
hippocampus. One possible explanation for this is that the number of recurrent excitatory synapses
may increase transiently during hippocampal development, only later to regress to numbers found
in the adult.

Recent studies of others suggest that activation of NMDA receptors may play a key role in
the maintenance of synapses during development (Cline et al., 1987; Kleinschmidt et al., 1987). In
this regard it is interesting that the characteristic of the NMDA receptor-ion channel complex on
immature and mature rat CA_3 hippocampal pyramidal cells appear to be quite different. These
differences may play a role in synapse formation and maintenance.

The role of NMDA receptors in LTP and learning (for review see Cotman et al., 1989; Lynch,
1986) are widely discussed. In other circles recurrent excitatory neuronal networks are thought to be
substrates for memory (Lynch, 1986; Haberly and Bower, 1989). Both NMDA receptors and
recurrent excitation are well represented in the CA_3 subfield of immature hippocampus. One
challenging area for future study will be the clarification of the interrelations between synaptogenesis and synaptic plasticity within neural networks of the developing hippocampus.

ACKNOWLEDGEMENTS

The studies reviewed here were supported by NIH grants NS18309 and NS23071.

REFERENCES

Ben-Ari, Y., Krnjevic, K, and Reinhardt, W., 1979, Hippocampal seizures and failure of
      inhibition, **Can. J. Physiol. Pharmacol.**, 57:1462-1466.
Ben-Ari, Y., Cherubini, E., and Krnjevic, K., 1988, Changes in voltage dependence of NMDA
      currents during development, **Neurosci. Letts.**, 94:88-92.
Brady, R.J. and Swann, J.W., 1984, Postsynaptic actions of baclofen associated with its
      antagonism of bicuculline-induced epileptogenesis in hippocampus, **Cell. Mol.
      Neurobiol.**, 4:403-408.
Brady, R.J. and Swann, J.W., 1986, Ketamine selectively suppresses synchronized afterdis-
      charges in immature hippocampus, **Neurosci. Letts.**, 69:143-149.
Brady, R.J. and Swann, J.W., 1988, Suppression of ictal-like activity by kynurenic acid does not
      correlate with its efficacy as an NMDA receptor antagonist, **Epilepsy Res.**, 2:232-
      238.
Brady, R.J. and Swann, J.W., 1988, The effects of extracellular calcium on the epileptiform
      activity and NMDA responses are different in mature and immature hippocampal
      slices, **Neurosci. Abstr.**, 14:239.
Cline, H.T., Debski, E.A., and Constantine-Paton, M., 1987, N-methyl-D-aspartate receptor
      antagonist desegregates eye-specific stripes, **Proc. Natl. Acad. Sci. USA**, 84:4342-
      4345.
Cowan, W.M., Fawcett, J.W., O'Leary, D.D.M., and Stanfield, B.B., 1984, Regressive events in
      neurogenesis, **Science**, 225:1258-1265.
Crepel, F., 1982, Regression of functional synapses in the immature mammalian cerebellum,
      TINS, 5:266-269.

Cotman, C.W., Bridges, R.J., Taube, J.S., Clark, A.S., Geddes, J.W., and Monaghan, D.T., 1989, The role of the NMDA receptor in central nervous system plasticity and pathology, J. NIH Res., 1:65-74.

Easter, S.S.Jr., Purves, D., Rakic, P., and Spitzer, N.C., 1985, The changing view of neural specificity, Science, 230:507-511.

Ganong, A.H., Lanthorn, T.H. and Cotman, C.W., 1983, Kynurenic acid inhibits synaptic and acidic amino acid-induced responses in the rat hippocampus and spinal cord, Brain Res., 272:170-174.

Grenningloh, G., Rienitz, A., Schmitt, B., Methfessel, C., Zensen, M., Beyreuther, K., Gundelfinger, E.D., Betz, H., 1987, The strychnine-binding subunit of the glycine receptor shows homology with nicotinic acetylcholine receptors, Nature, 328:215-220.

Haberly, L.B. and Bower, J.M., 1989, Olfactory cortex: model circuit for study of associative memory?, TINS, 12:258-264.

Huttenlocher, P.R., de Courten, C., Garey, L.J., and van der Loos, H., 1982, Synaptogenesis in human visual cortex - evidence for synapse elimmination during normal development, Neurosci. Letts., 33:247-252.

Huttenlocher, P.R., 1984, Synapse elimination and plasticity in developing human cerebral cortex, Amer. J. Mental Defic., 5:488-496.

Kleinschmidt, A., Bear, M.F., and Singer, W., 1987, Blockade of "NMDA" receptors disrupts experience-dependent plasticity of kitten striate cortex, Science, 238:355-358.

Lynch, G., "Synapses, Circuits, and the Beginnings of Memory," M.S. Gazzaniga, ed., The MIT Press, Cambridge (1986).

McCarren, M., and Alger, B.E., 1985, Use-dependent depression of ipsps in rat hippocampal pyramidal cells in vitro, J. Neurophysiol., 53:557-571.

Miles, R. and Wong, R.K.S., 1983, Single neurones can initiate synchronized population discharge in the hippocampus, Nature, 306:371-373.

Miles, R. and Wong, R.K.S., 1986, Excitatory synaptic interactions between CA3 neurons in the guinea-pig hippocampus, J. Physiol., 373:397-418.

Miles, R. and Wong, R.K.S., 1987(a), Latent synaptic pathways revealed after tetanic stimulation in the hippocampus, Nature, 329:724-726.

Miles, R. and Wong, R.K.S., 1987(b), Inhibitory control of local excitatory circuits in the guinea-pig hippocampus, J. Physiol., 388:611-629.

Mishina, M., Takai, T., Imoto, K., Noda, M., Takahashi, T., Numa, S., Methfessel, C., and Sakmann, B., 1986, Molecular distinction between fetal and adult forms of muscle acetylcholine receptor, Nature, 321:406-411.

Purpura, D.P., Prelevic, S., and Santini, M., 1968, Postsynaptic potential and spike variation in the feline hippocampus during postnatal ontogenesis, Exp. Neurol., 22:408-422.

Purves, D. and Lichtman, J.W., 1980, Elimination of synapses in the developing nervous system, Science, 210:153-157.

Rakic, P., Bourgeois, J.-P., Eckenhoff, M.F., Zecevic, N., Goldman-Rakic, P.S., 1986, Concurrent overproduction of synapses in diverse regions of the primate cerebral cortex, Science, 232:232-235.

Schneiderman, J.H., 1986, Low concentrations of penicillin reveal rhythmic, synchronous synaptic potentials in hippocampal slice, Brain Res., 398:231-241.

Schofield, P.R., Darlison, M.G., Fujita, N., Burt, D.R., Stephenson, F.A., Rodriguez, H., Rhee, L.M., Ramachandran, J., Reale, A.V., Glencourse, T.A., Seeburg, P.H., and Barnard, E.A., 1987, Sequence and functional expression of the GABA$_A$ receptor shows a ligand-gated receptor super-family, Nature, 328:221-227.

Schwartzkroin, P.A., 1982, Development of rabbit hippocampus: physiology, Dev. Brain Res., 2:469-486.

Schwartzkroin, P.A. and Haglund, M.M., 1986, Spontaneous rhythmic synchronous activity in epileptic human and normal monkey temporal lobe, Epilepsia, 27:523-533.

Stanfield, B.B., 1984, Postnatal reorganization of cortical projections: the role of collateral elimination, TINS, 7:37-41.

Smith, K.L., Turner, J. and Swann, J.W., 1988, Paired intracellular recordings reveal mono- and polysynatpic excitatory interactions in immature hippocampus, Neurosci. Abstr., 14:883.

Swann, J.W. and Brady R.J., 1984, Penicillin-induced epileptogenesis in immature rat CA3 hippocampal pyramidal cells, **Dev. Brain Res.,** 12:243-254.

Swann, J.W., Brady, R.J., and Martin, D.L., 1989, Postnatal development of GABA-mediated synaptic inhibition in rat hippocampus, **Neuroscience,** 28:551-561.

Traub, R.D., Miles, R., and Wong, R.K.S., 1989, Model of the origin of rhythmic population oscillations in the hippocampal slice, **Science,** 243:1319-1325.

A ROLE OF NMDA RECEPTORS AND Ca^{2+} INFLUX IN SYNAPTIC PLASTICITY IN THE

DEVELOPING VISUAL CORTEX

Tadaharu Tsumoto, Fumitaka Kimura and Ayahiko Nishigori

Department of Neurophysiology, Biomedical Research Center
Osaka University Medical School
Kitaku, Osaka, 530 Japan

INTRODUCTION

Acidic amino acids, such as glutamate (Glu) and aspartate (Asp), are suggested to be excitatory transmitters in the visual cortex (Clark and Collins, 1976; Baughman and Gilbert, 1981; Tsumoto et al. 1986). Recent studies have demonstrated that synaptic receptors for Glu/Asp can be classified into at least three types, on the basis of their most sensitive agonists, i.e., 1) N-methyl-D-aspartate (NMDA)-preferring receptors, 2) kainate-preferring receptors and 3) quisqualate-preferring receptors (see Watkins and Evans, 1981 for review). In CA1 area of the hippocampus, NMDA receptors have been shown to play a role in a form of synaptic plasticity, i.e., long-term potentiation (LTP) of synaptic efficacy probably by controlling entry of Ca^{2+} into postsynaptic sites (Collingridge et al., 1983; see Fagg et al., 1986; Bliss and Lynch, 1988; Cotman and Monagham, 1988 for review). In the cat's visual cortex, Tsumoto and his associates (Tsumoto et al., 1987, 1988; Hagihara et al., 1988) have reported evidence suggesting that receptors of kainate/quisqualate types (non-NMDA receptors) are involved in geniculo- cortical synaptic transmission while NMDA receptors may play a role in synaptic plasticity in the developing visual cortex.

Functional properties of visual cortical neurons can be changed by various experimental modifications of visual inputs during the critical period of postnatal development (see Fregnac and Imbent, 1984 for review). For example, if kittens are deprived of vision in one eye by lid sutures during this period, cortical neurons lose their responsiveness to visual inputs from that eye (Wiesel and Hubel, 1963). These changes are believed to be due to plasticity of synapses at afferents converging onto a cortical neuron from the two eyes; synapses from the open eye may functionally be potentiated and synapses from the sutured eye may be depressed as a consequence of "binocular synaptic competition" (Hubel and Wiesel, 1965). In fact, it was demonstrated that LTP of synaptic efficacy could most frequently be induced in the kitten visual cortex during the critical period by 2 Hz repetitive stimulation applied to the optic nerve (Tsumoto and Suda, 1979) and to the white matter underlying the cortex (Komatsu et al., 1981; Perkins and Teyler, 1988). In connection with the proposed role of NMDA receptors in the hippocampus, these results raise another question of whether or not the activation of NMDA receptors and subsequent influx of Ca^{2+} into postsynaptic neurons play a role in inducing LTP in the developing visual cortex. In the present study, we attempted to answer this question by administration of an NMDA receptor antagonist to visual cortical slices through perfusion medium and by direct intracellular injection of a Ca^{2+}-chelator.

Excitatory Amino Acids and Neuronal Plasticity
Edited by Y. Ben-Ari
Plenum Press, New York, 1990

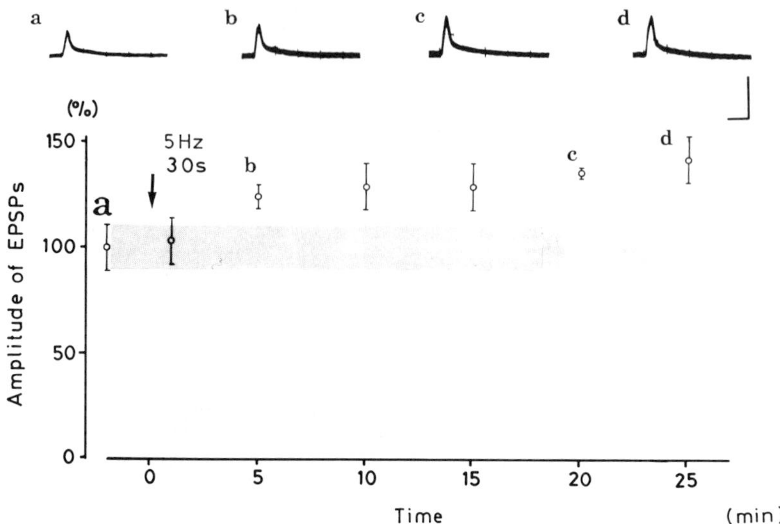

Fig. 1.   Time course of LTP of EPSPs recorded from a cell in layer II/III of the visual
cortex of a 29-day-old rat pup. Records a-d at top show potentials recorded at
corresponding time points in the graph. The upward deflections in the records
are EPSPs. Resting membrane potential of this cell was -58 mV. In the graph,
the amplitudes of EPSPs are plotted as a percentage of control value. Each
point and vertical bar represents mean ± standard deviation (S.D.) of 5 successive records. Scales at the right are 50 msce and 10 mV. The shaded horizontal band in the graph indicates mean ± S.D. of the control value. From
Kimura et al. (1989).

METHODS

Slices of the primary visual cortex were obtained from Sprague- Dawley rat pups aged 21-41
days. The animals were initially anaesthetized with urethane (i.p., 1.3 g/kg) and then infused
through the heart with cold oxygenated medium. They were then decapitated, and coronal sections (0.4-0.5 mm) were dissected from the visual cortex. Procedures for preparing the medium and
for incubating the slices in it have been described elsewhere (Kimura et al., 1989). The slices were
then transferred to a recording chamber. An arrangement of recording and stimulating electrodes
was essentially the same as described previously (Tsumoto et al., 1988).

For the activation of optic radiation fibers, a pair of tungsten wires, insulated except at the
tips and attached side by side with a tip separation of 0.2 mm, was placed in the white matter just
beneath layer VI of the cortex. Thses stimulating electrodes had mean resistances of about 1
Mohm. Stimulus intensity was within the range of 1 to 100 V for pulses of 0.1 msec duration. To
record field or intracellular potentials evoked by test stimulation of the white matter, glass
micropipettes filled with O.5 M sodium acetate containing 2% pontamine sky blue (resistance: < 10
Mohm) or 4 M potassium acetate (resistance: 60-150 Mohm), respectively, were inserted into layer
II/III of the cortex under visual control.

Intracellular signals were amplified by a high-input impedance amplifier, with built-in
current injection and bridge balance facilities (MEZ 8201, Nihon Kohden). Extracellular potentials
were filtered at 0.05-3kHz. During the experiments, the signals were displayed on a digitizing
oscilloscope (VC 11, Nihon Kohden), and recorded simultaneously on a data recorder (PC-108M,
Sony Magnescale) for later analysis. Initially, control responses to single shocks given to the

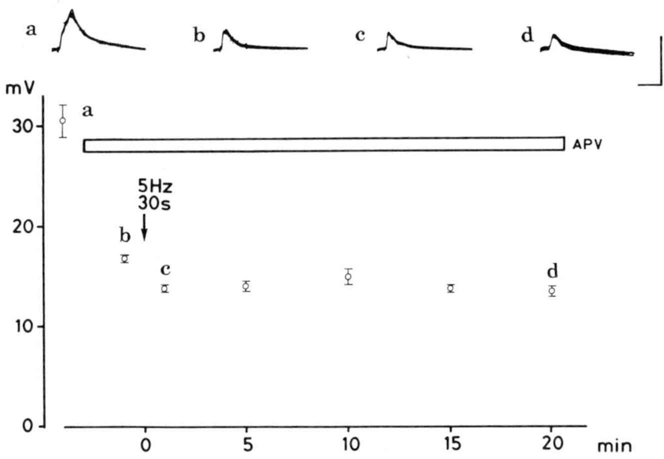

Fig. 2.   No induction of LTP during administration of APV. This cell was obtained
from a 38-day-old rat pup. Specimen records at the top (a-d) show EPSPs
recorded at the corresponding time in the graph. Scales at the right are 50
msec and 40 mV. In the graph, the amplitudes of the EPSPs are plotted
against time after tetany. Each point represents mean $\pm$ S.D. of 5 successive
responses. Resting membrane potential of this cell was -72 mV. Duration
during which APV was delivered is indicated by open, horizontal bar. From
Tsumoto et al. (1989).

underlying white matter at 0.1 Hz were recorded, and then repetitive stimulation (tetanus) was
applied to the white matter with pulses of the same amplitude but double the duration. The
parameters of the tetany were 5Hz for 30 or 60 sec. After tetany, test shocks at the same intensity
as the pre-tetanic control ones were given to the white matter at 0.1 Hz and the evoked responses
were further observed at least 20 min.

In some of the experiments, glass micropipettes for intracellular recordings were filled with
potassium acetate plus 0.2-0.5 M potassium ethylene glycol-bis-N, N, N', N'-tetraacetic acid
(EGTA) for intracellular injection. This electrode had resistances of 60-150 Mohm. Another
electrode for recording field potentials evoked by testing white matter stimulation was placed in
the site adjacent to the intracellular electrode. After neurons were impaled and cohtrol responses
were recorded, EGTA was injected by applying 0.5-2.0 nA hyperpolarizing pulses of 500-700 msec
width at 1 Hz for 8-15 min (see Results). Thereafter, testing and tetanic stimulation was given to
the white matter. EPSPs and field potentials were simultaneously recorded through the both
electrodes. Procedures for giving testing and tetanic stimulation were the same as described
previously (Kimura et al. 1989).

D,L-2-amino-5-phosphonovalerate (APV) and kynurenic acid (KYNA) were delivered to
the slices by switching the standard medium to one containing the drugs at known concentrations
through three-way switching taps so as not to cause an alteration in the rate of perfusion. The
standard and test solutions were exchanged conmpletely in the recording chamber within a few
minutes after switching the taps. D-APV is a selective antagonist for NMDA receptors (Davies et
al., 1981; Hagihara et al., 1988) while KYNA is a relatively non-selective antagonist for the
three types of EAA receptors (Perkins and Stone, 1982; Tsumoto et al., 1986). The concentration of
APV was usually 50 µM and that of KYNA was 0.5 - 1 mM. In the experiments in which removal of
gamma- aminobutyric acid (GABA)-mediated inhibition was attempted, bicuculline methiodide,
an antagonist of $GABA_A$ receptors was added to the medium at a concentration of 0.5 µM.

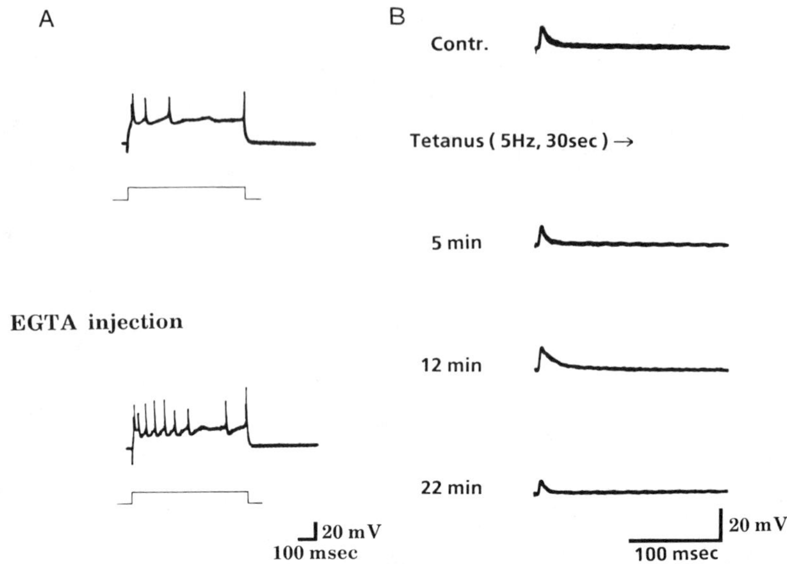

Fig. 3.   No induction of LTP after injection of EGTA. A; Reduction of accomodation of spike discharges by intracellular injection of EGTA. Top, records of response of a layer II/III cell to injection of depolarizing current pulses of 600 msec width. Bottom, records of responses to the current injection done 8 min after the initiation of EGTA injection. B; EPSPs which were not potentiated by tetanic stimulation given after the injection of EGTA. Time after stopping tetany is indicated at the left of each record.

RESULTS

Induction of LTP in the Visual Cortex

Excitatory postsynaptic potentials (EPSPs) evoked by white matter stimulation were recorded intracellularly from 23 cortical neurons located in layer II/III of the visual cortex. Cells which were lost before having recorded EPSPs at least 20 min after stopping the tetanic stimulation were discarded from the present analysis. Stimulus intensity was adjusted so that each shock elicited EPSPs which were just subthreshold for inducing spikes. In 8 of the 12 cells tested in the standard medium, the EPSPs were significantly enhanced after the tetanic stimulation, and this enhancement lasted as long as the recordings were continued (usually 30 min after the tetanus). The records in Fig. 1 show an example of LTP of the EPSPs observed in a 29-day-old rat pup. The test shocks given at 0.1 Hz to the underlying white matter evoked EPSPs with a mean peak amplitude of 6.0 mV (record a). Five minutes after tetanization these EPSPs were enhanced in amplitude significantly (record b). Thereafter, the amplitude was increased further and it became 8.5 mV, 25 min after the tetany (record d). The time course of this potentiation is shown graphically in Fig. 1. In some of the cells, slow EPSPs became manifest after the tetanic stimulation. Peak latencies of these slow EPSPs ranged from 30 to 110 msec. Details of the slow EPSPs have been described elsewhere (Kimura et al., 1989).

Blockade of LTP by APV Administration

The administration of APV prevented the induction of LTP in all the 11 cells, except one, from which stable intracellular recordings were done longer than 20 min after stopping the tetanus. An example for this is shown in Fig. 2. This cell was obtained from a 38-day-old rat pup. Before administration of APV, test shocks at an intensity just subthreshold for the generation of spikes evoked EPSPs having a mean peak amplitude of 30.5 mV (record a). Twenty min after the beginning of presentation of APV, test shocks at the same intensity elicited EPSPs with peak amplitudes of only 16.8 mV (record b). Following tetanic stimulation at 5Hz for 30 sec, the amplitude of EPSPs decreased further to 13.8 mV (record c), and this reduction of amplitude was maintained for a subsequent 20 min (record d).

As described above, APV reduced substantially the amplitude of EPSPs in most of the cells. There is a possibility, therefore, that the tetanic stimulation during the administration of APV might have been too weak to induce LTP. To test this possibility, the intensity of test shocks was increased during APV application in a few slices so that each shock could evoke EPSPs of about the same or higher amplitude than were obtained by pre-drug control stmulation. In no case such an intensified tetanus was effective in inducing LTP in the medium containing APV. The probability of LTP induction, assessed with measurements of the EPSP amplitudes, was only 9 % in the perfusate containing APV whereas it was 67 % in that without APV. This difference in the probability for the induction of LTP is statistically significant between the two perfusion conditions (X^2-test, $p<0.05$). These results strongly suggest that NMDA receptors play a crucial role in induction of LTP of synaptic transmission in the developing visual cortex.

Blockade of LTP by Intracellular Injection of EGTA

The intracellular injection of EGTA prevented the induction of LTP of EPSPs in layer II/III neurons, despite the fact that the same tetanic stimulation induced LTP of field potentials. As indication that EGTA had been injected sufficiently enough to chelate intracellular Ca^{2+}, we observed changes in accomodation of spike discharges on intracellular application of long, depolarizing current pulses, because the reduction in spike accomodation is known to be a Ca^{2+}-dependent event (Lynch et al., 1983; Madison and Nicoll, 1984; Malenka et al., 1988). An example for this is shown in Fig. 3A. When depolarizing current pulses of 600 msec duration was injected to this cell, it generated two spikes with a very short interval initially and then three spikes with successively longer intervals (Fig. 3A, upper record). The phenomenon that a cell cannot maintain its firing rate to a steady depolarizing current is called spike accomodation and believed to be due primarily to an increase in Ca^{2+}-activated potassium conductance (Madison and Nicoll, 1984). This accomodation of spikes was almost completely abolished after injection of EGTA for 7 min (Fig. 3A, bottom record). After having observed the disappearance of spike accomodation, EPSPs evoked by testing stimulation of the underlying white matter were recorded as control (Fig. 3B, top record) and then the tetanic stimulaion was given to the white matter. The tetanic stimulation was ineffective, however, in inducing potentiation of EPSPs (Fig. 3B, the second to fourth traces). Also, in another cell to which EGTA was injected, tetanic stimulation could not induce potentiation of EPSPs, although the same tetanus induced LTP of field potentials recorded simultaneously from the site adjacent to the cell.

DISCUSSION

The present study has demonstrated that tetanic stimulation of the underlying white matter induces LTP of EPSPs recorded from layer II/III neurons of the visual cortex of young rats. Also, the results have shown that the induction of LTP is blocked by the bath application of the NMDA receptor antagonist, APV. The concentration of APV was carefully adjusted so as to antagonize NMDA receptors selectively. Thus, it seems reasonable to conclude that NMDA receptors play a crucial role in inducing LTP in the developing visual cortex, as already reported in the hippocampus (Collingridge et al., 1983). In the hippocampus, NMDA receptors were shown to play such a role probably by permitting an entry of Ca^{2+} into neurons which initiates Ca^{2+}-dependent biochemical processes that underlie long-lasting changes in synaptic function (Mayer and Westbrook, 1987; Ascher and Nowak, 1988; see Fagg et al., 1986; Bliss and Lynch, 1988; Cotman and Monagham, 1988 for review). In the developing visual cortex, we have found that the injection of the Ca^{2+} chelator into postsynaptic neurons actually prevented the induction of LTP. This was not due to ineffectiveness of the tetanic stimulation, because it could actually induce LTP of field potentials recorded simultaneously from the site adjacent to the cell from which intracellular recordings were done. Thus, it is reasonable to conclude that the Ca^{2+} entry to postsynaptic site through the activated NMDA receptor-coupled channels during tetanic stimulation may play a crucial role in induction of LTP in the visual cortex. Such an activity-related entry of Ca^{2+} into postsynaptic sites may trigger Ca^{2+}-dependent, biochemical processes underlying long-lasting changes in synaptic efficacy in the developing visual cortex, as suggested previously (Tsumoto et al., 1988).

Relevance to Developmental Plasticity of Visual Cortex

Previous studies reported that LTP could be induced in the visual cortex with higher probability in young animals than in the adult (Tsumoto and Suda 1979; Komatsu et al. 1981; Perkins and Teyler 1988). From the present results it is possible to assume, therefore, that the activity of NMDA receptors may be high in the developing visual cortex and become relatively low as the cortex matures. In fact, Tsumoto et al. (1987) have found that NMDA receptors contribute more effectively to responses of visual cortical neurons to afferent inputs in kittens than in the adult cat. In line with this finding an autoradiographic study has demonstrated that APV-sensitive [3H]-L-glutamate binding is very high in the kitten visual cortex during the critical period of postnatal development, and thereafter become relativery low so as to reach the matured level (Bode-Greuel et al., 1989). In the matured visual cortex of the rat, however, Artola and Singer (1987) reported that an APV-sensitive component, i.e., NMDA receptor-mediated component is present in EPSPs recorded from layer II/III cells in response to white matter stimulation. Also, we observed that NMDA receptor mediated component of the EPSPs could be induced in some layer II/III neurons of the matured visual cortex of the cat (Shirokawa et al., 1989). In physiological conditions, however, the action of NMDA receptors seemed to be suppressed by the intracortical GABA-mediated inhibition (Shirokawa et al., 1989). Thus, we suggest that the intracortical inhibition may be involved in part in regulation of synaptic plasticity of the developing visual cortex. This interpretation agrees well with previous studies indicating that blockade of inhibition facilitates the induction of LTP in the visual cortex and hippocampus (Wigstrom and Gustafsson, 1985; Kimura et al., 1989) and that the intracortical inhibition matures relatively late in layer II/III of the kitten visual cortex (Komatsu, 1983).

ACNOWLEDGMENTS

We are very grateful to Dr. C. Yamamoto for providing the recording chamber for slice preparations. This work was supported by a Grant-in-Aid for Special Project Research of Plasticity of Neural Circuits from the Japanese Ministry of Education, Science and Culture to T.T.

REFERENCES

Artola, A., and Singer, W., 1987, Long-term potentiation and NMDA receptors in rat visual cortex, **Nature**, 330:649.
Ascher, P., and Nowak, L., 1988, The role of divalent cations in the NMDA responses of mouse central neurones in culture, **J. Physiol. (Lond.)**, 399, 247.
Baughman, R., and Gilbert, C.D., 1981, Aspartate and glutamate as possible neurotransmitters in the visual cortex, **J. Neurosci.**, 1:427.

Bliss, T.V.P., and Lynch, M.A., 1988, Long-term potentiation of synaptic transmission in the hippocampus: Properties and mechanisms, **in** "Long-Term Potentiation: From Biophysics to Behavior", Landfield, P.W. and Deadwyler, S.A. eds., Alan R. Liss, New York.

Bode-Greuel, K.M., and Singer, W., 1989, The development of N-methyl-D- aspartate receptors in cat visual cortex, **Develop. Brain Res.,** 46: 197.

Clark, R.M. and Collins, C.G.S., 1976, The release of endogeneous amino acids from the rat visual cortex, **J. Physiol. (Lond.),** 262:383.

Collingridge, G.L., Kehl, S.J., and McLennan, H., 1983, Excitatory amino acids in synaptic transmission in the Schaffer collateral- commissural pathways of the rat hippocampus, **J. Physiol. (Lond),** 334:33.

Cotman, C.W., and Monaghan, D.T., 1988, Excitatory amino acid neurotransmission: NMDA receptors and Hebb-type synaptic plasticity, **Ann. Rev. Neurosci.,** 11: 61.

Davies, J., Francis, A.A., Jones, A.W., and Watkins, J.C., 1981, 2-amino- 5-phosphonovalerate (APV), a potent and selective antagonist of amino acid-induced and synaptic excitation, **Neurosci. Lett.,** 21:77.

Fagg, G.E., Foster, A.C., and Ganong, A.H., 1986, Excitatory amino acid synaptic mechanisms and neurological function, **Trends Pharmacol. Sci.,** 7:357.

Fegnac, Y., and Imbert, M., 1984, Development of neuronal selectivity in primary visual cortex of cat, **Physiol. Rev.,** 64:325.

Hagihara, K., Tsumoto, T., Sato, H., and Hata, Y., 1988, Actions of excitatory amino acid antagonists on geniculo-cortical transmission in the cat's visual cortex, **Exp. Brain Res.,** 69:407.

Hubel, D.H., and Wiesel, T.N., 1965, Binocular interaction in striate cortex of the kittens reared with artificial squint, **J. Neurophysiol.,** 28: 1041.

Kimura, F., Nishigori, A., Shirokawa, T. and Tsumoto, T., 1989, Long-term potentiation and N-methyl-D-aspartate receptors in the visual cortex of young rats, **J. Physiol. (Lond.),** 414:125.

Komatsu, Y., Toyama, K,, Maeda, J., and Sakaguchi, H., 1981, Long-term potentiation investigated in a slice preparation of striate cortex of young kittens, **Neurosci. Lett.,** 26:269.

Komatsu, Y., 1983, Development of cortical inhibition in kitten striate cortex investigated by a slice preparation., **Develp. Brain Res.,** 8: 136.

Lynch, G,. Larson, J., Kelso, S., Barrionuevo, G., and Schottler, F., 1983, Intracellular injections of EGTA block induction of hippocampal long-term potentiation, **Nature,** 305:719.

Madison, D.V., and Nicoll, R.A., 1984, Control of the repetitive discharge of rat CA1 pyramidal neurones in vitro, **J. Physiol. (Lond.),** 354:319.

Malenka, R.C., Kauer, J.A., Zucker, R.S., and Nicoll, R.A., 1988, Postsynaptic calcium is sufficient for potentiation of hippocampal synaptic transmission, **Science,** 242:81.

Mayer, M.L., and Westbrook, G.L., 1987, Permeation and block of N- methyl-D-aspartic acid receptor channels by divalent cations in mouse cultured central neurones, **J. Physiol. (Lond.)** 394:501.

Perkins, N.M., and Stone, T.W., 1982, An iontophoretic investigation of the actions of convulsant kynurenines and their interaction with the endogenous excitant quinolinic acid, **Brain Res.,** 439:222.

Perkins IV, A.T., and Teyler, T.J., 1988, A critical period for long-term potentiation in the developing rat visual cortex, **Brain Res.,** 247:184.

Shirokawa, T., Nishigori, A., Kimura, F., and Tsumoto, T., 1989, Actions of excitatory amino acid antagonists on synaptic potentials of layer II/III neurons of cat visual cortex. **Exp. Brain Res.,** in press.

Tsumoto, T., and Suda, K., 1979, Cross-depression: an electro- physiological manifestation of binocular competition in the developing visual cortex, **Brain Res.,** 168:190.

Tsumoto, T., Masui, H., and Sato, H., 1986, Excitatory amino acid transmitters in neuronal circuits of the cat visual cortex, **J. Neurophysiol.,** 55:469.

Tsumoto, T., Hagihara, K., Sato, H., and Hata, Y., 1987, NMDA receptors in the visual cortex of young kittens are more effective than those of adult cat, **Nature,** 327:513.

Tsumoto, T., Kimura, F., Hagihara, K., Sato, H., and Sobue, K., 1988, A role of NMDA
        receptors and membrane-associated Ca^{2+}/calmodulin- binding proteins in synaptic
        plasticity in the developing visual cortex, **Biomed. Res. 9, Suppl. 2:43.**
Tsumoto, T., Kimura, F., Nishigori, A., and Shirokawa, T., 1989, Long-term potentiation and
        NMDA receptors in the developing visual cortex, **Biomed. Res.,** in press.
Watkins, J.C., and Evans, R.H., 1981, Excitatory amino acid transmitters, **Ann. Rev.
        Pharmac. Toxic.** 21:165.
Wiesel, T.N., and Hubel, D.H., 1963, Single-cell responses in striate cortex of kittens
        deprived of vision in one eye, **J. Neurophysiol.,** 26:1003.
Wigstrom, H., and Gustafsson, B., 1985, Facilitation of hippocampal long-lasting potentia-
        tion by GABA antagonists, **Acta Physiol. Scand.,** 125:159.

SPONTANEOUS AND EVOKED NMDA-RECEPTOR MEDIATED POTENTIALS IN THE

ENTORHINAL CORTEX OF THE NEONATE RAT *IN VITRO*

Roland S.G. Jones[1] and Uwe Heinemann[2]

[1]Division of Neuroscience John Curtin School of Medical
Research Australian National University Canberra
ACT2601 Australia.

[2]Institut fur Normale und Pathologische Physiologie
Der Universitat zu Koln Robert-Koch Strasse 39 D5000
Koln 41 West Germany

INTRODUCTION

It is becoming clear that amino acid receptors of the *N*-methyl-D-aspartate (NMDA)
subtype show changes in number or sensitivity during early development. During the first two
weeks of life in the rat there is a decline in sensitivity of the cell soma of CA1 neurones to NMDA
and a concurrent enhancement of dendritic responsiveness[1]. During the same period cerebellar
Purkinje cells show a progressive enhancement enhancement of sensitivity to NMDA which then
declines with further development[2,3]. Visual cortical cells in the cat also appear to show a
declining responsiveness to NMDA during early development[4].

Binding studies have now demonstrated that the physiological changes in sensitivity to
NMDA may result from changes in receptor density during the early development period. Thus, in
rat hippocampus, the number of NMDA binding sites at postnatal (P) day 4 is elevated compared
to adult levels, increases further up to P8-10 and thereafter declines to adult levels[5]. A transient
increased expression of NMDA-receptors is also seen in the human hippocampus with a peak
around foetal weeks 23-27[6].

We have demonstrated pronounced depolarizing synaptic potentials mediated by NMDA
receptors in layers IV/V of the adult rat entorhinal cortex (EC) *in vitro*[7,8]. More recent studies[9]
have shown that NMDA-mediated potentials are a much less evident component of synaptic
potentials of layer II cells in the EC evoked either by stimulation in the para-subiculum or by
activation of the collateral innervation from layer V. In view of the possibility that NMDA
receptors may be enhanced during development the present study sought to determine if NMDA-
mediated potentials could play a greater role in transmission in the superficial layers of the rat
EC during the first few weeks of life.

METHODS

Slices consisting of hippocampus, dentate gyrus, subicular complex and medial (MEC) and
lateral divisions of the entorhinal cortex were prepared from the brains of rats of various

Excitatory Amino Acids and Neuronal Plasticity
Edited by Y. Ben-Ari
Plenum Press, New York, 1990

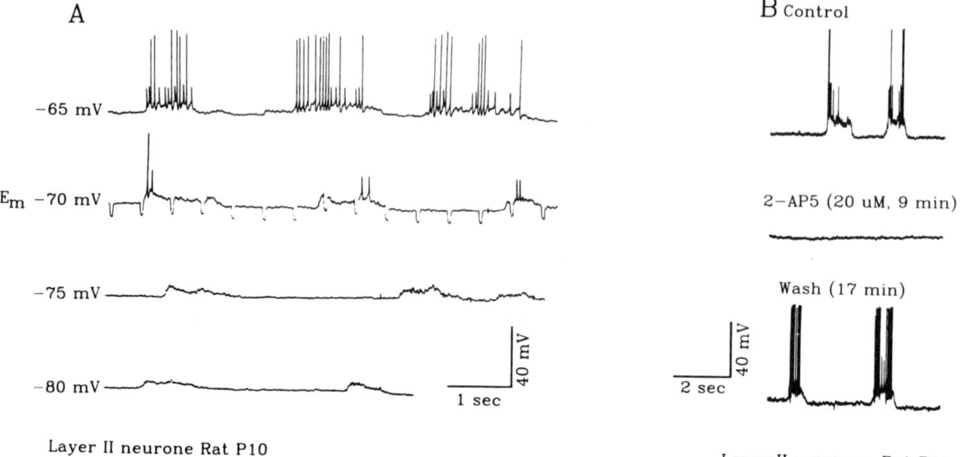

Fig. 1.   (A) Spontaneous depolarizing events recorded in a layer II cell in a slice from a rat P10. The traces are voltage recordings written out at slow speed from magnetic tape directly to a chart recorder. Spontaneous depolarizations at resting potential (E_m) gave rise to occassional spikes of two different amplitudes. Holding the membrane potential at hyperpolarized levels resulted in a decrease in amplitude and duration of the spontaneous events. At more positive levels the events were prolonged and the intensity of firing increased markedly. (B) Spontaneous events in another cell (P11) were abolished by 2-AP5.

postnatal ages using the approach previously detailed[8]. These were maintained *in vitro* in a recording chamber at the interface between a continuous perfusion of artificial CSF (see ref 7 for composition) and warmed moist carbogen gas (95% O_2-5% CO_2). Intracellular recordings were made from neurones in layers II or IV/V of MEC using potassium acetate (3M) filled electrodes. Synaptic potentials were elicited in layer II cells by stimulating in the para-subiculum or in layer V of MEC. The rats were arbitrarily divided into groups aged P5-7, P9-13 and P19-24 and compared to adults (> 70 days).

Artificial CSF which was nominally Mg^{++}-free was prepared by omitting $MgSO_4$. The NMDA-receptor antagonist, 2-amino-5-phosphonovalerate (2-AP5, 20-30 uM) and the non-NMDA (i.e. quisqualate/kainate) antagonist, 6-cyano-7-nitro-quinoxaline-2,3-dione (CNQX, 1-5 uM) were applied by bath perfusion.

RESULTS

Spontaneous depolarizing potentials

The great majority of layer II cells in the P5-7 and P9-13 groups displayed spontaneous depolarizing potentials or bursts of activity. These ranged in amplitude and duration from 0.5-15 mV and 0.2-6s respectively. Action potentials were associated with this depolarizing activity and these were often of two amplitudes with full blown spikes being interspersed with smaller amplitude (5-20 mV) spikes. Examples of this spontaneous activity are shown in figures 1 and 2. This type of spontaneous activity was much less marked in slices from the P19-24 rats with only a minority of cells displaying such events. Spontaneous depolarizations were not recorded in adult slices. In addition, the activity was restricted to layer II cells, with neurones in layer IV/V of the P9-13 day slices showing no such events.

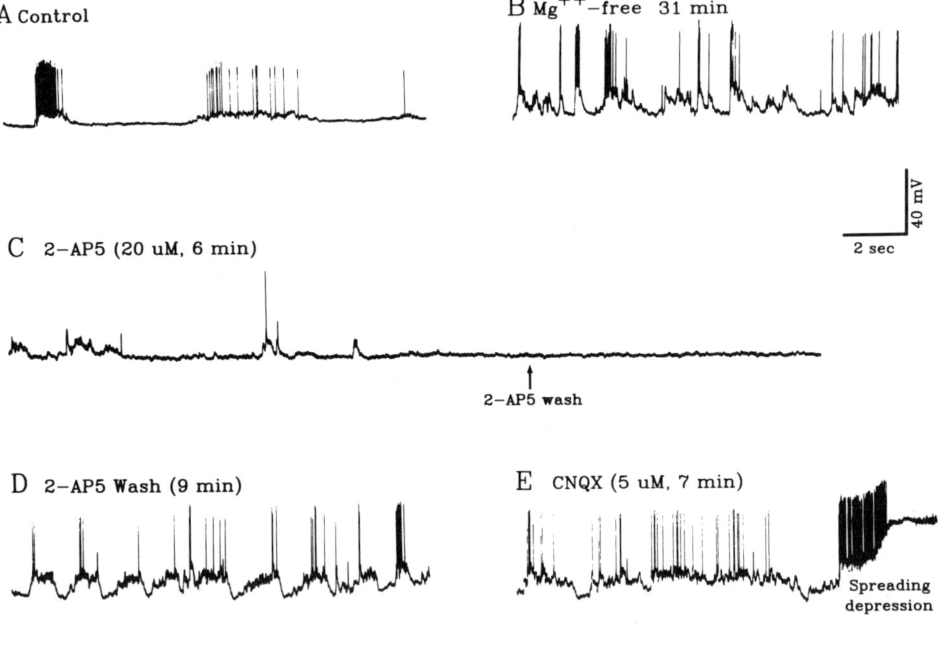

Fig. 2.   The spontaneous depolarizing events (A) in this cell were greatly enhanced by perfusion with Mg++-free medium (B). 2-AP5 abolished all activity (C). Following recovery (D), CNQX failed to affect the activity (E) and eventually a spreading depression occurred in the slice.

The voltage deflection elicited by hyperpolarizing current pulses showed a small increase (10-12 %) in amplitude during the bursts of activity in some cells, indicating an apparent decrease in conductance. However other cells showed the opposite effect and many showed no detectable change.

Figure 1 (A) shows that the voltage dependency of spontaneous activity in one cell. At rest, the activity consisted of depolarizing potentials which elicited occasional spikes of two different amplitudes. Holding the membrane potential at a more positive level resulted in an increase in duration of the events and a much increased intensity of firing associated with them. At more negative potentials the events clearly decreased in amplitude and possibly duration. While the majority of cells on which voltage dependency was tested showed similar effects, on a few, little change in the events was noted with either positive or negative shifts in membrane potential and others the events clearly decreased and increased in amplitude respectively with these manipulations of membrane potential.

Perfusion of slices with 2-AP5 very rapidly abolished all forms of spontaneous activity in cells in the P9-13 day slices. Other age groups have not yet been tested. In some cells there was a fairly rapid but progressive reduction in amplitude of the events before they disappeared. In other cells the disappearance of the spontaneous activity was practically all-or-none. One study is illustrated in figure 1 (B). Another effect noted during perfusion with 2-AP5 was a small (3-4 mV) hyperpolarization as the drug was washed in. On washing the antagonist, spontaneous activity recovered rapidly and the membrane potential repolarized.

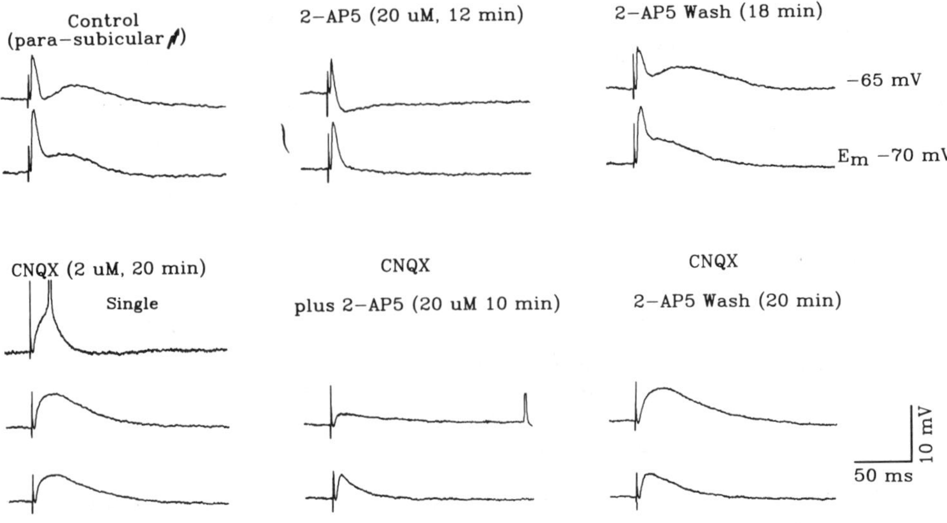

Fig. 3.   Synaptic potentials in a layer II cell evoked by para-subicular stimulation at resting potential (E_m, lower record in each panel) and with the cell held depolarized by 5 mV (upper record). The late, slow depolarization seen in the control records was blocked by 2-AP5. Following recovery, CNQX blocked the fast EPSP and IPSP leaving the slow, 2-AP5-sensitive EPSP in isolation.

In contrast to 2-AP5, CNQX did not abolish spontaneous activity. There was sometimes a reduction in the frequency of the events and also in amplitude, but the antagonist never prevented the spontaneous discharges completely.

The study in figure 2 shows the effects of depletion of $[Mg^{++}]_o$ on spontaneous depolarizing events in one cell. These were very rapidly enhanced in frequency and amplitude by perfusion with an Mg^{++}-free medium and the cell reached an almost continuous state of discharge. At this point, 2-AP5 quickly abolished all activity. Following washout of 2-AP5 and recovery of the spontaneous discharge, CNQX failed to markedly alter the activity and eventually a spreading depression phenomenon occurred in the slice.

Synaptic potentials

Layer II neurones in adult slices display complex synaptic potentials to stimulation of either the para-subiculum or the collateral input from layer V[9]. These are characterized by a fast monosynaptic EPSP which is CNQX sensitive, and pronounced fast and slow GABAergic inhibitory components. In many cells the evidence suggests that the fast IPSP is a monosynaptic GABAergic response while in others it appears to be mediated by recurrent interneurones[9]. In a minority of cells a slow NMDA-mediated depolarizing potential can be detected but this is weak and variable.

While it is obviously very difficult to quantify the contribution of NMDA-receptors to synaptic depolarizations, qualitative examination of studies to date suggest that NMDA-mediated potentials may be more pronounced in synaptic responses of layer II cells in immature rats.

Figure 3 shows synaptic responses evoked in a layer II cell in a slice from a P11 day rat following stimulation in the para-subiculum. At rest there was a fast depolarization which gave way to a more prolonged depolarization. A slow hyperpolarization then ensued which is not greatly evident in these records. At a more positive membrane potential (upper records) the fast EPSP decreased and a fast IPSP was shown to intervene between the two depolarizing phases of the response. 2-AP5 completely blocked the slow depolarization and reduced the fast event leaving the fast and slow IPSPs intact. Following recovery, CNQX perfusion blocked the fast EPSP and the hyperpolarizing responses leaving a prominent, slow depolarization which increased at more positive membrane potentials and was 2-AP5 sensitive. Similar studies in adult animals showed no, or much smaller, CNQX-resistant slow depolarizations in layer II cells.

DISCUSSION

The voltage dependent properties of the spontaneous events recorded in immature layer II cells resembled those described for neuronal responses mediated by NMDA-receptors. A block of the receptor associated ionophore by Mg^{++} is alleviated by membrane depolarization and enhanced by hyperpolarization and this results in an increase and decrease in receptor mediated potentials respectively[10,11,12]. The intimate involvement of NMDA receptors in the spontaneous depolarizations was confirmed by their abolition by 2-AP5. IN contrast, the non-NMDA receptor blocker CNQX was almost ineffective suggesting that the quisqualate/ kainate subtypes of receptor were little involved.

Further support for the role of NMDA receptors was indicated by a rapid and dramatic enhancement of the activity when $[Mg^{++}]_o$ was depleted. Again this enhanced activity was exquisitely sensitive to 2-AP5 but not CNQX. Finally, qualitative comparison indicated that synaptic potentials in layer II showed a greater contribution of NMDA-receptors during this early development period. It seems likely that these effects are dependant on an increased density of NMDA-receptors on layer II cells similar to that demonstrated in the hippocampus of the rat and human during development[5,6].

Whether this increased NMDA-dependant activity has a functional role in development is open to speculation. NMDA has a trophic effect on cells in culture[13,14] and the spontaneous activity in layer II could obviously be linked to promoting neuronal growth. It should also be remembered that neurones in layer II of the EC are the cells of origin of the perforant path input to the dentate gyrus. Increased impulse traffic in these axons could be involved in the high rate of synaptogenesis occurring in the dentate at this time[15].

A final note concerns epileptogenesis in the entorhinal area. Mg^{++}-free medium induces long lasting ictal-like discharges in mature EC cells[8]. However, this effect is increased greatly in severity in the immature EC. In addition, the epileptogenic effects of $GABA_A$-inhibition in the EC is also more severe in immature slices[16]. Both effects appear to be related to the increased expression of NMDA-receptors[15]. It is possible then that the increase in NMDA-receptors seen in the human brain[6] could be related to the high incidence of newborn and infantile convulsions.

REFERENCES

1. B. Hamon and U. Heinemann, Developmental changes in neuronal sensitivity to excitatory amino acids in area CA1 of the rat hippocampus, **Dev. Brain Res.** 38: 286 (1988)
2. J.-L. Dupont, R. Gardette and F. Crepel, Postnatal development of the chemosensitivity of rat cerebellar Purkinje cells to excitatory amino acids. An in vitro study, **Dev. Brain Res.** 34: 59 (1987).
3. G. Garthwaite, B. Yamini, and J. Garthwaite, Selective loss of Purkinje and granule cell responsiveness to N-methyl-D-aspartate in rat cerebellum during development, **Dev. Brain Res.** 36: 288 (1987).
4. T. Tsumoto, K. Hagihara, H. Sato and Y. Hata, NMDA receptors in the visual cortex of young kittens are more effective than those of adult cats, **Nature** 327: 513 (1987).
5. E. Tremblay, M.P. Roisin, A. Represa, C. Charriaut-Marlangue and Y. Ben-Ari, Transient increased density of NMDA binding sites in the developing rat hippocampus, **Brain Res.** 461: 393 (1989).

6. A.Represa, E. Tremblay and Y. Ben-Ari, Transient increased density of NMDA-binding sites in the human hippocampus during development, **Neurosci. Lett.** 99: 61 (1989)

7. R.S.G. Jones, Complex synaptic responses of entorhinal cortical cells in the rat to subicular stimulation in vitro: demonstration of an NMDA receptor-mediated component, **Neurosci. Lett.** 81: 209 (1987).

8. R.S.G. Jones and U. Heinemann, Synaptic and intrinsic responses of medial entorhinal cortical cells in normal and magnesium-free medium in vitro, **J. Neurophysiol.** 59 1476 (1988).

9. R.S.G. Jones, Synaptic responses of cells of origin of the perforant path in layer II of the rat entorhinal cortex in vitro, **J. Neurophysiol.** (submitted)

10. V. Crunelli and M.L. Mayer, Mg^{++} dependence of membrane resistance increases evoked by NMDA in hippocampal neurones, **Brain Res.** 311: 392 (1984).

11. M.L. Mayer, G.L. Westbrook and P.B. Guthrie, Voltage-dependent block by Mg^{++} of NMDA responses in spinal cord neurons, **Nature** 309: 261 (1984).

12. L. Nowak, P. Bregestovski, P. Ascher, A. Herbet and A. Prochiantz, Magnesium gates glutamate-activated channels in mouse central neurones, **Nature** 307: 462 (1984).

13. R. Balazs, N. Hack and O.S. Jorgensen, Stimulation of the N-methyl-D-aspartate receptor has a trophic effect on differentiating cerebellar granule cells, **Neurosci. Lett.** 87: 80 (1988).

14. I.A. Pearce, M.A. Cambray-Deakin and R.D. Burgoyne, Glutamate acting on NMDA receptors neurite outgrowth from cerebellar granule cells, **FEBS Lett.** 223: 143 (1987).

15. B. Crain, C. Cotman, D. Taylor and G. Lynch, A quantitative electron microscopic study of synaptogenesis in the dentate gyrus of the rat, **Brain Res.** 63: 195 (1973).

16. R.S.G. Jones, Enhanced susceptibility of the immature rat entorhinal cortex to epileptogenesis in vitro. **Neurosci. Lett.** (submitted)

LEARNING BY SEEING: N-METHYL-D-ASPARTATE RECEPTORS AND RECOGNITION

MEMORY

G. Horn and B.J. McCabe

University of Cambridge
Department of Zoology
Downing Street
Cambridge CB2 3EJ

Recognition memory and exposure learning

A common test of memory in human subjects involves first showing them a set of photographs, each photograph being presented alone. The experimenter then enlarges the set to include pictures that the subjects have not previously seen. The subject is subsequently shown a photograph, selected from the original set, or a novel photograph and asked whether he/she has seen it before (see Huppert and Piercy, 1976). If the response is correct the subject is considered to recognise the photograph with which, throughprevious exposure he/she is familiar Learning through exposure is probably a very common form of learning in humans and non- human primates (Nickerson, 1965; Overman and Doty, 1980) and indeed in a wide range of animals (see Sluckin, 1972). Exposure learning is not a form of 'developmental plasticity' for although children and the young of many species are capable of learning in this way, such learning is not restricted to the young : exposure learning occurs in adults, and even in a subgroup of patients with senile dementia of the Alzheimer type (Freed et al., 1989).

If an individual recognises an object we infer that the individual has a memory of that object. It has long been supposed - and certainly sinceWilliam James (1890) wrote his Principles of Psychology - that memories consist of 'traces' or 'engrams' left in the brain by previous experiences. If a stimulus leaves its 'mark' in the brain, where is the mark and what is its nature ?

Together with our colleagues, we have attempted to address these questions through a study of imprinting in the domestic chick. Imprinting is the process by which the young of animals (precocial) which show well-coordinated movements soon after birth learn to recognise their mother and other objects in their environment. Soon after hatching, young domestic chicks approach and follow their mother. This response is also elicited by a wide range of visually conspicuous objects especially if they are moving. After a period of exposure to such an object, the chick develops a social attachment to it. When this 'training' or 'imprinting' object is near, the chick emits contentment calls, and approaches it; if the object is moving away, the chick follows it. In addition, instead of approaching other conspicuous objects as it would have done in the naive state, the chick now avoids them (see Spalding, 1873; Lorenz 1937). As a result, when given a recognition test - for example a choice between a novel and a familiar object - the chick approaches the familiar one.

Because imprinting is usually measured by a locomotor approach response, this form of learning has most extensively been studied in precocial species. However, there is no requirement that the underlying learning process must be tied to a locomotor response (see Klopfer and Hailman, 1964; Baer and Gray, 1960). Indeed Sluckin (1972) considered the evidence to be against the view

Excitatory Amino Acids and Neuronal Plasticity
Edited by Y. Ben-Ari
Plenum Press, New York, 1990

that imprinting depends on the act of following. Sluckin and Salzen (1961) have emphasized the perceptual side of this learning process, as have Hinde (1962), Bateson (1966) and Kovach et al. (1966). Sluckin considered imprinting to be an instance of exposure learning, which "refers unambiguously to the perceptual registration by the organism of the environment to which it is exposed" (Sluckin, 1972, p.109; see also Thorpe, 1944). Such learning is not restricted to birds, but, as described above occurs in many animals; and far from being a special form of learning it may be a very common form. Furthermore when there is good performance in a recognition task we may infer that the memory for the previously experienced object, and accordingly the neural representation, is a highly specific one.

A specific brain region is crucially involved in imprinting

In the studies of imprinting described below chicks were trained by exposing them to a visually conspicuous object. Most experiments employed the following procedures (see McCabe et al., 1982). After hatching, chicks were reared in individual compartments in a dark incubator until they were between 15 and 30 h old. The chicks were then placed individually in running wheels, the centre of which was some 50 cm from the imprinting stimulus, the whole apparatus being contained within a large, black box. The chicks were exposed to the stimulus for between 1 and 4 h depending on the experiment. A chick's preferences were usually measured by exposing the chicks to the familiar object and to a novel object in succession and in balanced order. A measure of preference is given by the relative strength of the chick's approach to the familiar objects.

If the storage process underlying imprinting involves changes in the connections between neurones then changes in protein and RNA metabolism may be expected to occur in those regions in which storage takes place. Training was found to be associated with an increase in the incorporation of radioactive lysine into protein and of radioactive uracil into RNA in the dorsal part (forebrain roof) of the cerebral hemisphere (Bateson et al., 1972). The evidence provided by a number of control procedures suggested that the biochemical changes were closely related to the learning process because (i) when visual input was restricted to one cerebral hemisphere by dividing the supraoptic commissure and occluding one eye with a black patch, incorporation was higher in the forebrain roof of the 'trained' hemisphere than the 'untrained' hemisphere (Horn et al., 1973), (ii) the magnitude of incorporation was positively correlated with a measure of how much the chicks had learned, but was not correlated with a variety of other measures of the chicks' performance (Bateson et al., 1975), and (iii) the increase associated with training was not a result of some short-lasting effect of sensory stimulation (Bateson et al., 1973).

Using an autoradiographic technique (Horn and McCabe, 1978), an increased incorporation of radioactive uracil into RNA was found in the intermediate and medial part of the hyperstriatum ventrale (Horn et al., 1979), a region referred to as IMHV. Subsequently part or all of the same region of the medial hyperstriatum ventrale has been reported by other research groups to be involved in visual (Kohsaka et al., 1979) and auditory imprinting (Maier and Scheich, 1983) as well as in passive avoidance learning (Kossut and Rose, 1984; Davies et al., 1988). Bilateral destruction of IMHV prevents the acquisition of a preference for imprinting, and impairs the retention of an acquired preference (McCabe et al., 1981, 1982; see also Takamatsu and Tsukada, 1985). A variety of control procedures provided evidence that the impaired performance of the lesioned chicks in the preference test did not arise out of some disturbance of visuomotor coordination or motivation (McCabe et al., 1981, 1982; Johnson and Horn, 1986; Bolhuis et al., 1989); the most likely explanation of the behaviour of the IMHV-lesioned birds is that they are unable to recognise the training object.

Cellular and subcellular changes in IMHV following learning

Given the above evidence that IMHV plays a crucial role in the recognition process, probably in information storage, a series of studies was undertaken to determine whether or not imprinting led to changes in the structure or frequency of synapses in the region. Bradley et al. (1979, 1981) showed that imprinting led to an increase in the mean length of postsynaptic density (PSD) profiles of synapses in the left IMHV (Fig. la). The increase was restricted to axospinous synapses (Horn

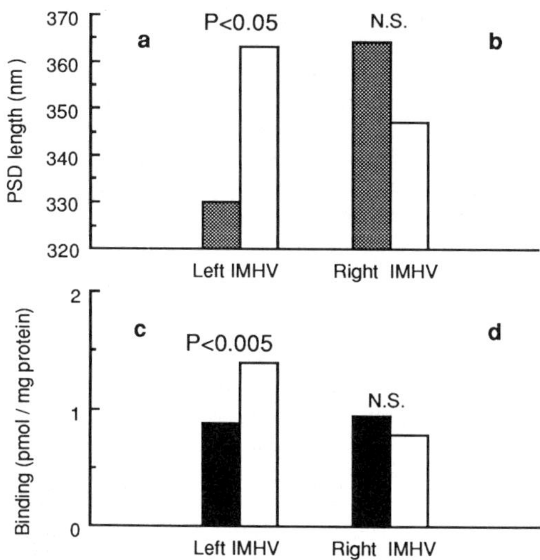

Fig. 1. Effects of training on the length of postsynaptic densities (PSDs) of synapses (a and b), and on NMDA-sensitive binding (c and d) in IMHV. Mean values are shown according to hemisphere and treatment. PSD length (a and b). Data for axospinous and axodendritic synapses are combined. Chicks were either exposed to the imprinting stimulus for 20 min (shaded bars, undertrained chicks) or 140 min (open bars, trained chicks). Further training led to a significant change in length of PSDs in the left IMHV only (After Bradley et al., 1981). NMDA-sensitive bindinq (c and d). Chicks were either dark-reared (filled bars) or trained (open bars) through exposure to the imprinting stimulus for 140 min. Training led to a significant change in NMDA-sensitive binding in the left IMHV only. (After McCabe and Horn, 1988).

et al., 1985). There was no significant effect of imprinting on the PSD length of synapses on dendritic shafts in the left IMHV nor on axospinous or dendritic shaft synapses in the right IMHV.

There is strong evidence that at least some axospinous synapses in the mammalian central nervous system are excitatory and possess receptors for L-glutamate (Nafstad, 1967; Errington et al., 1987). Accordingly we enquired (i) whether imprinting is associated with changes in the binding of L-[H]glutamate to membranes from the IMHV, (ii) whether imprinting is associated with 3changes in the N-methyl-D-aspartate (NMDA)-sensitive component of L-[H]glutamate binding and (iii) whether the left and right sides of the IMHV differ in this respect (McCabe and Horn, 1988).

Full details of the method of rearing and training chicks are given elsewhere (Bradley et al., 1981; see also first section above). Briefly the chicks were placed individually in running wheels and exposed to a rotating red box for a total of 140 min. All chicks were then assigned codes and held individually in darkness for at least 7 h before being decapitated. After coding all procedures were performed without knowledge of the chick's experimental history. Immediately after decapitation the left and right IMHV regions were dissected out (IMHV is a small piece of tissue, weighing approximately 2 mg). Details of the methods used for preparing membranes, the binding assays and the methods of statistical analysis are described by McCabe and Horn (1988).

L-glutamate sensitive binding was saturable with an apparent $Kd = 0.5 \pm 0.09$ μM. In this study only binding in the left IMHV was measured since our prior expectation, based on the electron microscope study, was that changes would be expected only in the left IMHV. L-glutamate-sensitive binding accounted for $61 \pm 1.7\%$ of total radioactivity. Binding in the trained chicks was $20 \pm 6.7\%$ greater than in the dark-reared chicks ($P < 0.01$). NMDA-sensitive binding was measured from the left IMHV of 96 trained and 99 dark-reared chicks. Measurements were made on samples derived from 3 chicks, and so were made on 32 and 33 samples for trained and dark-reared chicks respectively. Binding in the right IMHV was measured in a sub-sample of this population namely 22 samples from trained and 22 from dark-reared chicks. The mean NMDA-sensitive binding for all groups are shown in Fig. 1 c and d. NMDA-sensitive binding in the left IMHV was significantly higher, by 59 + 18%, in trained chicks than in dark-reared chicks. There were no signiflcant effects of training in the right IMHV. This pattern of results closely resembles that found in the first morphological studies of the effects of training on PSD lengths (Fig. la,b). In those studies (Bradley et al., 1979, 1981) one group of chicks was trained for 140 min, the other for 20 min. Exposure to the training object for the longer period of time leads to a significant preference for the training object, whereas training for the shorter period of time is inadequate to establish such a preference (Bateson, 1979). In a subsequent electron microscope study which involved a group of dark-reared birds as well as chicks trained for 20 min it was found that the mean PSD lengths of these 2 groups did not differ significantly. Thus, whereas the results shown in Fig. la and b are for PSD lengths pooled for both axospinous and axodendritic synapses, only the contributions for the axospinous synapses was found significantly to be affected by training (Horn et al., 1985).

Under the incubation conditions used in this study (50 mM Tris acetate buffer without Na+ Cl or Ca2+), the NMDA-sensitive binding sites correspond to physiological receptors (Fagg and Matus, 1984; Monaghan and Cotman, 1986; Foster and Fagg, 1987). This view is further supported by the finding of rapid association and disassociation of L-[3H]glutamic acid binding at the NMDA-sites, the pharmological profile of which was found to be characteristic of NMDA receptors (Olverman et al., 1984). The changes in NMDA-sensitive binding occur in a region of the brain which, as described above, plays a crucial role in the memory underlying imprinting (see also Cipolla-Neto et al., 1982; Horn, 1985). Accordingly there is an *a priori* hypothesis that the changes in NMDA-sensitive binding are related to the storage process. It remains possible, however, that these changes are some indirect consequence of the learning process especially since of the two groups that were compared, one was trained and hence visually experienced whereas the untrained controls were visually naive having been reared in darkness.

In a previous study of imprinting it was found that, in a group of 106 chicks all of which were exposed to the training object for a given, fixed length of time, there was a positive correlation between the strength of preference for the training object and the amount of [14C]-uracil incorporated into RNA in the forebrain roof (Bateson et al., 1975). To determine whether a similar correlation existed between preference score and NMDA-sensitive binding, 108 chicks were trained as described above. Approximately 8 h later the chicks were given a preference test after which they were killed, IMHV removed and MMDA-sensitive binding measured. Since the IMHVs of 3 chicks were combined in a sample, the 3 birds contributing to the sample were matched for preference scores. There was a significant positive correlation between preference score and NMDA-sensitive binding in the left IMHV ($r = 0.38$, $P < 0.012$, 1-tailed test). The corresponding correlation coefficient for the NMDA-sensitive binding in the right IMHV was not significant ($r = 0.10$). Approach during training is weakly correlated with preference score (Bateson and Jaeckel, 1974). It was possible therefore that the observed correlation between NMDA-sensitive binding and preference score reflected a relationship between binding and locomotor activity during training. Binding and preference scores were therefore corrected for training approach activity, using the method of partial correlation. Training approach activity was measured as the number of revolutions made by the running wheel as a chick attempted to approach the red box during the training period. There was a positive partial correlation between NMDA-sensitive binding in the left IMHV and preference score ($r = 0.41$; $P < 0.01$, 1-tailed test). The 36 values that contributed to thYis correlation were divided into 3 groups of equal size according to their ranked corrected preference scores. The mean corrected binding and preference scores together with the standard errors were calculated for each group of 12 values. The lowest mean corrected preference score (of the 'poor learning' group of chicks) was not significantly different from 50, the chance level of performance. The corresponding mean binding for this group (i) was not significantly different from that in the left IMHV of the dark-reared chicks of the previous study (see Fig. lc) and (ii) was significantly less than the group ('good learners') with the highest mean preference score. The partial correlation coefficient between NMDA-sensitive binding in the right IMHV and preference score was not significant ($r_{xyz} = 0.09$).

The apparent K_d of the NMDA-sensitive binding of L-glutamate was 0.1 µM. Accordingly, the binding would be expected to be virtually saturated at the concentration of the radioligand used in the above studies (l µM). The increase in binding in the left IMHV after imprinting (Fig. lc) is therefore likely to reflect an increase in the number of available binding sites, although we have not exluded the possibility that there is also a change in the affinity of these sites. These studies do not, of course, preclude the possibility of changes in L-glutamate receptors of the non-NMDA-type.

There are a number of reasons why the increase in NMDA-sensitive binding cannot simply be attributed to some side-effects of the training procedure or of accelerated maturation (Dudai, 1989). Firstly it is unlikely that the increase is some consequence of a general arousal response since such a response is likely to affect both the right and left IMHV. An effect of arousal, and indeed of accelerated maturation would also be expected to be expressed in behaviour. For example, the more aroused or the more developmentally mature the chicks, the more vigorously would they be expected to approach the red box during training. However the partial correlation coefficient between NMDA-sensitive binding and preference score was significant when the effect of approach activity during training was held constant. This latter finding also demonstrates that differences in loco-motor activity during training cannot account for the correlation between binding and corrected preference score. Secondly, light exposure as such does not account for the findings, since the corrected mean binding in the left IMHV of chicks that had been exposed to the red box for 140 min, but had not developed a preference for it, was closely similar to that in the the left IMHV of dark-reared chicks. Together these considerations suggest that NMDA-sensitive binding in the left IMHV is not influenced by maturation, arousal, light exposure or locomotor activity per se. Instead, the results suggest either (a) that birds with larger numbers of NMDA receptors learn better than birds with fewer NMDA receptors or (b) that binding increases as the chicks learn about the imprinting object and so form a preference for it. Certain considerations make it possible to distinguish between these hypotheses. The dark-reared birds were unselected for their learning abilities. It is reasonable to suppose, therefore, that this group contained poor learners which would if trained achieve a low preference score, as well as good learners which would if trained achieve a high preference score. Hence if hypothesis (a) is correct the variance of NMDA-sensitive binding in the left IMHV of the dark-reared chicks should be higher than that of both the poor learners and the good learners since these two groups are *ex-hypothesi* sub-groups of the unselected dark-reared chicks. The variances for the left IMHV binding from four groups of chicks, two dark-reared groups, the poor and the good learners, did not differ significantly (Bartlett's Test, x = 5.59; d.f. = 3; P > 0.1). Therefore hypothesis (a) may be rejected in favour of the view that learning leads to an increase in the number of NMDA-type receptors in the left IMHV. The functional significance of the increase is not known. Nor do we know whether the increase remains stable or is transient - enabling say a relatively large influx of Ca2+ to trigger intracellular events which are expressed as morphological changes in the postsynaptic cell - subsequently returning to a lower level.

Horn and McCabe (in prep.) have recently completed a study of the time-course of the change in NMDA-sensitive binding up to some 8.5 h after training. Chicks were trained for 140 min as before and killed approximately O or 3 or 6 or 8.5 h afterwards. As before, there was a significant increase in binding at 8.5 h; but there were no significant changes at 0, 3 and 6 h.

These results immediately pose problems for the view that the increase in number of NMDA receptor binding sites observed in the left IMHV some 8.5 h after training (McCabe and Horn, 1988) are crucial for learning. After all if chicks are given a preference test say 3 h after training, they prefer the training object to a novel one, just as they may do some 8 h after training (e.g. see control chicks in Cipolla-Neto et al., 1982). The new findings imply that the increase in NMDA receptor numbers observed in the left IMHV after imprinting cannot support the preference when it is tested 3 h after training, since there is no such increase; clearly a different mechanism must be invoked to account for the neural basis of a preference measured soon after training. The possibility that at least two mechanisms are involved in memory is not wholly surprising: different pharmacological agents have been found to disrupt retention when given at different times after training (for review see Andrew, 1980). A clue as to what might be happening comes from studies of long-term potentiation.

Recognition memory and long-term potentiation

Long-term potentiation (LTP) was first reported by Lømo (1971), Bliss and Lømo (1973) and Bliss and Gardner-Medwin (1973). Trains of shocks delivered to the perforant path lead to a long-lasting potentiation of transmission at the dentate gyrus. Although LTP has most extensively been studied in the hippocampus of various species, LTP-like changes in synaptic transmission have been demonstrated *inter-alia* in the cat medial geniculate body (Gerren and Weinberger, 1983), the rat neocortex (Artola and Singer, 1987; Bindman et al., 1987) and superior cervical sympathetic ganglion (Briggs et al., 1985), and at the neuromuscular junction of the crayfish opener muscle (Baxter et al., 1985). The diversity of structures in which LTP occurs makes it unlikely that it has a single function.

Since LTP ensues if a train of shocks is delivered to presynaptic fibres, the after-effect may in some sense be regarded as a 'memory' of the train of shocks. But whether LTP has a role to play in 'learning-memory' is another matter. The assumption that hippocampal LTP has a role in learning-memory rests, in part, on the finding that bilateral removal of the hippocampal region (with other parts of the medial temporal lobe) in humans, or degeneration of neurones in part of the region results in a profound anterograde amnesia with a variably short retrograde amnesia (B. Milner et al., 1968; Zola-Morgan et al., 1986). Such amnesics incidentally are very poor at recognition memory (see Baddeley 1982). Whilst it seems beyond dispute that the region plays **some** role in memory processes, just what that role is remains uncertain. There is no evidence that the hippocampus is a repository for long-established memories in humans (B. Milner, 1972; B. Milner et al., 1968; Zola-Morgan et al., 1986) or in rats (McNaughton et al ., 1986), so it is plausible to suppose that the brain region is somehow involved in processing information to be stored permanently elsewhere, possibly in the neocortex (see for example Griffith, 1966; Warrington and Weiskrantz, 1982; Mishkin, 1982; Bloch and Laroche, 1985; Teyler and DiScenna, 1986; P. Milner, 1989).

Hippocampal LTP may be established by delivering some 100-200 shocks given in one or more trains at frequences of 100-400 Hz (see Teyler and DiScenna, 1987; McNaughton, 1982). Such stimuli are likely to generate high frequency, synchronous volleys of impulses which may rarely occur under physiological conditions. Indeed, in one study of impulse activity recorded from units in various parts of the hippocampal region of unanaesthetised cats, the median background firing rate was 1.6 spikes/sec (Brown and Horn, 1977). The possibility must therefore be considered that the inferred massive discharge generated by the high frequency stimulation used for establishing hippocampal LTP may 'uncover' properties of that are not normally expressed in this region of the brain. These properties may, however, normally be expressed in other brain regions in which long-term storage takes place (see below).

Whatever the functions of hippocampal LTP prove to be in the economy of the nervous system and in behaviour the phenomenon is of great interest because it shows that transmission between neurones can be modified for relatively long periods of time, and because much is known of the cellular and subcellular mechanisms involved. Whilst these mechanisms may be different at different synapses the mechanisms involved in certain forms of hippocampal LTP are becoming clear (for review see Collingridge and Bliss, 1987; Teyler and DiScenna, 1987). Briefly, the **induction** of hippocampal LTP is considered to involve pre- and postsynaptic elements. Trains of impulses in the presynaptic fibres lead to the release of L-glutamate from their terminals. This neurotransmitter binds to two L-glutamate receptors those of the NMDA and quisqualate types. When activated in this way, the ion channels associated with the quisqualate receptors open leading to a depolarisation of the postsynaptic membrane. This depolarisation results in the opening 2°f the voltage-gated NMDA channels through which there is an influx of Ca + into the postsynaptic cell. In ways as yet unknown it is thought that a signal generated in the dendritic spine feeds back to the presynaptic bouton (cf. Griffith 1966; see p. 224 of Horn 1985; Bliss et al ., 1986) . The putative signal leads to changes in the presynaptic bouton such that, at the dentate gyrus, there is an increased release of L-glutamate when an impulse invades the terminal (see Bliss et al., 1986). This increased release of transmitter may also be implicated in the **maintenance** of LTP, although additional mechanisms are involved. Davies et al. (1989) studied hippocampal slices after the induction of LTP through stimulation of the Schaffer collateral-commissural pathway. Within 15 min after the induction of LTP the sensitivity of CAl neurones to iontophoretically-applied quisqualate receptor ligands increased, reaching a maximum after some 2 h. It is not known whether the increased sensitivity of the postsynaptic cells to these ligands is a consequence of an increased affinity and/or of an increased number of quisqualate receptors. There is no evidence that NMDA receptors play a role in the short-term maintenance of LTP; but the question of whether they become involved after, say, 8.5 h (cf imprinting) is an open one. Finally, LTP in the rat dentate gyrus leads to morphological changes in axospinous synapses (Desmond and Levy, 1986). These changes include an increase in the length of the postsynaptic density on certain dendritic spines.

There are some admittedly rather crude parallels between the cellular changes which follow the induction of hippocampal LTP and those in the left IMHV which follow imprinting. Both involve changes in the length of the PSDs of axospinous synapses. Imprin-ting leads to an increase in NMDA-sensitive binding which may reasonably be accounted for through an increased number of L-glutamate receptors of the NMDA type; LTP in the Shaffer collateral-commissural system leads to an increased sensitivity, possibly to an increased number of L-glutamate receptors, but of the quisqualate type. In both imprinting and LTP the changes in receptors do not occur immediately after the experience (that is training on the one hand or shocking the presynaptic fibres on the other), but appear gradually over a period of hours. If at the Schaffer collateral-commissural pathways the initial phase of LTP is maintained by an increased release of transmitter, it is plausible that the ligand itself may regulate the number of receptors. Such control is at least suggested by certain other studies. Thus Handleman et al. (1987) have shown that if rat pups are given daily injections of substance P on days 1 through 7 of postnatal life it is found, on the basis of autoradiographic studies, that there is an increased number of receptors for that peptide at the locus coeruleus and dorsal raphe nuclei when the rats are studied at the age of 2 40 days of age. These results suggests that for substance P the number of receptors may be determined by the amount of ligand present over a certain period of development. The results also raise the more general possibility that at synapses which are modifiable beyond that period, the transmitter released by the presynaptic terminal may upregulate the number of receptors in the postsynaptic membrane.

The parallels between the cellular changes associated with imprinting and LTP respectively raise the question of whether the early phases of memory in imprinting are sustained by presynaptic events, as they are in LTP. Although there is no direct evidence for this possibility in the case of imprinting, there is some evidence of a presynaptic involvement for another form of learning in the domestic chick. Stewart et al. (1984) undertook an electron microscope study of changes in IMHV following passive avoidance training, measuring aspects of synapse morphology similar to those measured by Bradley et al. (1981). In addition they counted the number of vesicles in synaptic boutons. Stewart et al. (1984) found PSD changes similar to those which follow imprinting. They also found that training was associated with a substantial and highly significant increase in the mean number of vesicles per synaptic bouton in the left IMHV. Such an increase would be consistent with an increased mobilisation of transmitter as a consequence of training, and raises the possibility that when such terminals are invaded by a nerve impulse more transmitter is liberated after training than before.

In the case of LTP the changed sensitivity of CA1 neurones increases and reaches a maximum after some 2 h. A clear increase in NMDA-sensitive binding occurs approximately 8.5 h after a 140 min period of exposure to the imprinting stimulus. Several factors may have to be taken into account when considering these different time-courses. Thus the LTP studies were conducted on brain slices *in vitro*, whereas the studies of imprinting were conducted on chicks that were trained before NMDA-sensitive binding was measured. In addition, with the stimuli used in our studies of imprinting, 2 30 min of training are required before the chick forms a preference for the imprinting object. Furthermore during the first few hours after hatching chicks appear to sleep a great deal of the time that they are in the incubator. When they see the imprinting object a whole variety of physiological processes must be engaged, as the chicks become aroused and attentive to the training object, and engaged for visual learning for the first time in the bird's life. In the light of all these factors it is not perhaps too surprising that the changes in NMDA-sensitive binding occur relatively slowly. These and other differences between hippocampal LTP and imprinting (e.g. changes in quisqualate receptors in the former and in NMDA receptors in the latter) are worth as much experimental attention as are the similarities, even though IMHV is upstream of the hippocampus.

Many workers in the field (see e.g. Bliss and Lømo, 1973; Teyler and DiScenna, 1987) would still probably agree that the significance of hippocampal LTP for learning requires further experimental clarification. Nonetheless, the cellular and molecular mechanisms of synaptic plasticity that the study of hippocampal LTP has revealed may be similar to those involved in the recognition memory of imprinting.

ACKNOWLEDGEMENTS

We are grateful to Barrie Puller for technical assistance, to Chris Percival for secretarial assistance and to the SERC for financial support.

REFERENCES

Andrew, R.J., 1980, The functional organisation of phases of memory consolidation, **Adv. Study Behav.**, 11:337-67.

Artola, A. and Singer, W., lg87, Long-term potentiation and NMDA receptors in rat visual cortex, **Nature Lond.**, 330:649-52.

Baer, D.M. and Gray, P.H., 1960, Imprinting to different species without overt following, **Percept. Mot. Skills**, 10:171-174.Baddeley, A.D., 1982, Implications of neuropsychological evidence for theories of normal memory, In The Neuropsycholoqy of Cognitive Function, D.B. Broadbent and L. weiskrantz, Eds., The Royal Society, London.

Bateson, P.P.G., 1966, The characteristics and context of imprinting, **Biol. Rev.** 41:177-220.

Bateson, P., 1979, Brief exposure to a novel stimulus during imprinting in chicks and its influence on subsequent preferences, **Anim. Learning Behav.**, 7:259-262.

Bateson, P.P.G., Horn, G. and Rose, S.P.R., 1972, Effects of early experience on regional incorporation of precursors into RNA and protein in the chick brain, **Brain Res.**, 39:449-65.

Bateson, P.P.G., Horn, G. and Rose, S.P.R., 1975, Imprinting: correlations between behaviour and incorporation of [l4C]uracil into chick brain, **Brain Res.**, 84:207-20.

Bateson, P.P.G. and Jaeckel, J.B., 1974, Imprinting: correlations between activities of chicks during training and testing, **Anim. Behav.** 22:899-906.

Bateson, P.P.G., Rose, S.P.R. and Horn, G., 1973, Imprinting: lasting effects on uracil incorporation into chick brain, **Science**, 181:576-78.

Baxter, D.A., Bittner, G.D. and Brown, T.H., 1985, Quantal mechanism of long-term synaptic potentiation, **Proc. Natl. Acad. Sci. U.S.A.**, 82:5978-82.

Bindman, L.J., Meyer, T. and Pockett, S., 1987, Long-term potentiation in rat neocortical neurones in slices, produced by repetitive pairing of an afferent volley with intracellular depolarizing current, **J. Physiol.**, 386:90P.

Bliss, T.V.P., Douglas, R.M., Errington, M.L. and Lynch, M.A., 1986, Correlation between long-term potentiation and release of endogenous amino acids from dentate gyrus of anaesthetised rats, **J. Physiol.**, 377:391-408.

Bliss, T.V.P. and Gardner-Medwin, A.R., 1973, Long-lasting potentiation of synaptic transmission in the dentate area of the unanaesthetised rabbit following stimulation of the perforant path. **J. Physiol.**, 232:357-74.

Bliss, T.V.P. and Lømo, T., 1973, Long-lasting potentiation of synaptic transmission in the dentate area of the anaesthetised rabbit following stimulation of the perforant path, **J. Physiol.**, 232:331-56.

Bloch, V. and Laroche, S., 1985, Enhancement of long-term potentiation in the rat dentate gyrus by post-trial stimulation of the reticular formation, **J. Physiol.**, 360:215-31.

Bolhuis, J., Johnson, M., Horn, G. and Bateson, P., 1989, Long-lasting effects of IMHV lesions on social preferences in domestic fowl. **Behav. Neurosci.**, 103:438-44.

Bradley, P., Horn, G. and Bateson, P., 1979, Morphological correlates of imprinting in the chick brain, **Neurosci. Lett. Suppl.** 3:S84.

Bradley, P., Horn, G. and Bateson, P., 1981, Imprinting: an electron microscopic study of chick hyperstriatum ventrale, **Exp. Brain Res.**, 41:115-20.

Briggs, C.A., Brown, T.H. and McAfee, D.A., 1985, Neurophysiology and pharmacology of long-term potentiation in the rat sympathetic ganglion, **J. Physiol.**, 359:503-521.

Brown, M.W. and Horn, G., 1977, Responsiveness of neurones in the hippocampal region of anaesthetised and unanaesthetised cats to stimulation of sensory pathways, **Brain. Res.**, 1977, 123:241-259.

Cipolla-Neto, J., Horn, G. and McCabe, B.J., 1982, Hemispheric asymmetry and imprinting: the effect of sequential lesions to the hyperstriatum ventrale, **Exp. Brain Res.**, 48:22-27.

Collingridge, S.L. and Bliss, T.V.P., 1987, NMDA receptors - their role in long-term potentiation, **Trends in Neurosci.**, 10:288-293.

Davies, D.C., Taylor, D.A. and Johnson, M.H. (1988), The effects of hyperstriatal lesions on one-trial passive avoidance learning in the chick. **J. Neuroscience** 8:4662-4666.

Davies, S.N., Lester, R.A.J., Reyman, K.G. and Collingridge, G.L., 1989, Temporally distinct pre- and post-synaptic mechanisms maintain long-term potentiation, **Nature**, 338:500-503.

Desmond, N.L. and Levy, W.B., 1986, Changes in the postsynaptic density with long- potentiation in the dentate gyrus, **J. comp. Neurol.**, 253:476-482.

Dudai, Y., (1989) **The Neurobiology of Memory. Concepts Findings Trends.** Oxford University Press.

Errington, M.L., Lynch, M.A. and Bliss, T.V.P., 1987, Long-term potentiation in the dentate gyrus: induction and increased glutamate release are blocked by D( )aminophonovalerate, **Neurosci.**, 20:279-84.

Fagg, G.E. and Matus, A., 1984, Selective association of N-methyl aspartate and quisqualate types of L-glutamate receptor with postysnaptic densities, **Proc. Natl. Acad. Sci. U.S.A.**, 81:6876-6880.

Foster, A.C. and Fagg, G.E., 1987, Comparison of L-[H]glutamate, D-[H]aspartate, DL-[3H]AP5 and [H]NMDA as ligands for NMDA receptors in crude postsynaptic densities from rat brain, **Eur. J. Pharmacol.**, 133:291-300.

Freed, D.M., Corkin, S., Growden, J.H. and Nissen, M.J., 1989, Selective attention in Alzheimer's disease: characterizing cognitive subgroups of patients, **Neuropsychologia**, 27:325-339.

Gerren, R.A. and weinberger, N.M., 1983, Long-term potentiation in the magnocellular medial geniculate body of the anaesthetised cat, **Brain Res.**, 265:138-42.

Griffith, J.S., 1966, A theory of the nature of memory, **Nature**, Lond., 211:1160-63.

Handelmann, G.E., Shults, C.W. and O'Donohue, T.L., 1987, A developmental influence of substance P on its receptor, **Int. J. Dev. Neurosci.**, 5 : 11-16.

Hinde, R.A., 1962, Some aspects of the imprinting problem, **Symp. Zool. Soc. Lond.** 8:129-38.

Horn, G., 1985, **Memory, Imprintinq and the Brain**, Clarendon Press, Oxford.

Horn, G., Bradley, P. and McCabe, B.J., 1985, Changes in the structure of synapses associated with learning, **J. Neurosci.**, 5:3161-3168.

Horn, G. and McCabe, B.J., 1978, An autoradiographic method for studying the incorporation of uracil into acid-insoluble compounds in the brain. **J. Physiol.**, London, 275: 2-3P.

Horn, G. and McCabe, B.J., 1989, Time-dependent changes in N-methyl-D-aspartate-sensitive binding following learning. In prep.

Horn, G., McCabe, B.J. and Bateson, P.P.G., 1979, An autoradiographic study of the chick brain after imprinting, **Brain Res.**, 168:361-373.

Horn, G., Rose, S.P.R. and Bateson, P.P.G., 1973, Monocular imprinting and regional incorporation of tritiated uracil into the brains of intact and 'split-brain' chicks. **Brain Res.**, 56:227-37.

Huppert, F.A. and Piercy, M., 1976, Recognition memory in amnesic patients: effect of temporal context and familiarity of material, **Cortex**, 12:3-20.

James, W., 1890, **Principles of Psychology**, Henry Holt, New York. Reprinted in 1950, Dover Publications, New York.

Johnson, M.H. and Horn, G. (1986), Dissociation of recognition memory and associative learning by a restricted lesion of the chick forebrain, **Neuropsychologia**, 24:329-40.

Klopfer, P.H. and Hailman, J.P., 1964, Perceptual preferences and imprinting in chicks, **Science**, 145:1333-1334.

Kohsaka, S, Takamatsu, K. Aoki, E. and Tsukada, Y. (1979), Metabolic mapping of chick brain after imprinting using [l4C]2-deoxyglucose technique, **Brain Res.** 172: 539-44.

Kossut, M. and Rose, S.P.R., (1984), Differential 2-deoxyglucose uptake into chick brain structures during passive avoidance training, **Neuroscience,** 12:971-77.

Kovach, J.K., Fabricius, E. and Fält, L., 1966, Relationship between imprinting and perceptual learning, **J. Comp. Physiol. Psychol.**, 61:449-454.

Lømo, T., 1971, Patterns of activation in a monosynaptic cortical pathway: the perforant path input to the dentate area of the hippocampal formation, **Exptl. Brain Res.**, 12:18-45.

Lorenz, K.Z., 1937, The companion in the bird's world, **Auk**, 57:245-273.

McCabe, B.J., Cipolla-Neto, J., Horn, G. and Bateson, P.P.G., 1982, Amnesic effects of bilateral lesions placed in the hyperstriatum ventrale of the chick after imprinting, **Exp. Brain Res.**, 48:13-21.

McCabe, B.J. and Horn, G., 1988, Learning and memory: regional changes in N-methyl-D-aspartate receptors in the chick brain after imprinting, **Proc. Natl. Acad. Sci. U.S.A.**, 85:2849-53.

McCabe, B.J., Horn, G. and Bateson, P.P.G., 1981, Effects of restricted lesions of the chick forebrain on the acquisition of filial preferences during imprinting, **Brain Res.** 205:29-37.

McNaughton, B.L., 1982, Long-term synaptic enhancement and short-term potentiation in rat fascia dentata act through different mechanisms, **J. Physiol.**, 324:249-262.

McNaughton, B.L., Barnes, C.A., Rao, G., Baldwin, J. and Rasmussen, M., 1986, Long-term enhancement of hippocampal synaptic transmission and the acquisition of spatial information, **J. Neurosci.**, 6:563-571.

Maier, V. and Scheich, H. (1983), Acoustic imprinting leads to differential 2-deoxy-D-glucose uptake in the chick forebrain, **Proc. Natl. Acad. Sci. U.S.A.**, 80:3860-64.

Milner, B., 1972, Disorders of learning and memory after temporal lobe lesions in man, **Clin. Neurosurq.**, 19:421-446.

Milner, B., Corkin, S. and Teuber, H.L., 1968, Further analysis of the hippocampal amnesia syndrome: 14-year follow-up study of H.M., **Neuropsychologia**, 6:215-234.

Milner, P.M., 1989, A cell assembly theory of hippocampal amnesia, **Neuropsychologia**, 27:23-30.

Mishkin, M., lg82, A memory system in the monkey, In **The Neuropsycholoqy of Cognitive Function**, D.E. Broadbent and L. Weiskrantz, eds., The Roya Society, London.

Monaghan, D.T. and Cotman, C.W., 1986, Identification and properties of NMDA receptors in rat brain synaptic plasma membranes, **Proc. Natl. Acad. Sci. U.S.A.**, 83:7532-36.

Nafstad, P.H.J., 1967, An electron microscope study of the termination of the perforant path fibers in the hippocampus and the fascia dentata, **Z. zellforsch. Mikrosk. Anat.**, 76:532-42.

Nickerson, R.S., 1965, A note on long-term recognition for pictorial material, **Psychon. Sci.**, 11:58.

Overman, Jr., W.H. and Doty, R.W., 1980, Prolonged visual memory in macaques and man, **Neurosci.**, 5:1825-1831.

Olverman, H.J., Jones, A.W. and Watkins, J.C., 1984, L-glutamate has higher affinity than other amino acids for [3H]-D-AP5 binding sites in rat brain membranes, **Nature**, 307:460-62.

Sluckin, W., 1972, **Imprinting and Early Learninq**, Methuen, London.

Sluckin, W. and Salzen, E.A., 1961, Imprinting and perceptual learning, **Ouart. J. exp. Psychol.**, 1961, 13:65-77.

Spalding, D.A., 1873, Instinct, with original observations on young animals, **Macmillan's Magazine**, 27:282-293. Reprinted in 1954, **Br. J. Anim. Behav.**, 2:2-11.

Stewart, M.G., Rose, S.P.R., King, T.S., Gabbot, P.L.A. and Bourne, R., 1984, Hemispheric asymmetry of synapses in chick medial hyperstriatum ventrale following passive avoidance training: a stereological investigation, **Devl. Brain Res.**, 12:261-69.

Takamatsu, K. and Tsukada, Y. (lg8S), Neurobiological basis of imprinting in chick and duckling, In **Perspectives on Neuroscience: From Molecule to Mind**, Y. Tsukada, ed., pp. 187-206, Springer Verlag, Berlin.

Teyler, T.J. and DiScenna, P., 1986, The hippocampal memory indexing theory, **Behav. Neurosci.**, 100:147-154.

Teyler, T.J. and DiScenna, P., 1987, Long-term potentiation, **Ann. Rev.** Neurosci., 10:131-161.

Thorpe, W.H., 1944, Some problems of animal learning, **Proc. Linn. Soc. Lond.**, 156:70-83.

Warrington, E.K. and Weiskrantz, L., 1982, Amnesia: a disconnection syndrome? **Neuropsycholoqia**, 20:233-48.

Zola-Morgan, S., Squire, L.R. and Amaral, D.G., 1986, Human amnesia and the medial temporal region: enduring memory impairment following a bilateral lesion limited to field CAl of the hippocampus, **J. Neurosci.**, 6:2950-2967.

THE ROLE OF THE NMDA RECEPTOR IN THE DEVELOPMENT OF THE FROG VISUAL

SYSTEM

Hollis T. Cline, Elizabeth A. Debski and Martha Constantine-Paton

Department of Biology, Yale University, New Haven, CT. 06511

Recent work has demonstrated an involvement of the NMDA receptor, a type of glutamate-sensitive receptor, in organizing the developing visual systems of frogs, fish and kittens (Cline et al, 1987; Kleinschmidt et al, 1987; Fox and Fraser, 1987; Tsumoto et al, 1987; Scherer and Udin, 1988; Schmidt, 1988; Cline and Constantine-Paton, 1989; Fox et al, 1989). It is thought that the NMDA receptor activation may be an initial cellular event in the experience-dependent phases of visual development, which include the formation of both topographic maps and ocular dominance columns. In this chapter, we will review our electrophysiological and anatomical data which demonstrate that NMDA receptor activation is crucial for the development of the retinotectal projection of the frog *Rana pipiens* and we will end with a discussion of the influence of the NMDA receptor in neuronal growth.

The frog visual system

The retina of the Rana pipiens tadpole sends an entirely crossed projection to the contralateral optic tectum, the primary visual area in the brain. The retinal projection to the tectum is organized topographically, as are the visual projections of every animal studied thus far. In a topographic projection the two-dimensional organization of the sensory cell bodies are reiterated in the organization of their axon terminals within the central nervous system (CNS). This means that neighboring regions of visual space activate neighboring ganlgion cells within the retina, which in turn project their axon terminals to neighboring spots within the tectal neuropil. Numerous studies using tetrodotoxin (TTX) to block afferent activity, or strobe light regimes to artificially synchronize the afferent activity, have demonstrated that the development or regeneration of the topographic projection requires normal patterns of electrical activity in the visual pathway (reviewed in Constantine-Paton et al, 1990). Further studies have shown that neighboring retinal ganglion cells (RGCs) of the same response type exhibit a high degree of temporally correlated activity compared to non-neighboring RGCs (Arnett, 1978; Mastronarde, 1983). Therefore, a possible means of communicating neighbor relations from the peripheral retinal ganglion cell bodies to the terminals in the optic tectum is through their patterns of electrical activity. It is thought that a high fidelity topographic projection would arise if coactive synapses of neighboring RGCs were selectively stabilized at the expense of non-coactive synapses from non-neighboring RGCs.

What cellular mechanism could detect coactivity in multiple afferent synapses and initiate a stabilization process selective to those coactive synapses?

It is widely known that the NMDA receptor is associated with a ligand-gated and voltage-sensitive channel which conducts the second messenger calcium only when two conditions

Excitatory Amino Acids and Neuronal Plasticity
Edited by Y. Ben-Ari
Plenum Press, New York, 1990

are met: neurotransmitter must bind to the receptor site at the same time that the postsynaptic membrane is significantly depolarized (Mayer & Westbrook, 1987, Ascher & Nowak, 1987 for reviews). Therefore the NMDA receptor/channel is capable of detecting afferent coactivity and transmitting an intracellular signal indicative of that coactivity.

Our experiments rely on the constant mobility of the RGC terminal arbor in the rapidly growing tadpole and the fact that there must be a mechanism to optimize the positions of new synapses with respect to the coactivity in neighboring synapses. The RGC terminals shift their positions along the tectum's rostrocaudal axis as the animal grows (Reh and Constantine-Paton, 1984). It is thought that the retinotectal synaptic connections are shortlived and that the arbors shift in a topographically ordered fashion. This means that the mechanisms that maintain topographic order as the synapses are broken and remade must be continuously active throughout the period of larval growth. In this context, our experiments have tested the ability of the growing retinotectal system to maintain coordinated migration of neighboring RGC terminals in the presence of NMDA receptor antagonists.

We have found that blocking the NMDA receptor, by chronically treating the optic tectum of tadpoles with NMDA receptor antagonists, disrupts the mechanism which sorts the afferent terminals according to their neighbor relations (Cline et al, 1987; Cline & Constantine-Paton, 1989). We have used two anatomical assays for retinotopic organization. One is retrograde labeling from the tectum to the retina in two-eyed tadpoles. In these experiments, we assessed the refinement of the retinotectal map from the degree of scatter of retinal ganglion cell bodies labeled retrogradely with horseradish peroxidase (HRP) from a controlled injection site in the tectum. With this procedure, a refined map is seen as a low degree of scatter in retrogradely labeled cell bodies, because ganglion cells from a local region of the retina terminate exclusively in a local region of the tectum. Conversely, a crude map is detected as a high degree of scatter in the labeled cell bodies because ganglion cells from a large portion of the retina are represented in each local region of the tectum.

Untreated tadpoles displayed a uniform dispersion of labeled calls, which was confined to about 5% of the retinal area. Chronic treatments with either DL-APV or a mixture of MK801 and NMDA, resulted in a widely distributed population of retrogradely labeled ganglion cells, covering 17% and 10% of the retinal area, respectively. Treatments with either L-APV, the inactive isomer of APV, or NMDA alone do not alter the fidelity of the retinotectal projection. It is possible that the disrupted topography was due to the enlargement of individual RGC terminals, comparable to the effect of TTX (Reh & Constantine-Paton, 1985). However, a detailed morphological analysis of the RGC terminals arbors indicated that neither APV- nor NMDA-treated arbors differed from untreated arbors with respect to their tangential area, branch number or branch density.

The second assay we used to monitor the activity-dependent sorting mechanism was the maintenance of eye-specific stripes in the three-eyed tadpole (Cline et al, 1987). A stereotyped striped pattern of eye-specific afferent terminal segregation can be induced by implanting a supernumerary eye primordium into the forebrain region of early *Rana* embryos (Law & Constantine-Paton, 1981). The projection of the supernumerary retina onto the tectal lobes, visualized by labeling the supernumerary optic nerve with HRP, forms a reproducible pattern of alternating rostrocaudally-directed stripes of afferent termination zones from the normal and supernumerary eyes. The stripe pattern is superimposed on a normal topographic retinotectal projection and is robust under a variety of experimental manipulations. Eye-specific segregation of RGC terminals appears to arise through the combined operation of the activity independent and activity-dependent mechanisms that are thought to produce the continuous retinotopic projection in normal animals (for review, see Udin and Fawcett, 1988).

We have found that exposure of the optic tectum of three-eyed tadpoles to either APV or MK801, in combination with NMDA, resulted in a desegregation of the eye-specific stripes. The only other agent which has been shown to cause desegregation is TTX (Reh and Constantine-Paton, 1984). Treatments with either L-APV or NMDA alone do not cause desegregation. In fact, NMDA treatment appears to potentiate the afferent segregation process.

It is interesting to note that exposure of the tecta of three-eyed tadpoles to MK801 in the absence of NMDA does not cause stripe desegregation. MK801, a non-competitive use-dependent NMDA channel blocker, blocks from within the NMDA channel and can only gain access to the binding site when the channel is open. The failure of MK801 alone to desegregate eye-specific stripes might be due to the infrequency of channel opening and therefore the poor access of the drug to the interior of the channel (Huettner and Bean, 1988).

How does the NMDA receptor organize the retinal afferents according to their activity patterns?

The role of the NMDA receptor in detecting coactivity can be seen by considering two neighboring synapses which arise from different RGCs. We assume that both the non-NMDA types of glutamate receptors, the quisqualate/kainate-sensitive receptors, and the NMDA receptors are colocalized at retinotectal synapses. When either one of the presynaptic terminals fires individually, the non-NMDA receptor type provides a rapid EPSP with each action potential in the afferent terminal. However, the postsynaptic depolarization in response to limited afferent activity is not sufficient to relieve the Mg^{+2} block of the NMDA channel. Therefore, even though transmitter binds to the NMDA receptor with each presynaptic event, the channel does not conduct ions because of the voltage-dependent Mg^{+2} block. However, when the two synapses fire synchronously, the EPSPs mediated by the quisqualate-type receptors sum so that the postsynaptic depolarization reaches the threshold for driving the Mg^{+2} out of the NMDA channel, thereby permitting conductance of calcium through the coactive synapses. We hypothesize that the calcium influx triggers cellular mechanism(s) which stabilize or increase the lifetime of those synapses. The stabilization probably includes a signal from the postsynaptic element to the presynaptic site, perhaps comparable to the post- to presynaptic signal invoked in long term potentiation.

This model predicts several features of the retinotectal projection. Most importantly, it suggests that the retinal ganglion cell transmitter would be an excitatory amino acid. We have receptor binding (McDonald et al, 1989), immunohistochemical and electrophysiological (Debski et al, 1987) evidence that retinotectal synaptic transmission is mediated by an excitatory amino acid, most likely glutamate.

Most of our electrophysiological experiments on retinotectal synaptic transmission have been done in a cannulated tadpole preparation, in which an anesthetized tadpole is cannulated through the heart, and the dorsal aspect of the brain is exposed and the nociceptive inputs and motor outputs are severed. This preparation allows the application of drugs to the brain through the vasculature and topically to the optic tectum. A postsynaptic evoked potential with both a positive and negative component is recorded from the tectal neuropil following stimulation of the optic nerve. We believe the positive and negative components of the potential result from the unusual cellular geometry of the tadpole neuropil, where neuronal cell bodies, which would act as sources, and nerve terminals, which would acts as sinks, are intermingled at the recording site in the superficial neuropil layers. We have used $CoCl_2$ treatment to determine which components of the evoked potential are postsynaptic to the RGCs. Application of kynurenic acid (1 mM), a broad range glutamate receptor antagonist, reversibly blocks the postsynaptic evoked potential (Debski et al, 1987). In contrast, APV (62 uM) does not block the postsynaptic potential. In fact, APV reversibly increases the amplitude of the positive component of the evoked potential to approximately 129% of control. In addition, APV increases the spontaneous single unit activity recorded in the tectum. It is interesting that NMDA application also increases single unit activity, but it blocks the evoked potential (Debski & Constantine-Paton, 1988). One possible explanation for these results is that NMDA receptors are concentrated on a population of inhibitory tectal neurons. Exposure to APV could initially decrease inhibition in the tectum and thereby increase the excitability of the tectum. Exposure of the tectum to the agonist NMDA would increase the activity of inhibitory tectal neurons, which in turn would decrease the tectal response to optic nerve stimulation. These data are consistent with a model of a circuit within the tectum, originally proposed to explain stripe segregation, in which lateral inhibition produces regions of excitation banded by regions of inhibition (reviewed in Debski et al, 1990).

How does the calcium transient increase the stability of coactive synapses?

One possible scenario is that a calcium transient through the NMDA channel activates calcium-sensitive protein kinase(s), such as protein kinase C (PKC) and/or calcium calmodulin-dependent protein kinase Type II (Cam K II), which in turn increases or prolongs the structural stability of the coactive synapses. We were interested to know if protein kinase activity might be involved in the formation and maintenance of topographic retinotectal maps. To test this hypothesis we chronically treated the optic tecta of three-eyed *Rana pipiens* tadpoles with drugs which alter protein kinase activity: phorbol esters, sphingosine and 1-[5-isoquinolinesulfonyl]-2-methyl piperazine (H-7). Chronic treatments with phorbol esters down regulate protein kinase C (Mathies et al, 1987). Sphingosine blocks PKC and Cam KII activity by competing with their activators, diacylglycerol and calmodulin, respectively, at the regulatory binding site (Hunnan et al, 1986). H-7 competes with ATP at the catalytic domain of several kinases including PKC and Cam KII (Hidaka et al, 1984).

We found that eye-specific stripes persist despite prolonged (up to 8 weeks) treatment with phorbol esters, sphingosine or H-7. Nevertheless, the drug treatments do result in alterations in retinal ganglion cell (RGC) terminal arbor morphologies and in the patterns of tectal cell proliferation, which provide convincing evidence that the drugs do reach the tissue in a biologically active form and in sufficient concentrations to exert an effect. Our data indicate that protein kinase activity may not be critical for the maintenance of eye-specific stripes. Further experiments to examine how these chronic treatments influence kinase activity are being conducted. We suggest that activity-dependent synapse stabilization in the retinotectal system may be mediated by an NMDA receptor-mediated protein kinase-independent mechanism which stabilizes synapses or increases synapse lifetime for a finite increment of time. Once established, the degree of coactivity of each synapse with its neighbors is assessed, according to the activation of the NMDA channel and the magnitude of the NMDA receptor-mediated calcium influx. If the coactivity is sufficient to elevate intracellular calcium levels to a threshold value, then the cellular machinery would be activated to prolong the lifetime of the synapse by a further increment of time. If the degree of coactivity of the synapses is too low to allow spatiotemporal summation of the EPSPs, and thereby relieve the magnesium block of the NMDA channel, then intracellular calcium levels would not rise and the cellular mechanisms which stabilize the synapse would not be triggered. Consequently, the life of the synapse would not be prolonged and it would retract. Precedense for such an NMDA receptor-mediated transient potentiation of synaptic transmission which is pharmacologically insensitive to protein kinase blockers has been demonstrated in the hippocampus (Collingridge et al, 1983; Malinow et al, 1988; Kauer et al, 1988).

Influence of the NMDA receptor on axon terminal growth

NMDA treatment changes the overall termination pattern of the retinal projection and also changes the growth patterns of individual RGC terminal arbors. Exposure of the tecta of three-eyed tadpoles to the agonist NMDA results in sharper stripe borders than in untreated animals. A closer examination of the stripe borders in treated and untreated animals has demonstrated that NMDA-treatment results in a 75% reduction in the number of axons crossing from a stripe to an inter-stripe zone. Thus, it appeared that NMDA treatments sharpened the stripe borders by altering the growth of the RGC terminals from the supernumerary eye into the stripe regions which were innervated primarily by the normal eye. To pursue this possibility, we drew individual HRP-labeled RGC terminals from untreated and NMDA-treated tecta and performed a detailed morphological analysis of the arbors. Terminals from NMDA-treated tecta have dramatically fewer branch tips, although the tangential extent of the arbor is not significantly different from the controls. In other words, NMDA-treatment reduced arbor branch density (branch tips/area) by approximately 50% compared to untreated terminals.

To understand how NMDA treatment might influence terminal growth, we considered our arbor morphology data from untreated and NMDA-treated two-eyed tadpoles and from untreated and NMDA-treated three-eyed tadpoles (Table 1). Terminal arbors from untreated two-eyed tadpoles differ from arbors from three-eyed tadpoles in one important respect: three-eyed arbors have about twice as many branch tips as two-eyed arbors confined to the same tangential area, and therefore have twice the branch density. Furthermore, two-eyed arbors do not show the decrease in branch density with NMDA treatment, seen in three-eyed arbors. We believe these differences in arbor morphologies are due to the difference in the density of afferent innervation in the singly- and doubly-innervated tecta and reflect an essential difference in endogenous NMDA receptor activity in the tecta from two- and three-eyed tadpoles.

Morphometric studies of the innervation density of singly- and doubly-innervated tecta have shown that striped tecta are innervated by approximately twice the normal number of retinal ganglion cells compressed onto a tectal neuropil which has expanded by only 40% (Constantine-Paton and Ferrari-Eastman, 1987). Therefore, within each striped termination zone RGCs from a larger retinal area converge on the same region of neuropil. This means that RGCs which would normally project to neighboring sites in the tectum are now forced to project to the same tectal site. In addition, electron microscopic studies have shown that despite the increased innervation density, doubly-innervated tecta have about the same number of synapses as singly-innervated tecta (Constantine-Paton & Norden, 1986). Together, these data suggest that RGC arbors in doubly-innervated tecta make fewer synaptic contacts than their counterparts in singly-innervated tecta. Furthermore, since each arbor from doubly-innervated tecta also has twice the branch number, each branch is supported by a severely reduced number of synapses compared to normal arbors.

These data bear on the degree of endogenous NMDA receptor activity within singly- and doubly-innervated tecta in the following way. We are assuming that activation of the NMDA receptor is triggered by converging, coactive retinal inputs. Since neighboring RGCs of the same response type have more highly correlated patterns of activity than non-neighboring RGCs, the increased retinal convergence in the doubly-innervated tecta should result in a decrease in coactivity and a relative decrease in the NMDA receptor activation compared to singly-innervated tecta. Based on these data we hypothesized that NMDA receptor activation may influence retinal terminal growth. In particular, we hypothesized that decreased NMDA receptor activation increases the probability of branch initiation and that a high degree of NMDA receptor activation correlates with synapse stabilization and branch maintenance. The balance between branch initiation and branch retraction in subregions of the arbor based on local correlated activity patterns would shape terminal morphology.

We suggest that increased NMDA receptor activation increases the stabilization of synapses and branch tips, so there would be a corresponding decrease in branch retraction. In addition, the increased stability of the synapses would have a inhibitory influence on local branch initiation. Conversely, the model predicts that under conditions of relatively low NMDA receptor activation there would be an increased branch initiation, but NMDA receptor activation would be too low to stabilize the synapses on the new branches and they would retract. If there is a lower limit of synapse density required to maintain a branch, then the terminal branchlets in doubly-innervated tecta are at greater risk of retraction than those in singly-innervated tecta. Electron microscopic analyses of single HRP-labeled RGC terminals in young frogs indicate that most of the synaptic contact between the retina and tectum is made through the fine branchtips that appear to be altered in the treatments we have employed (Yen and Constantine-Paton,1988; Yen, personal communication). Therefore, we feel it is justified to associate changes in the number of branchtips with branch survival, synapse density and synapse lifetimes, and to assume that the morphological alterations we observe reflect significant differences in retinotectal connectivity.

Before considering the effects of NMDA treatments on arbor morphology it is important to point out that electrophysiological investigations have shown that chronic NMDA treatment decreases the sensitivity of tectal NMDA receptors (Debski et al, 1989). As mentioned above, acute application of NMDA blocks the tectal evoked potential in a dose -dependent manner. Following chronic NMDA treatment comparable to that used for the anatomical studies, the dose-dependence curves of the NMDA block are shifted upwardso that a greater concentration is required to block the response.

The decreased branch density in arbors from three-eyed NMDA-treated animals may be due to a combined effect of drug treatment on branch survival and branch initiation. In drug-treated tadpoles where the NMDA receptors are less sensitive, increased branch retraction would occur in regions of the arbor where there is relatively little NMDA receptor activation. In addition, the chronic presence of the agonist would cause a simultaneous decrease in branch initiation in regions of the arbor where afferent coactivity is sufficient to activate NMDA receptors. Indeed, since NMDA treatment appears to increase the probability of NMDA channel opening (Cline and Constantine-Paton, 1989), chronic exogenous NMDA could simultaneously decrease the probability of new branch initiation from well stabilized regions of the arbor at the same time that it eliminates branches bearing poorly correlated synapses.

In singly-innervated tecta, where a smaller area of retina projects to a unit volume of tectum, the degree of correlated activity in any region of neuropil is higher than in doubly- innervated tecta. Therefore, even under conditions of lowered sensitivity, NMDA receptors throughout the arbor still have a high probability of being activated in NMDA-treated singly-innervated tecta and a net branch retraction is not observed. In addition, since each RGC arbor in a singly-innervated tectum makes more synapses than its counterpart in a doubly-innervated tectum, the survival of branch tips in normal terminals is less susceptible to variations in the stability of individual synapses, even under conditions of lowered NMDA receptor sensitivity. We believe these differences between singly- and doubly-innervated tecta account for the different responses of the terminal arbors to NMDa treatment.

SUMMARY

The importance of patterned retinal activity in visual system development has been recognized since Hubel and Wiesels early experiments (1963). The NMDA receptor is one cellular mechanism which can recognize patterned retinal activity and convey an intracellular message of that activity. It will now be of considerable interest to elucidate the cellular events involved in synapse stabilization subsequent to NMDA receptor activation.

REFERENCES

Arnett, D.W. 1978. Statistical dependence between neighboring retinal ganglion cells in goldfish. **Exp. Brain Res..** 32:49-53.

Ascher, P. & Novak, L. 1987 Electrophysiological studies of NMDA receptor. TINS 10: 284-288.

Cline, H.T., Debski, E., Constantine-Paton, M. 1987. NMDA receptor antagonist desegregates eye-specific stripes. **Proc. Natl. Acad. Sci.** 84:4342-4345.

Cline, H.T., Constantine-Paton, M. 1989. NMDA receptor antagonists disrupt the retinotectal topographic map. **Neuron** (in press).

Collingridge, G.L., Kehl, S.J., & McLennan, H (1983) Excitatory amino acids in synaptic transmission in the Schaffer collateral-commissural pathway in the rat hippocampus. **J. Physiol.** (Lond) 334: 33-43.

Constantine-Paton, M., Cline, H.T. and Debski, E., 1990 Patterned activity, synaptic convergence and the NMDA receptor in developing visual pathways. **Ann. Rev. Neuroscience** 13: in press.

Constantine-Paton, M. and Ferrari-Eastman, P. 1987. Pre- and Postsynaptic correlates of interocular competition and segregation in the frog. **J. Comp. Neurol.** 255, 178-195.

Constantine-Paton, M and Norden, J.J. (1986) Development of order in the visual system. **Cell and Developmental Biology of the Eye.** S.R. Hilfer and J.B. Sheffield (eds). Springer-Verlag, New York.

Debski, E.A., Cline, H.T., Constantine-Paton, M. 1987. Kynurenic acid blocks retinal-tectal transmission in *Rana pipiens*. **Proc. Soc. Neurosci.** 13:1691.

Debski, E.A., Constantine-Paton, M. 1988. The effects of glutamate receptor agonists and antagonists on the evoked potential in *Rana pipiens*. **Proc. Soc. Neurosci..** 14:674.

Debski, E.A., Cline, H.T., Constantine-Paton, M. 1989. Chronic application of NMDA or APV affects the NMDA sensitivity of the evoked tectal response in Rana pipiens. **Proc. Soc. Neurosci.** 15 (in press).

Debski, E.A., Cline, H.T., Constantine-Paton, M. 1990. Activity-dependent tuning and the NMDA receptor. **J. Neurobiol.** 21(1)(in press).

Fox, B.E.S., Fraser,S.E. 1987. Excitatory amino acids in the retino-tectal system of *Xenopus laevis..* **Proc. Soc. Neurosci.** 13:766.

Fox, K., Sato, H. and Daw, N. (1989) The location and function of NMDA receptors in cat and kitten visual cortex. **J. Neurosci.** 9: 2443-2454.

Fraser, S.E. 1985 Cell interactions involved in neuronal patterning: An experimental and theoretical approach._in : Molecular Bases of Neural Development, G.M. Edelman, W.E. Gall and W.M. Cowan, eds, John Wiley & Sons, New York, pp. 481-508.

Hannun, Y.A. and Bell, R.M. (1989) Functions of sphingolipids and sphingolipid breakdown products in cellular recognition. **Science** 243: 500-507.

Hidaka, H., Inagaki, M, Kawamoto, S., & Sasaki, Y. (1984) Isoquinolinesulfonamides, novel and potent inhibitors of cyclic nucleotide dependet protein kinase and protein kinase C. **Biochem,** 23: 5036-5040.

Hubel, D.H., Wiesel, T.N. 1965. Binocular interaction in striate cortex of kittens reared with artificial squint. **J. Neurophysiol.** 28:1041-1059.

Huettner, J. and Bean, B . 1988 Block of N-methyl-D-aspartate-activated current by the anticonvulsant MK801: Selective binding to open channels. **Proc. Natl. Acad. Sci.** USA 85: 1307-1311.

Kauer, J.A., Malenka, R.C. and Nicoll, R.A. (1988) NMDA application potentiates synaptic transmission in the hippocampus. **Nature** 334: 250-252.

Kleinschmidt, A., Bear, M.F., Singer, W. 1987. Blockade of NMDA receptors disrupts experience-dependent plasticity of kitten striate cortex. **Science** 238:355-358.

Law, M. I. and Constantine-Paton, M. (1981) Anatomy and physiology of experimentally produced striped tecta. **J. Neurosci.** 1: 741-759.

Malinow, R., Madison, V.D. and Tsien, R.W. (1988) Persistent protein kinase activity underlying long-term potentiation. **Nature** 335: 820-824.

Mathies, H.J.G., Palfrey, H.C., Hirning, L.D. and Miller, R.J. (1987) Down regulation of protein kinase C in neuronal cells: Effects on neurotransmitter release._J. Neurosci. 7: 1198-1206.

Mastronarde, D.N. 1983 Correlated firing of cat retinal ganglion cells. I. Spontaneously active inputs to X- and Y-cells. **J. Neurophysiol.** 49:303-324.

Mayer, M.L. and Westbrook, G.L. 1987 The physiology of excitatory amino acids in the vertebrate central nervous system. **Prog. Neurobiol.** 28: 197-276.

McDonald, J.W., Cline, H.T., Constantine-Paton, M., Maragos, W.E., Johnston, M.V., Young A.B. 1989. Quantitative autoradiographic localization of NMDA, quisqualate and PCP receptors in the frog tectum. **Brain Res.** 482: 155-159.

Reh, T., Constantine-Paton, M. 1984. Retinal ganglion cells change their projection sites during larval development of *Rana pipiens*. **J. Neurosci.** 4:442-457.

Reh, T.A., Constantine-Paton, M. 1985. Eye-specific segregation requires neural activity in three-eyed *Rana pipiens*. **J. Neurosci.** 5:1132-1143.

Scherer, W.S. and Udin, S.B. (1988) The role of NMDA receptors in the development of binocular maps in Xenopus tectum. **Proc. Soc. Neurosci.** 14: 272.16.

Schmidt, J.T. 1985 Formation of retinotopic connections: selective stabilization by an activity-dependent mechanism. **Cell. and Molec. Neurobiol.** 5: 65-84.

Schmidt, J.T. (1988) NMDA blockers prevent both retinotopic sharpening and LTP in regenerating optic pathway of goldfish. **Proc. Soc. Neurosci.** 14: 272.15.

Tsumoto, T., Hagihara, K., Sato, H., Hata, Y. 1987. NMDA receptors in the visual cortex of young kittens are more effective than those of adult cats. **Nature** 327:513-514.

Udin, S.B. and Fawcett, J.W. 1988. Formation of topographic maps. **Ann. Rev. Neurosci.** 11:289-327.

Yen, L-H, Constantine-Paton, M. 1988. EM analysis of single retinal ganglion cell terminals in developing *Rana pipiens*. **Proc. Soc. Neurosci.** 14:674.

COMMENTARY: EXCITATORY AMINO ACIDS AND IN VIVO DEVELOPMENT

G. Horn and E. Cherubini

The first part of this session was focused on the electrophysiological mechanisms underlying synaptic plasticity during development and the role of excitatory amino acids.

In the first paper Swann presented evidence that during the second postnatal week synaptic interactions between CA3 neurons are well developed and probably excitatory connections are more abondant than in adult neurones. At this period of development, network generated interictal bursts occur either spontaneously or evoked by orthodromic stimulation, in conditions of reduced inhibition (brief tetanization of the afferent pathway or in the presence of penicillin). In pair recordings from minislices more than 60 % of individual cells may initiate a population discharge (against 30 % of the mature neurones). Unitary epsps produced in one cell by a burst in another cell last longer than in adult neurones, suggesting that more synaptic contacts exist among pyramidal neurones.

The paper by Ben-Ari et al. concerned another form of synaptic plasticity present in CA3 pyramidal neurones which exhibit spontaneous giant depolarizing potentials. These are generated by the synchronous discharge of a population of neurones and are GABA mediated since they reverse at the same potential of exogenously applied GABA and are blocked by the GABA A antagonist bicuculline. Our interesting feature is that these GABA mediated events are controlled presynaptically by NMDA receptors (during this period of development the hippocampus is particularly enriched in NMDA receptors) and are facilitated by glycine, in a strychnine insensitive way. However glycine does not modify the NMDA induced inward current on the postsynaptic cell, suggesting that two glycine sites are present in neonatal rats : a high affinity one which will be saturated by the glycine present in the extracellular fluid and a low affinity, located on presynaptic GABAergic neurones and which will disappear during the second postnatal week, in relation to the synaptic rearrangement which will occur during this period. Although the role of these giant GABA mediated events is still not clear, it is possible that they represent an important signal in cell growth and differentiation. Thus calcium influx through NMDA activated channels constitutes an essential step in neuronal plasticity.

The following paper by Cherubini et al. shows that, during the first week of postnatal life, repetitive applications of NMDA in conditions in which synaptic transmission is blocked by TTX and potassium conductances are reduced by externast TEA, Cs or 4-AP, periodic inward currents are generated which persist for the life of the preparation. These currents are due to the pulsatile release of glutamate probably triggered by oscillations of intracellular calcium in presynaptic nerve terminals. It bear stressing that this form of developmental plasticity has many similarities with LTP, another form of synaptic plasticity triggered by NMDA in mature neurones. Like for LTP NMDA is necessary for the induction of the periodic inward currents, but is not necessary for their maintenance ; once established the periodic inward currents are insensitive to NMDA antagonists, but are blocked by non NMDA antagonists. Thus, during the immediate postnatal period, where electrical LTP can not be produced, fluctuations in intracellular calcium tirggered by NMDA receptors may provide a means of intracellular communication and therefore synaptic growth and consolidation before a more mature type of synaptic plasticity starts to operate.

In the last paper of this session, Jones and Heinemann describe the occurence of spontaneous depolarizing potentials in layer II of the entorhinal cortex of neonatal rats (P9-P13) which are blocked by NMDA antagonists. However, in contrats to GDPs observed in the hippocampus, these potentials were voltage dependent, suggesting that they were mediated by NMDA receptors. Moreover stimulation of the para-subiculum evokes in layer II cells of neonatal and not adult rats a fast non NMDA and a slow NMDA mediated depolarization (e.p.s.p.). Thus, although the functional role of NMDA mediated events in immature neurones is still open to speculation, it appears a general feature of mammalian CNS development.

The three papers presented in the second half of the session were concerned with the roles of excitatory amino acids in the developing brain and in learning in young animals. Tsumoto discussed plasticity in the visual cortex of the cat and the rat. He and his colleagues suggested that glutamate and aspartate may be trasmitters at lateral geniculate-cortical synapses: these amino acids are released in the visual cortex when the visual pathways are activated by direct electrical stimulation or by bars of light moved across the visual fields. In the mature visual cortex receptors appear primarily to be of the non-NMDA type, whereas NMDA receptors contribute substantially to synaptic transmission in the immature cortex during the sensitive period of postnatal development. Tsumoto also showed that long-term potentiation (LTP) of synaptic efficacy could be induced in the developing cortex and that the induction was blocked by APV as well as by the intracellular injection of EGTA. On the basis of these results he suggested that an influx of Ca^{2+}, probably through NMDA receptors, through postsynaptic membranes plays a key role in inducing long-term changes of synaptic efficacy in the developing visual cortex. Cline also considered that an influx of calcium through NMDA channels plays a role in synaptic plasticity in the optic tectum of tadpoles, stabilising synapses there. Cline and her colleagues investigated the role of NMDA receptors in the development of retinotectal connections in the frog. Exposure of the optic tectum of tadpoles to NMDA receptor antagonists, was found to disrupt retinal topography in normal animals and to cause desegregation of eye-specific stripes in 3-eyed animals. The loss of retinal topography was not due to an enlargement of retinal ganglion cell axon terminals. She referred to the work of McDonald and his colleagues who showed, using receptor binding techniques, that NMDA and quisqualate binding sites are present in the tectal neuropil. Cline described a series of electrophysiological experiments demonstrating that kynurenate blocked the evoked potential recorded in the tectum following optic nerve stimulation. APV increased single unit tectal activity and increased the amplitude of the evoked potential. NMDA also increased single unit activity but blocked the evoked potential. These data were interpreted as being consistent with a differentially high distribution of NMDA-receptors on inhibitory tectal neurones. Cline suggested that the intracellular calcium transient through NMDA receptor channels activates Ca^{2+}-sensitive protein kinases. To test this hypothesis she exposed the optic tecta of 3-eyed tadpoles to kinase blockers, but the treatment did not cause striped desegregation. The drugs were apparently reaching the tectum in a biologically active form because individual retinal ganglion cell terminals were decreased in size by the drug treatments. These data were interpreted to signify that protein kinases are involved in axon growth, but not in the selective stabilisation of co-active synapses. In a further set of experiments the optic tecta of 3-eyed tadpoles were treated with NMDA. This agonist was found to sharpen the stripe borders by decreasing the number of processes which grow from a striped zone to an interstriped zone. Cline concluded that NMDA receptors play two separate roles in retinotectal development: (i) the NlMIDA receptors on tectal neurones are implicated in the selective stabilisation of co-active retinal axons and (ii) presynaptic NMDA receptors are involved in the growth of the axon terminal.

In the last paper of the session evidence was presented that NMDA receptors play a role in the recognition memory that underlies visual imprinting in the domestic chick. Hrn, McCabe and their colleagues have shown that a restricted region of the chick forebrain - the intermediate and medial part of the hyperstriatum ventral (IMHV) - is crucial for imprinting and may be a site of information storage. Training with a visually conspicuous imprinting stimulus is associated with a significant increase in the mean area of postsynaptic densities of axospinous synapses in the left but not in the right IMHV. The binding of L-[3H]glutamate to membranes from IMHV were studied under conditions conductive to the labelling of synaptic L-glutamate receptors. Approximately 8 h after training NMDA-sensitive binding was significantly greater in trained than in dark-reared chics. The effect was, once again, found in the left but not in the right IMHV. The results were consistent with the change in binding being due to an increase in the number of NMDA receptors. Partial correlation analysis indicated that the effect was not attributable to locomotor activity during training, activity in sensory pathways, maturational processes or related side-effects. The increase in NMDA sensitive binding did not occur immediately after the enf of training. A study of the time-course of the increase showed that it was delayed until approximately 8 h after the end of the training period. Comparisons were made between the cellular mechanisms of hippocampal LTP and the effects of learning on IMHV synapses.

EXCITATORY AMINO ACIDS, GROWTH FACTORS, AND CALCIUM: A TEETER-TOTTER MODEL FOR NEURAL PLASTICITY AND DEGENERATION

Mark P. Mattson

Sanders-Brown Research Center on Aging and
Department of Anatomy & Neurobiology
University of Kentucky Medical Center, Lexington, KY 40536-0230

SUMMARY

This paper presents and examines the hypothesis that excitatory amino acids (EAAs) and growth factors (GFs) exert opposing actions on neuronal cytoarchitecture by influencing cellular Ca^{2+} homeostasis. This hypothesis is supported by experiments with cultured hippocampal pyramidal neurons in which EAAs induced dendritic regression and cell death, whereas fibroblast GF (FGF) promoted neurite outgrowth and cell survival. FGF protected against glutamate-induced neuronal degeneration by raising the threshold for the actions of this EAA. Pharmacological studies, and direct monitoring of intracellular Ca^{2+} levels, demonstrated that a sustained rise in intracellular Ca^{2+} levels was largely responsible for the degenerative actions of glutamate. FGF attenuated the Ca^{2+} response to glutamate. Experiments with glutamate, Ca^{2+} ionophore A23187, and Na^+-deficient culture medium provided evidence that FGF can enhance Na^+-dependent Ca^{2+} extrusion. These data suggest a model in which cell survival and neurite outgrowth in hippocampal neurons is regulated by the opposing actions of EAAs and GFs acting through the Ca^{2+} second messenger system. In this "teeter-totter" model the relative levels of input from EAAs and GFs determine whether a neuron lives or dies, and whether its outgrowth is in a progressive or regressive state. Interactions of EAAs and GFs may play important roles in: developmental events such as neurite outgrowth, synaptogenesis, and natural cell death; maintenance and plasticity of neural circuitry in the mature nervous system; and maladaptive neurodegeneration that occurs in aging and disorders such as Alzheimer's disease.

INTRODUCTION

Neuroarchitecture is defined by the neurons present, the shapes of their neuritic arbors, and the spatial arrangement of synaptic connections between the different neurons. A continuum of changes in neuroarchitecture occurs during a lifetime. The most dramatic changes are seen during development as axons and dendrites are elaborated, and synapses are formed, effectively establishing functional neural circuitry. In adult life, changes in neuroarchitecture also occur, and are believed to provide the nervous system with the ability to adapt to environmental demands. Such structural plasticity is likely to underlie important functional processes such as learning and memory (1). Unfortunately, adaptive change often does not continue through life, as degeneration of neurites and cell death occurs with aging or in specific disorders. Proper communication between different cells of the nervous system is vital for normal development and functioning. Two classes of intercellular signaling molecules that are critical to neuronal development and maintenance are GFs and neurotransmitters. Provocative examples of the potency of GFs and neurotransmitters include: the trophic effects of nerve GF (NGF) on sensory and sympathetic neurons (13); the neurite out

Excitatory Amino Acids and Neuronal Plasticity
Edited by Y. Ben-Ari
Plenum Press, New York, 1990

growth- and survival-promoting actions of FGF in CNS neurons (27,28,35); the neurotoxic action of EAAs on select groups of central neurons (5,19-25,33). In addition to the dramatic, life or death, effects of GFs and EAAs, these signals may also exert more subtle effects on neurite outgrowth and synaptogenesis (19,20). Thus, by influencing the presence of a neuron, its specific neuritic architecture, and its synaptic relations with other neurons, GFs and EAAs can act to sculpt neuronal circuitry.

Environmental signals such as EAAs and GFs regulate neuroarchitecture by activating (or inhibiting) specific second messenger pathways (16,20). Among the different second messengers, intracellular Ca^{2+} has risen to the forefront as an important regulator of neuroarchitecture. A provocative example is seen when hippocampal pyramidal neurons are exposed to increasing levels of glutamate, which acts through the Ca^{2+} system. The changes in neuroarchitecture range from an inhibition of dendritic growth cone motility, to a progressive regression of dendrites, to cell death (21). The importance of Ca^{2+} in neurons is evident from the presence of highly specialized cellular systems for the control of intracellular Ca^{2+} levels (i.e., Ca^{2+} channels, Ca^{2+} extrusion systems , and Ca^{2+} binding proteins (4). Recent evidence suggests that these different components of the neuronal systems for Ca^{2+} homeostasis are influenced by environmental signals, such as EAAs and GFs, that play important roles in the development and maintenance of the nervous system (22-27).

In addition to the importance of cellular signaling systems in neuronal development and plasticity, it is likely that they are intimately involved in the maladaptive neuronal degeneration that occurs in aging and specific disorders such as Alzheimer's disease. Abnormalities in GF systems (2), EAA systems (15,18,20,33) and Ca^{2+}-regulating systems (12) have been suggested to be involved in disorders ranging from Alzheimer's disease to stroke. Thus, it is important to consider that the signaling systems that provide the nervous system with the capability of developing and adapting to environmental demands, are the same systems that may be involved in pathological changes in neuroarchitecture (20).

Previous studies that examined the mechanisms regulating the formation and modification of neural circuitry focused on the actions of a single type of environmental signal. However, in the intact nervous system neurons are undoubtedly exposed to several different types of signals simultaneously. It is therefore important to understand how different signals interact on the same neurons. The present paper addresses this problem directly by presenting data from recent experiments which examined the interactive effects of FGF and glutamate on cultured hippocampal neurons. These findings lead us to forward a "teeter-totter" model for the regulation of neuroarchitecture by GFs and EAAs. This hypothesis states that *developmental and pathological changes in the structure of a neuron can be determined by the net level of activation of signaling pathways for trophic and regressive environmental signals.*

Location and Actions of FGF and EAAs Suggest Roles in Hippocampal Development

In order for a molecule to act as a morphogenic signal in development it must be available to its target cells, and the target cells must possess appropriate receptors and a functional signal transduction system. In the developing hippocampus, both FGF and glutamate appear to fulfill these criteria. Thus, FGF is present at high levels during development in the intact hippocampus (3,30), and cultured glial cells and neurons from embryonic hippocampus possess considerable FGF--like immunoreactivity (Fig. 1). FGF increases long-term neuronal survival in cultures of embryonic hippocampal cells (Fig. 2 and ref. 27). Basic FGF can also prevent naturally occurring neuronal death *in vivo* (9) suggesting that FGF may play a role in determining which neurons survive, and which are lost, during development. In addition, FGF has quite striking effects on neurite outgrowth; both neurite elongation rates and the frequency of neurite branching is increased greatly by FGF.

Glutamate is also present in developing hippocampal neurons, and is particularly concentrated in the pyramidal neurons (Fig. 1). The axons of developing glutamatergic neurons release glutamate in a calcium- and activity-dependent manner (26). In addition, developing hippocampal neurons possess glutamate receptors as determined by several different methods (20,31). In contrast to the trophic effects of FGF, EAAs exert negative effects on cell survival and neurite outgrowth in hippocampal pyramidal neurons. EAAs can induce neurodegeneration leading to cell death in cultured embryonic and postnatal pyramidal neurons (21,25). More subtle neuroarchitectural changes are seen in pyramidal neurons exposed to lower levels of glutamate (that do not cause cell death); dendritic outgrowth is selectively inhibited (21). Interestingly, glutamate may also act to promote synapse formation during hippocampal development (26).

Interactions Between EAAs and GFs in Development

A major effort in developmental neuroscience research has been given to the identification and characterization of individual factors that exert trophic or regressive effects on particular neurons. These studies have greatly increased our understanding of the kinds of signals present in the nervous system, and their influences on various aspects of neuronal development and function. However, in the intact nervous system, individual neurons are undoubtedly exposed to many signals simultaneously. In order to fully understand how the nervous system develops and functions it will therefore be necessary to examine the interactive effects of different environmental signals. We have begun this line of investigation by examining the interactive effects of EAAs and GFs on cell survival and neurite outgrowth in cultured embryonic hippocampal neurons.

When cultured hippocampal neurons are maintained in the presence of FGF, the dose-response curve for outgrowth-inhibiting and neurotoxic effects of glutamate is shifted such that higher levels of glutamate are required to elicit a given change in neuroarchitecture (27). Neurons maintained in the presence of FGF will continue dendrite outgrowth when exposed to concentrations of glutamate that normally inhibit dendritic outgrowth (Fig. 3). Perhaps most striking is the protective effect of FGF against glutamate neurotoxicity. Within a particular range of glutamate concentrations, pyramidal neurons grown in the absence of FGF were readily destroyed by glutamate, whereas pyramidal neurons in parallel cultures maintained in the presence of FGF were resistant to glutamate neurotoxicity and often continued their outgrowth . The ability of EAAs and GFs to interact in the regulation of cell survival and neurite outgrowth in cell culture suggests that such interactions might be involved in the normal development of neural circuitry. For example, the axon of a hippocampal neuron might be coaxed towards a brain region rich in FGF, while a dendrite might stabilize its outgrowth in a glutamate-rich region. Both glutamate and FGF can act locally on individual dendritic growth cones (20,21) suggesting that these signals have considerable potential to sculpt neuroarchitecture in quite specific ways. Indeed, glutamate released from growing afferent axons (from entorhinal cortex) can act locally on target hippocampal neurons to inhibit dendrite outgrowth and promote synapse formation (26). In regions where neurites encounter both an EAA and a GF, the outgrowth status might be determined by the relative levels of input from these two opposing signals.

Natural cell death is another arena of neuronal development in which GF-EAA interactions may play roles. It is well-known that, in many regions of the nervous system, there is an initial "overproduction" of neurons, and a subsequent dimunition in this initial neuronal pool as the axons of these neurons reach their target cells. It has been suggested that a competition of the different neurons for a limited supply of target-derived GF might determine which neurons survive this developmental phase (7). The results of the experiments above, demonstrating interactive effects of FGF and glutamate, provide a basis upon which to expand the working hypothesis for mechanisms of selective natural cell death to include interactions between EAAs and GFs. For example, suppose that two neurons receive equal levels of a GF from a common target. Further suppose that one of the neurons is being exposed to a greater level of EAA input. Under these conditions it might be expected that the neuron receiving greater excitatory input would degenerate. On the other hand, if two neurons were receiving equal levels of excitatory input, then differences in the levels of trophic input from a GF would determine which neuron survived. In this view, the "adjustment" of neuronal numbers during the formation of neural circuitry would arise from the actions of several different environmental inputs.

Intracellular Ca2+: A Common Mechanistic Focus for the Actions of EAAs and GFs ?

Within the last several years quite convincing evidence has been obtained supporting roles for intracellular Ca^{2+} in the regulation of neuroarchitecture in development, adult plasticity, and disease (16,20). An elevation in intracellular Ca^{2+} is a common correlate of the binding of EAAs to their receptors. The recent development of technologies for monitoring intracellular Ca^{2+} levels directly has allowed clear demonstrations of this effect of EAAs in several types of neurons (6,17, 23). In cultured embryonic hippocampal pyramidal neurons, glutamate induces Ca^{2+} influx through plasma membrane channels (22,23). Glutamate receptors linked to Ca^{2+} influx are localized in dendrites and the soma and are absent from the axon. This localized Ca^{2+} influx is largely responsible for the selective effects of glutamate on dendritic outgrowth (21). Similarly, the neuron-selective effects of EAAs on cell survival (see above) can be accounted for by the presence of Ca^{2+}-linked EAA receptors in the vulnerable neurons (22). Neurons which are vulnerable to glutamate neurotoxicity invariably show large and sustained Ca^{2+} responses to glutamate which precede degeneration. Other neurotransmitters that affect neurite outgrowth and cell survival also appear to act, at least in part, through their effects on intracellular Ca^{2+} (see ref. 20 for review).

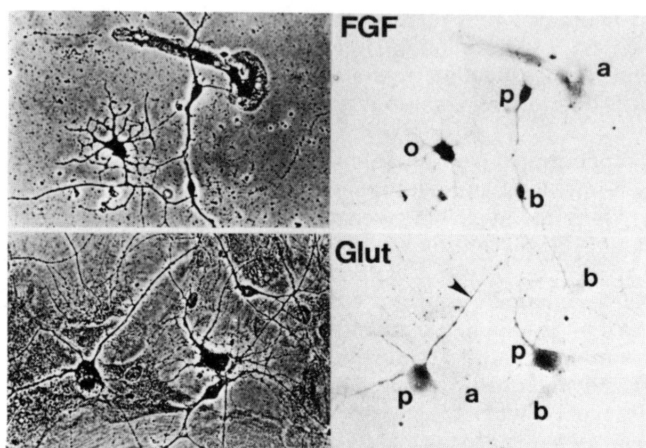

Fig. 1.   Localization of FGF and glutamate in cultured hippocampal cells. FGF and glutamate were localized by biotin-avidin-peroxidase immunocytochemistry using polyclonal antisera generously provided by Drs. M. Klagsbrun and A. Rustioni, respectively. Abbreviations: p, pyramidal neuron; b, bipolar neuron; o, oligodendrocye; a, astrocyte. Upper) Phase contrast (left) and bright-field (right) micrographs of neurons and glial cells immunostained for localization of FGF. Note that the pyramidal neuron and oligodendrocyte stain more intensely than the other cells. Lower) Phase contrast (left) and bright-field (right) micrographs of neurons and underlying glial cells immunostained for localization of glutamate. Note that the pyramidal neurons possess considerably more glutamate-like immunoreactivity than the other cell types. Scale bar, 25 μm.

Although GFs were identified as neuronal morphogens before neurotransmitters, the mechanisms of action of neurotransmitters have been easier to unravel. This is true because, as it happens, neurotransmitters use the same signal transduction pathways to elicit both electrophysiological and morphological responses. Might Ca^{2+} also be involved in the action of GFs? There is certainly precedence for this possibility in studies of nonneuronal cells. For example, fibroblasts show a rise in intracellular Ca^{2+} levels in response to platelet-derived GF, and this Ca^{2+} signal is apparently involved in the mitogenic response to this GF (32). Epidermal GF may also exert its effects on target cells by inducing Ca^{2+} influx (15). We recently examined the Ca^{2+} correlates of FGF actions in cultured hippocampal neurons (27). While FGF alone had no obvious effects on intracellular Ca^{2+} levels, FGF did attenuate the Ca^{2+} responses normally elicited by glutamate (Fig. 4). Further probing of FGF's effects on Ca^{2+} homeostasis suggested that FGF was improving the ability of the neurons to extrude Ca^{2+} in a Na^+-dependent manner (M. P. Mattson, unpublished). Thus, neurons grown in the presence of FGF were better able to reduce elevations in intracellular Ca^{2+}

214

induced by ionophore A23187 than were FGF-deprived neurons. However, when the culture medium lacked Na+, FGF-treated neurons were no better than untreated neurons in reducing a Ca2+ load.

Taken together, these data indicate that environmental signals that exert quite opposite effects on neuroarchitecture can act through a common intracellular messenger system. In this case, EAAs and FGF act at different points in the cellular system for Ca2+ regulation. EAAs induce Ca2+ influx through plasma membrane channels whereas FGF promotes Ca2+ extrusion, probably by enhancing the Na+-Ca2+ exchanger.

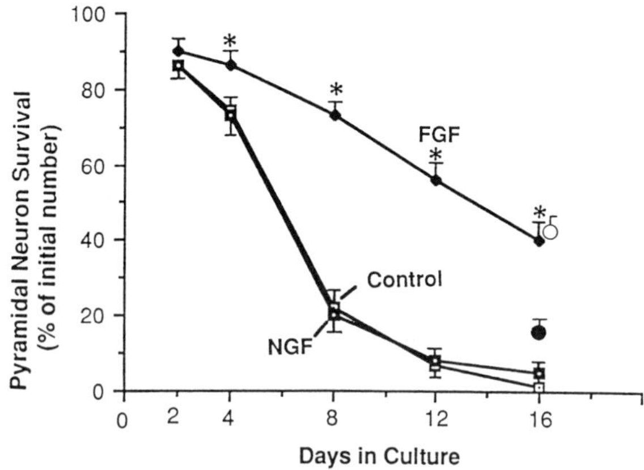

Fig. 2.   FGF promotes the survival of embryonic hippocampal neurons in cell culture. Cultures were maintained in the absence of growth factors (Control), or in the presence of 5 ng/ml FGF or 10 ng/ml NGF (growth factors were added on culture day 0 and every other day thereafter). Neuronal survival was determined by counting viable neurons in marked microscope fields (counts made on days 2-16 are expressed as a percentage of the initial number of neurons present on day 0). On day 16, the open and closed circles represent survival in cultures treated with 10 ng/ml and 1 ng/ml of FGF, respectively. Values represent the mean and SEM (n= 4-6 fields in two separate cultures).

Implications for Abnormal Neurodegeneration

A concept that has been emerging for some time now is that the maintenance of neural circuitry requires proper communication between the interacting neurons, and a breakdown in communication can lead to neuronal degeneration. Two previous hypotheses for the causes of abnormal neurodegeneration implicated GF deficiencies (2) or aberrant EAA activity (18,20,33). An example of data supporting the GF deficiency hypothesis comes from studies of the septo-hippocampal system. It has been suggested that a failure of basal forebrain cholinergic neurons to obtain or respond to NGF may result in the degeneration of these neurons in Alzheimer disease (10,11,14). Other trophic factors, such as EGF and FGF, may also be involved in neuronal maintenance in the mature nervous system. FGF, in particular, seems to influence neurons from a number of brain regions, and can prevent the death of septal neurons (29) and cultured hippocampal neurons (27). Future work should provide insight into the involvement of these and other GFs in abnormal neurodegeneration.

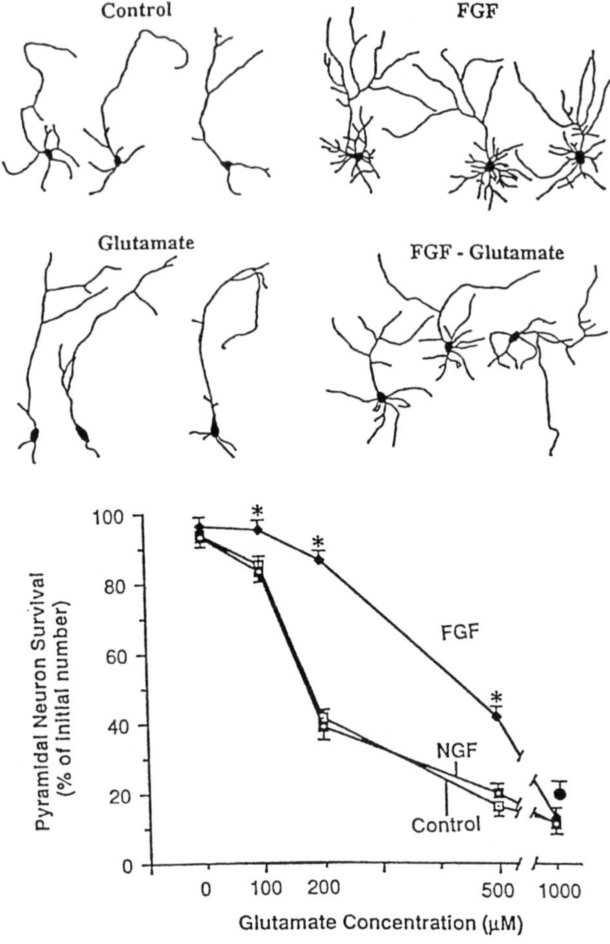

Fig. 3.   Interactive effects of FGF and glutamate on neurite outgrowth and cell survival in hippocampal pyramidal neurons.
Upper) Tracings of pyramidal neurons (5 days in culture) maintained under the indicated conditions. Control, no treatments; FGF (5 ng/ml added on days 0, 2, and 4; Glutamate, 50 µM glutamate added on culture day 2. Note that FGF enhanced the outgrowth of both dendrites and axons, glutamate reduced dendrite outgrowth selectively, and combined treatment with FGF and glutamate resulted in an intermediate level of neurite outgrowth.
Lower) In control cultures, glutamate caused a dose-dependent reduction in neuronal survival (glutamate was added on culture day 5 and survival was assessed 24 hr later). Neuronal survival in cultures maintained in the presence of FGF (5 ng/ml added on culture days 0, 2 and 4) and exposed to 100, 200, or 500 µM glutamate was significantly greater than in control cultures ($P<0.05$-0.001). NGF (10 ng/ml added on days 0, 2, and 4) did not affect glutamate - induced neuronal death. Values represent the mean and SEM (n=8 fields in two separate cultures.

216

Perhaps a more compelling case can now be made for the involvement of EAAs in disorders involving selective neurodegeneration. The data is quite convincing that EAAs are involved in the degeneration of hippocampal pyramidal neurons that often accompanies stroke and status epilepticus (see refs. 5 and 20 for reviews). Similar patterns of hippocampal cell death are seen following experimental ischemia or administration of EAAs, and the cell death can be prevented with EAA receptor antagonists. Interestingly, the hippocampal neurons most vulnerable in Alzheimer's disease (CA1 pyramidal cells) are also the most vulnerable to EAA neurotoxicity (25). In addition, hippocampal pyramidal neurons arising from a common progenitor cell in culture always showed similar sensitivities to glutamate neurotoxicity (22), suggesting that intrinsic neuronal properties acquired during development may contribute to selective vulnerability. Differences in the presence of glutamate receptors linked to Ca^{2+} influx can account for selective neuronal death (5,22,23). Finally, recent data from this laboratory indicates that glutamate can induce the expression of Alzheimer-associated antigens in cultured hippocampal neurons (M. P. Mattson, submitted), a finding consistent with a role for EAAs in Alzheimer's disease. Lines of investigation similar to those just described have also yielded data supporting a role for EAAs in Huntington's disease and amyotrophic lateral sclerosis (5,20,33). Thus, the selective degeneration of neurons in different brain regions, which lead to quite different functional impairments, may involve aberrations in similar cellular signaling systems.

The studies of EAA and FGF actions on cultured hippocampal neurons described above provide a basis upon which to extend and congeal the previous GF- and EAA-based hypotheses. These new findings indicate that GFs and EAAs can *interact* in neurodegenerative processes. Several important questions are raised by the demonstration of interactive effects of GFs and EAAs in neurodegeneration, not the least of which relates to preventative and therapeutic approaches to neurodegenerative disorders. Can the addition of exogenous GFs prevent or retard the progressive neurodegeneration that occurs in diseases such as Parkinson's, Alzheimer's and Huntington's? Will GFs promote recovery of functional neural circuits? Might EAA antagonists be employed to reduce neurodegeneration? EAA antagonists have, in fact, been shown to reduce neuronal damage in animal models of stroke (5,33). Interestingly, grafts of neonatal striatal tissue into the striatum of adult rats protected the host striatal neurons from destruction by EAAs (34). These kinds of beneficial effects of transplanted tissue (see also refs. 11,14) are entirely consistent with the hypothesis for GF-EAA interactions in abnormal neurodegeneration proposed herein. Although a major goal in the study of any disease is to understand its proximate cause, effective preventative and therapeutic strategies may be possible without this knowledge. Just as one need not know the cause of a forest fire in order to stop its progress, it may not be necessary to know the initiating causes of neurodegeneration in a disorder order to stop its progress. For example, suppose that EAA overactivity and calcium accumulation in neurons is involved in the cell death that accompanies disorders such as Huntington's or Alzheimer's diseases. Then it seems reasonable that blocking these steps in the neurodegenerative process would allay the progression of the disease. The importance of this approach is magnified when one considers the possibility of more than one initiating cause for some of these disorders. Abnormalities at any of several different points (e.g., increased release of EAAs, increased EAA receptor density, reduced inhibitory inputs, or faulty Ca^{2+} extrusion pumps) might lead to imbalances in cellular signaling, Ca^{2+} accumulation, and neuronal degeneration. Any of these abnormalities could, in principle, be allayed by treatments that reduce Ca^{2+} influx.

Interestingly, a recent report provided evidence that Ca^{2+} channel blockers can prevent age-associated declines in memory in rabbits (8). If Ca^{2+} is, in fact, ultimately responsible for the neuronal death that occurs in many of these neuropathological conditions, then similar Ca^{2+}-based preventative and therapeutic approaches might prove beneficial.

The Teeter-Totter Model

Taken together with the considerable prior literature on EAAs and GFs in neuronal development and degeneration, our findings suggest the following model for the regulation of neural circuitry in development, adult plasticity, and disease (Fig. 4). During the early stages of development (prior to arrival of EAA-releasing axons) a neuron is likely to be exposed to high levels of FGF compared to EAA; under these conditions intracellular Ca^{2+} levels are relatively low, and neurite outgrowth and cell survival are promoted. The action of the EAA comes into play as neurons interact;

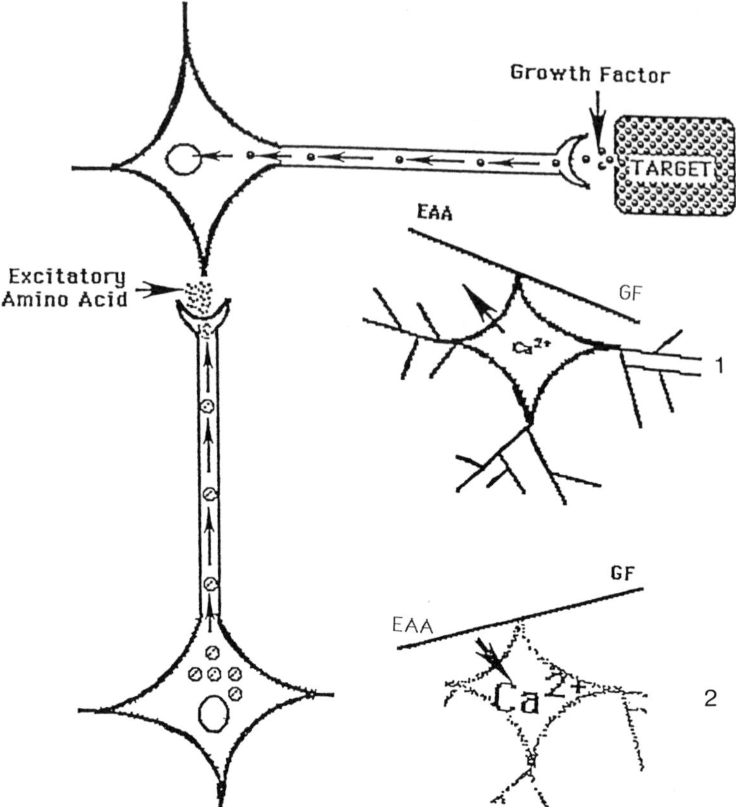

Fig. 4.   Teeter-totter model for the regulation of neuroarchitecture by EAAs and growth factors. This model is based upon studies of the interactive effects of glutamate and FGF in cultured hippocampal neurons. A neuron (upper left) receives input from both a target-derived growth factor (upper right) and an EAA-releasing neuron (lower left). Alone, the growth factor (GF) promotes cell survival and neurite outgrowth, whereas the EAA causes neuritic regression and cell death. The change in architecture of the neuron is determined by the net level of activation of each type of environmental signal. When levels of the growth factor are high or levels of the EAA are low, the teeter totter is tipped in favor of the growth factor and the neuron survives and grows (1). On the other hand, a deficiency in growth factor or an excess of EAA tips the teeter-totter to the EAA side and neurodegeneration occurs 2). See text for discussion.

dendritic outgrowth is inhibited and synpases begin to form (26). Some natural cell death may occur at this time as neurons compete for GF; variations in levels of EAA may also contribute to selective neuronal death during development.

Once a functional neural circuitry is established, the GF-EAA teeter-totter maintains a balance, and moderate changes in the levels of FGF and EAAs allow for adaptive plasticity. Again, intracellular calcium plays a key role as a mediator of synaptic plasticity. With aging, or the onset of specific neurodegenerative disorders, the GF-EAA teeter-totter may become unbalanced and shift towards the EAA side. This may result from large reductions in FGF levels, large elevations in EAA input, or perhaps even a combination of moderate reductions in FGF combined with moderate increases in EAAs. This shift in favor of the regressive (EAA) input would then lead to sustained rises in intracellular Ca^{2+} which, in turn, induces neuronal degeneration. Of course this model is still quite speculative, and is based largely on the data obtained from work in the hippocampal system. Additional in situ work will be necessary to verify such a scenario in the intact hippocampus, and the applicability of this model to other neural systems and other GFs will also need testing.

ACKNOWLEDGEMENTS

I thank Drs. P. B. Guthrie, S. B. Kater, and M. Murrain for their important contributions to the research. D. Giddings, J. Mattson, L. Nguyen, and B. Rychlic provided technical assistance. This work was supported by the NIH, the French Foundation for Alzheimer Research, and a Faculty Scholar Award to the author from the Alzheimer's Disease and Related Disorders Association.

REFERENCES

1. Alkon, D. L., 1989, Memory storage and neural system. **Sci. Amer.** 261:42-50.
2. Appel, S. H., 1986, A unifying hypothesis for the cause of amyotrophic lateral sclerosis, Parkinsonism, and Alzheimer disease. **Ann. Neurol.** 10:499-505.
3. Caday, C. G., A. Mirzabegian, J. Prosser, M. Klagsbrun and S. P. Finklestein., 1988, Fibroblast growth factor levels in developing brain. **Soc. Neurosci. Abstr.** 14:363.
4. Carafoli, E., 1987, Intracellular calcium homeostasis. **Ann. Rev. Biochem.** 56:395-433.
5. Choi, D. W., 1988, Glutamate neurotoxicity and diseases of the nervous system. **Neuron** 1:623-634.
6. Connor, J. A., W. J. Wadman, P. E. Hockberger and R. K. S. Wong., 1988, Sustained dendritic gradients of calcium induced by excitatory amino acids in CA1 hippocampal neurons. **Science** 240:649-653.
7. Cowan, W. M., J. W. Fawcett, D. D. M. O'Leary, and B. B. Stanfield. 1984. Regressive events in neurogenesis. **Science** 225 : 1258-1265.
8. Deyo, R. A., K. T. Straube, and J. F. Disterhoft. 1989. Nimodipine facilitates associative learning in aging rabbits. **Science** 243 : 809-811.
9. Dreyer, D., A. Lagrange, C. Grothe and K. Unsicker., 1989, Basic fibroblast growth factor prevents ontogenetic neuron death in vivo. **Neurosci. Lett.** 99:35-38.
10. Fischer, W. K. Wictorin, A. Bjorklund, L. R. Williams, S. Varon and F. H. Gage., 1987, Amelioration of cholinergic neuron atrophy and spatial memory impairment in aged rats by nerve growth factor. **Nature** 329:65-68.
11. Gage, F. H., D. M. Armstrong, L. R. Williams and S. Varon., 1988, Morphological response of axotomized septal neurons to nerve growth factor. **J. Comp. Neurol.** 269:147-155.
12. Gibson, G. E., and C. Peterson., 1987, Calcium and the aging nervous system. **Neurobiol. Aging** 8:329-343.
13. Green, L. A., and E. M. Shooter., 1980, The nerve growth factor. **Annu. Rev. Neurosci.** 4:353-402.
14. Hefti, F., A. Dravid and J. Hartikka., 1984, Chronic intraventricular injections of nerve growth factor elevate hippocampal choline acetyltransferase activity in adult rats with partial septo-hippocampal lesions. **Brain Res.** 293:305-311.
15. Hesketh, T. R., J. P. Moore, J. D. H. Morris, M. V. Taylor, J. Rogers, G. A. Smith and J. C. Metcalfe., 1985, A common sequence of calcium and pH signals in the mitogenic stimulation of eukaryotic cells. **Nature** 313:481-484.
16. Kater, S. B., M. P. Mattson, C. S. Cohan, and J. A. Connor., 1988, Calcium regulation of the neuronal growth cone. **Trends Neurosci.** 11:315-321.
17. Kudo, Y., K. Ito, H. Miyakawa, Y. Izumi, A. Ogura and H. Kato., 1987, Cytoplasmic calcium elevation in hippocampal granule cell induced by perforant path stimulation and L-glutamate application. **Brain Res.** 407:168-172.
18. Maragos, W. F., J. T. Greenamyre, J. B. Penney, and A. B. Young., 1987, Glutamate dysfunction in Alzheimer's disease: an hypothesis. **Trends Neurosci.** 10:65-68.
19. Mattson, M. P., 1988, Neurotransmitters in the regulation of neuronal cytoarchitecture. **Brain Res. Rev.** 13:179-212.

20. Mattson, M. P., 1989, Cellular signaling mechanisms common to the development and degeneration of neuroarchitecture. **Mech. Ageing Dev.** in press

21. Mattson, M. P, P. Dou, and S. B. Kater., 1988, Outgrowth-regulating actions of glutamate in isolated hippocampal pyramidal neurons. **J. Neurosci.** 8:2087-2100.

22. Mattson, M. P., P. b. Guthrie, B. C. Hayes, and S. B. Kater., 1989, Roles for mitotic history in the generation and degeneration of neuroarchitecture. **J. Neurosci.** 9:1223-1232.

23. Mattson, M. P., P. B. Guthrie, and S. B. Kater., 1988, Intracellular messengers in the generation and degeneration of hippocampal neuroarchitecture. **J. Neurosci. Res.** 20:447-464.

24. Mattson, M. P., P. B. Guthrie, and S. B. Kater. 1989. A role for Na+-dependent calcium extrusion in protection against neuronal excitotoxicity. **FASEB J.** in press.

25. Mattson, M. P., and S. B. Kater., 1989, Development and selective neurodegeneration in cell cultures from different hippocampal regions. **Brain Res.** 490:110-125.

26. Mattson, M. P., R. E. Lee, M. E. Adams, P. B. Guthrie, and S. B. Kater., 1988, Interactions between entorhinal axons and target hippocampal neurons: A role for glutamate in the development of hippocampal circuitry. **Neuron** 1:865-876.

27. Mattson, M. P., M. Murrain, P. B. Guthrie, and S. B. Kater., 1989, Fibroblast growth factor and glutamate: Opposing roles in the generation and degeneration of hippocampal neuroarchitecture. **J. Neurosci.** 9: in press

28. Morrison, R. S., R. F. Keating and J. R. Moskal., 1988, Basic fibroblast growth factor and epidermal growth factor exert differential trophic effects on CNS neurons. **J. Neurosci. Res.** 21:71-79.

29. Otto, D., M. Frotscher and K. Unsicker., 1989, Basic fibroblast growth factor and nerve growth factor administered in gel foam rescue medial septal neurons after fimbria fornix transection. **J. Neurosci. Res.** 22:83-91.

30. Pettmann, B., G. Labourdette, M. Weibel and M. Sensenbrenner., 1986, The brain fibroblast growth factor (FGF) is localized in neurons. **Neurosci. Lett.** 68:175-180.

31. Represa, A, E. Tremblay, and Y. Ben-Ari., 1989, Transient increase of NMDA-binding sites in human hippocampus during development. **Neurosci. Lett.** 99:61-66.

32. Rozengurt, E. and S. A. Mendoza., 1985, Synergistic signals in mitogenesis: Role of ion fluxes, cyclic nucleotides and protein kinase C in Swiss 3T3 cells. **J. Cell Sci.** 3 (suppl.):229-242.

33. Schwarcz, R., A. C. Foster, E. D. French, W. O. Whetsell, and C. Kohler., 1984, Excitotoxic models for neurodegenerative disorders. **Life Sci.** 35 : 19-32.

34. Tulipan, N., S. Luo, G. S. Allen and W. O. Whetsell., 1988, Striatal grafts provide sustained protection from kainic and quinolinic acid-induced damage. **Expl. Neurol.** 102:325-332.

35. Walicke, P. A., 1988, Basic and acidic fibroblast growth factors have trophic effects on neurons from multiple CNS regions. **J. Neurosci.** 8:2618-2627.

TROPHIC EFFECTS OF EXCITATORY AMINO ACIDS IN THE DEVELOPING NERVOUS

SYSTEM

R. Balázs and N. Hack

Netherlands Institute for Brain Research
Meibergdreef 33, 1105 AZ Amsterdam
The Netherlands

Cerebellar granule cells in culture exhibit specific survival requirements (e.g. Thangnipon et al., 1983; Kingsbury et al., 1985; Gallo et al., 1987a). These include chronic depolarization, usually effected by increasing the K^+ concentration of the medium above the physiological level (≥ 20 mM). The dependence on 'high' K^+ develops within a narrow window of time between 2 and 4 days *in vitro* (DIV) (Gallo et al., 1987a). It was also observed that the effect of depolarization on cell survival is mediated through stimulated Ca^{2+} influx via voltage sensitive Ca^{2+} channels (VSCC). On the basis of these findings we have put forward the hypothesis that the effect of high K^+ *in vitro* mimics the influence of the first innervation the immediately postmigratory granule cells receive from the mossy fibres (Balázs and Jo/rgensen, 1987; Gallo et al., 1987a). Many of these fibres are glutamatergic (Somogyi et al., 1986) and granule cells are endowed with glutamate (Glu) receptors including the N-methyl- D-aspartate (NMDA)-preferring subtype (Cull-Candy et al., 1988), which is known to be linked to an ion channel whose Ca^{2+} permeability, in comparison with the other Glu receptor subtypes, is high (Mayer and Westbrook, 1987). The hypothesis predicted therefore that the trophic influence of the mossy fibre innervation is mediated through Ca^{2+} influx via the NMDA receptor-ionophore complex.

This hypothesis was supported by the observation that the survival of granule cells can be promoted in culture by NMDA treatment (Balázs et al., 1988a,b). We have recently extended these studies to the examination of the possible involvement in the trophic effect of excitatory amino acid receptors other than the NMDA preferring subtype (Balázs et al., 1989b). Here we summarize our observations and discuss briefly certain implications as to the developing nervous system.

Cerebellar granule cell cultures

The studies have been conducted on granule cells cultured from rat cerebellum (Thangnipon et al., 1983; see legend to Fig. 1). By choosing the right age, about postnatal day 8, and employing certain manipulations of the growth conditions it is possible to obtain a culture which is relatively homogeneous in terms of both cellular composition (about 90% of the cells being granule neurons) and maturation: this facilitates the detection of transient and maturation stage-dependent effects, which could easily be masked *in vivo*.

Effect of NMDA on granule cell survival

As in the case of elevated K^+, the dependence of granule cells on NMDA develops with the maturation of cells in culture and becomes manifested during the period 2-4 days in culture (Balázs et al., 1988b).

Excitatory Amino Acids and Neuronal Plasticity
Edited by Y. Ben-Ari
Plenum Press, New York, 1990

The NMDA effect is both concentration-dependent and voltage- sensitive (Balázs et al., 1988b). Irrespective of the medium K+ concentration a 'saturation' of the effect of NMDA on cell survival is detectable: this occurs at about 100 µM of NMDA and cell survival remains unaltered at higher concentrations of NMDA, up to 1.1 mM (the highest concentration tested).

The level of NMDA induced cell survival is, however, dependent on the medium K+ concentration within a range (5-15 mM K+) at which K+ on its own has relatively little effect. Taking maximal survival at 5 mM K+ (K5) as 100%, the corresponding values in a medium containing 10 mM or 15 mM K+ (K10 or K15) are about 140% or 180%. This effect probably reflects the voltage dependence of the Mg^{2+} blockade of the ion channel linked to the NMDA receptor. In addition to the efficacy, the potency of NMDA in promoting cell survival is also a function of the K+ concentration approaching at 15 mM the reported affinity of NMDA in binding to brain membranes (Olverman et al., 1984).

NMDA promotes selectively the survival of nerve cells. Estimates of proteins which are relatively enriched in nerve cells (N-CAM, D3 and synaptin) (Bock et al., 1982; Jorgensen, 1983; Nybroe et al., 1985) showed a dose and voltage dependence which was similar to that of overall cell survival, while markers of astrocytes (the other major cell type present in the cultures) were unaffected by NMDA supplementation.

It seems that the effect of EAA is not mediated indirectly through stimulation of the release of trophic factors from astrocytes. Although astrocytes possess EAA receptors, these do not seem to include the NMDA preferring subtype (Bowman and Kimelberg, 1984; Kettemann and Schachner, 1985; Lehman and Hansson, 1988; Usowicz et al., 1989).

Pharmacological characterization clearly showed the involvement of 'conventional' NMDA receptors in the survival promoting effect (Balázs et al., 1989a). Selective antagonists, such as competitive inhibitors of the recognition site (APV, APH) and noncompetitive inhibitors acting on the ionophore (MK 801, dextromethorphan), proved to block the trophic effect, and their potencies were similar to those known from studies in which the binding of these substances to brain membrane preparations was determined (Foster and Fagg, 1984; Olverman et al., 1984; Wong et al., 1988). On the other hand, non-selective EAA inhibitors were relatively less potent.

The involvement of Glu - which, although not added to the medium, is released by the cultured cells - in promoting the survival of granule neurones through receptor stimulation was indicated by the observation that NMDA antagonists (APV or MK 801) reduced the limited survival detectable in low K+ media. This view was supported by the finding that continuous removal from the culture medium of Glu, by the addition of alanine aminotransferase and pyruvate, reproduced the effect of NMDA receptor blockers (Balázs et al., 1989a). Since Glu is a mixed agonist, we investigated next the effects of Glu analogues which are preferentially stimulating EAA receptors other than the NMDA-preferring subtype.

Effect of stimulation of Glu receptors other than the NMDA preferring subtype on granule cell survival

Quisqualic acid (QA) or a-amino-3-hydroxy-5-methyl-4-isoxazole- propionate (AMPA), studied over a wide range of concentrations, in 'low' K+ containing media, had no effect on granule cell survival. On the other hand, kainic acid (KA) had a significant and relatively complex influence (Balázs et al., 1989b) (Fig. 1). Cell survival was promoted at low concentrations, maximal effects being obtained at 25-50 µM. Further increase in [KA] resulted, at first, in a loss of the effect, while at high concentrations (>500 µM) KA became toxic to the cells. Estimation of proteins which are relatively enriched in nerve cells or astrocytes, such as N-CAM or glutamine synthetase, indicated that KA primarily rescues neurones.

The KA effect involved EAA receptors whose characteristics were clearly different from those of the NMDA receptors. While the promotion of granule cell survival by NMDA proved to be voltage-dependent (Balázs et al., 1988b), that effected by KA was voltage-independent (Balázs et al., 1989b) Furthermore, dinitroquinoxalinedione (DNQX), which is a relatively

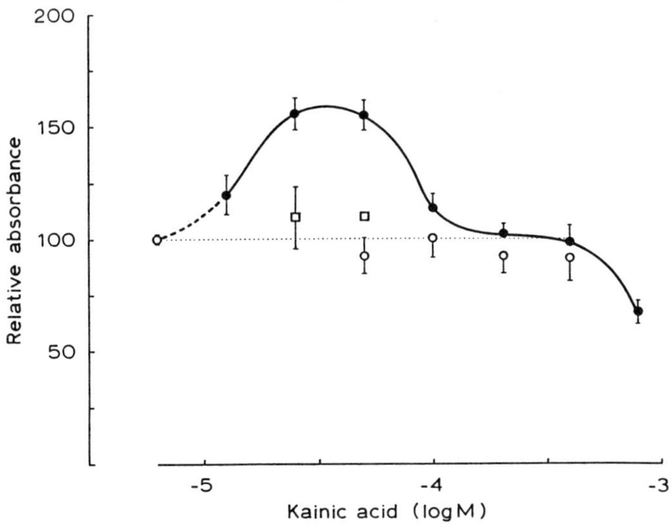

Fig. 1.   Cerebellar granule cells were grown, usually for 7-9 days, on poly-L-lysine coated plastic dishes in basal Eagle's medium containing 10% foetal calf serum supplemented with glutamine (2 mM) and genatmicin (100 µg/ml) (K5 medium). Non-neuronal cell proliferation was blocked with cytosine arabinoside (10 µM at about 18 h) (see e.g. Thangnipon et al., 1983). The Glu analogues, KA (o), QA ( ) or AMPA (o) were added at 2 DIV. Survival was monitored by estimating the reduction of a tetrazolium salt (MTT) to a coloured formazan by viable cells (see Balázs et al., 1988b). Data for the dose response to KA, QA and AMPA are derived from 13 and 2-3 independent experiments respectively. In each experiment, data were normalized by taking the optical density in the cultures in K5 (without added EAA) as 100. Estimates are means and the S.E.M. is indicated with a bar (unless it is too small to be shown). One-way ANOVA showed that only the effect of KA is significant; post-hoc Newman-Keuls test indicated that 25 and 50 µM KA significantly enhanced, whereas 800 µM compromised cell survival (p=0.05). Data are taken from Balázs et al. (1989b).

selective antagonist of the KA/QA receptors (Honoré et al., 1988), was found to be far the most potent inhibitor of the KA induced cell rescue (IC_{50} in the presence of 50 µM KA about 10 µM). The relative selectivity of the inhibition by DNQX was indicated by the observation that at a concentration of 50 µM, DNQX completely blocked the effect of KA, while cell survival promoted by NMDA appeared to be unaffected. Moreover, QA or AMPA, which had no influence on survival, inhibited potently the trophic effect of KA. These findings are consistent with other observations, showing that QA/ AMPA have high affinity for the KA- receptor (e.g. Simon et al., 1976; McCaslin and Morgan, 1987; Gallo et al., 1989).

It seems, therefore, that KA acting on a unique EAA receptor is responsible for the promotion of granule cell survival. Nevertheless, the effect may be secondary to KA-receptor stimulation, by being mediated through a depolarization induced activation of VSCC and thus increased Ca^{2+} influx. Nifedipine (NF) is a potent dihydropyridine antagonist of a subpopulation of VSCC, the L channels (Nowycky et al., 1985) and its effect on $^{45}Ca^{2+}$ influx and cell survival has also been demonstrated in cerebellar granule cell cultures (Kingsbury and Balázs, 1987: Gallo et al., 1987a).

Table 1.   Influence of the Calcium Antagonist, Nifedipine (NF), on the Trophic Effect
on Granule Cells of EAA Receptor Stimulation or Chronic Depolarization

a) Chronic depolarization

LDH activity/dish

Nifedipine (µM)	-	0.1	0.5	1.0
% of K25	100±1.3	54±3.9	37±2.2	39±4.0

b) Differential effect of nifedipine (NF) on the increase
in cell survival as a result of treatment with kainate or
NMDA

	50 µM KA		NMDA	
0.5 µM NF	-	+	-	+
%	(100)	1	(100)	97

In (a) the effect of NF on the survival of cells (grown in a medium containing 25 mM K^+, K25) was tested in two experiments by estimating LDH activities per dish: in each experiment the values after chronic NF treatment are expressed as a percentage of the mean LDH activity of the cells in the absence of NF. One way ANOVA followed by Newman-Keuls' procedure showed that NF interfered with cell survival in a concentration-dependent manner, exerting its full effect at 0.5 µM. In (b) treatment of cultures in K5 with 50 µM KA and in K10 with 140 µM NMDA increased LDH activity (units in µmol/h per dish) from 7.34±0.40 and 7.04±0.62 respectively by 1.07 and 4.00 units. The difference in activity in the presence of KA or NMDA + NF minus activity in K5 or K10 + NF was expressed as a percentage of the difference in the absence of NF. Data modified from Balázs et al. (1989c).

Table 2.   Effect of NMDA Receptor Blockade on the KA Induced Promotion of Granule
Cell Survival

% of K5

-	KA	MK 801	KA + MK 801
100±1.5	128±3.2	90±7.3	161±3.5

Pooled data from 6 independent cultures (7-9 DIV) in K5 medium (for each culture, 3-6 replications per condition). Concentration of supplements was 50 µM KA and 5 µM MK 801. The data were normalized by expressing LDH activities (as a marker of cell survival) as a percentage of the values in K5 without supplements. One way ANOVA showed significant effects ($p < 0.00001$), while the Newman-Keuls' procedure indicated that the addition of MK 801 to the cultures grown in the presence of KA increased survival. Data modified from Balázs et al. (1989c).

At a concentration of 0.5 µM NF blocked maximally the beneficial effect of elevated K+, but not that of NMDA, on cell survival (Table 1). In contrast to NMDA, the effect of KA was also completely blocked by 0.5 µM NF. These observations are apparently at variance with the findings of Holopainen et al. (1989), who reported that the KA elicited elevation of cytoplasmic free calcium concentration is not inhibited by 50 µM NF. However, in the latter studies NF was tested on cells exposed to 500 µM KA and it seems that at such a high concentration KA evokes Ca2+ influx through receptor gated channels and the elevation of $[Ca^{2+}]_i$ may also result from perturbation of the intracellular calcium homeostasis (Murphy and Miller, 1989).

It is known that KA can induced Glu release from cerebellar granule cells, the EC_{50} being in the concentration range (about 50 µM) where KA promotes cell survival (Nicoletti et al., 1986; Gallo et al., 1987b). It is therefore possible that the KA effect is mediated, at least in part, through the stimulation of NMDA receptors by Glu released in response to KA exposure. This possibility was tested by examining the effect on the KA-induced promotion of cell survival of selective inhibitors of the NMDA receptor. The non-competitive antagonist, MK 801, caused a significant potentiation of the effect of KA (Table 2). The implication of this observation is that upon supplementation with KA.the effect of a factor, presumably Glu, becomes manifested in the culture resulting in cell death mediated through NMDA receptor stimulation. In view of the promotion of cell survival by NMDA (see above), this finding was unexpected. Further experiments indicated that it is most unlikely that the potentiation of the effect of KA is due to side effects of MK 801, since APV, a competitive NMDA antagonist, had the same positive influence on cell survival (Table 3). If the potentiation of the effect of KA by MK 801 or APV was due to a blockade of the toxic effect of the released Glu on the NMDA receptor, the addition of an agonist should counteract the potentiation.

Table 3. The Potentiation of KA Effect by Antagonists Acting at the NMDA Receptor can be Blocked by Quinolinic Acid (QUIN), a Relatively Inert Agonist at this Receptor

KA	(50 µM)	+	+	+	+
APV	(50 µM)	-	+	+	+
QUIN	(mM)	-	-	2	4
%		100±2.	124±2.4	118±2.7	106±3.9

Cultures were grown for 7 DIV in K5 medium in the presence of 50 µM KA; additives are indicated. LDH activities per dish were normalized taking the mean value in the presence of KA as 100%. One way ANOVA showed significant effects and Newman-Keuls' procedure indicated that survival in the KA treated cultures was significantly increased by APV and this increase was blocked by 4 mM (although not by 2 mM) QUIN; 4 mM QUIN on its own had no significant effect on cell survival. Data modified from Balázs et al. (1989c).

Quinolinic acid is a very weak NMDA receptor agonist in the cerebellum (Stone and Burton, 1988): in our previous studies we could not detect significant effects on granule cell survival in culture even at the relatively high concentration of 4 mM (Balázs et al., 1989a), but a low potency toxic effect of QUIN observed in cerebellar slices does suggest that it is an agonist at the NMDA receptor (Garthwaite and Garthwaite, 1987). Table 3 shows that the APV induced potentiation of the effect of KA was indeed counteracted by QUIN in a concentration-dependent manner, thus supporting the view that in the presence of KA a released endogenous EAA does exert a toxic influence, which is mediated through an 'NMDA' receptor stimulation.

We tried to mimic this 'toxic' effect by adding different concentrations of NMDA to cultures grown in the presence of KA. However, when both EAA were present at suboptimal concentrations their effects were strictly additive, whereas at concentrations causing maximal promotion of cell survival, 50 µM KA did not increase cell survival beyond the level seen in the presence of 140 µM NMDA. It is therefore possible that the endogenous EAA liberated in the culture exerts receptor mediated effects which are not mimicked by the combined action of the Glu analogues tested.

In conclusion, it is worth to remember that it was not long ago that it was seriously debated whether Glu, a 'common' metabolite, could fulfil such a sophisticated role as required from a neurotransmitter. In the meantime this notion was completely accepted, while it also became evident that under certain conditions Glu, as an abundant excitatory transmitter, can also become toxic to nerve cells. Now we are putting forward evidence indicating that in addition, EAA exert a trophic effect on developing nerve cells. Our evidence concerning the trophic effect of EAA on cerebellar granule cells in culture together with observations on the toxic effect of EAA *in vivo* and tested in slices (e.g. Garthwaite and Garthwaite, 1986; see also Garthwaite in this volume) underline the fact that the trophic effect of EAA is maturational stage-dependent (for more detailed discussion of this point see Balázs et al., 1988b).

Evidence is forthcoming that indicates that the trophic influence of EAA is not unique to the cerebellum: such effects have also been detected recently in cultures of the spinal cord (Brenneman et al., 1989) and of the cerebral cortex (Ruijter and Baker, 1989). Furthermore, the trophic effect of EAA is not restricted to the promotion of cell survival. EAA seem to affect neurite emission and growth (Pearce et al., 1987; Mattson et al., 1988) differentially influencing the growth of dendrites and axons, thus shaping the architecture of the nerve cells and through this affecting the organization of neural networks (see Mattson, this volume). In addition, it seems that EAA can influence in a cell type specific manner neuronal differentiation (Moran and Patel, 1989; Patel et al., 1989).

Most of the evidence suggests that both the trophic and the toxic effects of EAA involve Ca^{2+} as the second messenger. However, we do not know yet the steps beyond the elevation of free cytoplasmic calcium activity, which ultimately lead to gene expression securing long-term survival. It is also possible that in addition to calcium other transduction systems linked to the relevant EAA receptors are also involved in the transient trophic effects. However, already at this initial stage of the research into the new trophic role of EAA, the observations have important implications. Thus, one of the characteristic features of the developing nervous system is an overproduction of nerve cells and their projections, including synaptic contacts, followed by a regression of the redundant elements (Hamburger and Oppenheimer, 1982; Levi-Montalcini, 1982; Cowan et al., 1984). According to common wisdom, this is the penalty paid for securing proper connectivity. However, it is possible that such redundant projections, for example from the glutamatergic pyramidal cells in the cerebral cortex, fulfil a role by providing, at the right time, a trophic influence ensuring the survival of their transient synaptic partners.

Furthermore, there is evidence implicating transmitters other than EAA exerting a transient trophic influence in the developing nervous system (e.g. Balázs et al., 1977; Busnikov, 1984 ; Lauder, 1988). It is possible, therefore, that during development transduction systems which are coupled to transmitter receptors may also trigger processes other than those involved in information coding, subserved by these systems in the fully differentiated cell. Such an effect would be determined by the maturational stage of the cell which receives the stimulus. The economy of such an arrangement is considerable in terms of significant sparing of genetic material.

ACKNOWLEDGEMENTS

We are indebted to Ms. Olga Pach for typing the manuscript. The support provided by NWO and the Van den Houten Foundation is gratefully acknowledged.

REFERENCES

Balázs, R., and Jørgensen, O. S., 1987, Trophic function of excitatory transmitter amino acids, **Neurosci.**, 22:S41.
Balázs, R., Patel, A. J., and Lewis, P. D., 1977, Metabolic influences on cell proliferation in the brain, in: "Biochemical Correlates of Brain Structure and Function", A. N. Davidson, ed., Academic Press, New York.

Balázs, R., Hack, N., and Jørgensen, O. S., 1988a, Stimulation of the N-methyl-D-aspartate receptor has a trophic effect on differentiating granule cells, **Neurosci. Lett.**, 87:80-86.

Balázs, R., Jørgensen, O. S., and Hack, N., 1988b, N-methyl-D-aspartate promotes the survival of cerebellar granule cells in culture, **Neurosci.**, 27:437-451.

Balázs, R., Hack, N., Jørgensen, O. S., and Cotman, C. W., 1989a, N-methyl-D-aspartate promotes the survival of cerebellar granule cells: pharmacological characterization, **Neurosci. Lett.**, 101:241- 246.

Balázs, R., Hack, N., and Jørgensen, O. S., 1989b, Selective stimulation of excitatory amino acid receptor subtypes and the survival of cerebellar granule cells in culture, **Neurosci.**, submitted.

Balázs, R., Hack, N., and Jørgensen, O. S., 1989c, Interactions in the effects of stimulation of excitatory amino acid receptors and survival of cerebellar granule cells in culture, **Int. J. Dev. Neurosci.**, submitted.

Bock, E., Divac, I., Norrild, B., Thorn, N. A., Thorp-Pedersen, C., and Treiman, M., 1982, Synaptic membrane proteins in mammalian brain, **Scand. J. Immun.**, 15:223-240.

Bowman, C. L., and Kimelberg, H. K., 1984, Excitatory amino acids directly depolarize rat brain astrocytes in primary culture, **Nature**, 311:656-659.

Brenneman, D. E., Yu, C., and Nelson, P. G., 1989, Multi-determinate regulation of neuronal survival: neuropeptides, excitatory amino acids and bioelectric activity, **Int. J. Dev. Neurosci.**, submitted.

Busnikov, G. A., 1984, The action of neurotransmitters and related substances on early embryogenesis, **Pharmac. Ther.**, 25:23-59.

Cowan, W. M., Fawcett, J. W., O'Leary, D. D. M., and Stanfield, B. B., 1984, Regressive events in neurogenesis, **Science**, 225:1258-1265.

Cull-Candy, S. G., Howe, J. R., and Ogden, D. C., 1988, Noise of single channels activated by excitatory amino acids in rat cerebellar granule neurones, **J. Physiol.**, 400:189-222.

Foster, A. C., and Fagg, G. E., 1984, Acidic amino acid binding sites in neuronal membranes: their characterization and relationship to synaptic receptors, **Brain Res. Rev.**, 7: 103-164.

Gallo, V., Kingsbury, A., Balázs, R., and Jørgensen, O. S., 1987a, The role of depolarization in the survival and differentiation of cerebellar granule cells in culture, **J. Neurosci.**, 7:2203-2213.

Gallo, V., Suergiu, R., Giovannini, C., and Levi, G., 1987b, Glutamate receptor subtypes in cultured cerebellar neurons: modulation of glutamate and gamma-aminobutyric acid release, **J. Neurochem.**, 49: 1801-1809.

Gallo, V., Giovanni, C., and Levi, G., 1989, Quisqualic acid modulates kainate responses in cultured cerebellar granule cells, **J. Neurochem.**, 52:10-16.

Garthwaite, G., and Garthwaite, J., 1986, In vitro neurotoxicity of excitatory amino acids analogues during cerebellar development, **Neurosci.**, 17:755-767.

Garthwaite, G., and Garthwaite, J., 1987, Quinolinate mimics neurotoxic actions of N-methyl-D-asparate in rat cerebellar slices, **Neurosci. Lett.**, 79:35-39.

Garthwaite, J., 1990, Mechanisms of excitatory amino acid neurotoxicity in rat brain slices, in: "Excitatory Amino Acids and Neuronal Plasticity", Y. Ben Ari, ed., Plenum Press, New York.

Hamburger, V., and Oppenheimer, R. W., 1982, Naturally occurring neuronal death in vertebrates, **Neurosci. Commun.**, 1:39-55.

Holopainen, I.,Enkvist, M. O. K., and Akerman, K. E. O., 1989, Glutamate receptor agonists increase intracellular Ca2+ independently of voltage-gated Ca2+ channels in rat cerebellar granule cells, **Neurosci. Lett.**, 98:57-62.

Honoré, T., Davies, S. N., Drejer, J., Fletcher, E. J., Jacobsen, P., Lodge, D., and Nielsen, F. E. (1988) Quinoxalinediones: potent competitive non-NMDA glutamate receptor antagonists, **Science**, 241: 701-703.

Jorgensen, O. S., 1983, D2-protein and D3-protein as markers for synaptic turnover and concentration, **J. Neur. Trans. Suppl.**, 18:245-255.

Kettemann, H., and Schachner, M., 1985, Pharmacological properties of Á- aminobutryic acid-, glutamate-, and asparate-induced depolarization in cultured astrocytes, **J. Neurosci.**, 5:3295-3301.

Kingsbury, A., and Balázs, R., 1987, Effect of calcium agonists and antagonists on cerebellar granule cells, **Eur. J. Pharmacol.**, 140: 275-283.

Kingsbury, A., Gallo, V., Woodhams, P., and Balázs, R., 1985, Survival, morphology and adhesion properties of cerebellar interneurones cultured in chemically defined and serum supplemented medium, **Dev. Brain Res.**, 17:17-25.

Lauder, J. M., 1988, Neurotransmitters as morphogens, **Progr. Brain Res.**, 73:365-387.

Lehman, A., and Hansson, E., 1988, Kainate-induced stimulation of amino acid release from primary astroglial cultures of the rat hippocampus, **Neurochem.Int.**, 13:557-561.

Levi-Montalcini, R., 1982, Developmental neurobiology and the natural history of Nerve Growth Factor, **A. Rev. Neurosci.**, 5:341-361.

McCaslin, P. P., and Morgan, W. W., 1987, Cultured cerebellar cells as an in vitro model of excitatory amino acid receptor function, **Brain Res.**, 417,:380-384.

Mattson, M. P., 1990, Interactions of excitatory amino acids and growth factors in the development and degeneration of hippocampal neuroarchitecture, in: "Excitatory Amino Acids and Neuronal Plasticity", Y. Ben Ari., ed., Plenum Press, New York.

Mattson, M. P., Dou, P., and Kater, S. B., 1988, Outgrowth-regulating actions of glutamate in isolated hippocampal pyramidal neurons, **J. Neurosci.**, 8:2087-2100.

Mayer, M. L., and Westbrook, G. L., 1987, The physiology of excitatory amino acids in the vertebrate central nervous system, **Prog. Neurobiol.**, 28:197-276.

Moran, J., and Patel, A. J., 1989, Stimulation of the N-methyl-D-aspartate receptor promotes the biochemical differentiation of cerebellar granule neurons and not astrocytes, **Brain Res.**, in press.

Murphy, S. N., and Miller, R. J., 1989, Regulation of Ca^{++} influx into striatal neurons by kainic acid, **J. Pharmacol.Exp. Ther.**, 249:184- 193.

Nicoletti, P., Wroblewski, J. T., Novelli, A., Alho, H., Guidotti, A., and Costa, E., 1986, The activation of inositol phospholipid metabolism as a signal-transducing system for excitatory amino acids in primary cultures of cerebellar granule cells, **J. Neurosci.**, 6: 1905-1911.

Nowycky, M. C., Fox, A. P., and Tsien, R. W., 1985, Three types of neuronal calcium channel with different calcium agonist sensitivity, **Nature**, 316:440-443.

Nybroe, O., Albrechtsen, M., Dahlin, J., Linnemann, D., Lyles, J. M., Moller, C. S., and Bock, E., 1985, Biosynthesis of the neuronal cell adhesion molecule: Characterization of polypeptide C, **J. Cell Biol.**, 10:1-6.

Olverman, H. J., Jones, A. W., and Watkins, J. C., 1984, L-glutamate has higher affinity than other amino acids for [3H]-D-AP5 binding sites in rat brain membranes. **Nature**, 307:460-462.

Patel, A. J., Hunt, A., and Sanfeliu, C., 1989, Cell-type specific effects of N-methyl-D-aspartate on biochemical differentiation of subcortical neurons in culture, **Int. J. Dev. Neurosci.**, submitted.

Pearce, I. A., Cambray-Deakin, M. A., and Burgoyne, R. D., 1987, Glu- tamate acting on NMDA receptors stimulates neurite outgrowth from cerebellar granule cells, **FEBS Lett.**, 223:143-147.

Ruijter, J. M., and Baker, R. E., 1989, The effects of potassium-induced depolarization, glutamate receptor antagonists and N-methyl-D- aspartate on neuronal survival in cultured neocortex explants. **Int. J. Dev. Neurosci.**, submitted.

Simon, J. R., Contrera, J. F., and Kuhar, M. J., 1976, Binding of [3H] kainic acid, an analogue of L-glutamate, to brain membranes, **J. Neurochem.**, 26: 141-146.

Somogyi, P., Halasy, K., Somogyi, J., Storm-Mathisen, S., and Ottersen, O. P., 1986, Quantification of immunogold labelling reveals enrichment of glutamate in mossy and parallel fibre terminals in cat cerebellum, **Neurosci.**, 19:1045-1050.

Stone, T. W., and Burton, N. R., 1988, NMDA receptors and ligands in the vertebrate CNS, **Prog. Neurobiol.**, 30:333-368.

Thangnipon, W., Kingsbury, A., Webb, M., and Balázs, R., 1983, Observations on rat cerebellar cells in vitro: Influence of substatum, potassium concentration and relationship between neurones and astrocytes, **Dev. Brain Res.**, 11:177-189.

Usowicz, M. M., Gallo, V., and Cull-Candy, S. G., 1989, Multiple conduc- tance channels in type-2 astrocytes by excitatory amino acids, **Nature**, 339: 380-383.

Wong, E. H. F., Knight, A. R., and Woodruff, G. N., 1988, [3H]MK-801 labels a site on the N-methyl-D-aspartate receptor channel complex in rat brain membranes, **J. Neurochem.**, 50:274-281.

MECHANISMS OF EXCITATORY AMINO ACID-INDUCED STIMULATION OF GABAERGIC

SYNAPTIC ACTIVITY IN CULTURES FROM THE RAT SUPERIOR COLLICULUS

M. Perouansky** and R. Grantyn*

*Department of Neurophysiology, Max Planck Institute
for Psychiatry, 8033 Martinsried, F. R. G.

**Present address: Department of Physiology, Hebrew
University, School of Medicine, P.O. Box 11 72
Jerusalem 91010, Israel

INTRODUCTION

Presynaptic autoreceptors appear to be a ubiquitous regulatory device in many systems
(Starke, 1981; Middlemiss, 1988). It is therefore not too surprising that excitatory amino acid receptors (EAARs) have been found on EAA-releasing terminals (Ferkany et al., 1982; Collins et al.,
1983; Ferkany and Coyle, 1983; Potashner and Gerard, 1983; Poli et al., 1985) or putative glutamatergic axons (Evans, 1980; Curtis et al., 1984) and terminals (Foster et al., 1981; Represa et al. 1987).
More controversial is the notion that EAARs can also modulate the release of the inhibitory neurotransmitter GABA (Pearce and Dutton, 1982; Drejer et al., 1987; Gallo et al., 1987; Harris and
Miller, 1989; Pin and Bockaert, 1989). Considering the major role that GABAergic inhibition plays
in primary visual brain structures it seems important to further explore the possibility that
EAARs participate in the regulation of GABA release. In particular, it should be clarified whether the EAA-stimulated release of GABA is indeed related to the generation of GABAergic synaptic activity. If so, it would be important to localize the respective EAARs and to characterize
these EAARs on the basis of the EAA-induced conductance changes.

Recent studies have clearly demonstrated that EAA can facilitate the release of GABA in
the presence of TTX (Perouansky and Grantyn, 1988; Harris and Miller, 1989; Pin and Bockaert,
1989), but it has been doubted that this GABA derives from a vesicular pool of the transmitter.
Instead, it was hypothesized that EAA-induced depolarization turns Na^+-dependent, tetanus
toxin-insensitive GABA-uptake into GABA-release (Harris and Miller, 1989; Pin and Bockaert,
1989). If reversed GABA-uptake was the only source of the EAA-stimulated GABA-release EAA
should have no influence on the frequency of discrete synaptic Cl- currents.

Previously described effects of EAA on the release of neurotransmitters (Giorguieff et al.,
1977 ; Pastuszko et al ., 1984; Roberts and Anderson, 1979; Krespan et al ., 1982; Ferkany et al.1982
; Collins et al., 1983; Ferkany et al., 1983; Poli et al., 1985; Errington et al ., 1987; Jones et al.,
1987) were characterized by a certain degree of variability, depending on the experimental
model. Even depression of GABA-release was seen, notably with lower concentration of KA
(McBean et al ., 1981; Poli et al ., 1985; Bray et al ., 1989) . As such, any further study on synaptic
activity modulated by direct application of EAA should test receptor-specific agonists over a
range of doses.

Excitatory Amino Acids and Neuronal Plasticity
Edited by Y. Ben-Ari
Plenum Press, New York, 1990

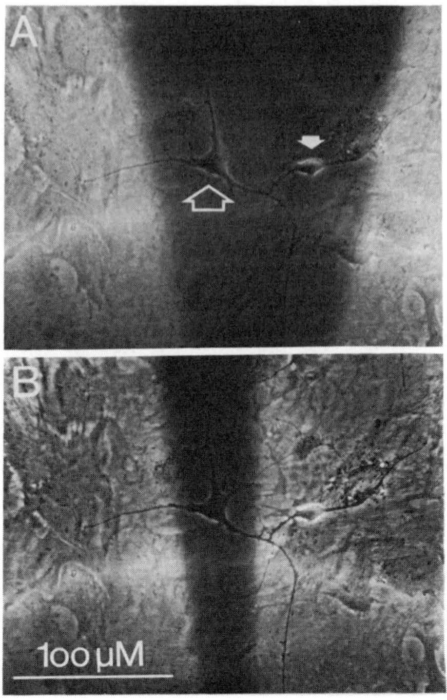

Fig. 1.   Photomicrographs with phase contrast illumination showing cultured
neurons from the rat superior colliculus after two weeks in vitro. Dark
shadow across the field: colored stream of salt solution between a superfusion
and a suction pipette. A: multiple cell superfusion. B: single cell superfusion.

The TTX-insensitivity of the EAA-stimulated transmitter release has been considered as
evidence for a presynaptic location of EAARs on axonal terminals (Ferkany et al ., 1982). However, this argument is not wholly acceptable. Notably, in sensory structures, such as the superior
colliculus (Tigges and Tigges, 1975) and the olfactory bulb (Jaffe and Cuello, 1980), GABA
releasing sites could also exist on dendrites. Moreover, a link between somatodendritic EAARs and
axonal release sites could be established by Ca^{2+} spikes (Dichter et al., 1983) or intracellular Ca^{2+}
signals (Kudo and Ogura, 1986; Connor et al., 1987; Hockberger et al ., 1989) propagated along the
axon. It is therefore important to directly test for the presence of presynaptic EAARs. With spatially restricted access of EAA to the axon terminals of a GABAergic neuron synapsing with the voltage clamped neuron in culture it should be possible to reveal a ligand-induced membrane potential
change in the presynaptic terminal by monitoring the frequency of spontaneous postsynaptic chloride currents. The KClinduced stimulation of transmitter release could serve as a control.

Here it will be shown that in cultures from the neonatal rat SC low concentrations of EAA do,
indeed, facilitate the generation of spontaneous synaptic Cl- currents. Therefore, an inversed
GABA uptake process cannot solely account for the EAA-evoked Ca^{2+} dependent release of GABA.
On the contrary, it is very likely that EAARs participate in the regulation of exocytotic GABA
release and thereby modulate GABAergic synaptic transmission in the superior colliculus. The
main objective of the present study was to determine the location of the respective EAARs.

METHODS

Cultured neurons from the stratum griseum superficiale (SGS) of the rat SC were studied with
patch-clamp techniques as described previously (Grantyn and Lux, 1988; Perouansky and Grantyn,

1989). In order to apply the test solutions in a spatially defined way we designed a push-pull type microsuperfusion system. This system consisted of a five-barrel common-outflow pipette (opening diameter 80 µm) and a suction pipette (opening 200 µm). The openings were positioned 250-300 µm apart from each other, with the investigated neuron between them. In the absence of a negative hydrostatic pressure through the suction pipette a wide stream of superfusion solution covered between 2 and 20 neurons, depending on culture density (multiple cell superfusion, Fig. 1A). If a wide bore superfusion pipette was placed at some distance from the voltage-clamped neuron, this cell could be maintained in a constant extracellular environment over the whole somadenritic area. By combining the superfusion with suction the stream of applied solutions could be made as narrow as 30-50 µm, thus contacting only one cultured cell (single cell superfusion, Fig. 1B). The width of the stream was regulated by changing either the hydrostatic pressure of the applied solution or the negative hydrostatic pressure of the suction. Selecting an appropriately shaped and positioned cell, the test solutions could be directed to particular areas of an individual neuron, i. e. either to the soma or to a neurite. The salt solution used for the single cell superfusion contained (in mMol): NaCl 140, KCl 3, $CaCl_2$ 10, HEPES 10, glucose 10, TTX $0.5 - 1 \times 10^{-6}$, pH 7.3. To visualize the superfusion stream, 0.3% fast Green (FG) was added in one barrel. The flow through the remaining barrels was adjusted to cover exactly the same area as the FG-colored solution. All records were taken at room temperature (22-24° C).

RESULTS

Properties of GABAergic spontaneous synaptic activity

The cultures were used for electrophysiological measurements between DIV 15 and 30. By this time *in vitro*, cultured SGS-derived neurons were contacted by numerous axon terminals that are visible even with phase contrast optics. Fig. 2A presents a photomicrograph from a culture processed to reveal immunoreactivity for glutamic acid decarboxylase (GAD), as reported elsewhere (Warton et al., 1990). It can be seen that a light GAD-negative neuron is covered with dark GAD-immunoreactive puncta. That these puncta truely represent presynaptic terminals was shown by electron microscopical examination of immunostained cultures (Warton et al., 1990).

Excitatory amino acid-induced stimulation of GABAergic synaptic activity

When cells were exposed to multiple cell superfusion with solutions containing high (10 mM) concentrations of $CaCl_2$ and low (1 mM) concentrations of $MgCl_2$ patch clamp recording revealed synaptic activity in nearly all neurons. The individual currents decayed with a time constant of 11 to 39 ms and displayed a slope conductance of 20 to 100 pS at a holding voltage of -50 mV. Spontaneous currents with these characteristics reversed at the Cl- equilibrium potential (Fig. 2B) and disappeared completely in the presence of 20 to 50 µM bicuculline methiodide (Fig. 2C). Their frequency decreased with decrease of extracellular Ca^{2+} concentration. The Cl--dependent spontaneous activity will be further referred to as $I_{Cl(GABA)SYN}$. It is assumed that spontaneous discrete $I_{Cl(GABA)SYN}$ result from GABA release, due to vesicle exocytosis from axon terminals. The properties of $I_{Cl(GABA)SYN}$ were similar to the previously described GABAergic currents in rat and chick cortical cultures (Dichter, 1980; Weiss et al., 1988), rat hippocampal cultures (Segal and Barker, 1984 a,b) and in synaptoneurosomes from the adult guinea pig brain stem (Drewe et al., 1988). The fact that spontaneous activity was generally absent in bicuculline-treated cultures (Fig. 2C) indicates that other types of synapses, if present, were not active under the present experimental conditions. This is clearly advantageous for a study aimed at identifying presynaptic receptors on GABAergic terminals.

Fig. 3A illustrates the common observation (more than 80 neurons tested) that superfusion of cultured neuronal networks with a salt solution containing EAA leads to an increase in the frequency of $I_{CL(GABA)SYN}$, despite the presence of 0.5-1 µM TTX. The stimulating effect of EAA was particularly convincing when the background activity was moderate, as is the case in Fig. 3A. The concentrations of EAA were varied around the previously determined EC_{50} for a current response in cultured SC-derived neurons (Perouansky and Grantyn, 1989) to find the EAA concentration that would, in the multiple cell superfusion paradigm, cause generation of $I_{CL(GABA)SYN}$ at clearly

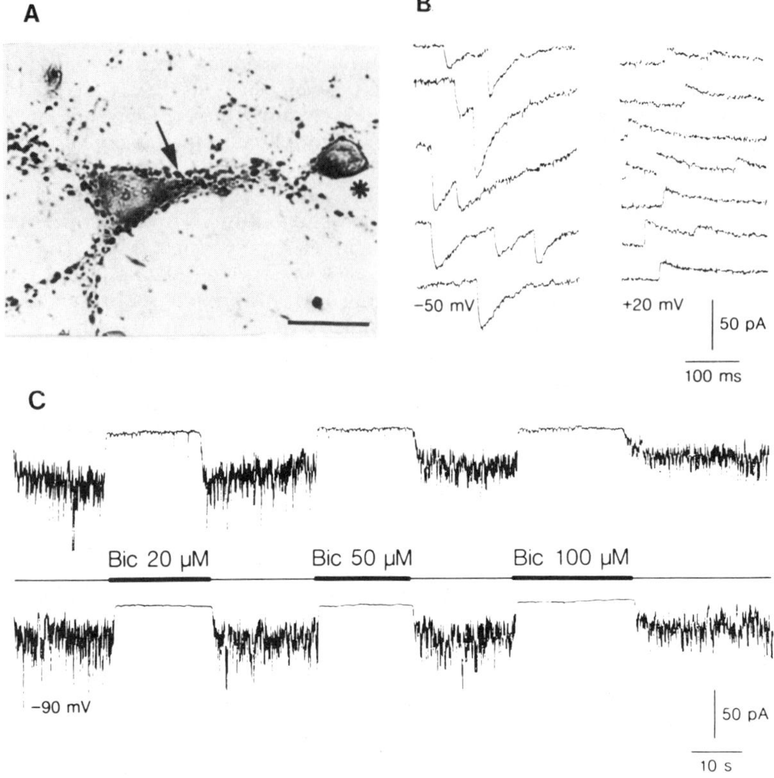

Fig. 2.   GAD-immunoreactive terminals fA) and spontaneous synaptic activity (B,C) in dissociated cell cultures from the rat superior colliculus. A: photomicrograph under oil immersion showing a larger GAD-negative and a smaller GAD-positive (asterisk) neuron from a SGS-derived culture after 27 DIV. The arrow points to a GAD-positive bouton. Bar: 20 µm. B: discrete spontaneous currents reversing at the Cl- equilibrium potential (-5 mV). The record was taken with an electrode containing 120 mM CsCl at holding voltages indicated below the bottom traces. Extracellular [Ca2+] : 10 mM. Mg2+ omitted. Multiple cell superfusion in the presence of 0.5 µM TTX. C: blockade of spontaneous Cl- currents with increasing concentrations of bicuculline methiodide. Records from two different cells after 16 DIV.

higher frequency than 80 mM KCl. It can be seen in Fig. 3A that the stimulating effect of 200 µM Glu or 30 µM KA was stronger than that of 80 mM KCl. At such concentrations of the EAA, synaptic activity was superimposed on a directly evoked current which had (with Cs-filled electrodes) an inward direction at negative voltages and reversed around O mV. A frequency increase of ICL(GABA)SYN during Glu- or KA-application was seen in 52 neurons. In 5 neurons with high background activity KA (5-30 µM) decreased the frequency of ICL(GABA)SYN.

To explore the possibility that dendritic release sites would account for the TTX-resistant EAAR-mediated stimulation of ICL(GABA)SYN multiple cell superfusion experiments were performed in cultures with very low neuronal density. Configurations were sought that allowed for superfusion of several solitary neurons. Under these conditions dendrodendritic contacts were excluded, yet superfusion with EAA stimulated the GABA release.

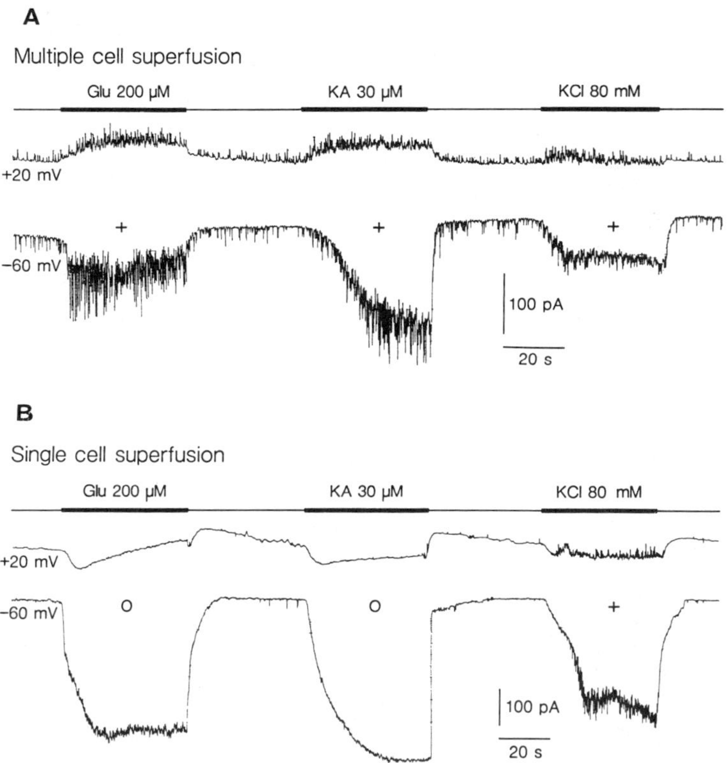

Fig. 3.   Effect of multiple cell and single cell superfusion with glutamate, kainate and elevated KCl in the presence of 0.5 µM TTX. Glu and KA increase the frequency of $I_{CL(GABA)SYN}$ only if the somatodendritic area of neurons in the neighbourhood of the voltage clamped neurones is covered with the stream of drug containing solution (A). When the access of EAA is restricted to the investigated cell with the attached axonal terminals no increase of $I_{CL(GABA)SYN}$ can be seen with Glu and KA, whereas elevated KCl continues to activate the generation of GABAergic synaptic activity (B) . $I_{CL(GABA)SYN}$ is superimposed on the direct current response of the postsynaptic cell. Filling solutions: 120 mM CsCl (A) and 120 mM N-methyl-glucamine in (B). Note that the synaptic currents are reversed at positive holding voltage (Ecl = -10 mV), while the directly evoked current response shown in B is still inward at +20 mV. Holding voltages are indicated on the left. Records from SGS-derived neurons after 15 (A) and 17 (B) DIV.

Absence of stimulatory effect on GABAergic synaptic activity during axonal application of glutamate and kainate

An experiment with single cell superfusion is presented in Fig. 3B. The access of EAA was in this case restricted to the axon terminals and the contacted soma and proximal dendrites of a bipolar cell. Between the periods of agonist application investigated neurons were held under permanent superfusion with nominally Mg^{2+} -free and high Ca^{2+} solution, not containing EAA or elevated KCl. To eliminate synaptic activity from the dendritic membrane outside the superfused area, the bath contained 50 µM bicuculline, in addition to 12 mM $MgCl_2$. Under such conditions, activation of putative presynaptic EAARs should cause axon terminal depolarization, GABA release, and an increase in the frequency of postsynaptic chloride currents. High $[K^+]_0$ would depolarize the terminals irrespective of the presence of presynaptic EAARs. The KCl-induced generation of $I_{CL(GABA)SYN}$ served as a control for the presence of functional synapses. Concentrations were adjusted in such a way that the effect of EAA on $I_{CL(GABA)SYN}$ surpassed, in the multiple cell superfusion paradigm, the effect of 80 mM KCl.

It is shown in Fig. 3B that under the condition of single cell superfusion $I_{CL(GABA)SYN}$ could not be stimulated with Glu and KA, although 80 mM KCl increased the frequency of the spontaneous $I_{CL(GABA)SYN}$ and a current response to Glu and KA was present in the postsynaptic neuron.

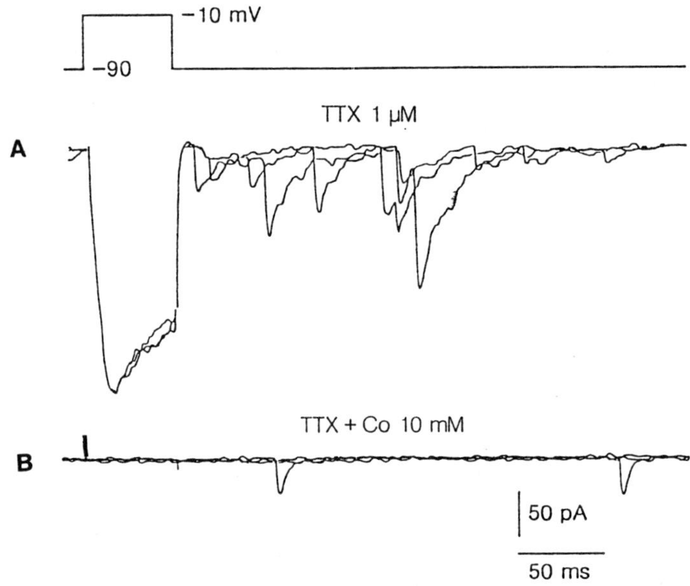

Fig.4.　Currents induced in a solitary SC-derived neuron by application of a depolarizing voltage step. Upper trace: voltage. A: high voltage activated Ca^{2+} current ($I_{Ca(HV)}$) and $I_{CL(GABA)SYN}$ recorded in a salt solution containing 10 mM Ca^{2+} and 1 µM TTX. B: the frequency of $I_{CL(GABA)SYN}$ is drastically reduced under the condition of completely blocked $I_{Ca(HV)}$

This test was repeated in a total of 27 neurons. Our sample included only neurons displaying $I_{CL(GABA)SYN}$ at a frequency of <2.5 events/s. In 22 out of 24 neurons 200 µM Glu and/or 30 µM KA failed to stimulate $I_{CL(GABA)SYN}$ although these cells responded to elevated KCl. In two cells KA reversibly reduced the frequency of $I_{CL(GABA)SYN}$. In three cases where the EAA were ineffective, synaptic activity wax also unaltered with KCl.

The absence of a change in the frequency of $I_{CL(GABA)SYN}$ to application of KA was not the consequence of a direct blocking action of KA at the GABAA receptor. This possibility was tested

234

by simultaneous application of 30 µM GABA and 30 µM KA. The compound current response to GABA and KA represented about the sum of the currents elicited with either of the agonists alone.

Depolarization-induced stimulation of GABAergic synaptic activity in single sc-derived neurons

To gain some insight into the mechanism by which somatodendritic EAARs may affect the axonal release of GABA we selected isolated cells that apparently formed autosynapses via recurrent axon collaterals. These cells were selected to test for the possibility that a voltage step would activate the mechanism that presumably links the somatodendritic EAARs with the axonal release sites.

A total of 9 neurons were investigated as illustrated in Fig. 4. The stimulation of $I_{CL(GABA)SYN}$ is quite obvious. In addition, it can be seen that the depolarization-induced stimulation of GABA release is absent if the high voltage-activated noninactivating Ca^{2+} current is blocked with Co^{2+}.

DISCUSSION

Our experiments have demonstrated that EAA do indeed modulate the spontaneous GABAergic synaptic activity. This result supports previous biochemical studies which revealed a change in Ca^{2+} dependent GABA release under the influence of EAA (Harris and Miller 1989). Additionally, our findings indicate that at least part of the GABA is released in a quantal manner, presumably by an exocytotic process.

From the TTX-resistant character of the EAA-stimulated GABAergic synaptic activity it follows that propagation of sodium action potentials along the axon is not a necessary condition for coupling the EAARs to the GABA release sites on the axon terminals. This fact could reflect three mutually nonexclusive circumstances: 1) A sufficient number of EAA-activated channels is located on the axon terminals; 2) Release sites are located on dendrites; 3) Somatodendritic EAARs are coupled to axonal release sites.

The present single cell superfusion experiments suggest that the EEA-induced depolarization of axon terminals, if present at all, is not large enough to activate exocytosis. EAA-stimulated GABA release is therefore unlikely to depend on presynaptic EAARs alone. This conclusion was reached also by Pastuszko et al. (1984) who investigated the effect of high doses of KA on rat brain synaptosomes.

GABAergic release sites on dendrites, too, cannot solely account for the EAA-induced activation of synaptic activity, as observed in the multiple cell superfusion mode, because the latter also occurred in solitary neurons that lacked dendrodendritic contacts. The results are, however, consistent with the possibility that a sufficiently strong coupling mechanism exists between somatodendritic EAARs and axonal GABA releasing sites.

This raises the question of how, in the absence of propagated Na^+ spikes, EAARs could possibly affect the distant axonal release sites. A reasonable assumption would be that the EAA-induced depolarization activates somatodendritic Ca^{2+} channels. The immature neurons in culture may then conduct a Ca^{2+} spike along the axon (Dichter et al., 1983) that causes an influx of Ca2t at any site, including the terminal membrane. This could lead to liberation of additional Ca^{2+} from intracellular stores (Kostyuk et al., 1989). Our limited sample of cells with presumed autosynapses gave us an opportunity to explore this possibility. It was found that a block of $I_{Ca(HV)}$ with Co^{2+} or Cd^{2+} prevented the depolarization-induced stimulation of GABA-release, indicating 1) that depolarization is not a sufficient condition to activate the transmitter release, and 2) that voltage activated Ca^{2+}-influx could account for the connection between somatodendritic receptors and axonal release sites.

It should be pointed out that our superfusion experiments were specifically designed to reveal a stimulatory effect of EEA on GABA release. Therefore neurons were chosen for analysis that displayed little spontaneous synaptic activity. The experiments cannot exclude a depressant action of EAA, as described in some previous studies with KA (McBean et al., 1981; Poli et al., 1985; Bray et al., 1989). In fact, some neurons of the present sample tended to reduce their spontaneous synaptic activity during application of 30 μM KA. It is conceivable that such an effect of Glu or KA is mediated via the type 4 EAARs where 2-amino-4-phosphonobutyrate (APB) is presumed to act presynaptically in some pathways (Cotman et al., 1986). A slight EAA-induced depolarization of the axonal membrane, due to a low density of respective receptors or channels, may result also in a decreased amplitude or inactivation of propagating spontaneous Ca^{2+} spikes. The contribution of the latter to the spontaneous synaptic activity is, however, unknown.

The present study focussed on the action of the receptor-specific ligand KA, because in SC-derived neurons the average number of KAactivated channels per cell is apparently higher than the number of channels activated with maximal concentrations of Glu, Nmethyl-D-aspartate (NMDA) or quisqualate (QA). Dividing, for instance, the average slope conductance during a KA-activated whole-cell current (1700 pS) by the average single channel conductance reported by others (Cull-Candy et al., 1988; Llano et al., 1988) (2 pS) predicts around 850 KA-activated channels per cell. In contrast, the respective predicted figure for NMDAactivated currents in SC-derived neurons is only 10 channels per cell. Such a low NMDA receptor density would not be sufficient to support an EAA modulation of transmitter release at each of the many axon terminals possessed by these cells.

Using KA has also the advantage that the lower amplitude of the current noise makes it easier to discriminate $I_{CL(GABA)SYN}$. Finally, it was found that KA-activated channels also appear in rather immature neurons (Perouansky and Grantyn, 1988). For these reasons the results with KA are considered most convincing. Nonetheless, the major conclusions of this study could also be valid for NMDA- and QA-selective receptors. First, it was shown that the compound current response to Glu involves also receptors for NMDA and for QA (Perouansky and Grantyn, 1989). Second, in some experiments 200 μM of NMDA or 3 μM of QA were used, instead of Glu. These drugs, too, stimulated $I_{CL(GABA)SYN}$ under condition of multiple cell superfusion, but they failed to have an effect with single cell superfusion.

Further studies will shed more light on the mechanism of coupling between the somadendritic EAARs and the transmitter release sites in GABAergic neurons.

ACKNOWLEDGEMENT

Similar devices for selective superfusion are being used in this department by Drs. K. Gottmann, J. Streit and H.D. Lux who made many helpful suggestions on the method. The authors wish to thank their collegues Dr. K. Kraszewski for participation in data analysis, Dr. C. Parsons for useful comments on the manuscript and Mrs. C. Pfitzner for excellent technical assistance.

REFERENCES

Bray, S., Bustos, G.A., Lee, P.H.K., Hong, J.S., Aitken, P.G., Nadler, J.V., Kainic acid inhibits cholecystokinin release from rat hippocampal slices, **Neurosci. Lett.**, 100 (1989) 313-318.
Collins, G.G.S., Anson J. and Surtees, L., Presynaptic kainate and N-methyl-D-aspartate receptors regulate excitatory amino acid release in the olfactory cortex, **Brain Res.**, 265 (1983) 157-159.
Connor, J.A., Tseng, H.-Y., Hockberger, P., Depolarization- and transmitter-induced changes in intracellular Ca^{2+} of rat cerebellar granule cells in explant cultures. **J. Neurosci.**, 7 (1987) 1384-1400.
Cotman, C.W., Flatman, J.A., Ganong, A.H., Perkins, M.N., Effects of excitatory amino acid antagonists on evoked and spontaneous excitatory potentials in guinea-pig hippocampus, **J.Physiol. (Lond.)** 378 (1986) 403-415.
Cull-Candy, S.G., Howe, J.R., Ogden, D., Noise and single channels activated by excitatory amino acids in rat cerebellar granule neurones, **J. Physiol. (Lond.)**, 400 (1988) 189-222.

Curtis, D.R., Headley, P.M. and Lodge, D., Depolarization of feline primary afferent fibres by acidic amino acids, **J. Physiol. (Lond.)**, 351 (1984) 461-472.

Dichter, M.A., Physiological identification of GABA as the inhibitory transmitter for mammalian cortical neurons in cell culture, **Brain Res.**, 190 (1980) 111-121.

Dichter, M.A., Lisak, J., Biales, B., Action potential mechanism of mammalian cortical neurons in cell culture, **Brain Res.**, 289 (1983) 99-107.

Drejer, J., Honore, T. and Schousboe, A., Excitatory amino acidinduced release of 3H-GABA from cultured mouse cerebral cortex interneurons, **J. Neurosci.**, 7 (1987) 2910-2916.

Drewe, J.A., Childs, G.V. and Kunze, D.L., Synaptic transmission between dissociated adult mammalian neurons and attached synaptic boutons, **Science**, 241 (1988) 1810-1813.

Errington, M.L., Lynch, M.A. and Bliss, T.V.P., Long-term potentiation in the dentate gyrus. Induction and increased glutamate release are blocked by D(-)aminophosphonovalerate, **Neurosci.**, 20 (1987) 279-284.

Evans, R.H., Evidence supporting the indirect depolarization of primary afferent terminals in the frog by excitatory amino acids, **J. Physiol.**, 298 (1980) 25-35.

Ferkany, J.W. and Coyle, J.T., Kainic acid selectively stimulates the release of endogenous excitatory acidic amino acids, **J. Pharm. exp. Ther.**, 225 (1983) 399-406.

Ferkany, J.W., Zaczek, R. and Coyle, J.T., Kainic acid stimulates excitatory amino acid neurotransmitter release at presynaptic receptors, **Nature**, 298 (1982) 757-759.

Fisher, R.S. and Alger, B.E., Electrophysiological mechanisms of Kainic acid-induced epileptiform activity in the rat hippocampal slice, **J. Neurosci.**, 4 (1984) 1312-1323.

Foster, A.C., Mena, E.E., Monaghan, D.T. and Cotman, C.W., Synaptic localization of Kainic acid binding sites, **Natllre**, 289 (1981) 73-75.

Gallo, V., Suergiu, R., Giovannini, C. and Levi, G., Glutamate receptor subtypes in cultured cerebellar neurons: Modulation of glutamate and -aminobutyric acid release, **J. Neurochem.**, 49 (6) (1987) 1801-1809.

Giorguieff, M.F., Kemel, M.L. and Glowinski, J., Presynaptic effect of L-glutamic acid on the release of dopamine in rat striatal slices, **Neurosci. Lett.**, 6 (1977) 73-77.

Grantyn, R. and Lux, H.D., Similarity and mutual exclusion of NMDA- and proton-activated transient Na+-currents in rat tectal neurons, Neurosci. Lett., 89 (1988) 198-203.

Harris, K.M. and Miller, R.J., Excitatory amino acid-evoked release of [3H]GABA from hippocampal neurons culture, **Brain Res.**, 482 (1989) 23-33.

Hockberger, P.E., Tseng, H.-Y., Connor, J.A., Fura-2 measurements of cultured rat purkinje neurons show dendritic localization of Ca2+ influx, **J. Neurosci.**, 9 (1989) 2272-2284.

Jaffe, E.H. and Cuello, A.C., Release of -aminobutyric acid from the external plexiform layer of the rat olfactory bulb: Possible dendritic involvement, **Neurosci.**, 5 (1980) 1859-1869.

Jones, S.M., Snell, L.D., Johnson, K.M., Inhibition by phencyclidine of excitatory amino-acid-stimulated release of neurotransmitter in the nucleus accumbens, **Neuropharm.**, 26 (1987) 173-179.

Kostyuk, P.G., Mironov, S.L., Tepikin, A.V., Belan, P.V., Cytoplasmic free Ca in isolated snail neurons as revealed by fluorescent probe Fura-2: Mechanisms of Ca recovery after Ca load and Ca release, **Membrane Biol.**, 110 (1989) 11-18.

Krespan, B., Berl, S. and Nicklas, W.J., Alteration in neuronalglial metabolism of glutamate by the neurotoxin kainic acid, **J. Neurochem.**, 38 (1982) 509-518.

Kudo, Y., Ogura, A., Glutamate-induced increase in intracellular Ca2+ concentration in isolated hippocampal neurones. **Br. J. Pharmacol.**, 89 (1986) 191-198.

Llano, I, Marty, A., Johnson, J.W., Ascher, P. and Gahwiler, B.H., Patch-clamp recording of amino acid-activated responses in 'organotypic' slice cultures, Proc. Natl. Acad. Sci., USA, 85 (1988) 3221-3225.

McBean, G.J. and Roberts, P.J., Glutamate-preferring receptors regulate the release of D-[3H]aspartate from rat hippocampal slices, **Nature**, 291 (1981) 593-594.

Middlemiss, D.N., Receptor-mediated control of neurotransmitter release from synaptosomes, **Trends Pharmac. Sci.**, 9 (1988) 83-84

Pastuszko, A., Wilson, D.F. and Erecinska, M, Effect of kainic acid in rat brain synaptosomes: the involvement of calcium, **J. Neurochem.**, 43 (1984) 747-754.

Pearce, B.R. and Dulton, G.R., L-Glutamate increases the spontaneous release of 3H-GABA from cultured cerebellar neurons, **Dev. Brain Res.**, 3 (1982) 492-496.

Perouansky, M., Grantyn, R., Mechanisms of glutamate excitation in visual neurons, **in:** E.A. Cavalheiro, J. Lehmann and L. Turski (Eds.), Frontiers in Excitatory Amino Acid Research, Alan Liss, New York, 167-170.

Perouansky, M. and Grantyn, R., Separation of quisqualate- and kainate-selective glutamate receptors in cultured neurons from the rat superior colliculus, **J. Neurosci.**, 9 (1989) 70-80.

Pin, J.-P. Pin, Bockaert, J., Two distinct mechanisms, differentially affected by excitatory amino acids. trigger release from fetal mouse striatal neurons in primary culture, **J. Neurosci.**, 9 (1989) 648-656.

Poli, A., Contestabile, A., Migani, P., Rossi, L., Rondelli, C., Virgili, M., Bissoli, R. and Barnabei, 0., Kainic acid differentially affects the synaptosomal release of endogenous and exogenous amino acid neurotransmitters, **J. Neurochem.**, 45 (1985) 1677-1686.

Potashner, S.J. and Gerard, D., Kainate-enhanced release of D-[^{3}H] aspartate from cerebral cortex and striatum : reversal by baclofen and pentobarbital, **J. Neurochem.**, 40 (1983) 1548-1557.

Represa, A., Tremblay, E., Ben-Ari, Y., Kainate binding sites in the hippocampal mossy fibers: localization and plasticity. **Neurosci.**, 20 (1987) 739-748.

Segal, M., Barker, J.L., Rat hippocampal neurons in culture: properties of GABA-activated Cl⁻ ion conductance. **J. Neurophysiol.**, 51 (1984) 500-515.

Segal, M., Barker, J.L., Rat hippocampal neurons in culture: voltage clamp analysis of inhibitory synaptic connections. **J. Neurophysiol.**, 52 (1984) 469-487

Tigges, M. and Tigges, J., Presynaptic dendrite cells and two other classes of neurons in the superficial layers of the superior colliculus of the chimpanzee, **Cell Tiss.Res.**, 162 (1975) 279-295.

Warton, S.S., Perouansky, M. and Grantyn, R., Development of GAGAergic synaptic connections in vivo and in cultures from the rat superior colliculus, **Dev. Brain Res.**, (in press).

Weiss, D.S., Barnes, E.M.jr. and Hablitz J.J., Whole-cell and single-channel recordings of GABA-gated currents in cultured chick cerebral neurons, **J. Neurophysiol.**, 59 (1988) 495-513.

THE ROLE OF TAURINE AND GLUTAMATE DURING EARLY POSTNATAL CEREBELLAR

DEVELOPMENT OF NORMAL AND *WEAVER* MUTANT MICE.[1]

E. Trenkner

Department of Pathology
Columbia University, Physicians & Surgeons
630 West 168 Street
New York, NY 10032

It is generally accepted that the stimulation of neurotransmitter receptors can lead to the differentiation of neuronal cells[1,2]. That astroglial cells might play an active role in this developmental process was suggested, but its molecular mechanism still remains elusive. This study analyzes the interaction between cerebellar granule cell neurons (GC) and astroglial cells based on an amino-acid-transmitter mediated regulation mechanism.

The neuroreactive free amino acids glutamate (excitatory), GABA, and taurine (both inhibitory) represent the majority of known transmitters in the developing mammalian cerebellum. Uptake and release of these transmitters and glutamate-receptor subclasses have been identified on both neurons and astroglial cells *in vivo*[3,4] and *in vitro*[5,6.] Astrocytes distinguish themselves from neurons in the expression of receptor subclasses (quisqualate and kainate receptors on cerebellar astrocytes, NMDA-sensitive receptors a.o. on neurons). They express higher uptake capacity for glutamate, suggesting a role in the control of excitatory amino acid concentrations in the extracellular space.

Taurine, 2-aminoethanosulfonic acid, has been found essential for the development and survival of neurons and has received clinical attention for its anticonvulsant properties[7,8]. The protective role of taurine was further confirmed after injections of kainate[9] into the cerebral cortex as well as in non-neuronal systems[10]. Excitatory amino acids and excitotoxins are potent stimulators of taurine release *in vivo*[11] and *in vitro*[12]. Although the mechanism of taurine's action is still obscure, more and more evidence supports the idea that taurine acts as an osmoregulator[13] and modulator of Ca^{2+} influx[14] as well as of Cl^- influx[15].

Next to glutamate, taurine is by far the most abundant free amino acid in mammals. Its concentrations in the brain reach the millimolar range, and the highest concentrations occur in newborn and neonatal brain, approximately 3 - 4 times higher than in mature brain in the same species. Although differing in their taurine concentrations, all brain regions show a similar decrease from birth to weaning, suggesting a possible role of taurine in development. The most compelling evidence for its essential role in development has been provided for cerebellar development of cat and mouse: female cats raised on ataurine-deficient diet produce offspring with severe cerebellar abnormalities[16]. This defect was reversible after their diet was supplemented with taurine or taurine precursors, but not other free amino acids. Similar defects to those described for cat were observed in the mouse mutation weaver (*wv/wv*)[17], both morphologically and with respect to taurine. Roffler-Tarlov and Turey[18] found, similar to taurine-deprived kittens, a significant reduction of glutamate and taurine concentrations (30-50% of normal mice) in

---

[1] Supported by NIH Grant NS 21457.

Excitatory Amino Acids and Neuronal Plasticity
Edited by Y. Ben-Ari
Plenum Press, New York, 1990

early postnatal *wv/wv* cerebellum, the critical period in the postnatal morphogenesis of mouse cerebellum; whereas GABA and aspartate were unchanged. *wv/wv* is, therefore, particularly relevant for analyzing the role of glutamate and taurine during neuron-glial interactions in cerebellar development.

Early postmitotic *wv/wv* GC degenerate in the external granule cell layer (EGL) without traversing a significant distance, possibly due to a defect in Bergmann glia fiber maturation, as well as due to an inherent defect of GC[19]. In normal mice on the other hand, GC neurons must play an active part in this interaction, because 6 - 8 generations of GC migrate along one set of Bergmann glial fibers (a specialized form of radial glial cells) from the EGL, through the molecular layer into the internal granule cell layer, their final destination where they home around protoplasmic and stellate astrocytes. Thus the maturation of GC is closely linked to the interaction with various astroglial cell types.

The approach taken in this study is based on our hypothesis that glutamate released from granule cell neurons stimulates astroglial cells to release taurine. Taurine is then taken up by granu-

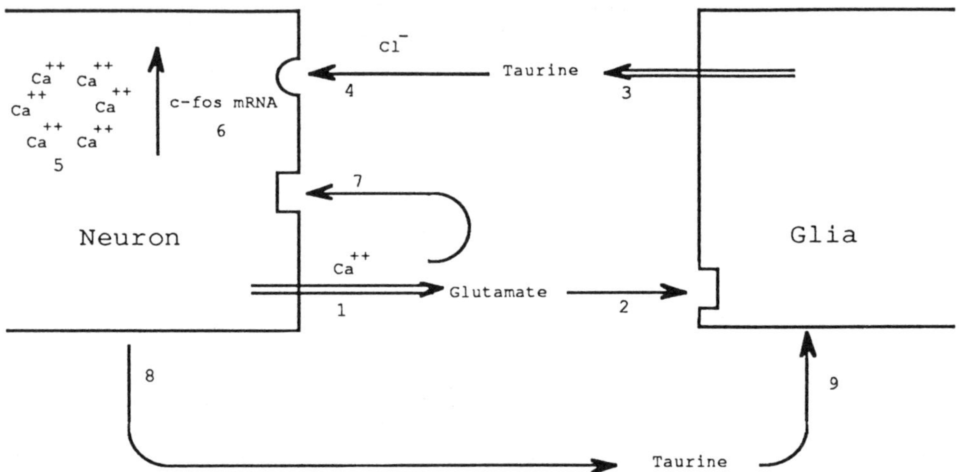

Fig. 1.　Model for the glutamate-taurine regulation of granule cell maturation

le cells and subsequently released following the induction via glutamate receptors on granule cells (Fig. 1). We propose that this balanced interaction between excitatory amino acids and their modulator taurine is established under normal conditions to control survival and function of neurons. If, however, this balance is perturbed, then excitotoxic reactions of glutamate released into extracellular space will destroy neurons, as has been observed in neurodegenerative diseases. Here we demonstrate the validity of this model in two systems, in the neurodegenerative mutation *wv/wv* and with excitotoxicity of cerebellar and hippocampal neurons in cultures.

ASSAY SYSTEMS :

1. The microwell tissue culture system[20] is based on cell-cell interactions and allows for the analysis of the dynamics of cell behavior in normal and mutant mouse development. Cells from trypsinized 7-day-old mouse cerebellum were maintained in BME and 10% dialyzed horse serum. The cells reaggregate within hours into clusters that later develop interconnections consisting either of sheets of migrating cells and cell processes, or of cables, consisting of parallel fiber bundles and astroglial processes, with cells migrating along their surfaces. Granule cells in several stages of

Glutamate is released from granule cells (1) and stimulates astroglial cells (2) to
release taurine (3). Taurine then stimulates granule cells independent of external Ca2+
(4) to release Ca2+ from internal pools (5) which leads to a significant stimulation of c-
fos mRNA (6). The induction of c-fos mRNA by glutamate at P1-P4 was insignificant.
Glutamate (7) or taurine (4) stimulate neuronal release of taurine (8), which is taken up
into glia (9).

differentiation, basket and/or stellate neurons, some larger neurons, as well as astrocytes and oligo-
dendroglia, were identified in reproducible, nonrandom cell patterns. Axonal and dendritic processes, as well as synapses, are generated in this system.

2. Monolayer cultures[21] were utilized to manipulate and observe defined cerebellar cell types of
both normal and *wv/wv* mice independently from a complex tissue structure, growing on 100 μg/ml
poly-D-lysine in BME and 10% horse serum.

RESULTS AND DISCUSSION

I. The mutant mouse *weaver* as a model for inherited neurodegenerative diseases *in vitro*.

Phenotypically the *wv/wv* defect was reproduced *in vitro*[22]. When 4 - 7 day old *wv/wv*
cerebellar cells were cultured for 2 - 3 days, GC degenerated within reaggregates and in monolayers
with time in culture. Significantly reduced numbers of extended fiberbundles and GC processes
reduced in length with only few migrating GC attached, are characteristic for the *wv/wv*-specific
cell pattern.

1. Following our hypothesis (Fig. 1), one possible explanation among several others for the
death of GC is that they suffer from NMDA receptor-mediated excitotoxicity of extracellular glutamate accumulated in the extracellular space. Since the *wv/wv* defect is inherent to GC, it is conceivable that, for still unknown reasons, *wv/wv* GC would be more vulnerable under such excitoto-

Table 1.   NMDA-antagonists rescue weaver cerebellar cells *in vitro*

additions	Surviving neurons [%]	
	wv/wv	*+/+*
medium	35 ± 4	87 ± 6
glutamate 10^{-4} M	14 ± 7	36 ± 7
APV 10^{-4} M	79 ± 9	83 ± 3
APV 10^{-5} M	86 ± 2	89 ± 5
CGS 19755 10^{-4} M	88 ± 7	86 ± 5
CGS 19755 10^{-5} M	84 ± 3	85 ± 5
CNQX 10^{-4} M	18 ± 5	79 ± 9

xic stress than GC of normal mice. If this assumption is correct, then NMDA receptor antagonists,
but not the quisqualate-receptor antagonist CNQX, should prevent *wv/wv* GC death. The results
of 5 experiments listed in Table 1 clearly demonstrate that this assumption is valid.

2. Extracellular taurine prevents the expression of the *wv/wv* phenotype *in vitro*. If the
reduction of the taurine concentrations in early postnatal *wv/wv* cerebellum[18] is affecting *wv/wv*
cerebellum development, as it was described for cat[16], then the reconstitution of the physiological
taurine concentrations should reverse the *wv/wv* phenotype. This was approached *in vitro*.
When *wv/wv* cerebellar cells were cultured in the presence of physiological concentrations of taurine, neuron survival as well as cell pattern formation and function were restored (Fig. 2). Glutamate, GABA and aspartate had no effect, suggesting taurine-specificity. Thus taurine could
prevent neurons from excitotoxicity in a concentration-dependent manner.

II.   Taurine prevents excitotoxic reactions *in vitro* :

Taurine has been linked in a variety of studies to tissue protection during tissue damage *in
vivo*[10]. We have confirmed this protective role *in vitro* by demonstrating that 10^{-2} M taurine
protected cerebellar and hippocampal neurons from kainate and quinolinate excitotoxicity[23].
Depending on the vulnerability of these tissues to excitotoxins, a defined ratio of taurine concentration to excitotoxin sensitivity appears to control the extent of this protection. For example,

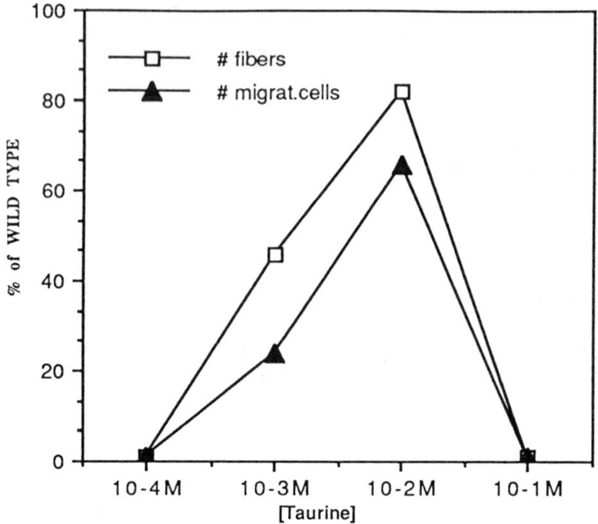

Fig. 2   Taurine reverses the *wv/wv* defect *in vitro*

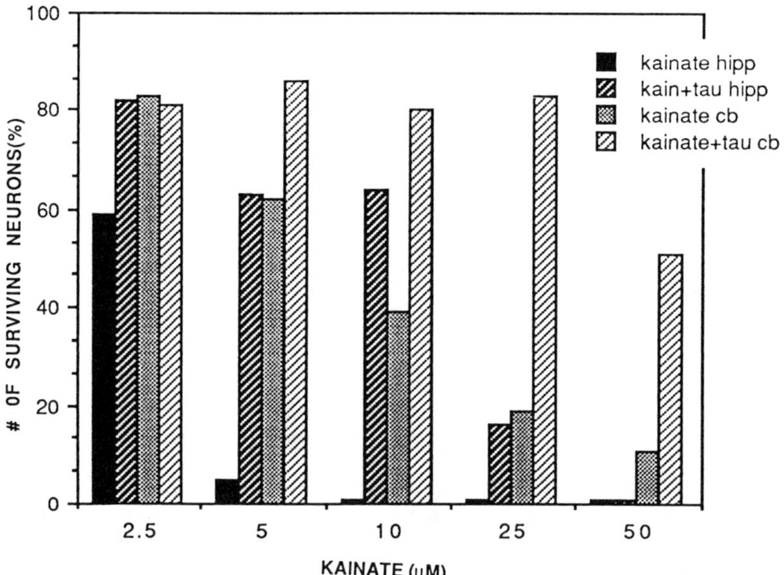

Fig. 3.   Survival of hippocampal and cerebellar neurons after
excitotoxicity in the presence of taurine.

Percent of neuronal survival after concentration-dependent kainate excitotoxicity in
cultures of mouse cerebellum (cb) and mouse hippocampus (hipp) in the absence and
presence of 10-2 M taurine. The results show that excitotoxicity can be prevented in the
cases where a balanced relationship between kainate and taurine concentrations is
maintained. This balance appears to be tissue specific. The data reflect an average of
three experiments with a variation of not more than 18%.

10^{-2} M taurine protected cerebellar cells effectively from up to 25 μM kainate; whereas, in hippocampal neurons, 10^{-2} M taurine protected significantly against only 5μM kainate (Fig. 3). Thus, it is conceivable that, in part, a balanced interaction between excitatory amino acids and their modulator taurine protects neurons from excitotoxicity under normal conditions and particularly in early postnatal stages of development.

III. Taurine release from cerebellar astrocytes

Cooperating astroglial cells are thought to play an important role in this balanced interaction. As hypothesized in Fig. 1, taurine when released from neurons is taken up into astrocytes via high-affinity uptake systems. It has been demonstrated in vitro that taurine release from astrocytic cell lines can be induced by norepinephrin agonists[24]. When applied to cerebellar astrocytes, glutamate, as well as kainate, quisqualate and AMPA but not NMDA, induced [3H]-taurine release[12]. The induction was blocked in the presence of 10 mM CNQX, suggesting that the induction of taurine release is modulated by quisqualate receptors. Taurine-release was also induced by living cerebellar granule cells (10 - 100 GC/glial cell); whereas, GC membranes, fixed GC, or glial cells of various origin did not induce taurine release[12]. These results support our hypothesis of a balanced interaction between glutamate and taurine during neuron-glial interaction. Once this balance is disturbed, as in inherited neurodegenerative diseases like *wv/wv* or after injury (i.e., ischemia), it will lead to neuronal death, unless extracellular taurine concentrations are increased[23].

Although the mechanism of taurine is still unknown, recent results[11] support the idea[13] that taurine acts as an osmoregulator, mediating water uptake during excitation via Na$^+$, Cl$^-$ or Ca^{2+} influx. This is supported by our finding that the protective role of taurine failed in the presence of tryptophan metabolites like 3-HAT and 3-Hydroxykynurenine, which irreversibly oxidize and permeabilize cell membranes[25].

IV. Taurine induces c-*fos* mrna expression at a narrow developmental window

In search for a regulatory function of taurine during early postnatal development, the induction of the nuclear proto-oncogenes c-*fos* and c-*myc* in cerebellar tissue slices were analyzed, because of their regulatory functions in gene activation (in collaboration with Drs.Chr. Ruppert and W. Wille, Cologne, FDR). Although preliminary the results in the S1 nuclease protection assay show undoubtedly that taurine, but not glutamate, stimulates c-*fos* mRNA in the first four postnatal days (Table 2). This induction was independent of extracellular Ca^{2+} but required Na$^+$ and Cl$^-$. In comparison, general depolarization with K$^+$ (50 mM) required extracellular Ca^{2+}. On the other hand, at seven days postnatally glutamate but not taurine stimulated c-*fos* mRNA in the presence of external Ca^{2+}, confirming the results of others[27]. Thus the induction of c-*fos* mRNA appears to be developmental-stage-specific and Ca^{2+} independent, suggesting a role in development.

Table 2.   Induction of *c-fos* mRNA in cerebellar slices by taurine and glutamate (in relative units)

	0.9 NaCl	10^{-2} M Taurine	10^{-4} Glutamate
P3	19 ± 2 U	80 ± 17	<2
P7	12 ± 6	<1	73 ± 21

Results of 4 experiments are expressed in relative units (u) obtained from gel scanning and normalized on the scanning values of constant amounts of β-microglobulin.

In summary, evidence is provided for a regulatory feedback mechanism that might control the extent of glutamate-excitotoxicity in a balanced relationship with taurine to prevent neuronal death. Once this balance is perturbed, however, as in *wv/wv*, the conditions for neurodegenerative diseases are fulfilled: excitotoxic reactions will destroy neurons unless sufficient concentrations of taurine, released from astroglial cells and/or neurons, are taken up into neurons and then function as an osmoregulator during excitation.

REFERENCES

1. H. Thoenen and D. Edgar, 1982, The regulation of neural gene expression, **Trends Neurosci.**, 5:311 313.
2. J. M. Lauder, 1983, Hormonal and humoral influences on brain development, **Psychoneuroendocrinology**, 8:121-155.
3. Cavalheiro (personnal communication)
4. J.A. Sturman, 1988, Taurine in development, **J. Nutr.**, 118:1169-11
5. H.K. Kimelberg and M.D. Norenberg, 1988, Astrocytes, **Scientific American**, April:66-76.
6. V. Gallo, C. Giovannini, R. Suergin, and G. Levi, 1989, Expression of excitatory amino acid receptorsby cerebellar cells of the type-2 astrocyte cell lineage, **J. Neurochem.**, 52:1-9.
7. R. J. Huxtable and K. Nakagawa, 1985, The anticonvulsant actions of two taurine derivatives in genetic and chemically induced seizures, **in**: "Taurine, Biological Actions and clinical Perspectives," pp. 435-448.
8. I.P. Lapin, 1981, Kynurenines and seizures, **Epilepsia**, 22:257.
9. P.R. Sanberg, W. Staines, and E.G. McGeer, 1979, Chronic taurine effects on various neurochemical indices in control and kainic acid-lesioned neostriatum, **Brain Res.**, 161:367-370.
10. R.E. Gordon, A.A. Shaked, and D.F. Solano, 1986, Taurine protects hamster bronchioles from acute NO_2 **AJP**, 125:586-600.
11. W. Walz and A.F. Allan, 1987, Evaluation of the osmoregulatory function of taurine in brain cells, **Exp. Brain Res.**, 68:290-298.
12. E. Trenkner, 1989, Possible role of glutamate with taurine in neuron-glia interaction during cerebellar development, **in** "Functional Biochemistry" H. Pasantes-Morales, R. Martin del Rio, W. Shain, Eds., in press.
13. N.M. van Gelder, 1989, Brain taurine content as a function of cerebral metabolic rate: osmotic regulation of glucose derived water production, **Neurochem. Res.**, 14:495-497.
14. N. Hori, F. French-Mullen, and D.O. Carpenter, 1985, Kainic acid responses and toxicity show pronounced Ca dependence, **Brain Res.**, 358:380-384.
15. K. Okamoto, H. Kimura, and Y. Sakai, 1983, Taurine induced increas of Cl-conductans of cerebellar Purkinje cell dendrites *in vitro*, **Brain Res.**, 259:319-323.
16. J.A. Sturman, R.C. Moretz, J.H. French, and H.M. Wisniewski, 1985b, Postnatal taurine deficiency in the kitten results in a persistence of the cerebellar external granule cell layer: correction by taurine feeding, **J. Neurosci. Res.**, 13:521-528.
17. R.L. Sidman, 1968, **in**: "Physiological and Biochemical Aspects of Nervous Integration", F.D. Carlson, Ed., Prentis Hall, N.J., pp. 163-193.
18. S. Roffler-Tarlov and M. Turey, 1982, The content of amino acids in developing cerebellar cortex and deep cerebellar nuclei of granule cell deficient mutant mice, **Brain Res.**, 247:65-73.
19. D. Goldowitz and R.J. Mullen, 1982, Granule cell as a site of gene action in the weaver mouse cerebellum: evidence from heterozygous weaver chimeras, **J. Neurosci.**, 2:1474-1484.
20. E. Trenkner and R.L. Sidman, 1977, Histogenesis of mouse cerebellum in microwell cultures: cell reaggregation and migration, fiber and synapse formation, **J. Cell Biol.**, 75:915-940.
21. M.E. Hatten, 1985, Neuronal regulation of astroglial morphology and proliferation *in vitro*. **J. Cell Biol.**, 100:384-396.
22. E. Trenkner, M. E. Hatten, and R.L. Sidman, 1978, Ether-soluble serum components affect *in vitro* behavior of immature cerebellar cells in weaver mutant mice, **Neuroscience**, 3:1092-1100.
23. E. Trenkner, A. Stern, and J.A. Dykens, 1989, Taurine moderates *in vitro* neurotoxicity of excitotoxic amino acids and tryptophan metabolites, submitted.
24. W.G. Shain, V. Madelian, D.L. Martins, H.K. Kimelberg, M. Perrone, and R. Lepore, 1986, Activation of b-ádrenergic receptors stimulates release of an inhibitory transmitter from astrocytes, **J. Neuroch.**, 46:1298-1303.
25. J.A. Dykens, A. Stern, and E. Trenkner, 1987b, Kainate-induced death of cerebellar neurons *in vitro* is mediated by xanthine oxidase, **J. Neurochem.**, 49:1222-1228.
26. M.R. Hanley, 1988, Proto-oncogenes in the nervous system, **Neuron**, 1:175-182.
27. A.M. Szekely, M.L. Barbaccia, and E. Costa, 1987, Activation of specific glutamate receptor subtypes increases c-fos proto-oncogene expression in primary cultures of neonatal rat cerebellar granule cells, **Neuropharm.**, 26:1779-1782.

REGULATION OF NEURITE OUTGROWTH FROM CEREBELLAR GRANULE CELLS IN CULTURE:

NMDA RECEPTORS AND PROTEIN KINASE C

Martin A. Cambray-Deakin and Robert D. Burgoyne

The Physiological Laboratory
University of Liverpool
P.O. Box 147, Liverpool L69 3BX, U.K.

INTRODUCTION

The timing of the initiation and the rate of axon and dendrite outgrowth are critically important for the formation of ordered connections within the developing nervous system. The extracellular signals and intracellular factors that regulate neurite outgrowth are largely unknown. Recent studies have begun to suggest a role for conventional neurotransmitters in the regulation of neurite growth through changes in the concentration of calcium in the neurite and growth cone (Kater et al, 1988; Robson and Burgoyne, 1989).

The developing cerebellar granule cell has been studied extensively since aspects of its development can be studied both in vivo and in cell culture (Burgoyne and Cambray-Deakin, 1988). During development newly divided granule cells each extend a single axon which bifurcates to form the parallel fibres. At the same time the cell body migrates down to the granular layer where several short dendrites are formed. The morphological changes that occur during the formation of axons and dendrites in vivo have been well described but little is known of the molecular basis of the control of these events. Dissociated cell cultures from early post-natal cerebellum consist predominantly of granule cells many of which rapidly extend neurites within minutes to hours after plating in culture (Cambray-Deakin et al, 1987a). We have used granule cell cultures to investigate the regulation of the initiation of neurite outgrowth by activation of N-methyl-D-aspartate (NMDA) receptors and begun to investigate the intracellular second messenger systems involved (Burgoyne et al, 1988; Pearce et al, 1987).

ACTIVATION OF NMDA RECEPTORS STIMULATES NEURITE OUTGROWTH FROM GRANULE CELLS IN CULTURE

Granule cells use glutamate as their major if not sole neurotransmitter and are themselves responsive to excitatory amino acids acting on the kainate/quisqualate- and the NMDA-type receptors since they receive a glutaminergic input from the mossy fibres. Following their last cell division and during the period of most rapid growth and differentiation of granule cells in the rat cerebellum these neurons exhibit a high level of sensitivity to NMDA which subsequently declines (Garthwaite et al, 1987). At this stage, however, granule cells newly arrived in the granular layer are not sensitive to the neurotoxic effects of NMDA (Garthwaite et al, 1986). These findings suggest that NMDA receptors on granule cells may play some role during development. Using short-term cultures of dissociated cerebellar cells we have asked the question of whether activation of NMDA receptors controls the initiation of neurite outgrowth from granule cells.

The culture medium that we used does not contain added glutamate but granule cells in culture release glutamate into the medium in reponse to depolarisation and also in a basal fashion (Pearce and Dutton, 1981). Therefore, it is possible that endogenously released glutamate could act on granule cells in culture. In our initial experiments (Pearce et al, 1987) we grew granule cells in culture for 8h either in a serum containing (serum$^+$ medium) or in defined medium without serum but with added insulin and sodium selenite (serum$^-$ medium). In both cases the cells were depolarised due to the presence of 25mM KCl in the medium. During an 8h culture period 20-30% of granule extend neurites in both serum$^+$ and serum$^-$ cultures (Fig.1) but it is clear from paired cultures that neurite outgrowth is more extensive at this time in serum$^+$ cultures (Pearce et al, 1987). An accumulation of released glutamate was detected in medium in serum$^-$ cultures (our unpublished observations).

Using serum$^+$ cultures no effect of added NMDA on neurite outgrowth was seen (Pearce et al, 1987). However, a stimulation of neurite outgrowth by NMDA is detectable in serum$^-$ cultures (see below). We examined the possibility that NMDA receptors were already being activated by endogenous glutamate by examining the effects of receptor antagonists on neurite outgrowth. The broad-spectrum glutamate receptor antagonist kynurenate inhibited neurite outgrowth by about 50% in both serum$^+$ and serum$^-$ cultures. This suggests that neurite outgrowth was stimulated by endogenous glutamate in the cultures. The inhibitory effect of kynurenate was specific since it could be reversed by addition of exogenous glutamate. The effect of glutamate appeared to be specifically through the NMDA-type receptors since the inhibitory effect of kynurenate could be reversed by addition of NMDA but not by kainate or quisqualate. In addition, a similar degree of inhibition of neurite outgrowth was found with the NMDA receptor antagonist D-2-amino-5-phosphonovalerate (APV). The inhibitory effect of APV was reversed by the addition of NMDA (Fig.2).

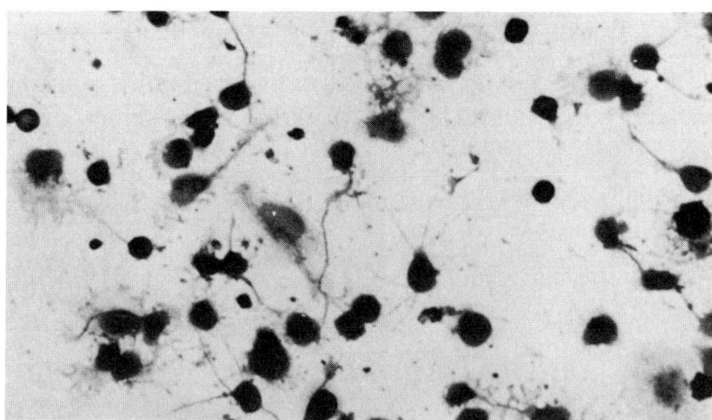

Fig.1. Micrograph showing cerebellar granule cells and their neurites after growth in serum$^-$ cultures for 8h. Cells were dissociated from the cerebellum of 6-8 day old rats and plated on poly-D-lysine coated coverslips. After 8h the cells were fixed and stained with Coomassie brilliant blue.

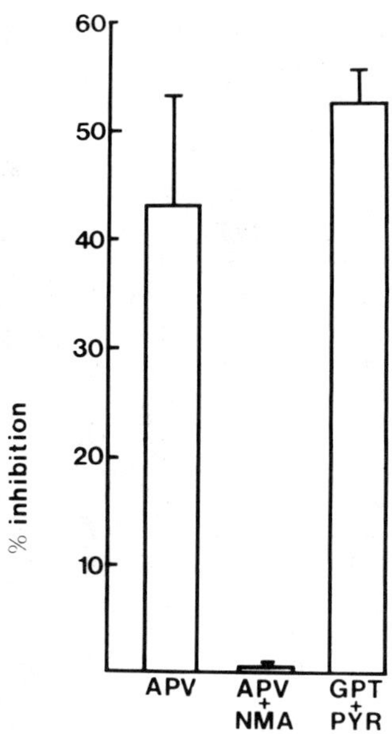

Fig.2. Effect of APV and glutamate pyruvate transaminase (GPT) + pyruvate on neurite outgrowth from
cerebellar granule cells in culture. Granule cells
were grown with no additions or in the presence of
30 µM DL-APV with or without 30 µM NMDA or with GPT
+ pyruvate. The data are expressed as percentage
inhibition of neurite outgrowth compared to controls.

The requirement for endogenous glutamate for neurite outgrowth was
also examined by using an enzymatic method to remove released glutamate
from the culture medium. Serum[+] cultures were grown with 2mM puruvate and
glutamate pyruvate transaminase (GPT) which will result in conversion of
glutamate to ketoglutarate. In these cultures neurite outgrowth was
inhibited to the same extent as seen with APV (Fig.2).

ACTIVATION OF NMDA RECEPTORS ELEVATES CYTOSOLIC CALCIUM CONCENTRATION IN
GRANULE CELLS

In order to begin to investigate possible second messenger pathways
that may mediate the stimulatory effect of NMDA receptor activation on
neurite outgrowth we examined the effect of NMDA on cytosolic calcium
concentration in granule cells in culture. Granule cells in short term
culture on coverslips were loaded with the calcium indicator fura-2 and
cytosolic calcium monitored in the cell population (Burgoyne et al, 1988).
Depolarisation with 25mM KCl raised cytosolic calcium concentration.
Quisqualate had no effect and kainate produced only a small rise in
cytosolic calcium concentration. In contrast to these excitatory amino acid
agonists, NMDA with or without prior depolarisation and even in the
presence of 1.3 mM $MgCl_2$ produced a marked elevation in cytosolic calcium
concentration. This was through a specific effect on the NMDA-type receptor
since the effect was reversed by APV (Fig.3). Of particular relevance to
the long term effect of NMDA receptor activation on neurite outgrowth was
the finding that the elevation in cytosolic calcium concentration elicited
by NMDA was sustained and remained at a plateau 100-200nM above basal for
as long as we monitored calcium concentration (up to 30 min after NMDA
addition). Therefore, it is clear that the NMDA receptors on immature
granule cells do not readily desensitise and activation of these receptors
over a prolonged time course could regulate events occurring over hours in
culture as a result of calcium entry through the NMDA channel.

In addition to elevating cytosolic calcium concentration (see above) activation of NMDA receptors of granule cells also results in the production of IP_3 (Nicoletti et al, 1987) and the translocation and activation of protein kinase C (Vaccarino et al, 1987). Therefore, one potential pathway through which NMDA receptor activation might stimulate neurite outgrowth is activation of protein kinase C by calcium and/or diacylglycerol and phosphorylation of growth related proteins and so we have examined the possibility that neurite outgrowth may be regulated by protein kinase C (Cambray-Deakin, Adu and Burgoyne, submitted for publication).

We examined the affect of activating protein kinase C on neurite outgrowth by growing granule cells in serum cultures in the presence of the phorbol ester TPA. As shown in Fig.4, TPA does have effects on neurite outgrowth and the dose-response curve is biphasic so that low concentrations (up to 1 nM) of TPA stimulate neurite outgrowth and

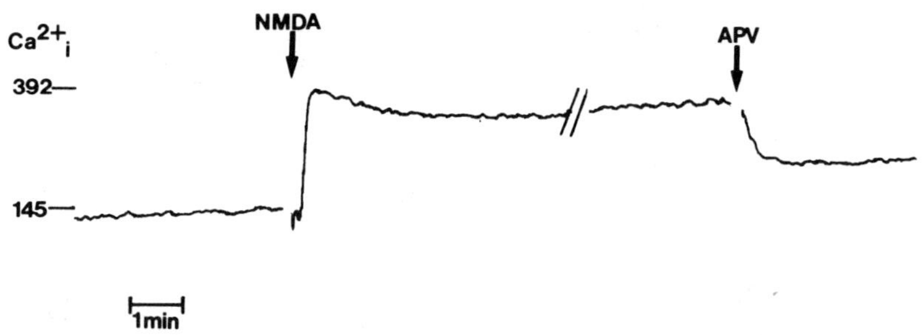

Fig.3. Effect of NMDA on cytosolic calcium concentration in granule cells in culture and reversal of the NMDA-effect by APV.

concentrations above 10nM result in an inhibition of neurite outgrowth. A phorbol ester that does not activate protein kinase C was used as a control: 4-α-phorbol-12,13-dibutyrate had no effect on neurite outgrowth. Cultures incubated with high levels of TPA appeared morphologically normal apart from the profound inhibition of neurite outgrowth. Treatment of cultures with 0.1 nM TPA stimulated a general increase in ^{32}P incorporation into most polypeptides resolved by one-dimensional polyacrylamide gel electrophoresis indicating that this low level of TPA was sufficient to activate protein kinase C. The higher levels of TPA that inhibited neurite outgrowth have been found in other cell types to down-regulate protein kinase C and treatment of granule cell cultures for 8h with 100nM TPA resulted in a 94% reduction in protein kinase C levels detected using ^{3}H-phorbol dibutyrate binding.

The results with TPA suggest that activation of protein kinase stimulates neurite outgrowth. In addition high levels of TPA that cause down-regulation of protein kinase C result in an almost complete inhibition of neurite outgrowth indicating that protein kinase C is <u>essential</u> for the initiation of neurite outgrowth from granule cells within 8h in the culture conditions that we used. In support of this interpretation we found that the protein kinase C inhibitors sphingosine (10 µM) and staurosporine (1 and 10 µM) also inhibited neurite outgrowth from granule cells.

THE NEURITE PROMOTING EFFECT OF NMDA REQUIRES PROTEIN KINASE C ACTIVITY

As indicated above NMDA added to serum[+] cultures had no effect on neurite outgrowth. In contrast, addition of 50 µM NMDA to serum[−] cultures

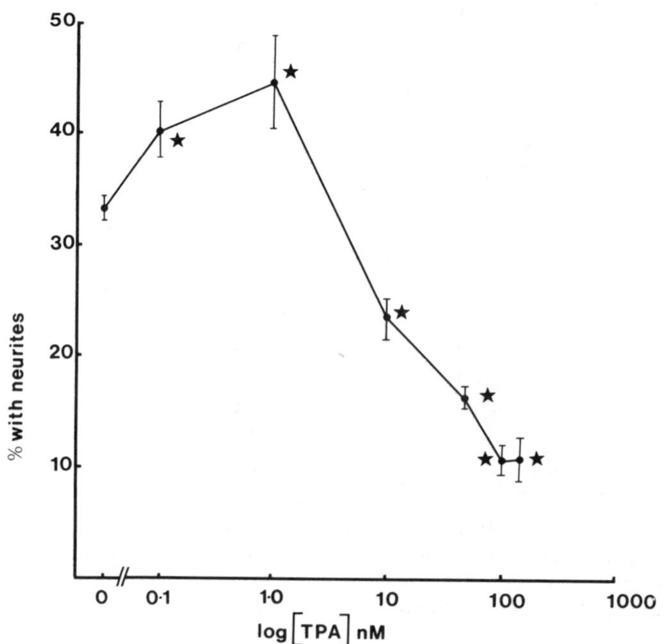

Fig.4. Effect of TPA on neurite outgrowth from granule cells. Granule cells were maintained in culture for 8h in the presence of the indicated concentration of TPA. The percentage of granule cells with neurites is shown. The star indicates a significant difference from control (p<0.001).

resulted in an increase in neurite outgrowth (Fig.5). The magnitude of this stimulation was similar to that seen with 1 nM TPA. To examine whether the stimulation of neurite outgrowth required protein kinase C we examined the effect of 50 µM NMDA on neurite outgrowth in granule cell cultures simultaneously incubated with 100 nM TPA to down regulate protein kinase C or with sphingosine to block protein kinase C. The results in Table 1 show that NMDA is unable to stimulate neurite outgrowth in conditions where protein kinase C activity is reduced or blocked. These data are consistent with the idea that activation of NMDA receptors stimulates neurite outgrowth through a pathway involving protein kinase C.

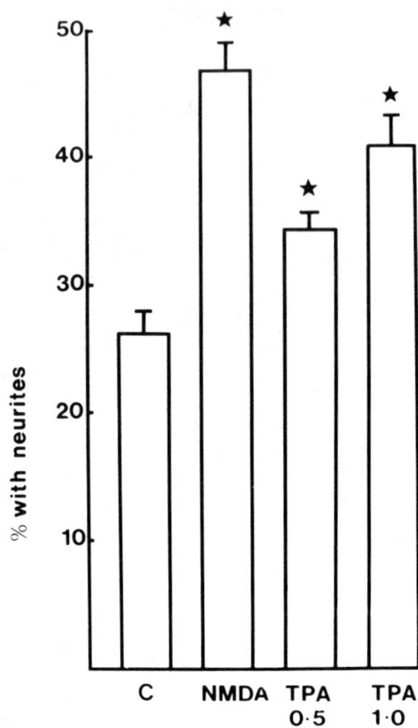

Fig.5. The effect of NMDA and low levels of TPA on neurite outgrowth from granule cells. Granule cells were maintained without additions or in the presence of 50 µM NMDA or 0.5 or 1.0 nM TPA. The stars indicate significant differences from control values (p<0.001).

Table 1. Effect of NMDA on neurite outgrowth from granule cells treated with high levels of TPA or with sphingosine.

Treatment	neurite outgrowth (% of control)
100 nM TPA	21.3 + 3.8
100 nM TPA + 50 µM NMDA	16.4 + 0.9
10 uM sphingo.	48.3 + 8.8
10 uM sphingo. + 50 µM NMDA	57.0 + 3.0

CONCLUSIONS

Our results have shown that glutamate acting through the NMDA receptor
can stimulate the initiation of neurite outgrowth from cerebellar granule
cells in culture. It is likely that this is mediated by prolonged calcium
entry through the NMDA receptor channel leading to sustained elevation of
cytosolic calcium concentration. This would in turn lead to activation of
protein kinase C either directly through calcium or following breakdown of
PIP_2 and generation of diacylglycerol. In many cell types phospholipase C
is activated by an elevation of cytosolic calcium and this may also be the
case in the granule cell following NMDA receptor activation and calcium
entry.

Results from other laboratories have shown that in longer term
cultures than those used by us NMDA can retard cell death in granule cell
culture (Balazs et al, 1988). In addition, treatment of granule cell
cultures between 2-5 days in culture with NMDA results in an increase in
the levels of glutaminase activity (Moran and Patel, 1989) the significance
of which is somewhat obscure. Thus it appears that activation of NMDA
receptors on immature granule cells can stimulate neurite outgrowth,
neuronal survival and the expression of a biosynthetic enzyme. Whether all
of these effects are mediated through protein kinase C is unknown. One
significant clue as to the mode of action of NMDA receptor activation in
long term effects on granule cells is the finding that activation of these
receptors results in rapid induction of c-fos mRNA expression (Skekely et
al, 1989). The increase in c-fos was specific to NMDA and not found with
kainate or quisqualate. c-fos is of particular interest since it is a
prototypical early onset gene which is believed to code for one of a class
of gene regulatory proteins that controls the transcription of genes
required for cellular differentiation or cellular plasticity. It will
clearly be of interest to determine which aspects of NMDA receptor
activation are mediated by direct second messenger effects and which
require genomic regulation. It is possible that even the shorter term (over
8h) effects on neurite outgrowth that we have identified could involve
changes in gene transcription and the long term effects on cell survival
surely do so.

Does activation of NMDA receptors control granule cell development in
vivo ? This is a crucial question for which there is a yet no convincing
experimental evidence. A stimulatory effect of NMDA receptor activation on
axon growth in vivo may be unlikely since at the time that granule cells
extend their axon in vivo they do not appear to possess detectable levels
of NMDA receptors (Cambray-Deakin, Foster and Burgoyne, unpublished data).
We have usually assumed that the neurites on granule cells in culture are
axonal in nature since they transport and accumulate synaptic vesicles in
varicosities. However, staining with an antibody against a synaptic vesicle
protein (Cambray-Deakin et al, 1987b) was not detectable in granules cells
after 24h in culture and only developed over the next three days. It is
possible the the initially formed neurites are actually dendritic in nature
since at early times in culture they do not express neurofilament
polypeptides which appear later (Cambray-Deakin et al, 1987a) but do
express the dendrite-specific protein MAP2. MAP2 persists for some days
throughout the neurites before becoming restricted to the proximal segments
of the neurites with neurofilament protein restricted to the long axon-like
distal segment of the neurite (Cambray-Deakin and Burgoyne, unpublished
data). This morphology is similar to the mature granule cell in vivo since
at least some granule cell axons emerge as branches from short dendrites
and is similar in some respects to the pattern of development of polarity
of hippocampal neurons in culture (Dotti and Banker, 1987). One possiblity
is that the processes that we see in culture result from NMDA receptor
dependent stimulation of <u>dendrite</u> outgrowth. In fact the neurites grown in

the first 8h are all rather short (about 30-50 µM) and there is relatively
little variation in length until some hours later when rapid growth of much
longer neurites   occurs. Only a sub-population of granule cells extend
neurites at early times in culture and it is not known whether this growth
represents development of post-mitotic neurons from the external germinal
layer or regeneration of granule cells from the granular layer following
cell dissociation and plating in culture. The idea that NMDA receptor
activation stimulates dendrite outgrowth would be consistent with the
developmental events in vivo in which granule cells only extend their short
dendrites when the cell body has migrated to the granular layer where the
cell bodies express NMDA receptors and, importantly, receive glutaminergic
input from the mossy fibres (Garthwaite and Brodbelt, 1989) which could
thus serve to stimulate the final stages of granule cell differentiation
and granule cell survival in the developing cerebellum.

REFERENCES

Balazs, R., Hack, N., and Jorgensen, O. S., 1988, Stimulation of the N-
methyl-D-aspartate receptor has a trophic effect on differentiating
cerebellar granule cells, Neurosci. Lett., 87:80.

Burgoyne, R. D. and Cambray-Deakin, M. A., 1988, The cellular neurobiology
of neuronal development: the cerebellar granule cell, Brain Res. Rev.,
13:77.

Burgoyne, R. D., Pearce, I. A. and Cambray-Deakin, M. A., 1988, N-methyl-D-
aspartate raises cytosolic calcium concentration in rat cerebellar granule
cells in culture, Neurosci. Lett., 91:47.

Cambray-Deakin, M. A., Morgan, A. and Burgoyne, R. D., 1987a, Sequential
appearance of cytoskeletal components during early stages of neurite
outgrowth from cerebellar granule cells in vitro, Dev. Brain Res., 37:197.

Cambray-Deakin, M. A., Norman, K. M. and Burgoyne, R.D., 1987b,
Differentiation of the cerebellar granule cell: expression of a synaptic
vesicle protein and the microtubule-associated protein MAP1A, Dev.Brain
Res., 34:1.

Dotti, C. G. and Banker G. A., 1987, Experimentally induced alterations in
the polarity of developing neurons, Nature, 330:254.

Garthwaite, J. and Brodbelt, A. R., 1989, Synaptic activation of N-methyl-
D-aspartate and non N-methyl-D-aspartate receptors in the mossy fibre
pathway in adult and immature rat cerebellar slices, Neurosci., 29:401.

Garthwaite, J., Garthwaite, G. and Hajos, F., 1986, Amino acid
neurotoxicity: relationship to neuronal depolarisation in rat cerebellar
slices, Neurosci., 18:449.

Garthwaite, J., Yamani, B. and Garthwaite, G., 1987, Selective loss of
Purkinje and granule cell responsiveness to N-methyl-D-aspartate in rat
cerebellum during development, Dev. Brain Res., 36:288.

Kater, S. B., Mattson, M. P., Cohan, C., and Connor, J. A., 1988, Calcium
regulation of the neuronal growth cone, Trends Neurosci., 11:315.

Moran, J. and Patel A. J., 1989, Stimulation of the N-methyl-D-aspartate
receptor promotes the biochemical differentiation of cerebellar granule
neurons and not astrocytes, Brain Res., 486:15.

Nicoletti, F., Wroblewski, J. T., and Costa, E., 1987, Magnesium ions inhibit the stimulation of inositol phospholipid hydrolysis by endogenous amino acids in primary cultures of cerebellar granule cells, J. Neurochem., 48:967.

Pearce, B. R. and Dutton, G. R., 1981, K^+-stimulated release of endogenous glutamate, GABA and other amino acids from neuron- and glia-enriched cultures of the rat cerebellum, FEBS Lett., 135:215.

Pearce, I. A., Cambray-Deakin, M. A. and Burgoyne, R. D., 1987, Glutamate acting on NMDA receptors stimulates neurite outgrowth from cerebellar granule cells, FEBS Lett. 223:143.

Robson, S. J., and Burgoyne, R. D., 1989, L-type calcium channels in the regulation of neurite outgrowth from rat dorsal root ganglion neurons in culture, Neurosci. Lett. in press.

Szekely, A. M., Barbaccia, M. L., Alho, H., and Costa, E., 1989, In primary cultures of cerebellar granule cells the activation of N-methyl-D-aspartate sensitive glutamate receptors induces c-fos mRNA expression, Mol. Pharm., 35:401.

Vaccarino, F., Guidotti, A., and Costa,E., 1987, Ganglioside inhibition of glutamate-mediated protein kinase C translocation in primary cultures of cerebellar neurons, Proc. Natl. Acad. Sci. USA, 84:8707.

REGULATION OF GABA$_A$ CURRENTS BY EXCITATORY AMINO ACIDS

Armin Stelzer

Department of Neurology, College of Physicians and Surgeons, 4-408
Columbia University 630 West 168th Street New York, N.Y. 10032 U.S.A.

INTRODUCTION

In the mammalian cortex, glutamate (1) amd γ-aminobutyric acid (GABA) (2) are the principal transmitters mediating excitatory and inhibitory synaptic events. Glutamate activates cation conductances that lead to membrane depolarization. This action is mediated by at least three distinct receptor subtypes defined by their main agonists as N-methyl-D-aspartate (NMDA), quisqualate and kainate receptors (1). GABA controls at least two conductances that produce hyperpolarization in cortical neurons: an early inhibitory synaptic potential mediated by chloride currents through GABA$_A$ receptors (2) and a late hyperpolarization mediated by potassium current through GABA$_B$ receptors (3).

The following chapters describe modifying actions of excitatory amino acids (EAA) in the regulation of GABAergic inhibitory potentials and GABA$_A$ currents in hippocampal pyramidal cells. The regulation of GABA currents by EAA is mediated by both, conventional glutamate receptors and novel glutamate binding sites that are not coupled to ion channels. In addition, some EAA actions on GABA currents are mediated through intracellular regulatory sites and second messenger systems. The discussion represents an attempt to classify the different EAA actions in the regulation of GABA currents and define the conditions under which the various modifications of GABA responses are expressed. The first two chapters summarize recently published data (4,5).

1) Activation of NMDA receptors blocks GABAergic inhibition in the hippocampal slice

The application of tetanic electrical stimuli to the stratum radiatum fiber pathway in the hippocampus in vitro produces an NMDA receptor-dependent enhancement of synaptic efficacy (4). Repeated administration of such stimuli leads to the genesis of spontaneous and stimulation-evoked epileptiform discharges (4,6,7). This model of tetanization-induced epilepsy was used to explore the cellular mechanisms underlying the developing enhancement of excitatory synaptic efficacy.

Experiments were performed on transverse slices (0.5 mM thick) of guinea-pig hippocampus. Intra-and extracellular recordings were made in the CA1 stratum pyramidale of slices in which the CA2/3 subfield was removed by dissection. Evoked responses were produced by stimulation of str. radiatum Schaffer collateral/commissural fibers. GABA and NMDA were applied through an extracellular, three-barrelled ionophoretic electrode to the edge of stratum pyramidale. Tetanic orthodromic stimulation (50 Hz, 2s, 5 trains every 10 sec, repeated every 30 min) resulted in a

Excitatory Amino Acids and Neuronal Plasticity
Edited by Y. Ben-Ari
Plenum Press, New York, 1990

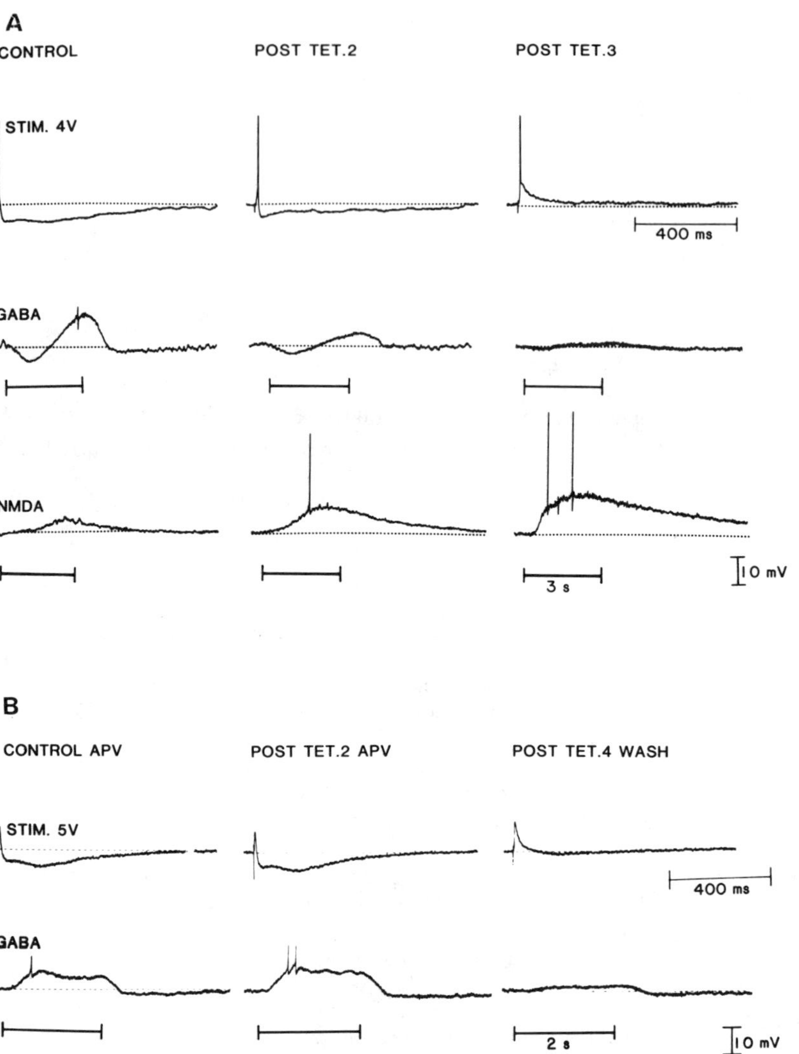

Fig. 1.   Effects of tetanic stimulation on postsynatpic potentials, and GABA and NMDA responses. A, Progressive fading of the response to ionophoretically applied GABA and the analogous decline of evoked IPSPs, the NMDA response was greatly enhanced. B, in presence of D-APV, IPSPs and GABA responses are preserved following tetanic stimulation. Further tetani in wash result in a reduction of both IPSPs and GABA responses (for experimental details see ref. 4).

progressive decrease in the magnitude of both phases of the orthodromically evoked IPSPs to less than 25% of control values after 4 tetani. Sponataneous IPSPs measured with 3M-filled microelectrodes were similarly reduced following tetanization. Increased EPSPs after the first tetanus were matched by a reduction of both, evoked and spontaneous IPSPs of about 50%. Epileptiform activity, including paroxysmal depolarization shifts (PDSs) occurred after IPSPs had been fallen below 10% of control IPSPs. The enhanced excitation and decreased inhibition induced by tetanization was always evident until the end of the recording (up to four hours). The lack of change in the reversal potential of either phase of the orthodromic IPSP indicates that the tetanus-induced

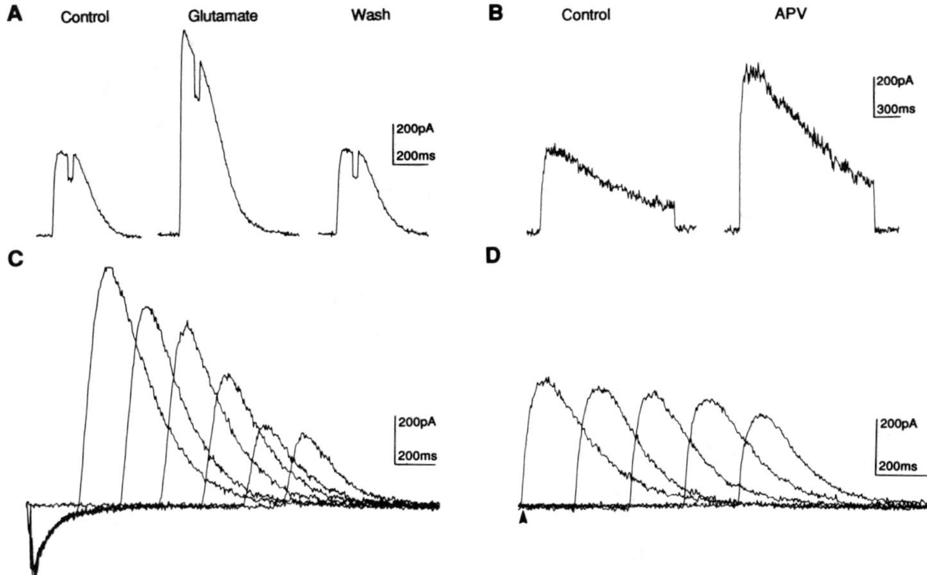

Fig. 2.   Glutamate enhances GABA-stimulated chloride currents. In A, 50 µM glutamate bath applied enhanced GABA chloride current in a reversible manner to about 250% compared to respective control currents. In B, GABA (10 µM) and GABA (10 µM) + D-APV (30 µM) were applied via a multibarreled inflow system. In C, glutamate and GABA were applied by different pressure electrodes. GABA pulses were triggered at five different points of time (200 ms up to 1000 ms, step interval 200 ms) following a glutamate pulse. The GABA response at the very right, was triggered without a preceding glutamate pulse and served as control GABA current. D depicts recordings using the same experimental protocol as described in C. The duration of glutamate ejection was adjusted so that the concentration of glutamate was below threshold for eliciting a glutamate stimulated inward current. The arrow indicates the time of glutamate application. GABA and glutamate concentrations as in C. Holding potential in all cells was -10 mV (for details see ref. 5).

reduction of IPSPs is not due to changes in the respective ionic equilibria of both phases. In the presence of the NMDA receptor antagonist D-APV (5-10 µM), tetanic stimuli did not affect excitatory or inhibitory PSPs. Following washout of APV, further tetani resulted in the reduction of IPSPs and an increase in EPSP amplitude. The involvement of NMDA receptors in the changes of PSPs is further demonstrated by an increased response to ionophoretically applied NMDA following high-frequency stimulation (Fig.1A). The response to ionophoretically applied GABA was reduced progressively by tetanization. In the presence of D-APV, however, IPSPs and the GABA response remained unaltered by high-frequency stimulation (Fig.1B). After washout of D-APV, further tetanization produced a progressive decrease of the GABA response, a reduction of orthodromically evoked IPSPs and an increase of EPSPs.

In conclusion, these results demonstrate that the progressive enhancement of excitatory synaptic efficacy and development of epileptiform activity produced by tetanic stimulation of afferent pathways to CA1 pyramidal cells in the hippocampus is mediated through an alteration of postsynaptic sensitivity to those neurotransmitters believed to be important in both inhibitory neurotransmission and synaptic plasticity. The results also indicate that the tetanus-induced enhancement of EPSPs and progressive reduction of IPSPs are mediated through postsynaptic NMDA receptor activation. The origin of these multiple effects produced by the activation of a

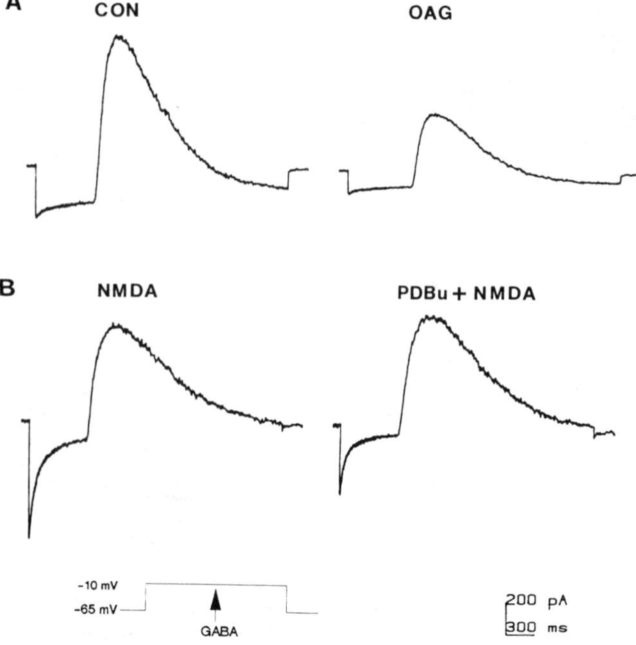

Fig.3.   Excitatory amino acids block the protein kinase C-mediated reduction of GABAA currents, but not the PKC mediated reduction of calcium currents. High voltage activated calcium currents and GABA chloride currents were recorded consecutively in a given neuron (experimental protocol at bottom). A, Bath application of 400 nM OAG resulted in a reduction of both currents measured. Phorbol ester (PDBu) produced similar results. B, When PKC stimulators were applied in presence of excitatory amino acids (here e.g. PDBu (500 nM) + NMDA (50 μM) only the calcium currents were suppressed, the GABA responses remained unaltered. The extracellular solution consisted of (in mM): NaCl, 140; CsCl, 5; TEACl 15, 4-aminopyridine, Hepes 10, Tetrodotoxin 0.4, CaCl$_2$ 10; MgCl$_2$ 1; 10; D-glucose, 25; pH, 7.4. The content of the intracellular recording pipette was identical with that described previously (5).

single class of postsynaptic receptors may lie in the influx of calcium ions through NMDA receptors activated by the tetanus (8,9). Increase in [Ca^{2+}]i has been demonstrated in acutely dissociated CA1 pyramidal cells to destabilize the GABA$_A$ conductance (10,11) most likely through the activation of a calcium/calmodulin-dependent phosphatase which dephosphorylates the GABA$_A$-receptor and renders it non-functional (11). These data raise the possibility that the reduction of GABA sensitivity after tetanization may stem from a change in the state of phosphorylation of the receptor (Figure 4.1). The link between NMDA-receptor mediated rises in [Ca$_{2+}$]i and GABA-receptor dephosphorylation remains to be demonstrated.

2) Excitatory amino acids potentiate GABA$_A$ responses in isolated hippocampal neurons

A variety of pharmacological and possibly endogenous agents can allosterically interact with the GABA$_A$/benzodiazepine receptor or the associated chloride channel and affect the flow of chloride currents (for review see ref. 12). These include anxiolytic (benzodiazepines), anxiogenic (ß-carbolines), anticonvulsant (barbiturates), convulsant (picrotoxin) and hypnotic (steroids) agents (12). The classical potentiators of GABA$_A$ currents, benzodiazepines and barbiturates, exert their action by increasing the receptor's affinity for GABA. The following chapter summarizes recently published data (5) that show that glutamate and other excitatory amino acids potentiate the GABAA conductance at concentrations below those required for their excitatory actions and without significantly altering GABA-binding affinity.   Slices of the hippocampus from the CA1

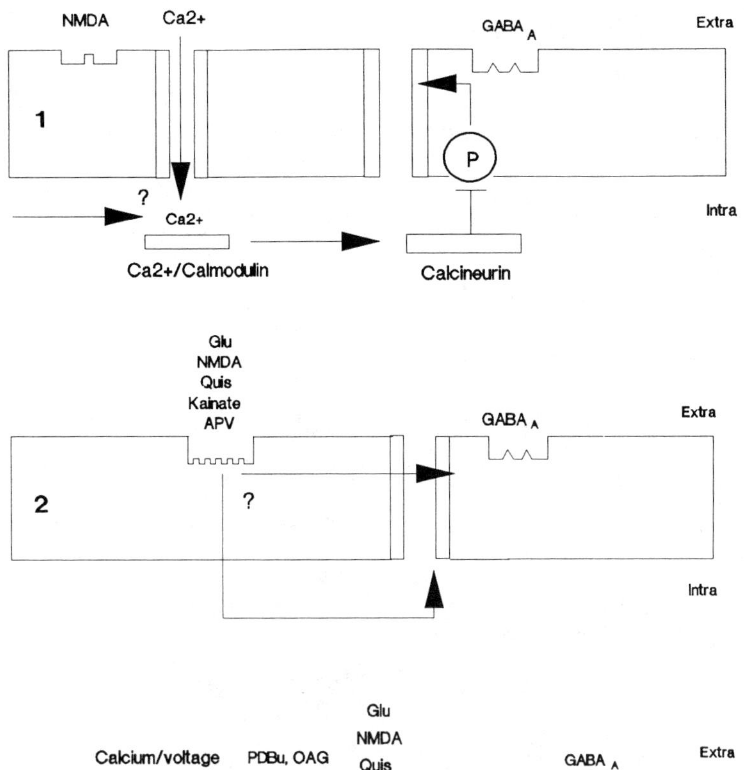

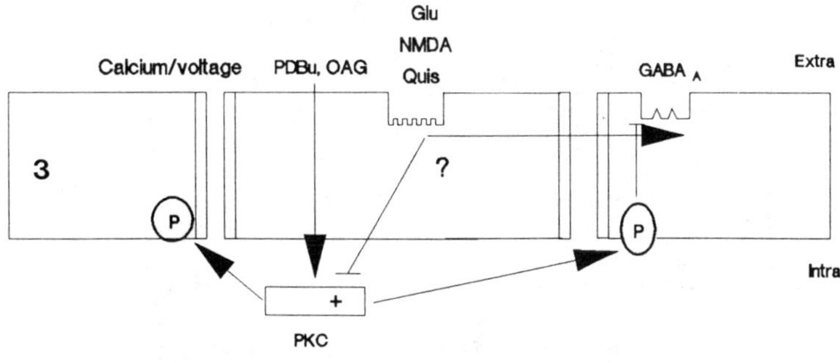

Fig.4.　Modulation of GABA$_A$-receptor/channel complex by excitatory amino acids. For the purpose of illustration, the transmitter binding sites, ion channels and modulatory proteins are depicted as separate functional units. Arrows denotes activation, T-like elements inhibiton. 1: Activation of NMDA receptors blocks GABA responses and rises in [Ca^{2+}]i suppresses GABA currents through the activation of a calcium/calmodulin-dependent kinase. "?" refers to possible sources of rises in [Ca^{2+}]i. 2: Excitatory amino acids potentiate GABA$_A$ responses. The two arrows depict alternative sites of action: excitatory amino acids may directly change GABA$_A$-receptor conformation (upper arrow). In this case the binding site for EAA may be an integral part of the GABA$_A$-receptor/channel complex. Alternatively EAA may trigger a most likely intracellular process that leads to the observed change of the GABA channel conductance state (lower arrow). 3: Excitatory amino acids block selectively the PKC-mediated reduction of GABA$_A$ currents. As in "2", two alternative sites of action are possible: EAA may block the actual phosphorylation process or may alter the conformation of the GABA$_A$ receptor thus rendering it immune to the PKC-mediated phosphorylation (upper arrow in the right part of diagram). The PKC-mediated reduction of voltage-gated calcium currents is not affected by EAA (left part of diagram). For further details see table 1 and text.

	1)	2)	3)
Agonist	NMDA, glutamate	glutamate, NMDA, quis, KA, APV	glutamate, NMDA, quis
Receptor	neurotransmitter	neuromodulator	neuromodulator
Antagonist	APV (other NMDA-receptor antagonists)	?	?
Selectivity	yes	no	partial
APV	antagonist	agonist	none
Ionotropic*	yes	no	no
Agonist (glu) Concentration	? (amount of synaptic release)	10^{-8} -10^{-4} M	M10^{-6} - 5 x 10^{-5} M (lower?)
Second messenger system	Ca^{2+} /calmodulin-dep. phosphatase	?	protein kinase C
[Ca^{2+}]$_i$	physiological	<10^{-8} M (10 mM BAPTA)	<10^{-8} M (10 mM BAPTA)
Preparation	Slice, CA1 hippocampus	Isolated neurons, CA1 hippocampus	Isolated neurons, CA1 hippocampus
Effect on GABA$_A$	conductance decreased	conductance increased	conductance maintained

Legend   List of properties of glutamate regulation of GABAA currents as described in more detail in chapters 1-3 (see above). "*" The term "ionotropic" describes those receptors whose recognition site is coupled to ion channels (effect 1).
Effects 2 and 3 are mediated by non-ionotropic receptors.

region were prepared for trypsin digestion in a bicarbonate free saline (13). Acutely dissociated cells from the CA1 hippocampal subfield of adult guinea-pigs were used to examine the effects of glutamate and other excitatory amino acids on neuronal GABA responses. Whole-cell clamp recordings (14) were performed to monitor the current response to GABA (0.1 to 1 mM in extracellular saline) applied with a picospritzer. Recordings in Fig. 2A illustrate that bath application of L-glutamate (50µM) resulted in a reversible increase of GABA-induced outward currents. When glutamate pulses, applied by sufficiently long pulse durations, were followed at varying intervals by a fixed pulse of GABA from a different pressure pipette, GABA responses within one second after glutamate application were potentiated (Fig. 2C).

When the amount of pressure applied glutamate was reduced by shortening the glutamate pressure pulse duration such that the concentration reaching the cell did not elicit any detectable inward current, enhancement of GABA responses was still observed at short intervals following the glutamate pulse (Fig. 2D). These data indicate that the potentiation of GABA responses may have a higher sensitivity to glutamate than the opening of inward currents.   In another series of experiments the effect of several glutamate analogues on GABA currents was examined. Quisqualate, kainate and NMDA potentiated GABA$_A$ responses. The potentiating effect was not blocked by conventional anagonists of glutamate receptor subtypes, including APV and γ-D-glutamyl-glycine. Surprisingly, the NMDA-receptor antagonist APV mimicked the effects of glutamate on GABA chloride currents (Fig. 2B).   Data from dose response curves show that glutamate produced a marked increase in the maximum peak of the GABA current. The threshold for GABA and the agonist concentration for half maximal response were not significantly altered indicating that glutamate does not affect GABA binding affinity. The potentiating effect became significant at a glutamate concentration of 10^{-8} (108.8 % + 2.0, n=14, compared to control = 100%, p<0.01, t-test), the KD of the curve was 10.2 µM. These values are significantly lower compared to those reported for glutamate-activated inward currents (10 µM for threshold, 1.1 mM for KD (15)).   There may be two ways excitatory amino acids modify GABA$_A$ receptor conductance (for illustration see Fig 4.2). Firstly, the glutamate analogues may activate their corresponding receptor subtypes and trigger a common, most likely intracellular process leading to the modification of the GABA response. Alternatively, the various agents may act at a common binding site at or closely associated with the GABA receptor complex to elicit this novel action. In either case the process underlying the GABAA conductance increase can be considered to be distinct from that leading to the opening of cationic conductances since the two responses differ considerably in their concentration requirement and in their selectivity towards agonists.

3) Excitatory amino acids block the protein kinase C-mediated reduction of GABA$_A$ responses in hippocampal neurons

Recent studies provide evidence for the notion that the GABA$_A$-receptor complex is suject to modification by protein kinase C (PKC). Activators of PKC decreased GABA responses in hippocampal pyramidal cells (16) and responses from receptors expressed in oocytes (17). Biochemical data show that the ß-subunit of the GABA$_A$/benzodiazepine receptor is phosphorylated by purified PKC (18).   Data presented here suggest that excitatory amino acids and the PKC system contribute in an intricate fashion to the regulation of GABA-mediated chloride currents. Calcium-/phospholipid-dependent protein kinase C is physiologically activated by the second messenger diacylglycerol (DAG)(19). Some synthetic diacylglycerols, such as 1-oleolyl-2-acetylglycerol (OAG) and tumor-promoting phorbol ester are readily intercalated into intact cell membranes and activate PKC directly (20,21). Bath application of the phorbol ester phorbol-12,13-dibutyrate (PDBu) in acutely isolated (13) CA1 hippocampal neurons of adult guinea pigs resulted in a steady, progressive reduction of GABA currents after a few minutes of experimental intervention. The amount of GABA current reduction was concentration-dependent, ranging from about 90% (compared with control = 100%) in 100 nM PDBu to about 55 % in 500 nM PDBu 5 min after PDBu application. Similar effects were observed in comparable OAG-containing solutions (Fig.3A). At low PDBu or OAG concentrations (up to 250 nM) the decline of the GABA-peak current amplitude progressed to a finite value of about 50% even in presence of the PKC-stimulating agent. At the high concentration ranges (400 and 500 nM, respectively) application of both types of PKC activators generally lead to a continuous reduction of the GABA response in presence of the PKC activating drug resulting in a complete abolition after sufficiently long administration. The decline could, however, be brought to a halt at any level by switiching back to control saline.

Bath applied glutamate prevented the PKC-mediated reduction of GABAA currents in a concentration-dependent fashion. 50 µM glutamate blocked PDBu effects completely, application of 10 µM glutamate resulted in a significantly smaller amount of reduction following 250 nM PDBu, whereas 1 µM extracellular glutamate did not have an effect in protecting GABA responses from protein kinase C action. 10 µM glutamate was sufficient to block the PKC reduction of GABA currents in 200 nM OAG. The block of the PKC action on GABA responses by glutamate was mimicked by other agonists including NMDA (Fig. 3B) and quisqualate. Like the NMDA-mediated potentiation of GABA-currents, the protection of GABA from PKC by NMDA was not prevented by the conventional NMDA-receptor blocker APV at all concentrations used (up to 100 µM D-L APV and 50 µM D-APV). Unlike the APV-mediated potentiation of GABA currents, however, APV did not mimick this novel excitatory amino acid action of blocking the PKC reduction of GABA currents: in presence of 50 μM D-APV, bath application of PDBu (250 nM) reduced the GABA current to a similar extent as without APV.

To examine whether glutamate blocks all PKC actions or whether the block is specific in the regulation of GABA currents, properties of the protein kinase C regulation of calcium currents were studied and compared with those of $GABA_A$ currents. In acutely dissociated hippocampal pyramidal cells, high-voltage-activated (N- and L-type) calcium currents are most markedly reduced by phorbol ester (22). These high-voltage activated calcium currents were studied for comparison. Bath application of both PKC stimulators resulted in a reduction of voltage activated calcium currents. Excitatory amino acids did not affect the reduction of calcium currents caused by PKC stimulators at any concentration and combination tested. Figure 3B illustrates that in the presence of NMDA (50 μM), addition of PDBu (500 nM) resulted in a decrease of voltage gated calcium currents, but did not affect the simultaneously recorded GABAA currents (Fig. 3B).

In summary, these data indicate that protein kinase C, or at least the part of the heterooligomeric enzyme that regulates calcium currents, is still activatable under conditions that block the PKC regulation of GABA currents.

DISCUSSION

Taken together, the results presented point to a complex regulation of GABAergic synaptic transmission by EAA and second messenger systems. In Table 1, the properties of three regulatory effects of glutamate on $GABA_A$-receptor currents and IPSPs are listed. Fig.4 illustrates these effects graphically.

Only effect 1, the NMDA-receptor dependent block of IPSPs is mediated by one of the classsical glutamate subtype receptors with glutamate acting as neurotransmitter and mediating the opening of NMDA-receptor linked channels. A recent study shows that GABA-mediated chloride currents are reduced by the activation of a Ca^{2+}/calmodulin dependent phosphatase which dephosphorylates the GABAA receptor and renders it non-functional (11). Although the exact link remains to be demonstrated, data from both studies support the notion that GABA currents are reduced by the influx of calcium ions through NMDA receptors activated by the tetanus. It remains to be be seen, whether other sources that increase $[Ca^{2+}]i$ (voltage-gated calcium currents, inositol 3-phosphate) are capable of triggering this process.

Effects 2 and 3, the potentiating effect of glutamate on GABA responses and the block of protein kinase C-mediated GABA-current reduction are both mediated by several glutamate analogues. Like the NMDA-mediated potentiation of GABA currents, the protection of GABA from protein kinase C by NMDA was not prevented by the conventional NMDA-receptor blocker APV. Unlike the APV-mediated potentiation of GABA currents, however, APV did not mimick this novel excitatory amino acid action of blocking the protein kinase C reduction of GABA currents. Both effects are not contingent upon the opening of ligand-gated inward currents. This fact together with the markedly different selectivity towards agonists (compared with conventional glutamate receptors) suggest that these effects are mediated by a different, so far unknwon class of glutamate receptors (Table 1).

Effects 2 and 3, the potentiating effect of glutamate on GABA responses and the block of PKC-mediated GABA-current reduction were examined under conditions where fluctuation of $[Ca^{2+}]i$ is minimized. The high sensitivity of glutamate in potentiating GABA responses may be significant in that it may allow the isolated expression of this effect without activation of cationic (particularly Ca^{2+} conductances (see Fig. 2D). Since excitatory amino acids inhibit the PKC-action of reducing GABA currents in a concentration-dependent manner, it is quite probable that under physiological conditions much lower concentrations of glutamate and analogues may be required than those needed to block the potent PKC-activation by PDBu or OAG (10-50μM). Both effects, the PKC protective action of glutamate and the potentiation of GABA currents occur at concentration ranges close to those reported for cerebrospinal fluid (1-10μM, ref. 23). It is therefore feasible that the composition of the extracellular environment represents a critical factor in the regulation of GABA currents in vivo.

ACKNOWLEDGEMENTS

Studies presented were supported by SFB, NIH, Epilepsy Foundation of America and the Klingenstein Foundation. I thank several colleagues, in particular Dr. R.K.S. Wong for continuous support.

REFERENCES

1. Watkins, J.C. & Evans,R.H., **Ann. Rev. Pharmacol. Tox.** 21 165-204 (1981).
2. Krnjevic,K., **Physiol.Rev.** 54, 418-540 (1974).3. Nicoll, R.A. & Alger,B.E. Science 212, 957-959 (1981).
4. Stelzer,A., Slater,N.T. & ten Bruggencate,G., Nature 326, 698-701 (1987).
5. Stelzer,A. & Wong,R.K.S., **Nature** 337, 170-173 (1989).
6. Slater,N.T., Stelzer,A., Galvan,M., **Neurosci. Lett.** 60, 25-31 (1985).
7. Stasheff,S.F., Bragdon,A.C. & Wilson, W.A., **Brain Res.** 344, 296-302 (1985).
8. MacDermott,A.B., Mayer,M.L, Westbrook,G.L., Smith,S.J. & Barker,J.L., **Nature** 321, 519-522 (1986).
9. Krnjevic,K, Morris,M.E. & Ropert,N., **Brain Res.** 374,1-11 (1986).
10. Stelzer, A., Kay, A.R. & Wong,R.K.S., **Science** 241, 339-341 (1988).
11. Chen,Q.X., Kay,A.R., Stelzer,A. & Wong,R.K.S., **J. Physiol.** (in press).
12. Olsen,R.W. & Venter, J.C., (eds.) Benzodiazepine/GABA Receptors and Chloride Channels: Structural and Functional Properties (Liss, New York, 1987).
13. Kay,A.R. & Wong,R.K.S., **J. Neurosci. Meth.** 16, 227-238 (1986).
14. Hamill,O.P., Marty,A., Neher,E., Sakmann,B. & Sigworth,F.J., Pfluegers **Arch. ges. Physiol.** 391, 85-100 (1981).
15. Kiskin,N.I., Krishtal,O.A. & Tsyndrenko,A.Y., **Neurosci. Lett.** 63, 225-230 (1986).
16. Stelzer,A. & Wong,R.K.S., **Soc. Neurosci. Abstr.** 369.2 (1988).
17. Sigel,E. & Baur,R., Proc. Natl. Acad.Sci. USA 86,2938-2942 (1988).
18. Browning et al., **Soc. Neurosci. Abstr.** (1989).
19. Nishizuka,Y., **Science** 233, 305-312 (1986).
20. Kaibuchi,K. et al., **Cell Calcium** 3, 323 (1982).
21. Castagna,M. et al., **J. Biol. Chem.** 257, 7847 (1982).
22. Doerner,D., Pitler,T.A. & Alger,B.E., **J. Neurosci.** 8(11), 4069-4078 (1988).
23. Ferraro,T.N. & Hare,T.A., **Brain Res.** 338, 53-60 (1985).

COMMENTARY : EXCITATORY AMINO ACIDS AND *IN VITRO* DEVLOPMENT

M. Mattson

The major events that occur during development of the nervous system include neurite outgrowth, synaptogenesis, and "natural" neuronal death. A major contribution to our understanding of the cellular and molecular mechanisms that regulated these events has come from *in vitro* studies. This session examined roles for excitatory amino acids (EAAs) and trophic factors in neuronal development in two regions of the mammalian brain, the hippocampus and the cerebellum. Three main questions were raised in the discussion of the presented papers: how do EAAs exert quite different effects on neurite outgrowth in different types of neurons?; what are the intracellular messenger systems involved in the effects of EAAs and fibroblast growth factor on neurite outgrowth and neuronal survival?; to what extent are developmental changes in the expression of EAA (and GABA) receptors involved in the formation of functional neural circuitry?

It was pointed out by R. Balazs that low levels of EAAs stimulate neurite outgrowth in cultured cerebellar granule neurons, whereas similar levels inhibit dendritic outgrowth in cultured hippocampal neurons. A probable explanation of this discrepancy was given by M. P. Mattson who described previous work that examined the involvement of intracellular calcium in regulating neurite outgrowth. It was suggested that, under the *in vitro* conditions used in the relevant studies, intracellular calcium levels in cerebellar neurons may be below the levels optimum for neurite outgrowth, whereas in hippocampal neurons levels are within the optimum range. Therefore, EAA-induced rises in intracellular calcium levels stimulate neurite outgrowth in the granule neurons, but reduce dendritic outgrowth in the hippocampal pyramidal neurons.

Another topic of interest concerned the data presented by M. P. Mattson which indicated that astrocyte-derived fibroblast growth factor (FGF) can reduce the neurite outgrowth-inhibiting and neurodegenerative effects of EAAs toward hippocampal neurons. The possibility that the observed protection (against EAA toxicity) of neurons contacting astrocytes resulted from a local uptake of glutamate by glia was raised. Several observations were pointed out which made such a scenario unlikely. These included: astrocytes protected neurons from kainate neurotoxicity (astrocytes do not have an uptake system for kainate); astrocytes also protected against calcium ionophore-induced neurodegeneration; a gradient of decreasing glutamate levels with increasing proximity to astrocytes would have to be present (this possibility seems unlikely given the lack of limits to diffussion in the cell culture system).

D. Choi was interested in the mechanism whereby FGF reduced EAA neurotoxicity. M. P. Mattson's findings indicated that FGF enhanced the ability of the neurons to handle a calcium load; it was suggested that FGF might improve the sodium-dependent calcium extrusion system since it had been shown that this system is important in protecting against glutamate neurotoxicity. D. Choi commented that he had also found that the sodium/calcium exchanger is important in protecting against N~DA neurotoxicity in cortical cell cultures.

Further comment on interactions between glia and neurons in relation to the actions of EAAs was raised by E. Trenkner's data that indicated that glutamate released from neurons can induce the release of taurine from glial cells, which may then afford protection against glutamate neurotoxicity. The importance of further studies of glia-neuron interactions was emphasized.

M. Cambray-Deakin's studies of the involvement of protein kinase C in the development of cerebellar granule neurons, taken together with the studies of intracellular calcium in hippocampal neurons, suggested that similar intracellular signaling mechanisms are used to regulate neurite outgrowth and cell survival. It was also noted that these same systems may be used in the mature nervous system in fundamental functional processes such as learning and memory.

The electrophysiological studies of EAA receptors by R. Grantyn in developing cultured neurons, and of GABA receptors by A. Stelzer in mature cultured neurons, illuminated the importance of *changes* in the presence and properties of receptors in neuronal plasticity. It was suggested by R. Balazs that such transient expression of EAA receptors may endow developing nervous systems the capacity for dynamic change in neuronal structure.

Finally, the results of the cell culture studies presented in this session provided support for the hypothesis that similar cellular signaling mechanisms are involved in brain development and function, and that EAAs and growth factors may be involved in the degeneration of neuroarchitecture that occurs in aging or specific disorders.

<u>SESSION VI</u> : EAA AND LTP

LONG-TERM POTENTIATION IN THE DENTATE GYRUS IN VIVO IS ASSOCIATED WITH A SUSTAINED INCREASE IN EXTRACELLULAR GLUTAMATE

T.V.P. Bliss, M.L. Errington and M.A. Lynch

Division of Neurophysiology and Neuropharmacology
National Institute for Medical Research
The Ridgeway, Mill Hill, London NW7 1AA, UK

There is general agreement concerning the NMDA receptor-mediated events underlying the induction of long-term potentiation (LTP) in two prominent hippocampal projections, the Schaffer-commissural input to CA1 pyramidal cells, and the perforant path input to granule cells of the dentate gyrus (Collingridge and Bliss, 1987; Gustafsson and Wigström, 1988). In contrast, the mechanisms by which the enhanced response is expressed and maintained[1] are not well understood, and there is disagreement as to whether the effect is (i) at least partly presynaptic (Bliss and Lynch, 1988; Routtenberg et al, 1985), (ii) wholly postsynaptic (Kauer et al, 1988; Muller and Lynch, 1988), (iii) not postsynaptic (Malinow et al, 1989), (iv) presynaptic initially, and at least partly postsynaptic after about the first 45 min (Davies et al, 1989) or (v) on the basis of morphological evidence, presynaptic (Applegate et al, 1987), postsynaptic (Greenough et al, 1986), or both (Lee et al. 1980; Chang and Greenough, 1984). We shall not attempt to produce a synthesis of these conflicting views in this chapter, and it is doubtful whether the time is ripe for such an attempt. Our less ambitious purpose is to present a summary of the results we have obtained in the last few years with one technique in one pathway: push-pull perfusion of the dentate gyrus in the rat. Contrary to the experience of Aniksztejn et al, 1989 (see Roisin et

---

[1]The terms 'maintenance' and 'expression' are often used interchangeably, but it is useful to distinguish between the terms by reference to agents which block pre-established LTP. When H7, an inhibitor of the catalytic subunit of protein kinase C, is bath applied to the hippocampal slice after the induction of LTP in the Schaffer-commissural pathway, the synaptic response declines to its pre-potentiated level; potentiation is restored when H7 is washed out, demonstrating the persistence of the mechanisms underlying the maintenance of LTP in circumstances in which its expression is blocked (Malinow et al, 1988). On the other hand, transient application of a monoclonal antibody raised against rat hippocampus has been reported to produce an irreversible block of pre-established LTP in vitro; this agent thus blocks the maintenance, as well as the expression of LTP (Stanton et al, 1987).

Excitatory Amino Acids and Neuronal Plasticity
Edited by Y. Ben-Ari
Plenum Press, New York, 1990

al, this volume), we have consistently observed an increase in the
concentration of glutamate in the perfusate during LTP in this pathway.
We argue that the most economical interpretation of these results is that
LTP is at least in part expressed as an increase in glutamate release from
potentiated terminals.

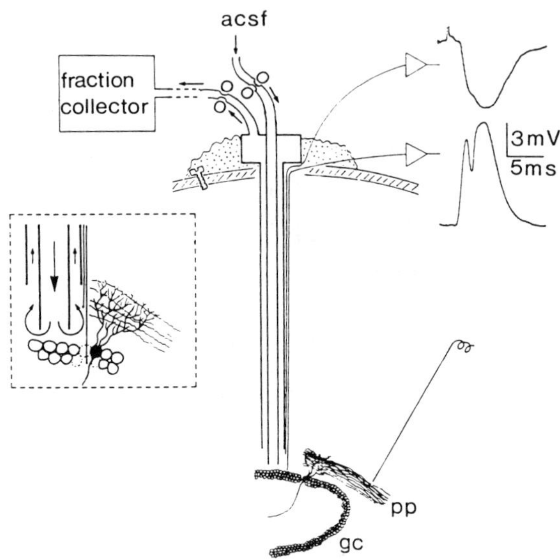

Fig 1. Diagram of push-pull cannula assembly in place in the molecular
layer of the dentate gyrus. Population responses evoked by stimulation of
the perforant path and recorded in the molecular layer (upper) and cell
body layer (lower) are shown. acsf artificial cerebrospinal fluid; pp
perforant path; gc granule cell layer.

Release of radiolabelled glutamate

The technique of push-pull perfusion, as adapted in our laboratory
for perfusing the dentate gyrus while recording responses evoked by
stimulation of the perforant path, is described in detail elsewhere
(Errington et al, 1983; see Fig 1). In our initial experiments with this
method, we examined the release of tritiated glutamate, newly synthesized
from tritiated glutamine which was administered as a bolus through the
cannula at the beginning of the experiment (Dolphin et al, 1982). We
showed first that the release of labelled glutamate was stimulus-
dependent (Fig 2B). In other experiments tetanic stimulation was given to
induce LTP. A sustained increase in glutamate release was observed in
those animals in which LTP was successfully induced; in another group of
animals in which, for unknown reasons, the attempt to induce LTP failed,
there was no increase in glutamate release (Fig 2C, D). These experiments

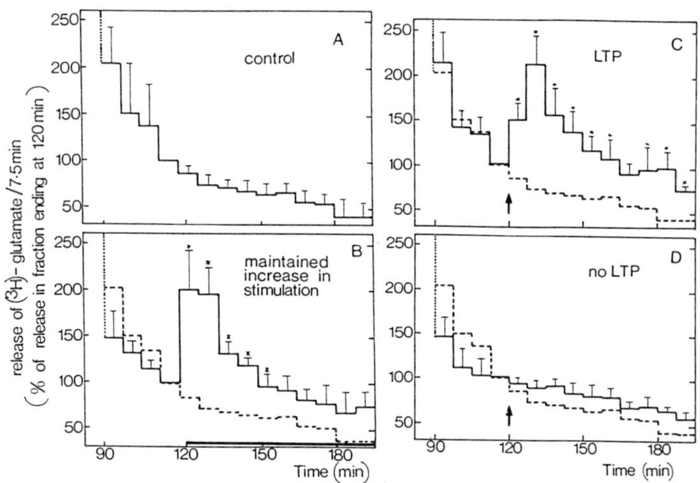

Fig 2. Efflux of ^{3}H-glutamate into push-pull perfusates in four groups of
rats anaesthetized with urethane, and injected through the cannula with a
bolus of ^{3}H glutamine. Perfusion was begun one hour later (t=0). A,
control group, receiving test stimuli to the perforant path at 0.1 Hz.
This curve is repeated (dashed lines) in B-D. B, activity-dependent
increase in ^{3}H-glutamate efflux. During the period indicated by the bar,
the rate of stimulation was increased to 3 Hz, and the stimulus intensity
was doubled. In B, C asterisks denote significant increase in
[3]H-glutamate efflux compared to corresponding period in control group
(p < 0.05, t-test). C, LTP-associated increase in ^{3}H-glutamate efflux.
LTP was induced in this group of animals by a single tetanus (arrow; 250
Hz for 0.5 sec). D, absence of increased efflux in a group of animals in
which the tetanus (arrow) did not produce LTP. (From Dolphin et al, 1982).

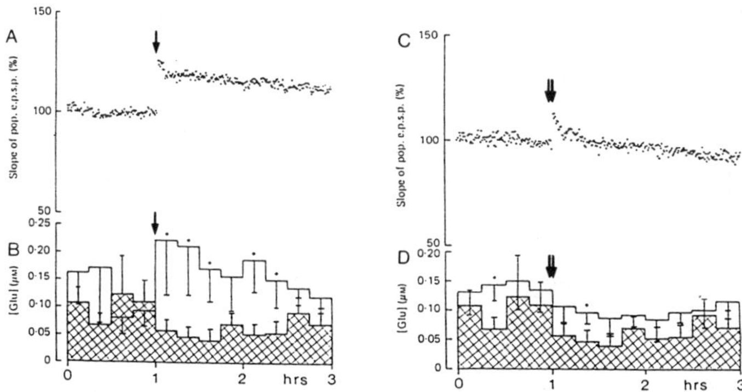

Fig 3. LTP is accompanied by an increased efflux of endogenous glutamate.
I. Commissural stimulation blocks both the induction of LTP and the
associated increase in glutamate release. A, LTP in a group of animals
receiving a single tetanus (250 Hz for 200 msec at the time indicated by
the arrow. B, histograms showing mean concentration of endogenous
glutamate in sequential push-pull fractions from a control group
receiving test stimulation only (hatched histograms in B, D) and the
group in which LTP was induced. Asterisks B, D mark significant
differences between control and experimental groups (p < 0.05).
C, D blockade of LTP (C) and of increase in glutamate release (D) by a
commissural tetanus given at the same time as the perforant path tetanus
(double arrow). (From Bliss et al, 1986).

showed that LTP in the dentate gyrus is associated with an increase in the
release of newly-synthesized, labelled glutamate, the putative transmitter
for the perforant path (Fonnum, 1984). However, because of the
possibility that recently synthesized transmitter is preferentially
release following LTP, it was not possible to decide whether or not there
was an absolute increase in glutamate release. To examine that point, and
to extend our observations to other potential transmitters, we went on to
measure the release of endogenous amino acids, using high-performance
liquid chromatography (HPLC) for separation of amino acids, combined with
fluorometric detection (Bliss et al, 1986).

Release of endogenous glutamate

The concentration of endogenous glutamate in push-pull perfusates
from the dentate gyrus is approximately 100nM. In anaesthetized animals,
we have often found that the concentration declines over a 3-4 hr
perfusion period. For this reason, most of our experiments have been
designed so that transmitter release in the group receiving tetanic
stimulation can be compared, at comparable times, with a control group
receiving only low-frequency test stimulation. Another principle of
experimental design has been to include a third group of animals in which
the induction of LTP has been blocked by various drugs or procedures.
Three such studies are presented in Figs 3-5. In the experiment shown in
Fig 3, the induction of LTP was blocked by coincident tetanic stimulation
of the commissural input to dentate gyrus (Douglas et al, 1982). This
manoeuvre also blocked the increase in glutamate release which in the
experimental group (Fig 3B) persisted at concentrations significantly
above corresponding control concentration for 90 min after tetanus.

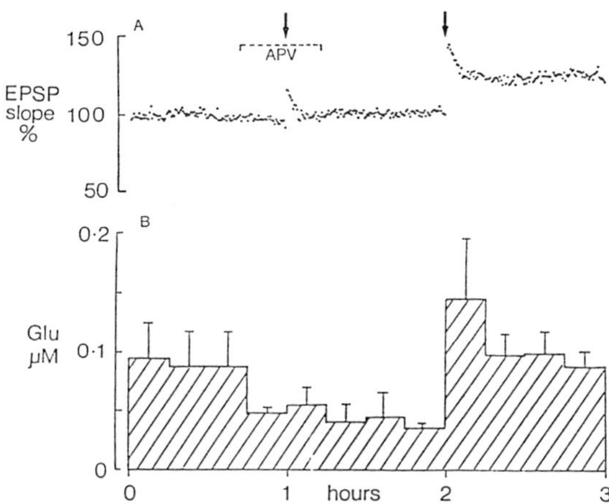

Fig 4. LTP is accompanied by an increased efflux of endogenous glutamate.
II. D(-)aminophosphonovalerate (APV) blocks both the induction of LTP,
and the associated increase in glutamate release. A, Addition of APV
(100µM) to acsf blocks the induction of LTP (first arrow). An identical
tetanus (250 Hz, 200 msec; second arrow) given after APV had been washed
out produced normal LTP. B, An increase in glutamate release occurs after
the second tetanus, which produces LTP, but not after the first, given in
the presence of APV to block LTP. (From Errington et al, 1987).

In the second experiment, we perfused the N-methyl-D-aspartate antagonist, D(-)aminophosphonovalerate (APV), into the dentate gyrus during tetanic stimulation (Fig 4). Induction of LTP was blocked, and there was no increase in glutamate release. APV was then washed out, and a second train of tetanic stimulation delivered. LTP was successfully induced, and the preceding decline in glutamate concentration sharply reversed, indicating a sustained LTP-associated increase in glutamate concentration in the perfusate.

In the third experiment we made use of the lipoxygenase inhibitor, nordihydroguaiaretic acid (NDGA), to block the induction of LTP (Lynch et al, 1989). The histograms paraded in Fig 5 include all the measurements of glutamate concentration in consecutive 15 min samples from each of the six animals in the three experimental groups (control, LTP and NDGA). The increase in glutamate release in the LTP group, which was sustained for at least 90 min after the tetanus in all six animals, and in four animals for the full two hours, is conspicuously absent in both the control and the LTP-blocked (NDGA) groups.

Thus, in all three sets of experiments, and in agreement with our earlier measurements of labelled glutamate release, a sustained increase in the release of endogenous glutamate occurred when, and only when, tetanic stimulation led to the successful induction of LTP. This strongly suggests that increased transmitter release is correlated with LTP rather than with some other consequence of tetanic stimulation.

Glutamate release in the unanaesthetised rat

We have recently measured endogenous glutamate release in rats with a chronically implanted push-pull cannula (Errington et al, 1988). In 3 out of 4 such animals, the induction of LTP was followed by an increase in glutamate release lasting for many hours. In the experiment, illustrated in Fig 6, LTP of the population EPSP and the increase in glutamate release both persisted for approximately 20 hours; in both cases a further tetanus produced only a transient enhancement (second arrow, Fig A,B). Interestingly, the concentration of glutamate in push-pull pefusates from unanaesthetised animals is significantly higher (range: 194-349nM) than in the anaesthetised preparation.

Glutamate release and calcium-induced LTP

It was reported by Turner et al in 1982 that brief exposure of hippocampal slices to high concentrations of calcium produces a long-lasting enhancement of synaptic transmission in CA1. A similar effect can be obtained in the dentate gyrus of the anaesthetized rat. The characteristics of calcium-induced LTP and tetanus-induced LTP in the dentate gyrus are not identical, the principal difference being the absence of spike potentiation in the case of calcium-induced LTP (Bliss et al, 1987; Melchers et al, 1987). Nevertheless, the fact that the two effects are mutually occlusive (maximal tetanus-induced LTP prevents calcium-induced LTP and vice-versa) suggests that both are mediated by similar cellular mechanisms. With this in mind, we examined the relationship between transmitter release and calcium-induced LTP using a push-pull cannula in the anaesthetized rat. As with tetanus-induced LTP, we found that calcium-induced LTP was accompanied by a sustained increase in glutamate release (Fig 7). Furthermore, increased release and enhanced transmission were both blocked by APV (Bliss et al, 1987).

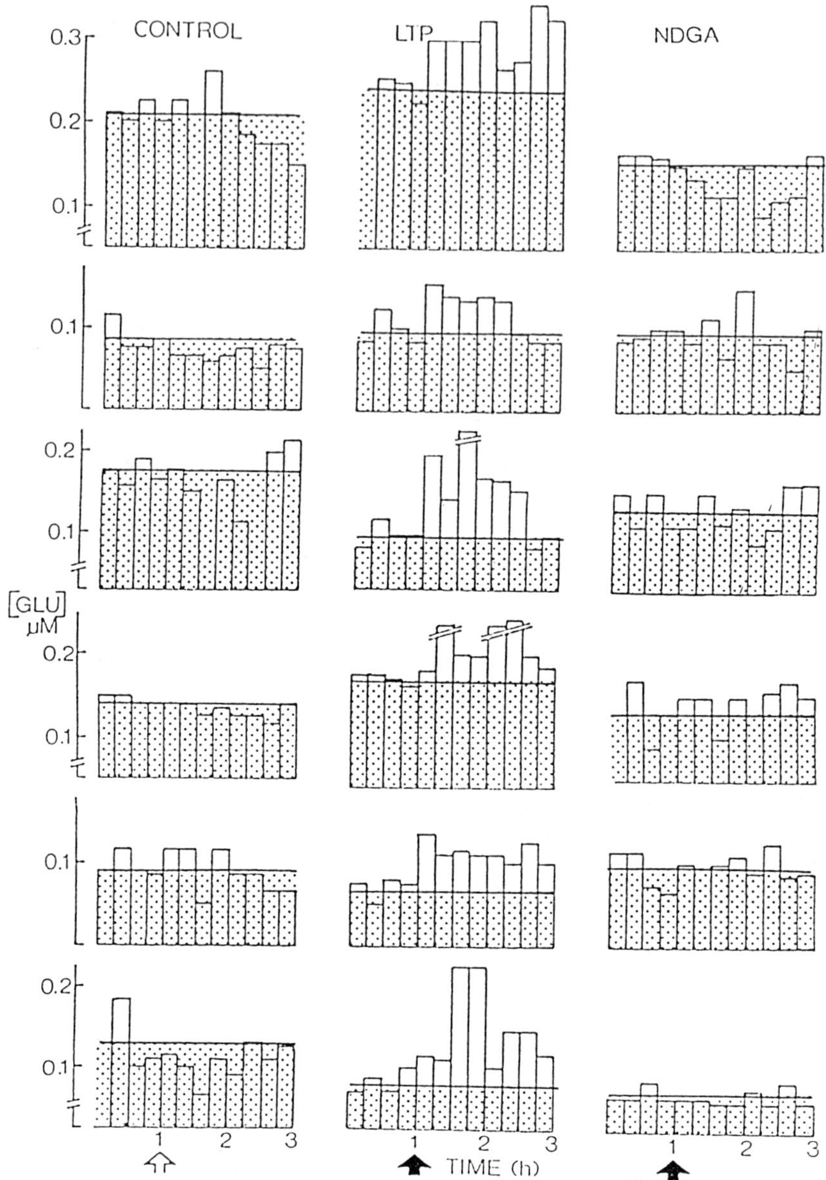

Fig 5. LTP is accompanied by an increased efflux of endogenous glutamate. III. Nordihydroguaiaretic acid (NDGA) blocks both the induction of LTP and the associated increase in glutamate release. Histograms depicting glutamate concentration in consecutive push-pull fractions in each of six animals in three groups; a control group (receiving only test stimulation at 0.033 Hz, increasing to 0.5 Hz for 5 min at the end of the first hour, open arrow, so that the total number of stimuli was the same in all three groups), an LTP group (receiving three 250 Hz, 200 msec tetani over 2 minutes, filled arrow) and an NDGA group in which NDGA was added to the pefusion medium during the tetani (bar) to block the induction of LTP. A consistent and sustained increase in glutamate release was seen only in the LTP group (From Lynch et al, 1989).

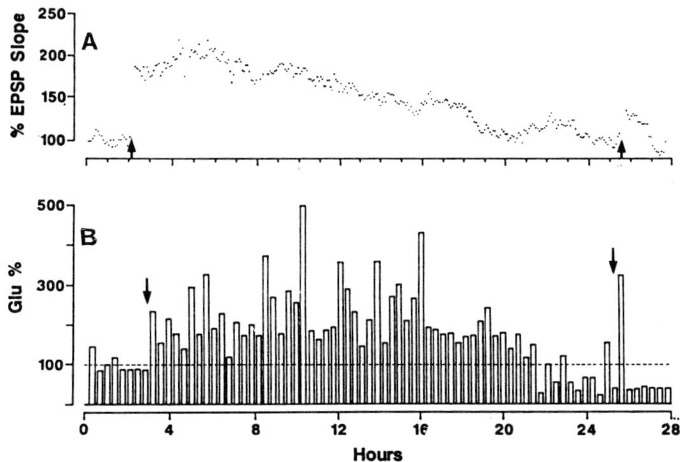

Fig 6. Plot of the slope of the EPSP (A) and the concentration of
glutamate in alternate 10 min push-pull fractions (B) from an animal with
an implanted push-pull cannula. Tetanic stimulation (five series of
trains [8 x (400Hz, 20msec) at 1Hz] at 1 minute intervals) was delivered
at the time indicated by the arrow. Note that LTP and the increase in
glutamate release both declined over the 24 hours after the first tetanus,
and that a second episode of tetanic stimulation produced only a transient
increase in both EPSP and in glutamate release. (From Errington et al.,
1989).

CONCLUSIONS

     In this chapter we have summarized our evidence that LTP in the
dentate gyrus is correlated with an increase in the efflux of glutamate
into push-pull pefusates. The reason for the discrepancy between our
results and those from Ben-Ari's laboratory, which were obtained largely
from CA1, are not clear, but on present evidence the possibility exists
that there are fundamental differences in the mechanisms mediating the
expression of LTP in the two regions. Two important questions which emerge
from our own experiments are: what is the source of the glutamate in
push-pull perfusates, and, more particularly, what is the source of the
increased glutamate in LTP? We have not been able to answer these
questions with any precision using the push-pull technique alone.
Repetitive antidromic stimulation of granule cell axons (mossy fibres)
does not result in detectable increases in glutamate release, so we can be
sure that the LTP-related increase in transmitter release does not derive
from granule cell axons (Bliss et al, 1986). The most likely source both
of basal and of LTP-related glutamate remains the perforant path. The
concentration of glutamate in the extracellular compartment will then be
determined by the balance between the rate of release from perforant path
terminals (and here it is necessary to distinguish between
activity-dependent quantal release, spontaneous quantal release, and
non-quantal release) and the rate of reuptake into glial and/or other
compartments. Changes in any combination of these sinks and sources could
contribute to LTP-associated changes in the glutamate content of push-pull
perfusates. It should be remembered, also, that a change in the amount of
transmitter released per action potential is not the only way in which a
change in extracellular glutamate concentration might be related to a
change in synaptic efficacy. Changes in the background level of
transmitter in the synaptic cleft may also modulate transmission by
altering the occupancy rate of multiple binding sites on receptors which
require more than one transmitter molecule to bind for receptor/channel

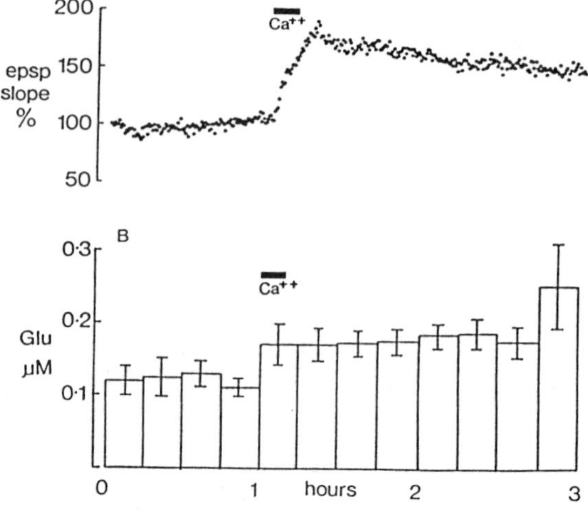

Fig 7. Calcium-induced LTP is associated with a sustained increase in glutamate efflux. A, B slope of the EPSP (A) and concentration of glutamate in successive push-pull fractions (B) from a group of animals in which the concentration of calcium was increased from 1.8mM for 10 min (bar). (From Bliss et al, 1987).

activation. For example, the efficacy of inhibitory inputs to the Mauthner cell in goldfish can be increased or decreased by experimentally modulating the background concentration of the inhibitory transmitter glycine; small increases potentiate transmission, large increases depress transmission, effects which can be attributed to the requirement for multiple binding of transmitter to the glycine receptor (Faber and Korn, 1987). These considerations raise the possibility that LTP is maintained (at least in part) by a small, but sustained increase in the background concentration of glutamate, brought about either by enhanced spontaneous release or by decreased uptake, while larger increases in extracellular glutamate account for the homo- and hetero- synaptic depression which is often engendered by intense tetanic stimulation.

REFERENCES

Aniksztejn L, Roisin MP, Amsellem R, Ben-Ari Y (1989) Long-term potentiation in the hippocampus of the anaesthetized rat is not associated with a sustained enhanced release of endogenous excitatory amino acids. Neuroscience 28, 387-392.

Applegate MD, Kerr DS, Landfield PW (1987) Redistribution of synaptic vesicles during long-term potentiation in the hippocampus. Brain Research 401, 401-406.

Bliss TVP, Douglas RM, Errington ML, Lynch MA (1986) Correlation between long-term potentiation and release of endogenous amino acids from dentate gyrus of anaesthetized rats. J Physiol (Lond.) 377, 399-408.

Bliss TVP, Errington MP, Lynch MA (1986) Calcium-induced long-term potentiation in the dentate gyrus is accompanied by a sustained increase in glutamate release. In: Excitatory Amino Acid Transmitters (eds) TP Hicks and D Lodge, pp 337-340, Alan R Liss, New York.

Bliss TVP, Lynch MA (1988) Long-term potentiation of synaptic transmission in the hippocampus: properties and mechanisms. In: Long-Term Potentiation (eds) SA Deadwyler and PW Landfield, pp 3-72, Alan R Liss, New York.

Chang F-LF, Greenough WT (1984) Transient and enduring morphological correlates of synaptic activity and efficacy change in the rat hippocampal slice. Brain Research 309, 35-46.

Collingridge GL, Bliss TVP (1987) NMDA receptors: Their role in long-term potentiation. Trends in Neuroscience 10, 288-293.

Davies SN, Lester RAJ, Reymann K, Collingridge GL (1989) Temporally distinct pre- and post- synaptic mechanisms maintain long-term potentiation. Nature 338, 500-503.

Dolphin AC, Errington ML, Bliss TVP (1982) Long-term potentiation of the perforant path is associated with increased glutamate release. Nature 297, 496-498.

Douglas RM, Goddard GV, Riives M (1982) Inhibitory modulation of long-term potentiation: Evidence for a postsynaptic locus of control. Brain Research 240, 259-272.

Errington ML, Dolphin AC, Bliss TVP (1983) A method for combining field potential recording with local perfusion in the hippocampus of the anaesthetized rat. J Neurosci Meths 7, 353-357.

Errington ML, Laroche S, Lynch MA, Bliss TVP (1989) Hippocampal long-term potentiation and glutamate release in freely-moving rats. Neurosci. Letts. Suppl. 36, S79.

Faber DS, Korn H (1988) Synergism at central synapses due to lateral diffusion of transmitter. Proc Natl Acad Sci 85, 8708-8712.

Fonnum F (1984) Glutamate: a neurotransmitter in the mammalian brain. J. Neurochem. 42, 1-11.

Greenough WT, Hwang H-M, Chang F-LF, Wallace C, Anderson B (1986) Changes in intramembranous particle aggregrates in neuronal membranes of CA1 pyramidal cells following induction of LTP in hippocampal slices. Soc. Neurosci. Abstr. 12, 505.

Gustafsson B, Wigström H (1988) Physiological mechanisms underlying long-term potentiation. Trends in Neuroscience 11, 156-162.

Kauer JA, Malenka RC, Nicoll RA (1988) A persistent postsynaptic modification mediates long-term potentiation in the hippocampus. Neuron 1, 911-917.

Lee KS, Schottler F, Oliver M, Lynch G (1980) Brief bursts of high frequency stimulation produce two types of structural change in rat hippocampus. J. Neurophysiol. 44, 247-258.

Lynch MA, Errington ML, Bliss TVP (1989) Nordihydroguaiaretic acid blocks the synaptic component of long-term potentiation and the associated increases in release of glutamate and arachidonate: an _in vivo_ study in the dentate gyrus of the rat. Neuroscience 30, 693-701.

Melchers BPC, Pennartz CMA, Lopes da Silva FH (1987) Differential effects
of elevated extracellular calcium concentration on field potentials in
dentate gyrus and CA1 of the rat hippocampal slice preparation. Neurosci.
Letts. 77, 37-42.

Muller D, Joly Y, Lynch G (1988) Contribution of quisqualate and NMDA
receptors to the induction and expression of LTP. Science 242, 1694-1697.

Routtenberg A, Lovinger DM, Steward O (1985) Selective increase in
phosphorylation of a 47 kDA protein (F1) directly related to long-term
potentiation. Behav Neural Biol 43, 3-11.

Stanton PK, Sarvey JM, Moskal JR (1987) Inhibition of the production and
maintenance of long-term potentiation in rat hippocampal slices by a
monoclonal antibody. Proc Natl Acad Sci 84, 1684-1688.

Turner RW, Baimbridge KG, Miller JJ (1982) Calcium-induced long-term
potentiation in the hippocampus. Neuroscience 7, 1411-1416.

LONG TERM POTENTIATION IS NOT ASSOCIATED WITH A SUSTAINED ENHANCED RELEASE OF GLUTAMATE IN THE RAT HIPPOCAMPUS *IN VIVO* AND *IN VITRO*

Marie - Paule Roisin , Laurent Aniksztejn and Y. Ben- Ari

INSERM U29, 123 boulevard de Port-Royal, 75014 PARIS

INTRODUCTION

Long term potentiation (LTP) first observed by Bliss and Lomo in 1973 is defined as a long lasting enhancement of synaptic transmission, produced by a high frequency train of electrical stimulation delivered to afferent pathways in the mammalian hippocampus. This LTP has been extensively studied by an experimental model of learning and memory (eg. review Teyler and DisCenna 1987) in particular in the hippocampus, a cortical structure which plays an important role in memory and learning. Different hypothesis have been suggested to explain the mechanism of induction and maintenance of LTP (Bliss and Lynch 1988, Wigström and Gustafsson 1988). LTP includes an initiation phase (20 min to 1 hour) and a maintenance phase which persists for several hours *in vitro* and days *in vivo* and which includes protein synthesis and post translational modification. It is for instance clear that a rise in Ca^{++} in the postsynaptic element plays a crucial role in initiating the sequence of events which will trigger the long lasting modification. This can be induced in certain synapses (the CA1 Schaffer collaterals system or perforant path-granules cell synapses) by the activation of the NMDA- channel complex (Bliss and Lynch, 1987). In other systems (Cf. the hippocampal mossy fibers synapse) it can be induced by the activation of voltage dependent Ca^{++} channels (Griffith et al. 1986). However the subsequent events and in particular the contribution of the presynaptic and postsynaptic elements of the potentiated signal have not been determined.There are several indications that a postsynaptic change underlines the maintenance of LTP. This includes i) the selective increase in the non NMDA component after LTP (Kauer et al., 1988 a) which cannot be readily reconciled with a presynaptic type of mechanism, ii) the delayed increase in quisqualate responses after LTP (Davies et al. 1989), suggesting in increase in the efficacy or density of post synaptic quisqualate receptors in keeping with the theory suggested initially by Baudry et al., in 1980, iii) the induction of LTP by direct post synaptic regulation of protein kinase C system (Hu et al., 1987, Malenka et al., 1989, Malinow et al., 1989) iiii) the increase in number of receptor sites for glutamate (Lynch et al., 1982). Other modifications have been reported including a change in the postsynaptic morphology (Fifkova and Van Harreveld 1977, Chang and Greenough 1984). The involvement of protein synthesis in the latter phases of LTP (Frey et al., 1988, Otani et al. 1989) as well as the release of new proteins (Duffy et al. 1981, Charriaut-Marlangue et al. 1988, Fazeli et al. 1988) may contribute to the persistent effects.

In contrast Bliss et al. in 1986 have measured transmitter release after LTP and suggest that LTP in the dentate gyrus is due to a sustained increase in the release of glutamate and aspartate. Using a similar approach (push-pull device *in vivo* and HPLC analysis) we have not found a persistent enhanced release of the transmitter candidates glutamate and aspartate (Aniksztejn et al. 1987, Aniksztejn et al. 1988 , Ben- Ari et al. 1989). In the present study, we have reexamined *in vivo* and *in vitro* the relationship between LTP and the release of endogenous glutamate and

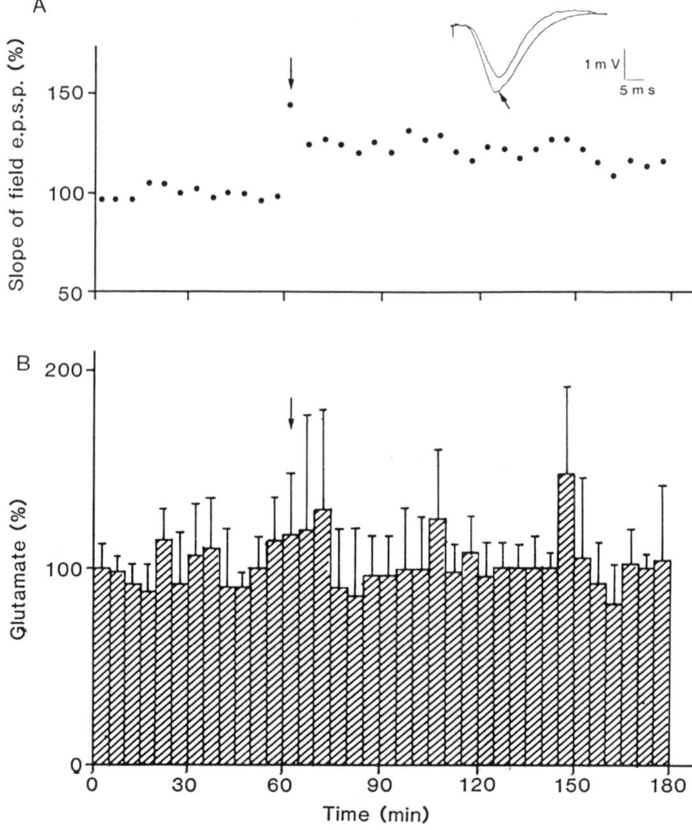

Fig. 1   Relationship between long term potentiation and the release of glutamate in the stratum radiatum of CA1.
A)The slope of the field EPSP evoked by the Schaffer -commisural stimulation (0.033 Hz, 0.05 ms duration) is expressed as a percentage of the 60-min control level. The traces represent the field EPSP just before the high frequency stimulation and the potentiated EPSP 1h after the stimulation (100Hz, 1sec). One field EPSP was digitized every minute and 5 consecutives EPSPs were sampled to obtain the average.
B) Release of glutamate expressed as a percentage of the 60-min control (% mean±S.D. of mean) (n=8). (Adapted from Fig.1 of Aniksztejn et al. 1989)

aspartate. We have used three paradigms to induce LTP: i) a train of electrical stimulation. ii) application of Phorbol esters. iii) application of a peptide, the mast cell degranulating peptide (MCD), which produces a particularly potent and persistent LTP (Cherubini et al.1987).

METHODS

Experiments were performed on adult male wistar rats as previously described. (Anikztejn et al. 1987, Anikztejn et al. 1988). Briefly, rats were anaesthetized with urethane and placed in a stereotaxic frame. A twisted bipolar electrode was implanted in the commisure to stimulate the commisural pathway or in the enthorinal cortex to stimulate the medial perforant patway. Test stimuli (0.05ms) were applied at a frequency of 0.033 Hz. A push pull cannula was introduced in one hippocampus and the position of the cannula was adjusted so as to record the typical field potential evoked by the electrical stimulation of the commisural pathway or the perforant pathway respectively. . For this purpose a recording monopolar electrode, which has the same length as the inner cannula was glued to the outer cannula and was twisted, so the end was very close to the inner cannula. Oxygenated artificial cerebrospinal fluid (ACSF) was perfused at a flow rate of 10μl/min. The perfusion was continued for a stabilization period of 1 hour. At the end of this period, samples were collected at 5 min intervals and the content of endogenous amino acids was determined by high performance liquid chromatography (HPLC) as previously described (Aniksztejn et al.,1987). The HPLC system consisted of two Beckman 114 M pumps controlled by a Beckman 421A programmer, a 7010 Rheodyne injection valve fitted with a 20 μl loop. The derivatization of amino acids was performed using one volume OPA for one volume of the perfusate collected and the mixture was injected 1 min. later using a Varian 9090 autosampler. A 4.6 X 75 mmcolumn packed with ultrasphere XL 3μ ODS (Beckman), an XL ODS precolumn were used for the separation of amino acids OPA derivatives.

RESULTS

1) Induction of LTP by electrical stimulation

As shown in Fig.1A in control conditions, stimulation of the commisural pathway evoked in the stratum radiatum of CA1 a negative field EPSP and the stimulation at 100 Hz enhanced the field EPSP. This potentiation persisted for 2 hours. The release of glutamate was expressed as a percentage of the control period and the mean values were pooled from 8 animals (Fig 1B). We found no significant change in the release of glutamate during 3 hours experiment. However in 2 animals, there was a small increase which persisted only during 10 min. after the train. Table 1 compare the results obtained in a control group. The values were pooled in 15 or 30 min samples after the train. Note the relative stability during the 3 hours of the control experiment. There was no significant difference between the control and LTP group. In the LTP group there was a small difference in glutamate release in the 15 min samples which followed the train. The release of glutamate was not different from the control period 1 and 2 hours after the train.

We have also studied the relationship of LTP in the dentate gyrus and the release of amino acids in 12 experiments. The same push pull device was used. Stimulation of the medial perforant pathway at 0.033 Hz evoked a negative population spike. The brief high frequency stimulation (100 Hz 1 sec) produces long term potentiation of the population spike in the fascia dentata. This effect persisted over 2 hours after the train. We used the same procedure to analyse the release of glutamate than for CA1. Again we found no significant increase in the release over time with a small increase of the release during the application of the train.The comparison in the Fascia dentata of the potentiated cases with control animals showed that as in CA1 there was a small (but not statistically significant) increase in the release of glutamate and aspartate in the first 15 min sample after the train.

Table 1   Release of glutamate in CA1

Time (min)	Control (n=6)	100 Hz LTP (n=8)	MCD LTP(n=6)	MCD Seizure (n=5)
0-60	100±33	100±14	100±35	100±24
60-70	83±25	121±50	538±320	590±435
70-80	108±56	90±22	98±49	215±100
90-120	103±36	102±23	92±37	175±52
120-150	106±38	107±17	100±36	130±38
150-180	130±53	98±24	116±19	126±41

The values of the release of glutamate in control animals, animals subjected to a brief high freequency, animals subjected to MCD application which induced LTP or interictal activity, are indicated as a percent of the 6O min control period.

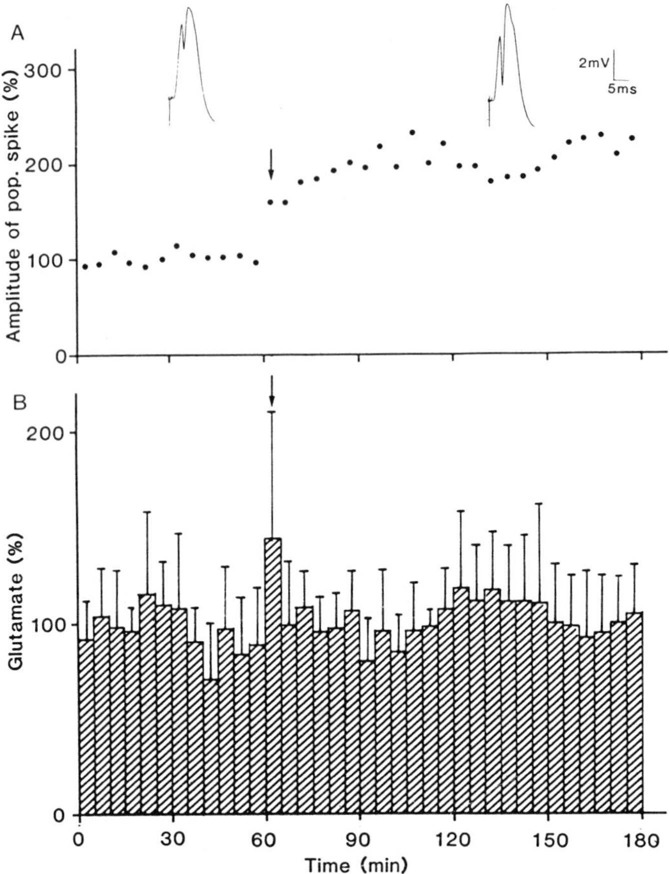

Fig. 2   Relationship between long term potentiation induced in the dentate gyrus and the release of glutamate. The brief high frequency stimulation of the medial perforant pathway (100Hz, 1sec.) produces long term potentiation of the population spike in the fascia dentata. A &B : same presentation as in Fig.1(n=6). (Adapted from Fig.2 of Aniksztejn et al. 1989).

2 Long lasting potentiation produced by a phorbol ester

Others forms of potentiation have been described *in vitro* . Malenka and coworkers (1986) have shown that in the hippocampal slice preparations, application of phorbol esters produces a long lasting enhancement of synaptic transmission. Phorbol esters are potent activators of Ca^{2+} and phospholipid -dependent protein kinase and the protein kinase C. We examined the relationship between long lasting enhancement of synaptic transmission produced by a phorbol-ester and the release of glutamate in the CA1 hippocampal of the anesthetized rat. The application of PDac during 5 min produced a sustained enhancement of the EPSP (Fig. 3). The EPSP was still enhanced 2 hours after the drug was washed out. In the same experiment, we determined the content of glutamate in 5 min samples. Note that the application of PDac produced a brief increase in the release of glutamate and aspartate, and 15 min later, the levels of glutamate and aspartate returned to control values. Similar observations were made in 8 rats. Clearly, the protein kinase C activator produced initially a highly significant increase in the release of glutamate (5 to 6 folds) and aspartate (3 folds).

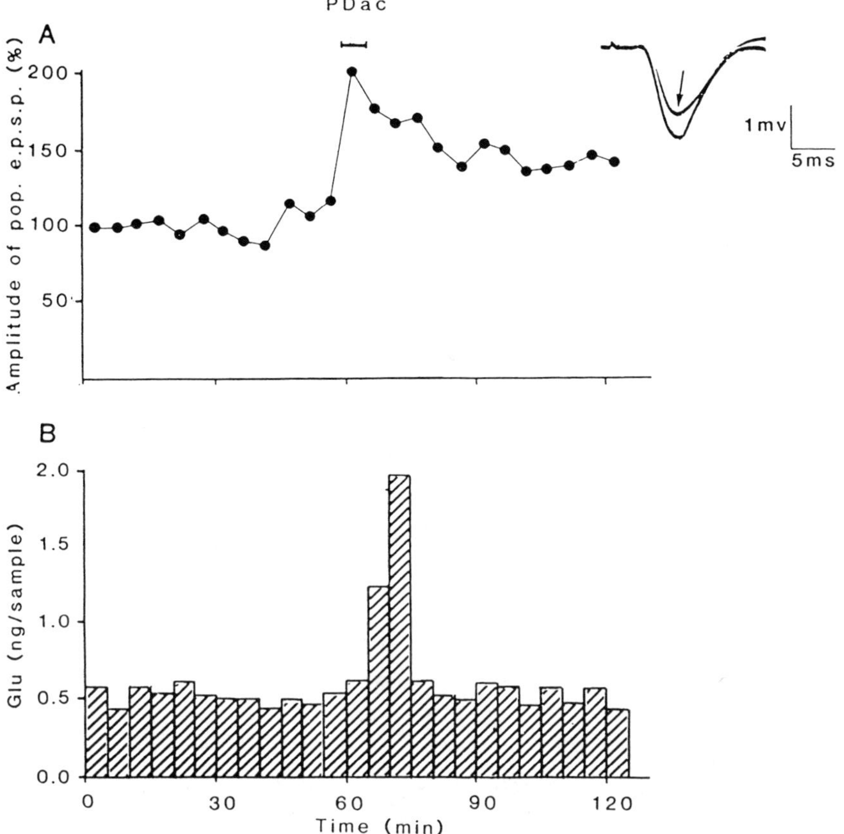

Fig. 3   Long-lasting potentiation induced by Phorbol 12-13 diacetate (PDac) and the release of glutamate in the stratum radiatum of CA1.
A) Long-lasting potentiation of the field EPSP (same presentation as in Fig.1). The traces represent the field EPSP just before the application of PDac (arrow) and the potentiated EPSP 1hr after the application of the drug (500μM, 5min).
B) Release of glutamate in the same experiment.(Adapted from Fig.1 of Aniksztejn et al. 1987)

3 Induction of LTP by MCD

We have recently studied the potentiation effect of another agent the mast cell degranulating peptide (MCD). MCD is a peptide of 22 amino acids with 2 disulfide bounds. MCD has been isolated from bee venom by Lazdunski and his group (Bidard et al.1987). It releases histamine from mast cells and produces convulsions upon central administration (Habermann 1977 and Bidard et al.1987). The effects of MCD are mediated by specific (K+) receptors located in cortical structures, notably in the hippocampus (Mourre et al. 1988). It has been suggested that MCD acts on a specific class of K+ channels.

A study in this laboratory have shown that in hippocampal slices, bath application of MCD induces in CA1 a long lasting enhancement of synaptic transmission during 2 to 6 hours (E.Cherubini et al. 1987). The potentiation induced by MCD was not associated with changes in the intrinsic membrane properties of the post-synaptic cell. It was not due to a reduction of the Cl-dependent

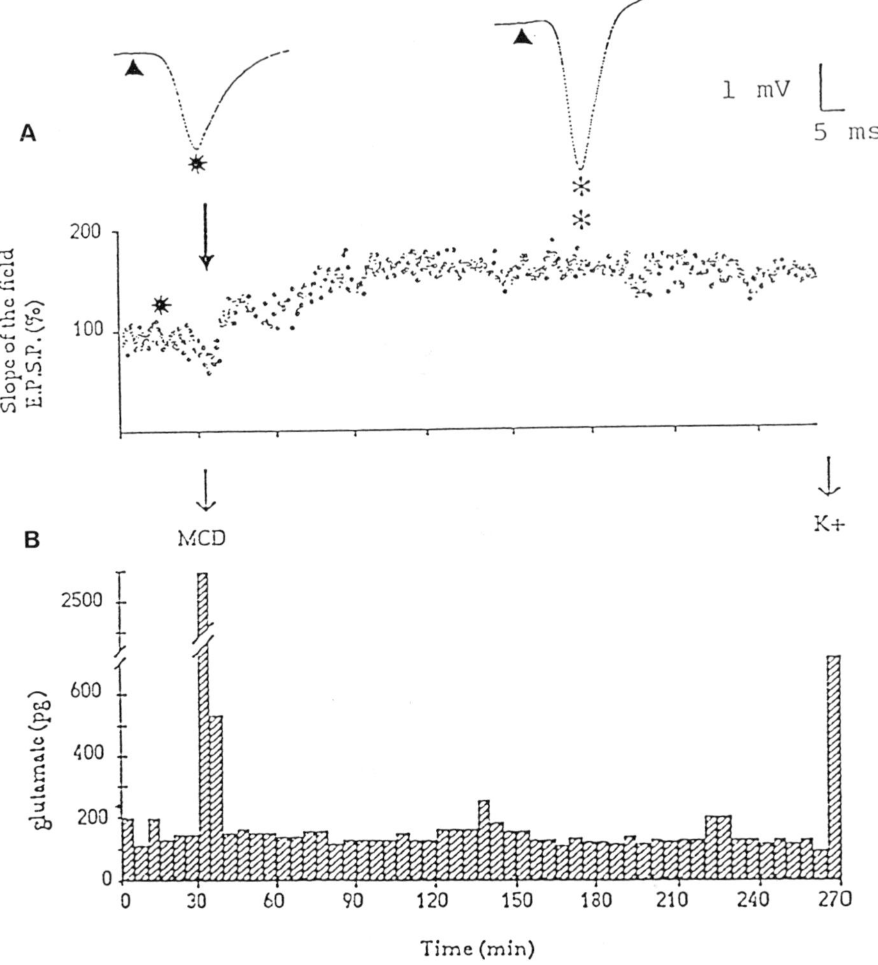

Fig. 4   Long term potentiation of in the stratum radiatum of CA1 produced by the mast cell degranulating peptide (MCD) and concomitant release of endogenous glutamate.A) The slope of the field EPSP evoked by the commisural Schaffer collaterals (stimulated at 0.033 Hz) is expressed as a percentage of the 30 minute control pre-drug level .) MCD produced a large enhancement of the field EPSP which persisted 3 hr after its application. The traces represent the field EPSP just before the MCD application (arrow) and the potentiated EPSP 1hr after the application of the drug . B) Release of glutamate from the same experiment. Note that the MCD produced a large,but transient , increase in the release of glutamate. At the end of the experiment a 5min pulse of K+ (30mM) was applied.

GABAergic inhibition. Once the synapses are potentiated by MCD, they cannot be further potentiated by a train of electrical stimuli, suggesting that the MCD and electrical LTP share common mechanisms.

The same push-pull device was used in the stratum radiatum of CA1 of anaesthetized rats. As shown in Fig. 4 , MCD 20 µM applied for 10 min induced a large enhancement of the field EPSP which persists 4 hours after its application.

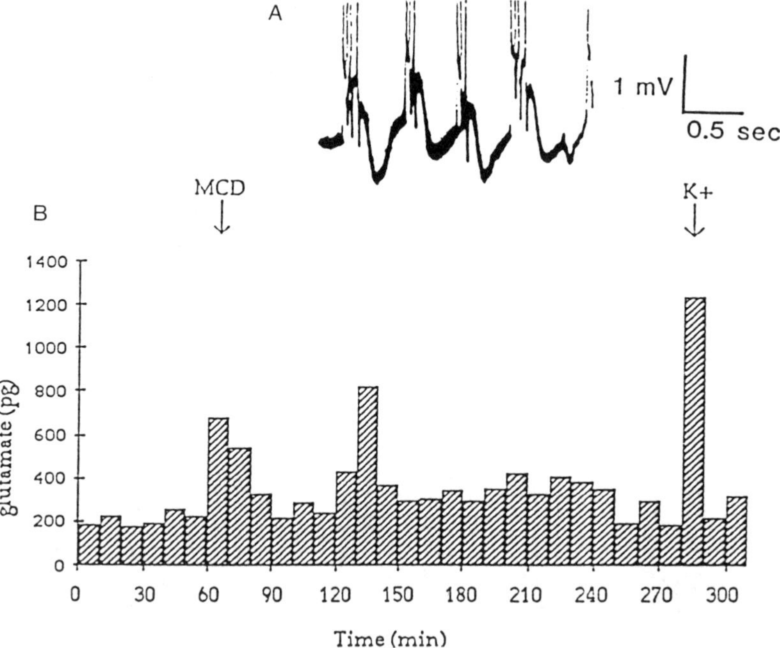

Fig. 5   Release of glutamate in a typical experiment in which MCD produced interictal activity.
A)The application of MCD (30μM,for 10 minutes) induced spontaneous paroxysmal discharges in the stratum radiatum of CA1.
B) Release of glutamate in 10 min samples. Note that the interictal activity was associated with a sustained increase of glutamate release

In the same experiment, we analyzed the release of glutamate and aspartate in 5 min samples and the results are expressed in pg/min. MCD produced a large increase of the glutamate and aspartate. The increase was maximal (30 folds) for aspartate (not shown) and (20 folds) for glutamate in the first 5 min samples. Then, at the end of the application of the drug, the glutamate values returned to the control values. In spite of some variation, the release of glutamate was not different from the controls. A K+ pulse (50 mM) applied at the end of the experiment increased the release of glutamate.A similar sequence of events was observed in 4 other rats. LTP produced by application of MCD was associated with only a transient increase in the release of aspartate and glutamate.

In 5 other rats, application of MCD for 10 min produced a sustained paroxysmal activity characterized by the presence of spontaneous discharges in the stratum radiatum of CA1 (Fig. 5A). This is probably due to an indirect activation of CA3 neurons. In fact, E. Cherubini et al. (1988) have shown that, MCD generates in CA3 interictal activity which propagates to CA1. In these experiments in which MCD induced seizures, there was an increase (Fig. 5B) in the release of glutamate during the application of the drug and a persistent and significant increase in glutamate release.Table I shows the pooled values from 6 rats in which MCD induced LTP and from 5 rats in which we observed an interictal activity. MCD application in these 2 cases produces a large increase in the release of glutamate.

In the LTP group, the concentration of glutamate return to control level. In contrast, in the other group, there was a stable enhancement of the release. This result shows that the technique allows to detect alteration in glutamate release.

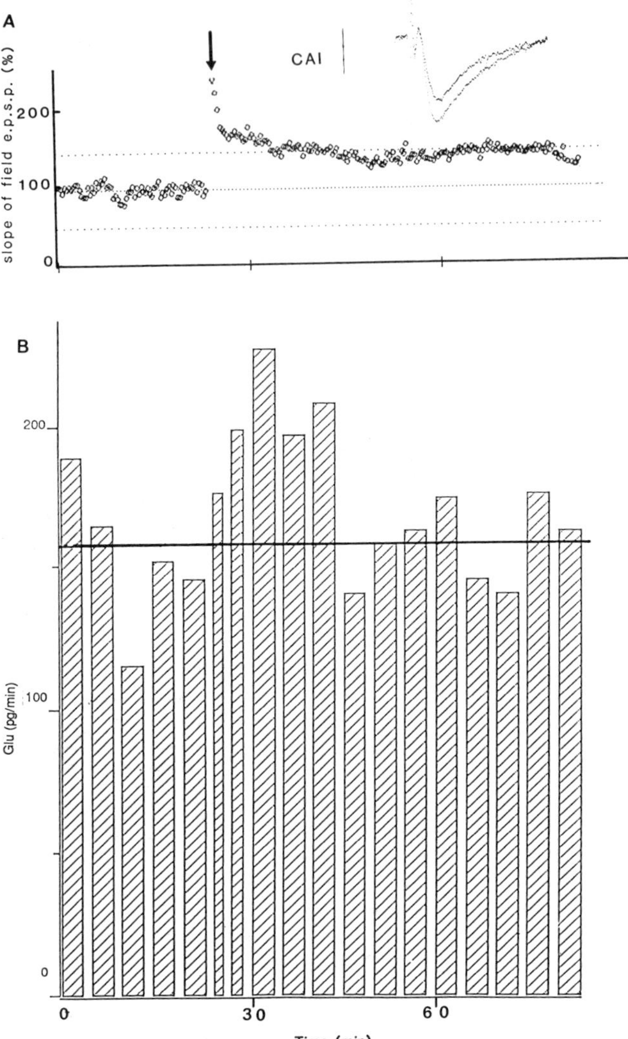

Fig. 6   Relationship between long term potentiation in the stratum radiatum of CA1 and *in vitro* release of glutamate . Hippocampal slices (500μm) were placed in a fully submerged chamber and superfused with ACSF at a reduced flow rate of 0.5 ml/min. A cannula was placed in stratum radiatum of CA1 and 5 min samples were collected at a rate of 25 μl/min. The stimulating and recording electrodes were placed under visual control in the Schaffer collateral. A)The slope of the field EPSP evoked by the Schaffer - commisural stimulation (0.033 Hz, 0.05 ms duration) is expressed as a percentage of the 30-min control level. The traces represent the field EPSP just before the high frequency stimulation and the potentiated EPSP 30 min after the stimulation (100 Hz,1sec). Signals were digitized and displayed on a digital oscilloscope and a computer driven chart recorder.
B) Release of glutamate in the same experiment is expressed in pg/min. The bar indicates the percentage of mean value obtained in control period. Note that the train produced transiently an increase (30%) of the release of glutamate.

4) Induction of LTP in slices preparation

To confirm these results, we have developped a new method (Brassart et al., 1989) which enables to measure in slices, the release of endogenous amino acids from localized regions combined with conventional electrophysiological experiments. With this method, stable levels of aspartate and glutamate were found in the stratum radiatum of CA1 and lower level in CA3. A pulse of K+ (50mM) produced a brief 4-6 fold increase in glutamate, aspartate and a detectable release of GABA The method is relatively sensitive since it enables to detect small changes in release such as these induced by an increase in the frequency of stimulation (Roisin et al. in preparation).Using this method, we have examined the correlation between long term potentiation and the release of glutamate in stratum radiatum of CA1.

As shown in Fig. 6, the stimulation at 100 Hz for 1 sec indicated by the arrow enhances the field of EPSP more than 200 % initially and a subsequent 50 % increase over the control which persisted for 1 hour. In the same experiment, we determined the content of glutamate in the perfusate. Application of the train produces a statistically significant increase of glutamate (30 %) release only during 17 min after train.

DISCUSSION

In conclusion, in the anaesthetized rats, using different models to induce LTP, phorbol ester or MCD and occasionaly with a train of electrical stimulation,there was during the initiation phase in CA1 an increase in the release of glutamate. The release of amino acids was relatively stable and there was no stastically significant differences between the release of glutamate in the maintenance phase versus the control period. We have also extended this observation to slice preparations using a newly developped local recuperation method.We conclude that the maintenance of LTP is not associated with a persistent enhanced release of glutamate.

What are the reasons for the differences with the experiments of Bliss et al. Although unlikely we cannot exclude a difference between the CA1 region (where most of our experiments were performed) and the fascia dentata (where Bliss' experiments have been made). In terms of experimental procedures the only major difference between our experiments design and that Bliss and coworkers are the type of push pull cannula device used. We have used an open system in which the speed of perfusion and recapture are identical. In Bliss' experiments local differences in perfusion speed between the two pumps could generate important changes in pressure and volume. In fact Bliss and coworkers have repeatedly seen local release of blood in the perfusate (e.g. Bliss et al. 1986) notably with perfusion with arachidonic acid (Williams et al. 1989), this could provide an important source of contamination. We have never found blood in our samples.

An additional source of discrepancy between the two laboratories concerns the method of analysis. Thus in the initial papers of Bliss (Bliss et al., 1986, Errington et al., 1987) we believe that the findings depicted do not allow to reach the conclusions drawn by the authors. The comparison was made between 8 naive animals and 8 animals which received a train of electrical stimulation and there was clearly no difference in the same group between the pre and post train period. The authors note that there was a continuous reduction of glutamate release in the control experiments and therefore this comparison could not be made but then i) this reduction and its level of significance has not been quantitavely evaluated ii) in the second paper (Errington et al., 1987), a stable release of glutamate during the 3 hours period of the experiment was observed. Furthermore, in the latter paper the authors examined the release in the same animal before and after the train in presence and following washout of the NMDA receptor antagonist D-APV. This agent reduced by more than 50% the release of glutamate (Fig, 2), this effect was not critically discussed. Why does the same application of APV in the next figure (Fig. 3) fail to produce such an effect? Furthermore, after washing APV, a train of electrical stimulation induced an LTP; the authors state that the train was associated with a sustained increase in glutamate release. However in fact since the APV reduced the release , it is more relevant to compare the post train period with the initial control period . When this comparison is made it is apparent that there was only a transient (15 min.) increase in the release. In a more recent paper however, Lynch et al. (1989) have found an enhanced release of glutamate in the post train period vs the pre train period in the same animals. The type of changes introduced in the experimental procedure has not been specified.

In conclusion, we propose that the initial phase (15 to 30 minutes) but not the maintenance phase (30 min to 3 hrs) is probably associated with an enhanced release of glutamate. The transition period between the initiation on and maintenance is crucial, since it may correspond to the activation of another type of mechanism which will cause the long lasting persistence of the potentiated signal. In fact several proteins appeared to be released during the tardive phase (Charriaut-Marlangue et al. 1988, Fazeli et al. 1988), we suggest that they are at least a part of the signal which will trigger the maintenance of the potentiation.

REFERENCES

Aniksztejn, L., Roisin, M.P.., Gozlan, H. and Ben-Ari, Y. (1987) Long lasting potentiation produced by a phorbol ester in the hippocampus of the anaesthetized rat is not associated with a persistent enhanced release of excitatory amino acids, **Neurosci. Lett.**, 81: 291-295.

Aniksztejn, L., Roisin, M.P., Amsellem, R. and Ben-Ari ,Y. (1989) Long term potentiation in the hippocampus of the anaesthetized rat is not associated with a sustained enhanced release of endogenous excitatory amino acids, **Neuroscience**, 28 : 387-392.

Baudry, M., Oliver, M., Creager, R., Wieraszko, A. and Lynch, G. (1980) Increase in glutamate receptors following repetitive electrical stimulation in hippocampal slices. **Life Sci.** 27, 325-330.

Ben-Ari, Y. Cherubini E., Aniksztejn, L., Roisin, M.P. and Charriaut-Marlangue, C (1989) Mechanism of induction of long term potentiation by the mast cell degranulating peptide. **Pharmacopsychiat.** 22, 107-110.

Bidard,J.N., Gondolfo, G., Mourre, C., Gottesman, C. and Lazdunski, M. (1987) The brain response to the bee venom peptide MCD. Activation and desensitization of an hippocampal target. **Brain Res.** 418, 235-244.

Bliss, T.V.P. and Lomo, T. (1973) Long lasting potentiation of synaptic transmission in the dentate area of the anaesthetized rabbit following stimulation of the perforant path. **J. Physiology** 232, 331-356.

Bliss ,T.V.P., Douglas, R.M., Errington, M.L. and Lynch, M.A., (1986) Correlation between long term potentiation and release of endogenous excitatory amino acids from dentate gyrus of anaesthetized rats. **J.Physiol.** (London), 377: 391-408.

Bliss, T.V.P. and Lynch, M.A. (1988) Long term potentiation of synaptic transmission in the hippocampus: properties and mechanisms. **In** Long term potentiation: From Biophysics to Behavior (Eds: Landfield, P.W. and Deadwyler, S.A.) Alan R. Liss, New-york. pp. 3-72.

Brassart, J.L., Charton, G., Roisin, M.P. and Ben-Ari, Y. (1989) Une nouvelle methode permettant la libération de transmetteurs endogènes sur coupes, associée à un enregistrement électrophysiologique. **Colloque National des Neurosciences**, pp.3194.

Chang, F.L.F. and Greenough, W.T. (1984) Transient and enduring morphological correlates of synaptic activity and efficacy change in the rat hippocampal slice. **Brain Res.** 309, 35-46.

Charriaut- Marlangue, C., Aniksztejn, L., Roisin, M.P. and Ben- Ari Y. (1988) Release of proteins during long term potentiation in the hippocampus of the anaesthetized rat. **Neuroscience Letters** 91, 308-314.

Cherubini E., Ben- Ari Y., Gho M., Bidart J.N. and Lazdunski M. (1987) Long term potentiation of synaptic transmission in the hippocampus induced by a bee venom peptide. **Nature (London)**. 328, 70-73.

Cherubini, E., Neuman, R., Rovira, C. and Ben-Ari, Y. (1988) Epileptogenic properties of the mast cell degranulating peptide in CA3 hippocampal neurones , **Brain Research**, 445: 91-100.

Davies, S.N., Lester, R.A.J., Reymann, K. G. and Collingridge, G.L. (1989) Temporally distinct pre and post synaptic mechanisms maintain long term potentiation. **Nature Lond.** 338, 500-502.

Duffy, C. J., Teyler, T.J. and Shashoua, V.E. (1981) Long term potentiation in the hippocampal slice: evidence for stimulated secretion of newly synthesized proteins. **Science.** 212, 1148-1151.

Errington, M.L., Lynch, M.A. and Bliss, T.VP. (1987) Long term potentiation in the dentate gyrus: induction and increased glutamate release are blocked by D (-) aminophosphonovalerate. **Neuroscience.** 20, 279-284.

Fazeli, M.S., Errington, M.L., Dolphin, A.C. and Bliss, T.V.P. (1988) Long term potentition in the dentate gyrus of of the anaesthetized rat is accompanied by an increase in protein efflux into push pull cannula perfusates. (1988) **Brain. Res.** 473, 51-59.

Fifkova, E. and Van Harreveld, A. (1977) Long lasting morphological changes in dentritic spines of dentate granular cells following stimulation of the enthorinal area. **J. Neurocytol.** 6, 211-230.

Frey, U., Krug, M., Reymann, K.G. and Matthies, H. (1988) Anisomycin, an inhibitor of protein synthesis, blocks late phases of LTP phenomena in the hippocampal CA1 region *in vitro*. **Brain Res.** 452, 57-65.

Griffith, W.H., Brown, T.H. and Johnston, D. (1986) Voltage - clamp analysis of synaptic inhibition during long term potentiation in hippocampus. **Journal of Neurophysiology** 55, 757-775.

Habermann, E. (1977) Neurotoxicity of apamin and MCD peptide upon central application. Naunyn-Schmiedeberg's **Arch. Pharmacol.** 300, 189-191.

Hu, G.Y., Hvalby. O., Walaas. S.I., Albert. K.A., Skjeflo, P., Andersen, P. and Greengard, P. (1987) Protein kinase C injection into hippocampal pyramidal cells elicits features of long term potentiation. **Nature.** 328, 426-429.

Kauer, J.A., Malenka, R.C. and Nicoll, R.A. (1988 a) NMDA application potentiates synaptic transmission in the hippocampus. **Nature Lond.** 334, 250-252.

Kauer, J.A., Malenka, R.C. and Nicoll, R.A. (1988) A persistent postsynaptic modification mediates long term potentiation in the hippocampus . **Neuron** 1, 911-917.

Lynch, M.A., Errington, M.L. and Bliss T.V.P. (1989) Nordihydroguaiaretic acid blocks the synaptic component of long- term potentiation and the associated increases in release of glutamate and arachidonate: an *in vivo* study in the dentate gyrus of the rat. **Neuroscience.** 30, 693-701.

Lynch, G., Halpain, S. and Baudry, M. (1982) Effects of high frequency synaptic stimulation on glutamate receptor binding studied with a modified *in vitro* hippocampal slice preparation. **Brain Res.** 244, 101-111.

Malenka,R.C., Madison, D.V. and Nicoll, R.A. (1986) Potentiation of synaptic transmission in the hippocampus by phorbol esters. **Nature** 319, 774-776.

Malenka,R.C., Kauer, J.A., Perkel, D.J., Mauk, M.D., Kelly, P.T., Nicoll, R.A. and Waxham, M.N. (1989) An essential role for postsynaptic calmodulin and protein kinase activity in long term potentiation. **Nature** 340, 554-557.

Malinow, R., Schulman, H. and Tsien, R.W. (1989) Inhibition of postsynaptic PKC or CaMKII blocks induction but not expression of LTP. **Science** 245, 862-866.

Mourre, C., Bidard,J.N., and Lazdunski, M. (1988). High affinity receptors for the bee venom peptide MCD. Quantitative autoradiographic localization at different stages of brain development and relationship with MCD neurotoxicity. **Brain Res.** 446, 106-112.

Otani, S., Marshall. C.J., Tate, W.P., Goddard, G.V. and Abraham, W.C. (1989) Maintenance of long-term potentiation in rat dentate gyrus requires protein synthesis but not mRNA synthesis immediately post-tetanization. **Neuroscience** 28, 519-526.

Roisin, M.P., Brassart, J.L., Charton, G., Crepel, V. and Ben-Ari, Y. A new method to measure endogenous transmitter release in localized regions of hippocampal slices. (submitted)

Teyler T.J. and DiScenna P. (1987) Long term potentiation. **Annu. Rev. Neurosci.** 10, 131-161.

Wigström, H., and Gustafsson B. (1988) Presynaptic and postsynaptic interactions in the control of hippocampal long-term potentiation in LTP: From Biophysics to behaviour P.W. Landfield, S.A. Deadwyler (Eds.) Alan R. Liss, Inc., New York 73-107.

Williams, J.H., Errington, M.L., Lynch, M.A., and Bliss T.V.P. (1989) Arachidonic acid induces a long - term activity - dependent of synaptic transmission in the hippocampus. **Nature.** 341, 739-742.

POSTSYNAPTIC MECHANISMS INVOLVED IN LONG-TERM POTENTIATION

Julie A. Kauer, Robert C. Malenka, David J. Perkel and Roger A. Nicoll

Departments of Pharmacology and Physiology, University of California, San Francisco, CA 94143

INTRODUCTION

Long-term potentiation (LTP) is a persistent enhancement of synaptic transmission observed at excitatory synapses in the mammalian hippocampus (Bliss and Lomo, 1973). This phenomenon is one of the most striking examples of synaptic plasticity in the vertebrate brain, and has been intensively studied as a model for learning and memory. LTP can be divided into two parts, the triggering or initiation events, and the long-lasting alteration in synaptic strength. In the past six years a considerable body of work has clarified some of the processes involved in triggering LTP, and it has become widely accepted that these processes are localized to the postsynaptic neuron. In contrast, the processes responsible for maintaining the potentiation over time are not as clearly understood, and the synaptic site of these processes remains controversial. This chapter will focus on work from our laboratory studying the cellular mechanisms involved in LTP at the Schaffer collateral-pyramidal cell synapse in the CA1 region of the hippocampal slice preparation.

RESULTS AND DISCUSSION

The presynaptic Schaffer collateral/commissural axon terminals release an excititatory amino acid, likely glutamate, which binds to receptors on the postsynaptic CA1 pyramidal cells to generate an EPSP consisting of two components (Coan and Collingridge, 1985; Forsythe and Westbrook, 1988; Wigstrom and Gustafsson, 1988; Kauer et al., 1988a). These components are mediated by two distinct receptor subtypes present at these synapses, the N-methyl-D-aspartate (NMDA) receptors and the quisqualate/kainate (Q/K) receptors. The Q/K receptors are coupled to ion channels that carry most of the current for EPSPs generated near the resting potential. In contrast, the NMDA receptors are blocked by physiological levels of $Mg2+$ at negative potentials, and so contribute to the EPSP only at potentials more positive than about -40 mV (Mayer et al., 1984; Nowak et al., 1984). LTP is usually experimentally triggered in the hippocampal slice in the following way. First, the afferent Schaffer collateral input is stimulated at low frequency (e. g., 0.1 Hz), and the postsynaptic EPSP is recorded, using an intracellular microelectrode, and/or an extracellular electrode which records from a population of cells. Next, a brief high-frequency train of stimuli (tetanus) is delivered to the afferents. Following the tetanus, the same low-frequency stimuli used initially now produces a much larger EPSP, and this enhancement lasts for many hours. There appear to be two phases of the potentiation following tetanic stimulation, 1) potentiation which decrements to baseline levels in 20-40 minutes, and 2) the persistent potentiation which is essentially non-decremental (LTP). Short decremental potentiation can be generated

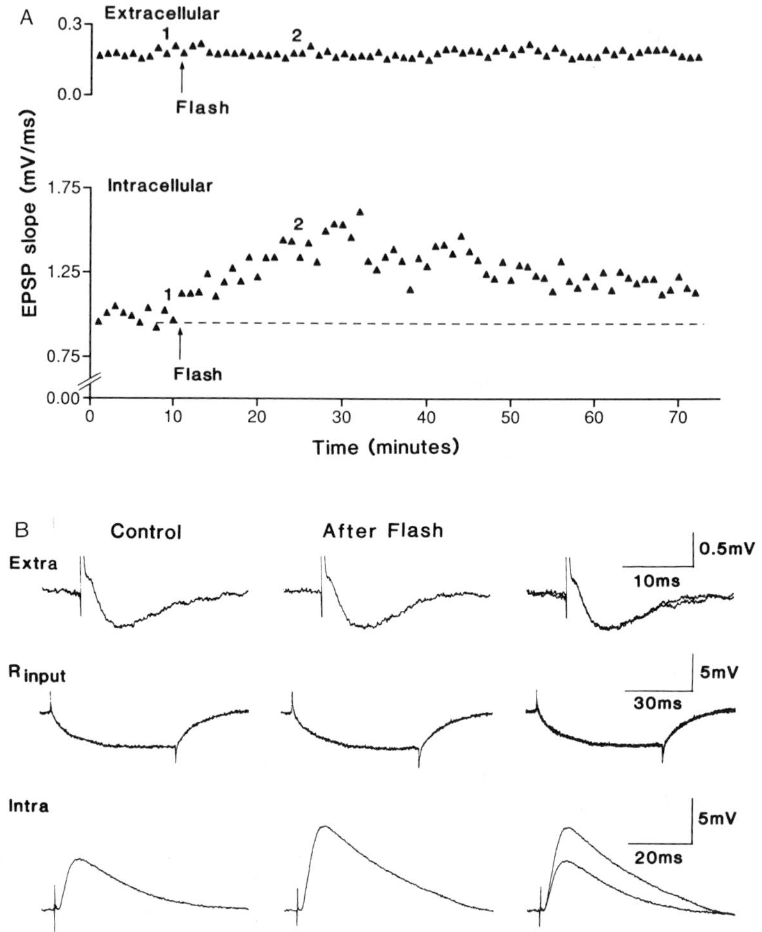

Fig. 1.   Photolysis of intracellularly injected nitr-5 enhances synaptic transmission. A) The slopes of the extracellular and intracellular EPSPs are graphed vs. time. The cell was penetrated 15 minutes before time 0 with an electrode containing 100 mM nitr-5 loaded with 50 mM Ca^{2+}. At the time marked by the arrows the slice was exposed to UV light for 25 sec. B) Sample records were obtained at times indicated by number in part A. Extra: extracellularly recorded EPSP; R_{input}: response to a current pulse (-0.11 nA); Intra: intracellularly recorded EPSP from the nitr-5 filled cell. (From Malenka et al., 1988)

without inducing LTP by ionophoresis of glutamate or NMDA into the CA1 dendritic region (Kauer et al., 1988b).

LTP can also be initiated by pairing several low-frequency afferent stimuli with direct depolarization of a postsynaptic cell (Kelso et al., 1986; Sastry et al., 1986; Gustafsson et al., 1987; Malenka et al., 1988). This manipulation provides the minimal requirements for LTP, i.e. afferent fiber stimulation during postsynaptic depolarization, and demonstrates that high-frequency firing of the presynaptic terminals is not necessary to induce LTP. Depolarization of the postsynaptic cell relieves the voltage-dependent Mg2+ block of the NMDA channel, allowing synaptically released glutamate to cause current flow through these channels. Activation of NMDA receptors was first shown to be essential to LTP induction in 1983, when it was found that the selective

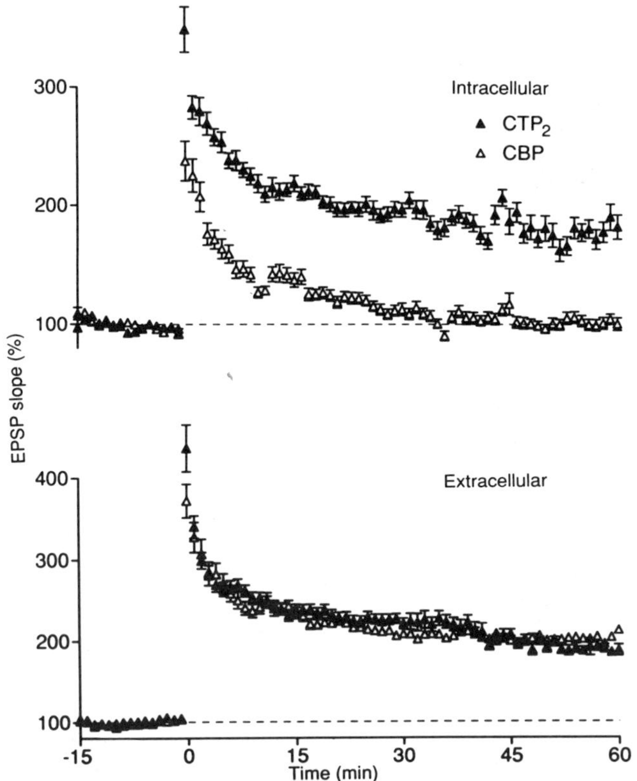

Fig. 2. Intracellular calmodulin-binding peptide (CBP) blocks LTP. The slopes of the EPSPs recorded intracellularly and extracellularly are plotted against time (test stimuli, 0.1 Hz). The intracellular electrodes were either filled with 190 uM CBP (open symbols, n=11) or with 190 uM control peptide (filled symbols, CTP2) (n=8). At time 0, the afferent pathway was stimulated twice at 100 Hz for 1 second, 20 seconds apart.
Sequence of CBP: MHRQETVDCLKKFNARRKLKGAILTTMLA.
Sequence of CTP2: ILTTMLATRNGSGGK. (From Malenka et al., 1989b)

NMDA receptor antagonist, 2-amino-5-phosphonovalerate (APV), prevents LTP (Collingridge et al., 1983).

Ca2+ entry through postsynaptic NMDA receptor channels is necessary and sufficient to potentiate synaptic transmission

In 1983, Lynch et al. found that loading the postsynaptic neuron with EGTA, a chelator of Ca2+, blocked the induction of LTP in that cell, suggesting that Ca2+ is an important second messenger for triggering LTP. Entry of Ca2+ through voltage-dependent channels does not appear to be sufficient to trigger LTP: strong depolarization of the membrane, prolonged Ca2+ action potentials (generated using intracellular cesium ions), or repetitive firing of the postsynaptic cell with tetanic stimulation in the presence of NMDA receptor antagonists all are ineffective in potentiation of synaptic transmission (Collingridge et al., 1983; Malenka et al., 1989a). Thus, the postsynaptic Ca2+ that triggers LTP must arise from an independent source. Unlike Q/K receptor channels, the NMDA channel is quite permeable to Ca2+, and is the likely conduit for the calcium required

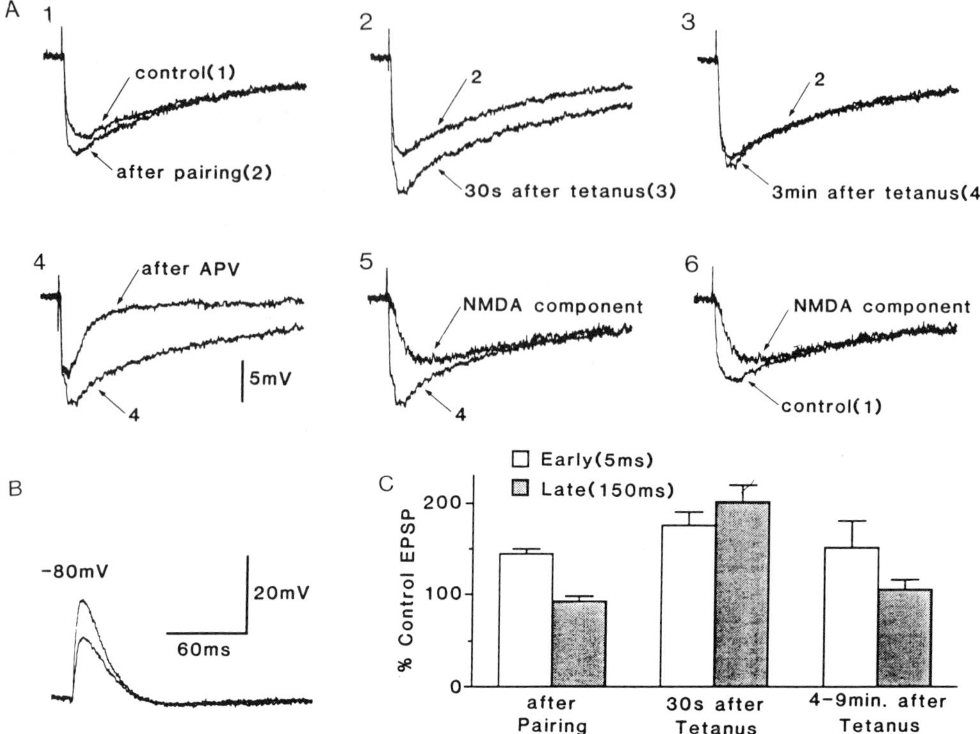

Fig. 3. The early component of the EPSP is selectively increased during LTP. The cell was held at -80 mV and an EPSP recorded in response to afferent stimuli (0.1 Hz). The neuron was then depolarized beyond the reversal potential, and stimulation was resumed. Each record is the average of 3 sweeps. A1) Superimposed records of the initial EPSP (control) and that recorded after an additional 4 min. of stimulation (after pairing). A2) EPSP just before and 30 s after tetanus (PTP). A3) EPSP just before and 3 min. after tetanus. A4) EPSP at 3 min. after tetanus and at 5 min. after 100 uM APV. A5-A6) The NMDA component of the EPSP was obtained from the traces in A4 by subtraction of the EPSP in the presence of APV from that before APV application. This component is superimposed on the EPSP recorded 3 min. post-tetanus (A5), and on the control EPSP (A6). B) The EPSP recorded while holding the neuron at -80 mV, at the start of the experiment and 15 min. after the tetanus. C) The bar graph compares each EPSP component (mean + S. E. M.) after pairing, 30 s after tetanus (PTP), and 4-9 min. after tetanus. The amplitudes of the EPSP at 5 ms (early) and at 150 ms (late) are plotted as a percentage of control size. Pairing: n=16; 30s and 4-9 min after tetanus: n=8. (From Kauer et al., 1988a)

to generate LTP (MacDermott et al., 1986; Jahr and Stevens, 1987; Mayer et al., 1987; Mayer and Westbrook, 1987; Ascher and Nowak, 1988).

We used the photolabile caged calcium compound, nitr-5, to test the effect of a rise in postsynaptic calcium on synaptic transmission (Malenka et al., 1988). Nitr-5 was loaded with Ca2+ and was introduced into the postsynaptic cell through the recording electrode. After recording a steady baseline value in reponse to afferent stimulation, the slice was exposed to UV light. This light flash reduces the affinity of nitr-5 for the ion, resulting in a sudden increase in free calcium. As shown in Figure 1, following the exposure to UV light the intracellular EPSP was strongly potentiated. A field electrode used to measure the population response of surrounding neurons recorded no persistent alteration in EPSP size with the light flash. In addition, cells filled with nitr-5 without calcium showed no alteration in EPSP size upon flashing the light, indicating that the release of calcium itself was responsible for synaptic potentiation (Malenka et al., 1988).

Like the earlier experiments using EGTA, experiments injecting unloaded nitr-5 (a rapid calcium chelator) into postsynaptic neurons blocked LTP induction (Malenka et al., 1988). These experiments confirm that a rise in intracellular Ca2+ is necessary for LTP, but they do not distinguish between a calcium increase resulting from influx of Ca2+ across the cell membrane and Ca2+ release from intracellular stores. In cultured spinal cord neurons, depolarization of the membrane more positive than +20 mV suppresses Ca2+ entry through the NMDA channel (Mayer et al., 1987). In CA1 pyramidal neurons, pairing low frequency stimuli with similar very strong membrane depolarization did not elicit LTP, while pairing the same stimuli with more modest depolarization produced potentiation in the same neuron (Malenka et al, 1988). Thus, suppressing Ca2+ entry across the cell membrane blocks LTP. Taken together, this set of experiments shows that a rise in postsynaptic Ca2+ is sufficient to produce synaptic potentiation and that postsynaptic Ca2+ influx is necessary for the induction of LTP.

Postsynaptic calmodulin and a postsynaptic kinase are required for the maintenance of LTP

The most likely means by which a transient Ca2+ signal in the postsynaptic cell may effect a prolonged cellular change is through the activation of one or more Ca2+-dependent enzyme systems. Three candidate enzymes have been proposed to play a role in LTP: Ca2+-dependent proteases, protein kinase C (PKC), and Ca2+/calmodulin dependent protein kinase II (CaM kinase II). A role for proteases was proposed since in vivo chronic administration of the protease inhibitor, leupeptin, blocks LTP (Staubli et al., 1988). Several studies have shown that bath application of kinase inhibitors (which have actions both on PKC and on CaM kinase II) prevent LTP (Lovinger et al., 1987; Malinow et al., 1988; Reymann et al., 1988). Interestingly, after LTP induction bath application of one of these, H-7, returned the EPSP to baseline levels, suggesting that persistent activity of a kinase is required to maintain LTP (Malinow et al., 1988). It is notable that protein kinase inhibitors that block LTP leave intact the short form of potentiation (Lovinger et al., 1987; Reymann et al., 1988; Malinow et al., 1988; Kauer et al., 1988b). This suggests that a protein kinase may not be required for the short decremental potentiation, but is necessary for development of LTP (or, possibly, conversion of the short form to LTP).

PKC has been implicated in LTP since LTP is associated with an increase in inositol phospholipid turnover (Lynch et al., 1988), translocation of PKC from cytosol to membrane (Akers et al., 1986), and an increased phosphorylation of a PKC substrate protein (Nelson and Routtenberg, 1985). Also, injection of PKC itself into the postsynaptic neuron enhances EPSPs (Hu et al., 1987). Considerable interest also exists in the possibility that CaM kinase II may be involved in LTP. This enzyme is a major component of the postsynaptic density (Kennedy et al., 1983; Kelly et al., 1984; Ouimet et al., 1984), and can undergo autophosphorylation which produces kinase activity that is independent of Ca2+/CaM (Saitoh and Schwartz, 1985; Miller and Kennedy, 1986). The properties of this kinase thus suit it nicely for action as a switch activated by a transient increase in postsynaptic Ca2+.

The protein kinase inhibitors have presynaptic as well as postsynaptic actions. We have therefore introduced various compounds directly into the postsynaptic cell via the intracellular microelectrode. We found that a tetanus delivered to the afferent pathway induced only short decremental potentiation in cells filled with the kinase inhibitor, H-7, while population EPSPs in the same slice exhibited LTP (Malenka et al., 1989b). This result implicates a protein kinase located postsynaptically in LTP.

A number of calmodulin (CaM) antagonists have been tested and shown to block LTP when bath applied. We tested the requirement for postsynaptic calmodulin in LTP using electrodes filled with calmidazolium, a calmodulin antagonist that is relatively selective at low concentrations. These cells exhibited only the short form of potentiation, but did not show LTP, whereas extracellular recordings demonstrated that the remaining cells in these slices did undergo LTP (Malenka et al., 1989b). This finding suggests that Ca2+ entering through NMDA receptors may act through postsynaptic calmodulin to initiate LTP.

Recently a synthetic CaM binding peptide (CBP) has been shown to bind with high affinity to CaM and to inhibit the CaM-dependent activation of CaM kinase II (Kelly et al., 1988). The sequence of this peptide was derived from the CaM-binding domain of CaM kinase II, and CBP is thus extremely selective for this kinase. We have compared effects of CBP with those of an inactive peptide of similar sequence and isoelectric point upon injection into postsynaptic CA1 cells (Malenka et al., 1989b). A tetanus delivered to cells recorded using CBP-containing electrodes elicited only short decremental potentiation, while cells recorded using electrodes filled with control peptide (CTP$_2$) exhibited LTP (Figure 2). The field EPSPs recorded simultaneously showed no observable difference in the LTP elicited in the two groups of slices. Control experiments showed that CBP has no observable effects on normal synaptic transmission. Additional experiments suggested that the ability of CBP to block LTP results from its antagonism of calmodulin dependent reactions, rather than solely from inhibition of CaM kinase II. Taken together, these data indicate that activation of calmodulin and a kinase, perhaps CaM kinase II, are required for LTP, but may not be involved in short decremental potentiation.

Site of the persistent change during LTP

As outlined above, considerable evidence favors the suggestion that the processes that underlie the induction of LTP occur in the postsynaptic cell. One of the most controversial points obscuring our understanding of LTP concerns the site of the changes responsible for the persistent increase in synaptic efficacy. Three major possibilities have been suggested: an increase in transmitter release from presynaptic afferents (Bliss et al., 1986), an alteration in the morphology of the synapse (Lee et al., 1980; Chang and Greenough, 1984), and a change in the sensitivity of the postsynaptic cell to transmitter (Lynch and Baudry, 1984). We have taken advantage of the dual nature of the EPSP at the synapse to examine these possibilities using electrophysiological methods (Kauer et al., 1988a). We compared the sizes of the NMDA and Q/K components of the EPSP before and after stimuli that trigger LTP. Upon the induction of LTP, an increase in the amount of transmitter released or a simple alteration in synaptic structure would increase both EPSP components simultaneously. In contrast, a selective increase in a single EPSP component would be most easily explained by an increase in the postsynaptic sensitivity to glutamate.

Both tetanic stimulation and the pairing of low-frequency stimulation with postsynaptic depolarization were used to induce LTP in these experiments. Because the NMDA component of the EPSP is suppressed at negative potentials due to block by Mg2+, we depolarized the neurons beyond the reversal potential (0 mV), where both EPSP components can be observed. The NMDA component is APV-sensitive and lasts hundreds of milliseconds, while the Q/K component is APV-insensitive and has a more rapid time course.

Figure 3 shows the reversed EPSPs from one experiment before and after the induction of LTP. Once the cell was depolarized, the first three EPSPs were taken as the control value. Since low frequency stimulation paired with depolarization triggers LTP, the subsequent EPSPs grew in size to a new stable level (Figure 3A1). However, only the early, Q/K-mediated component was enhanced. The afferent input was then tetanized, resulting in a transient increase in the EPSP throughout its time course. This increase represents post-tetanic potentiation (PTP), and results from an increase in transmitter release from presynaptic terminals. Thus, the effect seen in Figure 3A2 shows that it is possible to detect an increase in transmitter release under these experimental conditions. Within minutes, however, the late component of the EPSP returned to baseline levels (Figure 3A3). Treatment with APV selectively removed the late NMDA component of the EPSP. The NMDA component remains constant throughout the experiment (Figure 3A5 and 3A6), except during PTP when transmitter release is increased. The bar graph summarizes the data from a number of experiments, and shows that after pairing stimulation with depolarization, or several minutes after tetanus only the early, Q/K component is increased. During PTP, when transmitter release is increased, both EPSP components are enhanced.

Experiments in which we examined LTP in the presence of the selective Q/K antagonist, 6-cyano -
7-nitroquinoxaline-2,3-dione (CNQX), produced similar findings (Kauer et al., 1988a). In the presence of CNQX, the NMDA component is isolated from the Q/K component. Following pairing of
depolarization with low-frequency stimulation, there was no change in the EPSP. After a tetanus,
a transient, post-tetanic potentiation of the NMDA component was seen, but the EPSP then declined to baseline levels. APV abolished this EPSP entirely, showing that it is mediated through
the NMDA channels. Although there was no persistent change in the NMDA component following
LTP-inducing stimuli, after CNQX was washed out the Q/K component was considerably potentiated. Taken together these results suggest that only the Q/K component of the EPSP is increased
during LTP. Similar results have been reported recording population EPSPs, using low extracellular Mg2+ to bring out the NMDA component of the EPSP (Muller and Lynch, 1988; Muller et al.,
1988).

The selective increase in the Q/K component of the EPSP following LTP induction is unlikely to
result from either an increase in transmitter release or a simple morphological change in the synapse. Each of these changes would be expected to increase both components of the EPSP, as seen with
PTP. Instead, these findings are most easily explained by a modification of the postsynaptic
neuron which increases its sensitivity to transmitter. Our results combined with previous findings
have led us to the following hypothesis. The triggering of LTP requires a postsynaptic elevation
of calcium, which activates calmodulin and kinase systems. These calcium-dependent enzyme systems then act postsynaptically either to increase the number or modify the single-channel properties of the Q/K glutamate receptor/ion channels.

REFERENCES

Akers, R. E., Lovinger, D. M., Colley, P. A., Linden, D. J. and Routtenberg, A., 1986, Translocation of
        protein kinase C activity may mediate hippocampal long-term potentiation, **Science**
        231: 587.
Ascher, P. and Nowak, L., 1988, The role of divalent cations in the N-methyl-D-aspartate
        responses of mouse central neurones in culture, **J. Physiol.** 399: 247.
Bliss, T. V. P. and Lømo, T., 1973, Long-lasting potentiation of synaptic transmission in the dentate
        area of the anaesthetized rabbit following stimulation of the perforant path,
        J. Physiol. 232: 331.
Bliss, T. V. P., Douglas, R. M., Errington, M. L., and Lynch, M. A., 1986, Correlation between
        long-term potentiation and release of endogenous amino acids from dentate gyrus of
        anaesthetized rats, **J. Physiol.** 377: 391.
Chang, F.-F., and Greenough, W. T., 1984, Transient and enduring morphological correlates of
        synaptic activity and efficacy change in the rat hippocampal slice, **Brain Res.** 309:35.
Coan, E. J. and Collingridge, G. L., 1985, Magnesium ions block an N-methyl-D-aspartate
        receptor-mediated component of synaptic transmission in rat hippocampus, **Neurosci.**
        Lett. 53: 21.
Collingridge, G. L., Kehl, S. J. and McLennan, H., 1983, Excitatory amino acids in synaptic
        transmission in the Schaffer collateral-commissural pathway of the rat hippocampus,
        J. Physiol. 334: 33.
Forsythe, I. D. and Westbrook, G. L., 1988, Slow excitatory postsynaptic currents mediated by
        N-methyl-D-aspartate receptors on mouse cultured central neurones, **J. Physiol.** 396:
        515.
Gustafsson, B., Wigström, H., Abraham,W. C. and Huang, Y.-Y., 1987, Long-term potentiation in
        the hippocampus using depolarizing current pulses as the conditioning stimulus to
        single volley synaptic potentials, **J. Neurosci.** 7: 774.
Hu, G. -Y., Hvalby, O., Walaas, S.I., Albert, K. A., Skjeflo, P., Andersen, P., and Greengard, P.,
        1987, Protein kinase C injection into hippocampal pyramidal cells elicits features of
        long term potentiation, **Nature**, 328: 426.
Jahr, C. E. and Stevens, C. F., 1987, Glutamate activates multiple single channel conductances in
        hippocampal neurones, **Nature** 325: 522.
Kauer, J. A., Malenka, R. C. and Nicoll, R. A., 1988a, A persistent postsynaptic modification
        mediates long-term potentiation in the hippocampus, **Neuron** 1: 911.

Kauer, J. A., Malenka, R. C. and Nicoll, R. A., 1988b, NMDA application potentiates synaptic transmission in the hippocampus, **Nature** 334: 250.

Kelly, P. T., McGuinness, T. L., and Greengard, P., 1984, Evidence that the major postsynaptic density protein is a component of a Ca^{2+}/calmodulin-dependent protein kinase, **Proc. Natl. Acad. Sci. USA** 81: 945.

Kelly, P. T., Weinberger, R. P., and Waxham, M. N., 1988, Active site-directed inhibition of Ca^{2+}/calmodulin-dependent protein kinase type II by a bifunctional calmodulin-binding peptide, **Proc. Natl. Acad. Sci USA** 85: 4991.

Kelso, S. R., Ganong, A. H. and Brown, T. H., 1986, Hebbian synapses in hippocampus, **Proc. Natl. Acad. Sci. USA** 83: 5326.

Kennedy, M. B., Bennett, M. K., and Erondu, N. E., 1983, Biochemical and immunochemical evidence that the "major postsynaptic density protein" is a subunit of a calmodulin-dependent protein kinase, **Proc. Natl. Acad. Sci. USA** 80:7357.

Lee, K. S. Schottler, F., Oliver, M., and Lynch, G., 1980, Brief bursts of high-frequency stimulation produce two types of structural change in rat hippocampus, **J. Neurophysiol.** 44: 247.

Lovinger, D. M., Wong, K. L., Murakami, K. and Routtenberg, A., 1987, Protein kinase C inhibitors eliminate hippocampal long-term potentiation, **Brain Res.** 436: 177.

Lynch, G. and Baudry, M., 1984, The biochemistry of memory: a new and specifichypothesis, **Science** 224: 1057.

Lynch, G., J. Larson, S. Kelso, G. Barrioneuevo and F. Schottler, 1983, Intracellular injections of EGTA block induction of hippocampal long-term potentiation, **Nature** 305: 719.

Lynch, M. A., Clements, M., Errington, M. L., and Bliss, T. V. P., 1988, Increased hydrolysis of phosphatidylinositol-4,5-bisphosphate in long-term potentiation, **Neurosci. Lett.** 84: 291.

MacDermott, A. B., Mayer, M. L., Westbrook, G.L., Smith, S. J. and Barker. J. L., 1986, NMDA-receptor activation increases cytoplasmic calcium concentration in cultured spinal cord neurones, **Nature** 321: 519.

Malenka, R. C., Kauer, J. A., Zucker, R. S. and Nicoll, R. A. , 1988, Postsynaptic calcium is sufficient for potentiation of hippocampal synaptic transmission, **Science** 242: 81.

Malenka, R. C., Kauer, J. A., Perkel, D. J., and Nicoll, R. A., 1989a, The impact of postsynaptic calcium on synaptic transmission - its role in long term potentiation, **Trends Neurosci.**, in press.

Malenka, R. C., Kauer, J. A., Perkel, D. J., Mauk, M. D., Kelly, P. T., Nicoll, R. A., and Waxham, M. N., 1989b, Long-term potentiation: An essential role for postsynaptic calmodulin and protein kinase activity, **Nature**, in press.

Malinow, R., Madison, D. V. and Tsien, R. W., 1988, Persistent protein kinase activity underlying long-term potentiation, **Nature** 335: 820

Mayer, M. L., Westbrook, G. L. and Guthrie, P. B., 1984, Voltage-dependent block by Mg^{2+} of NMDA responses in spinal cord neurones, **Nature** 309: 262.

Mayer, M. L., MacDermott, A. B., Westbrook, G. L., Smith, S. J. and Barker, J. L., 1987, Agonist- and voltage-gated calcium entry in cultured mouse spinal cord neurons under voltage clamp, **J. Neurosci.** 7: 3230.

Mayer, M. L. and Westbrook, G. L., 1987, Permeation and block of N-methyl-D-aspartatic acid receptor channels by divalent cations in mouse central neurones, **J. Physiol.** 394: 501.

Miller, S. G. and Kennedy, M. B., 1986, Regulation of brain type II Ca^{2+}/calmodulin-dependent protein kinase by autophosphorylation: a Ca^{2+}-triggered molecular switch, **Cell** 44: 861.

Muller, D. and Lynch, G., 1988, Long-term potentiation differentially affects two components of synaptic responses in hippocampus, **Proc. Natl. Acad. Sci. USA** 85: 9346.

Muller, D., Joly, M., and Lynch, G., 1988, Contributions of quisqualate and NMDA receptors to the induction and expression of LTP, **Science**, 242: 1694-1697.

Nelson, R. B., and Routtenberg, A., 1985, Characterization of protein F1 (47kDa, 4.5 pI): a kinase C substrate directly related to neural plasticity, **Exp. Neurol.** 89: 213.

Nowak, L., Bregestovski, P. ,Ascher, P., Herbet, A., and Prochiantz, A., 1984, Magnesium gates glutamate-activated channels in mouse central neurones, **Nature** 307: 462.

Ouimet, C. C., McGuinness, T. L. and Greengard, P., Immunocytochemical localization of calcium/calmodulin-dependent protein kinase II in rat brain, **Proc. Natl. Acad. Sci. USA** 81: 5604.

Reymann, K. G., Frey, U., Jork, R. and Matthies, H., 1988, Polymixin B, an inhibitor of protien kinase C, prevents the maintenance of synaptic long–term potentiation in hippocampal CA1 neurons, **Brain Res.** 440: 305.

Saitoh, T. and Schwartz, J. H., 1985, Phosphorylation-dependent subcellular translocation of a Ca^{2+}/calmodulin-dependent protein kinase produces an autonomous enzyme in Aplysia neurons, **J. Cell Biol.** 100: 835.

Sastry, B. R., Goh, J. W. and Auyeung, A., 1986, Associative induction of posttetanic and long-term potentiation in CA1 neurons of rat hippocampus, **Science** 232: 988.

Staubli, U., Larson, J., Thibault, O., Baudry, M., and Lynch, G., 1988, Chronic administration of a thiol-proteinase inhibitor blocks long-term potentiation of synaptic responses, **Brain Res.** 444: 153.

Wigstrom, H. and Gustafsson, B., 1988, Presynaptic and postsynaptic interactions in the control of hippocampal long-term potentiation, in: "Long-term Potentiation: From Biophysics to Behavior". Landfield and Deadwyler ed. , Alan R. Liss, Inc., New York.

IDENTIFYING AND LOCALIZING PROTEIN KINASES NECESSARY FOR LTP

Roberto Malinow and Richard W. Tsien

Department of Molecular and Cellular Physiology
Beckman Center
Stanford University School of Medicine
Stanford California 94305

Long-term potentiation (LTP) of synaptic transmission follows a brief high-frequency stimulus delivered to afferent pathways (1). This activity-dependent enhancement of transmission, which can be studied at the behavioral, cellular and molecular level, is thought to play a role in learning and memory (2-4). Our recent work has centered on identifying and localizing intracellular signals responsible for the induction, maintenance and expression of LTP (5,6).

The triggering (induction) of LTP can be distinguished pharmacologically by agents which block LTP if present during the conditioning tetanus, but fail to affect established LTP (2-4). This profile is shown by APV (7,8), a specific antagonist of the NMDA-type glutamate receptor, and sphingosine (5), an agent which competes with kinase activators for kinase activation (9,10). Although both drugs block LTP, only APV blocks a slowly decaying potentiation which normally follows a conditioning tetanus (5). This shows that the effect of sphingosine is not by block of NMDA receptors. These results thus suggest that both NMDA-receptor activation and production of a kinase-activator inhibited by sphingosine are necessary to trigger LTP.

The maintenance and expression of LTP can be distinguished by agents which either eliminate (blocking maintenance) or reversibly inhibit (blocking expression of) potentiated transmission. Demonstrating that an agent blocks the maintenance of LTP is experimentally difficult because it requires showing i) that the drug inhibits established LTP and ii) that complete washout of the drug does not reverse the inhibition of established LTP. Such experiments have not been reported. Inhibition of the expression of LTP has been successful with two agents: H-7 (5), which blocks protein kinase activity by competing with ATP (11); and CNQX (12) (and DNQX (13)) which inhibit the non-NMDA subtypes of glutamate receptors (14). Inhibiting the expression of LTP can be demonstrated with experiments in which drug application selectively and reversibly blocks enhanced transmission, with no effect on transmission through an independent pathway which did not receive a conditioning tetanus and thus is assumed to be not potentiated (Fig. 1). It is important to recognize that although there are examples of such selective effects with H-7, this drug can also affect transmission through a pathway that was not delivered a tetanus. This effect on a non-tetanized pathway is always less than the effect on a tetanized pathway. More importantly, we never see an effect on the non-potentiated pathway without an effect on the potentiated pathway, whereas we often see an effect on the potentiated pathway with no effect on the non-potentiated pathway. We thus conclude that H-7 selectively blocks the mechanisms responsible for the expression of LTP. It is possible that the effects on a pathway which was not delivered a tetanus is due to prior potentiation of that pathway, either in the life of the rat or during the cutting of the slices.

CNQX (and DNQX) can be shown to inhibit the expression of LTP by their greater effect on potentiated over non-potentiated transmission (12,13). Furthermore, transmission through the

Excitatory Amino Acids and Neuronal Plasticity
Edited by Y. Ben-Ari
Plenum Press, New York, 1990

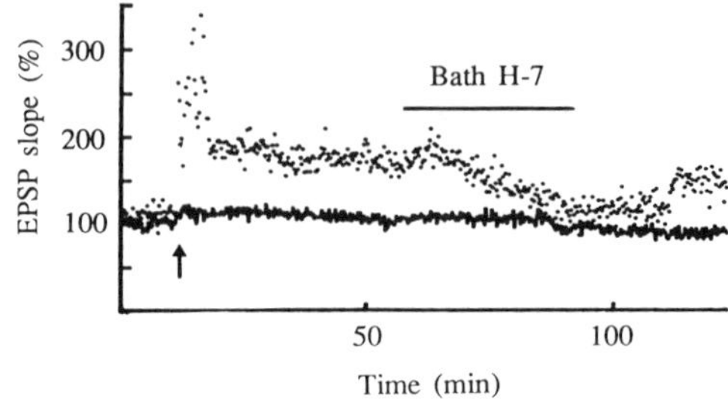

Fig. 1.   Bath applied H-7 selectively blocks expression of LTP with no effect on basal transmission. Transmission of two independent pathways in as measured extracellularly in CA1 stratum radiatum. Delivery of tetanus (arrow) to one pathway (unconnected dots) results in persistent synaptic enhancement. Bath applied H-7 (300μM) inhibits potentiated transmission with no effect on unpotentiated transmission. With wash of the drug potentiated transmission returns.

NMDA channel, as measured by the initial slope of the synaptic potential elicited in the presence of CNQX, does not change following a tetanus, thereby showing that the increase in transmission is expressed by an increase through the non-NMDA channel (12,13).

From these pharmacological studies one concludes that the induction of LTP requires NMDA receptor activation as well as activation of a sphingosine-sensitive kinase. Once LTP is established, its expression requires the persistent activity of an H-7-sensitive kinase which is responsible for an increase in the non-NMDA synaptic current. All of these studies (5,12,13) used bath applied drugs, and thus one cannot conclude anything about the location (pre- or postsynaptic) of these mechanisms without making critical assumptions (6). Furthermore, the kinase blockers used, sphingosine and H-7, affect several kinases, including protein kinase C (PKC) and the

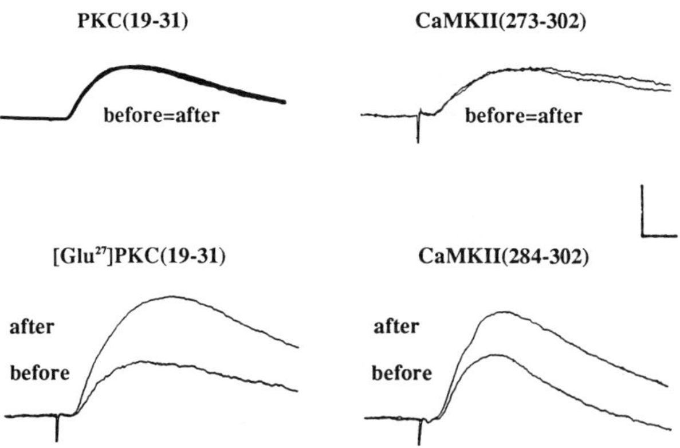

Fig. 2.   Postsynaptic PKC and CaMKII are necessary for LTP. Intracellular recordings with pipettes containing inhibitory or inactive peptides obtained 5 minutes before (before) or 60 minutes after (after) conditioning. With either 3 mM PKC (19-31) or 1 mM CaMKII(273-302) LTP is blocked. With inactive peptides [Glu27]PKC(19-31) (3 mM) or CaMKII(284-302) (1mM) LTP is not blocked. Each trace is average of 5 consecutive EPSPs. Scale bars 5.0 mM/ 12.5 ms.

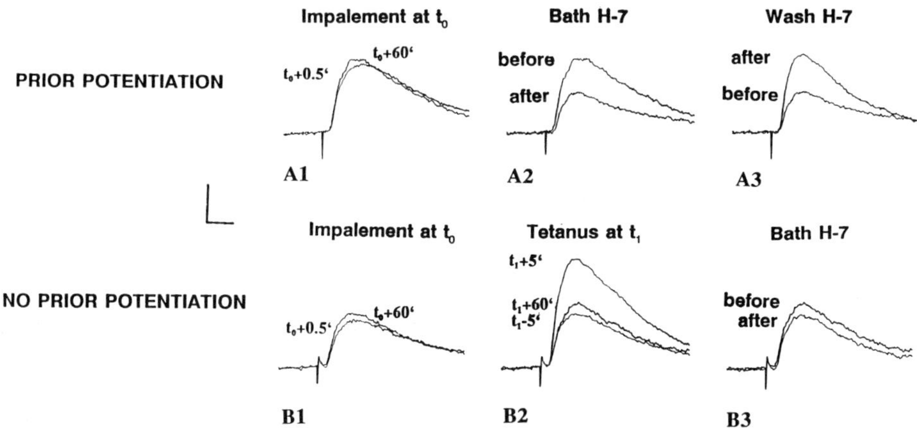

Fig. 3. Expression of LTP is not inhibited by postsynaptic delivery of PKC (19-31) and CaMKII(273-302). Intracellular synaptic monitoring of two independent pathways with an electrode containing 3mM PKC(19-31) and 1mM CaMKII (273-302). One pathway was potentiated with a conditioning tetanus prior to intracellular impalement (prior potentiation, A1, A2, A3), whereas the other pathway was not (no prior potentiation, B1 B2, B3). The previously potentiated pathway shows no synaptic decrement one hour after impalement with the inhibitor-containing microelectrode (A1), suggesting that postsynaptic PKC or CaMKII are not involved in maintaining or expressing enhanced transmission. Tetanic stimulation to the pathway not previously potentiated results in no LTP (B2), (despite significant potentiation as seen extracellularly; data not shown), showing that the inhibitors effectively reach postsynaptic targets and block the induction of LTP. Subsequent bath application of 300µM H-7 inhibits transmission in the previously potentiated pathway (A2) but not in the unpotentiated pathway (B3), showing that pathway A was in fact potentiated. With washout of H-7 potentiated transmission returns (A3).

multifunctional Ca/calmodulin-dependent protein kinase (CaMKII), thus precluding identification of specific enzymes.

To identify and localize the biochemical messengers mediating LTP we used intracellular recording techniques to deliver selective protein kinase inhibitors (6). We have tested the requirement of postsynaptic PKC activity with peptide fragments forming the pseudosubstrate region of the PKC regulatory domain. We have used PKC(19-31) and PKC(19-36), which show potent (IC_{50} < 0.2 µM) and selective (IC_{50} against CaMKII>50µM) inhibitory activity against PKC. As shown in Fig. 2, these agents blocked LTP when delivered to the postsynaptic cell with the recording intracellular microelectrode. To control for non-specific effects of the peptides we used [Glu^{27}]PKC (19-31), which has had a basic residue important for a substrate recognition substituted with glutamate. This peptide shows no inhibitory activity against PKC in vitro and postsynaptic delivery also fails to block LTP (Fig. 2). To investigate the role of postsynaptic CaMKII, we used a peptide fragment from the autoinhibitory region of the kinase, CaMKII(273-302). This peptide inhibits kinase activity with IC_{50}=1.0 µM, and fails to inhibit PKC (IC_{50}>200 µM). The inhibition of CaMKII cannot be overcome with increasing concentrations of calmodulin, showing that its mode of inhibition is not by competition with calmodulin. When delivered to postsynaptic cells through the recording microelectrode, CaMKII(273-302) blocks LTP (Fig. 2). To control for effects other that block of CaMKII, we used the shorter peptide CaMKII(284-302), which lacks a critical arginine necessary for inhibitory activity. This peptide does not block CaMKII activity (IC_{50}>80 µM) and also fails to block LTP. We thus conclude that postsynaptic PKC and CaMKII are both required to generate LTP.

To determine if these postsynaptic kinases are involved in the expression of LTP, we introduced H-7 to postsynaptic cells after a tetanus had been delivered to establish potentiated transmission (6). Postsynaptic delivery of H-7 did not inhibit established LTP, suggesting that postsynaptic kinases are involved in the induction but not the expression of LTP. We have repeated these experiments delivering a mixture of both inhibitors, PKC(19-31) and CaMKII(273-302), to postsynaptic cells after the establishment of LTP (see Fig. 3). With extracellular recording, a tetanus was delivered to one of two monitored pathways, potentiating transmission about 2-fold in one pathway (data not shown). Once the potentiated pathway stabilized, an intracellular recording was obtained with the inhibitor-containing microelectrode, and monitoring of transmission through the two pathways was commenced within 30 sec of impalement. If postsynaptic PKC or CaMKII activity were required for the expression of LTP, then as these inhibitors diffused into the postsynaptic cell one should see a decrement of transmission through the previously potentiated pathway. As shown in figure 3, this was not the case, as transmission through both pathways stayed constant (Fig 3A1). To test if the inhibitors had reached postsynaptic targets, we delivered a tetanus to the pathway which had not been potentiated. We saw a block of LTP through the intracellular electrode (Fig 3B2), despite LTP as measured with the extracellular electrode (data not shown), demonstrating that the peptides were effective. We subsequently applied H-7 in the bath and saw a decay of transmission through the previously potentiated pathway (Fig 3A2). This suggested that transmission through this pathway was in fact potentiated. Washout of the drug returned the potentiated transmission (Fig 3A3) showing that bath H-7 blocks the expression and not the maintenance of LTP.

These results imply that after the establishment of LTP, (intracellular) postsynaptic kinase activity is not necessary to express LTP. However, bath-applied H-7, which has access to both pre- and postsynaptic targets, is still effective in inhibiting the expression of LTP. This suggests that a presynaptic (or extracellular) H-7-sensitive process, possibly PKC activity, is required to express LTP.

REFERENCES

1. T. V. P. Bliss and T. Lomo, Long-lasting potentiation of synaptic transmission in the dentate area of the anaesthetized rabbit following stimulation of the perforant path. **J. Physiol.** (London) 232, 331 (1973).
2. T. V. P. Bliss and M. Lynch, in **Long-Term Potentiation: Mechanisms and Key Issues.** P.W. Landfield and S.A. Deadwyler, Eds (Liss, New York, 1988)
3. R. A. Nicoll, J. Kauer, R. C. Malenka, Current Excitement in Long-term Potentiation **Neuron** 1, 97 (1988)
4. T. H. Brown, P. F. Chapman, E. W. Kairiss, C. L. Keenan, Long-Term Synaptic Potentiation **Science** 242, 724 (1988).
5. R. Malinow, D. V. Madison, R. W. Tsien, Persistent Protein Kinase Activity Underlying Long-Term Potentiation **Nature** 335, 820 (1988).
6. R. Malinow, H. Schulman, R. W. Tsien, Inhibition of Postsynaptic PKC or CaMKII Blocks Induction but not Expression of LTP **Science** (in press)
7. G. L. Collingridge, S. J. Kehl, H. McLennan, Excitatory Amino Acids in Synaptic Transmission in the Schaffer Collateral-Commissural Pathway of the Rat Hippocampus. **J. Physiol. (London)** 334, 33 (1983).
8. E. W. Harris, A. H. Ganong, and C. W. Cotman Long-Term Potentiation in the Hippocampus Involves Activation of N-methyl-D-aspartate Receptors **Brain Res.** 323, 132 (1984).
9. A. B. Jefferson and H. Schulman, Sphingosine Inhibits Calmodulin-dependent Enzymes **J. Biol. Chem.** 263, 15241 (1988).
10. Y. A. Hannun and R. M. Bell, Functions of Sphingolipids and Sphingolipid Breakdown Products in Cellular Regulation **Science** 243, 500 (1989).
11. H. Hidaka, M. Inagaki, S. Kawamoto, Y. Sasaki, Isoquinolinesulfonamides, Novel and Potent Inhibitors of Cyclic Nucleotide dependent Protein Kinase and Protein Kinase C **Biochemistry** 23, 5036 (1984).

12. J. A. Kauer, R. C. Malenka, R. A. Nicoll, A Persistent Postsynaptic Modification Mediates Long-Term Potentiation in the Hippocampus **Neuron** 1, 911 (1988)
13. D. Muller, M. Joly, G. Lynch, Contributions of Quisqualate and NMDA Receptors to the Induction and Expression of LTP **Science** 242, 1694 (1988).
14. T. Honore, S. N. Davies, E. J. Fletcher, P. Jacobsen, D. Lodge, and F. E. Nielsen, Quinoxalinediones: Potent Competitive Non-NMDA Glutamate Receptor Antagonists, **Science** 241,701 (1988).

DELAYED ONSET OF POTENTIATION IN NEOCORTICAL EPSPS DURING LONG-TERM

POTENTIATION (LTP) - A POSTSYNAPTIC MECHANISM OR HETEROGENEOUS

SYNAPTIC INPUTS ?

Lynn J. Bindman and K.P.S.J. Murphy

Department of Physiology, University College London, Gower Street
London   WC1E 6BT   England

SUMMARY

Long-term potentiation (LTP) is an activity-dependent, long-lasting change in the efficacy of synaptic transmission, which is a possible neural substrate for learning and memory. We have induced LTP at some synapses in slices of rat sensorimotor cortex *in vitro*, by using either brief tetanic stimulation of afferent axons or an associative, postsynaptic method of induction. We were successful in inducing LTP of excitatory postsynaptic potentials (EPSPs) in 11 out of 53 neurones in layers III, V or VI.

In view of the importance of N-methyl-D-aspartate (NMDA) receptor activation for the induction of LTP in neocortex, as in regions of hippocampus, we hypothesized that the failures of induction of LTP might be due to a lack of NMDA-receptor mediated activity in the afferent pathway onto layer V and VI neurones. Hence we bath-applied the selective NMDA-receptor antagonist 2-amino-5- phosphonopentanoic acid (AP5) and examined its effect on composite PSPs evoked in layer V and VI neurones. We found that 94% of PSPs (44/47) were partly mediated by NMDA-receptor activity, so a lack of functioning NMDA receptors was not likely to be the reason for failures of induction of LTP by afferent tetanic stimulation.

When we induced LTP (in the absence of AP5) and applied AP5 15 to 20 min after the establishment of LTP, the expression of LTP was reversibly reduced: during LTP, synaptic responses consisted of an early AP5-insensitive component and a later AP5-sensitive component. This was seen in both EPSPs and field potentials in layers V and VI.

Although the depolarising slope of most of the neocortical EPSPs was increased during the expression of LTP, we found that in 9/11 short latency EPSPs the onset of the potentiated component of the EPSP was delayed by a few ms with respect to the onset of the EPSP. In 7 of the 9 EPSPs the LTP was induced postsynaptically by pairing an evoked EPSP repetitively with intracellular depolarising pulses. The explanation for the delayed onset of the potentiated component in these EPSPs could be either that a postsynaptic voltage-sensitive mechanism was involved, or that the afferent terminals on the neurones were heterogeneous. The afferents responsible for the shortest latency components of the neocortical EPSPs may not be capable of sustaining LTP.

Excitatory Amino Acids and Neuronal Plasticity
Edited by Y. Ben-Ari
Plenum Press, New York, 1990

INTRODUCTION

LTP of transmission at neocortical synapses has been demonstrated in several studies (for refs. see Bindman et al., 1988). The role of NMDA-receptor mediated activity in normal synaptic transmission in neocortex, and in the induction and expression of LTP, has been investigated using intracellular recordings by a number of laboratories (Thomson, 1986; Artola & Singer, 1987; Aram et al., 1987; Jones & Baughman, 1988; Bindman & Murphy, 1988; Sutor & Hablitz, 1989a, b; Kimura et al., 1989).

Our interest in excitatory amino acid neurotransmission and LTP was to elucidate why it was more difficult to induce associative LTP postsynaptically in layers V and VI of sensorimotor cortex (Baranyi & Szente, 1987; Bindman et al., 1988) than in hippocampal CA1 neurones (Wigstrom & Gustafsson, 1986). Associative LTP could be induced postsynaptically in only 15% and 28% of EPSPs respectively in these two studies of neocortical LTP compared with 80% of EPSPs in CA1. In many hippocampal regions, the activation of NMDA-receptor mediated activity is necessary for the induction of LTP [Collingridge et al. (1983); see also Bliss & Lynch (1988)] and preliminary results indicated this was true also in Layers II to IV of visual cortex (Artola & Singer, 1987).

We therefore studied the action of D-AP5 on PSPs and field potentials evoked in neocortical neurones during normal transmission, on the induction of LTP and on its expression. By comparing the time course of EPSPs before LTP was induced with that of potentiated EPSPs >10 min after the establishment of LTP, we have obtained evidence suggesting that the earliest part of the depolarising slope in layer III and V neocortical EPSPs is not potentiated.

METHODS

Neocortical slices from adult rat sensorimotor cortex were prepared and maintained as described in Bindman et al. (1988). Figure 1 shows the placement of tungsten stimulating electrodes on the subcortical white matter. In a few experiments we stimulated the grey matter instead. Microelectrodes, filled with 4MKAc for intracellular recording or 1.7 M NaCl for recording field potentials extracellularly, were positioned a measured distance from the pial surface and the border of the white/grey matter. Layer III recordings were taken from 25-30% of the depth of the grey matter perpendicularly below the pial surface, layer V from 40 to 60% and layer VI from 61 to 100%. Recording and stimulating techniques were as described in Bindman et al. (1988). Bicuculline methiodide (Sigma) 0.5 to 2 μM was used in about half the experiments. AP5 (Sigma) was bath-applied either as the D-isomer (10 to 20 μM) or the DL racemic mixture (17-40 μM). Kynurenate (Sigma) was bath-applied as a 1 to 2.5 mM solution.

RESULTS

In addition to field potential studies in 14 slices, a total of 66 neurones from 66 slices were used for intracellular studies of (1) the effect of AP5 on normal transmission, (2) the induction and expression of LTP and (3) of the effect of AP5 on the expression of LTP. The electrophysiological properties of the 66 neurones were: mean resting potential - 79.0 mV; mean action potential, 97.7 mV; mean firing threshold (mV above resting membrane potential) 23.6 mV; mean input resistance, 42.1 M . Of the 66 neurones, 5 were in layer III, 55 were in layer V and 6 in layer VI.

1. The Effect of AP5 on Normal Transmission in Layer V and VI Neurones

In 1.1. mM Mg^{++}, AP5 reduced the amplitude and/or duration of most PSPs in layer V and VI neurones. 47 composite PSPs were recorded from 26 neurones in layers V and VI and 44 of the 47 were partly mediated by NMDA-receptor activity. Bicuculline methiode was bath applied to 13 out of the 26 neurones > 1 h before the effects of AP5 were examined. Thus a GABA$_A$ antagonist was not essential to show the NMDA-receptor mediated component.

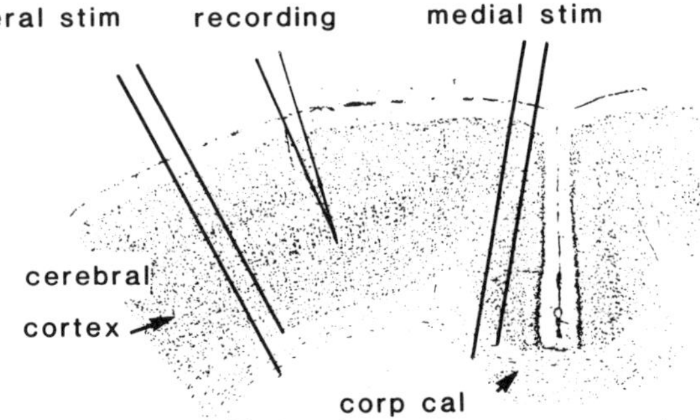

Fig. 1   Diagram showing the arrangement of recording and stimulating electrodes in slice preparation. Two pairs of stimulating electrodes were placed on the corpus callosum, one of the pairs laterally and one medially to the recording microelectrode sited in the grey matter of the sensorimotor cortex.

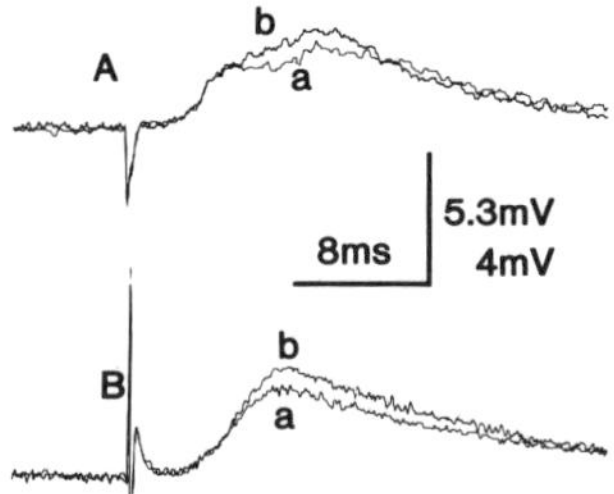

Fig. 2   Examples of EPSPs recorded from two layer III neurones (A and B) showing a delayed onset of the potentiated component of EPSP during LTP. Each trace represents the average of 8 successive EPSPs evoked by single shocks applied, for neurone A, to medial stimulating electrodes on the corpus callosum, and for neurone B to the adjacent grey matter. The lower traces, a, were taken during the control and the upper traces, b, during LTP >10 min after the end of the induction procedure. LTP was induced associatively by repetitively pairing EPSPs with intracellularly applied depolarizing current pulses (A; 31 pairings with 2.5 nA 100 ms pulses, onset of each pulse 10 ms after test shocks. B; 50 pairings with 2nA 200 ms pulses, onset 4.0 ms after test shocks). During LTP in A the latency of the averaged EPSP was 2.6 ms while the onset of the increased slope of potentiation was at 6.7 ms. During LTP in B the latency of the EPSP was 3.6 ms and that of the increased slope was 6.2 ms. Voltage calibration; 5.3 mV for A, 4 mV for B.

2. The Induction and Expression of LTP

We made > 100 attempts to induce LTP in 53 neurones and succeeded in eliciting LTP in 12 EPSPs in 11 neurones. Bicuculline methiodide was present in 6 out of 11 of the cells, so the presence of a $GABA_A$ antagonist was not essential for induction of LTP. Of the 42 cells in which we failed to induce LTP, bicuculline methiodide was present in two-thirds of the experiments. We failed to induce LTP by repetitive pairing of test shocks with intracellular current pulses in 36 neurones (34 of

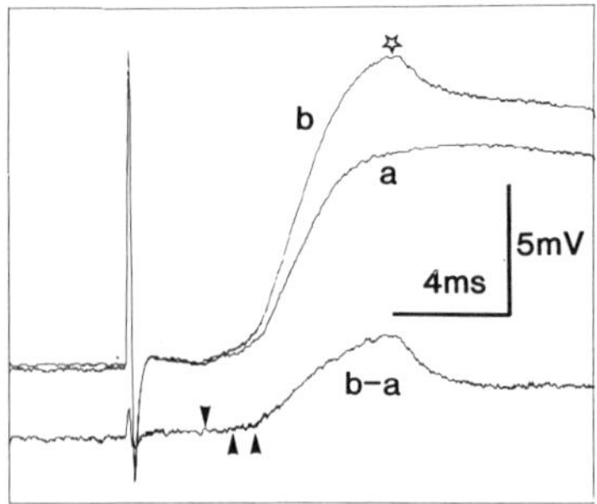

Fig. 3   Example of delayed onset of potentiation during LTP in a layer V neurone. Traces a and b are averages of 8 responses in control and LTP with EPSPs evoked by shocks to lateral stimulating electrodes on corpus callosum. Star above b shows onset of averaged (hence distorted) action potentials. LTP was induced associatively by pairing 30 current pulses (2nA 100 ms duration, onset 35 ms after test shocks). b-a shows electronic subtraction of averaged traces: the subtraction shows that while the onset of the EPSPs (at downward arrow) was at 2.8 ms, the first slight increase in slope (left upward arrow) was at 3.6 ms and the major increase in slope during LTP started at 4.5 ms (right upward arrow).

them in layers V and VI) in spite of repeated attempts with increasing current strength or duration.

The success rate for associative postsynaptic induction of LTP was approximately 15% i.e. 7 EPSPS in 6 neurones out of 42. Tetanic stimulation was applied to afferent pathways onto 25 cells (17 cases after unsucessful pairing attempts) so the success rate for induction of LTP by afferent tetani was 5 out of 25 neurones (20%).

Three examples of the changes in the magnitude and time course of the EPSPs once LTP was established are shown in Figs 2A, 2B & 3. In each neurone, LTP was induced postsynaptically as described in the figure legends. The time course of the induction and maintenance of the LTP in the neurone of Figure 2A is shown in the graph of Figure 4 of Bindman et al (1988). The increase in size of the EPSP in LTP was small but very highly significant. Each of the three cells in Figs. 2 & 3 shows a delayed onset of the increased depolarising slope of the EPSP during the expression of LTP. The mean latencies are given in the legends. A delayed onset such as this is not typically seen in LTP of EPSPs in the hippocampus, but was found in 9 of the 11 short latency EPSPS in this study.

The mean onset latency of the 11 EPSPS was 3.6ms (range 2.1 to 5.3ms) while the mean onset latency of the increased depolarizing slope of EPSPS during LTP was 9.9ms (range 0 to 35ms, n = 11). The mean difference of 6.3ms was significant at $p<0.05$ (paired t-test, 2 tailed). One additional EPSP in which LTP was induced was >40ms onset latency.

When kynurenate (2.5 mM) was bath applied it abolished EPSPs. When the microelectrode was withdrawn from other neurones, no sign of a field potential was detected following lateral or medial test shocks. We do not think that the non-potentiated onset of the EPSPs was due to an intracellular recording of an extracellular event.

We confirmed that the presence of 20 µM AP5 could block the induction of LTP. In field

310

potential studies, induction of LTP was prevented in 5 out of 7 slices. Following replacement of AP5 with csf, LTP was subsequently elicited.

3. The effect of AP5 on the expression of LTP

When AP5 was bath applied to the slice after LTP had been established, a reversible reduction in the amplitude of field potentials was produced in 13 out of 14 slices. The reduction was very highly significant ($p < 0.001$). An example was shown in Bindman & Murphy, 1988.

In EPSPs of layer V & VI cells after LTP had been established, AP5 reduced late portions of the EPSPs. The AP5 revealed that LTP occurred in an AP5-insensitive component that had shorter latency than the AP5-sensitive component. Similarly in field potentials of layers V & VI, the AP5-insensitive potentiation preceded the AP5-sensitive part of each response during LTP.

DISCUSSION

There are two features of LTP in the neocortex that differ from LTP in the CA1 area of hippocampus. One is that NMDA-receptor mediated activity participates in the expression of LTP in neocortex. EPSPs that are partly NMDA-receptor mediated can be recorded before LTP is induced in neocortical slices in 1.1mM Mg^{++} when evoked at 0.1 Hz and at normal resting potential (e.g. Thomson, 1986; Aram et al. 1987; Jones & Baughman, 1988). It is not surprising, therefore, that AP5 will also block components of the response in neocortex once LTP has been established. The effect of AP5 is reversible, so the NMDA-receptor mediated component is not necessary for the maintenance of LTP. In hippocampal CA1 cells, NMDA-receptor mediated activity is seen only when cells are more depolarised for example during higher frequency activation [Collingridge et al. (1988)]. It is of interest to consider whether the Ca^{++} entry into neocortical neurones via NMDA-receptor channels during synaptic activity (particularly during LTP) has functional consequences. The layer V & VI sensorimotor cortex neurones that we have studied and those in more superficial layers of frontal cortex studied by Sutor & Hablitz (1988 a,b) might differ from hippocampal cells either in the prolongation of firing during suprathreshold EPSPs and/or in some metabolic or structural change initiated by the increased Ca++ entry via NMDA-receptor activated channels during LTP.

The second feature of LTP in the neocortex that differs from LTP in the CA1 area is that the increased depolarizing slope of the potentiated EPSP is usually delayed by a few ms after the onset of the neocortical EPSPs. It is likely that the AP5-sensitive component in EPSPs during LTP in frontal cortex (Sutor & Hablitz, 1989 a,b) and visual cortex (Kimura et al. 1989) is polysynaptic. We found that an AP5-insensitive potentiated component precedes the NMDA- receptor mediated component of EPSPs and field potentials during LTP in layers V and VI of sensorimotor cortex. It is possible that even this early AP5-insensitive component of potentiated EPSPs may be a disynaptic response. From our studies of cells in layer III, V & VI in which LTP was induced postsynaptically, it seems that the shortest latency depolarisation of EPSPs in the neocortex (i.e. preceding the AP5-insensitive potentiated components seen in LTP) may not be capable of undergoing LTP. One possible explanation for the delay in onset of the enhanced depolarisation in EPSPs during LTP is that there are voltage-sensitive mechanisms involved in the expression of LTP in neocortical neurones. It could be that a certain level of depolarization, involving delay in the EPSP, has to be achieved before channels participating in the potentiation are opened. A second possibility is that the terminals onto the neurone are heterogeneous. By using a postsynaptic method of induction of LTP it is highly likely that the site of the prolonged change is at the synapses on the postsynaptic neurone. Although stimulation of the thalamic nucleus VPL evokes monosynaptic activity mainly in layer IV, a similarly short latency activation of layer V neurones occurs in somatosensory cortex (Armstrong-James et al. 1987). Perhaps monosynaptic inputs to neocortex do not participate in LTP.

REFERENCES

Aram, J.A. Bindman, L.J., Lodge, D. & Murphy, K.P.S.J., 1987, Glutamate receptor mediated e.p.s.p.s in layer V neurones in rat neocortical slices. **J. Physiol.(Lond)** 394: 117.

Armstrong-James, M., Britto, J.A., Fox, K. and Mercier, B.E., 1987, Short latency responses to vibrissa stimulation from layer Vb neurones in the rat barrel-field. **J. Physiol. (Lond)** 388:43P

Artola, A. and Singer, W., 1987, Long-term potentiation and NMDA receptors in rat visual cortex. **Nature** 330:649-652.

Baranyi, A. & Szente, M.B., 1987, Long-lasting potentiation of synaptic transmission requires postsynaptic modifications in the neocortex. **Brain Res.** 423: 378-384.

Bindman, L.J. and Murphy, K.P.S.J., 1988. NMDA-receptors participate in the maintenance of long-term potentiation of synaptic transmission in slices of rat neocortex *in vitro*. **J.Physiol. (Lond)** 406:176P.

Bindman, L.J., Murphy, K.P.S.J. and Pockett, S., 1988, Postsynaptic control of the induction of long-term changes in efficacy of transmission at neocortical synapses in slices of rat brain. J. **Neurophysiol.** 60:1053-1065.

Bliss, T.V.P. & Lynch, M.A., 1988, Long-term potentiation of synaptic transmission in the hippocampus: properties and mechanisms. In: **From Biophysics to Behavior**, pp3-72, Publ. Alan R. Liss, Inc. NY.

Collingridge, G.L., Kehl, S.J. & McLennan, H., 1983, Excitatory amino acids in synaptic transmission in the Schaffer collateral- commissural pathway of the rat hippocampus. J. **Physiol. (Lond)** 334:33-46.

Collingridge, G.L., Herron, C.E. & Lester, R.A., 1988, Frequency dependent N-methyl-D-aspartate receptor-mediated synaptic transmission in rat hippocampus. **J. Physiol. (Lond)** 339:301-312.

Jones, K.A. & Baughman, R.W., 1988, NMDA- and non-NMDA-receptor components of excitatory synaptic potentials recorded from cells in layer V of rat visual cortex. **J. Neurosci.** 8:3522-3534.

Kimura, F., Nishigori, A., Shirokawa, T. & Tsumoto, T., 1989, Long- term potentiation and N-methyl-D-aspartate receptors in the visual cortex of young rats. **J. Physiol. (Lond)** 414:125-144.

Sutor, B. and Hablitz, J.J., 1989a, EPSPS in rat neocortical neurons *in vitro* I. Electrophysiological evidence for two distinct EPSPs. **J. Neurophysiol.**, 61:607-620.

Sutor, B. and Hablitz. J.J., 1989b, EPSPS in rat neocortical neurons *in vitro* II. Involvement of N-methyl-D-aspartate receptors in the generation of EPSPS. **J. Neurophysiol.**, 61:621-634.

Thomson, A.M., 1986, A magnesium-sensitive potential in rat cerebral cortex resembles neuronal responses to N-methyl aspartate. **J. Physiol. (Lond)** 370: 531-549.

Wigstrom, H. & Gustafsson, B. 1986, Postsynaptic control of hippocampal long-term potentiation. J. **Physiol. (Paris)** 81:228- 236.

ACKNOWLEDGEMENTS

K.M. is an MRC Scholar. We thank The Wellcome Trust for financial support for this research.

LOCAL CIRCUIT CONNECTIONS MEDIATED BY NMDA AND NON-NMDA RECEPTORS IN

SLICES OF NEOCORTEX

Alex M. Thomson and Shahrzad Radpour

Department of Physiology
Royal Free Hospital School of Medicine
Rowland Hill Street, London NW3 2PF, UK

INTRODUCTION

The extent to which NMDA receptors contribute to simple information transfer has been much debated and may vary from region to region. For example, a significant proportion of sensory input to the thalamus, particularly that mediating noxious input (Eaton & Salt, 1989), is blocked by NMDA antagonists while thalamocortical inputs are sensitive to non-NMDA EAA antagonists, not to NMDA antagonists (P.L. Herrling & T.E. Salt, personal communication). Despite the high density of NMDA binding sites in cortical regions, such as neocortex and hippocampus (Greenamyre et al, 1985), only a relatively small proportion of epsps (excitatory postsynaptic potentials) are demonstrably mediated by these receptors. Indeed, in the CA1 region of hippocampus, the component of the input from CA3 that is purported to be NMDA receptor mediated was only revealed when non-NMDA EAA receptors were blocked and the input pathway stimulated at high strength (Kauer, this volume, Andreasen et al, 1988). Since evidence first began to accumulate that several types of synaptic plasticity are blocked by NMDA antagonists (Collingridge et al, 1983, Artola & Singer, 1987, Kleinschmidt et al, 1987), NMDA receptors have come to be regarded as mediators of plasticity, rather than as mediators of rapid information transfer. Whether or not this distinction is absolute, any discussion of the function of NMDA receptor mediated synapses should include their probable role as mediators of synaptic plasticity.

WHICH PRESYNAPTIC NEURONES ACTIVATE NMDA RECEPTORS ?

When electrical stimulation is employed to activate synaptic input, the precise identity and source of the active fibres must be in doubt. In neocortex, low intensity electrical stimulation of the white matter evokes either conventional non-NMDA receptor mediated or, less frequently NMDA receptor mediated epsps (Thomson, 1986), but whether the presynaptic fibres are afferents from eg. thalamus, transcallosal fibres or pyramidal axons activated antidromically is uncertain. Intermediate stimulus strengths evoke conventional compound epsps in which the NMDA receptor mediated component is obscured and high stimulus strengths result in an 'epsp/ipsp' sequence that may be a reflection of the protocol, rather than the physiology. When a large number of afferent fibres are activated together, inhibitory interneurones shared by the neuronal pool will also be activated, but it is still unclear whether inhibition necessarily follows physiological excitation in such a precise temporal sequence. It is quite possible that inhibition is normally arranged spatially, rather than temporally. The temporal sequence evident with electrical stimulus protocols has led several authors to propose that the slower time-course of NMDA receptor mediated epsps indicates that they will ordinarily be curtailed by inhibitory events. There is no evidence to support this.

To identify the presynaptic neurone acting upon NMDA receptors we employed spike-triggered-averaging in slices of neocortex (Thomson et al, 1988). The spike of a putative presynaptic neurone recorded either extra- or intra-cellularly, was used to trigger averages of the intracellular recording of a putative postsynaptic neurone. A single axon epsp was revealed in approximately 1 in 30 of the simultaneously recorded pairs, on averaging between 100 and 300 sweeps. Single axon epsps varied substantially in amplitude [from <0.1 to >1mV] and in time course, but the majority increased in amplitude and particularly in duration, as the postsynaptic cell was depolarized (see Figure 1.). The contribution to the epsp of NMDA and non-NMDA receptors was investigated by applying antagonists to the postsynaptic neurone from a third electrode (Thomson Girdlestone & West, 1989a). This multibarrel electrode was positioned so that maximal responses to brief electrophoretic pulses of NMDA were evoked. The selectivity and effectiveness of the antagonists was tested against neuronal responses to NMDA and quisqualate/AMPA before attempts were made to block the epsp. In all epsps tested, AP-5 or CPP reduced the epsp duration, leaving the rising phase relatively unaffected. In contrast, DGG reduced all phases of the epsp at doses that blocked responses to both NMDA and quisqualate. Lower doses that blocked only NMDA responses reduced only the later component of the epsp. This finding, that NMDA receptor mediated epsps result from activation of local circuit connections in neocortex, is consistent with recent in vivo studies in which the input from thalamus was insensitive to NMDA antagonists and sensitive to non-NMDA antagonists, while the epsps that resulted from antidromic activation of pyramidal tract neurones were sensitive to NMDA antagonists (P.L. Herrling & T.E. Salt, personal communication). More recently, using double intracellular recordings, we have been able to identify the presynaptic neurones as pyramidal, both by their electrophysiological properties and by their spiny dendrites, following intracellular injection of biocytin and subsequent histological processing for HRP (Radpour, Izzo & Thomson, unpublished results).

In neocortex therefore, NMDA receptors appear to be activated by local circuit connections. A prerequisite for LTP, if NMDA receptors must be activated, is therefore firing within the pyramidal cell pool. An input would not be potentiated, unless it coincided with activity in other pyramidal cells. Previous studies in neocortex in vivo are consistent with this suggestion since pairing stimulation of the thalamus with antidromic activation of the pyramidal tract led to potentiation of the thalamic input (Baranyi & Feher, 1981). If the same is true in CA1, it would help to explain the apparent need for powerful activation of CA1 pyramidal neurones during the induction of LTP. Recent results from this laboratory, indicate that pyramidal-pyramidal excitatory connections, although less common than in neocortex, do exist in the CA1 region and employ NMDA receptors (Radpour & Thomson unpublished results).

At the level of the input from one neocortical pyramid to another, we have so far obtained no evidence that a single presynaptic cell excites both another pyramid and an inhibitory interneurone presynaptic to that same cell. Some of the larger single axon epsps are curtailed by a hyperpolarization, when they are recorded from membrane potentials more depolarized than -65mV. This is however not an ipsp since it does not reverse at membrane potentials of >-85mV and is not apparent at membrane potentials close to rest (see Fig. 1). It is therefore only when large numbers of fibres are activated simultaneously that the temporal epsp/ ipsp relationship so often described, is induced. In CA1, a proportion of our paired recordings indicate that firing in one pyramid is correlated with long and variable latency activation of ipsps (probability 1 ipsp per 20 to 50 spikes) in another. However, none of these pairs was directly coupled. In the examples we have so far obtained of a single axon epsp

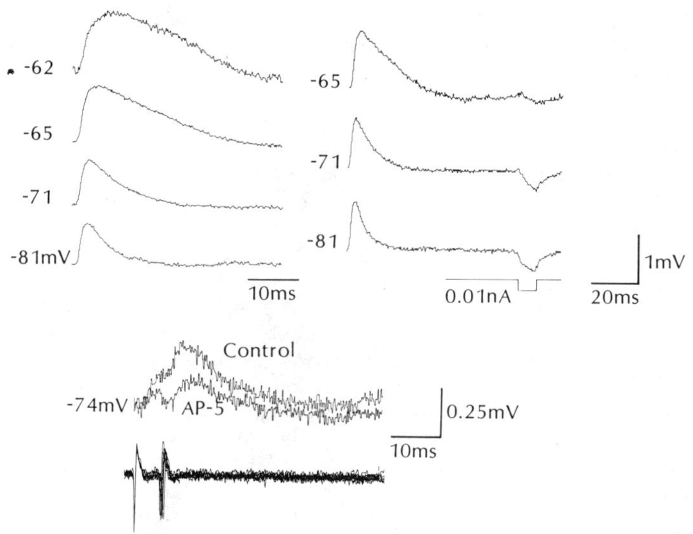

Figure 1. Single axon epsps recorded in neocortical neurones. Each record is an average of >500 sweeps. Top left: the epsp increased in amplitude and particularly in duration as the postsynaptic membrane was depolarized from -81mV to -62mV. Top right: the same epsp at 3 membrane potentials displayed at a slower sweep speed shows the hyperpolarization that curtails the epsp at -65mV. A brief current pulse (0.01nA) was injected into the postsynaptic neurone at the end of each sweep. Lower records illustrate an epsp recorded from a different pair of neurones, before (Control) and during application of AP-5. AP-5 reduces the peak and decay phase of the epsp. Note that the presynaptic neurone, spikes shown below (10 superimposed sweeps), fired in pairs of spikes and that the epsp exhibits summation.

generated in one CA1 pyramid by another, the epsp was not followed by an ipsp. The single axon epsps generated in CA1 neurones by CA3 neurones that have so far been studied also show little evidence of a systematic activation of ipsps after epsps, although feed-forward inhibition within the pool is apparent (Miles & Wong, 1987, Friedlander et al, 1989). These data seem to suggest that when a unitary epsp is activated, it is not automatically curtailed by an ipsp, but that neighbouring cells may be inhibited.

DOES THE VOLTAGE DEPENDENT BLOCKADE OF NMDA CHANNELS PRECLUDE THEIR ACTIVATION IN THE ABSENCE OF SUBSTANTIAL POSTSYNAPTIC DEPOLARIZATION ?

Several different experimental methods have been used to induce LTP, both in hippocampus and in neocortex. What most of these have in common is a synchronous activation of a relatively large number of input fibres, to evoke large compound epsps, either producing directly, or coincident with, depolarization of the postsynaptic cell. NMDA receptor/ channels are placed in a central position in many schemes proposed to explain the requirement for postsynaptic depolarization. The now famous, voltage dependent blockade of NMDA receptor/channels by Mg^{++} is proposed to prevent entry of Ca^{++} through these channels unless the blockade is released by postsynaptic depolarization. Since Ca^{++} entry, through the NMDA, rather than through voltage dependent Ca^{++} channels, appears to be a prerequisite for LTP (Kauer, this volume), potentiation would not occur if the immediately postsynaptic membrane was not sufficiently depolarized but, what constitutes a sufficient level of depolarization for activation of NMDA receptor/channels at a single synapse ?

Patch clamp studies (Nowak et al, 1984, Mayer & Westbrook, 1987) have shown that NMDA channels are so severely blocked by Mg^{++} at membrane potentials of -80mV that little response to NMDA receptor occupation could be expected to occur. This is apparently confirmed in recent studies in CA1 in which CNQX (a non-NMDA antagonist) blocked almost all synaptic response to Schaffer collateral stimulation at -80mV in the presence of Mg^{++}, but revealed a slow AP-5 sensitive component at -50 to -55mV (Kauer, this volume). However, in preparations in which an epsp mediated in part by NMDA receptors can be recorded under more normal conditions, a significant portion of even a single axon epsp can be shown to be AP-5 sensitive at -80mV (Thomson et al, 1989a). Somewhat surprisingly the entire epsp can be blocked by CNQX [unpublished results and see Wheal, this volume]. One explanation for this may be that CNQX blocks the high affinity glycine receptor whose activation is necessary for NMDA receptor/channel activation (Birch et al, 1988). Another is that the non-NMDA component of the epsp may depolarize the immediate postsynaptic site sufficiently to release some of the Mg^{++} blockade of the NMDA channel. In the absence of this 'priming' depolarization, NMDA receptor activation would be relatively ineffective unless a large depolarization were imposed from the soma.

The size of synaptic inputs to individual dendritic spines and the membrane characteristics of those spines are at present entirely a matter for speculation. Some tentative conclusions can perhaps be drawn to fit available experimental evidence. With the soma at -80mV, the immediate postsynaptic site is sufficiently depolarized during the rising phase of an epsp to allow activation of NMDA channels. As that epsp arrives in the soma, it may only be 0.1mV in amplitude, but might have reached tens of millivolts depolarization in a spine. Simulation of such a small epsp by injection of an equally small, standard sized current pulse into the soma cannot entirely reproduce the voltage relation of the epsp. Although the time course of the voltage response to the current pulse increases, as the apparent time constant of the soma rises with depolarization, it does not increase as much as that of the epsp. It becomes apparent therefore that the epsp, as seen at the soma, is generated by a larger and longer lasting event when the cell is depolarized. This could be due either to further release of Mg^{++} blockade of the NMDA channel, or to the activation of voltage dependent currents whose threshold has been exceeded. Perhaps prolongation of the epsp by activation of NMDA channels also facilitates the activation of local, time- and voltage-dependent currents, which facilitate the spread of synaptic current towards the soma.

DOES LTP INVOLVE AN INCREASE IN THE SIZE OF THE POSTSYNAPTIC CURRENT OR IN THE RESPONSIVENESS OF THE POSTSYNAPTIC MEMBRANE TO THE CURRENT ?

Previous studies of LTP in hippocampus have described the relative ease with which enhancement of the population spike can be achieved, compared with enhancement of the epsp itself (eg. Taube & Schwartzkroin, 1988). This has led to the suggestion that coupling between the depolarization at the subsynaptic membrane and the soma may be enhanced, in addition to, or even, in the absence of, any direct enhancement of the epsp. Any change in the coupling would have to be extremely selective, since gross changes in electrotonic or active properties of even a single dendritic branch would alter the effectiveness of many synaptic inputs.

Recently, elegant studies by Friedlander, Sayer and Redman (1989) have attempted to determine whether a single axon epsp from one CA3 to one CA1 pyramid can be potentiated by the standard stimulus protocols used to generate LTP in this pathway. Although LTP of larger, compound epsps (1.5 to 2.5mV in amplitude), recorded intracellularly, and field epsps recorded extracellularly, was achieved with the stimulus regimes used, LTP of single axon, or of small compound (<0.5mV) epsps was not achieved. Every attempt was made to ensure that the presynaptic CA3 cell was activated throughout the induction protocol. Whether this discrepancy indicates that a threshold for the expression, as well as for the induction of LTP must be reached, is uncertain, but if LTP is proposed to represent a physiological event, consolidation of single synapses by coactivation should be possible. It is therefore tempting to speculate that some change has occurred at each participating bouton/spine, but that transmission of that change to the soma still requires threshold for some other event [eg. activation of voltage-dependent conductances] to be exceeded. LTP could then involve either the enhancement of the synaptic current, so that it approaches threshold more closely, or a reduction in the threshold, of a very localized event. A previously subthreshold synaptic current could now exceed threshold. Prolongation of synaptic events by more slowly inactivating inward currents would be expected to facilitate current spread towards the soma.

WHY ARE NMDA RECEPTORS NOT REQUIRED FOR EXPRESSION OF LTP ?

It is now widely accepted that NMDA receptors are necessary for the induction, but not for the expression of LTP in CA1 neurones. In neocortex inputs from thalamus, which do not involve NMDA receptors can be potentiated if their activation is paired with activation of local circuit connections (Baranyi & Freher, 1981) that do. Expression of LTP in neocortex also involves NMDA receptors (see Bindman, this volume) although again, NMDA receptor activation is only essential for its induction. Following our recent finding, that CA1 neurones receive excitatory synapses from other CA1 cells, an explanation for the apparent difference between CA1 and neocortex arises. If, as in cortex, NMDA receptors are activated by local circuit axons and not directly from CA3, strong, high frequency stimuli used to induce LTP would activate pyramids and thereby their local connections. Single stimuli used to test for LTP might not activate them. Experimental tests for the expression of LTP in neocortex activate NMDA receptors, since here, typical test stimuli activate pyramids antidromically.

Some controversy surrounds the proposal that LTP involves an increase in the amount of transmitter released. One argument against the requirement for increased release of glutamate is that the expression of LTP does not

involve NMDA receptors in hippocampus, while the expression of post-tetanic and paired-pulse potentiation do (Kauer, this volume).
These latter two phenomena are traditionally accepted indications that transmitter release has been enhanced, by elevation of presynaptic Ca⁺⁺. If central synapses were totally analogous to the neuromuscular junction this argument would hold. However, there are significant differences. The very high affinity of glutamate for the NMDA receptor, which far exceeds its affinity for the quisqualate/kainate receptors, indicates that all postsynaptic NMDA receptors are already saturated if sufficient glutamate is released to activate non-NMDA receptors. In all cases studied so far, single axon synapses that involve NMDA receptors also involve non-NMDA receptors (Forsythe & Westbrook, 1988, Thomson et al, 1989a), though whether all presynaptic boutons from a given axon access both receptor types remains to be determined. Following this argument, if the available NMDA receptors are already saturated, increased transmitter release from any one bouton would not enhance NMDA receptor activation. To explain how LTP could involve enhanced glutamate release, without causing enhancement of NMDA receptor activation, it is only necessary to propose that LTP involves increased release from already active boutons, or that it occurs preferentially at synapses that do not involve NMDA receptors at all. In contrast the presynaptic Ca⁺⁺ loading, thought to underlie shorter-term potentiation, might bring previously silent boutons above threshold for transmitter release, thereby accessing more NMDA receptors. The only method presently available to answer these points would be to combine fluctuation analysis of single axon epsps (eg. Kullman et al, 1989) with pharmacology. This has yet to be attempted. Another possibility is that paired stimuli might activate pyramids within the pool and thereby NMDA receptor mediated synapses, while single stimuli do not.

GLYCINE FACILITATES THE ACTIVATION OF NMDA RECEPTOR/CHANNELS BY GLUTAMATE

Glycine, acting at a high affinity, strychnine insensitive binding site, augments and may even be essential for NMDA receptor/channel activation [see Ascher, Mayer and Dingledine, this volume]. Several previous studies had failed to demonstrate any augmenting effect of exogenous glycine, but in neocortical slices, electrophoretically applied glycine produces consistent and very large increases in neuronal responses to NMDA (Thomson et al, 1989b), homocysteic acid (HCA) and somewhat surprisingly, glutamate. Responses to AMPA and quisqualate were unaffected. Studies in this and several other labs (Knopfel et al, 1987, Thomson, 1985) had previously shown glutamate to be a poor activator of NMDA receptor/channels in intact tissue, despite its high affinity for the NMDA receptor and effectiveness in patch clamp studies. Our recent results (see Figure 2.) suggest that glutamate can be an effective NMDA agonist in neocortex, provided glycine levels are high enough. NMDA and HCA readily activate NMDA receptors in the presence of only endogenous glycine, while glutamate appears to require more glycine. Since epsps can be shown to include an NMDA receptor mediated component in the presence of only endogenous glycine, but that component can be augmented by additional exogenous glycine, glycine levels in our neocortical slices must be suprathreshold for enabling the natural transmitter to activate the receptor/channels, but submaximal. Modulation of NMDA receptor/ channel activity by glycine and by its endogenous antagonist, kynurenate, should therefore be taken into account in any scheme that involves NMDA receptors. Recently, exposure of hippocampal slices to a combination of NMDA, glycine and spermine has been shown to induce an enhancement of evoked field epsps that long outlasts their application (Thibault et al, 1989). Spermine and spermidine are endogenous polyamines that have been shown to augment NMDA receptor/channel activation via yet another binding site (Ransom & Stec, 1989). It is tempting to speculate on the roles that these compounds may play in LTP.

CONCLUSIONS

     Huge advances in our understanding of synaptic function have been made
using conventional techniques of electrical stimulation and recording of
gross evoked potentials. Similarly, a great deal of our present knowledge
of neuropharmacology comes from blockade, by relatively selective
antagonists, of compound psps and evoked field potentials. However, these
techniques may eventually restrict future advances if their limitations are
not fully grasped. Gross pathway stimulation combined with uncritical
antagonist application has led to the various suggestions, 1) that NMDA
receptors are not normally activated following a single presynaptic spike,
but require repetitive activation, possibly causing spill-over of
transmitter onto extra-synaptic sites, 2) that NMDA receptor mediated epsps
are so long in duration that they are necessarily and even selectively
depressed by ipsps, unless those ipsps are blocked pharmacologically, or

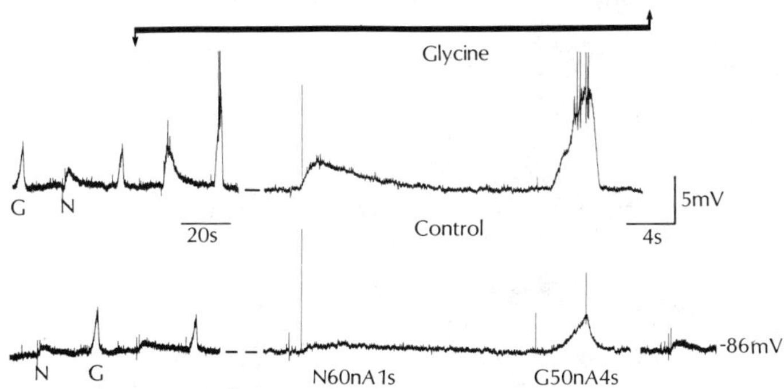

Figure 2.   Responses of a pyramidal neurone in a neocortical slice, to
electrophoretic pulses of glutamate and NMDA delivered alternately in an
electronically controlled cycle and recorded intracellularly. Top record:
electrophoretic application of glycine increases responses to both EAAs
(bar above gives duration of glycine application). Lower record: recovery
from glycine. In each trace one cycle is recorded at a faster speed.
Artifacts indicate the beginning and end of each electrophoretic pulse.

depressed during high frequency activation and 3) that almost pathological
postsynaptic depolarization must occur if the channels are to open. There
is no direct evidence to support any of these. At present the data
available about the functioning of single axon psps are very limited,
largely because maintaining dual recordings through manipulations of both
pre- and post-synaptic neurones is technically difficult. Fluctuation
analysis is one of the most powerful tools available, but must be used with
extreme caution and requires collection of several thousand unitary events
with the signal to noise ratios that are encountered in the cns. Setting
aside the inaccuracies that can be introduced by naive analysis, naive data
collection can also create interpretational problems. Double intracellular
recordings circumvent some of these, if capacitance-coupling artifacts are
understood and reduced, but can be difficult to obtain. For rapid sampling
of many pairs of cells in regions where the probability of

obtaining a connection is low (eg. CA1 to CA1, probability <1 in 100), we have activated the presynaptic neurone with glutamate and recorded from it extracellularly. This method ensures that the recording is from the soma/dendrites, rather than the axon of the presynaptic cell. However, if the glutamate is applied in brief pulses, a technique that has been employed elsewhere, several presynaptic neurones, as well as many that are not directly connected, are activated almost simultaneously. Since cells that are presynaptic for a given postsynaptic cell are frequently found in 'clusters', this is a strong possibility. Fluctuations may then be due to different numbers of presynaptic neurones being activated by each pulse. The recorded spike may not even be the cause of any part of the recorded epsp. Since it is impossible to place an electrophoretic pipette so that it activates only one neurone in a slice, it is essential to activate the cells asynchronously, by low continuous currents and to record many sweeps triggered by the recorded spike.

ACKNOWLEDGEMENTS

This work was supported by the Wellcome Trust and the Medical Research Council.

REFERENCES

Andreasen, M., Lambert, J.D.C. and Jensen, M.S. (1988) Direct demonstration of an N-methyl-D-aspartate receptor mediated component of excitatory synaptic transmission in area CA1 of the rat hippocampus. Neurosci. Lett. 93: 61-66.

Artola, A., and Singer, W. (1987) Long-term potentiation and NMDA receptors in rat visual cortex. Nature 330: 649-652.

Baranyi, A. and Feher, O. (1981) Synaptic facilitation requires paired activation of convergent excitatory pathways in the neocortex. Nature 290: 413-414.

Birch, P.J., Grossman, C.J. and Hayes, A.G. (1988) 6,7-dinitro-quinoxaline-2,3-dione and 6-nitro,7-cyano-quinoxaline-2,3-dione antagonise responses to NMDA in the rat spinal cord via an action at the strychnine-insensitive glycine receptor. Eur. J. Pharmacol. 156: 177-180.

Collingridge, G.L., Kehl, S.J. and McLennan, H. (1983) Excitatory amino acids in synaptic transmision in the Schaffer collateral-commisural pathway of the rat hippocampus. J. Physiol. 334: 33-46.

Eaton, S.A. and Salt, T.E. (1989) Synaptic responses to noxious physiological stimulation in the thalamus of the anaesthetised rat are not antagonized by the non-NMDA receptor antagonist, CNQX. J. Physiol. In the press.

Forsythe, I.D. and Westbrook, G.L. (1988) Slow excitatory postsynaptic currents mediated by N-methyl-D-aspartate receptors on cultured mouse central neurones. J. Physiol. 396: 515-533.

Friedlander, M.J., Sayer, R.J. and Redman, S.J. (1989) Evaluation of long term potentiation of small compound and unitary epsps at the hippocampal CA3-CA1 synapse. J. Neurosci. in the press.

Greenamyre, J.T., Olson, J.M.M. Penney, J.B. and Young, A.B. (1985) Autoradiographic characterization of N-methyl-D-aspartate-quisqualate- and kainate-sensitive glutamate binding sites. J. Pharm. & Exp. Ther. 233: 254-263

Kleinschmidt, A., Bear, M.F. and Singer, W. (1987) Blockade of "NMDA" receptors disrupts experience-dependent plasticity of kitten striate cortex. Science 238: 355-358.

Knopfel, T., Zeise, M.L., Cuenod, M. and Zieglgansberger, W. (1987) L-homocysteic acid but not L-glutame is and endogenous N-methyl-D-aspartic acid receptor preferring agonist in rat neocortical neurones in vitro. Neurosci. Lett. 81: 188-192.

Kullman, D.M., Martin, R.L. and Redman, S.J. (1989) Reduction by general aneasthetics of group Ia excitatory postsynaptic potentials and currents in the cat spinal cord. J. Physiol. 412: 277-296

Mayer, M.L. and Westbrook, G.L. (1987) Permeation and block of N-methyl-D aspartic acid receptor channels by divalent cations in mouse cultured central neurones. J. Physiol. 394: 501-527.

Miles, R. and Wong, R.K.S. (1987) Inhibitory control of local excitatory circuits in the Guinea pig hippocampus. J. Physiol. 388: 611-629.

Nowak, L., Bregestovski, P., Ascher, P., Herbert, A. and Prochiantz, A. (1984) Magnesium gates glutamate-activated channels in mouse central neurones. Nature 307: 462-464.

Ransom, R.W. and Stec, N.L. (1988) Cooperative modulation of [3H]MK-801[MK801] binding to the N-methyl-D-aspartate receptor-ion channel complex by L-glutamate, glycine, and polyamines. J. Neurochem. 51: 830-836.

Salt, T.E. (1987) Excitatory amino acid receptors and synaptic transmission in the rat ventrobasal thalamus. J. Physiol. 391: 499-510.

Sayer, R.J., Friedlander, M.J. and Redman, S.J. (1989) The time course and amplitude of epsps evoked at synapses between pairs of CA3/CA1 neurones in the hippocampal slice. J. Neurosci. in the press.

Taube, J.S. and Schwartzkroin, P.A. (1988) Mechanisms of long-term potentiation: A current-source density analysis. J. Neurosci. 8: 1645-55.

Thibault, O., Joly, M., Muller, D., Schottler, F., Dudek, S., and Lynch, G. (1989) Long-lasting physiological effects of bath applied N-methyl-D-aspartate. Brain Res. 476: 170-173

Thomson, A.M. (1986) A magnesium-sensitive post-synaptic potential in rat cerebral cortex resembles neuronal responses to N-methylaspartate. J. Physiol. 370: 531-549.

Thomson, A.M. (1985) Comparison of responses to transmitter candidates at an N-methylaspartate receptor mediated synapse in slices of rat cerebral cortex. Neurosci. 17: 37-47.

Thomson, A.M., Girdlestone, D. and West, D.C. (1988) Voltage-dependent currents prolong single-axon postsynaptic potentials in layer III pyramidal neurons in rat neocortical slices. J. Neurophysiol. 60: 1896-1907.

Thomson, A.M., Girdlestone, D. and West, D.C. (1989a) A local circuit neocortical synapse that operates via both NMDA and non-NMDA receptors Br.J.Pharmac. 96: 406-408.

Thomson, A.M., Walker, V.E. and Flynn, D.M. (1989b) Glycine enhances NMDA receptor mediated synaptic potentials in neocortical slices. Nature 338: 422-424.

MODULATION OF THE RESPONSIVENESS OF CEREBELLAR PURKINJE CELLS TO

EXCITATORY AMINO ACIDS

Francis Crépel and Mireille Krupa

Laboratoire de Neurobiologie du Développement. URA CNRS 1121

Université Paris-Sud. 91405 Orsay Cedex France

Introduction

Since the discovery and classification of receptor-types to excitatory amino acids (EAAs), (36) numerous studies have analyzed their role in synaptic plasticity, showing that, in particular in the hippocampus, n-methyl-D-aspartate (NMDA) receptors play a crucial role in long term potentiation of synaptic transmission (ref. dans 5).

The cerebellum of mammals appears to be another interesting region of the brain for such studies since the main excitatory afferents of Purkinje cells (PCs), i.e. parallel fibers (PFs) and climbing fibers (CFs), are likely to use glutamate (Glu) and aspartate (Asp) as neurotransmitters respectively (28, 37). Moreover, and in marked contrast with most other neuronal types in the brain, one know now that at least in the adult, PCs mainly bear non-NMDA receptors (8, 10), thus offering the opportunity to study synaptic plasticity in the absence of NMDA receptors.Indeed, the possible involvement of the cerebellum in motor learning was postulated by Brindley as early as 1964 (6). The idea was formalized in 1969 by Marr (27) who proposed the theory of the "external teacher" derived from the perceptron, a cybernetic machine invented by Franck Rosenblatt in 1962. According to the theory, during a motor learning involving the activation of PCs by PFs, coactivation of the cells by CFs is able to change durably the gain of the synaptic transmission between PFs and PCs, so that the cerebellar output is ajusted to the desired motor command. Later on, Albus (1) revisited the theory and proposed that this coactivation leads to a long term depression (LTD) of synaptic transmission at PF-PC synapses.

In an elegant series of in vivo experiments, Ito and coworkers (18, 20, 33) showed that coactivation of PCs by CFs and PFs leads indeed to a LTD of synaptic transmission between PFs and PCs. Moreover, they also showed that coactivation of the cells by CFs and by Glu and Asp also leads to a LTD of their responsiveness to glu, whereas that to Asp is unaffected. Finally, Kano and Kato (21, 22) demonstrated that this synaptic plasticity selectively involves quisqualate (QA) receptors of PCs, leading to the hypothesis that these receptors might be desensitized as a result of both their activation by Glu, and the Calcium (Ca) influx which occurs in PCs following stimulation of CFs (18, 19)

This led us to test more directly the role of Ca in the modulation of the responsiveness of PCs to EAAs, as well as to analyze the possible involvement of other intracellular mechanisms. In particular, one know now (31,35) that stimulation of QA receptors activates protein kinase C (PKC) by the production of diacylglycerol (DAG) (ref. in 16). One also knows that this enzyme is Ca dependent (30) and Alkon (2) has proposed that, at least in Hermissenda, the activation of PKC requieres a synergy between the production of DAG and a transiant rise of cytosolic Ca concentration due, for instance, to the opening of Ca channels.

Thus, it was temptating to speculate that the postulated desensitization of QA receptors of PCs involves a coactivation of PKC by DAG and by Ca, these second messengers being produced by the stimulation of the cells by Glu and by CFs respectively. Therefore, the present paper will deal both with the role of Ca and PKC in the modulation of the responsiveness of PCs to EAAs. The data presented here have been partly published elsewhere (11).

Methods

Experimental procedures were the same as described elsewhere (8). Briefly, cerebellar slices were prepared from 18 to 26 day old Sprague Dawlay rats and 2 month old animals, i.e near the end or after the end of cerebellar maturation (3). All PCs were extra- or intracellularly recorded at a somatic level in the vermian region of Larsell's lobules IX and X with glass micropipettes filled with 3M NaCl and 3M KCl respectively. They were identified by their typical climbing fiber responses (CFRs) (15) elicited by electrical stimulation of the white matter of the recorded folium. Glu, Asp and Qa were ejected in the dendritic field of the recorded cells through adjacent barrels of the same iontophoretic electrode. The individual barrels contained Glu (0.5·M, pH 6), Asp (0.5 M, pH 6), QA (0.1 M, pH 6) and 0.5 M NaCl for automatic compensation of ejecting currents. Quantitative estimates of EAA-induced responses of extracellularly recorded PCs were enabled by PSTHs of their unitary spike discharge constructed during 3 to 6 successive EAA applications, or by averaging PSTHs of different PCs tested with the same experimental protocole (see results). Purkinje cells were usually held at -65 mV in intracellularly recorded cells to better visualize the EAA-induced responses.

Results

1) Modulation of the responsiveness of PCs to EAAs by Ca

As described previously (11) the responses of all intracellularly recorded PCs (n=45) to Glu and Asp ejection in their dendritic field in 16 to 26 day old animals consisted of transient and dose dependent depolarizations which could reach the firing level of the cells when they were initially held at -65 mV (fig. 1). When these cells were coactivated at 4 hz for one minute by depolarizing currents pulses passed through the recording microelectrode and when the

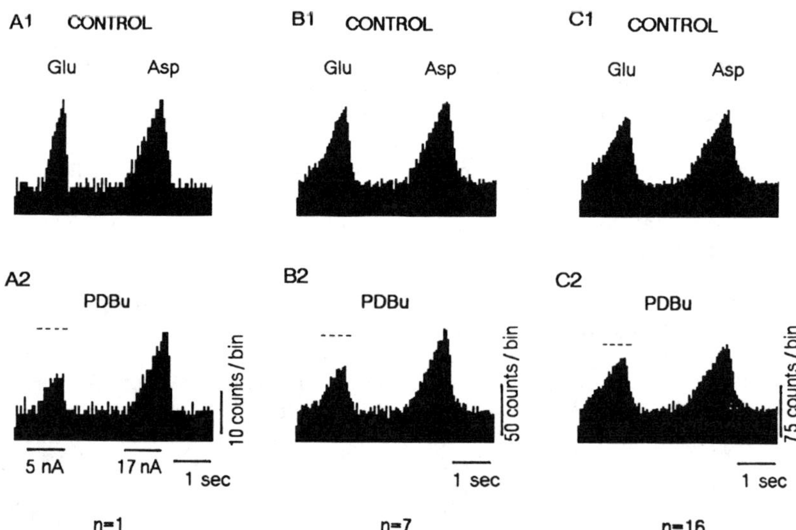

Figure 1. LTD of glutamate induced responses by direct activation of Purkinje cells.
A1: responses of an intracellularly recorded PC to Asp and Glu. A2: Sodium and Ca spikes induced in this cell by depolarizing current pulse (lower trace) passed through the recording microelectrode. A3: idem as A1, 20 min after the pairing procedure.
B1: idem as in A1 in another cell. B2, B3: idem as A2, except that depolarizing currents only evoked sodium spikes (B2), even during EAA application (B3). B4: idem as A3.

pairing procedure was repeated every 2 minutes, quite different results were obtained after 2 to 4 such pairings, depending on the nature of the responses elicited in PCs by stimulating currents (23, 24). When the current pulses were adjusted so that clearcut Ca spikes (fig. 1A 2) were elicited in the cells (n=23), the pairing procedure led to a clearcut depression of the responses of PCs to Glu in 5 of them, whereas those to Asp were left unaffected (fig.1A1, 1A3). This selective decrease of the responsiveness to Glu lasted as long as the cells were recorded, i.e. up to one hour (11). In marked contrast, when depolarizing currents were adjusted so that only Na spikes were evoked, even during Glu and Asp induced responses (fig.1B2, 1B3), no such LTD of Glu-induced responses was ever observed in the 22 cells tested in these conditions (fig.1B1, 1B4).

Therefore, these data strongly suggest that Ca is indeed involved in the modulation of the responsiveness of PCs to Glu (see also ref. 34), and, moreover, they also strongly suggest that this processes occurs at a postsynaptic level. Finally, it is noteworthy that the percentage of PCs exhibiting a decrease of their responsiveness to Glu in this experimental protocole is similar to that obtained by replacing the direct stimulation of the cells giving rise to Ca spikes by their activation by CFs during the periods of coactivation (11). This again suggests that CF inputs act only via the Ca influx which occurs in PCs during CF responses (32), as proposed by Ito (18, 19).

2) Effects of phorbol esters on the responsiveness of PCs to EAAs

The hypothesis of an involvement of PKC in the postulated desensitization of QA receptors of PCs was tested by studying the effects of bath application of 200 to 400 nM phorbol esters know to selectively activate PKC (7) on the responsiveness of the cells to EAAs. In these experiments, PCs were extracellularly recorded at a somatic level and their firing rate was quantified by the use of PSTHs. (fig. 2). In 19 of the 39 cells tested in 16 to 26 day old rats, this led again to a significant LTD of Glu induced responses, with an average decrease of 39 ± 5 %, whereas responses to Asp were not affected. In the other 20 PCs, Phorbol esters did not have any significant effect on their responses to EAAs. Similarly, a marked, selective, and persistent decrease of Glu induced responses was also induced by phorbol esters in 7 of the 16 PCs studied in adult animals. (fig. 2A, 2B), the other 9 cells being not affected, so that the mean decrease of Glu responsiveness induced by phorbol-esters in the 16 cells only averaged about 20% (fig. 2C2).

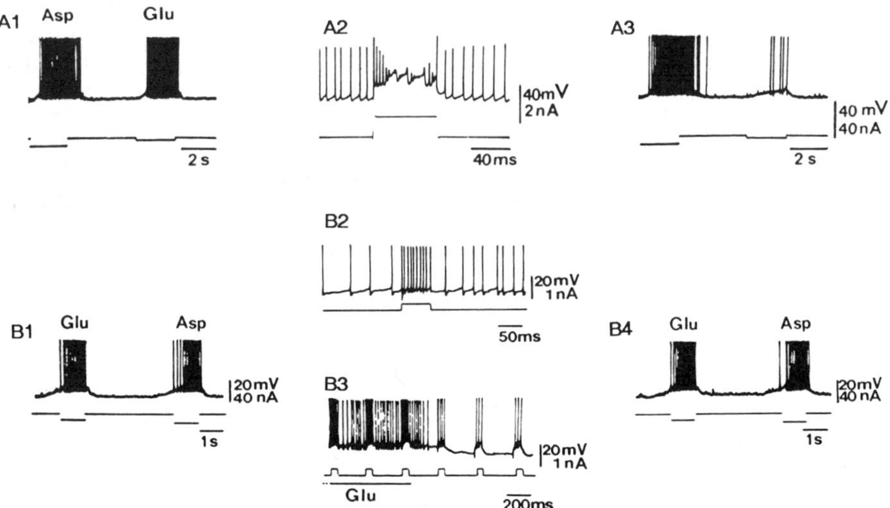

Figure 2 . Selective effect of phorbol esters on Glu induced responses.
A1, A2: PSTHs of Glu and Asp induced responses in one PC before (A1) and after (A2) bath application of 400 nM phorbol 12,13-dibutyrate (PDBu). B1, B2: average PSTHs of the 7 out of the 16 tested PCs for which PDBu induced a decrease of Glu-induced responses (B2) as compared to control one (B1). C1, C2: idem as in B1, B2 for the 16 tested PCs. Dotted lines in A2, B2, C2 represent control values of Glu-induced responses.

Furthermore, we also tested the effects of phorbol esters on QA induced responses in 22 other PCs obtained from 18 to 26 day old rats (Krupa and Crepel, in preparation). In 9 of them, there was a concomitant and long lasting (up to one hour) decrease of Glu and QA induced responses, those due to Asp being unaffected. In the remaining 13 cells, no effect of the drug was observed on their responsiveness to EAAs. Interestingly enough, the coactivation of these 22 cells by EAAs and CFs in the presence of phorbol esters was always ineffective in producing any further decrease of their responses to EAAs, whatever the former effect of phorbol esters was. These results suggest that the limiting step in the obtention of the LTD of Glu and QA induced responses by bath application of phorbol esters is not the activation of PKC itself but rather occurs in the subsequent cascade of events leading to the decreased responsiveness of the cells (see discussion).

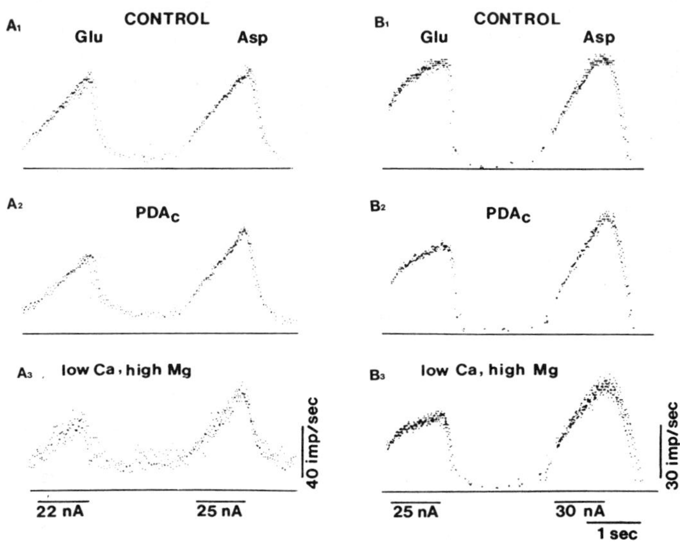

Figure 3 . Effects of phorbol esters on Glu-induced responses after blockade of synaptic transmission.

A1: Instantaneous frequency curve of the response of one PC to Glu and Asp application. A2: idem as in A1 after bath application of 400 nM phorbol 12,13-diacetate (PDAc). A3: idem as in A2 after replacing the standard bathing medium by low Ca, high magnesium solution. B1, B2, B3: as in A1, A2, A3 respectively in another PC.

Although these results also suggest that the depressant effect of phorbol esters on Glu and QA induced responses of PCs was induced at a post-synaptic level, this was checked more directly by studying the effect of phorbol esters on EAA-induced responses of PCs in the absence of synaptic transmission. In the 3 cells which were thus successfully tested, the selective depressant effect of phorbol esters on their responsiveness to Glu obtained in standard bathing medium was maintained after replacing this solution by a solution containing 0.1 mM Ca and 6 mM Mg (fig. 3).

Finally, the inactive analog 4-α-phorbol applied at the same concentration as β phorbols in the bath did not have any effect an EAA.-induced responses of PCs (n=20). Therefore these results strongly suggest that the depressant effect of phorbol esters on Glu and QA induced responses of PCs is indeed due to the activation of PKC of these neurones.

Discussion

On the whole, the present study strengthens our hypothesis that coactivation of PKC by Ca and DAG, resulting from the stimulation of PCs through CFs and QA receptors respectively, is requiered for the desensitization of the latter during generation of LTD. These results also strongly suggest that Asp activates PCs through a receptor-type distinct of the QA receptor, in keeping with the results of earlier studies (10, 11, 13).

In preliminary experiments (11), coactivation of PCs by CFs and EAAs was much less successful in adult rats than in immature ones in obtaining LTD of Glu-induced responses. The fact that direct activation of PKC by phorbol esters was as potent in the 2 groups in inducing such a LTD (see results) suggests that the limiting step in adult animals might be either a decrease in the coupling between QA receptors and the activation of phospholipase C as compared to younger animals (29, 31)), or a decrease in the number of NMDA receptors in adult PCs as compared to younger ones (13). Indeed, concerning the latter possibility, the activation of NMDA receptors by Glu during CFR induced depolarizations in young rats might contribute to Ca entry into PCs through NMDA coupled channels (4, 26), and thus reinforce the effects of the pairing procedure on the activation of PKC.

As already mentioned before (see results), phorbol esters failed to induce a LTD of Glu responsiveness of PCs in about 50 to 60% of the tested cells, and no further depressant effect was obtained in these neurons by their coactivation by EAAs and CFs. Although the sample of cells tested with this latter protocole is still rather small (see results), this suggests that in this case, the limiting step was not an insufficient activation of PKC by phorbol esters, due for instance to their poor penetration in the slices since, according to the scheme of Alkon (2), one would have expected to obtain additive effects of phorbol esters and of the pairing procedure in such circumstances. Rather, these results suggest that this limiting step occurs in the cascade of events linking the activation of PKC to the decrease of responsiveness of PCs to Glu. One hypothesis might be that PKC acts in synergy with other enzymes, the activation of which were not controled in the present experiments. The c-GMP dependent protein kinase might be such an enzyme also involved in the desensitization of QA receptors of PCs as recently proposed by Ito (19), in particular because this kinase is also highly concentrated in these neurons (25). In any case, our results are in keeping with previous observations showing that PKC may play a crucial role in feedback regulation of neuronal activation by several neurotransmitters including Glu, norepinephrine and acetylcholine (12, 17).

Finally, other sets of experiments (Krupa and Crepel, in preparation) show that phorbol esters induce a long term potentiation of synaptic transmission at PF-PC synapses, and this even when the intracortical inhibition was blocked by bicuculline, thus indicating that this processes probably affects the PF-mediated EPSP. Since, on the other hand, phorbol esters induce a LTD of the responsiveness of PCs to Glu (see before), this strongly suggests that phorbol esters also act at a presynaptic level to increase the evoked release of Glu by PFs, as already shown in other nerve fiber terminals (ref. in 5). Thus, it seems that the cerebellum which appeared as a rigid neuronal machine at the end of the sixties (14) can now be viewed as rather flexible, both at pre- and postsynaptic levels, and therefore as a good candidate for motor learning.

Acknowledgements — This work was supported by the CNRS, Université Paris XI and by an INSERM grant.

References

1. Albus, J.S. (1971) : A theory of cerebellar function. Math. Biosci. 10 : 25-61
2. Alkon, D.L. and Rasmussen, H. (1988) : A spatial-temporal model of cell activation. Science 239, 988-1005.
3. Altman, J. (1972) : Postnatal development of the cerebellar cortex in the rat. II.Phases in the maturation of Purkinje cells and of the molecular layer. J.Comp.Neurol. 145 : 399-464
4. Ascher, P. and Nowak, L. (1986) : Calcium permeability of the channels activated by N-Methyl-D-aspartate (NMDA) in isolated mouse central neurones. J.Physiol. (London) 377 43P
5. Bliss, T.V.P. and Lynch, M.A. (1988) : Long term potentiation of synaptic transmission in hippocampus : Properties and mechanisms. In long term potentiation from biophysics to behavior 3-72 Alan R. Liss. Inc.
6. Brindley, G.S. (1964) : The use made by the cerebellum of the information that it receives from sense organs. I.B.R.O. Bull. 3-80
7. Castagna, M.,Takai, Y.,Kaibuchi, K.,Sano, K.,Kikkawa, U. and Nishizuka,Y. (1982) : Direct activation of calcium-activated,phospholipid-dependent protein kinase by tumor-promoting phorbol esters. J.Biol Chem. 257 : 7847-7851

8. Crepel, F.,Dhanjal, SS. and Sears, T.A. (1982) : Effect of glutamate, aspartate and related derivatives on cerebellar Purkinje cell dendrites in the rat : An in vitro study. J. Physiol. (London) 329 : 297-317

9. Crepel, F., Dupont, J.L. and Gardette, R. (1982) : Connectivity and chemosensitivity of Purkinje cells in the immature cerebellum : an in vitro study. J. Physiol.(London) 332 : 62P

10. Crepel, F.,Dupont, J.L. and Gardette, R. (1983) : Voltage-clamp analysis of the effect of excitatory amino acids and derivatives on Purkinje cell dendrites in rat cerebellar slices maintained in vitro. Brain Research 279 : 311-315

11. Crepel, F. and Krupa, M. (1988) : Activation of protein kinase C induces a long-term depression of glutamate sensitivity of cerebellar Purkinje cells. An in vitro study. Brain Res. 458 : 397-401

12. Downing, J.E.G. and Role, L. (1987) : Activators of protein kinase C enhance acetylcholine receptor desensitization in synpathic ganglion neurons. Proc. Natl. Acad. Sci. USA. 84 : 7739-7743

13. Dupont, J.L.,Gardette, R and Crepel, F. (1987) : Postnatal development of the chemosensitivity of rat cerebellar Purkinje cells to excitatory amino acids. An in vitro study. Developmental Brain Research 34 : 59-68

14. Eccles, J.C., Ito, M. and Szentagothai, J. (1967) : The cerebellum as a neuronal machine. Springer-Verlag. New-York. Heidelberg.

15. Eccles, J.C.,Llinas, R. and Sasaki, K. (1966) : The excitatory synaptic action of climbing fibres on the Purkinje cells of the cerebellum. J. Physiol. (London) 182 : 268-296

16. Exton, J.H. (1985) : Mechanisms involved in alpha-adrenergic phenomena. Am. J. Physiol. 248 : E633-E647

17. Gonzales , R.A.,Greger, Jr, P.H.,Baker, S.P.,Ganz, N.I.,Bolden, C.,Raizada, M.K. and Crews, F.T. (1987) : Phorbol esters inhibit agonist-stimulated phosphoinositide hydrolysis in neuronal primary cultures. Develop.Brain   Research 37 : 59-66

18. Ito, M. (1987) : Characterization of synaptic plasticity in the cerebellar and cerebral neocortex. In Changeux, J.P. and Nonishi, M. Eds. The Neural and Molecular bases of Learning, John Wiley and Sons, LTD, 276-279

19. Ito, M. (1989) : Long term depression. Ann. Rev. Neurosci. 12 : 85-102

20. Ito, M., Sakurai, M. and Tongreach, P. (1982) : Climbing fibre induced depression of both mossy fibre responsiveness and glutamate sensitivity of cerebellar Purkinje cells. J. Physiol. (London) 324 : 113-134

21. Kano, M. and Kato, M. (1987) : Quiqualate receptors are specifically involved in cerebellar synaptic plasticity. Nature 325 : 276-279

22. Kano, M. and Kato, M. (1989) : Mode of induction of long term depression at parallel fibre-Purkinje cell synapses in rabbit cerebellar cortex. Neurosci. Res. 5 :544-556

23. Llinas, R. and Sugimori, M. (1980) : Electrophysiological properties of in vitro Purkinje cell somata in mammalian cerebellar slices. J. Physiol. (London) 305 : 171-195

24. Llinas, R. and Sugimori, M. (1980) : Electrophysiological properties of in vitro Purkinje cell dendrites in mammalian cerebellar slices. J. Physiol. (London) 305 : 197-213

25. Lohmann, S.M., Walker, U., Miller, P. E., Greengard, P., Camilli, P. D. (1981) : Immunohistochemical localization of cyclic GMP-dependent protein kinase in mammalian brain. Proc. Natl. Acad. Sci. USA. 78 : 653-657

26. MacDermott, A.B.,Mayer, M.L.,Westbrook, G.L.,Smith, S.J. and Barker, J.L. (1986) : NMDA-receptor activation increases cytoplasmic calcium concentration in cultured spinal cord neurones. Nature 321 : 519-522

27. Marr, D. (1969) : A theory of cerebellar cortex. J. Physiol. (Lond.)202 : 437-470

28. MC Bride, W.J.,Nadi, N.S.,Altman ,J. and Apprison, M.H. (1976) : Effects of selective doses of X-irradiation on the levels of several amino acids in the cerebellum of the rat. Neurochem. Res. 1 : 141-152

29. Nicoletti, F., Iadarola, M.J., Wroblews, J.T., and Costa, E.: (1986) Excitatory amino-acid recognition sites coupled with inositol phospholipid-metabolism. Developmental changes and interaction with alpha-1 adrenoceptors. Proc. Nat.   Acad. Sci. USA. 83 : 1931-1935

30. Nishizuka, Y. (1986) : Studies and perspectives in protein kinase C. Science 233 : 305-312

31. Recasens, M., Sassetti, I., Nourigat, A., Sladecze, F., and Bockaert, J. (1987) : Characterization of subtypes of excitatory amino-acid receptors involved in the stimulation of inositol phosphate synthesis in rat-brain synaptoneurosomes. Eur. J. Pharmacol. 141: 87-93

32. Ross, W. and Werman, R. (1987) : Mapping calcium transiants in the dendrites of Purkinje cells from the guinea-pig cerebellum in vitro. J. Physiol. (London) 389 : 319-336

33. Sakurai, M. (1987) : Synaptic modification of parallel fibre-Purkinje cell transmission in in vitro guinea-pig cerebellar slices. J. Physiol. (London) 394 : 463-480

34. Sakurai, M. (1988) : Depresion and potentiation of parallel fiber-Purkinje cell transmission in in vitro cerebellar slices. In Olivo cerebellar system in motor control, Ed. P. Strata, Berlin Springer-Verlag, in Press

35. Sladeczek, F., Pin, J.P.,Recasens, M., Bockaert, J. and Weiss, S.: (1985) Glutamate stimulates inositol phosphate formation in striatal neurones. Nature 314 : 717-719

36. Watkins, J.C. (1981) : Pharmacology of excitatory amino acid transmitters. In F.V. De Feudis and P. Mandel ,Eds. Advances in Biochemical Psychopharmacology, Vol. 29, Amino acid neurotransmitters, Raven, New-York, 205-212

37. Wiklünd, G., Toggenburger, G., Cuenod, M. (1982) : Aspartate : possible neurotransmitter in cerebellar climbing fibers. Science, 216 : 78-79

THE ROLE OF EPENDYMIN IN NEURONAL PLASTICITY AND LTP

Victor E. Shashoua

Ralph Lowell Laboratories, McLean Hospital

Harvard Medical School, Belmont, MA 02178, U.S.A.

INTRODUCTION

Ependymin is a brain extracellular glycoprotein that has been implicated in the formation of long-term synaptic changes after learning and neuronal regeneration. It was first identified as a brain protein that became more highly labeled after goldfish learned a new pattern of swimming behavior (Shashoua, 1976, 1977a, 1979). This type of change did not occur in a variety of experiments that controlled for the effects of stress, motor activity and physical strength of the animals (Shashoua, 1977b). Injections of anti-ependymin sera into the 4th ventricle of gold-fish brain at 8-24 hr after initiation of training blocked long-term (LTM) but not short-term memory (Shashoua and Moore, 1978). A blockade of LTM was also obtained for another training procedure in which goldfish learned to avoid a shock by escaping into another compartment in a shuttle box following the onset of a light in an avoidance conditioning experiment (Piront and Schmidt, 1988). More recent studies indicate that ependymin is also involved in classical (Pavlovian) conditioning (Shashoua and Hesse, 1989a). Goldfish were trained in a soundproof light-tight chamber (see Fig. 1) to associate the onset of light as the conditioning stimulus (CS) with the delivery of a shock as the US using the CS-US parameters established by Bitterman (1964). Measurements of the concentration of ependymin in goldfish brain extracts enriched with ECF proteins (Shashoua, 1981) using an ELISA procedure demonstrated that goldfish receiving the CS-US stimuli in a "paired" presentation showed changes whereas the controls receiving the same number of stimuli "unpaired" showed no changes (see Table I) in comparison to unstimulated animals. For the "paired stimulus" group there was an initial decrease at 3.5 hr after the start of the training followed by an increase at 52 hr, a time span that corresponds to the transcription and translation of messenger RNA in goldfish brain (Shashoua, 1970). This type of result was similar to that obtained for goldfish trained to swim with a float (R. Schmidt, 1986). Thus in two separate experiments it is found that ependymin is used up by the CNS as a consequence of the training procedure. The response of the brain is to synthesize and secrete more of the protein into the brain extracellular space. Moreover the quantity of ependymin that is depleted from ECF in a 70-mg goldfish brain is substantial, being about 7 μg for the acquisition of the classical conditioning experiment.

Ependymin is also used in mammalian brain during the consolidation process of learning. In experiments in which thirsty mice were trained to find water in one arm of a T-maze (Shashoua and Moore, 1980), a protein with the same molecular weight, subcellular localization and immunoreactivity properties as ependymin was the only brain component detectable by double-labeling methods that showed increased labeling. Controls did not show changes in any fraction including the cytoplasmic, synaptosomal and nuclear (Shashoua, 1982). Also double-labeling studies of the effects of long-term potentiation (LTP) of rat brain brain hippocampal slices showed that the increased labeling of the proteins released into the incubation medium occurred after LTP.

Table 1. Ependymin Concentration Changes After Classical Conditioning

Time[a] (hr)	Paired[b]	n	Unpaired[b]	n	%Change	P-value
3.5	115±3	8	142±24	7	-19	0.05*
5	175±18	24	139±13	24	+20	<0.0005*
25.5	141±18	8	145±11 5	- 3	n.s.	
Control (unstimulated)			137±3	5		

The paired groups received 25 stimuli in which the light was presented as the CS for 10 sec followed by a 0.5-sec shock (8 V at 60 Hz) as the US during the last 0.5 sec of the CS. The unpaired group received 25 presentations of the CS (10-sec light) and shock (0.5-sec duration) delivered separately 90 sec apart in a random order.

[a] Time after initiation of training; n = number of animals tested.
* = highly significant; n.s. = not significant, in a t-test.
[b] µg ependymin/mg brain ECF protein

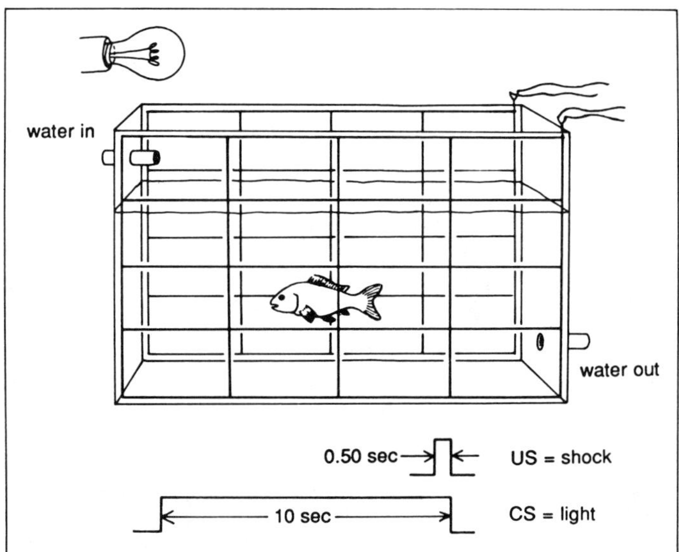

Fig. 1.   Apparatus for classical conditioning: Goldfish learn to associate light as the CS with the onset of an 8-V (60-Hz) shock as the US. The paired stimulus group received 25 CS-US paired stimuli delivered randomly at 90 or 180 seconds apart using the parameters shown above. The unpaired stimulus group, in a separate tank, received CS or US (25 each) in a random sequence at 90 seconds apart during a 90-minute period. The animals receiving the paired stimuli could recall the experience for at least 2 weeks.

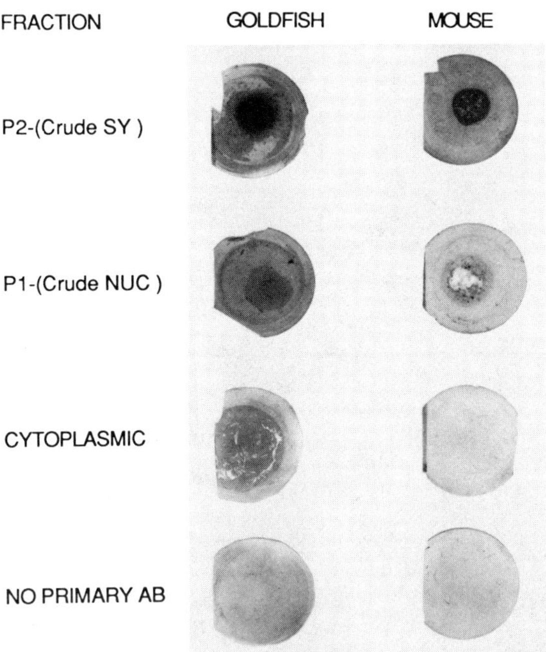

Fig. 2.   Fibrous insoluble polymer (FIP) aggregates isolated from subcellular fractions
of goldfish and mouse brain. Immunoreactivity vs. anti-ependymin IgG was
determined in a dot-blot assay. Note that the P2 fractions stain most heavily
and that there is no staining for the controls (FIP from P2 not exposed to
primary antibody).

No changes were found for the cytoplasmic or nuclear subcellular fractions. Thus three types of
behavioral experiments (swimming with a float, avoidance conditioning and classical
(Pavlovian) conditioning) in the goldfish and two experiments in mammalian brain (T-maze
learning in the mouse and LTP in rat hippocampal slices) implicate ependymin in the consolidation process of learning. How can an extracellular protein participate in such a phenomenon? Is
there a mechanism that may account for the relatively large amounts of the protein that appear
to be used up as a consequence of the learning? In an attempt to explore such questions we carried out
a detailed study of the molecular properties of ependymin.

Molecular Characteristics of Ependymin

Ependymin is a mixture of the α , $\alpha\beta$, and β_2 forms of disulfide linked dimers of two polypeptide chains (37 kD and 31 kD). In the goldfish brain about 50% of the protein is associated with
the extracellular fluid (ECF) and 50% with cytoplasmic fraction (R. Schmidt and Shashoua,
1981). Two-dimensional gel electrophoresis shows the presence of at least 7 α and 7 β subtypes of
ependymin, with isoelectric points at pH 5-5.7 indicating microheterogeneity and the possibility
of 49 types of dimers (Shashoua, 1988). Ependymin contains 10% carbohydrate as a N-aspargine
linked glycan (Shashoua et al., 1986). The microheterogeneity of the molecule persists after
cleavage of the carbohydrate fragment, indicating it is a function of the primary amino acid
sequence of the molecule and not principally due to its glycan fragments. It contains a glucuronic
acid sulfate (Shashoua et al., 1986) which confers a similar antigenicity to that reported for the
carbohydrate epitope of HNK-1, NCAM, J1, L1, MAG, and the glycolipid SGGL (Chou et al.,
1985).

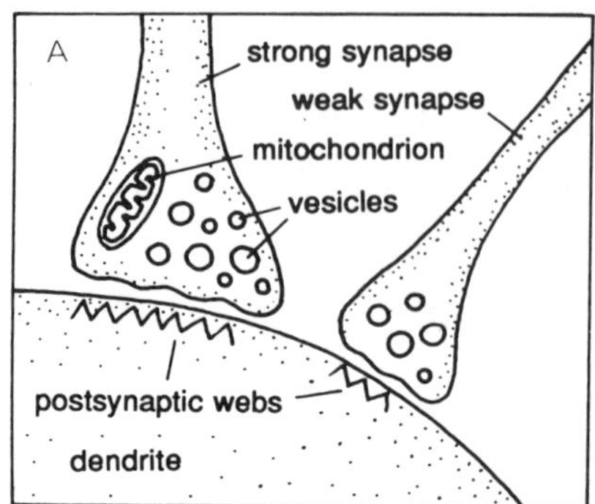

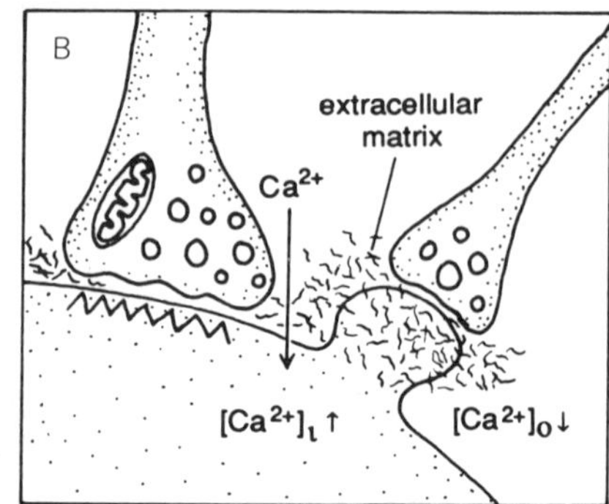

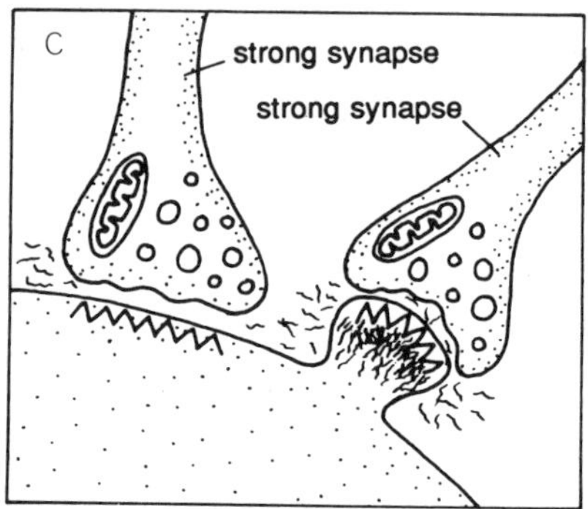

Fig. 3.   Ependymin polymerization hypothesis: In A, two inputs, a strong (large contact area) and a weak (small area of contact) converge onto a dendrite. During classical conditioning the paired stimulation of the two inputs results in a Ca++ depletion in the extracellular space and in increased Ca++ accumulation within the dendrite. In B, ependymin in the extracellular space polymerizes to form an extracellular matrix. The increased intracellular Ca++ activates proteases which break down intracellular cytoskeletons to cause the formation of a spine. Entry of ependymin in C into the spine produces polymers which form a new cyto-skeleton and the modification of the weak into a strong synapse.

Table 2 shows the amino acid sequence of one of the ependymin polypeptide chains as determined by cDNA methods (K.nigstorfer et al., 1989). We found the same partial sequence by analysis of peptides from proteolytic digests of the γ–ependymin chain of the molecule (Shashoua and Hesse, 1989b). The sequence shows no homologies to any known polypeptides; thus ependymin appears to be a unique brain macromolecule. Several characteristics of the molecule are confirmed by the sequence data. The N-terminal serine, microheterogeneity, the acidic nature of the molecule as well as the presence of an N-terminal leader sequence, which is consistent with the previous findings that show ependymin to be a secreted macromolecule (Majocha et al., 1982). A study of the phylogenetic distribution of "Ependymin-like" macromolecules shows that it is present in neural tissue of invertebrates and vertebrates including *Aplysia* and Hermissenda, goldfish, chick, mouse, rat, and human brain. Immunoprecipitation data confirm that antisera raised against the goldfish protein can precipitate mouse brain ependymin and those raised against the mouse can precipitate the goldfish brain ependymin (Shashoua et al., 1989a).

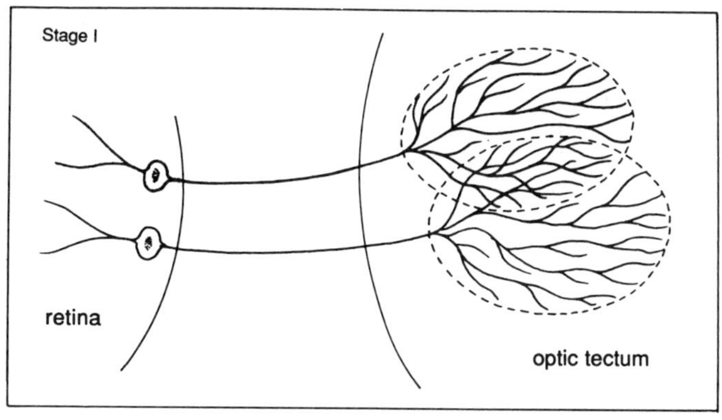

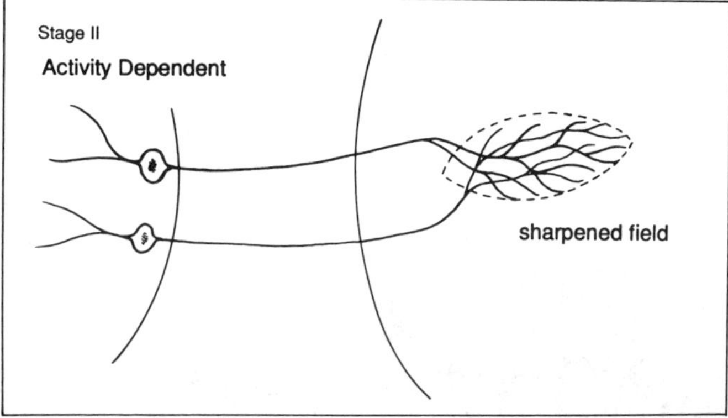

Fig. 4.   Regeneration of optic nerve axons: Stage I shows the regeneration pattern of a crushed optic nerve. A broad diffuse pattern of connections forms at the tectum. This includes overlapping regions from neighboring retinal ganglion cells. In stage II, an activitydependent process results in a preservation of the overlapping fields and an elimination of non-overlapping connections to form smaller but stable receptor fields.

Table 2. Amino Acid Sequence of γ-Ependymin

SSHRQPCHAPPLTSGTMKVVSTGGHDLESGEFSYDSKANKFRFVEDTAHA
SSHRQP HAPPLTSGTMKVVSTG   YDSKANKFRFVEDTAHA

NKTSHMDVLIHFEEGVLYEIDSKNESCKKETLQFRKHLMEIPPDATHESE
NKT   IHFEEGVLY   ETLQFR   MEIPPDATHESE

IYMGSPSITEQGLRVRVWNGKFPELHAHYSMSTTSCGCLPVSGSYHGEKK
IYMGSPSITE   VRVW   KK

DLHFSFFGVETEVDDLQVFVPPAYCEGVAFEEAPDDHSFFDLFHD
DLHFSFFGVETEVDDLQVFVPPAYCEGVAFEEAPDDHSFFDLFH

The table shows the total sequence data obtained by the cDNA method (Konigstorfer et al., 1989) and a partial sequence determined by analysis of proteolytic digests of the γ-polypeptide chain (shown underlined; Shashoua and Hesse, 198b).

Ependymin has the capacity to form fibrous aggregates in the absence of Ca++ . These polymers, once formed, could not be redissolved byraising the Ca++ concentration to normal ECF levels or even to 10 mM. The fibers were insoluble in boiling 2% SDS in 5 M urea, 100% acetic acid, saturated KCNS, CHCl /MeOH (2/1) and 100% trifluoroacetic acid. They were partially soluble in 80% formic acid; however, the polymer reprecipitated upon removal of the formic acid by dialysis or neutralization with NaOH. These properties appear to be distinct from those of other polymerizing macromolecules, such as tubulin, laminin, fibronectin and actin, which can be redissolved

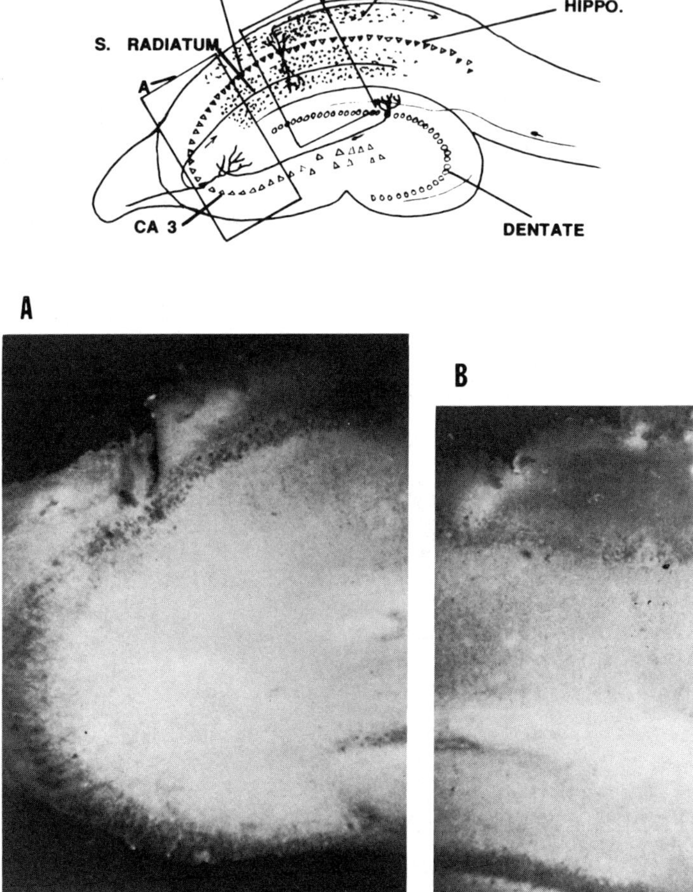

Fig. 5.   Effect of LTP on the development of anti-ependymin immunoreactive sites in rat hippocampal slices. The "dot-like" staining at S. radiatum and S. oriens occurs at the CA-1 where LTP was generated by stimulation but not at the unstimulated CA-3 region.

in 2% SDS in 5 M urea. Investigations of the mechanism of aggregation indicate that ependymin requires an activation step before its polymerization can be initiated by calcium-depleting conditions (i.e., EGTA addition) (Shashoua, 1988; Nolan and Shashoua, 1989). It appears that an ATP-dependent phosphorylation of the molecule may be an essential step in the process. These results suggest that the capacity to form fibrous aggregates may be important for its function in the CNS (Shashoua, 1985).

Evidence for the Polymerization of Ependymin *In Vivo*

We developed an assay procedure for detecting the presence of fibrous insoluble aggregates in brain tissue. The method is based on the fact that everything in the CNS should be soluble in 2% SDS in 5 M urea except ependymin polymers. Thus one can, in principle, dissolve away all brain tissue, leaving a residue of aggregates which should consist largely of polymerized ependymin fibers. Such solutions after sedimentation at 300 000 g for 3-4 hr yield pellets which, upon further purification and dialysis against distilled water, contain fibers that are retained onto Millipore filters (0.2-μ pore size). This assay showed that both goldfish and mouse brain contain about the same amount of insoluble fibrous aggregates (5-6 μg/mg brain protein) principally associated with the synaptosomal fraction (Shashoua et al., 1989c). Analysis of the immunoreactivity properties of the fibers using a dot blot type of assay indicates that the fibers readily interact with anti-ependymin IgG but not with preimmune serum or with HRP-coupled secondary antibody. Studies of the crossreactivity towards other fiber-forming proteins and components of the extracellular matrix indicate that anti-laminin, anti-fibronectin, anti-tubulin but not anti-mouse NCAM interact with ependymin. These same types of immunoreactivity properties were also obtained by fibers derived from pure ependymin, polymerized in vitro (Shashoua et al., 1989c). These cross-reactivity properties were quite surprising, since the primary amino acid sequence of ependymin did not reveal any significant homologies. However, when we restricted our search to small fragments 7 amino acids long, then homologies in segments that could function as suitable epitopes were found (Shashoua et al., 1989c). Preliminary results indicate that one such fragment can indeed be used to raise specific antisera that recognize goldfish brain ependymin (unpublished data).

A Hypothesis for the Role of Ependymin in Neuronal Function

Figure 3 depicts a proposed working hypothesis for the role of ependymin as a matrix-forming macromolecule in neuronal functions Shashoua, 1985). The diagram illustrates two pathways converging a on a single dendrite in a classical conditioning experiment. During the learning a simultaneous or a correct sequential firing of the strong ("large synaptic area") and the weak ("small contact area") results in a local depletion of the extracellular Ca^{++} in ECF and an activation of ependymin which then polymerizes to form an extracellular matrix. This process may also be accompanied by the formation of spines following the intracellular uptake of Ca^{++} and the breakdown of cytoskeletal elements by Ca-activated proteinases (Lynch and Baudry, 1984). Soluble locally activated ependymins present in ECF, upon entry into the spines, can immediately polymerize in the low intracellular Ca^{++} environment. Thus an extracellular matrix and a new stable cytoskeleton are formed (see Fig. 3C). Such structures essentially define the locus for formation of new or the modification of preexisting "weak" into "strong" synapses. The hypothesis has essentially four requirements: (1) The ECF must have a steady supply of ependymin. (2) The CNS must be able to generate a Ca^{++} local extracellular Ca^{++} depletion signal which is large (about 50%) during the learning as compared to the less than 10% that normally occurs during transmitter release functions. (3) An activation process must exist for converting ependymin into its polymerizable form. (4) The initiation Ca^{++} signal for forming an extracellular matrix (i.e., the Ca^{++} depletion signal) must be rapid, transient, and localized at specific sites. All these aspects of the hypothesis have some supporting evidence. In the goldfish the supply of ependymin in ECF is well regulated (Hesse, Hofsteinand Shashoua, 1984). Transient (lasting 2-3 min) ECF Ca^{++} depletions (up to 70%) have been demonstrated after stimulation of rat hippocampal slices (Krnjevic et al., 1982a,b) in vitro, goldfish optic tectum in vitro (Morris et al., 1986a), and cat hippocampus in vivo (Somjen, 1984). Also, studies of associative learning (Connor and Alkon, 1984) have shown that a transient Ca^{++} accumulation of Ca^{++} occurs in a post-synaptic cell of *Hermissenda*, if the two associating inputs were stimulated together but not when each was stimu-lated separately. The observed Ca^{++} influx could presumably result in a Ca^{++} depletion of extracellular Ca^{++} the local sites of the converging inputs. Recent studies of the mechanism of polymerization of ependymin (Nolan and Shashoua, 1989) suggest that phosphorylation by a kinase in the presence of ATP must occur as an activation step. Extracellular secretion of ATP following neuronal stimulation is a well known process which is estimated to generate ATP levels of over 100 μM (Gordon, 1986) at synapses. This has been proposed as a source for protein phosphorylation by an ectokinase (Ehrlich et al., 1988). Thus the required substances which control ependymin polymerization in vitro are also available in vivo and appear to be linked to effects of stimulation.

Regeneration as a Test for the Ependymin Polymerization Hypothesis

In order to test for the role of ependymin in the formation and alteration of synapses, it is essential to have a model system in which one knows when and where to look for the synaptic changes. The regenerating goldfish optic nerve provides such a model. A crushed optic nerve regenerates within a well-defined time course to arrive at its targets at the stratum opticum (Grafstein, 1969). It is a two-stage process. In stage I the regenerating optic nerve establishes a broad diffuse pattern of connections at the tectum at 20-22 days post-crush (J. Schmidt et al., 1983; Purves and Lichtman, 1980). At this stage visual stimulation of the retina produces a correct electrophysiological response, but the animals cannot see. In stage II an activity-dependent refinement process occurs at days 22 to about 36 postcrush and the goldfish regain vision and normal function. This process requires coincident, visually driven inputs that have a correlated activity at neighboring retinal ganglion cells (Ginsberg et al., 1984; J. Schmidt, 1985). Thus an initially diffuse connectivity pattern changes to a precisely organized tectal map. Figure 4 illustrates the transition process from stage I to II. The initially large multiple unit receptive field sizes (30-38°) are reduced by the elimination of inappropriate connections and the stabilization of specific synapses to give smaller receptive fields (sizes 10-12°). This apparent coincident activation of overlapping inputs from neighboring cells bears a formal resemblance to classical conditioning, and the requirements of the ependymin polymerization hypothesis (J. Schmidt and Shashoua, 1988). We therefore investigated the effects of anti-ependymin IgG on the pattern of regeneration of crushed optic nerve, using osmotic mini-pumps to continuously infuse anti-ependymin IgG into the 4th ventricle of goldfish brain. This had no effect on stage I of the regeneration process. Continuous infusion at days 22-35 post-crush (stage II), however, produced a blockade of the sharpening of the retinotectal projections. The multiple unit receptive fields, recorded by extracellular electrodes, remained large (30-32°), whereas control animals infused with pre-immune IgG, buffer or with another antibody (anti-S100 IgG) showed no effects, and no blockade of stage II occurred. The sizes of multi-unit receptor field decreased to 10.6-13.1°, with a time course comparable to untreated controls. This suggests that anti-ependymin may exert its effect by binding to the polymerized matrix that may be formed during the simultaneous stimulation of neighboring inputs. Such a blockade would presumably inhibit subsequent steps of synaptogenesis and would be consistent with the ependymin polymerization hypothesis. Support for such an interpretation is also provided by studies of the mechanism of reinnervation of the neuromuscular junction where regenerating axons have been found to seek and attach to specialized extracellular matrix sites (Sanes, 1983; McMahon et al., 1980).

LTP as a Test System for the Ependymin Polymerization Hypothesis

The hippocampal slice preparation has been widely investigated as a neurophysiological model system in studies of neuroplasticity (Bliss and Lomo, 1973; Anderson et al., 1980; Lynch and Schubert, 1980; Teyler, / 1980). It consists of a well-defined three-synaptic circuit. Afferent stimulation of the Schaffer collateral pathway to CA-1 dendrites by application of a brief tetanizing train (100 Hz for 1 sec) produces a post-synaptic field potential response that is 2to 10-fold larger and longer-lasting than that before tetanus. This long-term potentiation (LTP) of the response lasts for up to 10 hr in the in vitro preparation. Investigations of effects of LTP on the pattern of protein synthesis by the double-labeling method in our laboratory (Duffy et al., 1981) showed that the only protein changes detectable were those that were released into the extracellular medium. No changes were found in the cytoplasmic, nuclear, synaptosomal or membrane fractions. Additional experiments showed that ependymin was a major protein component of mouse and rat hippocampus (Shashoua, 1985; R. Schmidt et al., 1986). These findings raise the possibility that during LTP ependymin is used up in some process. The resultant decrease in ECF then stimulates its de novo synthesis and release into the extracellular medium. Recent studies have confirmed these findings by direct collection of the ECF-released ependymin (Hesse et al., 1984; Bliss et al., 1987; Fazeli et al., 1988; Charriaut-Malangue et al., 1988). These findings together with the reports that new synapse formation is detectable in stratum radiatum after LTP (Lee et al., 1980; Chang and Greenough, 1984) indicate that this model system may be suitable for looking for evidence of polymerized ependymin by immunocytochemical methods. Standard procedures were used to obtain LTP at CA-1 pyramidal cells in rat hippocampal slices by stimulation of Schaffer collaterals. At 1-2 hr post-tetanus, the slices which had a 2 to 3-fold potentiated response were incubated at 0°C for 3 hr to remove soluble ependymin from the

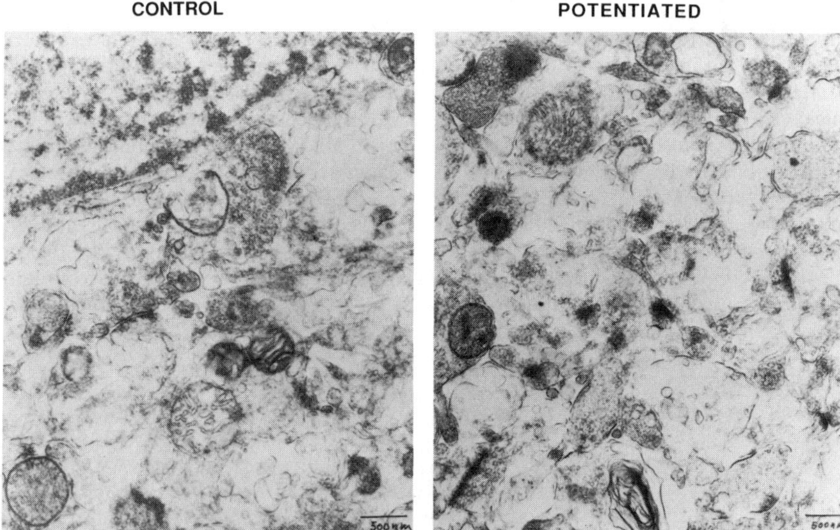

Fig. 6.   Electronmicrographs showing the immunocytochemical localization of polymerized ependymin in rat hippocampal slices after LTP. Note the difference in the features of the synaptic cleft and post-synaptic densities between the identically treated control and potentiated slices at the S. radiatum of the CA-1. Both sections were stained with anti-ependymin IgG and HRP-coupled goat anti-rabbit IgG. The sections were not treated with uranyl acetate or lead.

extracellular space of the slice (Shashoua, 1981) prior to its fixation with acrolein (King et al., 1983). This step was found to be essential for eliminating non-specific staining in the subsequent immunocytochemical localization of polymerized ependymin sites. Each slice was then cut into sequential 50μ thick sections on a vibratome, and immunohistochemical studies were carried out on the second and third sections. Figure 5 shows the pattern of staining with rabbit anti-ependymin (HIROKO) serum as visualized by the HRP reaction product with HRP-coupled goat anti-rabbit IgG. Both the control and experimental slices show the presence of intracellular ependymin in hippocampal pyramidal cells as well as in some cells of the protein in hippocampal neurons (R. Schmidt et al., 1986). The potentiated slices show the additional presence of "dot-like" staining in S. radiatum and S. oriens at the CA-1 region, which stops abruptly at CA-3 region. Thus a fine pattern of stained particles is obtained at a region where LTP was generated, i.e., CA-1, but not at a region where no stimulation was carried out (CA-3). Control unstimulated slices and potentiated slices that were not treated with primary antibody showed no such staining patterns (Shashoua et al., 1989d). These findings support the idea that ependymin polymerization occurs at the axo-dendritic regions of the slice where multiple pathways are stimulated.

Further evidence for these observations was obtained by electron microscopic studies of the slices (Shashoua et al., 1989d). Following the development of the HRP reaction product, each 50-μ slice was then fixed with glutaraldehyde and osmium and cut into thin sections for E.M. studies (Shashoua and Hesse, 1985; Shashoua et al., 1989c). Figure 6 shows the results for a control and a potentiated slice processed by an identical immunohistochemical protocol. No treatments with uranyl acetate and lead were used. The potentiated slices show the presence of an intense reaction product at local extracellular as well as post-synaptic sites. In the control slices synaptic regions appear as faintly staining regions because uranyl acetate was not used and no polymerized ependymin immunoreaction product is present. The results suggest that ependymin polymerization may have taken place at synaptic extracellular and postsynaptic regions.

Polymerization Recent studies indicate that the NMDA (N-methyl-D-aspartate) receptor has a key role in the initial events that take place during associative learning, developmental neuroplasticity, and neuronal regeneration. Detailed analyses of the electrophysiological and biochemical changes that follow activation of the NMDA receptor-channel complex by glycine andglutamate show that the influx of Ca++ (Cotman et al., 1989) causes, among other changes, a release of intracellular diacyl glycerol (DAG) inthe post-synaptic cell. The DAG and Ca++ then activate protein kinase C (PKC) and its translocation to form membrane-associated PKC (Nishizuka, 1984; Wolf et al., 1985; Murakami et al., 1986). Figure 7 shows a speculative hypothesis which proposes that the purpose of the membrane-bound PKC (i.e., the translocation process) is to deliver an enzyme to the cell surface and possibly release it to be used together with ATP to phosphorylate the ependymin present locally. The phosphorylated extracellular

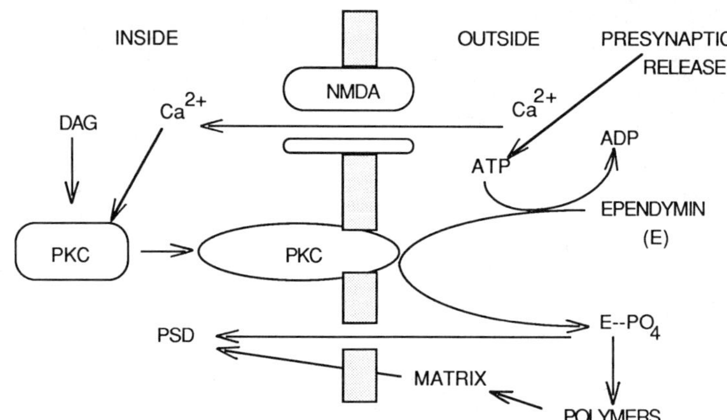

Fig. 7. A possible mechanism for a role for NMDA receptor activation and protein kinase C (PKC) translocation with the extracellular phosphorylation and polymerization of ependymin. DAG = diacyl glycerol; PSD = post-synaptic density. This mechanism proposes that PKC is translocated to membrane sites in order to function as an ectokinase which uses the ATP released by pre-synaptic stimulation to phosphorylate ependymin.

ependymin (E-P) can then polymerize to form an extracellularmatrix at the site of the NMDA activation if the extracellular Ca++ depletion is sufficiently large; in addition, E-P could enter the postsynaptic site to polymerize within the cell where calcium levels are low. The process depicted in Figure 7 provides a molecular mechanism for coupling short-term events to the process of ependymin polymerization and the generation of long-lasting synaptic changes (See Figure 3) that may occur after LTP and learning (Shashoua, 1985; Chang and Greenough, 1984; Morris et al., 1986b). Experiments investigating the validity of such a mechanism are in progress.

ACKNOWLEDGMENTS

This research was supported by a grant from the National Institutes of Health NINCDS, NS25748.

REFERENCES

Anderson, P., Sundberg, S. H., Swann, J. N., and Wigstrom, H., 1980, Possible mechanisms for
long-lasting potentiation of synaptic transmission in hippocampal slices from guinea
pigs, J. Physiol., 302:463-482.
Bitterman, M. E., 1964, Classical conditioning in the goldfish as a function of the CS-US interval,
J. Comp. Phys. Psychol., 58:359-366.
Bliss, T. V. P., Dolphin, A. C., Eryington, M. L., and Fazeli, M. S., 1987, Increases in the concentration of specific extracellular proteins during long-term potentiation in the dentate gyrus
of the rat, Proc. Physiol. Soc., 54P:Feb. 13.
Bliss, T. V. P., and Lomo, T., 1973, Long-lasting potentiation of/ synaptic transmission in the
dentate area of the anaesthetized rabbit following stimulation of the perforant path, J.
Physiol., 232L:1-356.
Chang, F. L., and Greenough, W. T., 1984, Transient and enduring morphological correlates of
synaptic activity and efficacy change in rat hippocampal slice, Brain Res., 309:35-46.
Charriaut-Malangue, C., Aniksztejn, L., Roisin, M. P., and Ben-Ari, Y., 1988, Release of proteins
during long-term potentiation in the hippocampus of anaesthetized rat, Neurosci. Let.,
91:308-314.
Chou, K. H., Ilyas, A. A., Evans, J. E., Quarles, R. H., and Jungalwala, F. B., 1985, Structure of a
glycolipid reacting with monoclonal IgM in neuropathy and with HNK-1, Biochem.
Biophys. Res. Comm., 128:383-388.
Connor, J. A., and Alkon, D. L., 1984, Lightand voltage-dependent increases of calcium ion
concentration in molluscan photoreceptors, J. Neurophysiol., 51:745-752.
Cotman, C. W., Bridges, R. J., Taube, J. S., Clarke, A. S., Geddes, J. W., and Monaghan, D. T., 1989,
The role of NMDA receptor in central nervous system plasticity and pathology, J. NIH
Res., 1:65-74.
Duffy, C., Teyler, T. J., and Shashoua, V. E., 1981, Long-term potentiation in the hippocampal
slice: Evidence for stimulated secretion of newly synthesized proteins, Science,
212:1148-1151.
Ehrlich, Y. H., Sinder, R. M., Kornecki, E., Garfield, M. G., and Lenox, R. H., 1988, Modulation of
neuronal signal transduction systems by extracellular ATP, J. Neurochem., 50:295-301.
Fazeli, M. S., Errington, M. L., Dolphin, A. C., and Bliss, T. V. P., 1988, Long-term potentiation in
the dentate gyrus of the anaesthetized rat is accompanied by an increase in protein
efflux into push-pull cannula perfusates, Brain Res., 473:51-59.
Ginsberg, K. S., Johnsen, J. A., and Levine, M. W., 1984, Common noise in the firing of neighboring
ganglion cells in goldfish retina, J. Physiol. (London), 351:433-444.
Gordon, J. L., 1986, Extracellular ATP: Effects sources and fate, Biochem. J., 233:309-319.
Grafstein, B., 1969, Changes in the morphology and amino acid incorporation of regenerating
goldfish optic neurons, Exp. Neurol., 23:544-560.
Hesse, G. W., Hofstein, R., and Shashoua, V. E., 1984, Protein release from hippocampus in vitro,
Brain Res., 305:61-66.
King, J. C., Lechan, R. M., Kugel, G., and Anthony, E. L. P., 1983, J. Histochem. Cytochem., 31:62-68.
Königstorfer, A., Sterrers, S., Eckerskorn, C., Lottspeich, F., Schmidt, R., and Hoffmann, W., 1989,
Molecular characterization of an ependymin precursor from goldfish brain, J.
Neurochem., 52:310-312.
Krnjevic, K., Morris, M. E., and Reiffenstein, R. J., 1982a, + Stimulation-evoked potentiation
changes in extracellular K andCa++ in pyramidal layers of the rat's hippocampus,
Can. J. Physiol. Pharmacol., 60:1643-1657.
Krnjevic, K., Morris, M. E., Reiffenstein, R. J. and Ropert, N., 1982b, Depth distribution and
mechanism of changes in extracellular K+ andCa++ concentrations in the hippocampus,
Can. J. Physiol. Pharmacol., 60:1658-1671.
Lee, K. S., Schottler, F., Oliver, M., and Lynch, G., 1980, Brief bursts of high frequency
stimulation produce two types of structural change in rat hippocampus, J.
Neurophysiol., 44:274-258.
Lynch, G., and Baudry, M., 1984, The biochemistry of memory: A new and specific hypothesis,
Science, 224:1057-1063.
Lynch, G., and Schubert, J., 1980, The use of in vitro brain slices for multidisciplinary studies of
synaptic functions, Annu. Rev. Neurobiol., 3:1-22.

Majocha, R. E., Schmidt, R., and Shashoua, V. E., 1982, Cultures of zona ependyma cells of goldfish brain: An immunological study of the synthesis and release of ependymins, J. Neurosci. Res., 8:331-342.

McMahon, U. J., Edgington, D. R., and Kuffler, D. P., 1980, Factors that influence regeneration of the neuromuscular junction, J. Exp. Biol., 89:31-42.

Morris, M. E., Ropert, N., and Shashoua, V. E., 1986a, Stimulus-evoked changes in extracellular calcium in optic tectum of goldfish: Possible role in neuroplasticity, Ann. N.Y. Acad. Sci., 481:375-377.

Morris, R. G. H., Anderson, E., Lynch, G. S., and Baudry, M., 1986b, Selective impairment of learning and blockade of long-term potentiation by an N-methyl-D-aspartate antagonist AP5, Nature, 319:774-766.

Murakami, K., Chan, S. Y., and Routtenberg, A., 1986, Protein kinase Cactivation by cis-fatty acid in the absence of Ca^{++} and phospholipids, J. Biol. Chem., 261:15424-15429.

Nishizuka, Y., 1984, The role of protein kinase C in cell surface signal transduction and tumour promotion, Nature, 308:693-698.

Nolan, P., and Shashoua, V. E., 1989, Mechanisms of ependymin phosphorylation in goldfish brain, Neurosci. Abstr., 15:383.35.

Piront, M. L., and Schmidt, R., 1988, Inhibition of long-term memory formation by anti-ependymin antisera after classical shock avoidance conditioning in goldfish, Brain Res., 442:53-62.

Purves, D., and Lichtman, J. W., 1980, Elimination of synapses in the developing nervous system. Science, 210:153-157.

Sanes, J. R., 1983, Roles of extracellular matrix in neural development, Annu. Rev. Physiol., 45:581-600.

Schmidt, J. T., 1985, Formation of retinotopic connections: Selective stabilization by an activity-dependent mechanism, Cell. Molec. Neurobiol., 5:65-84.

Schmidt, J. T., Edwards, D. L., and Stuermer, C. A. O., 1983, The reestablishment of synaptic transmission by regenerating optic axons in goldfish: Time course and effects of blocking activity by intraocular injection of tetrodotoxin, Brain Res., 269:15-27.

Schmidt, J. T., and Shashoua, V. E., 1988, Antibodies to ependymin block the sharpening of the regenerating retinotectal projection in goldfish, Brain Res., 446:269-284.

Schmidt, R., 1986, Biochemical participation of glycoproteins in memory consolidation after two different training paradigms in goldfish, in: "Learning and Memory: Mechanisms of Information Storage in the Nervous System, Advances in Bioscience," Vol. 59, H. Mattheis, ed., Pergamon Press, Oxford, pp. 213-222.

Schmidt, R., Loffler, F., Muller, H. W., and Seifert, W., 1986, Immunological cross-reactivity of cultured rat hippocampal neurons with goldfish brain protein synthesized during memory consolidation. Brain Res., 386, 245-257.

Schmidt, R., and Shashoua, V. E., 1981, A radioimmunoassay for ependymins ß and γ: Two goldfish brain proteins involved in behavioral plasticity, J. Neurochem., 36:1368-1377.

Shashoua, V. E., 1970, RNA metabolism in goldfish brain during acquisition of new behavioral patterns, Proc. Natl. Acad. Sci. USA, 65:160-167.

Shashoua, V. E., 1976, Brain metabolism and the acquisition of new behaviors. I. Evidence for specific changes in the pattern of protein synthesis, Brain Res., 111:347-344.

Shashoua, V. E., 1977a, Brain protein metabolism and the acquisition of new behaviors. II. Immunological studies of the α, ß and γ proteins of goldfish brain, Brain Res., 122:113-124.

Shashoua, V. E., 1977b, Brain protein metabolism and the acquisition of new patterns of behavior, Proc. Natl. Acad. Sci. USA, 74:1743-1747.

Shashoua, V. E., 1979, Brain metabolism and the acquisition of new behaviors. III. Evidence for secretion of two proteins into the brain extracellular fluid after training, Brain Res., 166:349-358.

Shashoua, V. E., 1981, Extracellular fluid proteins of goldfish brain: Studies of concentration and labeling patterns, Neurochem. Res., 6:1129-1147.

Shashoua, V. E., 1982, Molecular and cell biological aspects of learning: Towards a theory of memory, in: "Advances in Cellular Neurobiology," Vol. 3, 97-141.

Shashoua, V. E., 1985, The role of brain extracellular proteins in neuroplasticity and learning, Cell. Molec. Neurobiol., 5:183-207.

Shashoua, V. E., 1988, Monomeric and polymeric forms of ependymin: A brain extracellular glycoprotein implicated in memory consolidation processes, Neurochem. Res., 13:649-655.

Shashoua, V. E., Daniel, P. F., Moore, M. E., and Jungalwala, F. B., 1986, Demonstration of glucuronic acid on brain glycoproteins which react with HNK-1 antibody, Biochem. Biophys. Res. Comm., 138, 902-909.

Shashoua, V. E., Epstein, H., and Moore, M. E., 1989a, Enhanced synthesis of ependymin during rapid developmental periods. Submitted.

Shashoua, V. E., and Hesse, G. W., 1985, Role of brain extracellular proteins in the mechanism of long term potentiation in rat brain hippocampus, Neurosci. Abstr., 11:782.7.

Shashoua, V. E, and Hesse, G. W., 1989a, Classical conditioning leads to changes in extracellular concentrations of ependymin in goldfish brain, Brain Res., 484:333-339.

Shashoua, V. E., and Hesse G. W., 1989b, Determination of amino acid sequence of ependymin by analysis of proteolytic digests. Submitted.

Shashoua, V. E., Hesse, G. W., and Milinazzo, B., 1989c, Ependymin and neural function: Evidence for its polymerization in vivo. Submitted.

Shashoua, V. E., Hesse, G. W., and Paskevich, P., 1989b, Localization of ependymin in potentiated rat brain hippocampal slices. Submitted.

Shashoua, V. E., and Moore, M. E., 1978, Effect of antisera to ß and goldfish brain proteins on the retention of a newly acquired behavior, Brain Res., 148:441-449.

Shashoua, V. E., and Moore, M. E., 1980, Enhanced labeling of ECF proteins in mouse brain after training, Neurosci. Abstr., 6:290.4

Somjen, G. G., 1984, Acidification of interstitial fluid in hippocampal formation caused by seizures and by spreading depression, Brain Res., 311:186-188.

Teyler, T. J., 1980, Brain slice preparation: Hippocampus, Brain Res. Bull., 5:391-403.

Wolf, M., LeVine, H., III, Stratford, M., Cuatrecases, P., and Sahyoun, N., 1985, A model for intracellular translocation of protein kinase Cinvolving synergism between Ca++ and phorbol esters, Nature, 317:546-549.

THE ROLE OF PROTEIN KINASE C SUBSTRATE B-50 (GAP-43) IN NEUROTRANSMITTER

RELEASE AND LONG-TERM POTENTIATION

P.N.E. De Graan, L.H. Schrama, F.M.J. Heemskerk, L.V. Dekker, and
W.H. Gispen

Division of Molecular Neurobiology, Rudolf Magnus Institute, Laboratory for
Physiological Chemistry, and Institute of Molecular Biology and Medical
Biotechnology, Padualaan 8, 3584 CH Utrecht, NL

INTRODUCTION

Long-term potentiation (LTP) is a form of synaptic plasticity, which may be one of the events
underlying learning and memory. LTP is triggered by brief, high-frequency stimulation of afferents,
resulting in a long-lasting increase in the effectiveness of synaptic transmission (Bliss and Lomo,
1973; Bliss and Lynch, 1988; Brown et al., 1988; Matthies, 1989). Traditionally, LTP has been divided into two phases, the initiation phase and the maintenance phase. At present, three phases
can be distinguished: (i) an initiation phase, including several seconds after tetanic stimulation
when the events that trigger LTP begin; (ii) a transient phase, lasting about 30 min, during which
there is a slow decay of potentiation, which can be evoked by local transmitter application without stimulating the presynaptic terminal; and (iii) a maintenance phase, which can last for hours
(Kauer et al., 1988; Malinow et al., 1988a).

The molecular mechanisms underlying these three phases of LTP are only partly known. Most
synapses that exhibit LTP use the excitatory amino acid neurotransmitter glutamate, which acts
on two types of glutamate receptors, e.g. N-methyl-D-aspartate (NMDA) and quisqualate/ kainate. A transient activation of the NMDA receptor system, resulting in an influx of Ca^{2+} through the
receptor-linked channel, is essential for the initiation of LTP, whereas recent evidence points to a
role of quisqualate receptors during the transient phase of LTP (Davies et al., 1989). A major issue
is whether the molecular events during the different phases involve the pre-and/or the postsynaptic element. Recent evidence from several laboratories indicates that both elements are involved
and that cross-talk between the pre- and postsynaptic elements is a prerequisite for the development of LTP (Bliss and Lynch, 1988; Malinow et al., 1988a; Davies et al., 1989; Stevens, 1989;
Malinow et al., 1989). Stevens (1989) proposes that 4 interrelated signals are essential for LTP: 1.
an increase in glutamate release from the presynaptic terminals, 2. postsynaptic action of glutamate, resulting in an increased intracellular Ca^{2+} concentration $[Ca^{2+}]_i$, 3. an unidentified retrograde
message from the post- to the presynaptic element, augmenting glutamate release (probably by
increasing the presynaptic $[Ca^{2+}]_i$), and 4) a signal to the postsynaptic element inducing increased
sensitivity to glutamate (through kainate/quisqualate receptors). Thus, an increase in $[Ca^{2+}]_i$ in
both the pre- and post-synaptic element appears to be essential for LTP.

Three mechanisms for LTP invoking the long-term activation of Ca^{2+}-sensitive enzymes have
been proposed. Two of these hypotheses propose that protein kinases are involved, the third one,
which will not be discussed in this paper, concerns unmasking of glutamate receptors by the Ca^{2+}
activated protease calpain (Lynch and Baudry, 1984), an hypothesis which finds new support in
the recent findings showing that the expression of LTP is associated with an enhanced transmission mediated by kainate/quisqualate channels (Davies et al., 1989). Protein kinases are enzymes
which catalyze the transfer of phosphate from ATP to proteins (e.g. ion channels, receptors), the-

reby reversibly modifying their tertiary structure and hence their biological activity. The kinases
which have been implicated in the mechanism of LTP are the type II Ca^{2+}/calmodulin-dependent
kinase (CaMKII) and Ca^{2+}/phospholipid-dependent protein kinase C (PKC). CaMKII is a major
constituent of the post-synaptic density and is presumably activated by NMDA receptor-mediated
Ca^{2+} influx. Long-term activation of the enzyme has been observed *in vitro* through a multistep
mechanism involving autophosphorylation of the enzyme (Miller and Kennedy, 1986; Lisman and
Goldring, 1988; for a review see Schwartz and Greenberg, 1987). Recently, two laboratories have
independently provided strong evidence for the involvement of CaMKII in postsynaptic mechanisms underlying LTP by microinjecting synthetic peptides which interfere with CaMKII activity
in postsynaptic CA1 pyramidal cells. Malinow *et al.* (1989) injected a synthetic peptide (CaMKII
residues 273-302), containing the autoinhibitory domain of the kinase, which does not inhibit PKC
and does not compete with calmodulin. This peptide blocked the induction of LTP, but was ineffective with respect to established LTP. Malenka *et al.* (1989) injected synthetic peptides that are
potent calmodulin antagonists and inhibit CaMKII auto- and substrate phosphorylation. Their
findings demonstrate that postsynaptic activation of calmodulin and CaMKII are required for the
generation of LTP. These studies do not address a possible presynaptic role for CaMKII in LTP.

PKC has been implicated in LTP, because: 1. artificial activators of PKC, phorbol esters,
evoke LTP-like phenomena (Malenka *et al.*, 1986; Muller *et al.*, 1988), 2. PKC inhibitors block LTP
(Reymann *et al.*, 1988; Malinow *et al.*, 1988; Malinow *et al.*, 1989; Malenka *et al.*, 1989), 3. LTP is
parallelled by a translocation of PKC activity (Akers *et al.*, 1986), 4. PKC injection into hippocampal pyramidal cells elicits LTP-like phenomena (Hu *et al.*, 1987), and 5. increased phosphorylation of PKC substrates has been reported after tetanization (Lovinger *et al.*, 1985; Schrama *et al.*,
1986; Nelson *et al.*, 1989). Although these studies provide strong evidence for a role of PKC in LTP,
most of them either deal with postsynaptic mechanisms underlying LTP or do not distinguish between a pre- or postsynaptic involvement of PKC.

In this paper we will focus on the role of presynaptic PKC in LTP. Since it is not possible to separate pre- and postsynaptic PKC biochemically, we have developed a technique (De Graan *et
al.*, 1989) to monitor presynaptic PKC activity *in situ* by measuring the degree of phosphorylation
of a well-characterized presynaptic PKC substrate, the B-50 protein (Zwiers et al., 1980), also
known as GAP-43 and F1 (Benowitz and Routtenberg, 1987; Skene, 1989). B-50 is a nervous-tissue
specific protein (Kristjansson *et al.*, 1982), which is predominantly localized in the presynaptic
membrane (Sörensen *et al.*, 1981; Gispen *et al.*, 1985; Van Lookeren Campagne *et al.*, 1989a,b). The
protein is associated with the inner face of the membrane (Gispen *et al.*, 1985; Van Hooff *et al.*,
1989a) probably anchored by a thio-ester linkage to palmitate (Skene and Virag, 1989). B-50 is an
endogenous substrate of PKC (Aloyo *et al.*, 1983, Eichberg *et al.*, 1986; De Graan *et al.*, 1988a, 1989;
Van Hooff *et al.*, 1989b). It has been suggested that B-50 plays a role in the regulation of polyphosphoinositide metabolism (Gispen *et al.*, 1986; Van Hooff *et al.*, 1988) and/or calmodulin binding in
the presynaptic terminal (Andreasen *et al.*, 1983). Furthermore, the protein has been implicated in
the process of neurite outgrowth and in LTP (for reviews see Benowitz and Routtenberg, 1987; Skene,
1989). In this paper we summarize our evidence for a role of PKC-mediated B-50 phosphorylation
in neurotransmitter release, obtained in recent studies using K^+, phorbol esters and 4-aminopyridine
(4-AP) as secretagogues. In the second part we report on experiments designed to investigate long-
term changes in presynaptic PKC activity during LTP-like phenomena.

B-50 Phosphorylation and neurotransmitter release

A large number of studies provide evidence for the involvement of PKC in transmitter release. This evidence has mainly been obtained by studying the effects of agonists and antagonists of
PKC in cultured cells, tissue slices or synaptosomes on release of radiolabelled transmitter. Phorbol
esters, which are known to activate PKC, stimulate neurotransmitter release (Zurgil and Zisapel,
1985; Allgaier *et al.*, 1986; Versteeg and Florijn, 1986) and depolarization-induced neurotransmitter release can be attenuated by inhibitors of PKC or by down regulation of PKC by phorbol esters
(Allgaier and Hertting, 1986; Matthies *et al.*, 1987; Versteeg and Ulenkate, 1987; Bartmann *et al.*,
1989).

Since B-50 is a major substrate of PKC in the presynaptic membrane, we considered the possibility that PKC-mediated B-50 phosphorylation is essential in transmitter release. Indeed, in hippocampal slices a close correlation could be demonstrated between *in situ* B-50 phosphorylation

and neurotransmitter release (Dekker *et al.*, 1989a). K+ depolarization, phorbol esters and the combination of both concomitantly enhanced B-50 phosphorylation and release of D-aspartate. This increase in both parameters was dependent on extracellular Ca^{2+} and did not occur in the presence of the PKC inhibitor polymyxin B (Dekker *et al.*, 1989a). Subsequently, the mechanism underlying the depolarization-induced changes of B-50 phosphorylation was investigated in [32P]-labelled rat cortical synaptosomes (Dekker *et al.*, 1989b). In this preparation K+ depolarization induced a rapid transient increase in B-50 phosphorylation. The K+-induced increase in B-50 phosphorylation could be inhibited by several inhibitors of PKC. The trigger for the increase in B-50 phosphorylation seems to be an influx of Ca^{2+}, since extracellular Ca^{2+} is required and the effect of depolarization can partially be mimicked by the Ca^{2+}-ionophore A23187. Interestingly, as in hippocampal slices, the time course of the B-50 phosphorylation in synaptosomes closely parallels the K+-induced

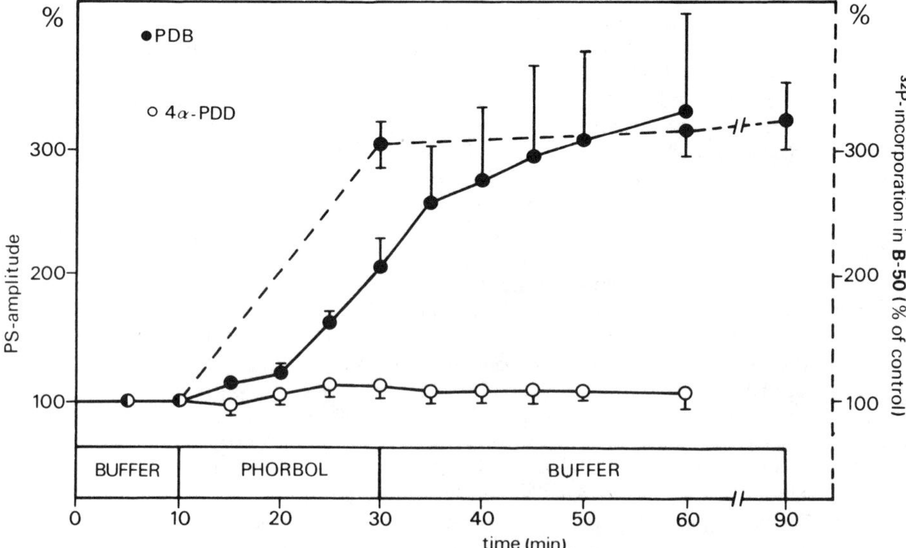

Fig. 1.   Long-lasting effects of a 20 min treatment of hippocampal slices with 4ß-phorbol 12,13-dibutyrate (PDB) on population spike (PS) amplitude and B-50 phosphorylation. After treatment with 10-6 M PDB (●) or 4 -PDB (o) slices were extensively washed or perfused with control buffer. Field potentials were recorded in the stratum pyramidale of CA1. The average PS amplitude (solid lines, mean ± SEM, n = 4) is expressed as % of the PS during the 10 min control period (pior to the phorbol treatment). *In situ* B-50 phosphorylation (broken lines, mean ± SEM, n = 4) was measured by quantitative immunoprecipitation from 32P-orthophosphate prelabeled slices, and is expressed as % of the value obtained for untreated control slices at each time point.

transmitter release (Cotman *et al.*, 1976; Nicholls *et al.*, 1987). K+ depolarization induced a concentration-dependent increase in noradrenaline (NA) release, which was Ca^{2+} dependent, and could be enhanced by phorbol esters. Phorbol esters did not stimulate release without depolarization. K+-induced NA release could be inhibited by the kinase inhibitors polymyxin B and staurosporin.

Transmitter release can also be induced by the convulsant drug 4-aminopyridine (4-AP). 4-AP stimulates release in a variety of brain regions *in vitro* and *in vivo* (Damsma *et al.*, 1988; Dole al and Tuçek, 1983; Tapia and Sitges, 1982; Tapia *et al.*, 1985; Tibbs *et al.*, 1989). Electrophysiological experiments have shown that 4-AP affects neuronal excitability by blocking fast-activating voltage-dependent K+ channels associated with Ca^{2+} influx (Thompson, 1982; Gustafsson *et al.*, 1982; Rogawski and Barker, 1983). *In situ* phosphorylation of B-50 in hippocampal slices was stimulated by 10-4 M 4-AP within 1 min and was parallelled by an increase in the release of NA (Heemskerk *et al.*, 1989a). The stimulation of B-50 phosphorylation by 4-AP was dependent on Na+ channel activity, since it could be blocked by tetrodotoxin (TTX; Heemskerk *et al.*, 1989b).

Similar results were obtained with 4-AP in synaptosomes, although the magnitude of the effect was smaller than in hippocampal slices (Heemskerk *et al.*, 1989c). A comparison between K+- and 4-AP-induced B-50 phosphorylation in synaptosomes reveals marked differences with respect to onset, durability and magnitude, 4-AP being the slower and longer acting of the two. The increase in B-50 phosphorylation induced by 4-AP seemed to be dependent on the state of depolarization, since the effect of 4-AP was largest under non-depolarizing conditions. Both the effects of 4-AP and K+ depolarization were abolished at low extracellular Ca^{2+} conditions ($< 10^{-7}$ M; Heemskerk *et al.*, 1989d).

4-AP enhances $^{45}Ca^{2+}$ entry either directly (Agoston *et al.*, 1983) or indirectly by blocking K+ channels (Segal *et al.*, 1984). This enhanced Ca^{2+} entry is observed within seconds after the addition of 4-AP as determined with the fluorescent dye Fura-2 (Gibson and Manger, 1988; Nicholls *et al.*, 1989; Heemskerk *et al.*, 1989c). Comparison of the effects of 4-AP and K+ depolarization on $[Ca^{2+}]_i$ measured with Fura-2 showed that K+ depolarization gives rise to a transient increase in $[Ca^{2+}]_i$ followed by a sustained elevation of $[Ca^{2+}]_i$, whereas 4-AP induced a rapid, sustained increase of Ca^{2+}. The effects of 4-AP and K+ depolarization on $[Ca^{2+}]_i$ were additive, again suggesting a different mechanism of action. The increase in $[Ca^{2+}]_i$ by 4-AP was found to be sensitive to TTX, whereas the K+-induced rise was not dependent on Na+ channel activity (Heemskerk *et al.*, unpublished results). The results suggest that the stimulation of B-50 phosphorylation by 4-AP is the result of the influx of Ca^{2+} through voltage-sensitive calciu channels (VSCC) by increased Na+ channel activity. High K+ exerts its effect on B-50 phosphorylation through a direct opening of VSCC, by depolarization the membrane below the Ca^{2+} threshold, a phenomenon not dependent on Na+ channel activity.

In the next series of experiments a more direct approach was used to investigate a causal relationship between the PKC substrate B-50 and NA release (Dekker *et al.*, 1989c). Anti-B-50 IgGs, which interfere with B-50 phosphorylation, were introduced into synaptosomes and their effect on Ca^{2+}-induced NA was investigated. Synaptosomes were permeabilized with streptolysin-O (SL-O), and K+-induced NA release was mimicked by increasing the Ca^{2+} concentration from 10^{-7} to 10^{-5} M. The NA is not released from a cytosolic pool, but from a vesicular pool, as Ca^{2+} did not induce efflux of the cytosolic marker protein lactate dehydrogenase. In SL-O permeabilized synaptosomes B-50 phosphorylation was complelety inhibited by anti-B-50 IgGs, whereas control IgGs were without effect. Under similar conditions Ca^{2+}-induced NA release was dose-dependently inhibited by anti-B-50 IgGs, whereas control IgGs and heat-inactivated IgGs were ineffective. Non of the IgGs tested affected basal NA release, showing that they did not affect Ca^{2+}-independent NA efflux. Thus, we have established a causal relationship between B-50 and NA release in synaptosomes (Dekker *et al.*, 1989).

In summary, three lines of evidence suggest that PKC-mediated B-50 phosphorylation is an essential step in the mechanism of NA release: 1. activation of PKC by phorbol esters induces the phosphorylation of B-50 and the release of a variety of transmitters; 2. depolarization and 4-AP-induced transmitter release from synaptosomes and hippocampal slices is closely correlated with the degree of B-50 phosphorylation; 3. anti-B-50 IgGs, which inhibit B-50 phosphorylation, completely block Ca^{2+}-induced release of NA from permeabilized synaptosomes. Since immunocytochemical studies have shown that B-50 is localized in synapses throughout the brain and does not appear to be colocalized with a particular neurotransmitter (Gispen *et al.*, 1985; Benowitz *et al.*, 1988), B-50 phosphorylation may be more generally involved in the mechanism of transmitter release, for instance of aminoacid transmitters. Indeed, there is a close correlation between B-50 phosphorylation and D-aspartate release from hippocampal slices (Dekker *et al.*, 1989a). This is an interesting possibility in view of the long-term changes in B-50 phosphorylation (Lovinger *et al.*, 1985,1986; Schrama *et al.*, 1986) and glutamate release (Dolphin *et al.*, 1982), which have been reported to occur during LTP.

Long-term changes in B-50 phosphorylation

To investigate a role of B-50 phosphorylation in LTP Routtenberg cf. (Routtenberg, 1985; Lovinger *et al.*, 1985; 1986; Linden *et al.*, 1988) elicited LTP by tetanization *in vivo* and subsequently measured B-50 phosphorylation in the homogenate of the extirpated tetanized brain region

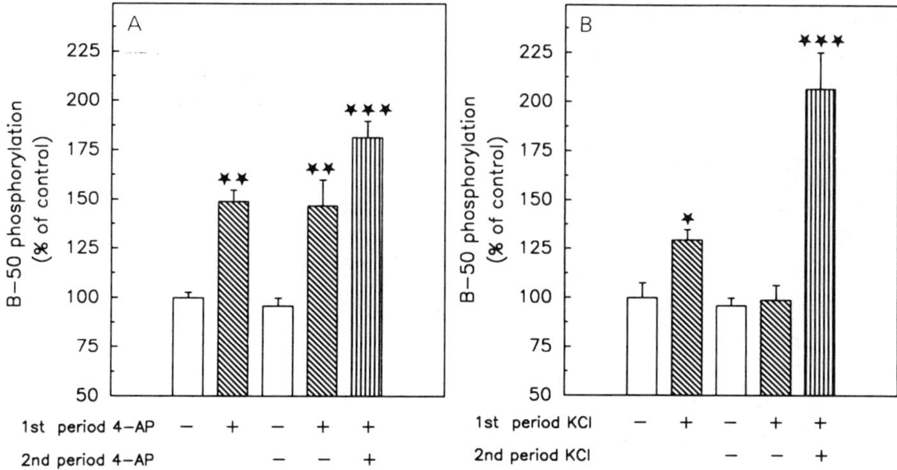

Fig. 2.   Effects of brief and prolonged incubation of hippocampal slices with 4-
aminopyridine (4-AP) or K+ on in situ B-50 phosphorylation.
A: Effects of 4-AP. Slices were treated for a first (5 min) and a second (25 min)
period with (hatched bars) or without (open bars) 10-4 M 4-AP. At the end
of the first period slices were extensively washed in buffer with (+) or
without (-) 4-AP.
B: Effects of K+. Slices were treated for a first (10 min) and a second (30 min)
period with high K+ buffer (30 mM K+, hatched bars) or control buffer (5
mM K+, open bars).
Significant differences are indicated: * 2p < 0.05; ** 2p <0.01; *** 2p < 0.001.

using [-^{32}P]ATP. In our initial studies, we have used a similar "posthoc" approach to monitor B-50 phosphorylation in synaptosomal fractions prepared from tetanized hippocampal slices (Schrama et al., 1986). In these studies precautions were taken to preserve the *in vivo* kinase activity as much as possible, but since the tissue is homogenized before the phosphorylation assay, the "post hoc" assay does not distinguish, between pre- and postsynaptic or glial PKC. Therefore we developed an *in situ* phosphorylation assay, which enables the measurement of B-50 phosphorylation in intact hippocampal slices (De Graan et al., 1989). Since tetanization of hippocampal slices only elicits LTP in a small population of synapses, we preferred to use phorbol esters or 4-AP, which induce a more generalized long-term enhancement of synaptic transmission.

A relatively short treatment of hippocampal slices with active phorbol esters elicits long-term synaptic enhancement which resembles that after tetanization (Malenka et al., 1986). Although there is some discussion about the nature and the properties (Gustafsson et al., 1988), phorbol- and tetanus-induced LTP are thought to share at least some molecular mechanisms. Application of the active phorbol ester PDB to hippocampal slices for 20 min followed by extensive washing with control buffer gave rise to both a sustained increase in the amplitude of the population spike and in B-50 phosphorylation, whereas the non-active phorbol ester 4 -PDD had no effect on either of the two parameters (Fig. 1; De Graan et al., 1988b). B-50 phosphorylation remained elevated for at least 60 min, despite continuous washing of the slices with control buffer, which removed more than 99% of the PDB (as measured with [^{3}H]PDB). This sustained increase in B-50 phosphorylation is most likely due to translocation of PKC from the cytosol to the membrane, a phenomenon which has also been reported to occur following tetanization (Akers et al., 1986).

After a brief incubation of slices with 4-AP (5 min), which induced a marked increase in B-50 phosphorylation (Fig. 2A), extensive washing and continued incubation in control medium for 25 min did not reduce B-50 phosphorylation. This increase in B-50 phosphorylation is smaller than that obtained when 4-AP is also present throughout the 25 min period used for the above

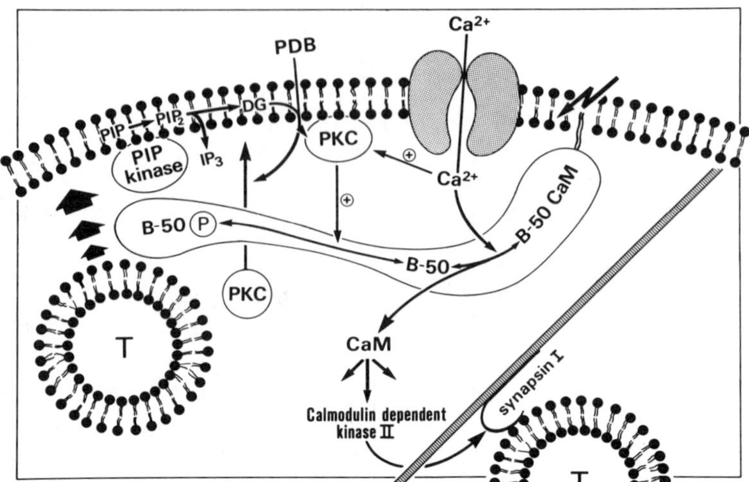

Fig. 3.   Model summerizing our working hypothesis concerning the role of B-50 (phosphorylation) in the regulation of neurotransmitter release. For explanation and abbreviations see text. (T) transmitter vesicle; (CaM) calmodulin.

mentioned washout procedure (Fig. 2A). Using a similar experimental protocol we measured the effect of brief 4-AP treatment on NA release. A 5 min incubation of slices with 4-AP increased the NA release to 7-fold the control after 5 min, and the release remained significantly increased until 20 min after removal of 4-AP from the superfusion medium (Heemskerk *et al.*, 1989a). To check the reversibility of the B-50 phosphorylation reaction in slices, we incubated slices for 10 min in depolarizing buffer (30 mM K+), followed by a washout period of 30 min in control medium. Depolarizing buffer enhanced B-50 phosphorylation and the effect was completely reversible within 25 min (Fig. 2B). Continued incubation in high K+ buffer further enhanced B-50 phosphorylation compared to the effect observed after 10 min depolarization (Fig. 2B).

Both active phorbol esters and 4-AP have been reported to stimulate the degree of tetanus-induced potentiation and both agents seem to be able to activate PKC and neurotransmitter release for periods extending their presence, indicating that they both induce a long-term activation of PKC (Routtenberg *et al.*, 1986; Haas and Greene, 1985; Lee *et al.*, 1986). Phorbol esters might exert their effect directly by translocating and activating PKC, and 4-AP probably facilitates LTP by a presynaptic calcium influx, which possibly results in activation of PKC.

Concluding remarks

Most of the evidence for the involvement of protein kinases in LTP points to a postsynaptic locus. However, as indicated in the introduction of this paper, in most of the studies only postsynaptic involvement could be investigated due to technical limitations (microinjection in the postsynaptic cell). The evidence for a role of presynaptic PKC in LTP can be summarized as follows: 1. the kinase inhibitor H-7 blocks expression of LTP when bath applied. However, H-7 does not inhibit enhanced synaptic transmission when injected postsynaptically after LTP has been established (Malinow *et al.*, 1989); 2. expression of LTP is mimicked by PKC activating phorbolesters (Malenka *et al.*, 1986), agents known to stimulate the release of various neurotransmitters (Zurgil and Zisapel, 1985; Kaczmarek, 1987); 3. selective presynaptic application of phorbol esters induces LTP-like phenomena (Malinow *et al.*, 1988b); 4. LTP is thought to involve an increase in transmitter release (Skrede and Malthe-Sorenssen, 1981; Dolphin *et al.*, 1982; Malenka *et al.*, 1987), a process which can be inhibited by various kinase inhibitors (Allgaier and Hertting, 1986; Versteeg and Ulenkate, 1987; Bartmann *et al.*, 1989), and by down-regulation of PKC activity in PC12 cells and sympathetic neurons (Matthies *et al.*, 1987); 5. phorbol ester-induced LTP is parallelled by a long-lasting increase in the phosphorylation of the nervous-tissue specific presynaptic PKC substrate B-50 (De Graan *et al.*, 1988b).

If presynaptic PKC is involved in LTP, B-50 is an interesting candidate to mediate the effect of sustained PKC activation, as we have provided evidence that B-50 is involved in the regulation of transmitter release. Figure 3 summarizes our working hypothesis concerning the role of B-50 in neurotransmitter release. In this model we try to describe the relationship between the different properties of B-50, which have been investigated more or less independently in different biological and *in vitro* systems. These properties are the affinity of B-50 for calmodulin at low Ca^{2+} concentrations (Andreasen *et al.*, 1983; Cimler *et al.*, 1987), the phosphorylation by PKC at residue 41 (Schrama *et al.*, 1988; Coggins and Zwiers, 1988), the modulatory effect of B-50 phosphorylation on phosphatydylinositol 4-phosphate kinase (PIP kinase) activity (Van Dongen *et al.*, 1985) and the involvement of B-50 in the mechanism of neurotransmitter release (Dekker at al., 1989c).

B-50 has calmodulin-binding properties as it is identical to the calmodulin-binding protein P-57 (Cimler *et al.*, 1987). This protein binds calmodulin *in vitro* under low calcium conditions and releases calmodulin when Ca^{2+} is raised (Andreasen *et al.*, 1983). Prephosphorylation of B-50 by PKC interferes with its calmodulin-binding capacities *in vitro* (Alexander *et al.*, 1987). These data are consistent with a model in which a depolarization-induced rise in intracellular Ca^{2+} *in vivo* would dissociate calmodulin and B-50, thus locally increasing the free calmodulin concentration (Fig.3). B-50, once liberated from calmodulin can be phosphorylated by PKC. PKC may be activated by Ca^{2+} entering the synaptic terminal after the depolarization. It is not clear whether such activation should involve the translocation of PKC from the cytosol to the membrane or whether Ca^{2+} directly activates membrane-bound (subtypes of) PKC. Alternatively, Ca^{2+} may activate phospholipase C generating DG which could induce activation of PKC (Fig. 3). PKC-mediated B-50 phosphorylation would subsequently facilitate fusion of transmitter vesicles with the presynaptic membrane through an as yet unresolved mechanism. PKC-mediated phosphorylation of B-50 could also prevent calmodulin from reassociating with B-50 when the local Ca^{2+} concentration is back to resting levels, thus prolonging the availability of calmodulin. This prolonged increase in calmodulin could activate a number of cellular pathways several of which are of importance to the release processes. For instance, calmodulin could activate CaMKII (Fig. 3), the kinase which phosphorylates synapsin I, a vesicle-associated protein, which is thought to link synaptic vesicle to each other and to the cytoskeleton (Benfenati *et al.*, 1989). Phosphorylation of Synapsin I would decrease its affinity for synaptic vesicles and cytoskeletal elements in the nerve terminal, thereby increasing the number of vesicles available for fusion with the plasma membrane (Benfenati *et al.*, 1989). Interestingly, calmodulin may also activate the phosphatase calcineurin, which has recently been shown to dephosphorylate B-50 (Liu and Storm, 1989; Schrama *et al.*, 1989). Calcineurin inhibits the activity of voltage-dependent Ca^{2+} channels (Armstrong, 1989) and may thus be involved in the down regulation of the Ca^{2+} signal after depolarization.

The degree of phosphorylation of B-50 has been reported to modulate the activity of the PIP-kinase (Fig. 3), the lipid kinase that phosphorylates PIP to PIP_2 (Van Dongen *et al.*, 1985). The availability of PIP_2 could be an important factor in the regulation of transmitter release processes. In the first place because depolarization-induced breakdown of PIP_2 by phospholipase C could generate DG, which can activate PKC. Secondly, PIP_2 itself may be an activator of PKC (O'Brian *et al.*, 1987). Thirdly, the amount of PIP_2 may regulate the activity of profilin and gelsolin, two proteins involved in the polymerization of G-actin, a process important for the regulation of the rigidity of the cytoskeleton and thus for vesicle mobility (Linstedt and Kelly, 1987; Forscher, 1989). Thus, regulation of the amount of PIP_2 through the phosphorylation state of B-50 could be of importance for the regulation of transmitter release.

Finally, yet unknown properties of B-50 (phosphorylation) could be involved in the regulation of transmitter release in the presynaptic nerve terminal. An intriguing, unexploited part of the B-50 molecule is the C-terminal domain, to which no specific function has been addressed yet. It is tempting to speculate that this area may direct the fusion of vesicles to the synaptic plasma membrane either by acting as a kind of anchoring molecule to synaptic vesicles or by acting as regulator of the fusion event itself.

REFERENCES

Akers, R.F., Lovinger, D.M., Colley, P.A., Linden, D.J. and Routtenberg, A., 1986, Translocation of protein kinase C activity may mediate hippocampal long-term potentiation, **Science** 231:587-589.

Alexander, K.A., Cimler, B.M., Meier, K.E. and Storm, D.R., 1987, Regulation of calmodulin binding to P-57, **J.Biol.Chem.** 262:6108-6113.

Allgaier, C. and Hertting, G., 1986, Polymyxin B, a selective inhibitor of protein kinase C, diminishes the release of noradrenaline and the enhancement of release caused by phorbol 12,13-dibutyrate, **Naunyn-Schmiedeberg's Arch. Pharmacol.** 334:218-221.

Allgaier, C., Von Kügelgen, O. and Hertting, G., 1986, Enhancement of noradrenaline release by 12-O-tetradecanoyl phorbol-13-acetate, an activator of protein kinase C. **Eur. J. Pharmacol.** 129:389-392.

Aloyo, V.J., Zwiers, H. and Gispen, W.H., 1980, Phosphorylation of B-50 by calcium-activated, phospholipid-dependent protein kinase and B-50 kinase, **J.Neurochem.** 41:649-653.

Andreasen, T.J., Luetje, C.W., Heideman, W. and Storm, D.R., 1983, Purification of a novel calmodulin binding protein from bovine cerebral cortex membranes, **Biochemistry** 22:4615-4618.

Bartmann, P., Jackisch, R., Hertting, G. and Allgaier, C., 1989, A role for protein kinase C in the electrically evoked release of [3H] -aminobutyric acid in rabbit caudate nucleus, **Naunyn-Schmiedeberg's Arch. Pharmacol.** 339:302-305.

Benfenati, F., Bahler, M., Jahn, R. and Greengard, P., 1989, Interactions of synaptsin I with small synaptic vesicles: distinct sites in synapsin I bind to vesicle phospholipids and vesicle proteins, **J.Cell Biol.**108: 1863-1872.

Benowitz, L.I. and Routtenberg, A., 1987, A membrane phosphoprotein associated with neural development, axonal regeneration, phospholipid metabolism and synaptic plasticity, **Trends Neurosci.** 10:527-532.

Benowitz, L.I., Apostolides, P.J., Perrone-Bizzozero, N., Finklestein, S.P. and Zwiers, H., 1988, Anatomical distribution of the growth-associated protein GAP-43/B-50 in the adult rat brain, **J.Neurosci.** 8:339-352.

Bliss, T.V.P. and Lomo, T., 1973, Long-lasting potentiation of synaptic transmission in the dendate area of the anaesthesized rabbit following stimulation of the perforant path. **J. Physiol.** (Lond.) 232:331-356.

Bliss, T.V.P. and Lynch, M.A., 1988, Long-term potentiation of synaptic transmission in the hippocampus: properties and mechanisms, **In:** Long-Term Potentiation: from Biophysics to Behaviour (eds. P.W. Landfield and S.A. Deadwyler), A. Liss, New York, pp. 3-72.

Brown, T.H., Chapman, P.F., Kairiss, E.W. and Keenan, C.L., 1988, Long-term synaptic potentiation, **Science** 242:724-728.

Cimler, B.M., Giebelhaus, D.H., Wakim, B.T., Storm, D.R. and Moon, R.T., 1987, Characterization of murine cDNAs encoding P-57, a neural-specific calmodulin-binding protein, **J.Biol.Chem.** 262:12158-12163.

Coggins, P.J. and Zwiers, H., 1988, Evidence for a single phosphorylation site in neuronal protein B-50, **Soc.Neurosci.Abstr.**14:452.7.

Cotman, C.W., Haycock, J.W. and White, W.F., 1976, Stimulus-secretion coupling processes in brain: analysis of noradrenaline and gamma-aminobutyric acid, **J.Physiol.(London)** 254:475-505.

Damsma, G., Biessels, P.T.M., Westerink, B.H.C., De Vries, J.B. and Horn, A.S., 1988, Differential effects of 4-aminopyridine and 2,4-diaminopyridine on the *in vivo* release of acetylcholine and dopamine in freely moving rats measured by intrastriatal dialysis, **Eur. J. Pharmacol.** 145:15-20.

Davies, S.N., Lester, R.A., Reymann, K. and Collingridge, G.L., 1989, Temporally distinct pre- and postsynaptic mechanisms maintain long-term potentiation. **Nature** 338:500-503.

De Graan, P.N.E., Dekker, L.V., De Wit, M., Schrama, L.H. and Gispen, W.H., 1988a, Modulation of B-50 phosphorylation and polyphosphoinositide metabolism in synaptic plasma membranes by protein kinase C, phorbol diesters and ACTH. **J. Rec. Res.** 8:345-361.

De Graan, P.N.E., Heemskerk, F.M.J., Dekker, L.V., Melchers, B.P.C., Gianotti, C. and Schrama, L.H., 1988b, Phorbol esters induce long- and short-term enhancement of B-50/GAP-43 phosphorylation in rat hippocampal slices, **Neurosci. Res. Commun.** 3:175-182.

De Graan, P.N.E., Dekker, L.V., Oestreicher, A.B., Van der Voorn, L. and Gispen, W.H., 1989a, Determination of changes in the phosphorylation state of the neuron-specific protein kinase C substrate B-50 (GAP-43) by quantitative immunoprecipitation. **J. Neurochem.** 52:17-23.

Dekker, L.V., De Graan, P.N.E., Versteeg, D.H.G., Oestreicher, A.B. and Gispen,W.H., 1989a, Phosphorylation of B-50 (GAP-43) is correlated with neurotransmitter release in rat hippocampal slices. **J. Neurochem.** 52:24-30.

Dekker, L.V., De Graan, P.N.E., De Wit, M., Hens, J.J.H. and Gispen, W.H., 1989b, Depolarization-induced phosphorylation of the protein kinase C substrate B-50 (GAP-43) in rat cortical synaptosomes, **J.Neurochem.**, in press.

Dekker, L.V., De Graan, P.N.E., Oestreicher, A.B., Versteeg, D.H.G. and Gispen, W.H., 1989c, Inhibition of noradrenaline release by antibodies to B-50 (GAP-43), **Nature** : in press.

Dole al, V. and Tuçek, S., 1983, The effects of 4-aminopyridine and tetrodotoxin on the release of acetylcholine from rat striatal slices. **Naunyn-Schmiedeberg's Arch. Pharmacol.** 323:90-95.

Dolphin, A.C., Errington, M.L. and Bliss, T.V.P., 1982, Long-term potentiation of the perforant path in vivo is associated with increased glutamate release, **Nature** 297:496-498.

Eichberg, J., De Graan, P.N.E., Schrama, L.V., and Gispen, W.H., 1986, Dioctanoylglycerol and phorbol diesters enhance phosphorylation of phosphoprotein B-50 in native synaptic plasma membranes, **Biochem.Biophys.Res.Communn.** 136:1007-1012.

Forscher, P., 1989, Calcium and phosphoinositide control of cytoskeletal dynamics, **Trends Neurosci.**, in press.

Gibson, G.E. and Manger, T., 1988, Changes in cytosolic free calcium by 1,2,3,4-tetrahydro-5-aminoacridine, 4-aminopyridine and 3,4-diaminopyridine, **Biochem. Pharmacol.** 37:4191-4196.

Gispen, W.H., Leunissen, J.L.M., Oestreicher, A.B., Verkleij, A.J. and Zwiers, H., 1985, Presynaptic localization of B-50 phosphoprotein: the (ACTH)-sensitive protein kinase substrate involved in rat brain polyphosphoinositide metabolism, **Brain Res.** 328:381-385.

Gispen, W.H., De Graan, P.N.E., Schrama, L.H. and Eichberg, J., 1986, Phosphoprotein B-50 and polyphosphoinositide-dependent signal transduction in brain. In: Phospholipids in the Nervous System: Biochemical and Molecular Pharmacology (Eds. L.A. Horrocks, L. Freysz and G. Toffano), Fidia Research Series, Vol. 4, pp. 31-41, Liviana Press, Padova.

Gustafsson, B., Galvan, M., Grafe, P. and Wigstrom, H., 1982, A transient outward current in a mammalian central neurone blocked by 4-aminopyridine, **Nature (Lond.)** 299:252-254.

Gustafsson, B., Huang, Y.Y. and Wigstrom, H., 1988, Phorbol ester-induced synaptic potentiation differs from long-term potentiation in the guinea pig hippocampus *in vitro*, **Neurosc.Lett.** 85:77-81.

Haas, H.L. and Greene, R.W., 1985, Long-term potentiation and 4-aminopyridine, **Cell. Mol. Neurobiol.** 5:297-301.

Heemskerk, F.M.J., Schrama, L.H., Gianotti, C., Spierenburg, H., Versteeg, D.H.G. and Gispen, W.H., 1989a, 4-Aminopyridine stimulates B-50 (GAP-43) phosphorylation and [3H]-noradrenaline release in rat hippocampal slices, **J. Neurochem.**, in press.

Heemskerk, F.M.J., Schrama, L.H. and Gispen, W.H., 1989b, Activation of protein kinase C by 4-aminopyridine dependent on Na+ channel activity in rat hippocampal slices, **Neurosci. Lett.**, in press.

Heemskerk, F.M.J., Schrama, L.H., De Graan, P.N.E., Ghijsen, W.E.J.M., Lopes da Silva, F.H. and Gispen W.H., 1989c, 4-Aminopyridine increases B-50 (GAP-43) phosphorylation and calcium levels in rat brain synaptosomes, **Soc. Neurosci. Abstr.** 15:189.15.

Heemskerk, F.M.J., Schrama, L.H., De Graan, P.N.E. and Gispen, W.H., 1989d, 4-Aminopyridine stimulates B-50 (GAP-43) phosphorylation in rat brain synaptosomes, **J. Mol. Neurosci.**, in press.

Hu, G.-Y., Hvalby, O., Walaas, S.I., Albert, K.A., Skjeflo, P., Andersen, P. and Greengard, P., 1987, Protein kinase C injection into hippocampal pyramidal cells elicits features of long term potentiation, **Nature**, 328:426-429.

Izumi, Y., Miyakawa, H., Ito, K. and Kato, H., 1987, Quisqualate and N-methyl-D-aspartate (NMDA) receptors in induction of hippocampal long-term facilitation using conditioning solution, **Neurosci. Lett.** 83:201-206.

Kaczmarek, L.K., 1987, The role of protein kinase C in the regulation of ion channels and neurotransmitter release. **Trends Neurosci.** 10:30-34.

Kauer, J.A., Malenka, R.C. and Nicoll, R.A., 1988, NMDA application potentiates synaptic transmission in the hippocampus, **Nature** 334:249-252.

Kikkawa, U. and Nishizuka, Y., 1986, The role of protein kinase C in transmembrane signalling. **Ann. Rev. Cell Biol.** 2:149-178.

Kristjansson, G.I., Zwiers, H., Oestreicher, A.B. and Gispen, W.H., 1982, Evidence that the synaptic phosphoprotein B-50 is localized exclusively in nerve tissue, **J. Neurochem.** 39: 371-378.

Lee, W.-L, Anwyl, R. and Rowan, M., 1986, 4-Aminopyridine-mediated increase in long-term potentiation in CA_1 of the rat hippocampus, **Neurosci. Lett.** 70:106-109.

Linden, D.J., Wong, K.L., Sheu, F.-S. and Routtenberg, A., 1988, NMDA receptor blockade prevents the increase in protein kinase C substrate (protein F1) phosphorylation produced by long-term potentiation, **Brain Res.** 458, 142-146.

Linstedt, A.D. and Kelly, R.B., 1989, Overcoming barriers to exocytosis, **Trends Neurosci.** 10:446-448.

Liu, Y. and Storm, D.R., 1989, Dephosphorylation of neuromodulin by calcineurin, **J. Biol. Chem.**, 264:12800-12804.

Lisman, J.E. and Goldring, M.A., 1988, Feasibility of long-term storage of graded information by the Ca^{2+}/calmodulin dependent protein kinase molecules of postsynaptic density. **Proc. Natl. Acad. Sci.** 85:5320-5324.

Lovinger, D.M., Akers, R.F., Nelson, R.B., Barnes, C.A., McNaughton, B.L. and Routtenberg, A., 1985, A selective increase in phosphorylation of protein F_1, a protein kinase C substrate, directly related to three day growth of long-term potentiation of long term synaptic enhancement, **Brain Res.** 343:137-143.

Lovinger, D.M., Colley, P.A., Akers, R.F., Nelson, R.B. and Routtenberg, A., 1986, Direct relation of long-term synaptic potentiation to phosphorylation of membrane protein F_1, a substrate for membrane protein kinase C, **Brain Res.** 399:205-211.

Lynch, M.A. and Baudry, M., 1984, The biochemistry of memory: a new and specific hypothesis. **Science** 224:1057-1063.

Lynch, M.A., Errington, M.L. and Bliss, T.V.P., 1985, Long-term potentiation of synaptic transmission in the dentate gyrus: increased release of [^{14}C]-glutamate without increase in receptor binding, **Neurosci. Lett.** 62:123-129.

Malenka, R.C., Madison, D.V. and Nicoll, R.A., 1986, Potentiation of synaptic transmission in the hippocampus by phorbol esters, **Nature (Lond.)** 321:175-177.

Malenka, R.C., Ayoub, G.S. and Nicoll, R.A., 1987, Phorbol esters enhance transmitter release in rat hippocampal slices, **Brain Res.** 403:198-203.

Malenka, R.C., Kauer, J.A., Perkel, D.J., Mauk, M.D., Kelly, P.T., Nicoll, R.A., and Waxham, M.N., 1989, An essential role for postsynaptic calmodulin and protein kinase activity in long-term potentiation, **Nature** 340:554-557.

Malinow, R., Madison, D.V. and Tsien, R.W., 1988a, Persistent protein kinase activity underlying long-term potentiation, **Nature** 335:820-824.

Malinow, R., Madison, D.V. and Tsien, R.W., 1988b, Selective activation of pre-synaptic protein kinase C enhances synaptic transmission in rat hippocampal slices, **Soc. Neurosci. Abstr.** 14:12.2

Malinow, R., Schulman, H. and Tsien, R.W., 1989, Inhibition of postsynaptic PKC or CaMKII blocks induction but not expression of LTP, **Science** 245:862-866.

Matthies, H., 1989, In search of cellular mechanisms of memory, Progr. Neurobiol. 32:277-349.

Matthies, H.J.G., Palfrey, H.C., Hirning, L.D. and Miller, R.J., 1987, Down regulation of protein kinase C in neuronal cells: effects on neurotransmitter release, **J. Neurosci.** 7:1198-1206.

Miller, S.G. and Kennedy, M.B., 1986, Regulation of brain type II Ca^{2+}/calmodulin-dependent protein kinase by autophosphorylation: a Ca^{2+}-triggered molecular switch. **Cell** 44:861-870.

Muller, D., Turnbull, J., Baudry, M. and Lynch, G., 1988, Phorbol ester-induced synaptic facilitation is different than long-term potentiation, **Proc. Natl. Acad. Sci. USA** 85:6997-7000.

Nelson, R.B., Linden, D.J., Hyman, C., Pfenninger, K.H. and Routtenberg, A., 1989, The two major phosphoproteins in growth cones are probably identical to two protein kinase C substrates correlated with the persistence of long-term potentiation, **J. Neurosci.** 9:381-389.

Nicholls, D.G., Sihra, T.S. and Sanchez-Prieto, J., 1987, Calcium-dependent and -independent release of glutamate from synaptosomes monitored by continuous fluorometry, **J.Neurochem.** 49:50-57.

Nicholls, D.G., Tibbs, G. and Barrie, A.P., 1989, Cytosolic free calcium in synaptosomes and its coupling to glutamate exocytosis, **J. Neurochem. Suppl.** 52:47D.

O'Brian, C.A., Arthur, W.L. and Weinstein, I.B., 1987, The activation of protein kinase C by the polyphosphoinositides phosphatidylinositol 4,5-diphosphate and phosphatilylinositol 4-monophophate, **FEBS Lett.** 214:339-342.

Reymann, K.G., Frey, U., Jork, R. and Matthies, H., 1988, Polymyxin B, an inhibitor of protein kinase C, prevents the maintenance of synaptic long-term potentiation in hippocampal CA1 neurons, **Brain Res.** 440:305-314.

Rogawski, M.A. and Barker, J.L., 1983, Effects of 4-aminopyridine on calcium action potentials and calcium current under voltage clamp in spinal neurons, **Brain Res.** 280:180-185.

Routtenberg, A., 1985, Protein kinase C activation leading to protein F1 phosphorylation may regulate synaptic plasticity by presynaptic terminal growth, **Behav.Neurobiol.** 44:186-200.

Routtenberg, A., Colley., P., Linden, D., Lovinger, D., Murakami, K. and Sheu, F.-S., 1986, Phorbol ester promotes growth of synaptic plasticity. **Brain Res.** 278:374-378.

Schrama, L.H., De Graan, P.N.E., Wadman, W.J., Lopes da Silva, F.H. and Gispen, W.H., 1986, Long-term potentiation and 4-aminopyridine-induced changes in protein and lipid phosphorylation in the hippocampal slice. Progr. Brain Res. 69:245-257.

Schrama, L.H., De Graan, P.N.E., Dekker, L.V., Oestreicher, A.B., Nielander, H., Schotman, P. and Gispen, W.H., 1988, Functional significance and localization of phsophosite(s) in the neuron-specific protein B-50/GAP-43, **Soc.Neurosci.Abstr.**14:197.15.

Schwartz, J.H. and Greenberg, S.M., 1987, Molecular mechanisms for memory: second messenger induced modification of protein kinases in nerve cells, **Ann. Rev. Neurosci.** 10: 459-476.

Segal, M., Rogawski, M.A. and Barker, J.L., 1984, A transient potassium conductance regulates the excitability of cultures hippocampal and spinal neurons, **J. Neurosci.** 4:604-609.

Skene, J.H.P., 1989, Axonal growth-associated proteins, **Ann. Rev. Neurosci.** 12:127-156.

Skene, J.H.P. and Virag, I., 1989, Posttranslational membrane attachment and dynamic fatty acylation of a neuronal growth cone protein, GAP-43, **J.Cell Biol.** 108:613-624.

Skrede, K.K. and Malthe-Sorenssen, D., 1981, Increased resting and evoked transmitter release following repetitive electrical tetanization of the hippocampus: a biochemical correlate to long-lasting synaptic potentiation, **Brain Res.** 208:436-441.

Sörensen, R.G., Kleine, L.P. and Mahler, H.R., 1981, Presynaptic localization of phosphoprotein B-50, **Brain Res. Bull.** 7: 57-61.

Stevens, C.F., 1989, Strengthening the synapses, **Nature**, 338, 460-461.

Tapia, R. and Sitges, M., 1982, Effect of 4-aminopyridine on transmitter release in synaptosomes. **Brain Res.** 250:291-299.

Tapia, R., Sitges, M. and Morales, E., 1985, Mechanism of calcium-dependent stimulation of transmitter release by 4-aminopyridine in synaptosomes. **Brain Res.** 361:373-382.

Thompson, S.H., 1982, Aminopyridine block of transient potassium current, **J. Gen. Physiol.** 80:1-18 (1982).

Tibbs, G.R., Dolly, J.O. and Nicholls, D.G., 1989, Dendrotoxin, 4-aminopyridine and ß-bungarotoxin act at common loci but by two distinct mechanism to induce calcium-dependent release of glutamate from guinea pig cerebrocortical synaptosomes, **J. Neurochem.** 52:201-206.

Van Dongen, C.J., Zwiers, H., De Graan, P.N.E. and Gispen, W.H., 1985, Modulation of the activity of purified phosphatidylinositol 4-phosphate kinase by phosphorylated and dephosphorylated B-50 protein, **Biochem.Biophys.Res.Commun.** 8:1219-1227.

Van Hooff, C.O.M., De Graan, P.N.E., Oestreicher, A.B. and Gispen, W.H., 1988, B-50 phosphorylation and polyphosphoinositide metabolism in nerve growth cone membranes. **J. Neurosci.** 8:1789-1795.

Van Hooff, C.O.M., Boonstra, J., Oestreicher, A.B., De Graan, P.N.E., Holthuis, J.C.M. and Gispen, W.H., 1989a, Nerve growth factor-induced changes in the intracellular localization of the protein kinase C substrate B-50 in pheochromocytoma PC12 cells, **J.Cell Biol.** 108:1115-1125.

Van Hooff, C.O.M., De Graan, P.N.E., Oestreicher, A.B. and Gispen, W.H., 1989b, Muscarinic receptor activation stimulates B-50 phosphorylation in isolated nerve growth cones, **J.Neurosci.**, in press.

Van Lookeren Campagne, M., Oestreicher, A.B., Van Bergen en Henegouwen, P.M.P. and Gispen, W.H., 1989a, Ultrastructural immunocytochemical localization of B-50/GAP43, a protein kinase C substrate, in isolated presynaptic nerve terminals and neuronal growth cones, **J. Neurocytol.**, in press.

Van Lookeren Campagne, M., Oestreicher, A.B., van Bergen Henegouwen, P.M.P. and Gispen, W.H., 1989b, Localization of B-50/GAP-43 and synaptophysin in the neonatal and adult rat brain, **Cell Diff.Devel.** 27 suppl.:S197.

Versteeg, D.H.G. and Florijn, W.J., 1986, Phorbol 12,-13-dibutyrate enhances electrically stimulated neuromessenger release from rat dorsal hippocampal slices *in vitro*, **Life Sci.** 40:1237-1243.

Versteeg, D.H.G. and Ulenkate, H.J.L.M., 1987, Basal and electrically stimulated release of [3H]-noradrenaline and [3H]-dopamine from rat amygdala slices *in vitro*: effects of 4ß-phorbol 12,13-dibutyrate, 4 -phorbol 12,13-didecanoate and polymyxin B, **Brain Res.** 416:343-348.

Zurgil, N. and Zisapel, N., 1985, Phorbol ester and calcium acts synergistically to enhance neurotransmitter release by brain neurons in culture, **FEBS lett.** 185:257-261.

Zwiers, H., Schotman, P. and Gispen, W.H., 1980, Purification and some characteristics of an ACTH-sensitive protein kinase and its substrate protein in rat brain membranes, **J. Neurochem.** 34:1689-1699.

DIFFERENT MECHANISMS AND MULTIPLE STAGES OF LTP

Hansjuergen Matthies[1,2], Uwe Frey[1], Klaus Reymann[1],
Manfred Krug[2], Reinhard Jork[1], and Helmut Schroeder[2]

Institute of Neurobiology and Brain Research , Academy of
Sciences[1],
Institute of Pharmacology, Medical Academy[2]
Leipziger Str.44 , 3090 Magdeburg (GDR).

MECHANISMS AND STAGES OF MEMORY FORMATION

On the basis of our own experimental data and results from other laboratories then available, we developed in 1972 a working hypothesis on neuronal
mechanisms of memory. We suggested that the assumed stages of short-term,
intermediate, and long-term memory, their different time course of origin
and decay, their biochemical correlates as well as their sensitivity to
interventions reflect the properties of the corresponding cellular mechanisms of a synaptic, synaptosomal and nuclear regulation of memory formation (Matthies,1972).
We could demonstrate that the formation of long-term memory (LTM) requires
the induction of protein synthesis and a subsequent posttranslational fucosylation of newly synthesized proteins.Their synthesis and processing seems
to be controled by transsynaptic action of mediators during learning-related
convergence of neuronal signals to integrating neurons . The learning model
used in our studies, a brightness discrimination on rats, revealed a biphasic occurrence of this macromolecular synthesis , an early stage during and
immediately after learning , producing mainly soluble proteins , and a late
stage about 4-8 hours after acquisition with an increased formation of membrane proteins.These changes could be most precisely differentiated in hippocampal regions, but occurred also in some neocortical structures. We
suggested that the early stage probably represents the synthesis of regulatory proteins which control the formation of target proteins finally remodelling the neuronal connectivity during the late stage of memory formation
(Popov et al.,1975, for review see Matthies,1989 a,1989 b).

MEMORY STAGES ALSO IN LTP ? METHODICAL ISSUES

Our efforts to separate and characterize more precisely the newly synthesized proteins were first of all not very successful with the methods
available at this time. Therefore, when posttetanic long-term potentiation
was discovered by Bliss and Lomo (1973),we applied also LTP as a model or
mechanism of long-term memory in our investigations by reason of its metho-
dical advantages . At this time, LTP was mainly investigated with electrophysiological methods and still regarded as an unitary phenomenon.Therefore,
the observation were mainly confined to the relatively short time of 30-60
minutes until the synaptic enhancement reached a certain stability.

However, by reason of our preceding results obtained with learning experiments and of the resulting conception, we assumed that different stages and corresponding mechanisms can be expected to occur also with LTP , if it actually represents a mnemonic device . But to verify this assumption, LTP should be investigated, even in vitro, during that time course of at least 8 - 10 hours which were found to be required for the protein synthesis-dependent formation of a more or less permanent memory trace.

PROTEIN SYNTHESIS AND FUCOSYLATION IS REQUIRED FOR LONG-LASTING LTP

First of all, we determined , whether the intraventricular application of anisomycin preventing the formation of long-term memory in several learning experiments also influences the long-term maintenance of LTP in the dentate gyrus. We could show that the presence of this inhibitor of protein synthesis during tetanization does not prevent the induction of LTP, but results in a gradual decrease of potentiation in the course of 5-7 hours (Krug et al.,1984) thus exhibiting a similar time course as the amnesic effect in our learning paradigm. This suppression of the late LTP by anisomycin was meanwhile confirmed by observations of Otani et al.(1989) . We obtained similar results also on CA1 cells of hippocampal slices (Frey et al.,1988) (Fig.3 C), indicating that this effect is not confined to a particular cell type. To obviate objections with regard to possible side effects of aniso-mycin not related to an inhibition of protein synthesis, we tetanized the Schaffer collaterals after separation of the CA1 dendrites from their cell bodies as the main site of neuronal protein synthesis. These isolated dendrites revealed a pronounced induction of EPSP-potentiaion as observed in intact slices. However, the potentiation gradually decreased in the course of 5-7 hours thus resulting in the same lack of the late maintenance as observed in complete CA1 neurons after inhibition of protein synthesis by anisomycin (Fig. 3 C). The observed decay of potentiation is not due to a functional impairment of the separated dendrites, because they can be again potentiated 8 hours after the first tetanization (Frey et al., 1989 a). The results altogether clearly indicate that a particular stage of long-lasting LTP-maintenance exists which depends on the intact protein synthesis during and immediately after tetanization . The evaluation of leucine incorporation into the dentate gyrus revealed a significant increase occurring immediately after tetanization of the perforant path , lasting about 2 hours (Loessner et al.,1987)(Fig.1 D). This transient enhancement of protein synthesis coincides with the time window after tetanization during which an inhibition by anisomycin prevents the late maintenance of LTP, suggesting that newly synthesized proteins are required (Abraham and Otani, 1988).
It was further shown in our studies on the formation of long-term memory that the fucosylation of proteins seems to be also critical (Jork et al., 1986). We demonstrated that dopamine is involved in the control of this fucosylation and that dopamine-induced improvement of long-term memory correlates with increased incorporation of fucose into hippocampal pro-teins. Particular glycoproteins are obviously completed by a posttransla-tional formation of a fucose 1-2 galactose linkage.This metabolic step can be prevented by a pretreatment with 2-deoxy-galactose. The false sugar is incorporeted into glycans instead of galactose, but can not serve for the fucose 1-2 linkage . Such pretreatment results, however, in a severe pre-vention of long-term memory . We could clearly prove that the deoxy-galac-tose-induced amnesia is related to the inhibited formation of those fuco-syl-glycoproteins which are increasingly synthesized after a training of brightness discrimination (Jork et al.,1989). Recently, we could show in our laboratory that a pretreatment of rats with 2-deoxygalactose also sup-presses the formation of the late, long-lasting maintenance without influ-encing the transmission at the glutamatergic synapses or the induction and early maintenance of LTP. (Fig.2). This result points to a role of fuco-syl-glycoproteins also in the late, long-lasting maintenance of LTP.

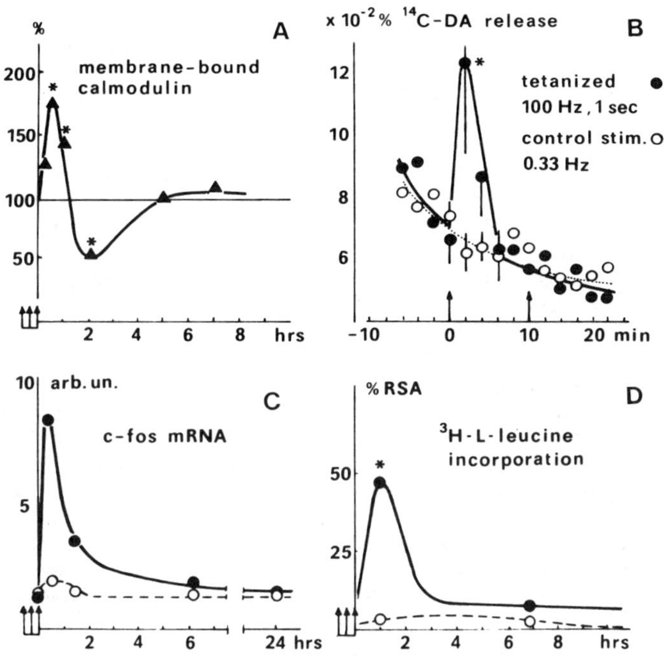

Fig.1   Biochemical correlates of hippocampal LTP
A. Translocation of calmodulin from the cytosol to the membran-bound fraction in rat hippocampal slices after tetanization (immunochemical determination)   (Popov et al.,1988)
B. Release of ¹⁴C-dopamine from superfused rat hippocampal slices by
   stimulation of Schaffer collaterals with 100 impulses (100 Hz)(filled
   circles) and by control stimulation with 4 pulses (o.33 Hz)(open circles).   Preloading with ¹⁴C-DA during 10 min. preincubation.
C. Accumulation of c-fos mRNA in the dentate area of freely moving rats
   after tetanic stimulation of the perforant path (4 trains of 3oo impulses, 200 Hz, arranged in 20 groups of 15 impulses each, separated
   by 5 sec . Intertrain interval 15 min). Tetanized rats (filled circles, control stimulation (o.5 Hz , the same number of impulses) open
   symbols. (Kaczmarek et al.,1989).
D. Incorporation of ³H-L-leucine into proteins of the dentate gyrus of
   freely moving rats at different times after tetanization (filled symbols) and after control stimulation (open symbols). % RSA : increase
   of relative specific activity as compared to unstimulated controls.
   (Loessner et al.,1988).

DIFFERENT RECEPTORS AND MESSENGERS MEDIATE EARLY MAINTENANCE

The question arised, which transmitter-mediated signal transduction indu-
duces the early and late maintenance . With regard to the well documented
role of glutamate and NMDA receptor-mediated calcium influx for the induction of LTP (for review see : Collingridge and Bliss,1987), we investigated the involvement of calmodulin using immunochemical methods. It was
shown that calmodulin is translocated from the cytosol to membranes within
30 min after tetanization and then redistributed during the following hour
(Popov et al.,1988 , Fig. 1 A).The involvement of the Ca-calmodulin system in the induction of LTP is confirmed by our finding that the presence
of calmidazolium, an inhibitor of calmodulin, during tetanization prevents
the initiation to a similar degree as the specific NMDA-antagonist D-APH
(Reymann et al.,1988 a, Fig.3 A, 3 D). Several experimental data suggested
a particular role of protein kinase C (PKC) in the initiation and maintenance of LTP, for instance the induction by phorbol esters (Malenka et al.,
1986) or by intracellular injection of PKC (Hu et al.,1987), the translocation to membranes after tetanization (Akers et al.,1986), as well as an
involvement of PKC-substrates (Routtenberg, 1987 , Gispen, 1986). Using
polymyxin B or other inhibitors of PKC, we could show that their presence
during tetanization of CA1 cells in vitro does not prevent the induction
and early maintenance of LTP. But in the course of 2 hours,the EPSP-potentiation gradually decreases to baseline (Reymann et al.,1988 b, Fig. 2 B).
Because the EPSP-potentiation after inhibition of protein synthesis decays
much slowlier in the course of 5-7 hours, an intermediate polymyxin B-sensitive stage of LTP-maintenance seems to exist obviously involving PKC-
dependent processes.To evaluate the receptor(s) probably mediating the signal which induces this intermediate stage,we tetanized CA1 neurons in vitro
in the presence of 50 uM L-APB. It was shown that this antagonist does not
prevent the induction of LTP, but gradually decreases the EPSP-potentiation
in the course of 2 hours thus eliminating both the intermediate and late
maintenance like PKC-inhibitors (Reymann and Matthies,1989 , Fig. 3 E).
L-APB is assumed to block a postsynaptic ibotenate-sensitive quisqualate
receptor differing from that activated by AMPA, and is probably coupled to
the phosphoinositide system (Nicoletti et al.,1988; Recasens et al.,1988;
Schoepp and Johnson.,1988) thus releasing calcium-mobilizing ITP and PKC
-activating diacylglycerol after occupation by glutamate. It was an open
question, whether this receptor represents a permanent subtype or a functional state which can be converted to an AMPA-sensitive quisqualate
receptor by phosphorylation after activation of PKC . It was shown that
15-30 min after tetanization , the sensitivity to AMPA slowly increases
in CA1 cells, reaching a peak about 2 hrs after induction of LTP (Davies
et al.,1989). This time course clearly corresponds to the anisomycin-
resistant, polymyxin-sensitive stage of intermediate maintenance suggested by our results. Recent observations demonstrated that the inhibition
of a K 252 b-sensitive protein kinase actually prevents the delayed post-
tetanic increase of AMPA responses (Reymann et al. 1989).

EARLY GENE (PROTO-ONCOGENE) EXPRESSION INVOLVED IN LATE LTP ?

The inhibition of PKC does not only suppress this intermediate stage, but
also prevents the occurrence of the late,anisomycin-sensitive LTP.We have
to consider that PKC is not a monofunctional enzyme and may also be involved in the control of genomic expression. Proto-oncogenes have been found
to play a role in the control of genomic expression in the nervous system
and particularly in neuronal plasticity (for review see: Hanley, 1988 ,
Chiarugi et al.,1989). C-fos proto-oncogene expression is induced in neurons by several mediators (Greenberg et al.,1986, Kaczmarek et al.,1988)
and not only by strong interventions like seizure production or brain damage.
Also the application of DOPA induces c-fos expression in the rat striatum

after ipsilateral dopaminergic denervation (Robertson et al.,1989),indi-
cating that dopaminergic signals can be involved in the control of genomic
functions. Therefore, we investigated the expression of c-fos mRNA by relatively specific hybridization techniques in the dentate gyrus of unanaes-
thetized freely moving rats after both control stimulation and tetanization
of the perforant path. It was shown that low frequency stimulation with
0.5 Hz for 45 min does not change the control values of c-fos mRNA. However,
a stimulation with 4 trains of 300 impulses (200 Hz) arranged in 20 groups
of 15 impulses , producing a pronounced long-lasting potentiation of the
population spike without generating seizure activity,results in a eightfold
increase of c-fos mRNA in the course of 45 min after the last tetanization.
This accumulation of the proto-oncogene mRNA already diappears in the following hour (Fig. 1 C).

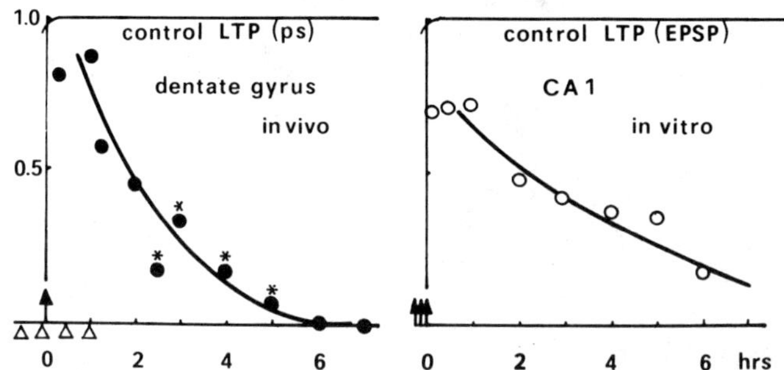

Fig. 2   Influence of 2-deoxygalactose (do-gal) on hippocampal LTP .
Right — Prevention of late maintenance of population spike potentiation
        in the dentate gyrus of freely moving rats. Four intracerebro-
        ventricular injections of 800 ug do-gal 60 and 30 min. before
        and 30 and 60 min after tetanization (5 trains with 15 impulses,
        200 Hz , train interval 5 sec.)
Left — Prevention of the late maintenance of EPSP- potentiation in CA 1 -
        neurons of rat hippocampal slices. Pretreatment of rats with bila-
        teral intracerebroventricular injection of 800 ug do-gal 90 and 30
        min before preparation of slices , and subsequent incubation of
        slices with do-gal for 180 min. Tetanization of Schaffer collate-
        rals: 3 impulse trains,100 impulses,100 Hz, train interval 10 min.

Our result differs from other studies using anaesthetized rats and antibodies for determination of c-fos-like proteins (Douglas et al.,1988), or
only weak tetanization probably not initiating a reliable long-lasting
potentiation (Bliss et al.,1988). Recently, Dragunow et al.(1989) observed
also an increase of c-fos mRNA and proteins in the dentate area after tetanization of the perforant path of non-anaesthetized , but not of anaesthetized rats. However, the c-fos induction did not strongly correlate with
the induction of LTP, when different schedules of tetanization were used.
Therefore, they suggest that c-fos induction is probably related to maintenance thus supporting our view of the existence of a late LTP-stage depending on a nuclear regulation and protein synthesis. Further studies are
necessary to elucidate the significance of the different experimental conditions, particularly of the frequency and the pattern of tetanic stimulation, influencing the induction of proto-oncogene expression, the subsequent formation of target proteins and the resulting persistence of LTP.

NOT ONLY GLUTAMATE IS REQUIRED FOR LATE LTP

The question arised, whether the different LTP-stages are induced by only
one transmitter and whether the activation of glutamate receptors is sufficient to induce protein synthesis and subsequent processing of macromolecules required for the late LTP, or if additional convergent transmitter-mediated signal(s) are necessary. The electrical stimulation of neuronal tissue used under experimental conditions obviously releases other
mediators,too.Therefore, considering our observations on the dopaminergic
control of fucosylation after learning, we investigated a possible role of
dopaminergic influences by tetanization of CA1 cells in vitro in the presence of different dopamine antagonists . It was shown that domperidone,
sulpirid and flupenthixol, respectively, in a concentration of 1 uM does
not influence the induction of LTP. But the potentiation slowly decreases
in the course of 5 - 7 hours in a similar manner, as after inhibition of
protein synthesis by anisomycin (Frey et al.,1989 b, Fig.3 F). This effect
was completely blocked by simultaneous presence of 1 uM apomorphin, indicating a specific involvement of dopamine receptors . Previous investigations did not detect these actions of the dopaminolytic drugs by reason of
a too short time of observations . To verify that dopamine is in evidence
during tetanization , we determined the release of the transmitter under
our experimental conditions after preloading the slice with ^{14}C-dopamine
during a preincubation period and measuring the efflux of specific radioactivity in 2 min-fractions obtained from a perfusion chamber.Control stimulation with 4 imp.(0.33 Hz) did not evoke any change of the efflux of
^{14}C-dopamine, but the tetanization with 100 impulses (100 Hz) results
in a significant short lasting release of dopamine (Fig.2 B). A second
tetanization 10 min later was ineffective, probably indicating the exhaus-
tion of the poorly labeled releasable pool . These results suggest that a
dopaminergic input to hippocampal CA 1 neurons at least contributes as a
necessary signal to the induction of the late LTP. Recently, Buzsaki and
Gage (1989) also suggested that the abundant release of glutamate onto
postsynaptic membranes of granule cells alone is not able to produce or
maintain LTP, but requires the coactivation of aminergic or other inputs
to the dentate gyrus.

CONCLUSIONS AND A SPECULATIVE VIEW

Summarizing our results also considering data of other laboratories, we
come to some conclusions and suggestions on the cellular processes underlying hippocampal posttetanic LTP :
1. LTP is not an unitary phenomenon, but reveals several subsequent stages
initiated by activation of different glutamate receptors and by coactivation of aminergic binding sites . The resulting involvement of different
second messenger systems modifies first of all functions of existing pre-
and postsynaptic proteins thus transiently enhancing synaptic efficacy.But
a final long-lasting LTP requires the synthesis of target proteins by a
change of genomic expression and/or translation.
2. Induction of LTP requires postsynaptic depolarization by occupation of
quisqualate receptors or other depolarizing inputs and a concomitant activation of NMDA-receptors resulting in an increased calcium influx.
Calmodulin , translocated to membranes by calcium influx , is pre- and/or
postsynaptically involved either in the control of transmitter release or
in the functional change of glutamate receptors. Presynaptic activation of
protein kinase C and subsequent phosphorylation of F1 (B50) protein probably initiate a remodelling of the terminal.
3. An intermediate stage of transient maintenance is probably induced by
occupation of an ibotenate-sensitive quisqualate receptor subtype coupled to the phosphoinositol-system and blocked by APB. The resulting release
of diacylglycerol and inositol-triphosphate activates protein kinase C and
mobilizes calcium from intracellular stores. This receptor itself or other

silent glutamate receptors are subsequently transformed by phosphorylation
to AMPA-sensitive binding sites thus transiently increasing the postsynaptic response to glutamate as a mechanism of intermediate maintenance.
4. The occupation of postsynaptic dopamine receptors obviously initiates an
additional second messenger system . In this way, either the expression of
an early gene (proto-oncogene) expression is induced, or the expression of

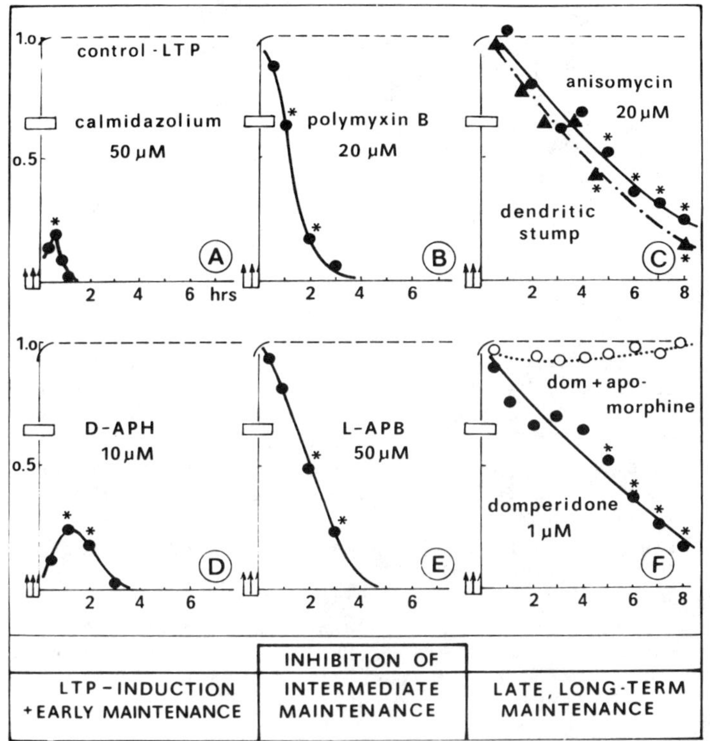

Fig.3   Evaluation of different stages and corresponding mechanisms of
         posttetanic LTP in the CA 1 region of rat hippocampal slices
         by the presence of various inhibitors during tetanization.
Tetanization :   3 impulse trains (100 impulses, bipolar, 0.2 msec per polarity, 100 Hz), train interval 10 min . Control EPSP-potentiation = 1.0,
* p < 0.05 versus control EPSP-potentiation

the target proteins is completeted by a competence signal.However,dopamine-
mediated influences also control posttranslational fucosylation of neuronal
proteins necessary for the formation of long-term memory traces . Even if
the tetanization of the perforant path results only in a weak increase of
fucose incorporation into hippocampal proteins (Pohle et al., 1987) , a
pretreatment with 2-deoxygalactose prevents the late stage of LTP, indica-

ting a critical role of a dopaminergic control of postranslational protein
fucosylation . These newly formed glycoproteins are transported into the
dendritic tree and preferably inserted into those postsynaptic structures
wich were previously labeled by mechanisms of transient LTP-maintenance.
Their insertion stabilizes synaptic facilitation thus accomplishing the
stage of the late, long-lasting maintenance.
This short representation of the scenario does not consider or explain all
important data obtained with LTP, for instance the release of arachidonic
acid or of proteins (Charriaut-Marlangue et al.,1988, Fazeli et al.,1988)
after tetanization. However, this conception is an approach to integrate
different results and interpretations, frequently presented as conflicting
alternative views. It makes clear the complex nature of posttetanic long-
term potentiation. In all probability, the enhancement of glutamatergic
transmission lasting several days or weeks can not be only attributed to
an activation of a glutamatergic terminal, but seems to require the coactivation of convergent inputs to postsynaptic principal cells. Considering
such a conception, the primary functional changes related to an enhanced
release of glutamate and resulting in stages of more or less transient potentiation can be only regarded as mechanisms of an operative or working
memory. The formation of a permanent memory trace obviously requires more
than the involvement of a single informational channel, but the cooperation
of several systems of the brain.

REFERENCES

Abraham,W.C., and Otani, S., 1988 , Maintenance of long-term potentiation
        in the dentate gyrus requires protein synthesized shortly after teta-
        nization in the anaesthesized rat, J.Physiol.(London) 407 : 50 P
Akers,R.F., Lovinger,D.M., Colley,P.A., Linden,D.J., and Routtenberg,A.,
        1986, Translocation of protein kinase C activity may mediate hippo-
        campal long-term potentiation , Science , 231 : 587
Bliss,T.V.P., Errington,M.L., Evan,G., and Hunt,S.P., 1988 , Induction of
        c-fos like protein in rat hippocampus following electrical stimula-
        tion , J.Physiol.(London), 398 : 96 P
Buzsaki,G. and Gage,F.H., 1989 , Absence of long-term potentiation in the
        subcortically deafferented dentate gyrus , Brain Res. 484 : 94
Charriaut-Marlangue,C.,Aniksztein,L., Roisin,M.P., and Ben-Ary,Y.,1988,
        Release of proteins during long-term potentiation in the hippocampus
        of the anaesthetized rat, Neurosci.Lett. 91 : 308
Chiarugi,V.P., Ruggiero,M., and Corradetti,R., 1989 , Ontogenes, protein
        kinase C, neuronal differentiaation and memory, Neurochem.Int 14 : 1
Collingridge,G.L., and Bliss,T.V.P., 1987 , NMDA-receptors - their role
        in long-term potentiation, Trends Neurosci. 10 : 288
Douglas,R.M., Dragunow,M., and Robertson,H.A.,1988, High-frequency dis-
        charge of dentate granule cells, but not long-term potentiation,
        induces c-fos protein, Mol.Brain.Res. 4 : 259
Dragunow,M., Abraham,W.C., Goulding,M., Mason,S.E., Robertson,H.a., and
        Faull,R.L.M., 1989, Long-term potentiation and the induction of c-fos
        mRNA and proteins in the dentate gyrus of unanaesthetized rats,
        Neurosci.Lett. ??? : 274
Fazeli,M.S., Errington,M.L., Dolphin,A.C., and Bliss,T.V.P.,1988 , Long-
        term potentiation in the dentate gyrus of the anaesthetized rat is
        accompanied by an increase in protein efflux into push-pull cannula
        perfusates, Brain Res. 473 : 51
Frey,U., Krug,M., Reymann,K., and Matthies, H., 1988 , Anisomycin, an inhi-
        bitor of protein synthesis, blocks late phases of LTP phenomena in
        the hippocampal CA 1 region in vitro, Brain Res. 452 : 57
Frey,U., Krug,M., Broedemann,R., Reymann,K.G.,and Matthies,H., 1989, Long-
        term potentiation induced in dendrites separated from their CA1 pyra-
        midal somata does not establish a late phase, Neurosci.Lett. 97 : 135

Frey,U., Hartmann,S., and Matthies,H.,1989, Domperidon, an inhibtor of the
     D2 - receptor, blocks a late phase of electrically induced long-term
     potentiation in the CA1 region in rats, Biomed.Biochem.Acta 48 : 473
Gispen, W.H., 1986 , Synaptic protein phosphorylation and long-term poten-
     tiation, in: Learning and Memory : Mechanisms of information storage
     in the nervous system, H.Matthies,ed., Pergamon Press, Oxford , p. 25
Greenberg,M.E., Ziff,E.B., and Greene.,L.A., 1986 , Stimulation of neuro-
     nal acetylcholine receptors induces rapid gene transcription, Science
     234 : 80

Hanley, M.R., 1988, Proto-oncogenes in the nervous system, Neuron 1 : 175
Hu, G.Y., Hvalby, O., Walaas,S.I., Albert, K.A., Skjeflo,P., Andersen,P.,
     and Greengard,P.,1987, Protein kinase C injection into hippocampal
     pyramidal cells elicits features of long-term potentiation,Nature
     328 : 426

Jork,R., Grecksch,G, and Matthies,H.,1986, Impairment of glycoprotein fuco-
     sylation in rat hippocampus and the consequences on memory formation,
     Pharmacol.Biochem.Behav. 25 : 1137
Jork,R.,Schnurra,I.,Smalla,K.H.,Grecksch,G.,Popov,N.,and Matthies,H.,1989,
     Deoxy-galactose mediated amnesia is related to an inhibition of trai-
     ning-induced increase in rat hippocampal glycoprotein fucosylation,
     Neurosci.Res.Commun. , 5 : 3
Kaczmarek,L., Siedlecki,J.A., and Danysz,W.,1988, Proto-oncogene c-fos
     induction in rat hippocampus, Mol.Brain.Res. 3 : 183
Kaczmarek,L.,Nikolajew,E.,Jakulski,D.,Krug,M.,Tischmeyer,W.,and Matthies,H.,
     1989, Induction of c-fos protooncogene expression in rat hippocampus
     ocurring in long-term potentiation and learning, submitted
Krug.,M., Loessner,B., and Ott,T., 1984 , Anisomycin blocks the late phase
     of long-term potentiation in the dentate gyrus of reely moving rats,
     Brain Res.Bull. 13 : 39
Loessner,B., Schweigert,C., Pchalek,V., Krug,M., Frey,S., and Matthies,H.,
     1987 , Training and LTP-induced changes of protein synthesis in rat
     hippocampus, Neurosci. 22 : S 512
Malenka,R.C., Madison,D.V., and Nicoll,R.A.,1986, Potentiation of synaptic
     transmission in the hippocampus by phorbol esters, Nature 321 : 175
Matthies,H., 1972 , Pharmacological influence on teaching and memoriza-
     tion processes Farmakol. i.Toxikol. 35 : 259
Matthies,H., 1989 a, Neurobiological aspects of learning and memory , Annu.
     Rev.Psychol. 40 : 381
Matthies,H., 1989 b , In search of cellular mechanisms of memory , Progr.
     Neurobiol. 32 : 277
Nicoletti,F., Wroblewski,J.T., Fadda,E., and Costa,E.,1988, Pertussis
     toxin inhibits signal transduction at a specific metabolotropic glu-
     tamate receptor in primary cultures of cerebellar granule cells,
     Neuropharmacol. 27 : 551
Otani,S., Marshall,C.J., Tate,W.P., Goddard,G.V., and Abraham,W.C., 1989 ,
     Maintenance of long-term potentiation in rat dentate gyrus requires
     protein synthesis but not messenger RNA synthesis immediately post-
     tetanization, Neurosci. 28 : 519
Pohle,W., Acosta,L., Ruethrich,H., Krug,M., and Matthies,H.,1987, Incor-
     poration of ^{3}H-fucose in rat hippocampal structures after condi-
     tioning by perforant path stimulation and after LTP-producing teta-
     nization, Brain Res. 410 : 245
Popov,N., Schulzeck,S., Schmidt,S., and Matthies,H.,1975, Changes in labe-
     ling of soluble and solubilized rat brain proteins using ^{3}H-leucine
     as precursor during a learning experiment, Acta biol.med.Germ.
     34 : 583
Popov,N.S.,Reymann,K.G.,Schulzeck,K., Schulzeck,S., and Matthies,H.,1988,
     Alterations in calmodulin content in fractions of rat hippocampal
     slices during tetanic and calcium-induced long-term potentiation,
     Brain Res. Bull. 21 : 201

Recasens,M., Guiramand,J., Nourigat,A., Sassetti,I., and Devilliers,G.,
     1989, A new quisqualate receptor subtype (sAA2) responsible for the
     glutamate-induced inositol phosphate formation in rat brain synap-
     toneurosomes, Neurochem.Int. 13 : 463
Reymann,K.G., Broedemann,R., Kase,H., and Matthies,H., 1988 a, Inhibitors
     of calmodulin and protein kinase C block different phases of hippo-
     campal long-term potentiation, Brain Res. 461 : 388
Reymann,K.G., Frey,U., Jork,R., and Matthies,H., 1988 b , Polymyxin B,
     an inhibitor of protein kinase C, prevents the maintenance of synap-
     tic long-term potentiation in hippocampal CA1 neurons, Brain Res.
     440 : 305
Reymann,K.G., and Matthies,H.,1989 a, 2-Amino-4-phosphonobutyrate selecti-
     vely eliminates late phases of long-term potentiation in rat hippo-
     campus, Neurosci.Lett., 98 : 166
Reymann,K.G., Davies,S.N., Matthies,H., Kase,H., and Collingridge,G.L.,
     1989 b, Activation of a K 252 b -sensitive protein kinase is neces-
     sary for a postsynaptic phase of long-term potentiation in area CA 1
     of rat hippocampus, Eur. J. Neurosci., submitted
Robertson,G.S.,Herrera,D.G.,Dragunow,M., and Robertson,H.A., 1989, L-DOPA
     activates c-fos in the striatum ipsilateral to a 6-hydroxydopamine
     lesion of the substantia nigra, Eur.J.Pharmacol. 159 : 99
Routtenberg,A., 1987 , Phospholipid and fatty acid regulation of signal
     transduction at synapses : potential role for protein kinase C in
     information storage, J.Neural Transm. 24 (Suppl.) : 239
Schoepp,D.D., and Johnson,B.G., 1988 , Excitatory amino acid agonist-anta-
     gonist interactions at 2-amino-4-phosphonobutyric acid-sensitive
     quisqualate receptors coupled to phospho-inositide hydrolysis,
     J.Neurochem. , 50 : 1605

EXTRACELLULAR PROTEASES AND S100 PROTEIN IN LONG-TERM POTENTIATION IN THE

DENTATE GYRUS OF THE ANAESTHETIZED RAT

M.S. Fazeli[+], M.L. Errington[*], A.C. Dolphin[+] and T.V.P. Bliss[*]

[+] Department of Pharmacology & Clinical Pharmacology, St. George's
Hospital Medical School, Cranmer Terrace, London SW17 ORE.

[*] Division of Neurophysiology and Neuropharmacology, National Institute
for Medical Research, The Ridgeway, Mill Hill, London NW10.

INTRODUCTION

The present chapter summarizes the results of experiments aimed at identifying macromolecules whose extracellular concentration is increased during long-term potentiation (LTP). The main
approach has been sampling of extracellular fluid, using a push-pull cannula, before and after the
induction of LTP.

Increased efflux of proteins following the induction of LTP

Push-pull cannula perfusates were analyzed on 5-20% gradient polyacrylamide gels using a
sensitive silver staining procedure. The perfusates were collected via a cannula implanted in the
molecular layer of the dentate gyrus of the anaesthetized rat (4.2mm posterior to bregma, 2.5mm
lateral to midline). The stimulating electrode was placed in the angular bundle (4.4mm lateral to
lambda). Extracellular evoked potentials were recorded via electrodes attached to the outside of
the cannula, and test shocks were delivered every 30sec. (Errington et al., 1983). Perfusion was
with artificial CSF at 6.5 1/min.; fractions were collected every 15min. Typical experiments were
of 4 hours duration, LTP being induced one hour after the beginning of the experiment.

The tetanic induction of LTP (3 trains of 250Hz, 200msec., every 30sec.), is followed by a
delayed (45-60min.), but sustained (3hours) increase in protein efflux into push-pull cannula perfusates (Fazeli et al., 1988a). The increase occurs across a wide spectrum of proteins (Fig. 1) and is
blocked by the prior perfusion of the selective NMDA-receptor antagonist, 5-D-amino-phosphono-
valeric acid (APV; 100 μM), which also blocks the induction of LTP. Charriaut-Marlangue et al.
(1988) have also demonstrated a non-selective increase in protein efflux.

Induction of an LTP-like effect by perfusing 10mM Ca^{2+}-CSF (normal Ca^{2+}-concentration 1.2
mM), also leads to an increase in protein efflux (Fazeli et al., 1988a). This increase, however, is
apparent within 15-30min. of the induction, in contrast to the longer delay observed when LTP is
induced tetanically. Moreover, the increase is not sustained and protein efflux returns to basal
levels in the second hour after the induction of LTP. The Ca^{2+}-induced increase in protein efflux, as
well as the synaptic potentiation, is also blocked by 100 μM APV.

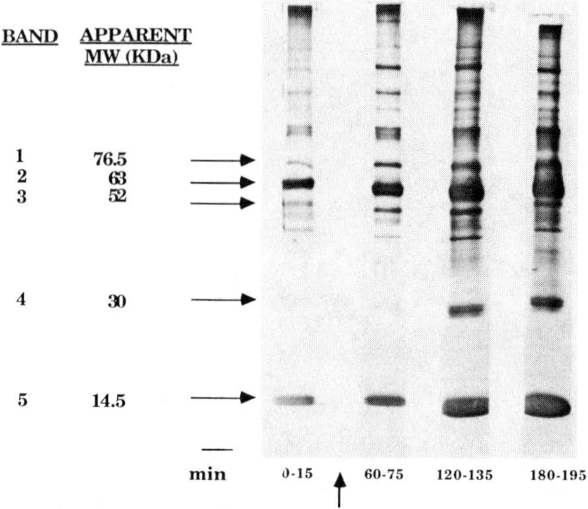

Fig. 1   Polyacrylamide gel electrophoresis of push-pull perfusates. The four lanes show the silver stained protein patterns of perfusates obtained during the first 15min. of each hour of a four hour experiment. Five protein bands (B1 to B5) were densitometrically scanned for quantitative analysis. The induction of LTP (arrow) was followed by a delayed increase in protein efflux.

The above results relate the increases in protein efflux to an NMDA-dependent process. As an initial step towards identifying the possible mechanisms of this increase we chose to characterize protein B5 (Mw 14.5kD), one of the proteins which increased following the induction of LTP by both stimuli. The main reason for selecting B5 was the presence in rat CSF of a protein of similar molecular weigh; the absence of this protein from serum suggested that it may be brain-specific (Fazeli et al., 1988a). We considered, and rejected, two proteins which on the basis of their molecular weights were candidates for B5 : i) S100; immunoreactivity to S100 antibodies was present in the perfusates at 11.5kD, clearly separable from the B5 protein band, ii) Nerve-growth factor; no NGF-like activity was found in the perfusates, before or after the induction of LTP (Fazeli et al., 1988b). The possibility that B5 may be haemoglobin came to light when samples of haemolyzed blood were found to contain high levels of a protein with similar electrophoretic mobility to B5.

B5 is similar to haemoglobin

One of the now established methods for the comparison of two or more proteins, was introduced by Cleveland et al. (1977). In this technique, the electrophoretic mobilities of the peptide fragments produced from a protein by limited proteolytic digestion, are compared to those of a known protein. Staphylococcus Aureus V8 protease, which cleaves peptide bonds at glutamic and aspartic acid residues, was used for this purpose. Fig.2 depicts the densitometric scans of the peptide digests of rat haemoglobin (a) and B5 (b). There are 8 peptide fragments with similar electrophoretic mobilities, which are not present in the lane containing protease alone (c). Additionally, a comparison was made between the spectrophotometric absorbance of two samples of perfusate containing different amounts of B5 (as judged by silver staining) and a solution of rat haemoglobin. The haem group present in haemoglobin has a characteristic absorbance spectrum between 380 and 600nm. The perfusates contain haemoglobin absorbance peaks, the amplitudes of which are

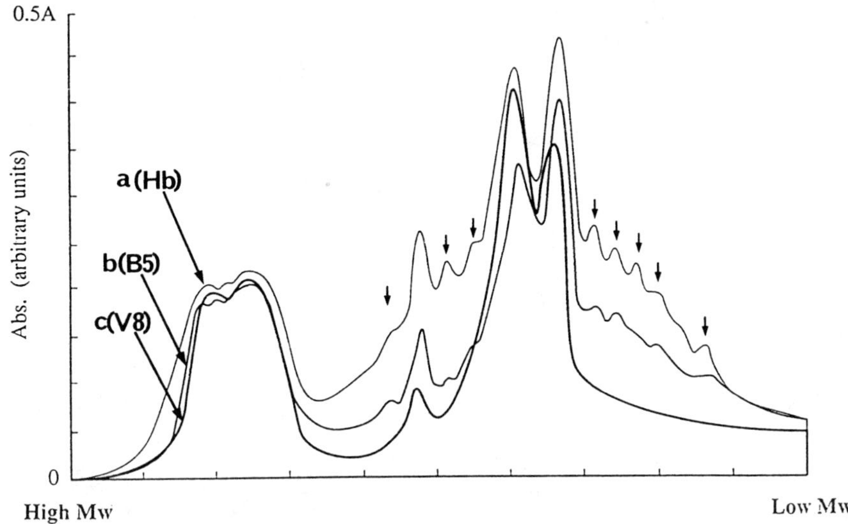

Fig. 2   One-dimensional *in situ* peptide mapping of a) haemoglobin and b) protein B5. A 15min. perfusate fraction and a sample of rat haemoglobin were run on a 15% polyacrylamide gel and then excised following coomassie blue staining. The excised bands were loaded into the sample wells of another gel (also 15%) and co-electrophoresed in the presence of S. Aureus protease V8. S. Aureus protease, run in the absence of any other proteins is depicted in (c). The resulting peptide fragments (in a & b) have similar electrophoretic mobilities (arrows).

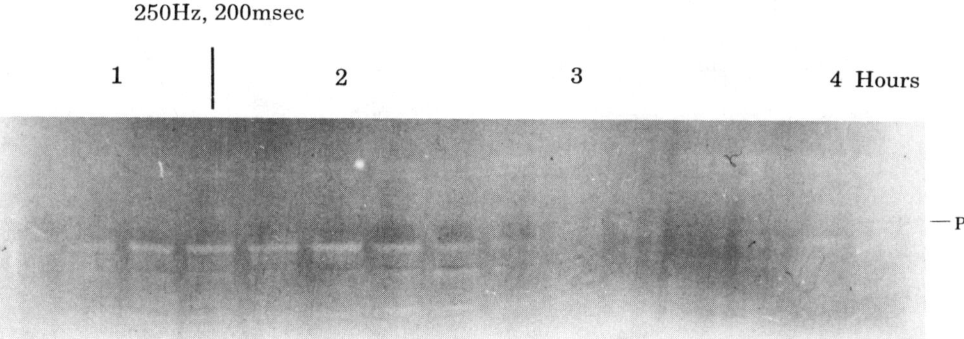

Fig. 3   LTP is associated with an increase in extracellular protease activity as monitored by SDS-PAGE zymography. The proteolytic activity present in a sequence of perfusates, collected over 15min. (corresponding to a total of 3 hours) from one experiment, is shown here. There is a delayed increase in proteolytic activity (P) at approximately 80kD, following the induction of LTP (arrow). Similar results were also obtained in three other experiments.

greater in the sample containing more B5. Moreover, on two-dimensional gels, protein B5 demonstrates a similar microheterogeneity to haemoglobin. Finally, protein B5 is immunoreactive to haemoglobin antibodies in immunoblotting experiments (Fazeli M.S., unpublished observations).

It is not clear why the induction of LTP is associated with an increase in haemoglobin efflux into push-pull perfusates. A possible explanation is that, following the induction of LTP, there is an increased haemolysis of clots formed when the cannula was implanted. Since haemolytic breakdown of clots is a proteolytic event, we speculated that the induction of LTP may be accompanied by the release or activation of extracellular proteases which digest blood clots present in the vicinity of the cannula, and thus give rise to an increase in protein efflux, including haemoglobin (B5).

LTP is associated with an increase in extracellular proteolytic activity

The technique of SDS-PAGE zymography was chosen for protease detection, since it is suitable for a wide range of proteases, including acetylcholinesterase (AchE (80kD)), which has recently been shown to possess trypsin-like proteolytic activity, independent of its esteratic activity (Small & Simpson, 1988).

It is apparent, from the example shown in Fig.3, that the induction of LTP by high-frequency stimulation, is associated with a delayed increase in extracellular protease activity (n=4). The bands on these gels are faint and the low contrast level does not allow densitometric scanning. Nevertheless, there are clear differences between control, where no protease activity was seen after the second or third fraction (n=3, data not shown), and LTP experiments. The most consistent change is seen in the protease activity migrating at approximately 80kD (Fig.3, P). There are also occasional increases seen in protease bands migrating at other molecular weights. When LTP was blocked by APV, in three out of four experiments, the increase in protease did not occur (data not shown).

As mentioned above, S100 protein immunoreactivity was detected in the perfusates. S100 is a highly acidic Ca^{2+}- and Zn^{2+}-binding protein of 23kD. Three S100 species have been identified, $S100a_o$ (two alpha-subunits), S100a (one alpha and one beta subunit) and S100b (two beta subunits). In the rat CNS only S100-beta mRNA has been detected, and the protein is exclusively localized to glial elements (for review see Donato, 1986). A number of experiments on S100 proteins have focused on the role of this protein in the control of behavioural activity, particularly learning and memory processes. Only one study has investigated their involvement in LTP (Lewis and Teyler, 1986). In that experiment, the induction of LTP in the CA1 area of hippocampal slices was found to be blocked by previous application of S100 antibodies. The following is a summary of our experiments on the involvement of S100 in LTP *in vivo*.

Effect of s100 antibodies on the induction/maintenance of ltp *in vivo*

We have studied the effect of S100 antibodies on LTP. Since the experiments were conducted acutely, the delivery of the antibody posed a problem. The major requirement was for a high concentration of antibodies to be present at the recording site. For this purpose, the tip of the recording electrode was broken to facilitate diffusion of antibody from the electrode. S100 antibodies and normal rabbit IgG (NRIgG, control) were diluted (1+3) in the electrode solution (0.5M sodium acetate and a trace of pontamine sky blue).

Stimulating electrodes were positioned bilaterally in the angular bundle, and recording electrodes were placed in the dentate hilus. Following electrode placements and maximization of responses, a rest period of 30min. followed. The diffusion of electrode solution into the neuropil did not affect evoked potentials. Test stimuli were delivered every 30sec. and LTP was induced on both sides after 1 hour of baseline recording, using the same tetanization protocol as above.

Fig. 4 shows the effect of high-frequency stimulation in the presence of S100 antibody (filled squares) and NRIgG (empty squares). Results are plotted as the percentage mean (averaged every 5min., n=5) of the average response in the 10min. just prior to tetanic stimulation. Tetanic stimulation in the presence of NRIgG results in LTP, which is sustained for 3 hours. The average increase in

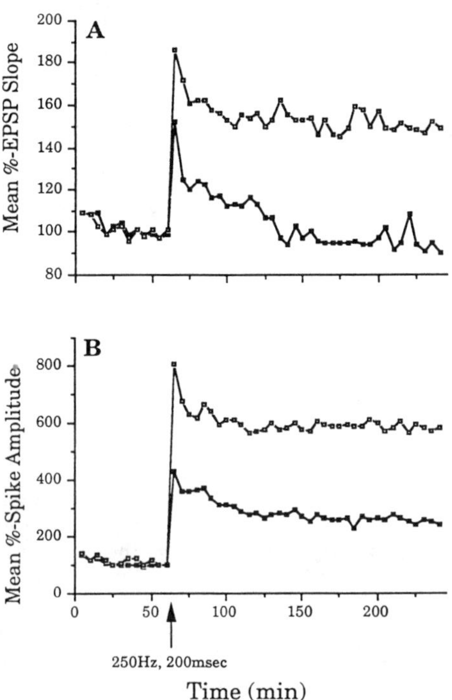

Fig. 4   High-frequency stimulation normally capable of inducing LTP *in vivo*, is
ineffective in the presence of S100 antibodies.
S100 antibodies were diluted (1+3) in electrode solution and delivered to the
dentate gyrus via large bore electrodes, used for extracellular recording. In
control experiments, normal rabbit IgG was used. Test shocks were presented
to both sides, at a rate of 0.03Hz, via stimulating electrodes positioned in the
angular bundle, at a rate of 0.03 Hz. Results are presented as percentages of
the basal response (the mean of the responses 10min. before tetanization),
averaged every 5min.

A) The initial potentation of the EPSP slope induced in the presence of S100
antibody (full squares) is less than in the presence of NRIgG (empty squares)
and potentiation is maintained for less than 1h. (n=5).

B) Initial potentiation of the spike amplitude is reduced on the antibody-
treated side compared to the control side, but it is maintained for the
duration of the experiment, in contrast to EPSP potentiation.

the initial EPSP slope and spike amplitude, measured over the last hour of the experiment, were 51.1% and 479.3% respectively (Fig. 4). In the presence of S100 antibody, the initial potentiation of the EPSP slope, measured 5min. after the tetanus, was less than in controls (25.6% compared to 72.5%, and by the second hour after the tetanus the response had declined to pre-tetanus values (96% of basal). On the other hand, the spike amplitude remained potentiated, though its magnitude, during the last hour, was significantly less than in controls. Such differential effects on LTP of the EPSP and population spike have been reported for protein synthesis inhibitors and inhibitors of PKC activity and inhibitors of arachidonic acid metabolism (Fray et al., 1988; Malinow et al., 1988 and Williams & Bliss, 1989).

In conclusion, the presence of S100 antibodies either masks or blocks the maintenance of LTP of the EPSP slope, in the dentate gyrus of the anaesthetized rat. One possible role for S100 proteins in LTP, may be an effect on transmitter release properties from nerve terminals. In cultured cerebellar granule cells, pre-incubation with S100 (5ug/ml) potentiates subsequent K+-stimulated release (in the absence of exogenous S100) of ^{3}H-glutamate newly synthesized from ^{3}H-glutamine (Fazeli et al., 1989).

DISCUSSION

We have described changes in extracellular protein composition associated with LTP. The findings can be divided into two categories :

i) Involvement of extracellular proteases in LTP.

Changes in synaptic morphology accompanying LTP have been previously described (e.g. Desmond & Levy, 1986). Morphological changes to pre- and/or post-synaptic structures imply corresponding changes to the extracellular matrix. We hypothesize that the extracellular proteases of the kind seen here are candidates for the task of extracellular restructuring. The release of an activating factor from potentiated synapses may trigger the proteolytic breakdown of extracellular matrix proteins, with consequent changes in synaptic morphology. Intracellular proteases may also play a role in LTP (Lynch & Baudry, 1984). Proteases such as plasminogen have previously been implicated in the control of neuronal growth and morphology (Krystosek & Seeds, 1981). A soluble form of acetylcholinesterase (AchE) is released by various stimuli in a number of CNS structures, including the hippocampus (Appleyard & Smith, 1987). In addition, the protein has a proteolytic activity associated with it. In our hands, AchE (Sigma) has proteolytic activity present at approximately 80kD. This is similar to the molecular weight of the major protease (Fig. 3, P) present in our perfusates. The possibility that AchE is released from hippocampal neurones following the induction of LTP is currently being investigated.

ii) A possible role for S100 proteins in the maintenance of LTP.

The involvement of S100 in LTP suggests a role for astroglial cells in activity-dependent changes in synaptic efficacy, since S100 (present in our perfusates, Fazeli et al., 1988b) probably originates from surrounding glia. The mechanism which links the induction of LTP to the release of S100 from astrocytic elements is unknown. Muller & Best (1989) have demonstrated that resupplimentation of cat visual cortex with immature astrocytes reinduces ocular-dominance plasticity in adult animals.

In this chapter, we have presented evidence suggesting that LTP in the dentate gyrus is associated with the release of proteases into extracellular space. This may be a significant factor in the remodelling of synaptic structure which has been observed in LTP. Finally, we have shown that antibodies to the glial protein S100, block a late (>1 h.) component of synaptic LTP. The role of S100 in modulating synaptic transmission in hippocampus is unknown. One possibility is that it regulates glutamate release.

REFERENCES

Appleyard M.E. & Smith A.D.   (1987). **Neurochem. Int.** , 11(4) : 397-406.
Charriaut-Marlangue C., Aniksztejn L., Roisin M.P. & Ben-Ari Y.   (1988). **Neurosci. Lett.**, 91 : 308-
        314.

Cleveland D.W., Fischer S.G., Kirschner M.W. & Laemli U.K.   (1977). **J. Biol. Chem.**, 252 : 1102-1106.

Desmond N.L. & Levy W.B. (1986). **J. Comp. Neurol.**, 253 : 466-475.

Donato R.   (1986). **Cell Calcium**, 7 : 123-145.

Errington M.L., Dolphin A.C. & Bliss T.V.P.   (1983). **J. Neurosci. Meth.**, 7 : 353-357.

Fazeli M.S., Errington M.L., Dolphin A.C. & Bliss T.V.P.   (1988a). **Brain Res.**, 473 : 51-59.

Fazeli M.S., Errington M.L., Dolphin A.C. & Bliss T.V.P.   (1988b). **Eur. J. Neurosci.** (Suppl. No.1) : 20.7.

Fazeli M.S., Errington M.L., Dolphin A.C. & Bliss T.V.P.   (1989). **Eur. J. Neurosci.** (Suppl. No.2) : 67.6.

Fray U., Krug M., Reymann K.C. & Matthies H.   (1988). **Brain Res.**, 452 : 57-65.

Krystosek A. & Seeds N.W.   (1981). **Science**, 213 : 1523-1534.

Lewis D. & Teyler T.J.   (1986). **Brain Res.**, 383 : 159-164.

Lynch G. & Baudry M. (1984). **Science** 343 : 137-143.

Malinow R., Madison D.V. & Tsien R.W.   (1988). 335 : 820-824.

Muller C.M. & Best J. (1989). **Nature** 342 : 427-430.

Small D.H. & Simpson R.J.   (1988). **Neurosci. Lett.**, 89 : 223-228.

Williams J.H. & Bliss T.V.P.   (1989). **Neurosci. Letts**, 88(1) : 81-85.

MODULATION OF THE INDUCTION OF LONG-TERM POTENTIATION IN THE HIPPOCAMPUS

B. R. Sastry, H. Maretic, W. Morishita and Z. Xie

Neuroscience Research Laboratory
Department of Pharmacology & Therapeutics
The University of British Columbia
Vancouver, B. C., Canada, V6T 1W5

LONG-TERM POTENTIATION

In 1894, in his Croonian lecture to the British Royal Society, Ramon y Cajal suggested that learning induces morphological changes at junctions between neurons and that these structural changes form the basis for memory. Hebb (1949) postulated that synaptic strengthening occurs during coincident activity in pre- and post-synaptic elements and that this strengthening forms the basis for learning and memory. This hypothesis of Hebb or a modification of it is generally used to describe the cellular mechanisms of learning and memory. Bliss and Lømo (1973) showed that a transient tetanic stimulation of afferents in the hippocamppus resulted in a post-tetanic long-term potentiation (LTP) of the postsynaptic responses to the same afferent stimulation. LTP is generally thought to be associated with learning and memory although there is no clear evidence in support of this idea.

COOPERATIVITY AND ASSOCIATIVITY

Tetanic stimulations of hippocampal afferents with stimulation strengths that induce only weak responses in the postsynaptic cells are not usually adequate to induce LTP (McNaughton, 1982; Lee, 1983; McNaughton and Barnes, 1983). Tetanic stimulations with strengths that evoke large trans-synaptic responses, however, reliably induce LTP. A cooperativity between two separate excitatory inputs to presumably the same population of postsynaptic neurons to induce LTP has also been shown (McNaughton, 1982; McNaughton and Barnes, 1983). This dependence of LTP induction on the co-activation of several afferents is interpreted by these authors to be due to the involvement of the postsynaptic cells in the induction of LTP. Tetanic stimulation of some afferents in stratum radiatum is known to influence the excitability of other untetanized afferents (Goh and Sastry, 1985). The induction of this interaction between afferents is dependent on the presence of extracellular calcium ions. Low frequency stimulation of stratum radiatum that does not induce LTP of population EPSPs in the CA1 area at this frequency, does induce LTP if the stimulation is paired with a tetanic stimulation of stratum oriens. Depolarization of CA1 neurons with intracellular current injections during a low frequency stimulation of stratum radiatum causes LTP (Gustafsson and Wigstrom, 1986; Sastry et al., 1986; Kelso et al., 1986). In these studies, however, neither the stimulation of stratum radiatum at low frequency nor the depolarization of CA1 neurons alone could induce LTP. It, therefore, appears that activity in the presynaptic terminals is necessary for the induction of LTP.

N-METHYL-D-ASPARTATE RECEPTORS AND LTP

The interesting study of Collingridge et al. (1983) which showed that 2-amino-5-phosphonovalerate (APV), an agent that blocks the actions of N-methyl-D-aspartate (NMDA), prevents the onset of tetanus-induced LTP led the way for further studies that examined the involvement of NMDA receptors in LTP. Magnesium ions interfere with the proper activation of NMDA receptor-coupled channels (MacDermott et al., 1986). This interference by magnesium ions is shown to be reduced if the recorded neuron is depolarized. It has, therefore, been postulated that in the studies involving the pairing of postsynaptic depolarization and presynaptic activation to induce LTP (Gustafsson and Wigstrom, 1986; Sastry et al., 1986; Kelso et al., 1986), the postsynaptic depolarization is minimizing the interference by magnesium ions with ionic channels coupled to NMDA receptors. Presynaptic activation during this depolarization allows the opening of NMDA receptor-coupled channels allowing further events necessary for the induction of LTP to take place (Gustafsson and Wigstrom, 1988; Malinow and Miller, 1986).

GLIAL CELLS AND LTP

As discussed earlier, LTP in the hippocampus could be induced by pairing a depolarization of CA1 neurons with the activation of stratum radiatum (eg., Sastry et al., 1986). In these studies, however, the depolarization of CA1 neurons produced by intracellular current injections are unusually large (often adequate to reverse the polarity of EPSPs). During the tetanic stimulations of stratum radiatum, that induce LTP, CA1 neurons are often depolarized only marginally even when the recording electrode is positioned in the apical dendrites of the CA1 neurons (Sastry, 1988). Tetanic stimulations of stratum oriens during a low frequency stimulation of stratum radiatum results in LTP of the stratum radiatum induced CA1 responses (Kelso and Brown, 1986; Sastry et al., 1986). In the above experiment, when the stimulation of stratum oriens is adjusted so that no depolarization can be detected with electrodes placed in the proximal apical dendrites of CA1 neurons, LTP of the stratum radiatum stimulation-induced EPSP can still be induced in the same neurons (Sastry, 1988). Presumably, any depolarization produced by tetanization of stratum oriens has electrotonically spread mainly from the basal dendrites through the soma to the apical dendrites to influence the Schaffer collateral-CA1 synapses. If so, it is unlikely that the subsynaptic membranes in the apical dendritic area are depolarized by the above stimulation of stratum oriens. It is, therefore, clear that in the experiments in which CA1 neuronal depolarization (with intracellular current injections) is paired with stratum radiatum stimulation to induce LTP (eg., Sastry et al., 1986), excessive depolarization of CA1 neurons is necessary. These results cast doubt on a direct involvement of CA1 neuronal depolarization in the induction of LTP. Voltage- and current-clamps of CA1 neurons as well as injections of chemicals into these neurons in hippocampal slices have been reported in literature to influence the ability to induce LTP by tetanic stimulation of the afferents. In these studies, the interpretations are based on the assumption that such intracellular insults influence the membrane properties of only the recorded neuron. The density of synaptic contacts in the apical dendritic area of CA1 pyramidal neurons is high and the extracellular space is quite limited. It is, therefore, not illogical to think that dramatic interference with the membrane electrical properties of CA1 neurons can alter ionic concentrations in the extracellular space

378

(eg., potassium ions) and affect the presynaptic terminals as well as the surrounding glia.
Moreover, injections of chemicals into the CA1 neurons may interfere with the retrograde
interaction between these neurons and the presynaptic terminals, with CA1 neuronal interactions with surrounding glia, or even with the concentrations of extracellular ions (eg., if the
injected chemicals decrease intracellular free calcium ions, calcium-mediated potassium
currents can be diminished). Another presumption in the literature has been that the
postsynaptic depolarization is necessary to unblock the magnesium ion's interference with
NMDA channels. It is, however, possible that the availability of magnesium ions at the
NMDA channel can be reduced by a release into synaptic clefts of agents that sequester the
ions thereby minimizing the need for postsynaptic depolarization to induce LTP. We,
therefore, examined the involvement of factors other than CA1 neuronal depolarization in
the induction of LTP (Sastry, 1988a, b, c).

Tetanic stimulations of hippocampal afferents (or even single stimulations) cause
large and prolonged depolarization of glial cells in the CA1 area of the hippocampus (Alger
et al., 1983; Casullo and Krnjević, 1987). Electrical coupling between the CA1 neurons and
glia has also been suggested to occur (Alger et al., 1983; Casullo and Krnjević, 1987). It is,
therefore, possible that large depolarizations of CA1 neurons can affect the nearby glia. The
possibility that glial depolarization may contribute to the development of LTP is, therefore,
examined (Sastry et al., 1988c). The depolarization of non-spiking cells (presumed to be
glia) in the CA1 apical dendritic area produced by tetanic stimulation of stratum radiatum is
smaller if the tetanus is given when the slices are incubated in a medium without added
calcium or in a medium containing APV (the involvement of other glutamate receptors is
not yet tested). Both these treatments are known to block the induction of LTP. A depolarization of glial cells with intracellular current injections during (but not without) the activation of stratum radiatum results in a subsequent LTP. This LTP is found not only at single
neuronal level but also at the population spike level. Perhaps, glial depolarization spreads to
several other glial cells (because of electrical coupling to each other) and then influences
synapses on several CA1 neurons. The input specificity of LTP is preserved in this case
because only activated synapses are potentiated. How a depolarization of glial cells can
induce LTP, is unclear. Perhaps, the depolarization releases substances from glia that then
find access to synapses. Since only activated synapses are affected, the endocytosis of the
substances into neurons requires this activation, or the action of these substances on neuronal membranes is voltage dependent. It is also possible that the substances released from
glial cells are involved in allosteric modulation of NMDA receptor-coupled channels (as in
the case of glycine) since LTP induction by pairing glial depolarization with afferent stimulation is greatly diminished if APV is present during the pairing (Sastry, 1988). It is possible that glial depolarization results in the release of potassium ions that depolarize nearby
neuronal elements and then help in the induction of LTP (by unblocking the effects of
magnesium ions on NMDA channels). Further investigation is necessary to examine this
possibility.

PEPTIDES, GROWTH-ASSOCIATED MACROMOLECULES AND LTP

Eccles (1983) raised the possibility that substances released from the postsynaptic
dendrites could act on the presynaptic terminals inducing further changes necessary for
LTP. If peptides are involved in the induction of LTP, they can be released from the dendrites. If this occurs, a replenishment of the releasable pool for subsequent release should
not be difficult because peptides are known to be synthesized in the cellbody. In the brain,
the levels of nerve growth factor (NGF) are highest in the hippocampus and the cerebral

cortex, and the corresponding mRNA is also detected in these two areas, suggesting that the growth factor is locally synthesized (Korsching et al., 1985). NGF in the hippocampus is known to be released following injury to hippocampal pathways (Nieto-Sampedro and Cotman (1985). The peptide is believed to be released from the postsynaptic neurons and taken up by the presynaptic terminals (although a release from the glial cells is also thought to occur); it causes sprouting of boutons, brings about biochemical changes in the presynaptic terminals, and helps establish synaptic contacts (Hendry et al., 1974; Springer and Loy, 1985). Moreover, the growth factor has been proposed to help Alzheimer's patients by helping the survival and the synaptic contacts of cholinergic neurons in the brain (Will and Hefti, 1985; Hefti and Weiner, 1986). High frequency stimulation of hippocampal afferents causes alterations in synaptic morphology as well as LTP (Van Harreveld and Fifkova, 1975; Fifkova and Van Harreveld, 1977; Fifkova and Anderson, 1981; Harris et al., 1987; Racine et al., 1983; Lee et al., 1981; Chang et al., 1986; Geinisman et al., 1988). We, therefore, decided to examine whether LTP- and neurite-inducung substances are released by tetanic stimulations.

Tetanic stimulations in guinea pig hippocampus in vivo causes a release of collectable substances that can induce LTP when applied on hippocampal slices (Chirwa and Sastry, 1987). The effects arere, however, small and the collection procedure is extremely tedious. LTP is known to occur in the neocortex (Lee, 1982) and the surface of the cerebral cortex is accessible for sample collection. Fluids were, therefore, collected from large surface areas on rabbit cortical surface during and without tetanic stimulation of the cortical surface. When these samples are applied on guinea pig hippocampal slices, samples collected during neocortical tetanization but not those that are collected without tetanic stimulation induce LTP of the CA1 population spike (Sastry et al., 1988a, 1988b). This effect of the applied sample occurs only if stratum radiatum is stimulated during the application of the sample. It, therefore, appears that presynaptic activity is necessary for the agents to act. The intracellularly recorded EPSP is also potentiated subsequent to the application of the neocortical sample. During or after the application of the sample the resting membrane potential and the input resistance of the hippocampal CA1 neurons are not altered significantly (Morishita and Sastry, 1989). It is interesting that if the samples are preheated and cooled, their LTP - inducing effect is greatly diminished. It is unlikely that the LTP - inducing agent is either glutamate or acetylcholine (Sastry et al., 1988a, 1988b). It is clear that heat-sensitive substances that induce LTP are released by the tetanic stimulation.

Substances collected from rabbit neocortex during tetanic stimulation also induce growth of neurites in PC-12 cells in culture (Sastry et al., 1988a, 1988b). As in the case of LTP, neocortical samples collected without tetanic stimulation, or samples collected with tetanic stimulation but are heated and cooled prior to application, do not induce any significant neurite growth in PC-12 cells. Saccharin, which interferes with the growth of neurites induced by NGF (Ishii, 1982), also blocks the neurite inducing action on PC-12 cells of the neocortical sample (Sastry et al., 1988a, 1988b). It is, therefore, thought that NGF-like substances may be involved in the induction of neurites in PC-12 cells and in LTP. Further studies are needed to clarify the involvement of NGF-like agents in the induction or establishment of LTP.

Proteins ranging from 14 to 86 kD have been shown to be released into extracellular fluid during LTP (Duffy et al., 1981; Hesse et al., 1984). Similarly, Charriaut-Marlangue et al. (1987) report that 90 min after the induction of LTP in the rat hippocampus, the release of peptides with 16-64kD molecular weight is increased and that a new peptide of 19 kD is observed. Nystrom and associates (1986) report that application of NMDA into the hippocampus causes a release of peptides with molecular weights ranging from 14 to 66 kD. Whether any of these peptides are involved in LTP induction is not clear. While protein synthesis inhibitors (Stanton and Sarvey, 1984) and antibodies to proteins (Lewis and Teyler, 1986; Stanton et al., 1987) are known to interfere with LTP, certain proteins have been shown to produce LTP when applied (Cherubini et al., 1987). Charriault-Marlangue et al. (1987) found that APV, which blocks LTP generation, decreased the presence of the 19 kD peptide discussed above. It is possible that activation of NMDA receptors that occurs during the inductive phase of LTP results in the release of peptides that are involved in further events leading to LTP. Gel electrophoresis (1 D and 2 D) of the samples collected from the tetanized neocortex shows a release of peptides with molecular weights >50 as well as <10 kD. We also fractionated the samples to find the approximate molecular weights of the substances that induce LTP in guinea pig hippocampal slices (Sastry et al., 1989a). Table 1 summarizes the results.

Table 1. Effects of various fractions of samples collected from the rabbit neocortical surface on the CA1 population spike in guinea pig hippocampal slices.

Fraction	CA1 population spike as a % of control expressed as mean $\pm$ SEM		
	30 min post-application	60 min post-application	n
<3 kD	110.00 $\pm$ 9.68	127.75 $\pm$ 10.65*	10
3-10 kD	116.88 $\pm$ 6.67	136.88 $\pm$ 9.11*	8
10-25 kD	98.63 $\pm$ 17.58	93.13 $\pm$ 20.45	8
<25 kD	152.25 $\pm$ 14.60*	164.38 $\pm$ 12.29*	8
25-50 kD	97.13 $\pm$ 9.69	101.38 $\pm$ 9.99	8
>50 kD	127.75 $\pm$ 13.37*	155.25 $\pm$ 10.25*	8
<25 kD (MK-801)**	94.50 $\pm$ 6.44	121.17 $\pm$ 16.20	6
>50 kD (MK-801)**	98.00 $\pm$ 8.88	78.20 $\pm$ 18.39	5
TNS (MK-801) Tet.***	194.50 $\pm$ 19.86*	174.83 $\pm$ 19.43*	6

** The rabbits were injected with MK-801 (0.5 mg/kg, i.p.) one hour prior to sample collection.

***The sample collected during tetanic stimulation of the rabbit neocortex (TNS) pretreated with MK-801, was applied during the tetanic stimulation of the guinea pig hippocampal slices (100 Hz, 5 s).

*The increases of responses were statistically significant at P<0.05 using Duncan's multiple comparison's test.

It is clear from Table 1 that fractions containing substances with molecular weights of <3, 3-10 and >50 kD can produce LTP and that samples collected from rabbits pretreated with MK-801 can not induce LTP. It, therefore, appears that NMDA receptor activation is necessary to cause the release of the LTP-inducing substances into the extracellular fluids. The fraction containing substanses with molecular weights <3 kD appears to contain LTP-inducing substances that are not heat sensitive. An examination of the presence of amino acids in the samples collected from the rabbit neocortex shows that the levels of released glycine increase by at least 4 fold during tetanic stimulation of the neocortex (Sastry et al., 1989b). Identification of a variety of other amino acids released is currently underway. The presence of glycine in the collected samples is interesting because it has recently been shown by Forsythe et al. (1988) that added glycine potentiates the NMDA component of hippocampal CA1 neuronal EPSPs.

GLIA AS A SOURCE OF LTP-INDUCING SUBSTANCES

Astrocytes in primary culture are well known to release substances into the medium that promote the growth and survival of hippocampal neurons in culture (Banker, 1980). Subsequent to an injury to the pathways to the hippocampus, trophic factors are released in the lesioned areas (Gage et al., 1984; Needels et al., 1986). Extracts of hippocampus pro-duce neurotrophic effects on medial septal cells (Ojika and Appel, 1984). It has, therefore, been suggested that the target cells in the hippocampus have trophic effect on the input cells from the septum. It is, however, possible that the neurotrophic agents come from hip-pocampal glial cells. Carlstedt (1985) report that regenerating axons in peripheral nerves develop synaptic boutons (but not functional synaptic contacts with other neurons) when they are surrounded by astrocytes. That astrocytes release not only NGF but also other growth related macromolecules is well known (Lindsay, 1979). While glia appear to have trophic influences on neurons, neurons are also thought to have such actions on glia (see review by Varon and Somjen, 1979). It is generally believed that glial cells release growth related macromolecules that are taken into neuronal presynaptic terminals by endocytosis and that these agents alter the intracellular biochemistry and eventually get transported towards the neuronal somata. The biochemical changes that take place inside the presynap-tic terminal include the activation of protein kinase C and the phosphorylation of B50 (Varon and Somjen, 1979; Cremins et al., 1986; Perrone-Bizzozero et al., 1986). In this connection it is interesting that in the goldfish, learning induces the release of two glial proteins into the extracellular medium (Shashoua, 1979).

SUMMARY AND CONCLUSIONS

High frequency stimulation in the guinea pig hippocampus or on the rabbit cerebral cortical surface results in a release of substances that produce, when applied, LTP in the CA1 area of guinea pig hippocampal slices. The substances collected from the neocortex also induce neurite growth in PC-12 cells. The samples collected during the tetanic stimula-tion of the neocortex contained increased concentrations of glycine, and various other amino acids that are being identified, as well as peptides. Whether the release of the substances is from neurons or from glia is being investigated.

Tetanic stimulations of stratum radiatum in the guinea pig hippocampus that induce LTP in CA1 neurons also cause large and prolonged depolarization of glial cells in the CA1 apical dendritic area. Artificial depolarization of glial cells during the activation of stratum radiatum results in LTP of the CA1 neuronal EPSP. It is, therefore, suggested that glial depolarization is involved as one of the steps in the induction of LTP. We speculate that the depolarization results in the release of substances into the extracellular space and that these substances are involved in directly or indirectly modulating the NMDA receptor-coupled channels as well as in producing trophic effects to induce structural changes in the synapses that are thought to be associated with the establishment and maintenance of LTP.

REFERENCES

Alger, B. E., McCarren, M. and Fisher, R. S., 1983, On the possibility of simultaneously recording from two cells with a single microelectrode in the hippocampal slice, Brain Res., 270:137-141.

Bliss, T. V. P. and Lømo, T., 1973, Long-lasting potentiation of synaptic transmission in the dentate area of the anaesthetized rabbit following stimulation of the perforant path, J. Physiol. (Lond.), 232:331-356.

Carlstedt, T., 1985, Regenerating axons from nerve terminals at astrocytes, Brain Res., 347:188-181.

Casullo, J. and krnjevic, K., 1987, Glial potentials in hippocampus, Can. J. Physiol. Pharma col., 65:847-855.

Chang, F.-L. F., Hwang, M., Isaacs, K. R. and Greenough, W. T., 1986, Morphological effects of LTP studied with rapid freezing and freeze substitution in rat hippocampal slice, Soc. Neurosci. Abstr., 12:506.

Charriault-Marlangue, C., L. Aniksztejn, L., Roisin, M. P. and Ben-Ari, Y., 1988, Release of proteins during long-term potentiation in the hippocampus of the anaesthetized rat, Neurosci. Lett., 91:308-314.

Cherubini, E., Ben Ari, Y., Goh, M., Bidard, J. N. and Lazdunski, M., 1987, Long-term potentiation of synaptic transmission in the hippocampus induced by a bee venom peptide, Nature, 328:70-73.

Chirwa, S. S. and Sastry, B. R., 1987, Chemical and post-synaptic modulation of the induc tion of synaptic potentiation in guinea pig hippocampus in vitro., J. Physiol. (Lond.), 386:29P.

Collingridge, G. L., Kehl, S. J. and McLennan, H., 1983, Excitatory amino acids in synaptic transmission in the Schaffer collateral-commissural pathway of the rat hippocampus, J. Physiol. (Lond.), 334:33-46.

Cremins, J. Wagner, J. A. and Halegoua, S., 1986, Nerve growth factor action is mediated by cyclic AMP- and Ca-phospholipid-dependent protein kinases, J. Cell Biol., 103:887-893.

Crutcher, K. A. and Collins, F., 1986, Entorhinal lesions result in increased nerve growth factor-like growth-promoting activity in medium conditioned by hippocampal slices, Brain Res. , 399:383-389.

Duffy, C. J., Teyler, T. J. and Shashoua, V. E., 1981, Long-term potentiation in the hip pocampal slices: evidence for stimulated secretion of newly synthesized proteins, Science, 212:1148-1151.

Eccles, J. C., 1983, Calcium in long-term potentiation as a model for memory, Neuros cience, 10:1071-1081.

Fifkova, E. and Anderson, C. L., 1981, Stimulation-induced changes in dimensions of stalks of dendritic spines in the dentate molecular layer, Exp. Neurol., 74:621-627.

Fifkova, E. and Van Harreveld, A., 1977, Long-lasting morphological changes in dendritic spines of dentate granule cells following stimulation of the entorhinal region, J. Neurocytology, 6:211-230.

Forsythe, I. D., Westbrook, G. L. and Mayer, M. L, 1988, Modulation of excitatory synaptic transmission by glycine and zinc in cultures of hippocampal neurons, J. Neurosci., 8:3733-3741.

Gage, F. H., Bjorklund, A. and Steveni, U., 1984, Denervation releases a neuronal survival factor in adult rat hippocampus, Nature, 308:637-639.

Geinisman, Y., Morrell, F. and deToledo-Morrell, 1988, Modelling of synaptic architecture during hippocampal "kindling", Proc. Natl. Acad. Sci. USA, 85:3260-3264.

Goh, J. W. and Sastry, B. R., 1985, Interactions among presynaptic fiber terminals in the CA1 region of the rat hippocampus, Neurosci. Lett., 60:157-162.

Gustafsson, B. and Wigstrom, H., 1986, Hippocampal long-lasting potentiation produced by pairing single volleys and brief conditioning tetani evoked in separate afferents, J. Neurosci., 6:1575-1582.

Gustafsson, B. H. and Wigstrom, H., 1988, Physiological mechanisms underlying long-term potentiation, Trends in Neurosciences, 11:156-162.

Harris, K. M., Jensen, F. E. and Tsao, B., 1987, Development of hippocampal synapses, spines and LTP, Soc. Neurosci. Abstr., 13:1429.

Hebb, D. O., 1949, The organization of behavior, John Wiley & Sons.

Hefti, F. and Weiner, W. J., 1986, Nerve growth factor and Alzheimer's disease, Ann. Neurol., 20:275-281.

Hendry, I. A., Stokel, K., Thoenen, H. and Iversen, L. L., 1974, The retrograde axonal transport of nerve growth factor, Brain Res., 68:103-121.

Hesse, G. W., Hofstein, R. and Shashoua, V. E., 1984, Protein release from hippocampus in vitro, Brain Res., 305:61-66.

Ishii, D. N., 1982, Effect of the suspected tumor promoters saccharin, cyclamate, and phenol on nerve growth factor binding and response in cultured embryonic chick ganglia, Cancer Res., 42:429-432.

Kelso, S. R. and Brown, T. H., 1986, Differential conditioning of associative synaptic enhancement in hippocampal brain slices, Science, 232:85-87.

Kelso, S. R., Ganong, A. H. and Brown, T. H., 1986, Hebbian synapses in hippocampus, Proc. Natl. Acad. Sci. USA, 83:5326-5330.

Korsching, S., Auburger, G., Heumann, R., Scott, J. and Thoenen, H., 1985, Levels of nerve growth factor and its mRNA in the central nervous system of the rat correlate with cholinergic innervation, Eur. Med. Biol. Organization J., 4:1389-1393.

Lee, K. S., 1982, Sustained enhancement of evoked potentials following brief, high-fre quency stimulation of the cerebral cortex in vitro, Brain Res., 239:617-623.

Lee, K. S., 1983, Cooperativity among afferents for the induction of long-term potentiation in the CA1 region of the hippocampus, J. Neurosci., 3:1363-1372.

Lee, K., Oliver, M., Schottler, F. and Lynch, G., 1981, In: Electrophysiology of isolated mammalian CNS preparations, Eds. G.A. Kerkut and H.V. Wheal, Academic press NY, pp. 189-211.

Lewis, D. and Teyler, T. J., 1986, Anti-S-100 serum blocks long-term potentiation in the hippocampal slice, Brain Res., 383:159-164.

Lindsay, R. M., 1979, Adult rat brain astrocytes support survival of both NGF dependent and NGF-insensitive neurons, Nature, 282:80-82.

MacDermott, A. B., Meyer, M. L., Westbrook, G. L., Smith, S. J. and Barker, J. L., 1986, NMDA-receptor activation increases cytoplasmic calcium concentration in cultured spinal cord neurons, Nature, 321:519-522.

Malinow, R. and Miller, J. P., 1986, Post-synaptic hyperpolarization during conditioning reversibly blocks induction of long-term potentiation, Nature, 320:529-530.

McNaughton, B. L., 1982, Long-term synaptic enhancement and short-term potentiation in rat fasia dentata act through different mechanisms, J. Physiol. (Lond.), 324:249-262.

McNaughton, B. L. and Barnes, C. A., 1977, Physiological identification and analysis of dentate granule cell responses to stimulation of the medial and lateral perforant pathways in the rat, J. Comp. Neurol., 175:439-454.

Morishita, W. and Sastry, B. R., 1989, Unpublished observations.

Needels, D. L., Nieto-Sampedro, M. and Cotman, C. W., 1986, Induction of a neurite-promoting factor in rat brain following injury or deafferenation, Neuroscience, 18:517-526.

Nieto-Sampedro, M. and Cotman, C. W., 1985, Growth factor induction and temporal order in central nervous system repair. In: Synaptic Plasticity, Ed. C.W.Cotman, Guilford press, NY. pp. 407-456.

Nystrom, B. Hamberger, A. and Karlsson, J.-O., 1986, Changes of extracellular proteins in hippocampus during depolarization, Neurochem. Int., 1:55-59.

Ojika, K. and Appel, S. H., 1984, Neurotrophic effects of hippocampal extracts on medial septal nucleus in vitro, Proc. Natl. Acad. Sci. USA, 81:2561-2571.

Perrone-Bizzozero, N. I. Finklestein, S. P. and Benowitz, L. I., 1986, Synthesis of a growth-associated protein by embryonic rat cerebrocortical neurons in vitro, J. Neurosci., 6:3721-3730.

Racine, R. J., Milgram, N. W. and Hafner, S., 1983, Long-term potentiation phenomena in the rat limbic forebrain, Brain Res., 260:217-231.

Sastry, B. R., 1988, Unpublished observations.

Sastry, B. R., Goh, J. W. and Auyeung, A., 1986, Associative induction of posttetanic and long-term potentiation in CA1 neurons of rat hippocampus, Science, 232:988-990.

Sastry, B. R., Chirwa, S. S., May, P. B. Y., Maretić, H., Pillai, G., Kao, E. Y. H. and Sidhu, S. D., 1988a, Are nerve growth factors involved in long-term synaptic potentiation in the hippocampus and spatial memory? In: Synaptic plasticity in the hippocampus, Eds. H. Haas and G. Buzsaki, Springer-Verlag, Berlin, pp. 102-105.

Sastry, B. R., Chirwa, S. S., May, P. B. Y. and Maretić, H., 1988b, Substances released during tetanic stimulation of rabbit neocortex induce neurite growth in PC-12 cells and long-term potentiation in guinea pig hippocampus, Neurosci. Lett., 91:101-105.

Sastry, B. R., Goh, J. W., May, P. B. Y. and Chirwa, S. S., 1988c, The involvement of non-spiking cells in long-term potentiation of synaptic transmission in the hippocampus, Can. J. Physiol. Pharmacol., 66:841-844.

Sastry, B. R., Maretić, H., Morishita, W. and Xie, Z., 1989a, Unpublished observations.

Sastry, B. R., Hansen, S., Maretić, H., Morishita, W. Perry, T. L. and Xie, Z., 1989b, Unpublished observations.

Shashoua, V. E., 1979, Brain metabolism and the acquisition of new behaviors III. Evidence for secretion of two proteins into the brain extracellular fluid after training, Brain Res., 166:349-358.

Springer, J. E. and Loy, R., 1985, Intrahippocampsl injections of antiserum to nerve growth factor inhibit sympathohippocampal sprouting, Brain Res. Bull., 15:629-634.

Stanton, P. K. and Sarvey, J. M., 1984, Blockade of long-term potentiation in rat hippocampal CA1 region by inhibitors of protein systhesis, J. Neurosci., 4:3080-3088.

Stanton, P. K., Sarvey, J. M. and Moskal, J. R., 1987, Inhibition of the production and maintenance of long-term potentiation in rat hippocampal slices by a monoclonal antibody, Proc. Natl. Acad. Sci. USA, 84:1684-1688.

Van Harreveld, A. and Fifkova, E., 1975, Swelling of dendritic spines in the fascia dentata after stimulation of the perforant fibers as a mechanism of post-tetanic potentiation, Exp. Neurol., 49:736-749.

Varon, S. S. and Somjen, G. G., 1979, Neuron-glia interactions, In: Neurosciences Programme Bulletin, 17:1-239, MIT press.

Will, B. and Hefti, F., 1985, Behavioral and neurochemical effects of chronic intravenous injections of nerve growth factor in adult rats with fimbria lesions, <u>Behav. Brain Res.</u>, 17:17-24.

ROLES OF METABOTROPIC AND IONOTROPIC GLUTAMATE RECEPTORS IN

THE LONG-TERM POTENTIATION OF HIPPOCAMPAL MOSSY FIBER SYNAPSES

Hiroyuki Sugiyama, Isao Ito[*] and Daisuke Okada

Department of Cellular Physiology, National Institute for
Physiological Sciences, Okazaki 444, and [*]Fujigotemba
Research Labs, Chugai Pharmaceutical Company, Gotemba 412,
Japan

INTRODUCTION

Glutamate (Glu) is thought to be a major excitatory neurotransmitter in the central nervous system, and its receptors are believed
to play important roles in brain functions such as learning and memory.
For instance, Glu receptors of N-methyl-D-aspartate (NMDA) type are
critically important in a certain form of synaptic long-term potentiation
(LTP), which is a cellular or molecular model for memory.   In other
cases, however, in hippocampal mossy fiber synapses for instance, NMDA
receptors may not be involved in the formation of LTP.   The type of
receptors involved there is not known.

We have been trying to characterize different subtypes of Glu
receptors using a simpl model system, $\underline{Xenopus}$ oocytes injected with rat
brain mRNA, and to apply the knowledge obtained there for the analysis of
molecular mechanism of brain functions.   In this chapter, we shall
briefly summerize the recent results of such analyses.

PROPERTIES OF GLUTAMATE RECEPTORS ANALYZED IN $\underline{XENOPUS}$ OOCYTES

Injection of mRNA isolated from rat brains has been shown to induce
different subtypes of Glu receptors in $\underline{Xenopus}$ oocytes (Fig.1).
Functional and pharmacological analyses indicated that these subtypes of
receptors could be classified into two major categories, metabotropic and
ionotropic Glu receptors (Sugiyama et al., 1989a)

Metabotropic Glu Receptors

Glu receptors of this category function by activating G proteins
and cause inositol phospholipid metabolism and intracellular Ca^{2+}
mobilization (Sugiyama et al., 1987).   The receptors are activated by
Glu, quisqualate (QA), ibotenate, and homocystein sulfinate (Fig.1A)
(Sugiyama et al., 1989a).   Antagonist pharmacology indicated that the

metabotropic Glu receptor is very different from the ionotropic subtypes
in that none of the ionotropic Glu receptor antagonists, either NMDA
type, non-NMDA type or non-selective type, is effective on this receptor
(Sugiyama et al., 1989a).

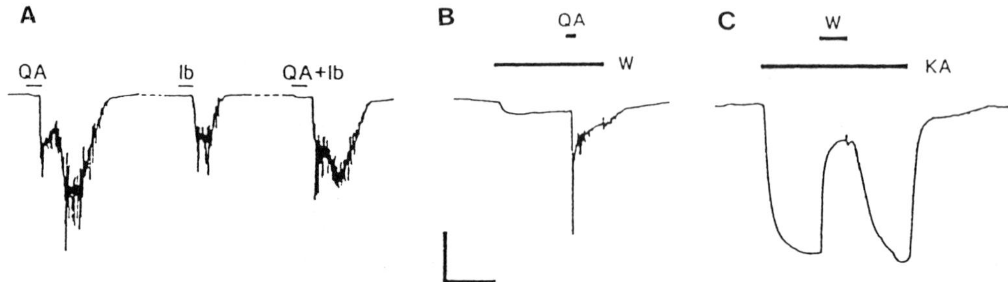

Fig.1. Typical examples of current responses of _Xenopus_ oocytes
injected with rat brain mRNA.   Metabotropic Glu receptors are
activated by QA or ibotenate (Ib).   Willardiine (W) functions
as a pure ionotropic QA agonist, as AMPA does.   It evokes
small smooth responses (B), and suppresses KA responses (C).
The membrane potential was clamped at -60 mV.   Calibrations:
40 nA and 2 min for A, 20 nA and 2 min for B and C.   From
Sugiyama et al., 1989a, and Sugiyama et al., 1989b.

Ionotropic Glu Receptors

Messenger RNA-injected oocytes show ionotropic responses to three
typical glutamatergic agonists, NMDA, QA and kainate (KA).   The
responses evoked by NMDA are considered to be mediated by Glu receptors
of NMDA type (Verdoorn et al., 1987).   However, clear discrimination
between receptors mediating QA responses and KA responses (which are
usually several to more than ten fold larger than QA responses) has been
hampered due to the lack of selective antagonists.   Moreover, KA
responses are suppressed by QA or QA-type agonists in a competitive
manner when applied simultaneously (Fig.1C) (Hirono et al., 1988;
Verdoorn and Dingledine, 1988; Rassendren et al., 1989; Sugiyama et al.,
1989b).   This observation has led to an interpretation that KA and QA
activate the same receptor, the efficacy of KA being much larger than
that of QA (Verdoorn and Dingledine, 1988; Rassendren et al., 1989).

Although this problem has not been resolved conclusively, we made
the following two observations which seem to be more readily consistent
with the interpretaion that QA and KA activate their own separate
receptors.

Effects of 7-L-glutamyl glycine (LGG)   This dipeptide was first
introduced by Watkins and his collaborators to distinguish QA and KA
responses (Davies and Watkins, 1981).   We confirmed that LGG suppressed
KA responses but not responses evoked by QA or α-amino-3-hydroxy-5-
methyl-4-isoxazolepropionic acid (AMPA), a specific ionotropic QA agonist
(Hirono et al., 1988).   Instead, LGG potentiated AMPA responses to a
small extent.   In contrast to LGG, however, γ-D-glutamyl glycine (DGG),
a reportedly more potent dipeptide than its L form isomer, did not
distinguish between them (Table I).

<u>Table I</u>. Effects of the dipeptides on KA and AMPA responses of <u>Xenopus</u> oocytes injected with rat brain mRNA.

Peptide	KA responses	AMPA responses
LGG	67 $\pm$ 12 (3)	135 $\pm$ 25 (4)
DGG	19 $\pm$ 13 (2)	33 $\pm$ 12 (2)

The values are relative current responses (%) to 100 μM KA or 10 μM AMPA in the presence of 1 mM LGG or DGG compared to its absesnce, mean $\pm$ SEM (n). From Sugiyama et al.,1989b.

<u>Effects of aniracetam</u>. This compound, reported to affect learning and memory processes (Cumin et al., 1982), was found to potentiate ionotropic QA or AMPA responses to 2-3 times the control (Fig.2). The potentiating effects were specific for ionotropic QA type responses, and were not exerted on the responses evoked by NMDA, KA or GABA (Table II).

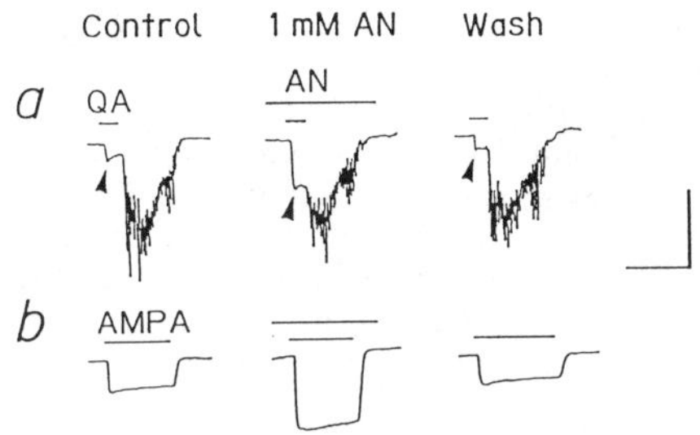

<u>Fig.2</u>. Potentiating effects of aniracetam (AN) on current responsses mediated by QA receptors in <u>Xenopus</u> oocytes injected with rat brain mRNA. <u>a</u>, QA evoked ionotropic responses (arrow heads) were followed by the oscillatory responses mediated by the metabotropic Glu receptors. The former was selectively potentiated by AN. <u>b</u>, Pure ionotropic responses evoked by AMPA were potentiated by AN. Concentrations of QA and AMPA were 20 μM. Calibration: 10 nA and 2 min.

ROLES OF GLUTAMATE RECEPTORS IN HIPPOCAMPAL MOSSY FIBER SYNAPSES

The synaptic transmission of mossy fiber-CA3 pyramidal cells is sensitive to non-NMDA antagonists such as kynurenate (Harris and Cotman, 1986) or 6-cyano-7-nitroquinoxaline-2,3-dione (CNQX) (Neuman et al., 1988). Since these antagonists block both ionotropic QA and KA

receptors, however, the problem still remains which of the non-NMDA
receptors, QA type or KA type, is involved. In addition, very little is
known about the receptors that trigger the formation of LTP in this
synapse. Thus we analyzed pharmacolgical properties of mossy fiber
synapses by field potential measurement using rat hippocampal slices.

Receptor subtypes involved in the synaptic transmission

It has been reported that the mossy fiber termination zones in CA3
regions are particularly dense in kainate binding sites compared to AMPA
binding sites, whereas AMPA sites are highly concentrated in CA1 regions
(Cotman et al., 1987). This led to a hypothesis that mossy fiber
synapses are mainly mediated by KA receptors whereas Schaffer/commissural
synapses by ionotropic QA receptors. However, the direct functional
evidence for this hypothesis is still missing due to the lack of specific
antagonists which can distinguish these two receptor subtypes.

Since aniracetam can distinguish, at least in oocytes, ionotropic QA
and KA responses, we tried to apply this compound as a pharmacological
tool to analyse the properties of hippocampal synapses.

Fig.3 shows examples of field potentials recorded in stratum lucidum
of CA3 regions of rat hipocampal slices (inset). Aniracetam (1 mM)
reversibly potentiated mossy fiber synaptic transmission. This suggests
that a subtype of Glu receptor equivalent to ionotropic QA receptors
expressed in oocytes is involved, at least partly, in the mossy fiber
synaptic transmission. The average potentiation was 29 $\pm$ 8 % (n=7).
Similar results were obtained from the analyses performed on field
potentials recorded in stratum radiatum of CA1 regions in response to the
stimuli given to Schaffer/commissural fibers. Aniracetam (1mM)
reversibly potentiated CA1 population EPSP by 26 $\pm$ 3 % (n=5). These
results may be more readily consistent with a notion that pharmacological
properties of mossy fiber transmission and Schaffer/commissural
transmission may not be very much different, than with a hypothesis
assuming the differential roles of ionotropic QA and KA receptors in
these two types of synapses. At least, the results suggest that Glu
receptors equivalent to ionotropic QA receptors in oocytes contribute to

Table II. Specificity of the aniracetam effects on amino acid
responses in mRNA-injected oocytes.

Amino acids	Relative amplitudes
QA	356 $\pm$ 16 (15)
AMPA	211 $\pm$ 10 (10)
KA	96 $\pm$ 8 (5)
NMDA	92 $\pm$ 5 (5)
GABA	90 $\pm$ 5 (5)

The amplitudes of current responses to various amino acids in
the presence of 1 mM aniracetam are shown relative to its
absence (%), as mean $\pm$ SEM (n). Concentrations are: QA and
AMPA, 20 μM; KA and GABA, 100 μM; NMDA, 100 μM together with 10
μM glycine.

a similar extent in both of these synapses. More extensive
characterization of ionotropic QA and KA receptors in the oocytes and
intracellualar analyses of EPSPs in slices are required on one hand, but
re-examinations of the significance and identity of the binding sites, on
the other hand, may also be necessary to settle the issue.

<u>Receptors involved in the mossy fiber LTP</u>

As shown in Fig.3, aniracetam potentiated the mossy fiber
transmission both before and after the formation of LTP to a similar
extent. The average ratio of aniracetam potentiation after the
formation of LTP to that before LTP was 1.01 $\pm$ 0.10 (n=7). The results
suggest that pharmacological properties of the transmission, at least the
contribution of the aniracetam-sensitive Glu receptor subtype, may not be
changed drastically by the formation of LTP. This is expected if the
mossy fiber LTP is maily due to the increase in the release of
transmitters, as suggested by Yamamoto and his collaborators (Kamiya et
al., 1989).

Although NMDA receptors are crucial for the formation of CA1 LTP,
NMDA antagonists do not affect CA3 LTP. Instead, we have shown that
mossy fiber LTP was suppressed by pertussis toxin which is known to
inactivate certain types of G proteins, suggesting that some kind of
neurotransmitter receptors coupled to G proteins are most likely

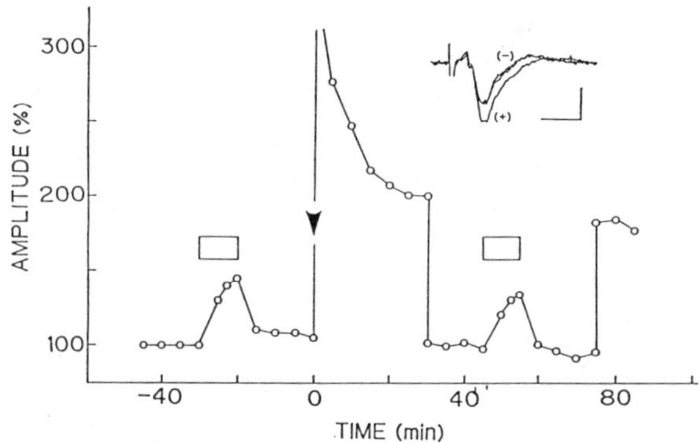

<u>Fig.3</u>. Time courses of changes in the amplitudes of field
potentials recorded in stratum lucidum of CA3 regions of rat
hippocampal slices stimulated to the granular cell layer. At
the arrowhead, a tetanic stimulation of 100 Hz for 1 sec was
given twice (20 sec apart), with the same strength as test
pulses. The open boxes indicate the periods during which
aniracetam (1 mM) was perfused. During the period of t=30 to
75 min, the strength of stimulation was reduced so that the
amplitude of the field potential was approximately the same as
the pre-tetanic original level. <u>Inset</u>: Examples of the field
potentials. The traces marked (-) were recorded before the
application of aniracetam and after 30 min wash, and the trace
marked (+) was recorded when aniracetam (1 mM) was perfused for
10 min. Calibrations: 0.5 mV and 5 ms.

involved.   One of the possible candidates is the metabotropic Glu
receptor which is known to be coupled to a G protein (Sugiyama et al.,
1987).   This receptor is unaffected by either NMDA antagonists such as
2-amino-5-phosphonovaleric acid (AP5) or non-NMDA antagonists such as
CNQX, as described above.   Thus we examined the effects of these
antagonists on the mossy fiber LTP.

As described previously (Neuman et al., 1988), mossy fiber synapses
are blocked by CNQX.   However, when a tetanic stimulation was given at
100 Hz in the presence of CNQX, we consistently observed a shift of the
field potential baseline which is accumulated during the tetanus.   To
minimize the shift, we adopted the stimulation pattern similar to the
theta pattern used by Lynch and his collaborators (Larson et al., 1986).
An example of such stimulation is shown in Fig.4.   Under such
conditions, the tetanic stimulation caused only negligible responses
(Fig.4).   However, when the antagonists were washed out after such
tetanic stimulation, the field potential reappeared, and its amplitude
reached a level which is substantially larger than the pre-tetanic level.
After 50 min washing, the amplitude was 154 $\pm$ 5 % of the pre-tetanic
amplitude (mean $\pm$ SEM, n=6).   This value of the potentiation was
essentially the same as the normal LTP induced by the identical tetanic
stimulation give in the absence of the antagonists (161 $\pm$ 4 %, mean $\pm$
SEM, n=6).   Thus, CNQX (or CNQX plus AP5) did not prevent the formation
of LTP, although the transmission itself was almost completely blocked
(Neuman et al., 1988).   These results suggest that transmitter receptors
which trigger the mossy fiber LTP are insensitive to CNQX and AP5.

We have shown that G proteins are involved in the mossy fiber LTP.
Since G proteins can promote phospholipid metabolism by activating either
phospholipase C   or phospholipase A$_2$, we examined the effects of

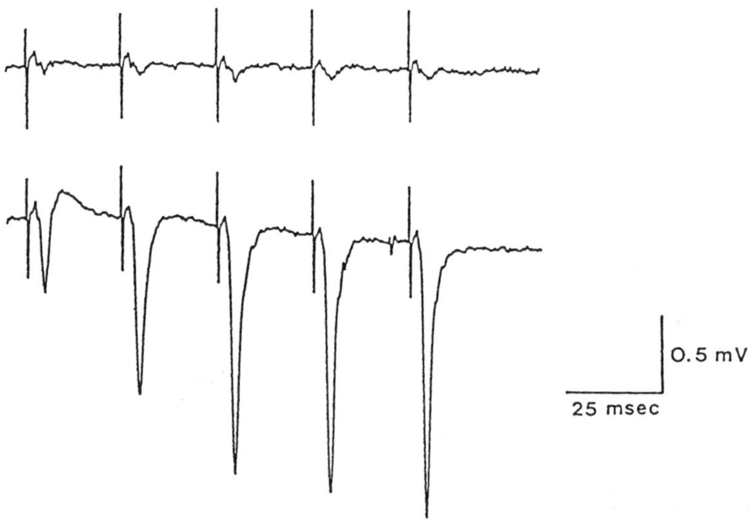

<u>Fig.4</u>. Examples of field potential responses recorded during
tetanic stimulations in stratum lucidum of CA3 regions of rat
hippocampal slices.   The tetanic stimulation was given to the
granular cell layer in the presence (upper trace) or absence
(lower trace) of 20 μM CNQX and 50 μM AP5, as bursts of 5
pulses at 40 Hz, repeated 10 times at 5 Hz.   The strength of
each pulse was the same as test pulses.

phospholipase inhibitors.   As shown in Fig.5, nordihydroguaiaretic acid (NDGA), an inhibitor of phospholipase A_2 and lipoxygenase, suppressed the CA1 LTP, confirmaing the results of Williams and Bliss (1988).   In addition, we observed that bromophenacyl bromide (BPB), an inhibitor of phospholipase A_2, also suppressed the CA1 LTP.   In contrast, however, neither of them affected the mossy fiber LTP (Okada et al., 1989). Since G protein-mediated reactions are involved in the mossy fiber LTP, this result would suggest that phospholipase C reactions may more likely be involved in it than phospholipase A_2 reactions, whereas the latter reactions may be involved in the CA1 LTP (Williams and Bliss, 1988).

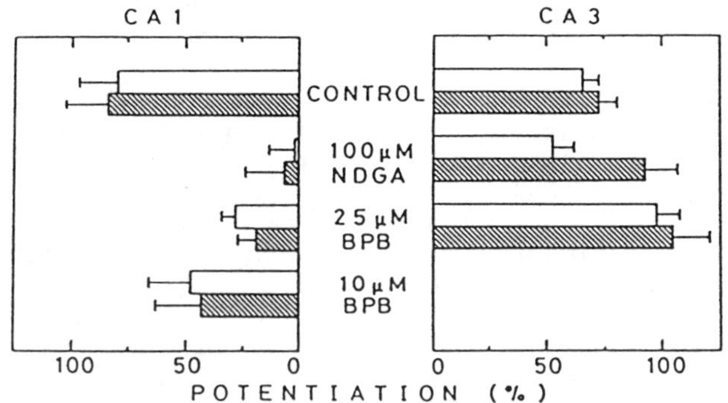

Fig.5. The differential effects of NDGA and BPB on LTP in the CA1 and CA3 subfields.   The bars show the SEM.   The open and hatched columns represent the percent increases above the pre-tetanic controls in the field potential amplitude and initial slope, respectively.   From Okada et al., 1989.

Taken together, these results suggest that the metabotropic Glu receptor is one of the plausible candidates which mediates the formation of mossy fiber LTP, because the receptor is insensitive to the known glutamatergic antagonists, functions by activating G proteins, and promotes inositol phospholipid metabolism.

REFERENCES

Cotman, C.W., Monaghan, D.T., Ottersen, O.P. and Storm-Mathisen, J., 1987, Anatomical organization of excitatory amino acid receptors and their pathways, Trends Neurosci., 10:273.
Cumin, R., Bandle,E.F., Gamzu, E. and Haefely, W.E., 1982, Effects of the novel compound aniractam (Ro 13-5057) upon impaired learning and memory in rodents, Psychopharmacology, 78:104.
Davies, J. and Watkins, J.C., 1981, Differentiation of kainate and quisqualate receptors in the cat spinal cord by selective antagonism with gamma-D (and L)-glutamylglycine, Brain Res., 206:172.
Harris, E.W. and Cotman, C.W., 1986, Long-term potentiation of guinea pig mossy fiber responses is not blocked by N-methyl D-aspartate antagonists, Neurosci. Lett.,70:132.

Hirono, C., Ito, I., Yamagishi, S. and Sugiyama, H., 1988, charactyerization of glutamate receptors induced inXenopus oocytes after injection of rat brain mRNA, Neurosci. Res., 6:106.

Kamiya, H., Sawada, S. and Yamamoto, C., 1989, Long-term potentiation in the mossy fiber-CA3 synapse of the guinea pig, International Workshop on "Plasticity of synaptic transmission", March 1989, Kanazawa.

Larson, J., Wong, D. and Lynch, G., 1986, Patterned stimlation at the theta frequency is optimal for the induction of hippocampal long-term potentiation, Brain Res., 368:347.

Neuman, R.S., Ben-Ari, Y., Gho, M. and Cherubini, E., 1988, Blockade of excitatory synaptic transmission by 6-cyano-7-nitroquinoxaline-2,3-dione (CNQX) in the hippocampus in vitro, Neurosci. Lett., 92:64.

Okada, D., Yamagishi, S. and Sugiyama, H., 1989, Differential effects of phospholipase inhibitors in long-term potentiation in the rat hippocampal mossy fiber synaposes and Schaffer/commissural syanpses, Neurosci. Lett., 100:141.

Rassendren, F.A., Lory, P., Pin, J.P., Bockaert, J. and Nargeot, J., 1989, A specific quisqualate agonist inhibits kainate responses induced in Xenopus oocytes injected with rat brain RNA, Neurosci. Lett., 99:333.

Sugiyama, H., Ito, I. and Hirono, C., 1987, A new type of glutamate receptorlinked to inositol phospholipidmetabolism, Nature, 325:531.

Sugiyama, H., Ito, I. and Watanabe, M., 1989a, Glutamate receptor subtypes may be classified into two major categories: a study on Xenopus oocytes injected with rat brain mRNA, Neuron, 3:129.

Sugiyama, H., Watanabe, M., Taji, H., Yamamoto, Y. and Ito, I., 1989b, Pharmacological properties of quisqualate- and kainate-preferring glutamate receptors induced in Xenopus oocytes by rat and chick brain mRNA, Neurosci. Res., 7: in press.

Verdoorn, T.A., Kleckner, N.W. and Dnigledine, R., 1987, Rat brain N-methyl0D-aspartate receptors expressed in Xenopus oocytes, Science, 238:1114.

Verdoorn, T.A. and Dingledine, R., 1988, Excitatory amino acid receptors expressed in Xenopus oocytes: agonist pharmacology, Molec. Pharmacol., 34:298.

Williams, J.H. and Bliss, T.V.P., 1988, Induction but not maintenance of calcium-induced long-term potentiation in dentate gyrus and area CA1 of the hippocampal slices is blocked by nordihydroguaiaretic acid, Neurosci. Lett., 88:81.

EFFECTS OF A NEUROTROPHIC FACTOR (FGF) ON DEVELOPMENT, REGENERATION

AND SYNAPTIC PLASTICITY OF CENTRAL NEURONS

W. Seifert, F. Förster, B. Flott and H. Terlau

Dept.of Neurobiology
Max-Planck-Institut f. biophys. Chemie
3400 Göttingen, W-Germany

INTRODUCTION

Fibroblast growth factor (FGF) was originally isolated from bovine brain and pituitary and described as a mitogenic factor for fibroblasts and other cell types of mesodermal origin (1,2). In recent years two astroglial growth factors (AGF1 and AGF2) were purified from bovine brain and demonstrated to be identical with basic and acidic fibroblast growth factor (3). In the following the term FGF is used only for the basic factor (bFGF) of molecular weight 16.000.

FGF stimulates the proliferation and maturation of astrocytes, as demonstrated by the work of M. Sensenbrenner and collaborators(3). They also showed by immunohistochemistry that FGF is localized in neurons of the brain and cerebellum *in-vivo* and is found also in dissociated primary cultures in-vitro. Thus FGF would appear to be an astroglial growth factor which is produced by the neurons of the CNS.

On the other hand, it has been also demonstrated in recent years that FGF can exhibit neurotrophic functions on central neurons (3). We can confirm this observation by studies of the effects of FGF on neurons in our hippocampal cell culture system (5), where FGF clearly stimulates both survival and neurite outgrowth (6). Therefore FGF may also function as a neurotrophic factor, although this function has not yet been demonstrated for the in-vivo stituation.

In addition to the possible roles of FGF during development of the nervous system, we have evidence to suggest also a role for FGF during regeneration after lesions ("wounding") and a role in the modulation of synaptic plasticity, as shown by experiments on LTP in the hippocampal slice.

RESULTS AND DISCUSSION

I. Role of FGF in development: experiments in a dissociated cell culture system of fetal rat hippocampus

We have studied the effect of FGF on neuronal cells in a hippocampal cell culture system derived from 18 day old rat fetuses. These cultures consist mainly of pyramidal neurons (60%) and basket neurons (30%) as well as astrocytes (10%). The granular neurons of the area dentata proliferate only during the first postnatal week and therefore are still absent in these cultures .

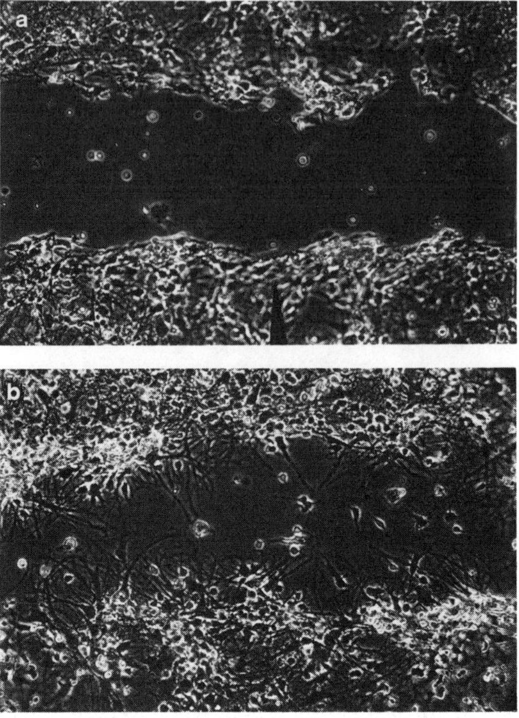

Fig. 1.   Cell cultures from rat hippocampus (E 18) or from rat cortex of newborn rats
were prepared in serum-free hormone- supplemented medium as previously
described (8,9). For the "wound experiment" shown here, cells were plated at
high density on polylysine coated coverslips. After the initial attachment
phase of 45 min in DMEM medium containing 5% horse serum and 5% fetal
calf serum, the coverslips were gently washed and placed into serumfree
medium supplemented with a mixture of hormones. These cultures were
medium- changed every 2 days and used for experiments after 10 or 14 days.
After scratching a wound into the cell layer on the coverslip, the cells and
their processes were observed by phasecontrast microscopy, by microcinema-
tography and by fluorescence microscopy following immunofluorescence staining with anit-GFAP for the astrocytes and anti-neurofilament for the neurons. Fibroblast growth factor (FGF) was purchased from Sigma Col., Munich.

Fig. 1a.   Phasecontrast photograph of a dense cortex culture immediately after
scratching the wound.

Fig. 1b.   The same wound culture system at 24 h after wounding.

Table 1)

Length (µm) of Neurites/Cell		Average No. of Neurites/Cell	Survival (% living cells of after 72 h in culture)	
Con.: 81	4.1	2.4	19.4%	0.4
FGF : 147	6.6	3.6	32.5%	6.7

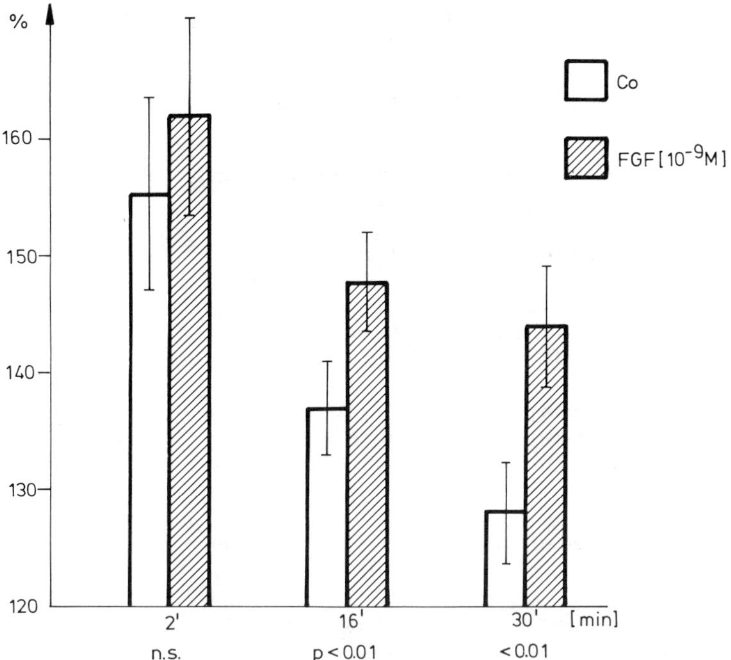

Fig. 2.   Effect of FGF on Long-Term Potentiation (LTP): Stimulation was done in the Schaffer-commissural pathway and extracellular recording of the field EPSP in the dendritic layer of the CA1 region (stratum radiatum). 100% = average field EPSP slope 10 min before tetanic stimulation. The stimulus strength before the tetanus was so adjusted that the slope of the field EPSP in control slices and FGF-treated slices was quite similar (about 3.00 mV/ms; n = 20 pairs of slices). FGF at 10^{-9} M was added to the bath medium 20 min before the tetanus was given (1 x 100 Hz for 1 sec).

In serum-free culture the postmitotic hippocampal neurons develop processes within 1-2 days and form neuronal networks with synapses after 8-10 days in culture.

When added to these cultures at nanomolar concentrations (10-9 M), FGF exhibits the typical effects of a neurotrophic factor : it stimulates neurite outgrowth (neuritogenic effect) and it improves the survival of these neurons (neuronotrophic effect). We have quantitatively measured these effects under various conditions. The following data give as an example for the neuritogenic effect the results of a 48 h time point and for the survival effect the results of a 72 h time point.

Thus FGF is not only a mitogenic growth factor for astrocytes, but also a neurotrophic factor for central neurons. Whether the fibroblast growth factor has this double function also under physiological conditions in the *in-vivo* brain, remains to be investigated. But it is tempting to speculate that FGF controls both the development of neurons and of astrocytes. If it is only localized in neurons, then these cells would provide the neurotrophic factor for neuronal-glial communication in the CNS, and FGF would be the first factor to act in a concerted action on two different types of cells: neurons and astrocytes.

II. Role of FGF in regeneration: experiments with an in-vitro lesion system of cortex or hippocampus cultures

We have developed and characterized in our laboratory a "wound culture system" (7) which we consider a suitable model to study lesions under in-vitro conditions. By scratching a wound in a dense cortex culture derived from newborn rats, it is possible to observe the various phases of process extension and cell migration into the wound by neuronal cells and glial cells under well controlled conditions. This system is used in our laboratory to study the effects of neurotrophic factors such as FGF and of neurotrophic agents such as gangliosides on the regenerative processes induced by the wounding.

Fig. 1 shows this "wound culture system" at time 0 after scratching the wound into the culture (1a) and at 24 h after wounding (1b).

If FGF is added to this wound culture system, we observed a significant increase in the length and rate of astrocytes processes stretching out into the open space. By measuring GFAP-stained cultures, we can clearly distinguish between neuronal processes and astrocyte processes. There are also glial cells migrating into the wound. These processes mimick the well-established "reactive gliosis" which occurs in-vivo after lesion or injury to the brain.

From our studies we like to propose a role for FGF during the regenerative processes following lesions: such damage to the nervous tissue will destroy neurons and in this way the released FGF could then activate the proliferation and migration of glial cells in and around the wound.

III. Role of FGF in synaptic plasticity: experiments on long term potentiation (LTP) in the hippocampal slice

Long term potentiation is a widely studied example of synaptic plasticity which was first established in the hippocampus in 1973 by Bliss and Lomo (8). Since then the mechanism of LTP has been studied extensively in the hippocampal slice preparation. It seems now well established that the induction of LTP is caused on the post-synaptic side by activation of the NMDA-receptor provided the postsynaptic neuron is sufficiently depolarized to remove the Mg-block and thus allow the influx of Ca-ions. There is also good evidence for a sustained increase of glutamate release from the presynaptic side during the maintenance phase of LTP. However, it is not known by which mechanism the postsynaptic side with its NMDA-receptors informs the presynaptic side of its activation. Therefore - according to a hypothesis by T. Bliss - there should be a retrograde messenger molecule operating from the post- to the pre-synaptic side. Apart from glycolipids, also peptides could be considered as possible candidates.

Therefore we started to look for the possible effects of neurotrophic factors such as EGF or FGF on the LTP process. We first studied the effect of EGF (9). These experiments showed no effect for single pulse or paired pulse stimulation in the CA1 field of the hippocampal slice. However after tetanic stimulation of the Schaffer-commissural pathway in the presence of 10^{-8} M EGF, we observed a significant increase in the amount of LTP, measured as size of the population spike in the CA1 somatic field. Intracellular recordings showed no change in the membrane resting potential (H. Terlau and W. Seifert, 1989).

In a similar series of experiments we have now studied the effect of FGF in the same paradigm: stimulation of the Schaffer-commissural pathway and recording from the CA1 region of the hippocampal slice. **Fig. 2** shows the result of tetanic stimulation in presence of 10^{-9} M FGF: there is a significant increase in LTP, measured as slope of the dendritic field EPSP, especially at later times (30 min).

As in the case of EGF we observed no effect of FGF on single pulse or paired pulse stimulation. More details of this investigation will be published elsewhere (H. Terlau and W. Seifert, manuscript in preparation).

Since FGF is definitely localized in the pyramidal neurons of the hippocampus, as demonstrated for example by immunohistochemical studies in the laboratory of M. Sensenbrenner, it seems at least feasible that during the course of tetanic stimulation or after activation of the NMDA-receptor this neurotrophic factor might be released from the presynaptic side or the post-synaptic side and in some way modulate the size or the time course of LTP.

Although the physiological significance of our observations remains to be established, the fact that both EGF and FGF - neurotrophic factors for central neurons and astrocytes - modulate synaptic plasticity in the hippocampal slice, adds a new facet to the multiple roles of neurotrophic factors in the nervous system. As a new hypothesis we propose that neurotrophic factors such as FGF which are specific for the glutamatergic pyramidal neurons of the hippocampus and cortex are modulators of synaptic plasticity in the adult brain.

We like to acknowledge that B.R. Sastry proposed that NGF or related growth factors are released by tetanic stimulation of cortex or hippocampus and could induce LTP (B.R. Sastry et al. pp. 102-105 in "Synaptic plasticity in the hippocampus eds. H.L. Haas and G. Buzsaki, Springer-Verlag 1988".

REFERENCES

1. D. Gospodarowicz, J. Cheng, G. M. Lui, A. Baird, and P. Bohlent, Isolation of brain fibroblast growth factor: identity with pituitary fibroblast growth factor **Proc. Natl. Acad. Sci.** USA 81:6963 (1984).
2. P. S. Rudland, W. Seifert, and D. Gospodarowicz, Growth control in cultured mouse fibroblasts: Induction of the pleiotypic and mitogenic response by a purified growth factor, **Proc. Natl. Acad. Sci.** USA, 71:2600 (1974).
3. B. Pettmann, C.Gensburger, M. Weibel, F. Perraud, M.Sensenbrenner, and G.Labourdette, Isolation of two astroglial growth factors from bovine brain; comparison with other growth factors; cellular localization, **in:** "Glia- Neuronal Communication in Development and Regeneration," NATO ASI Series H: Cell Biology, Vol. 2, H.Althaus and W.Seifert, ed., Springer-Verlag, Berlin (1987).
4. P. Walicke, W.M. Cowan, N. Meno, A. Baird, and R.Guillemin, Fibroblast growth factor promotes survival of dissociated hippocampel neurons and enhances neurite extension, Proc Natl Acad Sci USA 83:3012 (1986).
5. W. Seifert, B. Ranscht, H.J. Finck, F. Förster, S. Beckh, H.W. MÅller, Development of hippocampal neurons in cell culture: a molecular approach, **in:** "Neurobiology of the Hippocampus," W.Seifert ed., Academic Press, London (1983).
6. F.Förster, and W.Seifert, Influence of ganglioside and growth factors in hippocampus and cortex cellcultures, **J. Neurochemistry** 52, Supplement, ISN-Meeting 52:S 214A (1989).
7. W. Seifert, and F. Förster, Effects of gangliosides and fibroblast growth factor on central neurons in cell cultures of rat hippocampus and cortex during development and regeneration in-vitro, **in:** "Neural development and regeneration," A. Gorio et al., ed., Springer-Verlag, Berlin (1988).
8. T.V.P. Bliss, and T. Lomo, Long-lasting potentiation of synaptic transmission in the dentate area of the anaesthetized rabbit following stimulation of the perforant path, **J. Physiol.** 232:331 (1973).
9. H. Terlau, and W. Seifert, Influence of epidermal growth factor on long-term potentation in the hippocampal slice, **Brain Research** 484:352 (1989).

TRANS-SYNAPTOPHOBIA

Aryeh Routtenberg

Northwestern University

"These considerations suggest that the change must occur directly, not indirectly, at or near the synapse. It occurs just when or after the surface membrane on the two sides of the synaptic cleft has been disturbed (depolarized) and must involve both cells. This seems to suggest strongly that something passes between cell A and cell B, that this something can only pass through the two membranes when both have been recently disturbed, and that the "something" forms the basis of learning and memory. Let us call this "something" F. F passes between A and B and then alters the synapse, either by causing it to grow or atrophy or perhaps by affecting the enzyme-controlled production of transmitter. Which way does F pass? There are several reasons for thinking it more likely to be from B to A."

Griffith, <u>Nature,</u> 1966, p. 1160

In this chapter I briefly review certain hoary issues in the study of synaptic plasticity. These comments may be generalized to the study of information storage in the nervous system, hopefully to provide some focus to discussion of these pervasive issues.

1. Pre- vs. Post-Synaptic Localization of Mechanism

Perhaps more than any other issue, there is the continued debate as to whether the fundamental change occurring at the synapse that underlies memory is on one side or the other. It is unfortunate in this writer's view that the issue has taken on the mien of a political discussion: it must be either pre- or post-synaptic, and the possibility that it is both, is not an acceptable solution. Anyone who follows American politics will know that often the candidates are equally good, or embarassingly, equally bad. Often one candidate has certain strengths not possessed by the other and one wonders if it were possible to have a pan-political solution: a Repubocrat. I suggest that unlike politics we are not forced to choose. Our task is to locate the particular molecular events in space and in time. It would appear that, on logical grounds alone, it necessary to conclude as Hebb (1949), Griffith (1966) and Schmidt et al. (1976) did, that synaptic modification involves both sides of the synapse.

Despite this logical conclusion, theories have been proposed that provide examples of an unusual disorder "transynaptophobia" - fear of the other side of the synapse. For example, Goelet et al. (1986) provide a theory of presynaptic terminal change replete with retrograde genomic regulation, but there is no post-synaptic membrane even depicted in their summary figure. Lynch and Baudry (1984) propose a cytoskeletal rearrangement of the postsynaptic region, and like Goelet et al. render the other side of the synapse invisible. Both theories appear to claim that the mechanism proposed is universal and fundamental to information storage processes. Buth then one would be forced to conclude that only one of them could be right, or one of these theories is dead wrong. For a recent example of trans-synaptophobia the reader is referred to Olds et al. (1989).

2. Alternatives to an exclusive uni-sided synaptic theory: evidence from NMDA receptor blockade

Without placing too fine a point on it, the position taken here is that both sides of the synapse participate in the process of enhanced synaptic communication. Given the evidence on communication in both directions between synaptic sides over the past two decades (see Schmidt et al., 1976, and theories of Hebb (1949) and Griffith (1966) for earlier statements of this position) it is difficult to imagine that alterations occurring in one side would not have an impact on the other.

An example of this comes from two studies which have shown that blockade of the NMDA receptor influences the presynaptic terminal (Bliss, Lynch et al., 1988; D. Linden et al., 1988). If one assumes that the blocker, APV, is acting on the post-synaptic receptor - which all hippocampal slice and LTP studies indicate - then this would suggest a novel "retrograde" mechanism for synaptic plasticity. But to come back to the major theme, it emphasizes the point that synaptic plasticity involves the cooperative interaction between the pre- and postsynaptic sides of the junction. That the temporal order of interaction may involve first post- then pre- is an interesting particular that Bliss and Lynch (1988) and Linden and Routtenberg (1989) discuss in recent reviews. Hopeful signs that the phobia is lessening may be viewed in the reports by Malinow et al. (1989) and Collingridge et al. (1988).

3. Is there a Plasticity Kinase?

It appears that alterations of the signal transduction mechanism could play a cental role facilitating synaptic transmission. This mechanism, as reviewed elsewhere (Nishizuka, 1986) has a focal point in the activation of protein kinase C (PKC). Although the present author has emphasized the critical nature of this enzyme in the process of synaptic enhancement, it was not meant to exclude the involvement of other kinases such as calmodulin II and cyclic AMP kinases. Indeed, as described in Nelson et al. (1989), Malinow et al. (1989) and Malenka et al. (1989), there is a growing body of evidence to suggest their involvement. The evidence for PKC is at this writing more extensive than for the other kinases, since activators enhance, inhibitors block, and synaptic enhancement or learning itself activates the enzyme such that the extent of enhancement correlates with extent of enzyme activation. The next year should provide more evidence, I suspect, for the extent and nature of the involvement of the other kinases, particular calmodulin kinase II.

Having indicated that several mechanisms are likely to participate in the plasticity process, one should not lose sight of the SAMITO principle: "Some are more important than others." That is, in the transduction network certain molecular events represent a focal point around which modulations occur. It is becoming increasingly obvious that signal transduction and transmembrane signalling mechanismsinvolve both amplifying cascades as well as negative and positive feedback loops. Thus, perturbation of the network at any point could have a significant effect on the outcome. This issue will apply to both presynaptic terminal and postsynaptic region where signal transduction and transsynaptic signalling will take place. The solution to this issue may be significantly aided by the recent discovery that certain key enzymes, PKC for example, exist as a family with distinctive subcellular distributions and protein substrate selectivity. There are now recognized to be at least 7 subtypes of the enzyme with differential localization in the central nervous system (Nishizuka, 1988). The possibility exists that different subtypes are present within the presynaptic vs. the postsynaptic domains. Compelling evidence for this perspective does not as yet exist, though immunocytochemical and in situ hybridization results provide an initial basis for such surmises (e.g., Linden and Routtenberg, 1989).

4. Concluding Remarks

The present period of research in neurobiology is exciting for a variety of reasons not the least of which is the technological developments that facilitate what once were very difficult experiments. The availability of molecular biological techniques will provide us with insights that will stretch our perpectives to inter-relate genomic events to cellular events to information storage processes. In this period of discovery, a euphemism for ignorance to be sure, it would be wise to emulate the process we study, and keep synapses flexible, on both sides!

REFERENCES

Bliss, T.V.P. and Lynch, M.A. Long-term potentiation of synaptic transmission in the hippocampus: properties and mechanisms. In P.W. Landfield and S.A. Deadwyler (Eds.), Long-Term Potentiation: **From Biophysics to Behavior**, Liss, New York, 1988, 3-72 .

Collingridge, G.L and Lester, R.A.J. The sensitivity of CA neurones to quisqualate following high-frequencey stimulation in rat hippocampus in vitro. **J. Physiol.** (Lond), 1988, **398**, 22.

Davies, S.N., Lester, R.A.J., Reymann, K.G. and Collingridge, G.L. Temporally distinct pre- and postsynaptic mechanisms maintain long-term potentiation. **Nature**, 1989, **338**, 500-503 .

Goelet, P., Castellucci, V.F., Schacher, S. and Kandel, E.R. The long and the short of long-ter memory - a molecular framework. **Nature**, 1986, **322**, 419-422.

Griffith, J.S. A theory of the nature of memory. **Nature**, 1966, **211**, 1160-1163.

Hebb, D.O., **The Organization of Behavior**, Wiley, New York, 1949.

Linden, D.J., Wong, K.L., Sheu, F.-S. and Routtenberg, A. NMDA receptor blockade prevents the increase in protein kinase C substrate (protein Fl) phosphorylation produced by long-term potentiation. **Brain Res.**, 1988, **458**, 142-146.

Linden, D.J. and Routtenberg, A. The role of protein kinase C in long-term potentiation: a testable model. **Brain Res. Rev.**, 1989, **14**, 279-296.

Lynch, G., Gribkoff, V.K. and Deadwyler, S.A. Long-term potentiation is accompanied by a reduction in dendritic responsiveness to glutamic acid. **Nature** (Lond), 1976, **263**, 151-153.

Lynch G. and Baudry, M. The biochemistry of memory: a new and specific hypothesis. **Science**, 1984, **224**, 1057-1063.

Malenka, R.C., Kauer, J.A., Perkel, D.J., Mauk, M.D., Kelly, P.T., Nicoll, R.A. and Waxham, M.N. An essential role for postsynaptic calmodulin and protein kinase activity in longterm potentiation. **Nature**, 1989, **340**, 554-556.

Malinow, R., Schulman, H. and Tsien, R. Inhibition of postsynaptic PKC or CaMKII blocks induction but not expression of LTP. **Science**, 1989, **245**, 862-866.

Nelson, R.B., Linden, D.J., and Routtenberg, A. Phosphoproteins localized to presynaptic terminal linked to persistence of long-term potentiation (LTP): quantitive analysis of two-dimensional gels. **Brain Research**, 1989, **497**, 30-42.

Nishizuka, Y. Studies and perspectives of protein kinase C, **Nature** (Lond.), 1986, **233**, 305-312.

Nishizuka, Y. The molecular heterogenicity of protein kinase C and its implications for cellular regulation. **Nature** (Lond.), 1988, **334**, 661-665.

Olds, J.L., Anderson, M.L., McPhie, D.L., Staten, L.D. and Alkon, D.L. Imaging of memory-specific changes in the distribution of protein kinase C in the hippocampus. **Science**, 1989, **245**, 866-869.

Schmitt, F.O., Dev, P., Smith, B.H. Electronic processing of Information by brain cells. **Science**, 1976, **193**, 114-120.

KINDLING, PRENATAL EXPOSURE TO ETHANOL AND POSTNATAL DEVELOPMENT

SELECTIVELY ALTER REPONSES OF HIPPOCAMPAL PYRAMIDAL CELLS TO NMDA

J. V. Nadler, D. Martin, M. A. Bowe, R. A. Morrisett
and J. O. McNamara

Departments of Pharmacology, Neurobiology and Medicine
(Neurology), Duke University Medical Center, Durham, North
Carolina 27710, USA

INTRODUCTION

Research that built upon the discovery by Collingridge and his colleagues (1983) that the induction of long-term potentiation (LTP) requires
activation of the NMDA receptor has led to the realization that many higher
brain phenomena depend on activation of the NMDA receptor. LTP is regarded
as one of the key cellular events that links experience to memory and
learning. Thus several groups have sought and obtained evidence that the
NMDA receptor must be activated for certain types of associative learning to
occur (Morris et al., 1986; Lincoln et al., 1988; Mondadori et al., 1989).
Activation of the NMDA receptor appears to be a prerequisite also for experience-dependent modifications of neuronal response properties in the developing visual cortex (Kleinschmidt et al., 1987) and for the induction of
stimulus train-induced bursting, a cellular homologue of epileptogenesis
(Anderson et al., 1987). It can even influence the morphological (Pearce et
al., 1987; Brewer and Cotman, 1989) and biochemical (Moran and Patel, 1989)
differentiation of neurons. These effects are believed to result from a
substantial entry of Ca^{2+} through the NMDA receptor channel, which
occurs if the plasma membrane is sufficiently depolarized at the time of receptor activation (Ascher and Nowak, 1987). Although these forms of neuronal plasticity depend for their induction on activation of the NMDA
receptor, there is as yet no evidence of any role for this receptor in the
endurance or permanence of the change. In fact, LTP at the Schaffer collateral-commissural synapse in hippocampal area CA1 has been shown to result
from a long-lasting increase in the quisqualate receptor-mediated component
of the EPSP, with little or no change in the NMDA receptor-mediated component (Kauer et al., 1988; Muller and Lynch, 1988; Davies et al., 1989).

We have studied two stimuli that produce changes in the rat brain which
appear to last a lifetime: namely, kindling and prenatal exposure to
ethanol. Each is considered to model some aspects of a human disease, epilepsy in the case of kindling and fetal alcohol syndrome in the case of prenatal ethanol. The neurobiological changes associated with these stimuli
are subtle, in that animals so treated can function relatively normally.
Yet the kindling-induced propensity toward seizures and the cognitive
deficits brought about by exposure to ethanol in utero can be readily
demonstrated with appropriate tests. Considering the known properties of
the NMDA receptor and its involvement in other forms of neuronal plasticity,

Excitatory Amino Acids and Neuronal Plasticity
Edited by Y. Ben-Ari
Plenum Press, New York, 1990

we reasoned that the lifelong effects of kindling and prenatal exposure to ethanol might be related to long-lasting modifications of NMDA receptor function. When investigations of these phenomena yielded positive findings, we sought to determine whether similar changes in the neuronal response to NMDA might occur during normal development.

KINDLING

Kindling is an animal model of epilepsy in which periodic administration of an initially subconvulsive electrical stimulus eventually results in a generalized electrographic and behavioral seizure (Goddard et al., 1969). Once developed, this enhanced sensitivity to electrical stimulation seems to endure for the life of the animal. NMDA receptor antagonists markedly suppress development of the kindled state (McNamara et al., 1988) and, at higher doses, attenuate kindled seizures (Peterson et al., 1984; McNamara et al., 1988). The finding that higher doses of NMDA receptor antagonists are required to attenuate kindled seizures than to retard development of the kindled state could be explained by an enhancement of NMDA receptor function during the kindling process. Indeed electrophysiological studies demonstrated that NMDA receptors on dentate granule cells are more readily activated by perforant path stimulation in hippocampal slices from kindled rats than in slices from control rats (Mody and Heinemann, 1987; Mody et al., 1988). However, kindling could have brought this change about by several different mechanisms, not all of which involve the receptor directly. To investigate the possibility that kindling is associated with enhanced NMDA receptor function, we utilized two experimental approaches that test the ability of excitatory amino acids to depolarize hippocampal neurons.

We first asked whether kindling altered the ability of NMDA to inhibit the carbachol-induced stimulation of phosphatidyl inositol (PI) turnover in rat hippocampal slices (Morrisett et al., 1989a). This effect of NMDA depends on neuronal depolarization (influx of Na^+) and not on the influx of Ca^{2+} through the channel (Baudry et al., 1986). Amygdaloid kindling increased the inhibitory effect of NMDA. When studied 1 day after attainment of the behavioral criterion for kindling, NMDA more potently depressed carbachol-stimulated PI turnover in slices from kindled rats. The response to 10 μM NMDA increased from 23 ± 4% in slices from implanted controls to 54 ± 4% in slices from kindled rats (means ± S.E.M., P < 0.025 (Wilcoxon test)). This enhancement persisted for at least another 4-5 weeks without further stimulation. However, kindling affected neither basal nor carbachol-stimulated PI turnover. NMDA in the absence of carbachol little affected PI turnover in slices from control or kindled rats. Finally, kindling did not alter the ability of kainate or a phorbol ester to inhibit carbachol-stimulated PI turnover. Thus the effect of kindling appeared selective for the response to NMDA.

The ability of prototypic amino acid excitants to depolarize specific populations of hippocampal neurons was assessed with use of newly-developed grease-gap preparations. To study responses of CA3 pyramidal cells (relative to their axons in area CA1), transverse hippocampal slices were prepared and the dentate gyrus was dissected away. Each slice was arranged in an individual superfusion chamber such that area CA3a,b and area CA1 were in different compartments separated by a grease barrier (Martin et al., 1989). Various concentrations of excitant were tested by replacing the medium superfusing the CA3 compartment with 5.7 compartment volumes of medium that contained the excitant and their effects were determined on the DC potential differentially recorded between the two compartments. The potential difference produced by each excitant concentration was taken as a global measure of neuronal depolarization.

Angular bundle kindling shifted the NMDA concentration-response curve
in area CA3 to the left, when compared with the CA3 area from implanted
control rats studied at the same postsurgical survival times (Fig. 1). The
EC_{50} for NMDA was reduced, whether the studies were performed in the
presence or absence of Mg^{2+} and either 1 day or 1-2 months after the
last evoked seizure (Table 1). Differences in NMDA potency were greater at
the longer survival time. However, this resulted from a loss of NMDA
potency in the control CA3 area during this period. We are investigating
whether this change reflected the 1-2 month difference in age between the
two control groups or the presence of implanted electrodes in the brain for
different lengths of time. Either way, NMDA more potently depolarized CA3
pyramidal cells in slices from kindled rats. Preliminary studies suggest
that the effect of kindling is selective for responses to NMDA, because they
have shown no difference between CA3 pyramidal cells from kindled and
control rats in their response to AMPA or glutamate.

These results agree well with findings from our study of PI turnover.
A number of factors can influence measures of NMDA potency obtained with a
grease-gap preparation. The most likely explanation for the effect of

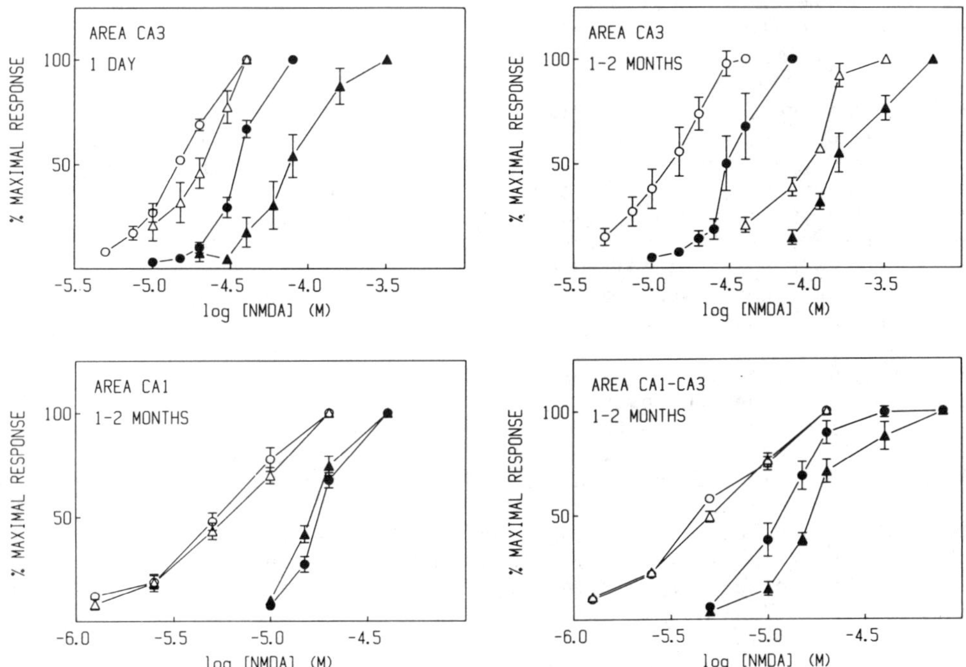

Fig. 1. Effects of angular bundle kindling on NMDA concentration-
response relationships in rat hippocampal regions. (Δ) control,
0 Mg^{2+}; (o) kindled, 0 Mg^{2+}; ($\blacktriangle$) control, 1 mM Mg^{2+};
($\bullet$) kindled, 1 mM Mg^{2+}. Experiments were performed either 1
day or 1-2 months after the last evoked seizure. Values are means
$\pm$ S.E.M. for the number of slices given in Table 1. Because the
amplitude of excitant-induced voltage deflections obtained in
grease-gap studies varies with the preparation and with the resistance of the grease seal, responses were normalized before results
from different slices were averaged. Each response was expressed
as a percentage of the voltage deflection evoked in the same slice
by a maximal concentration of NMDA.

Table 1. EC_{50} Values for NMDA after Angular Bundle Kindling

Hippocampal region	$[Mg^{2+}]$ (mM)	EC_{50} (μM)	
		Control	Kindled
Area CA3			
1 day	0	20.3 ± 3.0 (4)	14.2 ± 0.4 (5)
1–2 months	0	93.9 ± 5.3 (7)	13.7 ± 2.2 (6)
1 day	1	80.0 ± 16.3 (5)	35.0 ± 1.2 (6)
1–2 months	1	173 ± 15 (5)	40.5 ± 5.1 (7)
Area CA1			
1–2 months	0	5.8 ± 0.6 (5)	5.4 ± 0.4 (6)
1–2 months	1	16.1 ± 0.8 (9)	17.8 ± 0.4 (8)
Area CA1–CA3			
1 day	0	5.3 ± 0.5 (9)	5.8 ± 0.3 (13)
1–2 months	0	5.4 ± 0.3 (6)	5.0 ± 0.3 (8)
1 day	1	19.3 ± 1.6 (9)	19.6 ± 1.1 (11)
1–2 months	1	17.5 ± 1.1 (7)	11.8 ± 0.9 (8)

Responses to NMDA were studied either 1 day or 1–2 months after attainment of the behavioral criterion for kindling. EC_{50} values were calculated by regression of the linear portion of the semi-logarithmic concentration-response curve and are expressed as means ± S.E.M. for the number of experiments in parentheses.

kindling, however, is that kindling increases the expression of NMDA receptors by CA3 pyramidal cells. We have found a long-lasting increase in the density of antagonist-preferring NMDA receptors ([3H]CPP binding sites; Monaghan et al., 1988) on hippocampal membranes from kindled rats (Yeh et al., 1989; McNamara et al., this volume), although there was no increase in the density of agonist-preferring receptors (NMDA-sensitive [3H]glutamate binding sites) (Okazaki et al., 1989). An increase in NMDA receptor density would be expected to enhance agonist potency in the presence of a receptor reserve (Kenakin, 1984).

To study responses of CA1 pyramidal cells (relative to their axons in the subiculum), longitudinal slices of the hippocampal formation were prepared. Regio inferior and the fascia dentata were dissected away, leaving just area CA1 and the retrohippocampal area including the subiculum (Martin et al., 1989). In this case the grease barrier was placed across the CA1-subiculum border. In contrast to its effects on CA3 pyramidal cells, kindling produced no long-lasting change in the response of CA1 pyramidal cells to NMDA (Fig. 1, Table 1). Thus kindling does not necessarily change the properties of all neurons in the circuit activated by the stimulus, at least not in the same way.

In our early grease-gap studies with CA1-subiculum slices, we found indications that kindling enhanced NMDA receptor function in CA1 pyramidal cells by depressing the ability of Mg^{2+} to regulate channel opening (Martin et al., 1988). Subsequent investigations established that this result could be obtained only by leaving a small segment of the CA2-CA3 area attached to the slice. In such a preparation, kindling enhanced the potency of NMDA only in the presence of added Mg^{2+} and only after more than a day had passed since the last evoked seizure (Fig. 1, Table 1). Depolarizing responses to AMPA and glutamate were enhanced to a much lesser extent. This effect of kindling differs from its effect on "pure" responses of CA1 or CA3 pyramidal cells to NMDA. The presence of CA3 pyramidal cells in the preparation opens the possibility of a synaptic component in the depolarizing response of CA1 pyramidal cells. In addition to depolarizing the CA1 pyramidal cells, NMDA would also depolarize CA3 pyramidal cells, perhaps

thereby activating excitatory synapses made by these neurons in area CA1.
Release studies suggest that synaptic terminals of CA3 pyramidal cell axons
express NMDA receptors whose activation enhances the evoked release of glutamate and aspartate (Nadler et al., 1990). It may be these NMDA terminal
autoreceptors that are less readily blocked by Mg^{2+} after kindling.
Possibly the enhanced depolarization of CA3 pyramidal cells through NMDA receptor activation eventually leads to a rather different form of NMDA receptor plasticity at the axon terminals of these cells.

Our results suggest a plausible mechanism to account, at least partially, for the maintenance of kindled epilepsy. Enhancement of NMDA
agonist potency would be expected to unmask normally dormant NMDA receptors
on CA3 pyramidal cells and thus to augment and prolong the EPSP at Schaffer
collateral-commissural synapses (Collingridge et al., 1988; Forsythe and
Westbrook, 1988). This effect might also indirectly modify the actions of
neurotransmitters other than excitatory amino acids (Stanton et al., 1987;
Stelzer et al., 1987). Increasing the influence of NMDA terminal autoreceptors over the release of glutamate and aspartate might account for the
finding of Geula et al. (1988) that elevated K^{+} evoked the release of
more glutamate from rat hippocampal slices after kindling. These kindling-
induced changes in NMDA receptor function are in the right direction and
occur at the right times to underlie the establishment and maintenance of a
hyperexcitable state. Kindling is the first instance in which an environmental stimulus was found not only to depend for its effect on NMDA receptor
activation, but also to modify properties of the NMDA receptor in the
process.

It is interesting that kindling and LTP involve the enhancement of responses to different excitatory amino acids. This finding is consistent with
recent work that emphasizes the probable mechanistic differences between LTP
and kindling, despite their formal similarities (Cain, 1989).

PRENATAL EXPOSURE TO ETHANOL

Children born to mothers who consumed even moderate amounts of ethanol
during pregnancy may suffer learning disabilities, as well as disturbances
of behavior and cognition (Abel, 1984). Similarly, animals exposed to
ethanol before birth evidence long-lasting deficits in a number of learning
and memory tasks (Abel, 1979; Riley et al., 1979). Because of the role
played by NMDA receptors in associative learning, it seemed reasonable to
investigate the possibility that prenatal exposure to ethanol reduces the
sensitivity of CNS neurons to NMDA (Morrisett et al., 1989b). These
studies utilized the CA1-subiculum grease-gap preparation from which CA2-CA3
pyramidal cells were carefully excluded.

Pregnant rats were fed a liquid diet that contained 3.35% ethanol (v/v)
throughout gestation. Peak blood concentrations were only 30-40 mg/dl.
Control dams were fed an isocaloric liquid diet, in which a maltose-dextrin
mixture was substituted for ethanol. After the pups were born, they were
assigned to surrogate dams and were never again exposed to ethanol. Grease-
gap studies were performed at 70-90 days of age.

Prenatal exposure to ethanol markedly reduced the sensitivity of CA1
pyramidal cells to NMDA, but only when the tests were performed in the presence of added Mg^{2+} (Fig. 2). In slices from pair-fed controls,
Mg^{2+} concentrations of 1 and 3.16 mM shifted the NMDA concentration-
response curve to the right, but did not reduce the maximal response to NMDA
(see also Harrison and Simmonds, 1985; Martin et al., 1989). In slices from
experimental animals, however, 1 mM Mg^{2+} depressed the maximal response
to NMDA by 38 ± 9% and 3.16 mM Mg^{2+} depressed it by 60 ± 5% (means ±

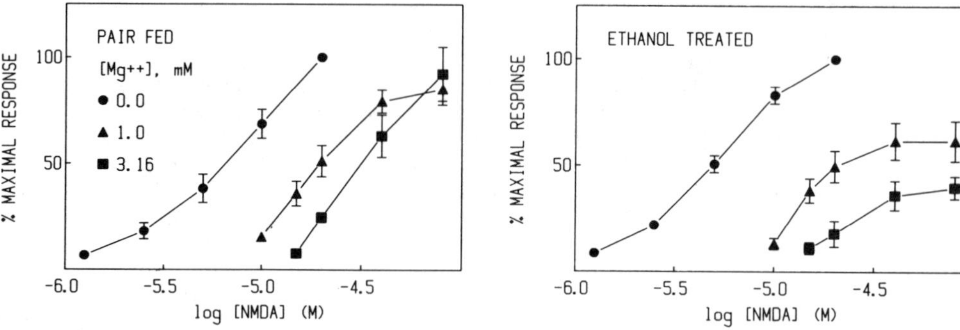

Fig. 2. Effects of prenatal exposure to ethanol on NMDA concentration-
response relationships in area CA1. Values are means ± S.E.M.
for 4-6 experiments. Before results from different slices were
averaged, the responses were expressed as a percentage of the
voltage deflection evoked in the same slice by a maximal concentration of NMDA in the absence of added Mg^{2+}. From Morrisett
et al. (1989b).

S.E.M.). These effects reversed when Mg^{2+} was removed from the superfusion medium. Prenatal ethanol exposure did not alter the sensitivity of
CA1 pyramidal cells to AMPA or kainate in either the presence or absence of
added Mg^{2+}.

These results suggest that low level exposure to ethanol _in utero_
results in a long-lasting and probably permanent loss of responsiveness to
NMDA on the part of CA1 pyramidal cells. The effect appears to result from
an enhancement of Mg^{2+} regulation over channel opening. Mg^{2+}
regulation could have been augmented through a post-translational modification of the receptor, biosynthesis of a receptor with a different comple-
ment of subunits or hyperpolarization of the CA1 pyramidal cells. The blood
ethanol level of the pregnant rats in this study was below that necessary to
produce learning deficits in the offspring. Thus a long-lasting depression
in the ability of NMDA to depolarize hippocampal neurons may at least partly
explain these deficits. It may also account for reduced LTP of the Schaffer
collateral-commissural synapse in area CA1 (Swartzwelder et al., 1988) and
for the retarded development of kindling (Savage and Reyes, 1985) in these
animals.

NORMAL DEVELOPMENT

Because NMDA receptor activation appears crucial to at least some forms
of associative learning, it seems reasonable that mechanisms might exist
that facilitate activation of this receptor during the period of life when
most learning takes place, that is, during the developmental period. Bode-
Greuel and Singer (1989) reported that the density of NMDA receptors in the
visual cortex is greatest during the critical period for experience-
dependent plasticity in that region and suggested that changes in NMDA receptor density might, in part, determine the onset and decline of plasticity. In light of our findings in the previous studies, we predicted that
either the sensitivity of hippocampal neurons to NMDA would be greater in
developing rats than in adults or that Mg^{2+} would be less able to
regulate channel opening. Our studies of this issue utilized the CA1-subi-
culum grease-gap preparation free of CA2-CA3 pyramidal cells. Unfortunately, our method proved unsuited to the study of rats less than 10 days
old.

Table 2. Developmental Changes in NMDA-induced Depolarization
         of CA1 Pyramidal Cells

	10-15 days	60+ days
EC_{50} of NMDA (0 Mg^{2+})	6.4 ± 0.2 µM (39)	6.1 ± 0.3 µM (18)
Concentration-Ratio (1 mM Mg^{2+})	2.7 ± 0.1 (21)	3.5 ± 0.2 (9)
IC_{50} of Mg^{2+} (against 10 µM NMDA)	397 ± 30 µM (6)	157 ± 4 µM (4)
Depression of Maximal Response by 10 mM Mg^{2+}	4 ± 6% (6)	48 ± 7% (4)
Mg^{2+} Increased Maximal Response to NMDA	24/46	2/26
Spontaneous Activity in 0 Mg^{2+}	19/71	2/22
EC_{50} of AMPA	4.7 ± 0.2 µM (4)	4.1 ± 0.3 µM (6)
Mg^{2+} Increased Maximal Response to AMPA	0	0

Values are either means ± S.E.M. for the number of experiments in parentheses or the observed incidence of the event.

Our results suggest that Mg^{2+} gains greater control over NMDA
channel opening as development proceeds. A comparison of CA1 pyramidal cell
responses from 10-15 day old and adult rats revealed no difference in sensitivity to NMDA when Mg^{2+} was omitted from the superfusion medium, but
CA1 pyramidal cells from the younger animals were significantly more sensitive to NMDA in the presence of 1 mM Mg^{2+} (Table 2). Analysis of
NMDA concentration-response curves generated in the presence of various
Mg^{2+} concentrations (0.1-10 mM) provided more detailed information
(Fig. 3). Mg^{2+} differentially affected responses to NMDA in at least
three ways. First, Mg^{2+} was less potent in slices from 10-15 day old
rats. The concentration required to depress responses to 10 µM NMDA was
more than double that needed for a similar degree of inhibition in slices

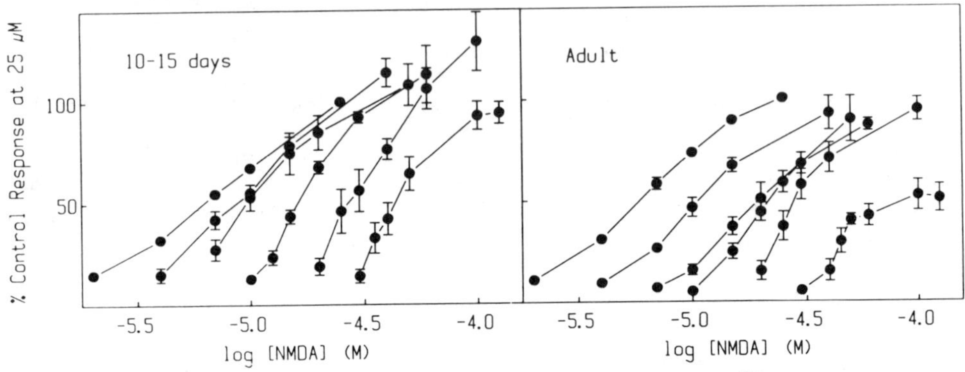

Fig. 3. Developmental differences in the effects of Mg^{2+} on NMDA
        concentration-reponse relationships in area CA1. The CA1 area
        was exposed to various concentrations of NMDA in the presence of
        (from left to right) 0, 0.1, 0.316, 1.0, 3.16 and 10 mM added
        Mg^{2+}. Values are means ± S.E.M. for 4-39 determinations.
        Responses were normalized before averaging as in figure 2. In
        slices from both 10-15 day old and adult rats, 25 µM NMDA evoked
        a maximal response in the absence of added Mg^{2+}.

413

from adult rats (Table 2). Second, Mg^{2+} was less efficacious in the
younger rats. Mg^{2+} did not depress the maximal response to NMDA within
the range of concentrations tested, whereas in slices from adult rats 10 mM
Mg^{2+} depressed the maximal response by about half. Finally, Mg^{2+}
(0.1-3.16 mM) actually increased the maximal response to NMDA by more than
10% in about half the slices from 10-15 day old rats that were tested. This
effect was obtained about as frequently with 0.1 mM Mg^{2+} as with 3.16
mM Mg^{2+}. In slices from adult rats Mg^{2+} enhanced the maximal response to NMDA in only two instances, a frequency that probably reflects the
normal variability of the grease-gap method. Also, exposing area CA1 of
developing rats to nominally Mg^{2+}-free medium commonly elicited spontaneous activity that was suppressed by adding Mg^{2+}. The occurrence of
Mg^{2+}-sensitive spontaneous activity in other preparations has been attributed to the enhanced activation of NMDA receptors by excitants that are
normally present in the tissue (Horne et al., 1986). This response was
rarely observed in CA1 slices from adult rats unless some CA2-CA3 pyramidal
cells were present. No such developmental differences were found in the
response of CA1 pyramidal cells to AMPA.

Lesser Mg^{2+} regulation of NMDA channel opening during development
might facilitate associative learning by permitting the channel to open at
less depolarized membrane potentials. This mechanism could also at least
partially explain the observations that responses of CA3 pyramidal cells to
NMDA are less voltage-dependent in developing rats (Ben-Ari et al., 1988),
that NMDA is more neurotoxic in developing rats (McDonald et al., 1988),
that the magnitude of LTP is greatest in 15 day old rats (Harris and Teyler,
1984) and that the kindling process evolves more rapidly in rat pups than in
adult rats (Moshe and Ludvig, 1988). However, these findings may also be
explained, in part, by the enhanced expression of NMDA receptors during the
developmental period (Tremblay et al., 1988; Bode-Greuel and Singer, 1989).

SUMMARY AND CONCLUSIONS

Our studies suggest that treatments, such as kindling and exposure to
ethanol _in utero_, which produce irreversible pathological changes in brain
function also selectively alter neuronal responses to NMDA. We have identified increases in agonist potency, which may result from enhanced NMDA receptor expression, as well as both positive and negative modifications of
Mg^{2+} regulation. There is no proof as yet that NMDA receptor plasticity accounts for the kindling phenomenon, for the cognitive deficits associated with fetal exposure to ethanol or for developmental events. However,
all the effects we obtained are in the appropriate direction. The same
functional parameters altered by kindling and prenatal exposure to ethanol
are also modified during the normal development of the brain. It is
tempting to speculate that these pathological stimuli produce their effects
through the same mechanisms that would normally be activated by developmental signals.

ACKNOWLEDGMENTS

We wish to acknowledge the participation of Drs. H.S. Swartzwelder,
D.D. Savage and W.A. Wilson in the prenatal ethanol studies and the technical assistance of Ms. L. Butler. These studies were supported by NIH
grants NS 16064 and NS 17771.

REFERENCES

Abel, E.L., 1979, Prenatal effects of alcohol on adult learning in rats,
    Pharmacol. Biochem. Behav., 10:239.

Abel, E.L., 1984, Prenatal effect of alcohol, Drug Alcohol Depend., 14:1.

Anderson, W.A., Swartzwelder, H.S. and Wilson, W.A., 1987, The NMDA receptor antagonist 2-amino-5-phosphonovalerate blocks stimulus train induced epileptogenesis but not epileptiform activity in the rat hippocampus, J. Neurophysiol., 57:1.

Ascher, P. and Nowak, L., 1987, Electrophysiological studies of NMDA receptors, Trends Neurosci., 10:284.

Baudry, M., Evans, J. and Lynch, G., 1986, Excitatory amino acids inhibit stimulation of phosphatidylinositol metabolism by aminergic agonists in hippocampus, Nature, 319:329.

Ben-Ari, Y., Cherubini, E. and Krnjevic, K., 1988, Changes in voltage dependence of NMDA currents during development, Neurosci. Lett., 94:88.

Bode-Greuel, K.M. and Singer, W., 1989, The development of N-methyl-D-aspartate receptors in cat visual cortex, Dev. Brain Res., 46:197.

Brewer, G.J. and Cotman, C.W., 1989, NMDA receptor regulation of neuronal morphology in cultured hippocampal neurons, Neurosci. Lett., 99:268.

Cain, D.P., 1989, Long-term potentiation and kindling: how similar are the mechanisms?, Trends Neurosci., 12:6.

Collingridge, G.L., Kehl, S.J. and McLennan, H., 1983, Excitatory amino acids in synaptic transmission in the Schaffer collateral-commissural pathway of the rat hippocampus, J. Physiol., 334:33.

Collingridge, G.L., Herron, C.E. and Lester, R.A.J., 1988, Synaptic activation of N-methyl-D-aspartate receptors in the Schaffer collateral-commissural pathway of the rat hippocampus, J. Physiol., 399:283.

Davies, S.N., Lester, R.A.J., Reymann, K.G. and Collingridge, G.L., 1989, Temporally distinct pre- and postsynaptic mechanisms maintain long-term potentiation, Nature, 338:500.

Forsythe, I.D. and Westbrook, G.L., 1988, Slow excitatory postsynaptic currents mediated by N-methyl-D-aspartate receptors on cultured mouse central neurones, J. Physiol., 396:515.

Geula, C., Jarvie, P.A., Logan, T.C. and Slevin, J.T., 1988, Long-term enhancement of K^+-evoked release of L-glutamate in entorhinal kindled rats, Brain Res., 442:368.

Goddard, G.V., McIntyre, D.C. and Leech, C.K., 1969, A permanent change in brain function resulting from daily electrical stimulation, Exp. Neurol., 25:295.

Harris, K.M. and Teyler, T.J., 1984, Development of long-term potentiation in area CA1 of the rat hippocampus, J. Physiol., 346:27.

Harrison, N.L. and Simmonds, M.A., 1985, Quantitative studies on some antagonists of N-methyl-D-aspartate in slices of rat cerebral cortex, Brit. J. Pharmacol., 84:381.

Horne, A.L., Harrison, N.L., Turner, J.P. and Simmonds, M.A., 1986, Spontaneous paroxysmal activity induced by zero magnesium and bicuculline: suppression by NMDA antagonists and GABAmimetics, Eur. J. Pharmacol., 122:231.

Kauer, J.A., Malenka, R.C. and Nicoll, R.A., 1988, A persistent postsynaptic modification mediates long-term potentiation in the hippocampus, Neuron, 1:911.

Kenakin, T.P., 1984, The classification of drugs and drug receptors in isolated tissues, Pharmacol. Rev., 36:165.

Kleinschmidt, A., Bear, M.F. and Singer, W., 1987, Blockade of 'NMDA' receptors disrupts experience-dependent plasticity of kitten striate cortex, Science, 238:355.

Lincoln, J., Coopersmith, R., Harris, E.W., Cotman, C.W. and Leon, M., 1988, NMDA receptor activation and early olfactory learning, Dev. Brain Res., 39:309.

Martin, D., Bowe, M.A., McNamara, J.O. and Nadler, J.V., 1988, Kindling depresses magnesium regulation of depolarizing responses to amino acid excitants, Soc. Neurosci. Abstr., 14:865.

Martin, D., Bowe, M.A. and Nadler, J.V., 1989, A grease-gap method for studying the excitatory amino acid pharmacology of CA1 hippocampal pyramidal cells, J. Neurosci. Meth., 29:107.

McDonald, J.W., Silverstein, F.S. and Johnston, M.V., 1988, Neurotoxicity of N-methyl-D-aspartate is markedly enhanced in developing rat central nervous system, Brain Res., 459:200.

McNamara, J.O., Russell, R.D., Rigsbee, L. and Bonhaus, D.W., 1988, Anticonvulsant and antiepileptogenic actions of MK-801 in the kindling and electroshock models, Neuropharmacology, 27:563.

Mody, _. and Heinemann, U., 1987, NMDA receptors of dentate gyrus granule cells participate in synaptic transmission following kindling, Nature, 326:701.

Mody, I., Stanton, P.K. and Heinemann, U., 1988, Activation of N-methyl-D-aspartate receptors parallels changes in cellular and synaptic properties of dentate gyrus granule cells after kindling, J. Neurophysiol., 59:1033.

Monaghan, D.T., Olverman, H.J., Nguyen, L., Watkins, J.C. and Cotman, C.W., 1988, Two classes of N-methyl-D-aspartate recognition sites: differential distribution and differential regulation by glycine, Proc. Natl. Acad. Sci. USA, 85:9836.

Mondadori, C., Weiskrantz, L., Buerki, H., Petschke, F. and Fagg, G.E., 1989, NMDA receptor antagonists can enhance or impair learning performance in animals, Exp. Brain Res., 75:449.

Moran, J. and Patel, A.J., 1989, Stimulation of the N-methyl-D-aspartate receptor promotes the biochemical differentiation of cerebellar granule neurons and not astrocytes, Brain Res., 486:15.

Morris, R.G.M., Anderson, E., Lynch, G.S. and Baudry, M., 1986, Selective impairment of learning and blockade of long-term potentiation by an N-methyl-D-aspartate receptor antagonist, Nature, 319:774.

Morrisett, R.A., Chow, C., Nadler, J.V. and McNamara, J.O., 1989a, Biochemical evidence for enhanced sensitivity to N-methyl-D-aspartate in the hippocampal formation of kindled rats, Brain Res., in press.

Morrisett, R.A., Martin, D., Wilson, W.A., Savage, D.D. and Swartzwelder, H.S., 1989b, Prenatal exposure to ethanol decreases the sensitivity of the adult rat hippocampus to N-methyl-D-aspartate, Alcohol, in press.

Moshe, S.L. and Ludvig, N., 1988, Kindling, in: "Recent Advances in Epilepsy, Volume 4," T.A. Pedley and B.S. Meldrum, eds., Churchill Livingstone, Edinburgh, p. 21.

Muller, D. and Lynch, G., 1988, Long-term potentiation differentially affects two components of synaptic responses in hippocampus, Proc. Natl. Acad. Sci. USA, 85:9346.

Nadler, J.V., Martin, D., Bustos, G.A., Burke, S.P. and Bowe, M.A., 1990, Regulation of transmitter release from the Schaffer collaterals and other projections of CA3 hippocampal pyramidal cells, in: "Understanding the Brain Through the Hippocampus: The Hippocampal Region As a Model for Studying Brain Structure and Function," J. Storm-Mathisen, J. Zimmer and O.P. Ottersen, eds., Elsevier, Amsterdam, in press.

Okazaki, M.M., McNamara, J.O. and Nadler, J.V., N-Methyl-D-aspartate receptor autoradiography in rat brain after angular bundle kindling, Brain Res., 482:359.

Pearce, I.A., Cambray-Deakin, M.A. and Burgoyne, R.D., Glutamate acting on NMDA receptors stimulates neurite outgrowth from cerebellar granule cells, FEBS Lett., 223:143.

Peterson, D.W., Collins, J.F. and Bradford, H.F., 1983, The kindled amygdala model of epilepsy: anticonvulsant action of amino acid antagonists, Brain Res., 275:169.

Riley, E.P., Lochry, E.A. and Schapiro, N.R., 1979, Lack of response inhibition in rats prenatally exposed to ethanol, Psychopharmacology, 62:47.

Savage, D.D. and Reyes, E., 1985, Prenatal exposure to ethanol retards the development of kindling in adult rats, Exp. Neurol., 89:583.

Stanton, P.K., Jones, R.S.G., Mody, I. and Heinemann, U., 1987, Epileptiform activity induced by lowering extracellular $[Mg^{2+}]$ in combined hippocampal-entorhinal cortex slices: modulation by receptors for norepinephrine and N-methyl-D-aspartate, Epilepsy Res., 1:53.

Stelzer, A., Slater, N.T. and ten Bruggencate, G., 1987, Activation of NMDA receptors blocks GABAergic inhibition in an in vitro model of epilepsy, Nature, 326:698.

Swartzwelder, H.S., Farr, K.L., Wilson, W.A. and Savage, D.D., 1988, Prenatal exposure to ethanol decreases hippocampal plasticity in the adult rat, Alcohol, 5:121.

Tremblay, E., Roisin, M.P., Represa, A., Charriaut-Marlangue, C. and Ben-Ari, Y., 1988, Transient increased density of NMDA binding sites in the developing rat hippocampus, Brain Res., 461:393.

Yeh, G.-C., Bonhaus, D.W., Nadler, J.V. and McNamara, J.O., 1989, NMDA receptor plasticity in kindling: quantitative and qualitative alterations in the NMDA receptor/channel complex, Proc. Natl. Acad. Sci. USA, in press.

SPROUTING OF MOSSY FIBERS IN THE HIPPOCAMPUS OF EPILEPTIC HUMAN AND RAT

Alfonso Represa, Evelyne Tremblay and Yehezkel Ben-Ari

INSERM U29, 123 Bd Port Royal 75014 Paris, France

Epilepsy is a neurological disorder characterized by the presence of recurrent seizures and during the interictal periods, synchronous activation of a neuronal cell group. Numerous reports have suggested that there is common underlying mechanisms with long term potentiation; i.e. the kindling model of epilepsy induced by stimulation of the entorhinal cortex is associated with a synaptic potentiation at the perforant pathway-granule cell synapse (Goddard et al. 1969); moreover, as in LTP induction, these synaptic potentiation involves the activation of NMDA receptors (Mody and Heinemann 1987; Ben-Ari and Gho 1987). The hypothesis that the morphological substrate of epilepsy resides in the establishment of new aberrant connections has been frequently considered. Morphological evidence has recently been obtained for the hippocampal mossy fibers which originate in the dentate granule cells and project to the giant pyramidal neurons of CA3 area. Thus, mossy fibers sprout and establish aberrant connections both in kindling epilepsy (Sutula et al. 1988, Represa et al 1989) and in childhood epilepsy (Represa et al 1989 b).

Since the mossy fibers have an unusually high Zn^{2+} content (Frederickson et al. 1983), the Timm method can be used as a marker to follow the trajectory of mossy fibers (Haug 1973). In human postmortem material it is however difficult to obtain a quantitative estimate with the Timm method. A more useful marker of the mossy fibers is provided by high affinity binding sites for KA, which are particularly enriched both in rat (Unnerstall and Wamsley 1983, Berger and Ben-Ari 1983) and human (Tremblay et al. 1985) in the mossy fiber terminal region (stratum lucidum). Lesion experiments indicate that these sites are predominantly located presynaptically on the mossy fiber terminals (Represa et al. 1987) where they could modulate transmitter release (Ferkany et al. 1982). Kainate binding sites persist for 24 h in post-mortem conditions (Tremblay et al. 1985) and image analysis can readily be used to quantify changes in the density or distribution of the mossy fibers. Employing Timm staining and quantitative autoradiography we report that plastic changes occur in the CA3 region and fascia dentata of epileptic children and that these changes are similar to those detected in kindled rats.

Kindling model of epilepsy in rats

In control Wistar rats, in agreement with earlier observations (Unnerstall and Wamsley 1983, Berger and Ben-Ari 1983) a distinct single band of high density KA binding sites was present in the stratum lucidum of CA3 (Fig 1 B); this correlated with the dark Timm positive stain which depicts the trajectory of the mossy fibers (Fig 1 A). In Amygdala kindled rats (Represa et al. 1989a) there was a significant increase in the density of KA binding sites in this region as compared to control rats (59±4 and 70±5 fmol/mg tissue S.E.M. in control and amygdala kindled rats respectively; p < .01 ; Fig 2). In addition, a new band of Timm deposits appeared in the infrapyramidal layer of the CA3 region of kindled rats (Fig 1) suggesting that mossy fibers had sprouted

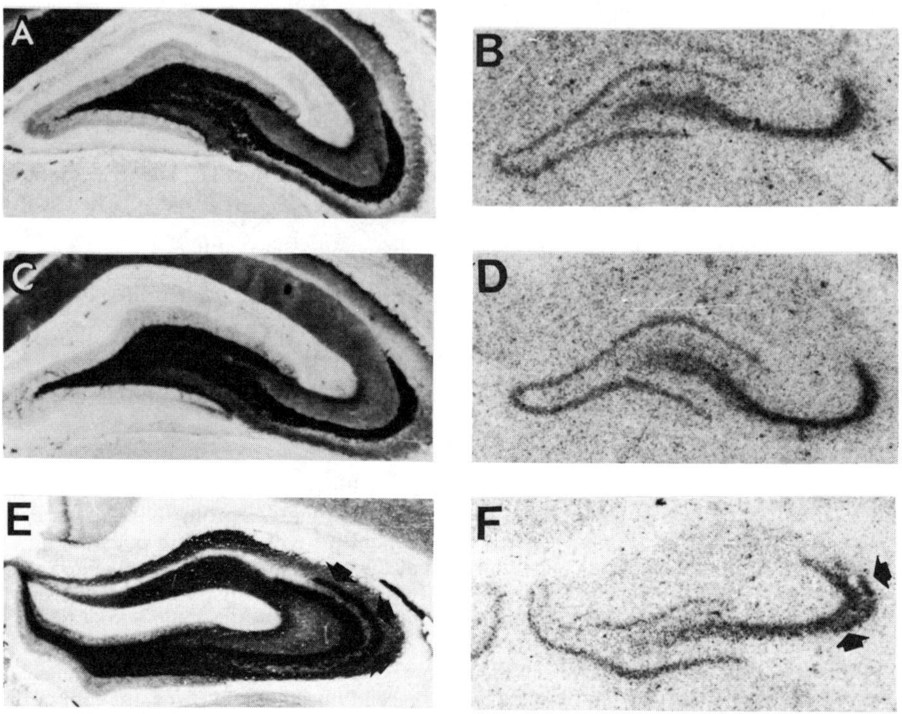

Fig. 1.   Photomicrographs depicting the distribution of Timm stained mossy fiber (A, C, E) and [^{3}H] - KA- binding sites (B, D ,F) in the CA3 region of control (A,B) entorhinal (C,D) and amygdala (E,F) kindled rats. Arrows shown the aberrant infrapyramidal band of mossy fiber in amygdala kindled rats.

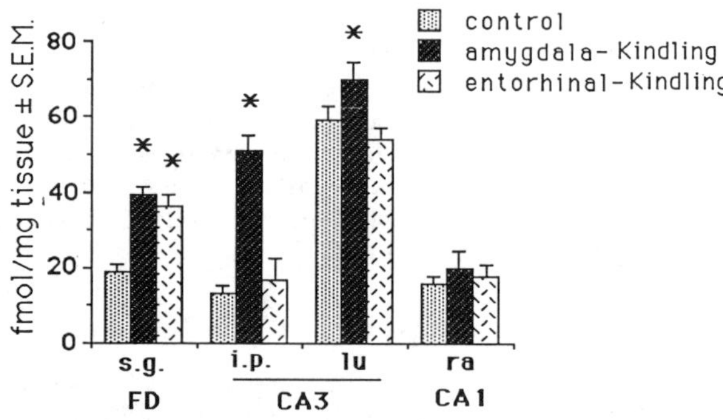

Fig. 2.   Quantitative measurements of high affinity KA binding sites in control and epileptic rats. Abbreviations: sg supragranular layer; i.p. and lu infrapyramidal layer and stratum lucidum of CA3; ra stratum radiatum of CA1. * p< 0.01.

and established aberrant synaptic contacts in this zone; this was associated with the appearance of a new band of high affinity KA binding sites (Fig 1 D and 2). In agreement with a recent study (Sutula et al. 1988), a similar sprouting occurred in the supragranular region of amygdala kindled rats; these authors have found in kindled rats the presence of Timm deposits in axons terminals forming synapses with supragranular dendrites. Figure 2 also shows that mossy fiber sprouting is associated in the fascia dentata with a significant increase in KA binding sites; therefore we can conclude that in the fascia dentata of kindled animals there is an establishment of new synapses between mossy fiber and granular cells. As shown in Figure 2 this reactive synaptogenesis in fascia dentata was also associated with a significant increase in the density of high affinity KA binding sites (19 ± 2 and 39 ± 2 fmol/mg tissue in control and amygdala kindled hippocampi respectively).

Figures 1 and 2 also show that in contrast to the amygdala kindling the mossy fibers do not sprout after kindling of entorhinal cortex in the CA3 field; Timm deposits and KA binding sites increases exclusively in the supragranular layer of these animals.

Childhood epilepsy

In control children (n= 5, who died without clinical evidence of neurological disease) the pattern of Timm stained mossy fibers and KA binding sites in hippocampal sections was very similar to that observed in control rats; the stratum lucidum and the pyramidal layers of CA3 were heavily labelled. A distinct narrow band was also conspicuous in the supragranular layer of fascia dentata. In epileptic children (n=6; Table I for details of the neurological findings; Represa et al. 1989b), despite the absence of conspicuous morphological lesions we found that the density of Timm deposits increases in the pyramidal and oriens strata of CA3 (Figure 3). This is associated with a highly significant increase in the density of KA binding sites both in the pyramidal-lucidum layer of CA3 and in the supragranular layer of the fascia dentata (Fig 3). Thus, in CA3 the mean density of KA binding sites (fmol/mg tissue $\pm$ S.E.M.) was 49.6 ± 7 in controls and 127 ± 12 in epileptics cases (p< .005); in the supragranular layer the values were of 24 ± 3 in controls and 58 ± 15 in epileptic cases (p< .01). In contrast, in CA1 there was no significant difference between the two groups (31 ± 5 and 22 ± 5 fmol/mg tissue in epileptics and controls respectively). The present changes likely represent a increase in the density of binding sites without change in the affinity (the K_D of KA binding sites in epileptic CA3 area was about 17 nM; Represa et al. in preparation).

Table 1   Clinical information on epileptic cases studied

AGE	NEUROLOGICAL FINDINGS
4 m	Hemimegalencephaly. Frequent generalized seizures since birth (pm = 4h)
8 m	Hemimegalencephaly. Frequent generalized seizures starting 11 days after birth. (pm 17h)
2 y	Seizures since 6 months old. (pm< 24h)
2 y	Agenesia of the corpus callosum. frequent generalized seizures since birth. (pm<24h)
10 y	Cytomegalic encephalitis. Myoclonic seizures in the last week of life. (pm< 24h)
15 y	Subacute sclerosing penencephalitis. Myoclonic seizures during the last 6 months. (pm< 24h)

m and **y** age in months and years respectively; **pm** postmortem delay in hours.

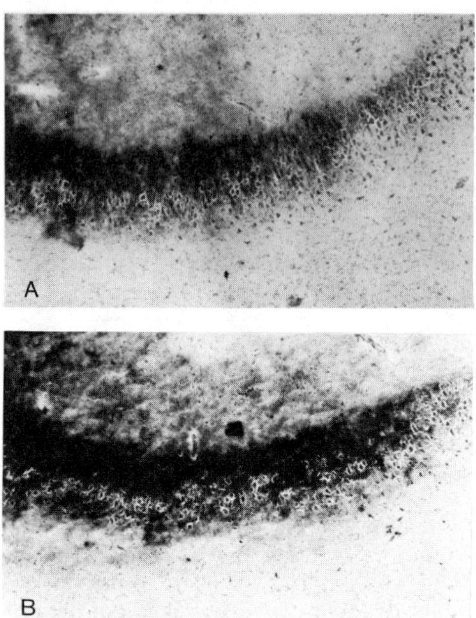

Fig. 3.   Photomicrographs depicting the distribution of Timm stained mossy fiber in the CA3 field of control (A) and epileptic (B) children. Note in B the increased density of Timm deposits in the pyramidal and oriens layers .

Therefore, in epileptic children as in kindled rats there is an increased density of KA binding sites exclusively in the regions innervated by the mossy fibers. This most likely reflects the sprouting of mossy fibers which establish abnormal projections with aberrantly located KA receptors. In a recent report Lanerolle et al. (1989) have found that dynorphine-A and somatostatine immunoreactivity increase in the inner third of molecular layer in the fascia dentata of cryptogenic epileptics (also see Hauser et al., 1989. The authors have also found a extensive loss in the hilus of neuropeptide Y and somatostatine neurons. A similar loss of hilar interneurons after KA treatment has been suggested to induce a sprouting of mossy fiber, which aberrantly innervate granular cells (Frotscher and Zimmer 1983); the extension of dynorphine immunoreactivity observed in human epileptics is consistent with a sprouting of mossy fibers collaterals in epileptic patients.

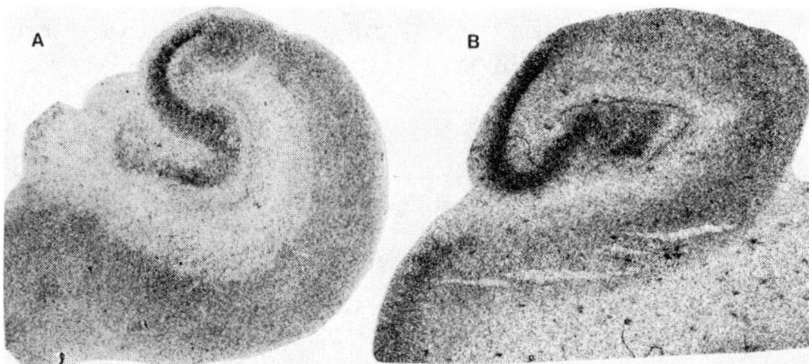

Fig. 4 .   Changes in high affinity kainic acid binding sites in the hippocampus of epileptic childrens. The autoradiography obtained from epileptic children (B) shows a increase of labelling in both, the supragranular layer of fascia dentata and the stratum lucidum-pyramidal of CA3 as compared to control one (A).

The mechanism of mossy fiber sprouting and its underlying selectivity of reinnervation is presently unknown. However, it is clear that the increased density of KA receptors will promote further excitability of the hippocampal circuitry. The epileptogenic properties of KA are due to a its powerful excitatory effects (Robinson and Deadwyler 1981; Westbrook and Lothman 1983; Ben-Ari and Gho 1988) and to a reduction of both GABA A and B mediated potentials and several voltage dependent K+ conductances (notably I_Q and I_{AHP}; Gho et al. 1986; Rovira et al 1989; Fisher and Alger 1984). KA also induces long lasting hyperexcitability of the synaptic responses, since several hours after a brief (2-3 min) bath application of KA (50-200 nM), the stimulation of the mossy fibers or other synaptic inputs to CA3 neurons generates, instead of the typical epsp-ipsp sequence, a giant depolarizing potential with a burst of action potentials (Ben-Ari and Gho 1987). This effect involves the activation of NMDA receptors since prior application of NMDA antagonists which did not block the spontaneous burst induced by kainate, prevented the long-lasting effects on the synaptic responses (ibid. Ben-Ari and Gho). Therefore NMDA antagonists enable to differentiate between the induction of seizure activity and its maintenance; NMDA receptors are clearly involved in the former procedure but not in the latter. This provides direct evidence that NMDA receptors play an important role in epileptogenesis as in long-term potentiation of synaptic transmission (LTP) produced by electrical stimulation.

To conclude we suggest that the mechanisms underlying 2 types of neuronal plasticity, i.e. LTP and sprouting and establishment of new synaptic connections may play a role in epileptogenesis. The establishment of new aberrant connections with high affinity KA binding sites will contribute to hyperexcitability. In fact, in slices from KA treated rats the aberrant excitatory granule-granule cell connection is functional since a single stimulus of mossy fibers induces repetitive discharge in these neurons (Tauck and Nadler 1985).

REFERENCES

Ben-Ari Y. (1985) Limbic seizures and brain damage produced by kainic acid: mechanisms and relevance to human temporal lobe epilepsy. **Neuroscience** 14: 375-403.

Ben-Ari Y. and Gho M. (1988) Long-lasting modification of the synaptic properties of rat CA3 hippocampal neurones induced by kainic acid . **J. Physiol.** 404: 365-384.

Berger M. and Ben-Ari Y. (1983) Autoradiographic visualization of [3H] kainic acid receptor subtypes in the rat hippocampus. **Neurosci. Lett.** 39: 237-242.

Fisher R.S. and Alger B.E. (1984) Electrophysiological mechanisms of kainic acid- induced epileptiform activity in the rat hippocampal slice. **J. Neurosci.** 4: 1312-1323.

Frotscher M. and Zimmer J. (1983) Lesion induced mossy fibers to the molecular layer of the rat fascia dentata; identification of postsynaptic granule cells by the golgi-EM technique. **J.Comp.Neurol.** 215: 299-311.

Gho M. , King A.E., Ben-Ari Y. and Cherubini E. (1986) Kainate reduces two voltage dependent potassium conductances in rat hippocampal neurons in vitro. **Brain Res.** 385: 411-414.

Goddard G.V., McIntyre D.C. and Leech C.K. (1969) A permanent change in brain function resulting from daily electric stimulation. **Exp. Neurol.** 25: 295-330.

Haug F.M.S. (1967) Electron microscopical localization of the zinc in the hippocampal mossy fibre synapses by a modified sulfide silver procedure.**Histochemie** 8: 355-368.

Hauser, C.R., J.E. Miyashiro, B.E., Swartz, G.O., J.R. Rich and A.V. Delgado-Escueta (1990) Altered patterns of dynorphin immunoreactivity suggest mossy fiber reorganization in human hippocampal epilepsy. **J. Neurosci.** In press.

Lanerolle N.C., Kim J.H., Robbins R.J. and Spencer D.D. (1989) Hippocampal interneuron loss and plasticity in human temporal lobe epilepsy. **Brain Res.** 495: 387-395.

Nadler J.V. (1981) Kainic acid as a tool for the study of temporal lobe epilepsy. **Life Sci.** 29: 2031-2042.

Represa A. Tremblay E. and Ben-Ari Y.(1987) Kainate binding sites in the hippocampal mossy fibers : localization and plasticity. **Neurosci.** 20: 739-748.

Represa A. Le Gall La Salle G. and Ben-Ari Y. (1989a) Hippocampal plasticity in the kindling model of epilepsy in rats. **Neurosci. Lett.** 99: 345-350.

Represa A. , Robain O., Tremblay E. and Ben-Ari Y. (1989b) Hippocampal plasticity in childhood epilepsy. **Neurosci. Lett.** 99: 351-355.

Robinson J.H. and Deadwyler S.A.(1981) Kainic acid produces depolarization of CA3 pyramydal cells in the in vitro hippocampal slice. **Brain Res.** 221: 117-127.

Rovira C., Gho M. and Ben-Ari Y. (1989) Block of $GABA_B$ -activated K^+ conductance by kainate and quisqualate in rat CA3 hippocampal pyramidal neurones. **Eur. J. Physiol.** in press.

Sutula T. Xiao-Xian H., Cavazos J. and Scott G. (1988) Synaptic reorganization in the hippocampus induced by abnormal functional activity. **Science** 239: 1147-1150.

Tauck D.L. and Nadler J.V. (1985) Evidence for functional mossy fiber sprouting in hippocampal formation of kainic acid treated rats. **J.Neurosci.** 5: 1016-1022.

Tremblay E., Represa A. and Ben-Ari Y. (1985) Autoradiographic localization of kainic acid binding sites in the human hippocampus. **Brain Res.** 343: 378-382.

Unnerstall J.R. and Wamsley J.K. (1983) Autoradiographic localization of high-affinity [3H] kainic acid binding sites in the rat forebrain. **Eur. J. Pharmacol.** 86: 361-371.

Westbrook G.L. and Lothman E.W. (1983) Cellular and synaptic basis of kainic acid- induced hippocampal epileptiform activity. **Brain Res.** 273: 97-109.

MOLECULAR MARKERS OF REACTIVE PLASTICITY

James W. Geddes[1], Michael C. Wilson[3], Freda D. Miller[4],
and Carl W. Cotman[2]

[1]Div. Neurosurgery and [2]Dept. Psychobiology
Univ. Calif., Irvine, CA 92717; [3]Dept. Neuropharmacology
Scripps Clinic & Res. Foundation, La Jolla, CA 92037
and [4]Dept. Anatomy & Cell Biology, Univ. Alberta
Edmonton, Canada

INTRODUCTION

Reactive plasticity, including axonal and dendritic sprouting and reactive synaptogenesis, has been proposed to contribute to the pathogenesis of several neurological disorders. We have obtained evidence suggestive of plasticity in Alzheimer's disease and temporal lobe epilepsy. In each of these disorders, an altered distribution of excitatory amino acid receptors, particularly of the kainic acid subtype, was observed in the hippocampal formation (Geddes et al., 1985; Cahan et al., 1987; Geddes et al., submitted). Altered distribution of kainic acid binding sites has also been observed in other forms of childhood epilepsy (Represa et al., 1989). Additional markers of plasticity in human neurological disorders have included intensification of acetylcholinesterase staining in Alzheimer's disease (Geddes et al., 1985), and supragranular Timm's staining in temporal lobe epilepsy (Babb et al., 1988). Although these results are suggestive of sprouting, this interpretation is open to question. The increase in receptor density could simply be the result of receptor upregulation. The intensification of Timm's staining could result from increased zinc in existing terminals, and intensification of acetylcholinesterase staining is also relatively nonspecific. In addition to their lack of specificity, many of the morphological methods used to demonstrate sprouting in the rodent brain are unsuitable for use in postmortem human tissue. For example, Timm staining requires perfusion with a sulfide solution for optimal results.

Recently, molecular genetic analysis has provided another approach to identify and characterize proteins that may participate in the sprouting response. Sprouting requires the extension of neuronal processes and the rebuilding of the dendritic spine, both of which require the synthesis of cytoskeletal proteins. Following the contact of the axon with the target, synaptic proteins must be synthesized. Hippocampal protein synthesis, monitored by [3H] leucine incorporation, is increased at 6 and 12 days following lesions of the entorhinal cortex (Fass and Steward, 1983). Similarly, β - tubulin and actin mRNAs increase in the hippocampus 6 days following an entorhinal lesion (Philips et al., 1987. These results suggest that the synthesis of specific proteins is increased in response to deafferentation. Immunochemical identification of, or examination of the expression of mRNAs corresponding to, the proteins involved in axonal and dendritic growth and in synapse formation may provide molecular markers of plasticity and enable examination of the events which underlie the sprouting response.

In this report, we present the results obtained using two such markers. SNAP-25 was selected using *in situ* hybridization to identify cDNA clones of mRNA species with unique patterns of expression in the murine brain (Branks and Wilson, 1986; Oyler et al., in press). Tubulin-α1 (Tα1) mRNA was identified as an mRNA expressed at 10 fold higher abundance in the developing as compared to the adult rat brain (Miller et al., 1987). We examined the distribution of SNAP-25m RNA and protein, and Tα1 mRNA, following entorhinal lesions. We also examined the distribution of SNAP-25 mRNA and protein following selective destruction of CA3 pyramidal neurons and dentate gyrus granule cells. SNAP-25 immunoreactivity was maintained in projections to the lesioned region, and enhanced in areas adjacent to those deafferented by the lesions. The expression of both SNAP-25 and Tα1 mRNA was increased in deafferented neurons, and in those neurons participating in the sprouting response. The results demonstrate that examination of SNAP-25 and Tα1 mRNA expression and of SNAP-25 immunoreactivity may be useful in investigating alterations in major hippocampal circuits in animal lesion studies and in human neurological disorders such as Alzheimer's disease and temporal lobe epilepsy.

SNAP-25: A novel presynaptic marker of hippocampal neurons

SNAP-25 is a novel 25 kD protein associated with the synaptosomal fraction of hippocampal membranes (Oyler et al., in press) and encoded by a neuronal-specific mRNA (MuBr8, Branks and Wilson, 1986). Based on the predicted amino acid sequence, SNAP-25 consists of 206 amino acids and there are four closely spaced cysteine residues which is characteristic of metal binding proteins (Oyler et al., in press). Within the rodent hippocampus, SNAP-25 mRNA is widely expressed, but is particularly abundant in CA3 pyramidal neurons. The mRNA levels in CA3 are 3-4 fold greater than those in CA1 and the dentate gyrus. Using a polyclonal antisera raised against a synthetic peptide corresponding to carboxyl 12 residues (194-205) of predicted amino acid sequence, the distribution of the SNAP-25 protein has been characterized in the mouse brain (Oyler et al., in press). SNAP-25 immunoreactivity is greatest in the mossy fibers and there is also a band of immunoreactivity associated with the inner molecular layer of the dentate gyrus. We examined SNAP-25 immunoreactivity and mRNA expression in the rat hippocampus following lesions of the pre- and postsynaptic components of the hippocampal mossy fibers, the commissural/associational pathway, the Schaffer collaterals, and the perforant path. These studies focused on two regions the dentate gyrus molecular layer and the CA3 region of the hippocampus. The results demonstrate that SNAP-25 protein is a unique presynaptic marker in these major hippocampal pathways. The results further suggest that examination of the distribution of SNAP-25 immunoreactivity and SNAP-25 mRNA expression may provide a useful marker of plasticity in neurological disorders (Geddes et al., 1989; Geddes et al., submitted).

Dentate Gyrus Molecular Layer

There are two major inputs to the granule cells of the dentate gyrus. The perforant path, originating in the entorhinal cortex, terminates on the outer 2/3 of granule cell dendrites while the commissural/associational (C/A) pathway, originating in contralateral and ipsilateral polymorphic hilar neurons respectively, terminates on granule cell dendrites in the inner one-third of the molecular layer. Following lesions of the entorhinal cortex, which denervates the outer molecular layer, this region undergoes a major reorganization of its inputs. The remaining afferent fibers sprout to form new synapses on the denervated target cells. The C/A terminal zone expands from the inner 1/3 of the molecular layer until it occupies the inner half (see Cotman and Anderson, 1987, for a review). The changes in the distribution of SNAP-25 immunoreactivity following entorhinal lesions parallel this reorganization. There is a clearing of immunoreactivity from the outer molecular layer, corresponding to the loss of perforant path innervation, and an intensification of immunoreactivity in the inner molecular layer, consistent with the sprouting of the C/A fibers (Fig. 1). This is evident as early as 4 days postlesion, the time of initial afferent outgrowth, and is still present as late as one year postlesion. In addition, the expression of SNAP-25 mRNA appears increased in the granule cells deafferented by the lesion, and in the polymorphic hilar neurons, which provide the C/A afferent fibers (Figs, 2a and 2b). These results indicate a presynaptic localization of SNAP-25 in the perforant and C/A pathways and further suggest that the synthesis of SNAP-25 is increased in association with a sprouting response.

Bilateral kainic acid lesions destroy the polymorphic hilar neurons and remove the C/A fiber system (Nadler et al., 1980). Kainic acid lesion is accompanied by a clearing of SNAP-25

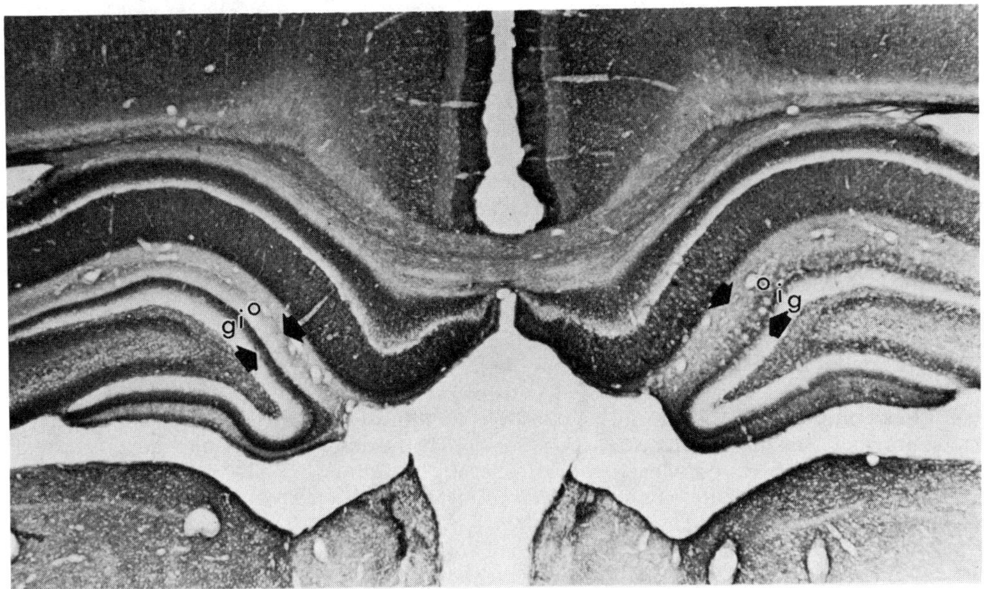

Fig. 1   Immunocytochemical localization of SNAP-25 in a coronal section of adult
rat hippocampus. The section was stained with a polyclonal antisera raised against a synthetic peptide corresponding to the carboxyl 12 amino
acids of SNAP-25 (Oyler et al., in press). The hippocampus on the right is
contralateral to an entorhinal lesion, and the pattern of SNAP-25 immunoreactivity is similar to that observed in control animals. Immunoreactivity is apparent in the mossy fibers and in the inner molecular layer of the
dentate gyrus. In the hippocampus ipsilateral to a unilateral entorhinal
lesion, shown on the rleft, intensification of immunoreactivity is observed
in the inner molecular layer of the dentate gyrus and a clearing of immunoreactivity in the middle to outer molecular layer. Abbreviations used:
g, granule cells of the dentate gyrus; i, inner molecular layer; m, middle
molecular layer; o, outer molecular layer.

immunoreactivity from the innermost region of the molecular layer and an intensification of
immunoreactivity in the outer molecular layer, presumably due to sprouting of the perforant path
input to this region (Geddes et al., 1989).. There did not appear to be aberrant sprouting of mossy
fiber collaterals into the supragranular zone, as has been observed previously using Timm's stain
(Nadler et al., 1980; Laurberg and Zimmer, 1981). Moreover, the expression of SNAP-25 mRNA
was not increased in the granule cells following the kainic acid lesions. The difference between the
SNAP-25 immunoreactivity observed in the present study and the previous Timm's stain results
may be the result of the shorter survival times used in the present study.

Colchicine injections into the hippocampus, at the proper dosage, are selectively toxic to
granule cells (Goldschmidt and Steward, 1980). Ten days following the destruction of dentate
gyrus granule cells by colchicine, SNAP-25 immunoreactivity is more diffuse and more intense in
the molecular layer of the dentate gyrus. This may result from sprouting of the major inputs to this
region following loss of the target granule cell dendrites (see Geddes et al., 1989).

Hippocampal area CA3

The mossy fiber axons of the dentate gyrus granule cells terminate on CA3 pyramidal neurons.
The removal of granule cells by colchicine is accompanied by elimination of the mossy fibers, as
revealed by Timm staining for heavy metals (Goldschmidt and Steward, 1980). In the present

study, colchicine lesion resulted in the disappearance of SNAP-25 immunoreactivity in CA3 stratum lucidum, and in an increased expression of SNAP-25 mRNA in the denervated CA3 pyramidal neurons. Kainic acid lesion-induced destruction of CA3 pyramidal neurons did not remove the SNAP-25 immunoreactivity in CA3 stratum lucidum, but did result in a loss of staining in CA1 stratum radiatum, the target of the Schaffer collateral projection, and in increased expression of SNAP-25 mRNA in CA1 pyramidal neurons. SNAP-25 immunoreactivity was enhanced, in CA1 stratum lacunosum-moleculare and stratum oriens following kainic acid lesion. This likely represents sprouting of the perforant path fibers as an attempt to compensate for the loss of C/A input to the molecular layer.

The results obtained by examining SNAP-25 immunoreactivity and mRNA expression following lesions of the pre- and post-synaptic components of the hippocampal mossy fiber, Schaffer collateral, commissural/associational, and perforant pathways are consistent. Selective destruction of hippocampal neurons does not diminish SNAP-25 immunoreactivity in the dendritic fields. In contrast, lesioned neurons exhibit an extensive loss of immunoreactivity at the site of their axonal projections. These results support the identification of SNAP-25 as a novel presynaptic protein in the major hippocampal pathways. The results also indicate that expression of SNAP-25 mRNA is increased in deafferented neurons. In addition, SNAP-25 immunoreactivity appears to be increased in afferent fibers which project to the lesioned neurons, and in the areas adjacent to the region deafferented by the lesion.

Examination of SNAP-25 mRNA expression and immunoreactivity may provide a useful marker of hippocampal plasticity in neurological disorders such as Alzheimer's disease and temporal lobe epilepsy. In preliminary results obtained with hippocampal tissue obtained post-mortem from Alzheimer's disease patients, the pattern of SNAP-25 immunoreactivity was similar to that observed in the rodent brain. In addition, SNAP-25 immunoreactivity was observed in neuritic plaques in subiculum, CA1, and in the molecular layer of the dentate gyrus. The presynaptic localization of SNAP-25, the intensification of SNAP-25 immunoreactivity in association with sprouting, and the presence of SNAP-25 immunoreactivity in neuritic plaques are consistent with the hypothesis that an aberrant sprouting response may contribute to neuritic plaque formation (Geddes et al., 1986; Geddes and Cotman, 1989).

Sprouting as a replay of development: Tubulin-α1

Several investigators have hypothesized that CNS sprouting may involve a reactivation of mechanisms that operated during development (Shepherd, 1988; Crutcher, 1987; Cotman and Neito-Sampedro, 1984) but this hypothesis has never been examined directly. There are many similarities between development and sprouting. Both sprouting and development require the extension of neuronal processes toward, and presumably competition for, vacant postsynaptic sites. In both cases, glial proliferation precedes neuronal extension and may provide trophic support and chemotactic guidance (for review see Cotman and Neito-Sampedro, 1984). Sprouting also results in a loss and reacquisition of dendritic spines and it is hypothesized that spine formation in both development and sprouting is dependent upon the arrival and normal function of axon terminals. Following synaptogenesis, both developmental and reactive, there appears to be communication between the afferent and the target, such that the phenotype of the innervating neurite can influence properties, including receptor distribution, in the target (Geddes et al., 1985). Both sprouting and development involve axonal growth, synaptogenesis, and differentiation. Thus, it would not be surprising if at least some of the molecular mechanisms associated with sprouting represent a recapitulation of developmental patterns of gene expression.

There are several gene families which are under developmental control and are differentially expressed in a developmental stage and/or tissue specific manner. Many examples of developmentally regulated genes include cytoskeletal proteins. A developmentally-regulated form of α-tubulin, referred to as Tα1 in the rat (Miller et al., 1987) and bα1 in the human brain (Cowan et al., 1983) is expressed at high levels in the fetal brain during the developmental expression of neuronal processes, but is greatly reduced in abundance in mature mammalian neurons (Miller et al., 1987). Following axonal injury in the peripheral nervous system, Tα1 mRNA is rapidly reinduced, is maintained at high levels during the period of axonal outgrowth, and is subsequently down-

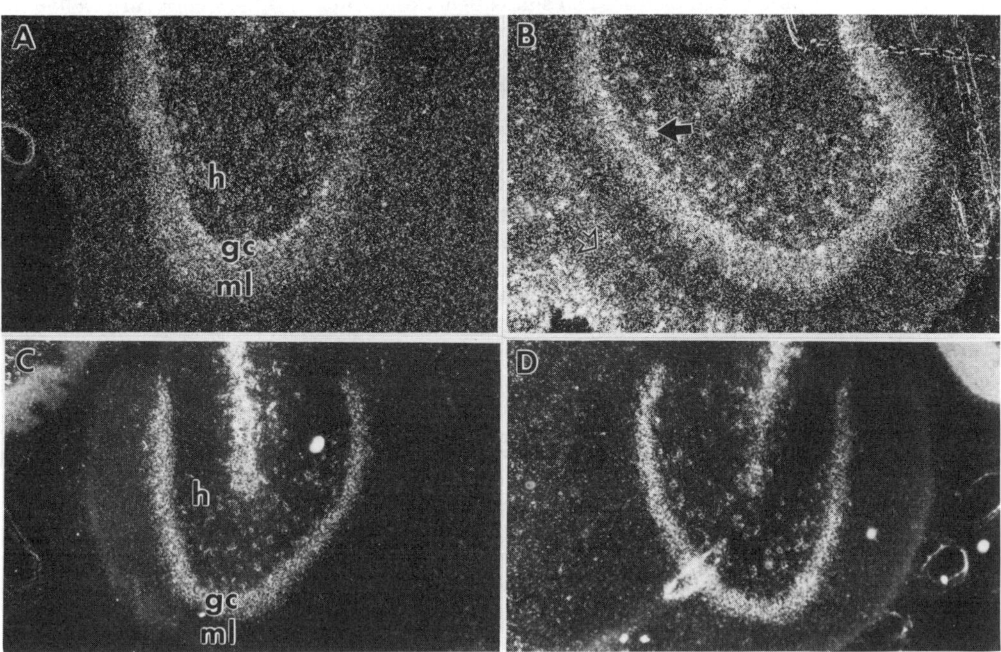

Fig. 2.   Darkfield photomicographs of emulsion-coated coronal sections of adult rat hippocampus hybridized with a ^{35}S-labelled riboprobe complementary to (A,B) SNAP-25 mRNA, and (C,D) the 3' untranslated region of tubulin-α1 mRNA. The rat sections were obtained from the contol animals (A,C) and from the ipsilateral hippocampus 7 days following a unilateral lesion of the entorhinal cortex (B,D).   Entorhinal lesion resulted in increased expression of both SNAP-25 and tubulin-α1 mRNA in granule cells and polymorphic hilar neurons.

regulated to control levels around the time of target contact (Miller et al., 1989). To determine whether sprouting central mammalian neurons also recruit the major embryonic form of α-tubulin mRNA, we used *in situ* hybridization to examine the expression of Tα1 mRNA in the rat hippocampus following unilateral electrolytic lesions of the entorhinal cortex.

An increase in the expression of Tα1 mRNA in dentate gyrus granule cells, denervated by the entorhinal lesion, was evident as early as one day postlesion. Elevated Tα1 mRNA expression was also observed in polymorphic hilar neurons, in pyramidal neurons of the CA3 and CA1 regions, and in cortical neurons located near the site of the lesion. An increase in Tα1 mRNA expression was also evident in the hippocampus contralateral to the lesion.and similar results were obtained following knife cut transections of the angular bundle. The maximal increase in Tα1 mRNA expression in the various hippocampal neuronal populations was at 3-5 days following surgery and by postlesion day 14, the expression of Tα1 mRNA had begun to return to control levels (see Fig 2c and 2d). The rapid lesion-induced increase and subsequent decline in Tα1 mRNA correlates well with previous studies on sprouting and synapse turnover triggered by entorhinal cell loss (See Cotman and Anderson, 1987 for review). The neurons which exhibited increased Tα1 mRNA expression include those which are predicted to extend axon collaterals (polymorphic hilar neurons and contralateral entorhinal neurons) and also neurons which may have to rebuild dendritic spines and postsynaptic densities following the loss of entorhinal input (granule cells of the dentate gyrus, pyramidal neurons in CA1 and CA3).

The observed elevation in levels of Tα1 mRNA during sprouting of hippocampal neurons in the rat supports the hypothesis that sprouting in the CNS involves a selective reactivation of developmental mechanisms. The results further demonstrate that increased expression of Tα1 mRNA is a useful marker of deafferented neurons, and of neurons participating in the sprouting response. The increase in Tα1 mRNA expression in the contralateral hippocampus, is surprising in view of the selective ipsilateral increase in tubulin immunoreactivity (Kwak and Matus, 1989), but consistent with a previously described cycle of synapse loss and replacement in the contralateral dentate gyrus following unilateral lesions of the entorhinal cortex (Hoff et al., 1981).

Alzheimer's disease also results in a loss of entorhinal input to the hippocampus (Hyman et al., 1984). In hippocampal sections obtained from five Alzheimer's disease patients, as compared to five age-matched controls, analysis of the X-ray film autoradiograms demonstrated a consistent elevation of bα1 mRNA in the dentate gyrus and CA3 region of the AD hippocampus (Geddes et al., in press). Analysis of emulsion coated sections at higher resolution demonstrate increased numbers of grains over neurons of the dentate gyrus, the hilus, and the CA3 regions in AD vs. controls. In most cases, the occasional neuron in CA1 exhibits an increased signal for bα1 mRNA, although in some AD patients, almost all of the neurons in CA1 exhibit increased expression of bα1 mRNA.

The results suggest that some of the embryonic markers observed in the AD brain appear in association with an aberrant sprouting response, as a consequence of the reinduction of developmental patterns of gene expression Recent results indicate that both mature and embryonic tau mRNA sequences are expressed in AD (Kosik et al., 1989). In addition, an antigen recognized by the monoclonal antibody Alz-50 is present in the embryonic and AD brain, but not in the normal adult brain (Wolozin et al., 1988). The presence of embryonic forms of cytoskeletal proteins, including tubulin, tau, and other MAPs, including MAP 2(see Papandrikopoulou et al., 1989) may impact on the polymerization, phosphorylation, and stability of the neuronal cytoskeleton in Alzheimer's disease (see Nunez, 1988). Thus, the inappropriate production of select developmentally-regulated cytoskeletal proteins, in association with an aberrant sprouting response, may contribute to the etiology of neuropathological disorders such as Alzheimer's disease.

SUMMARY AND CONCLUSIONS

Sprouting has been studied extensively using morphological markers, but relatively little is known regarding the molecular and biochemical events which underlie the sprouting response. Moreover, due to the lack of suitable markers, it has been difficult to examine changes in the major hippocampal pathways in animal lesion models and in neurological disorders. We utilized two markers, obtained through molecular genetic analysis, to examine alterations in hippocampal circuits following partial deafferentation.

SNAP-25, a neuronal specific protein, is located presynaptically in the perforant path, Schaffer collaterals, mossy fibers, and commissural/associational pathways. Selective destruction of CA3 pyramidal neurons, dentate gyrus granule cells, and entorhinal cortical neurons resulted in a loss of SNAP-25 immunoreactivity at the site of the axonal projections of the lesioned neurons. SNAP-25 immunoreactivity was maintained in projections to the lesioned region, and enhanced in areas adjacent to those deafferented by the lesions. Expression of SNAP 25 mRNA was increased in denervated regions, and in neurons which would be expected to participate in the sprouting response.

Tubulin-α1 is an mRNA which is expressed at high levels in the fetal brain during periods of neurite outgrowth, but at low levels in the adult brain. The expression of Tα1 mRNA was increased in the rat hippocampal neurons following entorhinal lesions, and in the human hippocampus in patients with Alzheimer's disease. These results demonstrate that sprouting in the CNS may result in the replay of developmental patterns of gene expression.

Examination of SNAP-25 and Tα1 mRNA expression and of SNAP-25 immunoreactivity may be useful in investigating alterations in major hippocampal circuits in a variety of conditions such as learning, hypoxia, hypoglycemia, and also in neurological disorders such as temporal lobe epilepsy and Alzheimer's disease.

REFERENCES

Babb, T.L., Kupfer, W.R., and Pretorius, J.K., 1988, Synaptic reorganization of mossy fibers into inner molecular layer in human epileptic Fascia dentata. **Soc. Neurosci.** Abstr. 14: 881.

Branks, P.L., and Wilson, M.C., 1986, Patterns of gene expression in the murine brain revealed by *in situ* hybridization of brain-specific mRNAs. **Mol. Brain Res. 1**: 1-16.

Cahan, L.D., Geddes, J.W., Choi, B.H., and Cotman, C.W., 1987, Alterations in excitatory amino acid receptors in temporal lobe epilepsy.**Soc. Neurosci.Abstr. 13**.

Cotman, C.W. and Anderson, K.J. ,1987, Synaptic plasticity and functional stabilization in the hippocampal formation: possible role in Alzheimer's disease. **Adv. Neurol. 47**: 313-35

Cotman, C.W., and Nieto-Sampedro, M., 1984, Cell biology of synaptic plasticity. **Science 225**: 1287-1294.

Cowan, N.J., Dobner, P.R.,Fuchs, E.V. and Cleveland, D.W., 1983, Expression of human alpha-tubulin genes: interspecies conservation of 3' untranslated regions, **Mol. Cell Biol. 3** : 1738-1745.

Crutcher, K.A., 1987, Sympathetic sprouting in the central nervous system: a model for studies of axonal growth in the mature mammalian brain. **Brain Res. Rev. 12**: 203-233.

Fass, B. and Steward,O., 1983, Increases in protein-precursor incorporation in the denervated neuropil of the dentate gyrus during reinnervation. **Neurosci. 9**: 653-664.

Geddes, J.W., Monaghan, D.T., Cotman, C.W., Lott, I.T., Kim, R.C. and Chui, H.C., 1985, Plasticity of hippocampal circuitry in Alzheimer's disease. **Science 230**: 1179-1181.

Geddes, J.W., Cahan, L.D., Cooper, S.M., Choi, B.H., Kim, R.C., and Cotman, C.W. Altered distribution of excitatory amino acid receptors in temporal lobe epilepsy. (submitted).

Cahan, L.D., Geddes, J.W., Choi, B.H., and Cotman, C.W., 1987, Alterations in excitatory amino acid receptors in temporal lobe epilepsy.**Soc. Neurosci. 13** .1079.

Geddes, J.W., Hess, E.J., Hart, R.A., Kesslak, J.P., Cotman, C.W., and Wilson, M.C., 1989, Expression of SNAP-25 protein and mRNA following hippocampal lesions. **Soc. Neurosci._Abstr. 14**: 1246.

Geddes, J.W., Anderson, K.J., and Cotman, C.W., 1986, Senile plaques as aberrant sprout stimulating structures. **Experimental Neurology 94**: 767-776.

Geddes, J.W. and Cotman, C.W., 1989, Plasticity, pathology, and Alzheimer's disease (Commentary). **Neurobiol. Aging 10**: 571-573.

Geddes, J.W., Hess, E.J., Hart, R.A., Kesslak, J.P., Cotman, C.W., and Wilson, M.C. Lesions of hippocampal circuitry define synaptosomal-associated protein-25 as a novel presynaptic marker. (submitted).

Geddes, J.W., Wong, J., Choi, B.H., Kim, R.C., Cotman, C.W., and Miller, F.D. Increased expression of an embryonic, growth-associated, mRNA in Alzheimer's disease.**Neurosci. Lett.** (in press)..

Goldschmidt, R.B. and Steward, O., 1980, Preferential neurotoxicity of colchicine for granule cells of the dentate gyrus of the adult rat. **Proc. Natl. Acad. Sci. USA.** 77: 3047-3051.

Hoff, S.F., Scheff, S.W., Kwan, A.Y., and Cotman, C.W., 1981, A new type of lesion-induced synaptogenesis: I. Synaptic turnover in non-denervated zones of the dentate gyrus in young adult rats. **Brain Res.** 222: 1-13.

Hyman, B.T., Van Hoesen, G.W. Damasio, A.R. and Barnes, C.L., 1984, Alzheimer's disease: cell-specific pathology isolates the hippocampal formation. **Science** 225: 1168-1170

Kosik, K.S., Orecchio, L.D. Bakalis, S. and Neve, R.L., 1989, Developmentally regulated expression of specific tau sequences. **Neuron,** 2: 1389-1397.

Kwak, S. and Matus, A., 1988, Denervation induces long-lasting changes in the distribution of microtubule proteins in hippocampal neurons. **J. Neurocytol..** 17: 189-195.

Laurberg, S. and Zimmer, J., 1981, Lesion-induced sprouting of hippocampal mossy fiber collaterals to the fascia dentata in developing and adult rats. **J. Comp. Neurol.** 200: 433-459.

Lewis, S.A. Lee, M.G.-S and Cowan, N.J., 1985, Five mouse tubulin isotypes and their regulated expression during development. **J. Cell Biol.** 101:852-861.

Miller, F.D., Naus, C.C.G., Durand, M., Bloom, F.E. and Milner, R.J, 1987, Isotypes of α-tubulin are differentially regulated during neuronal maturation. **J. Cell Biol.** 105: 3065-3073.

Miller, F.D., Tetzlaff, W. Bisby, M.A. Fawcett, J.W., and Milner, R.J., 1989, Rapid induction of the major embryonic α-tubulin mRNA, Tα1, during nerve regeneration in adult rats. **J. Neurosci.** 9 : 1452-1463.

Nadler, J.V., Perry, B.W., Gentry, C., and Cotman, C.W., 1980, Degeneration of hippocampal CA3 pyramidal cells induced by intraventricular kainic acid.**J. Comp. Neurol.** 192: 333-359.

Nunez, J., 1988, Immature and mature variants of MAP2 and tau proteins and neuronal plasticity. **Trends Neurosci.** 11: 477-479

Oyler, G.A., Higgins, G.A., Hart, R.A., Battengerg, E., Bloom, F.E., and Wilson, M.C. The sequence and characterization of a neuronal specific mRNA encoding a novel synaptosomal associated protein, SNAP-25. **J. Cell Biol.**(in press)

Papandrikopoulou, A., Doll, T., Tucker, R.P., Garner, C.C., and Matus, A. (1989) Embryonic MAP2 lacks the cross-linking sidearm sequences and dendritic targeting signal of adult MAP2. **Nature** 340: 650-652.

Phillips, L.L., Chikaraishi, D.M., and Steward, O., 1987, Increases in messenger RNA for actin and tubulin within the denervated neuropil of the dentate gyrus during lesion-induced synaptogenesis. **Soc. Neurosci Abstr.** 13: 1428.

Represa, A., Duyckaerts, C., Tremblay, E., Hauw, J.J., and Ben-Ari, Y., 1988, Is senile dementia of the Alzheimer type associated with hippocampal plasticity? **Brain Res.** 457: 355-359.

Represa, A., Robain, O., Tremblay, E., and Ben-Ari, Y., 1989, Hippocampal plasticity in childhood epilepsy. **Neurosci. Lett.** 99: 351-355.

Shepherd, Gordon M., 1988, "Neurobiology", 2nd edition, Oxford University Press, New York.

Wolozin, B.L., Scicutella, A. and Davies, P., 1989, Re-expression of a developmentally regulated antigen in Down Syndrome and Alzheimer disease. **Proc. Natl. Acad. Sci. USA** 85:6202-6206.

TRANSPLANTATION OF DEVELOPING HIPPOCAMPAL NEURONS TO ISCHEMIC AND

EXCITOTOXIC LESIONS OF THE ADULT RAT HIPPOCAMPUS

Niels Tønder, Torben Sørensen, [1]Flemming Fryd Johansen and
Jens Zimmer

PharmaBiotec, Institute of Neurobiology, University of Aarhus
Denmark and [1]PharmaBiotec, Institute of Neuropathology
University of Copenhagen, Denmark

INTRODUCTION

Intracerebral transplantation of immature brain tissue is now a well-
established technique for the study of the development and regenerative
capacity of neurons and nerve connections in the central nervous system.
With this purpose we have previously grafted hippocampal and fascia dentata
tissue to newborn and adult rats (Sunde and Zimmer, 1983; Sunde et al., 1984;
Zimmer and Sunde, 1984; Zimmer et al., 1985, 1986, 1987, 1988a,b,c; Tønder
et al., 1986, 1988; Sørensen and Zimmer, 1988a,b). One set of studies included grafting to rats with induced maldevelopment of the hippocampal
neuronal circuitry and demonstrated that dentate granule cells damaged by
neonatal X-irradiation can be partly replaced by homotypic transplants of
immature fascia dentata (Sunde et al., 1984, 1985). Another observation was
that the exchange of hippocampal graft-host nerve connections clearly de-
pended on the age of the recipient animal. The exchange of nerve connections
was much more extensive in animals X-irradiated at birth and grafted immediately thereafter than in other irradiated rats where the grafting was
delayed into early adulthood (Sunde et al., 1985). In accordance with this,
the number of graft neurons with commissural projections is much higher
after grafting to normal newborn rats (Tønder et al., 1988) than after
grafting to normal adult rats (unpublished results). In this chapter we
shall, however, review some of our recent data, demonstrating that extensive
reciprocal innervation of the graft and the host brain can take place also
in the adult brain, provided that the grafting is performed into so-called
"axon-sparing" lesions. In these types of lesions the neurons in the lesion
area degenerate while the extrinsic afferent nerve fibers are spared and
remain in the area for several months (Coyle and Schwarcz, 1983).

AXON-SPARING LESIONS

One week prior to transplantation, adult recipient rats were subjected
to axon-sparing lesions, induced either by direct intrahippocampal injection
of the glutamic acid analogue, ibotenic acid (Tønder et al., 1989b,c) or by
transient cerebral ischemia (Tønder et al., 1989a) (Fig. 1).

Excitatory Amino Acids and Neuronal Plasticity
Edited by Y. Ben-Ari
Plenum Press, New York, 1990

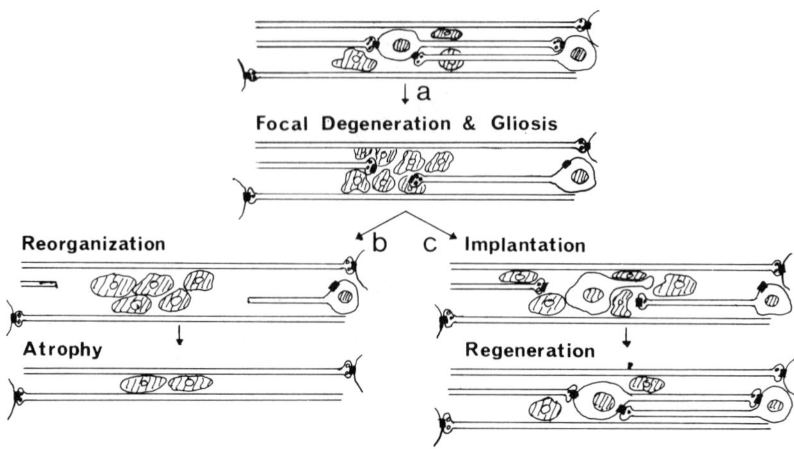

Fig. 1. Schematic illustration of the axon-sparing lesion-
transplantation model. - a: Ibotenic acid-/ischemia
induced neuronal degeneration and reactive gliosis.
- b: Long term reorganisation and atrophy of the
lesioned area. - c: Transplantation and connective
integration of homotypic neural tissue with the
host brain. Modified from Coyle and Schwarcz (1983).

Ibotenic acid lesion

Intrahippocampal injection of ibotenic acid (IA) in a dose of 5 µg in
0.5 µl phosphate buffer (pH 7.4) results in a total loss of neurons at the
injection site (Schwarcz et al., 1979; Köhler et al., 1979; Tønder et al.,
1989b,c). The axon-sparing property of the IA lesion has been demonstrated
by neurochemical methods, showing persistence of cholinergic, noradrenergic
and serotonergic afferents from septum, locus coeruleus and the raphe
nuclei, respectively (Schwarcz et al., 1979). In the present studies IA was
injected into either the dorsal fascia dentata (FD) or the dorsal CA3 area.
The persistence of cholinergic septohippocampal fibers was demonstrated by
histochemical staining for acetylcholinesterase, and spared dentate and
hippocampal afferents were demonstrated by Timm staining, silver staining of
induced anterograde axonal degeneration or immunohistochemistry for neuropeptides (Köhler, 1983; Tønder et al., 1989b,c).

Ischemic lesion

The ischemic lesions were induced by transient interruption of the
cerebral blood flow, using the 4-vessel occlusion technique described by
Pulsinelli and Brierley (1979). The ischemia results in a selective lesion,
which is most likely excitotoxic in nature (Jørgensen and Diemer, 1982;
Benveniste et al., 1984), of the pyramidal cells in CA1 (Pulsinelli and
Brierley, 1979; Jørgensen and Diemer, 1982) and the somatostatin-immunoreac-
tive neurons in the dentate hilus (Johansen et al., 1987). The axon-sparing
property of the ischemic CA1 lesion has previously been shown by persistence
of presynaptic structures at the electron microscopical level as well as by
preservation of high-affinity glutamate uptake 4 days after ischemia
(Johansen et al., 1984). One week and longer after ischemia, histochemical
stains have demonstrated the persistence of AChE-positive, cholinergic
septohippocampal fibers and Timm-positive and immunohistochemically reactive
hippocampal afferents in the CA1 lesions (Tønder et al., 1989a).

<u>The axon-sparing lesion as a transplantation site</u>

Although the excitotoxic and ischemic insults do not initially damage
the afferent nerve fibers, some long-term reorganization of these does take
place, just as the lesioned areas do undergo long-term atrophy (Fig. 1)
(Coyle and Schwarcz, 1983; Nadler et al., 1981; Tønder et al., 1989a,b,c).
After one week, at which time the transplantation was performed (see below),
the two main characteristics of the lesions were a dense reactive gliosis
and the persistence of afferent nerve fibers originating from intact neurons
located outside the lesion area. These afferents were accordingly deprived
of their normal target cells, but they had not actually been damaged or
transected. It might therefore be anticipated that they, one week after the
lesion would be in a state, where they would be able to enter and innervate
transplants of immature hippocampal tissue more readily than both intact and
acutely transected nerve fibers.

TRANSPLANTATION

In order to test the above mentioned possibility we grafted neonatal
dentate tissue or late fetal hippocampal tissue into one week old ischemic
or IA-induced lesions (Tønder et al., 1989a,b,c). Into the dentate IA-
lesions we grafted blocks of fascia dentata from newborn rats. Into CA1
ischemic lesions and CA3 IA-lesions we grafted cell suspensions made from
embryonic (E18-20) CA1 and CA3 tissue, respectively. After 6 weeks or more
the morphology of the grafts and their structural and connective integration
with the host brain were examined using ordinary cell staining, AChE and
Timm staining, immunohistochemical stains for neuropeptides and astroglia,
and fiber tracing by retrograde axonal transport of the injected fluorescent
dye, Fluoro-Gold, or light- or electronmicroscopy of induced anterograde
axonal degeneration.

TRANSPLANT SURVIVAL AND HISTOLOGICAL ORGANIZATION

The transplants survived well in both the IA- and the ischemic lesions.
After 6 weeks or more they displayed a near normal immunoreactivity for the
astroglial marker, glial fibrillary acidic protein (GFAP), in contrast to
the heavily gliotic, surrounding parts of the damaged host brain. The host
brain gliosis was accordingly not harmfull to the transplants. The gliosis
might in fact have beneficial effects on the grafts, considering that
several studies have shown that reactive glial cells produce neurotrophic
factors (Lindsay et al., 1982; Nieto-Sampedro et al., 1984; Cotman and
Nieto-Sampedro, 1985; Whittemore et al., 1985; Heacock et al., 1986;
Kesslak et al., 1986).

Regarding the intrinsic cytological and histological organization, the
transplants contained neurons typical of the grafted tissue. The FD transplants were thus dominated by dentate granule cells (Fig. 2), and the CA3
and CA1 transplants by the respective pyramidal cells (Fig. 3). In addition,
non-pyramidal neurons, characterized by their immunoreactivity against the
peptides cholecystokinin and somatostatin, were present in all transplants,
as also previously shown by Zimmer and Sunde (1984) in grafts of solid
pieces of hippocampal and fascia dentate tissue to normal newborn or adult,
and by Mudrick et al. (1987) after grafting of CA1 cell suspensions to CA1
ischemic lesions. The solid FD transplants were organotypically organized,
with clearly distinguishable granule cell and molecular layers and a hilar
area (Fig. 2) (cp. Sunde and Zimmer, 1983; Zimmer and Sunde, 1984). In the
CA1 (Fig. 3) and CA3 transplants the pyramidal cells were arranged in clusters or small bands, demonstrating a tendency of the suspended cells to
reaggregate.

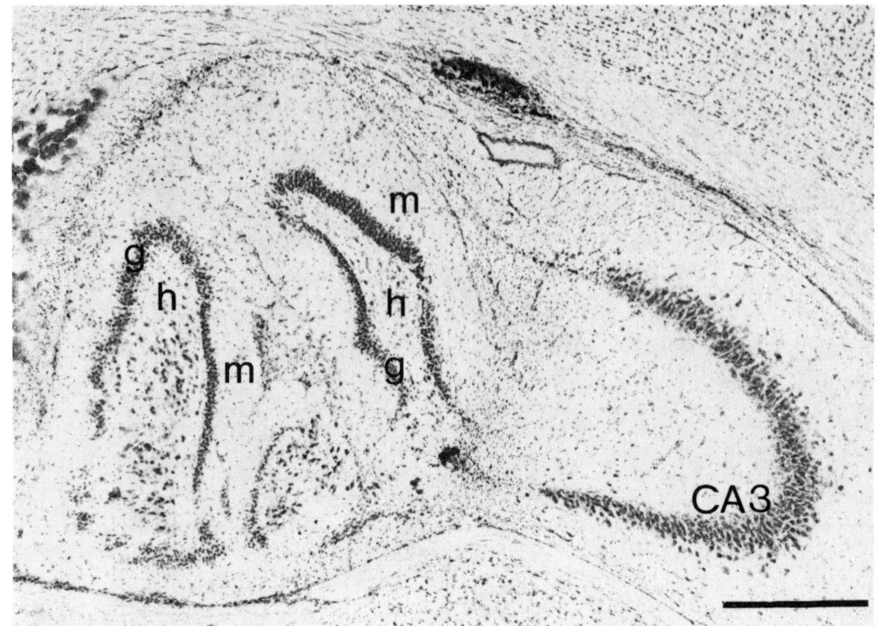

Fig. 2. Cell stain of dentate transplant 6 weeks after grafting
into one week old IA-lesion. The transplant has grown to
occupy almost the entire IA-lesion area. Note the organotypic organization of the graft, with a characteristic
dentate granule cell layer (g), molecular layer (m) and
hilar area (h). Bar = 500 µm.

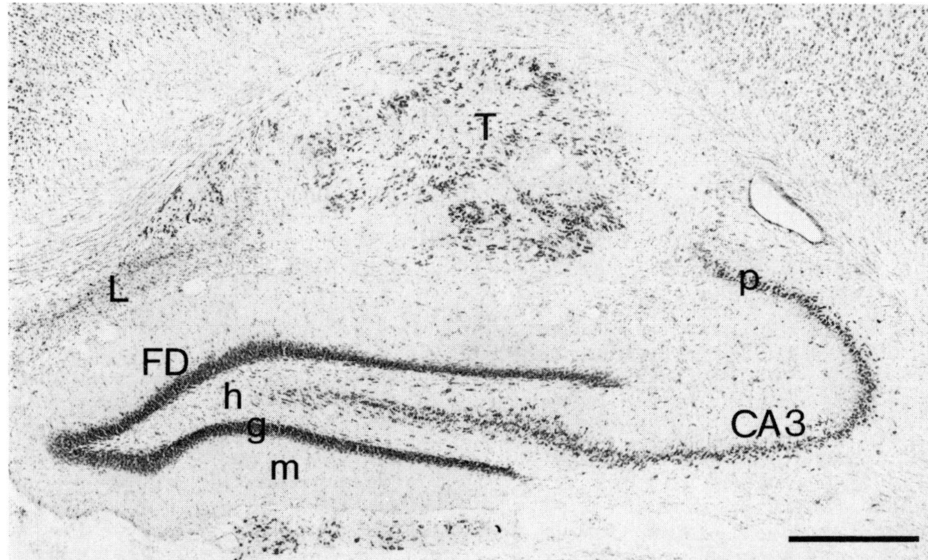

Fig. 3. Cell stain of CA1 suspension transplant (T) 6 months after
grafting to a one week old ischemic lesion. The graft is located in the lesioned CA1 and displays clusters and bands of
cells. FD: fascia dentata; g: granule cell layer; h: dentate
hilus; L: lesioned host CA1; m: molecular layer; p: pyramidal
cell layer. Bar = 500 µm.

NERVE CONNECTIONS

Fascia dentata transplants

The transplants were innervated by cholinergic AChE-positive, host
septohippocampal fibers. Histochemical staining for AChE has previously been
used to monitor the ingrowth of septohippocampal cholinergic fibers in hippocampal and fascia dentata transplants (Sunde and Zimmer, 1983; Zimmer et
al., 1985, 1987, 1989), and as in these studies the most dense AChE staining
was present around the cell layers, indicating a homotypic ingrowth of cholinergic fibers. When the transplants were innervated by host perforant path
fibers (see below), the AChE-positive fibers adapted to the laminar distribution of these main afferent projections, as previously shown by Zimmer et
al. (1986). The ingrowth of cholinergic, host septohippocampal fibers was
not surprising, as these fibers belong to a global, and in functional terms
"level setting" system, known to have a high regenerative capacity even in
the adult brain (Björklund and Stenevi, 1984; Sunde and Zimmer, 1983;
Zimmer et al., 1986, 1989).

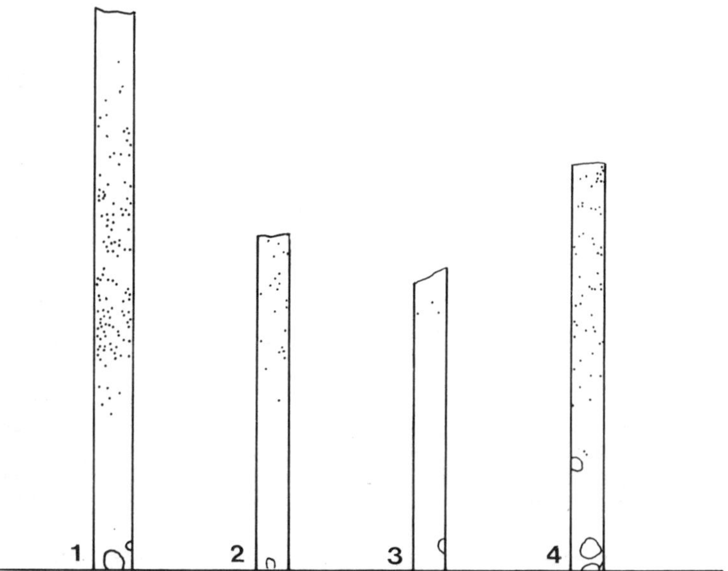

Fig. 4. Density and distribution of degenerating perforant pathway terminals in columns of electron
micrographs taken across the dentate molecular
layer of normal (1) and grafted (2-4) fascia
dentata 3 days after entorhinal lesions. 1-3:
Normal adult rat with predominantly medial
perforant path degeneration (1). Dentate graft
placed in intact adult fascia dentata, showing
host perforant pathway innervation next to
graft-host interface (2), but a significant
decrease just 200 μm away (3). 4: Dentate
graft placed in axon-sparing, ibotenic acid
lesion of the adult fascia dentata. Note the
increased number of degenerating terminals
after grafting to the axon-sparing lesion (cf.
Tønder et al., 1989b).

Perforant path fibers projecting from the entorhinal cortex to the
dentate molecular layer belong to one of the so-called "point-to-point"
systems, and normally they display very low regenerative capacity in the
mature brain. When, however, pieces of neonatal FD were grafted homotypically into dentate IA-lesions, these transplants became heavily innervated
by perforant path fibers from the host entorhinal cortex. This was demonstrated both by Timm staining and tracing by anterograde axonal degeneration
and Fink-Heimer silver staining or electron microscopy. In well-placed
grafts, the host perforant path fibers occupied the outer 1/2 to 2/3 of the
graft molecular layer, resembling the normal distribution of perforant path
afferents. When this host fiber ingrowth was compared with the ingrowth
observed after grafting to normal adult rats without other lesions than the
one caused by the actual insertion of the grafts (Sunde et al., 1984;
Zimmer et al., 1985, 1987), it was evident that the ingrowth after a preceding IA-lesion was both much more extensive and more dense (Fig. 4). It
resembled the ingrowth observed after grafting to normal, newborn rats, or
early postnatal rats with irradiation induced reduction of granule cells
(Sunde et al., 1984, 1985; Sørensen et al., 1986, 1987). Our observations
accordingly very strongly indicate that a preceding lesion of the axon-
sparing type can enhance the growth of adult nerve fibers of the point-to-
point type, monitored by their capability to innervate grafts of immature
neural tissue.

Outgrowth of dentate graft mossy fibers to the adult host CA3 region
has earlier been shown both by Sørensen et al. (1986) after grafting to
normal adult rats, and by Pohle (1987) after grafting to colchicine lesions
of the FD of adult rats. A similar outgrowth of dentate graft mossy fibers
was not found after grafting to IA lesions, but these transplants were not
always oriented with their hilar parts towards the deafferented host CA3,
and gliotic areas of host dentate or CA3 tissue were often present between
the grafts and the intact parts of the host CA3 pyramidal cell layer
(Tønder et al., 1989c).

<u>CA3 transplants</u>

The CA3 transplants exchanged several types of connections with the
host brain. These connections are schematically shown in figure 5, together
with the main connections to and from the normal and the IA-lesioned CA3. A
dense AChE-staining, located in particular around the cell clusters, was
found in all CA3 grafts, indicating a homotypic ingrowth of cholinergic,
host septohippocampal nerve fibers, as also shown after grafting of solid
pieces of hippocampal tissue (see above).

In Timm stained sections, large characteristically stained mossy fiber
terminals were seen in the transplants. At the ultrastructural level this
ingrowth of mossy fibers from the host fascia dentata was shown to follow a
homotypic pattern. Typical large mossy fiber terminals thus formed multiple
asymmetric synapses with dendritic shafts and complex spines corresponding
to the proximal parts of the dendrites of the CA3 pyramidal cells. As the
mossy fiber projection is a good example of a point-to-point system, these
observations provide yet another example of the increased and specific regenerative capacity, which can be induced by a preceding axon-sparing lesion.

Ingrowth of commissural fibers from the contralateral host hippocampus
was demonstrated by anterograde axonal degeneration and electron microscopy,
showing terminals of commissural origin forming asymmetric synaptic contacts
with dendritic spines and shafts of the graft CA3 pyramidal cells.

In recent studies we have demonstrated that hippocampal CA3 and CA4
neurons grafted to the newborn (Tønder et al., 1988) and the adult rat hippocampus (unpublished results) can establish efferent commissural projections

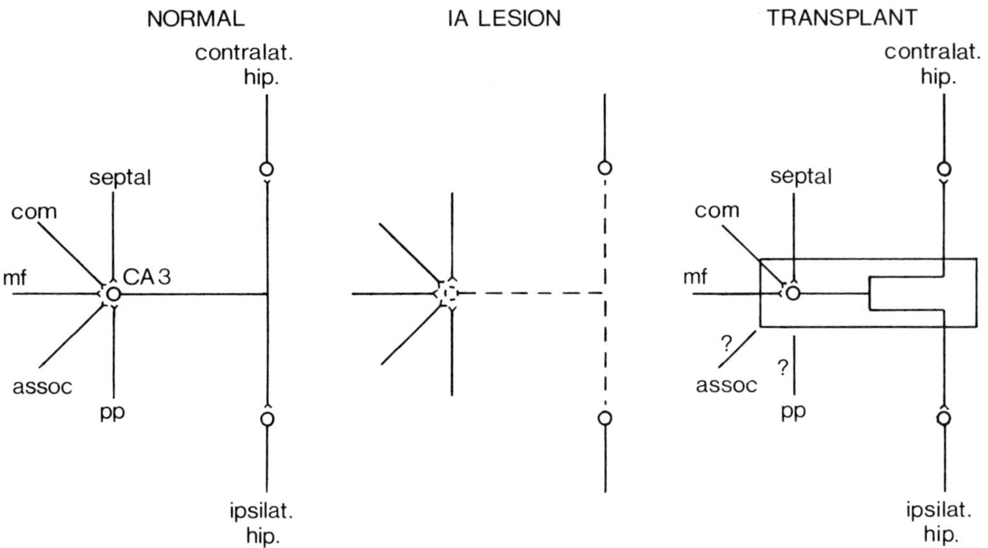

Fig. 5. Schematic illustration of the connective integration of CA3
transplants in IA-induced, axon-sparing lesions of the adult
rat CA3 region. assoc: CA3 associational fibers; com: commissural fibers; mf: mossy fibers; pp: entorhinal perforant path
fibers; septal: cholinergic septohippocampal fibers.

to the contralateral hippocampus, although the number of projecting neurons
clearly were limited after grafting to the intact adult brain. Using retrograde axonal transport of Fluoro-Gold, CA3 graft neurons placed in IA-lesions
were found to project to normal CA3 target areas in the host brain, like the
intact parts of the ipsilateral hippocampus (Schaffer collaterals) as well
as the contralateral hippocampus (commissural fibers). The ipsilateral projections correspond in part to those demonstrated by Zhou et al. (1985)
by immunohistochemical staining of mouse hippocampal transplants grafted
to normal adult mice. The trajectory in the host brain of the projecting
graft fibers is not known. In order to reach the contralateral hippocampus
the fibers must, however, have traveled for some distance through white
matter in the host fimbria and hippocampal commissures. This is noteworthy,
considering that other studies have shown that white matter impedes fiber
outgrowth (Caroni et al., 1988).

CA1 transplants

The CA1 transplants exchanged several types of connections with the
host brain (Fig. 6). The AChE staining of the CA1 transplants showed the
same pattern as in the CA3 transplants, indicating that also CA1 transplants
were homotypically innervated by host cholinergic, septohippocampal fibers.

Ingrowth of Schaffer collaterals from the ipsilateral and commissural
fibers from the contralateral hippocampus were demonstrated by silver
staining of anterograde axonal degeneration after acute lesion of the ipsilateral CA3 and the contralateral fimbria, respectively. At the electronmicroscopical level, electron dense, degenerating nerve terminals were found
even deep within the transplants in synaptic contact with dendritic spines,
following transection of host hippocampal commissural fibers in the contralateral fimbria.

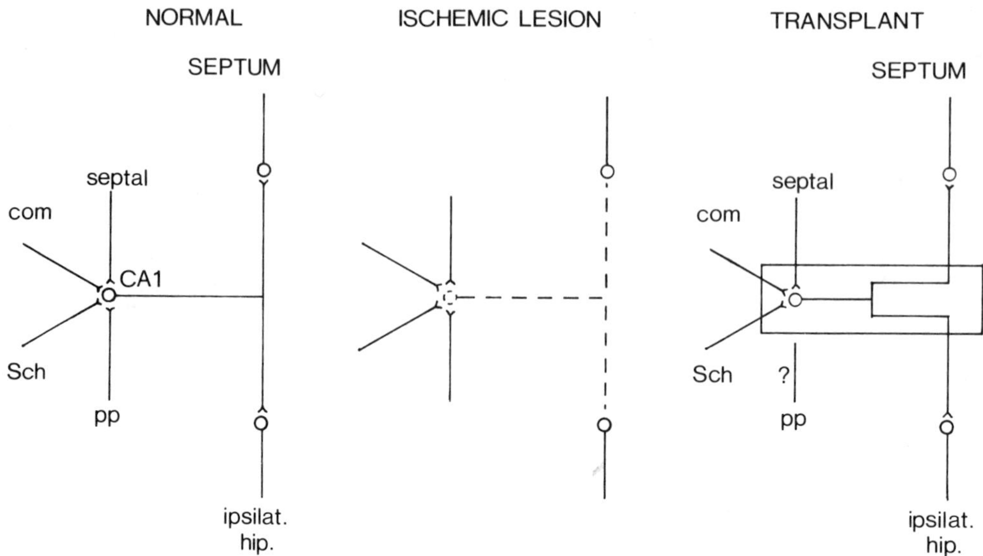

Fig. 6. Schematic illustration of the connective integration of CA1 transplants in ischemia-induced axon-sparing lesions of the adult rat
CA1 region. com: commissural fibers; pp: entorhinal perforant
path fibers; Sch: CA3 Schaffer collaterals; septal: cholinergic
septohippocampal fibers.

Using retrograde axonal transport of Fluoro-Gold efferent projections
arising from graft CA1 pyramidal cells were demonstrated to normal CA1 projection areas, like the undamaged, ipsilateral host CA1/subiculum at levels
temporal to the grafts, and the septum (Tønder et al., 1989a).

CONCLUDING REMARKS

The results reviewed here extend and support the observations made in
other studies showing that excitotoxic and ischemic brain lesions provide
suitable or even favourable environments for the survival and growth of
developing neurons (Pritzel et al., 1986; Wictorin et al., 1988; Pechanski
and Isacson, 1988; Nothias et al., 1988; Mudrick et al., 1988, 1989). The
lesion-induced reactive gliosis might thus be beneficial for the grafted
neurons through the supply of glial derived growth factors (Lindsay et al.,
1982; Nieto-Sampedro et al., 1984; Cotman and Nieto-Sampedro, 1985;
Whittemore et al., 1985; Heacock et al., 1986; Kesslak et al., 1986). The
axon-sparing properties of the ischemic and ibotenic acid lesions, moreover,
significantly enhance the growth into the grafts of adult host fibers of
the point-to-point type. This is likely to be due to a priming effect of
the preceding lesion on the host fibers, allowing them to compete more successfully with the immature and developing intrinsic graft fibers for available synaptic sites in the grafts. The formation of long, efferent projections from CA3 and CA1 grafts to normal target areas in the host brain
demonstrates the axonal growth capacity of grafted projection neurons. In
addition, it allows us to conclude that white matter in the mature brain
does not form an absolute barrier for growing nerve fibers, although in
vitro studies have shown that white matter impedes neurite outgrowth
(Caroni et al., 1988). At present we do not, however, know whether the
graft fibers are growing along lesioned host fibers or through intact white
matter.

At present little is known about the possible functional effects of
the dentate and hippocampal transplants placed in IA lesions, including
possible improvement of behavior deficits in the lesioned animals. Mudrick
et al. (1988, 1989) have, however, shown that efferent projections from CA1
neurons grafted to ischemic CA1 lesions in the host brain are electrophysiologically functional. Studies examining the behavioral effects of these
as well as the CA3 and dentate transplants are in progress.

Acknowledgements

These studies were supported by the Aarhus University Research Foundation, the Danish MRC, the Lundbeck Foundation and the NOVO Foundation.

Reference

Benveniste, H., Drejer, J., Schousboe, A., and Diemer, N.H., 1984, Elevation
of the extracellular concentrations of glutamate and aspartate in rat
hippocampus during transient cerebral ischemia monitored by intracerebral microdialysis, J. Neurochem., 43:1369.
Björklund, A., and Stenevi, U., 1984, Intracerebral neural implants:
Neuronal replacement and reconstruction of damaged circuitries, Ann.
Rev. Neurosci., 7:279.
Caroni, P., Savio, T., and Schwab, M.E., 1988, Central nervous system re-
generation: oligodendrocytes and myelin as non-permissive substrates
for neurite growth., Prog. Brain Res., 78:363.
Cotman, C.W., and Nieto-Sampedro, M., 1985, Progress in facilitating the
recovery of function after central nervous system trauma, Ann. NY Acad.
Sci., 457:83.
Coyle, J.T., and Schwarcz, R., 1983, The use of excitatory amino acids as
selective neurotoxins, in: "Handbook of Chemical Neuroanatomy. Vol. 1:
Methods in Chemical Neuroanatomy", A. Björklund and T. Hökfelt, eds.,
Elsevier Science Publishers B.V.
Heacock, A.M., Schonfeld, A.R., and Katzman, R., 1986, Hippocampal neuro-
trophic factor: characterization and response to denervation, Brain
Res., 363:299.
Johansen, F.F., Jørgensen, M.B., Ekström von Lubitz, D.K.J., and Diemer,
N.H., 1984, Selective dendrite damage in hippocampal CA1 stratum
radiatum with unchanged axon ultrastructure and glutamate uptake after
transient cerebral ischaemia in the rat, Brain REs., 291:373.
Johansen, F.F., Zimmer, J., and Diemer, N.H., 1987, Early loss of somato-
statin neurons in dentate hilus after cerebral ischemia in the rat
precedes CA-1 pyramidal cell loss, Acta Neuropathol. (Berl.), 73:110.
Jørgensen, M.B., and Diemer, N.H., 1982, Selective neuron loss after cerebral ischemia in the rat: Possible role of transmitter glutamate, Acta
Neurol. Scand., 66:536.
Kesslak, J.P., Nieto-Sampedro, M., Globus, J., and Cotman, C.W., 1986,
Transplants of purified astrocytes promote behavioral recovery after
frontal cortex ablation, Exp. Neurol., 92:377.
Köhler, C., 1983, Neuronal degeneration after intracerebral injections of
excitotoxins. A histological analysis of kainic acid, ibotenic acid
and quinolinic acid lesions in the rat brain, in: "Wenner-Gren International Symposium Series. Vol. 39: Excitotoxins", K. Fuxe, P. Roberts
and R. Schwarcz, eds., Macmillan Press, London.
Köhler, C., Schwarcz, R., and Fuxe, K., 1979, Hippocampal lesions indicate
differences between the excitotoxic properties of acidic amino acids,
Brain Res., 175.366.
Lindsay, R.M., Barber, P.C., Sherwood, M.R.C., Zimmer, J., and Raisman, G.,
1982, Astrocyte cultures from adult rat brain. Derivation, characterization and neurotrophic properties of pure astroglial cells from corpus
callosum, Brain REs., 243:329.

Mudrick, L.A., Baimbridge, K.G., and Miller, J.J., 1987, Subpopulations of fetal hippocampal neurons can be localized immunohistochemically after transplantation into the hippocampal CA1 region irreversibly damaged by cerebral ischemia, Can. J. Physiol. Pharmacol., 65(5):Axxiv.

Mudrick, L.A., Leung, P.P.-H., Baimbridge, K.G., and Miller, J.J., 1988, Neuronal transplants used in the repair of acute ischemic injury in the central nervous system, Prog. Brain Res., 78:87.

Mudrick, L.A., Baimbridge, K.G., and Peet, M.J., 1989, Hippocampal neurons transplanted into ischemically lesioned hippocampus: electroresponsiveness and reestablishment of circuitries, Exp. Brain Res., 76:333.

Nadler, J.V., Evenson, D.A., and Cuthbertson, G.J., 1981, Comparative toxicity of kainic acid and other acidic amino acids toward rat hippocampal neurons, Neuroscience, 6:2505.

Nieto-Sampedro, M., Whittemore, S.R., Needels, D.L., Larson, J., and Cotman, C.W., 1984, The survival of brain transplants is enhanced by extracts from injured brain, Proc. Natl. Acad. Sci. USA, 81:6250.

Nothias, F., Onteniente, B., Geffard, M., and Peschanski, M., 1988, Rapid growth of host afferents into fetal thalamic transplants, Brain Res., 463:341.

Peschanski, M., and Isacson, O., 1988, Fetal homotypic transplant in the excitotoxically neuron-depleted thalamus: Light microscopy, J. Comp. Neurol., 274:449.

Pohle, W., 1987, Transplantation of dentate granular cells after lesion of the dentate area by colchicine, Biomed. Biochem. Acta, 46:795.

Pritzel, M., Isacson, O., Brundin, P., Wiklund, L., and Björklund, A., 1986, Afferent and efferent connections of striatal grafts implanted into the ibotenic acid lesioned neostriatum in adult rats, Exp. Brain Res., 65:112.

Pulsinelli, W.A., and Brierley, J.B., 1979, A new model of bilateral hemispheric ischemia in unanesthetized rat, Stroke, 10:267.

Schwarcz, R., Köhler, C., Fuxe, K., Hökfelt, T., and Goldstein, M., 1979, On the mechanism of selective neuronal degeneration in the rat brain: Studies with ibotenic acid, Adv. Neurol., 23:655.

Sunde, N.A., and Zimmer, J., 1983, Cellular, histochemical and connective organization of the hippocampus and fascia dentata transplanted to different regions of immature and adult rat brains, Dev. Brain Res., 8:165.

Sunde, N., Laurberg, S., and Zimmer, J., 1984, Brain grafts can restore irradiation-damaged neuronal connections in newborn rats, Nature, 310:51.

Sunde, N., Zimmer, J., and Laurberg, S., 1985, Repair of neonatal irradiation-induced damage to the rat fascia dentata. Effects of delayed intracerebral transplantation, in: "Neural Grafting in the Mammalian CNS", A. Björklund and U. Stenevi, eds., Elsevier, Amsterdam.

Sørensen, T., Jensen, S., Møller, A., and Zimmer, J., 1986, Intracephalic transplants of freeze-stored rat hippocampal tissue, J. Comp. Neurol., 252:468.

Sørensen, T., Finsen, B., and Zimmer, J., 1987, Nerve connections between mouse and rat hippocampal brain tissue: Ultrastructural observations after intracerebral xenografting, Brain Res., 413:392.

Sørensen, T., and Zimmer, J., 1988a, Ultrastructural organization of normal and transplanted rat fascia dentata: I. A qualitative analysis of intracerebral and intraocular grafts, J. Comp. Neurol., 267:15.

Sørensen, T., and Zimmer, J., 1988b, Ultrastructural organization of normal and transplanted rat fascia dentata: II. A quantitative analysis of the synaptic organization of intracerebral and intraocular grafts, J. Comp. Neurol., 267:43.

Tønder, N., Gaarskjaer, F.B., Sunde, N.A., and Zimmer, J., 1986, Neonatal hippocampal neurons, retrogradely labeled with Granular Blue, survive intracerebral grafting and explantation to tissue culture, Exp. Brain Res., 65:213.

Tønder, N., Sørensen, J.C., Bakkum, E., Danielsen, E., and Zimmer, J., 1988, Hippocampal neurons grafted to newborn rats establish efferent commissural connections, Exp. Brain Res., 72:577.

Tønder, N., Sørensen, T., Zimmer, J., Jørgensen, M.B., Johansen, F.F., and Diemer, N.H., 1989a, Neural grafting to ischemic lesions of the adult rat hippocampus, Exp. Brain Res., 74:512.

Tønder, N., Sørensen, T., and Zimmer, J., 1989b, Enhanced host perforant path innervation of neonatal dentate tissue after grafting to axon-sparing, ibotenic acid lesions in adult rats, Exp. Brain Res., 75:483.

Tønder, N., Sørensen, T., and Zimmer, J., 1989c, Grafting of fetal CA3 neurons to excitotoxic, axon-sparing lesions of the hippocampal CA3 area in adult rats. Prog. Brain Res., in press.

Whittemore, S.R., Nieto-Sampedro, M., Needels, D.L., and Cotman, C.W., 1985, Neuronotrophic factors for mammalian brain neurons: Injury induction in neonatal, adult and aged rat brain, Dev. Brain Res., 20:169.

Wictorin, K., Isacson, O., Fischer, W., Nothias, F., Peschanski, M., and Björklund, A., 1988, Connectivity of striatal grafts implanted into the ibotenic acid-lesioned striatum - I. Subcortical afferents, Neuroscience, 27:547.

Zhou, C.-F., Raisman, G., and Morris, R.J., 1985, Specific patterns of fibre outgrowth from transplants to host mice hippocampi, shown immunohistochemically by the use of allelic forms of Thy-1, Neuroscience, 16:819.

Zimmer, J., and Sunde, N., 1984, Neuropeptides and astroglia in intracerebral hippocampal transplants: An immunohistochemical study in the rat, J. Comp. Neurol., 227:331.

Zimmer, J., Sunde, N., Sørensen, T., Jensen, S., Møller, A.G., and Gähwiler, B.H., 1985, The hippocampus and fascia dentata. An anatomical study of intracerebral transplants and intraocular and in vitro cultures, in: "Neural Grafting in the Mammalian CNS", A. Björklund and U. Stenevi, eds., Elsevier, Amsterdam.

Zimmer, J., Laurberg, S., and Sunde, N., 1986, Non-cholinergic afferents determine the distribution of the cholinergic septohippocampal projection: A study of the AChE staining pattern in the rat fascia dentata and hippocampus after lesions, X-irradiation, and intracerebral grafting, Exp. Brain Res., 64:158.

Zimmer, J., Finsen, B., Sørensen, T., and Sunde, N., 1987, Hippocampal transplants: Synaptic organization, their use in repair of neuronal circuits and mouse to rat xenografting, in: "NATO ASI Series H, Vol. 2: Glial-Neuronal Communication in Development and Regeneration", H.H. Althaus and W. Seifert, eds., Springer-Verlag, Heidelberg.

Zimmer, J., Finsen, B., Sørensen, T., and Poulsen, P.H., 1988a, Xenografts of mouse hippocampal tissue. Exchange of laminar and neuropeptide specific nerve connections with the host rat brain, Brain Res. Bull., 20:369.

Zimmer, J., Finsen, B., Sørensen, T., and Poulsen, P.H., 1988b, Xenografts of mouse hippocampal tissue. Formation of nerve connections between the graft fascia dentata and the host rat brain, Prog. Brain Res., 78:271.

Zimmer, J., Finsen, B., Sørensen, T., Sunde, N.A., and Poulsen, P.H., 1988c, Brain grafts: A survey with examples of repair and xenografting of hippocampal tissue, in: "Recovery of Function in the Nervous System, Fidia Research Series, 13", F. Cohadon and J. Lobo Antunes, eds., Liviana Press, Padova.

Zimmer, J., Tønder, N., and Sørensen, T., 1989, Hippocampus and fascia dentata transplants: Anatomical organization and connections, in: "The Hippocampus, New Vistas", V. Chan-Palay and C. Köhler, eds., Alan R. Liss, Inc., New York.

CHANGES IN SYNAPTIC TRANSMISSION IN THE KINDLED HIPPOCAMPUS

U. Heinemann, H. Clusmann, J. Dreier, and J. Stabel

Inst. Neurophysiology, University Cologne
Robert Kochstr. 39, D 5000 Köln 41, FRG

INTRODUCTION

The entorhinal cortex- hippocampus complex has particular features not only with respect to synaptic plasticity but also with respect to the readiness to develop progressive epilepsy. It might well be that this particular feature depends on the impressive plastic properties of the involved structures. A model for the progression of epilepsy is the kindling epilepsy. It is caused by stimuli (1 s, 50 Hz) of a type which also elicits long term potentiation [1]. Such stimulation induces generalized seizures when the stimuli are repeated in daily intervals [4]. The number of stimulations required to induce the kindling epilepsy is dependent on structure and varies between roughly 20 and more than 100.

Interestingly, kindling can also be induced by a number of drugs including anxiety inducing agents such as pentetrazol and the ß carbolines as well as agents such as cocain [17]. For this reason the kindling phenomenon is also of interest to psychiatrists. Once established the alterations remain more or less stable. It is possible after intervals of some months to evoke the same seizure by a single stimulus which originally was at best inducing local afterdischarges [17]. Since epilepsy is defined as a situation where many cells are more or less synchroneously active, the kindling epilepsy offers itself for studies on mechanisms underlying long term changes in synaptic connectivity and neuronal hyperexcitabillity.

A particular problem involved in the progression of epileptiform activity in the temporal lobe is illustrated in Fig 1. It shows epileptiform activity in a slice preparation which contains large parts of the temporal lobe: i.e. the rhinal, perirhinal and entorhinal cortex, the subicular regions and the hippocampal formation. The activity was induced by lowering $[Mg^{2+}]_o$. The figure is based on three different experiments and shows that epileptiform activity is rather different in the areas of interest. While entorhinal cortex and dentate gyrus lack interictal activity [2,13], area CA1 and CA3 discharge with a frequency of about 20/min [21]. These epileptiform discharges last for 40 to 80 ms and hence have the appearance of interictal events.

The rhinal cortex is also capable to produce short epileptiform discharges, but these precede seizure like events and hence represent true interictal discharges. The seizure like events can be of considerable duration (20 s and more). They are characterized by tonic and clonic electrographic activity in field potential recordings and by large ionic changes (Dreier et al, in prep). The entorhinal cortex and the subiculum develop exclusively seizure like events which last for often more than 20 s. In spite of the intense epileptiform activity in the entorhinal cortex, the dentate gyrus - is not recruited into active participation in the seizures. This would be indicated by large negative slow field potentials in the stratum granulare and by significant alterations of the ionic microenvironment [8,10,11]. Also pyramidal cells in area CA1 and CA3 do not actively participate in the seizures although they receive considerable input from the alveus which is likely relayed from the subiculum. Active participation of cells in the stratum oriens is indicated by the large negative field potentials and ionic changes in this layer. The weak participation of PC cells may find an explanation in the inhibitory role which most cells have in this layer. The fact that seizure activity from EC is notz relayed through the dentate gyrus suggests that the dentate gyrus imposes a powerful filter which normally prevents spread of epileptiform activity into the hippocampus. Studies in vitro suggest that active participation of the hippocampus in temporal seizures can be induced. Hence the filter is adjustable.

The filter function of the hippocampus appears to critically depend on slow IPSPs [3]. It is well known [7,18] that paired pulse and repetitive stimulation of the perforant path which con-

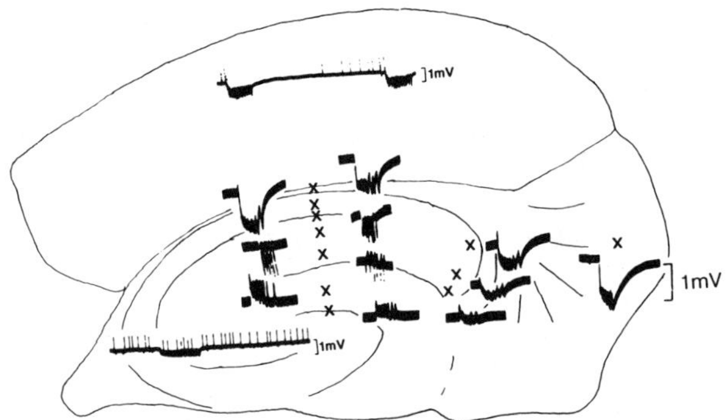

Fig. 1.   Effects of lowering extracellular [Mg]o on spontaneous activity in the temporal lobe. Combined figure from different experiments. Note difference in epileptiform activity in dentate gyrus, area CA1 and CA3 and in the entorhinal cortex.

nects the EC with the dentate gyrus leads to a decrease in synaptic responses (frequency habituation). When slow IPSPs are reduced in efficacy by application of the GBAB-B-receptor agonist baclofen it is noted that frequency habituation is removed. Consequently inhibition of the hilar inhibitory neurons which mediate these slow IPSPs also removes the inhibitory filter. This can be shown by application of baclofen to the hilus, which leads to blockade of synaptic inhibition in the dentate gyrus and consequently to epileptiform discharges upon stimulation [19].

Another way to overcome the filter function of the dentate gyrus is application of norepinephrine. NE blocks afterhyperpolarisations and frequency accomodation, i.e. the decrease in discharge frequency during prolonged cellular depolarisation induced by intracellular current injection. Indeed, NE can also reverse frequency habituation into frequency potentiation [30].

A third possibillity to overcome frequency habituation is the reduction of slow IPSPs by ionic changes in the cellular microenvironment. Such alterations are typically evoked by seizures and

446

repetitive stimulation. Relevant for the removal of the slow IPSP are decreases in [Ca]o and increases in [K]o. The slow IPSP is dependent on activation of potassium conductances [5,26, Rausche et al, submitted] and hence very sensitive to extracellular potassium accumulation. Indeed, elevation of [K]o by one or two mM can reverse frequency habituation into frequency potentiation. During active seizure generation [K]o can increase by up to 9 mM which would certainly be sufficient. In our experiments we never noticed any major ionic changes in dentate gyrus unless we directly stimulated the perforant path or the mossy fibers. Likewise, the slow IPSPs can be blocked by lowering extracelluar calcium [12,27]. This will lead to reduced transmitter release from presynaptic endings. Since disynaptic pathways are more sensitive to such treatment than monosynaptic pathways, lowering of [Ca]o indeed reduces and finally blocks fast and slow IPSPs.

Neuronal changes during kindling plasticity

Long term alteration in synaptic and cellular properties are induced during the kindling epilepsy which probably affects the filter function of the DG. We compared so far four types of kindling: namely, pentylenetetrazole kindling, kindling of area CA1, kindling of the commissural fibers and kindling of the amygdala. While in all conditions with electrical kindling a number of alterations are found in the hippocampal formation we detected sofar no changes during pentylenetetrazole kindling. Data from the laboratory of M. Gutnick suggest however alterations with this kindling model in the neocortex, some of which might resemble the alterations observed with the other kindling models in the hippocampus. All three kindling models affecting the hippocampus had in common that stimulus induced changes in $[Ca^{2+}]_o$ were augmented particularly in the synaptic input layers of the hippocampus [11,33] (Fig. 2). This effect was more prominent in area CA1 when this structure was kindled while the effects were more drastic in dentate gyrus during kindling of the amygdala and commissural fibers. In case of the amygdala, however, changes in dentate gyrus became apparent only when more then ten type five seizures were induced.

Studies on aspartate and DL-homocysteic acid induced [Ca]o changes in the CA1 model had suggested that changes in the laminar distribution of excitatory amino acid induced Ca^{2+} signals were associated with this kindling model . Laminar profiles of NMDA and Quis induced [Ca]o changes showed that cellular Ca uptake mediated through NMDA receptor activated ionophores is strongly augented while the laminar profiles of Quis induced ionic changes were unaltered on a long term scale [9]. This suggested some enhancement in the utilization of NMDA receptors. The possibility that GABAergic inhibition which may be diminished in kindled area CA1 [14,15] accounts for the alteration of NMDA induced laminar profiles was tested [6]. It was found that bicuculline augmented [Ca]o decreases induced by NMDA but also those induced by quisqualate. Also the laminar profiles of Quis or NMDA induced decreases in $[Ca^{2+}]_o$ were similarly altered [6]. Since in kindled slices the effect remained restricted to the NMDA induced ionic signals we conclude that disinhibition is not the sole mechanism underlying the kindling epilepsy.

Indeed, kindling leads in the dentate gyrus both to an augmentation of the early as well as the late IPSP [25]. Nevertheless also in this structure stimulus induced changes in $[Ca^{2+}]_o$ are augmented [23]. Intracellular recordings revealed that lateral perforant path induced EPSPs which show little participation of NMDA receptors in control recordings, now do. This statement needs some qualification. Recently a number of reports appeared indicating an NMDA component in synaptic transmission from the perforant path [16]. In order to clarify the issue we determined the laminar distribution of NMDA receptors in the dentate gyrus by measuring Quis and NMDA induced decreases in $[Na]_o$ (Stabel et al, submitted). The findings indicated that the density of NMDA receptors is largest at a distance of 50 to 100 m from the SG with a steep decline in receptor density towards outer SM.

We subsequently investigated the effects of lowering [Mg]o on synaptic responses evoked by lateral and medial perforant path stimulation as well as on comissural fiber and antidromic stimulation induced responses. While antidromic responses are little affected by lowering $[Mg^{2+}]_o$, medial perforant path and commissural fiber evoked responses are strongly augmented. These effects are very sensitive to ketamine (20 M) and 2APV (30 M). Lateral perforant path evoked responses also show some, although weaker, enhancement [20,21]. It appears always later during Mg^{2+} washout and it is less sensitive to NMDA antagonists (Clusmann, Stabel and Heinemann, in prep). In kindled animals this is markedly different. There stimulation of the lateral perforant path results in EPSPs which are strongly augmented soon after onset of Mg^{2+} washout [20,23]. This effect is readily reversed by 2APV. As expected for NMDA dependent signals the amplitude of the EPSPs increases with depolarization (Fig. 2), an effect which is reversed to the control situation by 2APV [20]. Thus we conclude that in a preparation with enhanced inhibition nevertheless there can be enhancement of NMDA receptor utilization. Since autoradiographic studies show no enhancement of binding sites [29] it appears that other mechanisms must account for the augmented NMDA receptor utilization. This could be due to an enhancement in receptor phosphorylation [22], but other mechanisms might contribute as well.

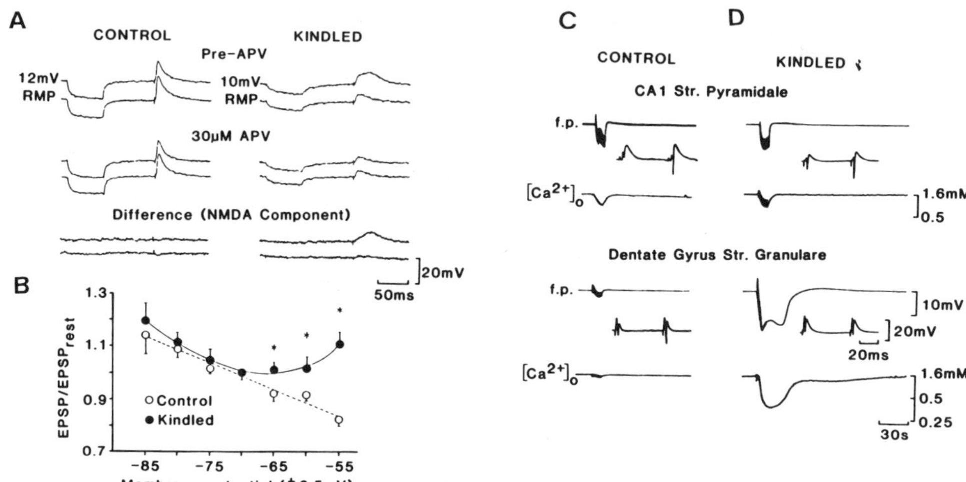

Fig. 2.   Alterations in cellular responses and in stimulus induced decreases in [Ca^{2+}]$_o$ following kindling. A. Intracellualr responses to hyperpolarizing current injection and to stimulation of lateral perforant path fibers before and after application of 2 APV in a control cell and in a cell from kindled dentate gyrus. Note decrease of amplitude of the EPSP upon depolarisation in control but increase upon depolarisation in a kindled cell. Note effect of 2 APV onto the two responses. Subtraction of recordings in kindled animals reveals a clear NMDA component not present in the recordings from normal animals. B. Voltage dependance of stimulus induced EPSPs in normal and kindled animals. Stars mark statistical significant differences. C, D. Stimulations which produce about equally sized field potentials produce similar sized decreases in extracellular calcium concentration in area CA1 after kindling of commissural fibers or the amygdala but a large enhancement of stimulus induced decreases in extracellular calcium concentration in dentate gyrus.

One such mechanism could be the enhanced input resistance of kindled cells [23]. It would augment synaptic responses and thereby increase contribution of NMDA receptors to synaptic transmission. Sprouting of perforant path fibers and consequently enhanced transmitter release with subsequent activation of extrasynaptic NMDA receptors is another possibility, presently investigated in our lab (see also [28]. We have sofar studied the possibility that the Ca^{2+} dependence of transmitter release is altered by measuring the level at which synaptic transmission is blocked. The statistical evaluation of the data in area CA1 suggests no statistical significant difference.

It should be noted that these are not the sole differences in the kindling model. The observation that LTP is much more difficult to elicit lead us to study typical norepinephrine (NE) and ß receptor mediated responses. We found that block of afterhyperpolarisations and frequency accomodation mediated by ß receptor agonists was lost in cells from kindled animals [31,32]. Also the receptor dependent reduction of changes in [Ca]o was lost, thus indicating a reduction in the number of NE receptors. Meanwhile there is also evidence for a change in the production of opioid peptides which is defined on the level of the mRNA [24]. Thus many of the cellular and synaptic properties change during kindling and it remains to be seen which are involved in the larger ease by which the EC can propagate seizure activity to the dentate gyrus.

In conclusion, the data particularly on dentate gyrus granule cells suggest that this normally habituating synaptic connection can change its coupling behaviour profoundly. We suggest that such alterations are important in the chronic progression of temporal lobe epilepsies which are often difficult to treat. It is well possible, although far from proven, that similar alterations in synaptic connections underly also processes of memory formation in this area.

REFERENCES

1-Bliss, T.V.P. and Lomo, T., Long-lasting potentiation of synaptic transmission in the dentate area of the anaesthetized rabbit following stimulation of the perforant path, **J. Physiol.** 232 (1973) 331-356.

2-Collingridge, G.L. and Bliss, T.V.P., NMDA receptors - their role in long-term potentiation, **Trends in Neuroscience** 10 (1987) 288-293.

3-Cotman, C.W. and Monaghan, D.T., Excitatory amino acid neurotransmission: NMDA receptors and Hebb-type synaptic plasticity, **Ann. Rev. Neurosci.** 11 (1988) 61-80.

4-Goddard, G.V., McIntyre, D.C. and Leech, C.K., A permanent change in brain function resulting from daily electical stimulation, **Exp. Neurol.** 25 (1969) 295-330.

5-Hablitz, J.J. and Thalmann, R.H., Conductance changes underlying a late synaptic hyperpolarization in hippocampal CA3 neurons, **J. Neurophysiol.** 58 (1987) 160-179.

6-Hamon, B. and Heinemann, U., Effects of GABA and bicuculline on N-methyl-D-aspartate- and quisqualate-induced reductions in extracellular free calcium in area CA1 of the hippocampal slice, **Exp. Brain Res.** 64 (1986) 27-36.

7-Harris, E.W. and Cotman, C.W., Effects of synaptic antagonists on perforant path paired-pulse plasticity: Differentiation of pre- and post-synaptic antagonism, **Brain Res.** 334 (1985) 348-353.

8-Heinemann, U., Basic mechanisms of the epilepsies. **In** A.M. Halliday, S.R. Butler and R. Paul (Eds.), A textbook of clinical neurophysiology, John Wiley & Sons, Chichester, New York, Brisbane, Toronto, Singapore, 1987, pp. 497-534.

9-Heinemann, U., Hamon, B., Konnerth, A. and Wadman, W.J., Stimulus-dependent synaptic plasticity in area CA1 of the in vitro hippocampal slice of rats, **Exp. Brain Res.** 14 (1986) 291-299.

10-Heinemann, U. and Jones, R.S.G., Neurophysiology of epilepsy. **In** L. Gram and M. Dam (Eds.), Perspectives of epilepsy, Raven Press, New York, 1989, pp. in press.

11-Heinemann, U., Konnerth, A., Pumain, R. and Wadman, W.J., Extracellular calcium and potassium concentration changes in chronic epileptic brain tissue. **In** A.V. Delgado--Escueta, A.A. Ward, D.M. Woodbury and R.J. Porter (Eds.), Advances in neurology / Basic mechanisms of the epilepsies, Raven Press, New York, 1986, pp. 641-661.

12-Jones, R.S.G. and Heinemann, U., Abolition of the orthodromically evoked IPSP of CA1 pyramidal cells before the EPSP during washout of calcium from hippocampal slices, **Exp. Brain Res.** 65 (1987) 676-680.

13-Jones, R.S.G. and Heinemann, U., Synaptic and intrinsic responses of medial entorhinal cortical cells in normal and magnesium-free medium in vitro, **J. Neurophysiol.** 59 (1988) 1476-1497.

14-Kamphuis, W., Huisman, E., Wadman, W.J. and Lopes da Silva, F.H., Decrease in GABA immunoreactivity and alteration of GABA metabolism after kindling in the rat hippocampus, **Exp. Brain Res.** 74 (1989) 375-386.

15-Kamphuis, W., Lopes da Silva, F.H. and Wadman, W.J., Changes in local evoked potentials in the rat hippocampus (CA1) during kindling epileptogenesis, **Brain Res.** 440 (1988) 205-215.

16-Lambert, J.D.C., Jones, R.S.G., Andreasen, M., Jensen, M.S. and Heinemann, U., The role of excitatory amino acids in synaptic transmission in the hippocampus, **Comp. Biochem. Physiol. [A]** 93A (1989) 195-201.

17-McNamara, J.O., Bonhaus, D.W., Shin, C., Crain, B.J., Gellman, R.L. and Giacchino, J.L., The kindling model of epilepsy: a critical review, **Crit Rev. Clin. Neurobiol.** 1 (1985) 341-391.

18-McNaughton, B.L. and Barnes, C.A., Physiological identification and analysis of dentate granule cell responses to stimulation of the medial and lateral perforant pathways in the rat, **J. comp. Neurol.** 175 (1977) 439-454.

19-Misgeld, U., Klee, M.R. and Zeise, M.L., Blockade of hippocampal GABA-ergic inhibition by baclofen. In E.-J. Speckmann, H. Schulze and J. Walden (Eds.), Epilepsy and Calcium, Urban & Schwarzenberg, Muenchen,Wien,Baltimore, 1986, pp. 17-33.

20-Mody, I. and Heinemann, U., NMDA receptors of dentate gyrus granule cells participate in synaptic transmission following kindling, **Nature (London)** 326 (1987) 701-704.

21-Mody, I., Lambert, J.D.C. and Heinemann, U., Low extracellular magnesium induces epileptiform activity and spreading depression in rat hippocampal slices, **J. Neurophysiol.** 57 (1987) 869-888.

22-Mody, I., Salter, M.W. and MacDonald, J.F., Requirement of NMDA receptor/channels for intracellular high energy phosphates and the extent of intraneuronal calcium buffering in cultured mouse hippocampal neurons, **Neurosci. Lett.** 93 (1988) 73-78.

23-Mody, I., Stanton, P.K. and Heinemann, U., Activation of N-methyl-D-aspartate receptors parallels changes in cellular and synaptic properties of dentate gyrus granule cells after kindling, **J. Neurophysiol.** 59 (1988) 1033-1054.

24-Morris, B.J., Feasey, K.J., ten Bruggencate, G., Herz, A. and Höllt, V., Electrical stimulation in vivo increases the expression of prodynorphin mRNA in rat hippocampal granule cells, **Proc. Nat. Acad. Sci.** 85 (1988) 3226-3230.

25-Oliver, M.W. and Miller, J.J., Alterations of inhibitory processes in the dentate gyrus following kindling-induced epilepsy, **Exp. Brain Res.** 57 (1985) 443-447.

26-Rausche, G., Igelmund, P. and Heinemann, U., Effects of changes in extracellular potassium, magnesium and calcium concentration on synaptic transmission in area CA1 and the dentate gyrus of rat hippocampal slices, **Pflügers Arch.** in press.

27-Rausche, G., Sarvey, J.M. and Heinemann, U., Lowering extracellular calcium reverses paired pulse habituation into facilitation in dentate granule cells and removes a late IPSP, **Neurosci. Lett.** 88 (1988) 275-280.

28-Represa, A., Le Gall La Salle, G. and Ben-Ari, Y., Hippocampal plasticity in the kindling model of epilepsy in rats, **Neurosci. Lett.** 99 (1989) 345-350.

29-Stabel, J., Arens, J., Lambert, J.D.C. and Heinemann, U., Effects of lowering $[Na^+]_o$ and $[K^+]_o$ and of ouabain on quisqualate-induced ionic changes in area CA1 of rat hippocampal slices, **Neurosci. Lett.** (1989) in press.

30-Stanton, P.K., Mody, I. and Heinemann, U., Mechanisms of action of norepinephrine in dentate gyrus granule cells: implications for long-term neuronal plasticity, **Exp. Brain Res.** submitted.

31-Stanton, P.K., Mody, I. and Heinemann, U., A role for N-methyl-D-aspartate receptors in norepinephrine-induced long-lasting potentiation in the dentate gyrus, **Exp. Brain Res.** 77 (1989) 517-530.

32-Stanton, P.K., Mody, I. and Heinemann, U., Downregulation of noradrenaline receptors during kindling, **Brain Res.** 476 (1989) 367-372.

33-Wadman, W.J. and Heinemann, U., Laminar profiles of $(K^+)_o$ and $(Ca^{2+})_o$ in region CA1 of the hippocampus of kindled rats. In M.et al. Kessler (Ed.), Ion Measurements in Physiology and Medicine, Springer-Verlag, Berlin, Heidelberg, 1985, pp. 221-228.

NMDA RECEPTOR PLASTICITY IN THE KINDLING MODEL

J. O. McNamara[1,2,3], G. Yeh[2], D. W. Bonhaus[1,3], M. Okazaki[2], and J. V. Nadler[2]

Departments of [1]Medicine (Neurology) and [2]Pharmacology
Duke University Medical Center and
[3]Veterans Administration Medical Center
Durham, North Carolina

INTRODUCTION

The N-methyl-D-aspartate (NMDA) subtype of excitatory amino acid receptor serves a critical role in the development and stabilization of synapses in the developing nervous system (Cline et al., 1987) and in plasticity of the adult nervous system, particularly with respect to formation of some forms of learning and memory (Morris et al., 1986; Mondadori et al., 1989) Its role in these processes almost certainly derives from two unique features of this ionotropic neurotransmitter receptor: 1) its regulation by magnesium which results in its sensitivity to membrane voltage, thereby endowing it with associative properties (MacDonald et al., 1982; Flatman et al., 1983; Nowak et al., 1984; Mayer et al., 1984); and 2) its permeability to calcium (MacDermott et al., 1986), a second messenger capable of controlling a host of calcium sensitive enzymes.

The primary goal of this laboratory is to elucidate the mechanisms of an animal model of neuronal plasticity and epilepsy, kindling. Not surprisingly, activation of NMDA receptors appears to be critically involved in formation of kindling. Unexpectedly, formation of kindling somehow transforms NMDA-receptive neurons so that they exhibit a long-lasting enhanced sensitivity to NMDA. This chapter will briefly review pharmacologic and electrophysiologic studies of NMDA receptors in this model and describe evidence documenting the enhanced sensitivity to NMDA. The main focus will be on studies aimed at elucidating a molecular explanation of the enhanced sensitivity to NMDA. Upregulation of multiple components of the NMDA receptor channel complex appears to explain part of the enhanced responsiveness to NMDA.

Pharmacologic and electrophysiologic studies of NMDA receptors in the kindling model

Kindling is a phenomenon in which repeated application of initially subconvulsive electrical stimulations eventually results in long lasting and widely propagated electrical seizures accompanied by behavioral manifestations of limbic and clonic motor seizures (Goddard et al., 1969). Once established, this enhanced sensitivity to electrical stimulation appears to persist for the life of the animal.

Enhancement of excitatory synaptic transmission could contribute to this enduring hyperexcitability. Excitatory amino acids appear to be the principal excitatory neurotransmitters in mammalian brain. Among subtypes of excitatory amino acid receptors, the involvement of NMDA receptors in particular is suggested by pharmacologic studies. NMDA antagonists profoundly inhibit (McNamara et al., 1988; see McNamara et al., 1989 for review) and may actually prevent (Roundoin et al., 1989) the development of kindling, suggesting that activation of NMDA receptors is an absolute prerequisite for induction of kindling. NMDA receptor antagonists partially inhibit the propagation of kindled seizures (Peterson et al., 1983; McNamara et al., 1988), suggesting that activation of NMDA receptors contributes to the enhanced seizure propagation in kindled animals. The dose of NMDA antagonists required to inhibit seizure propagation appears to exceed that required to inhibit development of kindling (McNamara et al., 1988), thereby raising the possibility that kindling somehow enhances the function of synapses using NMDA receptors.

Electrophysiologic studies demonstrated that kindling does enhance the function of at least some synapses using NMDA receptors. Mody et al. (1988) performed intracellular recordings from dentate granule cells in hippocampal slices isolated from animals kindled by stimulation of the amygdala or hippocampal commissure and sacrificed at intervals from one day to six weeks after the last seizure. They reported the presence of an NMDA component of the EPSP induced by perforant path stimulation in granule cells of kindled but not control animals. These findings suggested that enhancement of excitatory synaptic transmission contributed to the increased seizure propagation observed in these animals. Among the variety of questions raised by these findings, we were particularly interested in understanding the molecular basis of the altered physiology.

A change intrinsic to NMDA-receptive neurons is one possible explanation for the enhanced participation of NMDA receptors in synaptic transmission. To test this idea, we implemented means of assessing NMDA receptor-mediated depolarization and compared the responses of hippocampal slices from control and kindled animals. These studies are reviewed in greater detail elsewhere in this volume (Nadler et al.). The electrophysiologic approach used the grease gap preparation to measure depolarizations of CA3 pyramidal neurons. The biochemical approach measured NMDA-mediated inhibition of muscarinic cholinergic receptor-stimulated phosphoinositide hydrolysis in transverse slices (Morrisett et al., submitted). An increased sensitivity to NMDA manifested as a steepening and/or a leftward shift of the dose-response curve was detected with both measures in animals sacrificed either 24 hours or 4-6 weeks after the last kindled seizure (Morrisett et al., 1989; Nadler et al. this volume). We concluded that an alteration of NMDA-receptive neurons themselves accounted for at least part of the electrophysiologic findings.

Table 1. [^{3}H] glutamate binding to NMDA receptors in kindling

	Side	1 day		28 days	
		Control	Kindled	Control	Kindled
Dorsal Hippocampal Formation					
Area CA1s.radiatum	I	713±57	689±50	738±43	666±52*
	C	691±43	676±48	760±40	676±56*
Area CA1s.oriens	I	686±52	681±47	702±37	649±54
	C	668±40	655±41	719±38	660±60
Area CA3s.radiatum	I	439±37	447±32	344±26	333±22
	C	424±30	444±32	357±22	351±26
Area CA3s.oriens	I	407±35	418±32	314±24	305±22
	C	382±27	408±36	330±19	315±22
Dentate molecular	I	597±47	608±46	599±39	592±42
layer	C	569±42	585±42	600±37	593±45

Rats were either kindled or implanted but not stimulated. Kindled rats were sacrificed for receptor autoradiography either 1 or 28 days after attainment of the behavioral criterion. Values are means ± S.E.M. for 8 animals and are expressed as fmol/mg protein. I, ipsilateral; C, contralateral (with respect to the stimulating electrode). *P <0.025

Biochemical and radiohistochemical analyses of the NMDA receptor channel complex in kindling

We next sought to understand the molecular basis of the long lasting enhanced sensitivity to NMDA detected in hippocampal neurons of kindled animals. To test this idea, our initial studies examined the binding of L-[3H]glutamate to NMDA receptors with a quantitative radiohistochemical method (Okazaki et al., 1989). Kindling was produced by stimulation of the angular bundle and animals were sacrificed either 24 hours or 28 days after the last seizure. No significant changes were detected in any region one day after the last seizure. Measurements in animals sacrificed 28 days after the last seizure disclosed small (7-11%) but significant declines of NMDA receptor binding in stratum radiatum of the dorsal hippocampal area CA1 (Table 1), in both deep and superficial layers of the motor cortex and in layers I-IV of the somatosensory cortex. No significant changes were detected in other brain regions including stratum granulosum or moleculare of the dentate gyrus. At the time, we concluded that the enhanced sensitivity to NMDA can not be explained by an increased expression of NMDA receptors.

The discovery of additional components of the NMDA receptor channel complex led us to investigate the possibility of kindling-related alterations in recognition sites allosterically linked to the NMDA receptor recognition site. Glycine potentiates the opening of the NMDA receptor-gated cation channel (Johnson and Ascher, 1987) and is required for the action of NMDA (Kleckner and Dingledine, 1988); these actions are mediated through a receptor which is coupled to the NMDA receptor and the NMDA channel. We therefore measured ligand binding to the glycine site with [3H] glycine. We also measured the binding of [3H]TCP, an NMDA channel blocker, to quantify the number of NMDA receptor-gated channels (Bonhaus et al., in press). Finally, we used the competitive NMDA receptor antagonist, [3H] CPP, to measure NMDA receptor recognition sites. These measurements were performed with membranes isolated from the hippocampus of control and amygdaloid kindled animals sacrificed 28-32 days after the last kindled seizure.

The B_{max} of [3H]glycine, [3H] CPP, and [3H] TCP binding increased by 42%, 47%, and 25% respectively (Fig. 1 and Table 2) (p <.02 in each case) (Yeh et al., in press). No significant changes in ligand affinity were found. Surprisingly, no alterations in either the number or affinity of [3H] glycine binding sites were detected in animals sacrificed one day after the last kindled seizure (Table 2).

Table 2. Binding of NMDA Receptor Channel Complex in Hippocampal Membranes of Kindled Rats

A. Kindled animals 28-32 days after the last kindled seizure.

	B_{max} (pmol/mg protein)		K_d (nM)	
	Control	Kindling	Control	Kindling
[3H]Gly	2.6±0.3 (N=7)	3.7±0.3* (N=7)	210±50	270±80
[3H]CPP	1.7±0.2 (N=6)	2.5±0.2* (N=6)	90±20	120±20
[3H]TCP	1.6±0.1 (N=8)	2.0±0.1* (N=8)	9±1	11±1

B. Kindled animals one day after the last kindled seizure.

[3H]Gly	2.7±0.6 (N=5)	2.8±0.2 (N=5)	260±110	240±40

These values represent the means ± the S.E.M. for each group. * Means the difference between kindled and control groups was statistically significant (P < 0.02, unpaired Student's T-test). Numbers in parentheses refer to the number of animals tested.

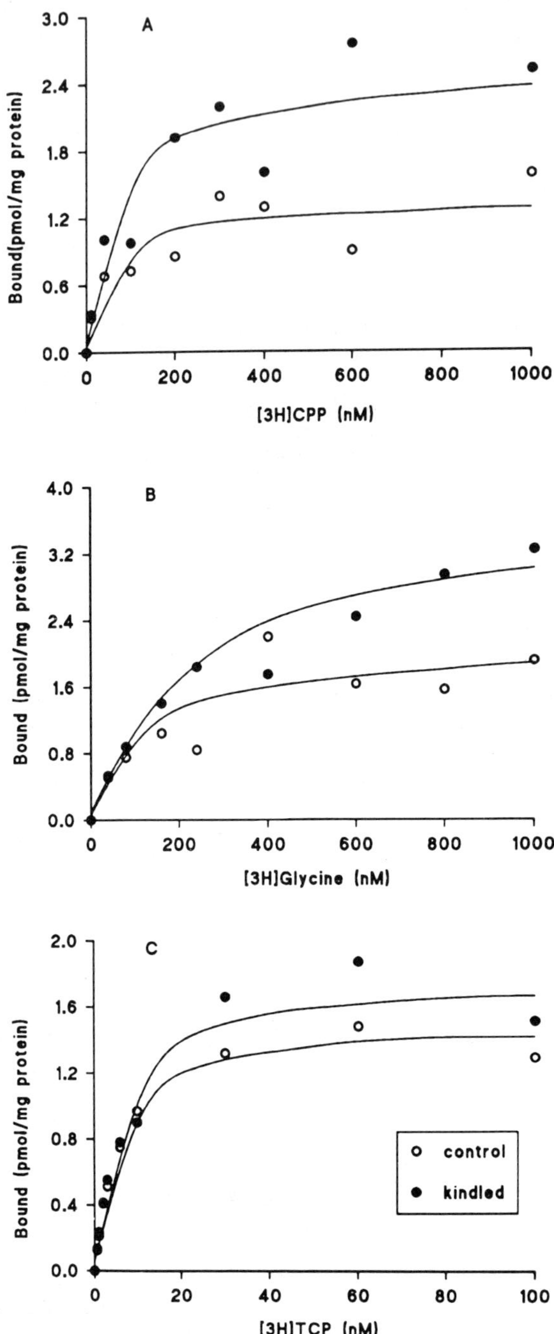

Fig. 1.   Representative [3H]CPP (A), [3H]glycine (B), and [3H]TCP (C) binding isotherms for representative control and kindled animals sacrificed one month after the last kindled seizure.

DISCUSSION

Together these results demonstrate that kindling is associated with a delayed upregulation of hippocampal NMDA receptors as evidenced by the binding of ligands to three distinct sites on this receptor/channel complex. The absence of detectable increase in glycine binding site density in animals sacrificed one day after the last kindled seizure indicates that these increases in receptor density appear to develop some time after establishment of the kindled state.

The precise nature of these alterations of the NMDA receptor is presently unclear. The increase in density of each of the three principal components of this complex suggests an increase in the number of complete, functional NMDA receptor channel complexes. Importantly, this view fails to explain our previous findings of a 7-11% **decrease** in [3H]L-glutamate binding to NMDA receptors in the hippocampal formation measured autoradiographically one month after completion of angular bundle kindling (Okazaki et al., 1989). We do not think that differences in site of kindling (amygdala versus angular bundle) explain these discrepancies, since equivalent alterations of GABA A, benzodiazepine, and muscarinic cholinergic receptor binding occur with kindling from amygdala and entorhinal cortex, the origin of afferents carried in the angular bundle (Valdes et al., 1982; Shin et al., 1985; Dasheiff and McNamara, 1982; Savage et al., 1983). We also doubt that differences in the method of measurement (membranes versus slide-mounted sections for radiohistochemistry) account for the discrepancies, since the radiohistochemical method has been at least as sensitive as membrane measurements for kindling-induced alterations of benzodiazepine and muscarinic cholinergic receptors (McNamara et al., 1980, Valdes et al., 1982; Byrne et al., 1980; Savage et al., 1983).

We favor the idea that the differences in binding may be explained by the ligands used in these experiments; specifically we propose that kindling selectively upregulates a subpopulation of **antagonist-preferring** NMDA receptors. The suggestion of distinct subtypes of NMDA receptors distinguished by differential binding of agonists and antagonists was initially made by Monaghan et al. (1988); a differential anatomic distribution of radiolabelled antagonist and agonist binding in radiohistochemical studies of rat brain formed the basis of this suggestion. This was supported by a preliminary report using high energy radiation inactivation analysis to determine the functional target size of molecules in rat brain membranes (Honore and Drejer, 1988); these studies detected a target size of 125,500 and 209,000 daltons for [3H]L-glutamate (agonist) and [3H] CPP (antagonist) binding respectively.

The ideas that subtypes of amino acid receptors could be defined based on different affinities for agonists and antagonists arose initially from the work of Olsen and Snowhill (1983) with the GABA A receptor. The existence of at least four subtypes of GABA A receptors has been suggested based on preferential binding of agonists and antagonists in radiohistochemical and biochemical studies of different regions of rat brain. The NMDA and GABA A receptors almost certainly share some similarities since both are amino acid ligand-gated ion channel receptors with multiple regulatory sites. In contrast to the NMDA receptor in which knowledge of its structure is based almost entirely on inferences from electrophysiologic studies, the structure of the GABA A receptor has been elucidated by molecular cloning and expression. This is a multiple subunit receptor in which the GABA and benzodiazepine recognition sites reside on separate subunits (Schofield et al. 1987). Interestingly, distinct genetic isoforms of the GABA A receptor have been identified which can result in dramatic differences in efficacy and potency of GABA-mediated ion conductances (Pritchett et al., 1988). The question thus arises whether the antagonist and agonist preferring subtypes of GABA A receptor represent distinct genetic isoforms or post-translational modifications of a single polypeptide. Available evidence suggests that distinct genetic isoforms are at least partly responsible for these subtypes; this is based on correlative studies of radioligand binding in different anatomic regions of rat brain combined with antibodies directed at distinct polypeptides and cDNAs selective for distinct mRNAs (Olsen et al., in press).

It seems likely that agonist and antagonist preferring subtypes of both GABA A and NMDA receptors will exhibit functional differences. Such differences may relate to the efficacy and/or potency of endogenous agonists, rate of desensitization, agonist selectivity, etc. Our data suggest that the antagonist-preferring subtype of NMDA receptor will somehow underlie increased function since its apparent presence in the kindled hippocampus correlates with enhanced sensitivity to NMDA-mediated depolarization (Morrisett et al., in press; Nadler et al., this volume) and enhanced participation of NMDA receptors in synaptic transmission (Mody et al., 1988). Correlation of the anatomic distribution of these biochemical and electrophysiologic changes at a cellular level will be required to further assess these possibilities.

Importantly, these findings are not sufficient to explain establishment of the kindled state, because animals are kindled before these changes occur. We suspect that some alteration of this receptor channel complex is present at the time of establishment of the kindled state; this could explain the enhanced NMDA receptor-mediated responses measured with both biochemical and electrophysiologic methods in hippocampus one day after completion of kindling. Such alterations might consist of a post-translational modification of the receptor channel complex or a change in genetic isoforms resulting in enhanced NMDA receptor function yet escaping detection by our methods.

The present findings stand in sharp contrast to alterations in a variety of other neurotransmitter receptors previously described in the kindling model. Alterations in the content of other receptors have occurred shortly after establishment of the kindled state and have dissipated within one to fourteen days (McNamara, 1978; Shin et al., 1985; Savage et al., 1984a; Savage et al., 1984b; Crain et al., 1987). Moreover, the direction of these alterations has been consistent with an increase in inhibition rather than an increase of excitability. We have interpreted these alterations in receptor content to reflect adaptive responses to the seizures.

One key question arising from these data is the functional significance of the long-lasting enhanced sensitivity to NMDA with respect to kindling itself. One possibility is that activation of NMDA receptors themselves underlies the enhanced propagation of seizures characteristic of these animals. If so, multiple mechanisms must be operative because NMDA antagonists only partially inhibit propagation of kindled seizures (McNamara et al., 1988). Another possibility is that the enhanced sensitivity to NMDA is indirectly responsible for whatever changes in brain structure and/or function are required for expression of a kindled seizure. This could be accomplished in a variety of ways, two of which are considered in the following paragraphs.

Persistent enhancement of NMDA receptor activation could regulate other molecules which themselves control neuronal excitability. Such changes could be effected within existing neuronal architecture. For example, Stelzer et al. (1987) found that NMDA receptor activation was required for a stimulus train induced reduction in the sensitivity to iontophoretically applied GABA in CA1 pyramidal neurons; the mechanism of this rapidly developing effect could involve post-translational modification (e.g. phosphorylation) of the GABA A receptor resulting in its altered function. Numerous other possibilities exist including NMDA receptor directed alterations in gene expression (Szekely et al., 1989).

Parallel changes in NMDA receptor function in the developing nervous system and kindled hippocampus suggest another mechanism by which NMDA receptor activation may indirectly effect the expression of a kindled seizure. A striking similarity with respect to NMDA receptor function has emerged in three settings: the hippocampus of a kindled adult rat, the hippocampus of a normal rat early in development, and the visual cortex of kittens during the critical period of postnatal development characterized by robust synaptic plasticity. Each condition is characterized by enhanced responsiveness to NMDA (Nadler et al., this volume; Tsumoto et al., 1987) enhanced NMDA receptor binding (Bode-Greul and Singer, 1989; Tremblay et al., 1989), and evidence consistent with enhanced participation of NMDA receptors in synaptic transmission (Tsumoto et al., 1987; Mody et al., 1988) (direct evidence for the last point has not been established for developing rat hippocampus). This suggests that the functions subserved by NMDA receptors in the developing nervous system may be enhanced in the hippocampus of a kindled adult. A key function of NMDA receptors in the developing retinal-tectal system is the development and stabilization of specific synaptic connections (Cline et al., 1987). Remodelling of neuronal architecture in the normal adult autonomic nervous system has recently been demonstrated to be a remarkably dynamic process (Purves et al., 1988); preliminary data from this same group suggests that similar processes may occur in the adult central nervous system. If so, enhanced NMDA receptor function of a kindled animal may underlie an elevated rate of change in synapse formation and stabilization required for the expression of a kindled seizure. The recent demonstration that NMDA promotes survival and enhances neurite outgrowth and branching in cultured cerebellar and dentate granule cells supports this idea (Pearce et al., 1987; Balazs et al., 1988; Brewer and Cotman, 1988). Interestingly, long-lasting sprouting of mossy fiber axons of dentate granule neurons has been identified in kindled animals (Sutula et al., 1988). Whether such changes are induced by NMDA receptor activation and whether these or other NMDA receptor directed morphologic rearrangements are required for the expression of a kindled seizure remains to be determined.

ACKNOWLEDGMENTS

This work was supported by N.I.H. grants NS24448, NS17771, NS27311, and by two Merit Review Grants awarded by the Veterams Administration. Appreciation is extended to Ms. Rena Wethington for her assistance in the preparation of this manuscript.

REFERENCES

Balazs, R., Jorgensen, O.S., and Hack, N., 1988, N-methyl-D-Aspartate promotes the survival of cerebellar granule cells in culture, **Neuroscience** 27:437.

Bode-Greul, K. and Singer, W., 1989, The development of N-methyl-D-aspartate receptors in cat visual cortex, **Dev. Brain Res.** 46:197.

Bonhaus, D.W., Yeh, G.C., Skaryak, L. and McNamara, J.O., Glycine regulation of the NMDA receptor-coupled ion channel in hippocampal membranes, **Mol.Pharm.**, in press.

Brewer, G.J. and Cotman, C., 1988, NMDA receptor regulation of neuronal morphology in cultured hippocampal neurons, **Neurosci. Lett.** 99:268.

Byrne, M.C., Gottlieb, R. and McNamara, J.O., 1980, Amygdala kindling induces muscarinic cholinergic receptor declines in a highly specific distribution within the limbic system, **Exp. Neurol.** 69:85.

Cline, H., Debski, E. and Constantine-Paton, M., 1987, NMDA receptor antagonist desegregates eye specific stripes, **Proc. Natl. Acad. Sci. U.S.A.** 84:4342.

Crain, B.J., Chang, K.J., and McNamara, J.O., 1987, An in vitro autoradiographic analysis of mu and delta opioid binding in the hippocampal formation of kindled rats, **Brain Res.** 412:311.

Dasheiff, R.M. and McNamara, J.O., 1982, Electrolytic entorhinal lesions cause seizures, **Brain Res.** 231:444.

Flatman, J.A., Schwindt, P.C., Crill, W.E., Strafstrom, C.E., 1983, Multiple actions of N-methyl-D-aspartate on cat neocortical neurons in vitro, **Brain Res.** 266:169.

Goddard, G.V., McIntyre, D.C., Leech, C.K., 1969, A permanent change in brain function resulting from daily electrical stimulation, **Exp. Neurol.** 25:295.

Honore, T. and Drejer, J., 1988, Binding characteristics of Non-NMDA receptors, in "Excitatory Amino Acids in Health and Disease," D. Lodge, ed., Wiley, London, pp. 91-106.

Johnson, J.W., Ascher, P., 1987, Glycine potentiates the NMDA response in cultured mouse brain neurons, **Nature** 325:529.

Kleckner, N.W., Dingledine, R., 1988, Requirement for glycine in activation of NMDA-receptors expressed in Xenopus oocytes, **Science** 241:835.

Mayer, M.L., Westbrook, G.L., Gutherie, P.B., 1984, Voltage-dependent block by Mg^{2+} of NMDA responses in spinal cord neurones, **Nature** 309:261.

MacDermott, A.B., Mayer, M.L., Westbrook, G.L., Smith, S.J., Barker, J.L., NMDA-receptor activation increases cytoplasmic calcium concentration in cultured spinal cord neurones, **Nature** 321:519.

MacDonald, J.F., Porietis, A.V., Wojtowicz, J.M., 1982, L-aspartic acid induces a region of negative slope conductance in the current voltage relationship of cultured spinal cord neurons, **Brain Res.** 237:248.

McNamara, J.O., 1978, Muscarinic cholinergic receptors participate in the kindling model of epilepsy, **Brain Res.** 154:415.

McNamara, J.O., Bonhaus, D.W., Nadler, J.V. and Yeh, G.C., N-methyl-D-aspartate (NMDA) receptors and the kindling model, in "Kindling 4," J. Wada, ed., Plenum Pub. Co., New York, in press.

McNamara, J.O., Peper, A.M., Patrone, V., 1980, Repeated seizures induce long term elevation of hippocampal benzodiazepine receptors, **Proc. Natl. Acad.Sci., U.S.A.** 77:3029.

McNamara, J.O., Russell, R.D., Rigsbee, L.C., and Bonhaus, D.W., 1988, Anticonvulsant and antiepileptogenic actions of MK-801 in the kindling and electroshock models, **Neuropharmacology** 27:563.

Mody, I., Stanton, P.K. and Heinemann, U., 1988, Activation of N-methyl-D-aspartate receptors parallels changes in cellular and synaptic properties of dentate gyrus granule cells after kindling, **J.Neurophysiol.** 59:1033.

Monaghan, D.T., Olverman, H.J., Nguyen, L., Watkins, J.C. and Cotman, C.W., 1988, Two classes of N-methyl-D-aspartate recognition sites: differential distribution and differential regulation by glycine, **Proc. Natl. Acad. Sci. U.S.A.** 85:9836.

Mondadori, C., Weiskrantz, L., Buerki, H., Petschke, F. and Fagg, G.E., 1989, NMDA receptor antagonists can enhance or impair learning performance in animals, **Exp. Brain Res.**, 75:449.

Morris, R.G.M., Anderson, E., Lynch, G.S. and Baudry, M., 1986, Selective impairment of learning and blockade of long-term potentiation by an N-methyl-D-aspartate receptor antagonist, **Nature**, 319:774.

Morrisett, R.A., Chow, C., Nadler, J.V. and McNamara, J.O., Biochemical evidence for enhanced sensitivity to N-methyl-D-aspartate in the hippocampal formation of kindled rats, **Brain Res.**, in press.

Morrisett, R.A., Chow, C., Sakaguchi, T., Shin, C., and McNamara, J.O., Inhibition of muscarinic-coupled phosphoinositide hydrolysis by N-methyl-D-aspartate is dependent upon depolarization via channel activation, submitted for publication.

Nowak, L., Bregestovski, P., Ascher, P., Herbet, A., Prochiantz, A., 1984, Magnesium gates glutamate-activated channels in mouse central neurons, **Nature** 307:462.

Okazaki, M.M., McNamara, J.O., and Nadler, J.V., 1989, N-methyl-D-aspartate receptor autoradiography in rat brain after angular bundle kindling, **Brain Res.** 482:359.

Olsen, R.W., Bureau, M., Khrestchatisky, M., MacLennan, A.J., Ciang, M.Y., Tobin, A.J., Xu, W., Jackson, M., Sternini, C., and Brecha, N. Isolation of pharmacologically distinct GABA-benzodiazepine receptors by protein chemistry and molecular cloning, **in** "GABA and Benzodiazepine Receptor Subtypes: from Molecular Biology to Clinical Practice, Raven Press, New York, in press.

Olsen, R.W. and Snowman, A.M., 1983, [3H]Bicuculline methochloride binding to low-affinity gamma-aminobutyric acid receptor sites, **J.Neurochem.** 41:1653.

Pearce, I., Cambray-Deakin, M., and Burgoyne, R., 1987, Glutamate acting on NMDA receptors stimulates neurite outgrowth from cerebellar granule cells, **FEBS Lett.** 223:143.

Peterson, D.W., Collins, J.F. and Bradford, H.F., 1983, The kindled amygdala model of epilepsy: anticonvulsant action of amino acid antagonists, **Brain Res.** 275:169.

Pritchett, D.B., Sontheimer, H., Gorman, C.M., Kettenmann, H., Seeburg, P.H., Schofield, P.R., 1988, Transient expression shows ligand gating and allosteric potentiation of $GABA_A$ receptor subunits, **Science** 242:1306.

Purves, D., Snider, W.D., and Voyvodic, J.T., Trophic regulation of nerve cell morphology and innervation in the autonomic nervous system, **Nature** 336:123.

Rondouin, G., Involvement of excitatory amino acids in the mechanisms of kindling, **in** "Kindling 4," J. Wada, ed., Plenum Press, New York, in press.

Savage, D.D.S., Dasheiff, R., and McNamara, J.O., 1983, Kindled seizure induced reduction of muscarinic cholinergic receptors in rat hippocampal formation: evidence for localization to dentate granule cells, **J. Comp. Neurol.** 221:106.

Savage, D.D.S., Werling, L.L., Nadler, J.V., and McNamara, J.O., 1984a, Selective and reversible increase in the number of quisqualate-sensitive glutamate binding sites on hippocampal synaptic membranes after angular bundle kindling, **Brain Res.** 307:332.

Savage, D.D.S., Nadler, J.V., and McNamara, J.O., 1984b, Reduced kainic acid binding in rat hippocampal formation after limbic kindling, **Brain Res.** 323:128.

Schofield, P.R., Darlison, M.G., Fujita, N., Burt, D.R., Stephenson, F.A., Rodriguez, H., Rhee, L.M., Ramachandran, J., Reale, V., Glencorse, T.A., Seeburg, P.H., and Barnard E.A., 1987, Sequence and functional expression of the $GABA_A$ receptor shows a ligand-gated receptor super-family, **Nature** 328:221.

Shin, C., Pedersen, H.B., and McNamara, J.O., 1985, Gamma-aminobutyric acid and benzodiazepine receptors in the kindling model of epilepsy: a quantitative radiohistochemical study, **J.Neurosci.** 5:2696.

Stelzer, A., Slater, N.T. and ten Bruggencate, G., 1987, Activation of NMDA receptors blocks GABAergic inhibition in an in vitro model of epilepsy, **Nature** 326:698.

Sutula, T., Xiao-Xian, H., Cavazos, J. and Scott, G., 1988, Synaptic reorganization in the hippocampus induced by abnormal functional activity, **Science** 239:1147.

Szekely, A.M., Barbaccia, M.L., Alho, H. and Costa, E., 1989, In primary cultures of cerebellar granule cells the activation of N-methyl-D-aspartate-sensitive glutamate receptors induces c-fos mRNA expression, **Mol. Pharm.** 35:401.

Tremblay, E., Roisin, M., Represa, A., Charriaut-Marlangue, C., and Ben-Ari, Y., 1989, Transient increased density of NMDA binding sites in the developing rat hippocampus, **Brain Res.** 461:393.

Tsumoto, T., Hagihara, K., Sato, H., and Hata, Y., 1987, NMDA receptors in the visual cortex of young kittens are more effective than those of adult cats, **Nature** 327:513.

Valdes, F., Dasheiff, R.M., Birmingham, F., Crutcher, K.A., and McNamara, J.O., 1982, Benzodiazepine receptor increases following repeated seizures: evidence for localization to dentate granule cells, **Proc. Natl. Acad. Sci. U.S.A.** 79:193.

Yeh, G.C., Bonhaus, D.W., Nadler, J.V., and McNamara, J.O., NMDA receptor plasticity in kindling: quantitative and qualitative alterations in the NMDA receptor/channel complex, **Proc. Natl. Acad. Sci. U.S.A.**, in press.

NMDA-RECEPTORS ARE INVOLVED IN SYNAPTIC PLASTICITY FOLLOWING PARTIAL

DENERVATION OF CA1 HIPPOCAMPAL CELLS

Howard Wheal

Department of Neurophysiology
University of Southampton
Bassett Crescent East
Southampton, SO9 3TU. UK

INTRODUCTION

Our original motive for partially denervating the CA1 pyramidal cells in the hippocampus
was to try to separate the two major excitatory afferents onto these cells, the Schaffer collateral
and commissural fibers, in order to explore whether they would independently supported long -
term potentiation, LTP (Bliss et al, 1983; Wheal et al,1983). The method that we used was that
described by Nadler et al (1978), where the neurotoxin kainic acid was injected unilaterally into
the lateral ventricle of the rat. This produced a specific loss of CA3/CA4 pyramidal cells in the
hippocampus ipsilateral to the injection site (Nadler et al (1978), Lancaster andWheal,1982). In
addition to the excitatory afferents onto the surviving CA1 pyramidal cells showing potentiation,
the cells themselves fired with graded bursts and also appeared to have lost some of their inhibitory responses (Lancaster andWheal,1984).

Consequences of the kainic acid (KA) lesion.

Following the degeneration of the excitatory Schaffer collateral terminals onto CA1 pyramidal cells, there was a chronic failure of components of synaptic inhibition (Lancaster andWheal,
1984; Ashwood, Lancaster and Wheal, 1986; Cornish and Wheal,1989; Wheal,1989). This included a loss of the early $GABA_A$ -receptor mediated IPSP as well as some reduction in the amplitude of the late K+ IPSP that follows activation of the $GABA_B$ -receptor. Although some recovery
of inhibition has been reported (Frank and Schwartzkroin,1985) we have recorded failure of this
inhibition 16 weeks post lesion (Cornish and Wheal, 1989). Histochemical studies have shown
that the numbers of possible inhibitory interneurones in the CA1 area that are immunopositive to
somatostatin significantly reduced 1-4 weeks post lesion (Bruce, Cornish and Wheal,1989). We
believe that the changes in both the excitatory and inhibitory circuitry contribute to the epileptiform bursting activity that has been recorded from the surviving CA1 pyramidal cells (Wheal,
1988;1989). The relationship between these two mechanisms is discussed in more detail below.

Reactive synaptogenesis

Associated with the loss of cells in the CA3/CA4 areas of the rat hippo- campus following
the intracerebroventricular injection (ICV) of KA (Lancaster and Wheal,1982) many degeneration
processes can be seen in the CA1 area. Abnormal morphology is clearly observed one day after the
lesion, with electron dense and lucent degeneration profiles, increased glial activity, damaged
myelin sheaths and swollen axons.

Excitatory Amino Acids and Neuronal Plasticity
Edited by Y. Ben-Ari
Plenum Press, New York, 1990

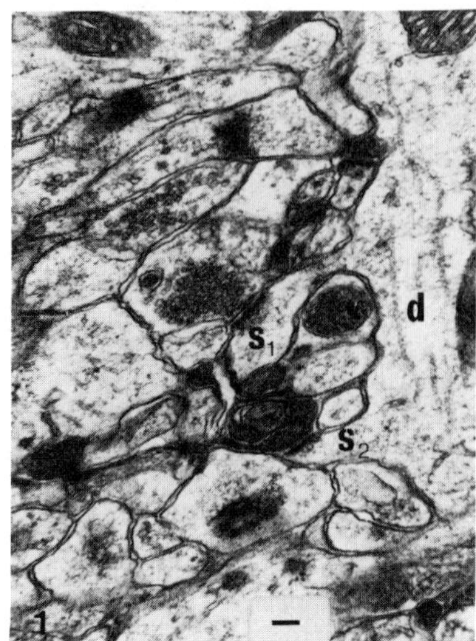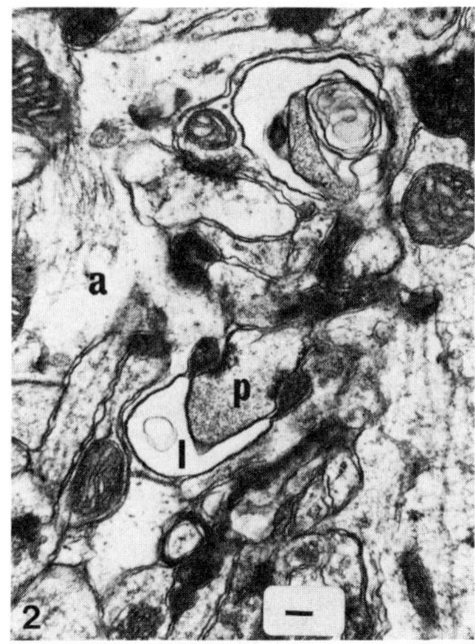

Figs. 1 & 2.　　　EM photomicrographs of the stratum radiatum of CA1 in the KA lesioned hippocampus, 3 days post injection (mag 40,670). Scale bars are 0.1μ.

Fig. 1.　　　shows dendrite(d) with two spines (S_1 and S_2). Both spines make asymmetric synaptic contacts: S_1 with a normal vesicle containing bouton, S_2 with a degenerating bouton.

Fig. 2.　　　shows two lucent areas and several electron dense profiles. Lucent profile (l) appears to be continuous with two dark profiles, and to be partly surrounded by an astrocytic process(a). The dark profiles are in asymmetric contact with a post-synaptic element(p), probably a spine. (Phelps, Wheal and Mitchell, unpublished observations)

This damage appears maximal at 3 days. At this stage postsynaptic thickenings are frequently apposed to the dark profiles, in which vesicles and mitochondria may still be detected, and are in invariably asymetric (see Figure 1). In terminals with no apparent signs of degeneration, the vesicles may vary more greatly in size and form larger, more dense clumps than normal. The crescent shape of the lucent areas may appear continuous with small dark profiles as shown in figure 2, however they rarely appose postsynaptic thickenings. The damage to the myelin sheaths is most commonly seen as a single continuous split between the lamina, often filled with a dark material, possibly oligodendrocytoplasm or microglial cytoplasm. At 3 days post injection of KA no changes were seen in the shape of the spines, except for that produced by pre-terminal shrinkage (Phelps, Wheal and Mitchell, unpublished observations).

A considerable range of patterns of synaptic regeneration have been reported following lesions of pathways in different CNS nuclei. In the cerebellum, Chen and Hillman (1982) have reported the extension of spines to reach functional terminals as well as the growth of double headed spines. Cells in the dentate gyrus also show considerable reactive plasticity with growth of intact afferents and proliferation of boutons (Parnavelas,1974; Lee et al, 1977), as well as elongation of spines and a consequential increase in the synaptic contact areas (Hillman and Chen,1988). Complex synapses are common in the dentate, and in development their growth continues after completion of synapse formation (Cotman et al 1973).

In the hippocampus, studies 3-7 days after bilateral ICV KA injections have shown a reduction in the number of synapses in the stratum radiatum from 35 to 5 per 100u^2 (Nadler et al 1980). 6-8 weeks after the lesion the number of synapses in the stratum radiatum had recovered to normal. Similar reactive synaptogenesis in the CA1 area was also reported following a unilateral lesion of the fimbria (Goldowitz et al 1979). However, this group also reported an increase in the number of complex synapses, or synaptic contacts per bouton, in the hippocampus. They suggested that this might produce a compensatory increase in the efficiency of intact synapses.

Role of NMDA receptors in synaptic function

The excitatory amino acid receptors that are responsible for normal synaptic function at the terminals of the Schaffer collateral and commissural afferents onto the CA1 cell are thought to be of a non-NMDA subtype. The evidence is primarily based on the use of the specific NMDA-receptor antagonist, D-2-amino-5-phosphonovalerate (D-APV), to which the normal monosynaptic potentials recorded in CA1 were insensitive (Koerner and Cotman,1982; Collingridge et al,1983). These observations were particularly surprising in the light of the very high density of NMDA receptors found at the termination sites of these afferent systems in the stratum radiatum and stratum oriens of CA1 (Monaghan et al,1983). The densities of NMDA-receptors found in these regions are infact the highest in the brain (Cotman and Monaghan,1987).

A functional role for the NMDA-receptors in the hippocampus has been found in that they are involved in the induction of LTP (Collingridge et al,1983; Harris et al, 1984; Wigstrom and Gustafsson, 1984; Collingridge and Bliss, 1987) or in the presence of low extracellualr Mg^{++} (Herron et al, 1985) which removes a voltage-dependent Mg^{++} block of the channels linked to the NMDA receptors (Nowak et al, 1984; Mayer et al, 1984). NMDA-receptors also appear to contribute to epileptiform activity. Various NMDA antagonists have been found effective in blocking seizures in several *in vivo* animal models of epilepsy (Croucher et al, 1982; Meldrum et al, 1983). D-APV has also been shown to reduce acute convulsant induced epileptiform activity in the *in vitro* hippocampus (Herron et al, 1985; Dingledine et al, 1986).

Suprathreshold stimulation of excitatory afferents in the KA lesioned hippocampus evoked a synchronised graded burst of population spikes from the surviving CA1 pyramidal cell population. This epileptiform activity was substantially attenuated by perfusion of the slices with 20uM D-APV (Ashwood and Wheal, 1986). Similar experiments were also carried out intracellularly where D-APV was found to have no direct effect on a range of membrane and cell properties. While D-APV (20uM) considerably reduced the number of action potentials in the epileptiform burst, it had little or no action on control excitatory or inhibitory synaptic potentials (Ashwood and Wheal,1987). It was clear from the subtraction of the suprathreshold events, before and after the application of D-APV, that the underlying NMDA-receptor dependent potential was early in onset but of long duration.

In these experiments the NMDA-receptor mediated burst was recorded with a concentration of 1mM Mg^{++} in the perfusion medium (ACSF). Even with this concentration of Mg^{++} present it is easy to speculate that the suprathreshold burst of somatic action potentials would depolarise the cell sufficiently to allow the NMDA-receptor activated conductance to contribute to the epileptiform event. However, from these data one could not tell whether the NMDA-receptors were contributing to synaptic function at a more discrete level. Further intracellular studies were therefor carried out on small subthreshold EPSPs recorded from CA1 pyramidal cells in lesioned hippocampi. Slices were selected that fired bursts of 2-4 action potentials in response to afferent stimulation in the stratum radiatum , at twice threshold intensity for cell firing. Subthreshold EPSPs were analysed for peak magnitude, time course and fluctuation characteristics (Andersen et al, 1987; Turner, 1988).

EPSPs were found to have a voltage sensitive variable late component, shown in figure 3, that was not present in control CA1 pyramidal cells (Wheal and Turner,1988; Turner and Wheal, 1988; Turner et al,1989). This late component was sensitive to D-APV (1-20uM) and therefore appeared to be mediated by NMDA- receptors. An early EPSP component was observed in all neurones, averaging 0.92 ± 0.51mV (mean $\pm$ SD) in amplitude with a 10-90% risetime of 8.1 ± 3.1ms and halfwidth of 30.9 ± 9.1 ms (n=23). The time course of the variable late EPSP component could occasionally be separated from the early EPSP by subtracting adjacent responses. The subtracted responses (n=41) averaged 2.93 ± 1.0mV in peak amplitude and had 10-90% risetime of 23.7 ± 11.9 ms and halfwidth of 42.9 ± 11.7ms (Turner and Wheal,1988). Thus the late subthreshold component that was seen at the normal resting potential for these cells(-63mV), exhibited significantly

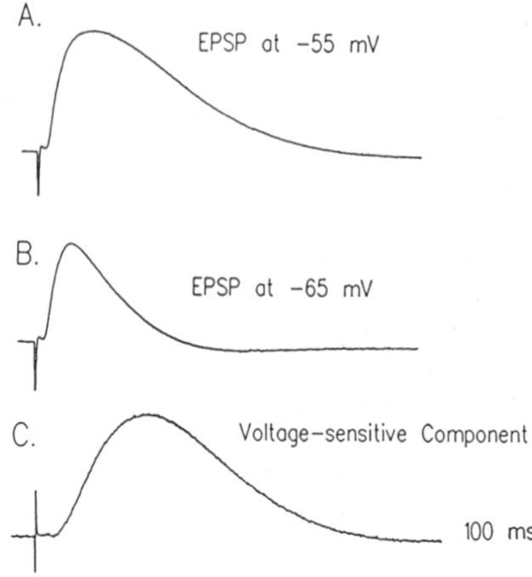

Fig. 3.   Evoked subthreshold EPSPs recorded from a CA1 pyramidal cell in the KA
lesioned hippocampus. The traces illustrate the voltage sensitivity of the
late NMDA receptor-mediated component of the EPSP. They also show the
slow time course of the the late EPSP. (With permission from Turner et
al,1989).

longer 10-90% risetime and halfwidth than the early EPSP (P<0.001 Students t test). This domi-
nating late NMDA-receptor mediated component to these EPSPs has led us to suggest that there
had been a shift in the receptor population following denervation of the CA1 pyramidal cell dendrites (Wheal and Turner,1988).

Slow time course of the NMDA-receptor mediated EPSP

The slow timecourse of the NMDA-receptor mediated EPSP described here is in common
with other observations of synaptic events in the hippocampus, neocortex and thalamus
(Thomson, 1986; Thomson et al, 1989; Wigstrom and Gustafsson, 1984 ; Dingledine et al 1986;
Salt,1986; Forsythe and Westbrook,1988). Another common property is that the NMDA-component is always preceded by a fast non-NMDA EPSP, even in single axon preparations (Forsythe
and Westbrook, 1988; Thomson et al,1989). The simplest explanation of the latter results is that
the neurotransmitter released at the excitatory synapses activates both sets of receptors, particularly if like glutamate the neurotransmitter had a higher affinity for the NMDA-receptor.

Many mechanisms have been proposed to explain the early and late components of these
EPSPs. However we are still short of information on the spatial relationship between synapses
and the distribution of the different receptor types on the spines and dendrites. However, we
would like to suggest that as a first step , changes in the presynaptic elements described above
may produce a variety of responses postsynaptically including failures of the late NMDA-component to the EPSP. Following denervation, the NMDA-receptor sites on the dendrites may prove to
be a preferred location for new synapse formation, whether that be as a result of spine elongation
or afferent sprouting. A dia- grammatic representation of this process is shown in figure 4.

The precise molecular or cellular mechanism that is responsible for the long timecourse of
the D-APV sensitive depolarisation is also a matter of some debate. However like Forsythe and
Westbrook (1988), our modelling studies have also excluded an extreme distal location on the
dendrites as being a simple solution (Turner et al,1989; Wheal and Turner, unpublished observations). In order to model the waveform parameters of the late EPSP , even assuming location at a
distal dendritic site, required a significantly larger and longer conductance transient than the
early EPSP (Turner et al,1989).

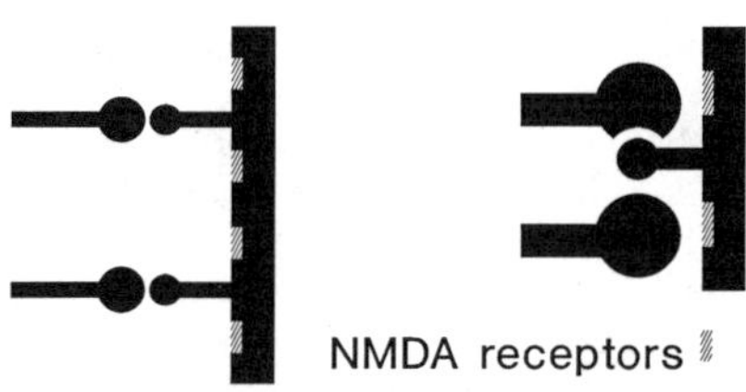

Fig. 4.　Diagrammatic representation of the change in synaptic morphology following partial denervation of the CA1 pyramidal cell. Growth of synapses into a different configuration might result in the recruitment of the NMDA receptors that mediate the epileptiform activity.

Relationship between the failure of inhibition and the expression of the NMDA-receptor mediated EPSP

There is a temptation to claim that the failure of inhibition in the KA model is the cause of the expression of the NMDA-receptor mediated event. However in our hands D-APV only had a very slight attenuating affect on the bicuculline-induced bursts in CA1 pyramidal cells when compared with its effects on similar bursts from lesioned slices(Ashwood and Wheal,1986). Furthermore, intracelluar studies by others have shown that D-APV at similar concentrations blocks only the last action potential in bicuculline induced epileptiform bursts (Herron et al,1986; Dingledine et al,1986). In contrast, more than 50% of the evoked burst of action potentials in CA1 cells in lesioned slices was blocked by D-APV. Whilst it has been suggested that $GABA_A$ -Cl- mediated IPSPs may hyperpolarise and shunt the dendritic membrane, thus controlling the NMDA channels (Dingledine et al,1986), a more effective control might be induced by the $GABA_B$ -K+ conductance (Wheal,1987).

Modulation by glycine receptors

An alternative mechanism for the major NMDA-component of the pyramidal cell EPSP in the KA lesioned hippocampus is is modulation of the NMDA receptor/channel comple by glycine (see review by Thomson,1989). Johnson and Ascher (1987) were the first reported that low concentrations of glycine augmented NMDA and glutamate activated currents in cultured neurones. Additional glycine was also shown to augment NMDA receptor-mediated components in slices of neocortex (Thomson et al, 1989). The distributions of strychnine- insensitive glycine binding sites and NMDA-receptors in the human hippocampus are very similar (Jansen et al,1989). Furthermore, especially high levels of glycine have been found associated with epileptogenic foci in the human brain (Van Gelder et al,1977). Thus changes in synapse and receptor distribution together with changes in the absolute concentration of glycine may all contribute to the role of the NMDA channel in the genesis of epileptiform activity.

Use of the non-NMDA receptor antagonist CNQX

CNQX (6-Cyano-7-nitroquinoxaline-2,3-dione) is a competitive antagonist of non-NMDA receptors (Drejer and Honore,1988) and as such it has been used to demonstrate an NMDA receptor component in the normal EPSP in control CA1 pyramidal cells (Andreasen et al,1989). In their hands 5uM CNQX blocked the early component of the EPSP, leaving a residual late D-APV sensitive component. In contrast, the same concentration of CNQX has been found to block both the non - NMDA as well as the NMDA receptor components to the epileptiform burst (Williamson and Wheal,1989). A possible explanation for these observations is that CNQX also blocks the glycine site on the NMDA receptor complex, represented in figure 5. Differences in the actions of CNQX have been found in other preparations (see Thomson,1989).

Comparisons between the mechanisms underlying LTP and epileptiform activity

It has been suggested that there might be similarities in the mechanisms underlying LTP and and the epileptogenic process of kindling (Collingridge and Bliss,1987), they both have a role for NMDA receptors. However as these authors mentioned, the important difference is that NMDA receptors are involved in the chronicity of epileptiform activity, whilst they do not contribute to the maintenance of LTP in the hippocampus. The synaptic changes associated with LTP are subtle, those associated with some epilepsy models are not. The data obtained from the KA acid lesioned hippocampus provides a good example of the differences between these two synaptic processes. As stated in the introduction, the afferents that survive the lesion can undergo LTP (Bliss et al,1983; Wheal et al,1983). Yet at the same time the reactive synaptogenesis that occures in the CA1 region is possibly recruiting the NMDA receptors that generate the epileptiform activity.

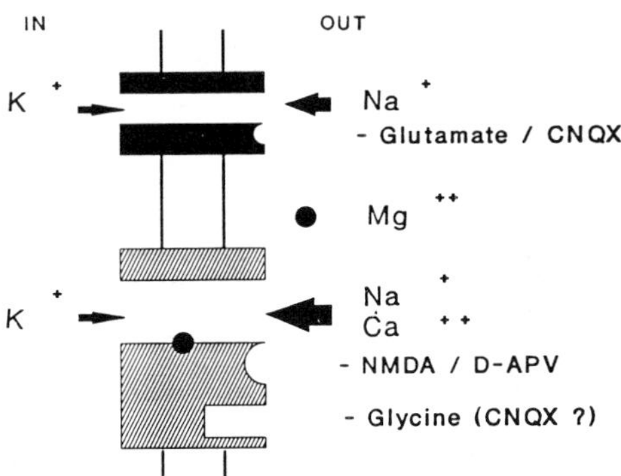

Fig. 5.   Diagrammatic representation of the membrane mechanisms associated with epileptiform activity in CA1 pyramidal cells.

CONCLUSIONS

The mechanisms underlying the epileptiform activity in surviving CA1 pyramidal cells in the KA lesioned hippocampus is dominated by the role of NMDA receptors and their channel. Although these receptors contribute very little to the normal EPSPs recorded in these cells, following the lesioning a prominent late voltage sensitive component to the EPSP was observed (Wheal and Turner, 1988; Turner et al,1989). Furthermore this late component was totally blocked by a range of concentrations of the NMDA receptor antagonist D-APV(1-10uM).

It should be restated that although the late EPSP component was voltage sensitive it was voltage sensitive it was contributing sigificantly to cell activity at the normal resting potential of these cells (-63mV), and that the recordings were made in an ACSF with $[Mg^{++}]_o$ of 1mM.

In order to explain this change in synaptic activity it has been suggested that following the partial denervation of the CA1 pyramidal cells, there is some form of reactive synaptogenesis that recruits the NMDA receptor sites on the dendritic membrane for new synapse formation. It is unclear at the moment which of the surviving excitatory afferents would provide the terminals for reinnervation, however local recurrent excitatory afferents in CA1 (Knowles and Schwartzkroin, 1981; Christian and Dudek, 1988) must be prime candidates. The resulting reinforcement of recurrent excitation would further contribute to the hyperexcitability of the CA1 pyramidal cells.

ACKNOWLEDGEMENTS

This work was supported by the Wellcome Trust, MRC, Action Research and Wesse Medical School Trust.

REFERENCES

Andersen, P., Storm, J. and Wheal, H.V., 1987, Thresholds of action potentials evoked by synapses on the dendrites of pyramidal cells in the rat hippocampus in vitro. **J. Physiol.** 383 :509-526.

Andreason, M., Lambert, J.C.D. and Jensen, M.S. 1988. Direct demonstration of an N-methyl-D-aspartate receptor mediated component of excitatory synaptic transmission in area CA1 of the rat hippocampus. **Neuro Sci Lett.** 93:61-66.

Ashwood, T.J., Lancaster, B. and Wheal, H.V. 1986. Intracellular electrophysiology of CA1 pyramidal neurones in slices of the kainic acid lesioned hippoampus. **Exp. Brain Res.** 62: 189-198.

Ashwood, T.J. and Wheal, H.V. 1986. Extracellular studies on the role of N-methyl-D-aspartate receptors in epileptiform activity recorded from the kainic acid-lesioned hippocampus. **Neurosci. Letts.** 67 :147-152.

Ashwood, T.J. and Wheal, H.V. 1987. The expression of N-methyl-D-aspartate receptor- mediated component during epileptiform synaptic activity in the hippocampus. **Br. J. Pharmac.** 91 : 815-822.

Bliss, T.V.P., Lancaster, B. and Wheal, H.V. 1983. Long-term potentiation in commissural and Schaffer projections to hippocampal CA1 cells: an *in vivo* study in the rat. **J. Physiol.** 341 :617-626.

Bliss, T.V.P. and Collingridge, G.L. 1987. NMDA-receptors and their role in long-term potentiation. **TINS** 10 :288-293.

Bruce, R.C., Wheal, H.V. and Cornish, S.M. 1989. Somatostatin-like immunoreactive interneurones in the kainic acid lesioned rat hippocampus. **Neurosci Lett Suppl.** 36: S43.

Chen, S., and Hillman, D.E. 1982. Marked reorganisation of Purkinje cell dendrites and spines in adult rat following vacation of synapses due to deafferentation. **Brain Res.** 245:131-135.

Christian, E.P. and Dudek, F.E. 1988. Electrophysiological evidence from glutamate micropipette applications for local excitatory circuits in the CA1 area of rat hippocampal slices. **J. Neurophysiol.** 59:110-123.

Collingridge, G.L., Kehl, S.J. and McLennan, H. 1983. Excitatory amino acids in synaptic transmission in the Schaffer collateral-commissural pathway of the rat hippocampus. **J.Physiol.** 334: 33-46.

Cornish, S.M. and Wheal, H.V. 1989. Long term failure of synaptic inhibition in the kainic acid lesioned rat hippocampus of the rat. **Neurosci.** 28:563-571.

Cotman, C.W. and Monaghan, D.T. 1987. Organisation of excitatory amino acid receptors : Functional implications. **In:** P, Hicks., D. Lodge and H. McLennan ed., "Excitatory amino acid transmission". Liss, NY.pp 325-332.

Cotman, C.W., Taylor, D and Lynch, G.S. 1973. Ultrastructural changes in synapses in the dentate gyrus of the rat during development. **Brain Res.** 63: 205-213.

Croucher, M.J., Collins, J.F. and Meldrum, B.S. 1982. Anticonvulsant action of excitatory amino acid antagonists. **Science.** 216: 899-901.

Dingledine, R., Hynes, M.A. and King, G.L. 1986. Involvement of N-methyl-D-aspartate receptors in epileptiform bursting in the rat hippocampal slice. **J.Physiol.** 380: 175-189.

Drejer, J. and Honore, T. 1988. New quinoxalinediones show potent antagonism of quisqualate responses in cultured mouse cortical neurons. **Neuro. Sci. Lett.** 87: 104-108.

Forsythe, I.D. and Westbrook, G.L. 1988. Slow excitatory postsynaptic currents mediated by N-methyl-D-aspartate receptors on cultured mouse central neurones. **J. Physiol.** 396: 515-533.

Franck, J.E. and Schwartzkroin, P.A. 1985, Do kainate-lesioned hippocampi become epileptogenic? **Brain Res.** 329: 309-313.

Goldwidtz, D., Scheff, S.W. and Cotman, C.W. 1979. The specificity of reactive synaptogenesis: a comparative study in the adult rat hippocampal formation. **Brain Res.** 170: 427-441.

Harris, E.W., Ganong, A.H. and Cotman, C.W. 1984. Long term potentiation involves activation of N-methyl-D-aspartate receptors. **Brain Res.** 323:132-137.

Herron, C.E., Williamson, R. and Collingridge, G.L. 1985, A selective N-methyl-D-aspartate antagonist depresses epileptiform activity in rat hippocampal slices. **Neurosci. Lett.** 61: 225-260.

Hillman, D.E. and Chen, S. 1988. Early plasticity in the dentate gyrus following surgical lesions to the perforant path._Soc Neurosci. 14:1133.

Jansen, K.L.R., Dragunow, M. and Faull, R.L.M. 1989. [3H] Glycine binding sites, NMDA and PCP receptors have similar distributions in the human hippocampus: an autoradiographic study. **Brain Res.** 482: 174-178.

Knowles, D. and Schwartzkroin, P.A. 1981. Local circuit synaptic interactions in hippocampal brain slices J. **Neurosci.** 1 : 318-322.

Koerner, J.F. and Cotman, C.W. 1982. Response of Schaffer collateral CA1 pyramidal cell synapses of the hippocampus to analogues of acidic amino acids. **Brain Res.** 251 : 105-115.

Lancaster, B. and Wheal, H.V. 1982. A comparative histological and electrophysiological study of some neurotoxins in the rat hippocampus. **J. Comp. Neurol.** 211 : 105-114.

Lancaster, B. and Wheal, H.V. 1984. Chronic failure of inhibition in the CA1 area of the hippocampus following kainic acid lesions of the CA3/CA4 area. **Brain Res.** 295 : 317-324.

Lee, K.S., Stanford, E.J., Cotman, C.W. and Lynch, G.S. 1977. Ultrastructural evidence for bouton proliferation in the partially deafferented dentate gyrus of the adult rat. **Exp. Brain Res.** 29: 475-485.

Mayer, M.L., Westbrook, G.L. and Guthrie, P.B. 1984. Voltage dependent block by Mg++ of NMDA responses in spinal cord neurones. **Nature.** 309:261-263.

Meldrum, B.S., Croucher, M.J., Badman, G. and Collins, J.F. 1983. Antiepileptic action of excitatory amino acid antagonists in the photosensitive baboon *Papio papio*. **Neurosci. Lett.** 39 :101-104.

Monaghan, D.T., Holets, V.R., Toy, D.W. and Cotman, C.W. 1983. Anatomical distributions of four pharmacological distinct 3H-L-glutamate binding sites. **Nature.** 306: 176-179.

Nadler, J.V. Perry, B.W. and Cotman, C. 1978. Intraventricular kainic acid preferentially destroys hippocampal pyramidal cells. **Brain Res.** 205: 676-677.

Nadler, J.V., Perry, B.W., Gentry, C. and Cotman, C.W. 1980. Loss and reaquisition of hippocampal synapses after selective destruction of CA3-CA4 efferents with kainic acid. **Brain Res.** 191: 387-403.

Nowak,L., Bregestovski, P., Ascher, P., Herbet, A. and Prochiantz, A.1984. Magnesium gates glutamate-activated channels in mouse central neurons. **Nature.** 307: 462-465.

Parnavelas, J.G., Lynch, G.S., Brecha, N., Cotman, C.W. and Globus, A. 1974. Spine loss and regrowth in hippocampus following deafferentation. **Nature.** 248: 71-73.

Salt, T.E. 1986. Mediation of thalamic sensory input by both NMDA and non-NMDA receptors. **Nature.** 322: 263-265.

Thomson, A.M. 1986. A Mg++ sensitive excitatory postsynaptic potential in slices of rat cerebral corte resembles responses to NMA. **J.Physiol.** 370: 531-549. Thomson, A.M. 1989. Glycine modulation of the NMDA receptor/channel complex. **TINS.** 12: 349-353.

Thomson, A.M., Girdlestone, D. and West, D.C. 1989. A local circuit neocortical synapse that operates via both NMDA and non-NMDA receptors **Br.J.Pharmacol.** 96: 406-408.

Thomson, A.M., Walker, V.E. and Flynn, D.M. 1989. Glycine enhances NMDA-receptor mediated synaptic potentials in neocortical slices. **Nature** 338: 422-424.

Turner, D.A. 1988. Waveform and amplitude characteristics of evoked responses to dendritic stimulation of CA1 guinea-pig pyramidal cels. **J. Physiol.** 395: 419-439.

Turner, D.A. and Wheal, H.V. 1988. Components of subthreshold synaptic potentials in CA1 pyramidal neurones from kainic acid lesioned rat hippocampus in vitro.**J. Physiol.** 400 : 50P.

Turner, D.A., Wheal, H.V., Stockley, E and Cole H. 1989. Three-dimensional Reconstructions and analysis of the cable properties of neurones. **in:** "Cellular Neurobiology" H.V. Wheal and J. Chad, ed., IRL-OUP, Oxford.

Van Gelder, N.M., Sherwin, A.L. and Rasmussen, T. 1977. Amino acid content of epileptogenic human brain: Focal versus surrounding regions. **Brain Res** 40: 385-397.

Wheal, H.V. 1988. Changes in synaptic function that underlie epileptiform activity in hippocampal pyramidal cells. **In** Synaptic Plasticity in the Hippocampus Ed. Buzsaki and Haas, Springer Verlag, Berlin and Heidelberg, pp,140-143.

Wheal, H.V. 1989. Function of synapses in the CA1 region of the hippocampus: their contribution to the generation or control of epileptiform activity. **Comp Biochem Physiol.** 93A: 211-220.

Wheal, H.V., Ashwood, T.J. and Lancaster, B. 1984. A comparative in vitro study of the kainic acid lesioned and bicuculline treated hippocampus: chronic and acute models of focal epilepsy. **In**: Electrophysiology of Epilepsy ed. Schwartzkroin and Wheal, Academic Press, London.

Wheal, H.V., Lancaster, B. and Bliss, T.V.P. 1983. Long-term potentiation in Schaffer collateral and commissural systems of the hippocampus: in vitro study in rats pretreated with kainic acid. **Brain Res.** 272 : 247-253.

Wheal, H.V. and Turner, D.A. 1988. Early and late EPSP components in CA1 pyramidal neurons from kainic acid lesioned rat hippocampus. **Soc Neurosci** 14: 790.

Wigstrom, H. and Gustafsson, B. 1984. A possible correlate of the postsynaptic condition for long-lasting potentiation in the guinea-pig hippocampus in vitro. **Neurosci Lett.** 44: 327-332.

Williamson, R. and Wheal, H.V. 1989. CNQX blocks synaptic responses in the kainic acid lesioned hippocampus. **Neurosci Lett Suppl.** 36, S23.

COMMENTARY - EXCITATORY AMINO ACIDS AND REACTIVE PLASTICITY

V. Nadler and J. Zimmer

This session covered three topics that fall under the broad heading of reactive plasticity: namely, (1) the ability of developmental and environmental events to modulate properties of the NMDA receptor, (2) markers for axon sprouting in relation to excitotoxic neuropathology and (3) the use of transplanted neurons to repair excitotoxic lesions.

NMDA RECEPTOR PLASTICITY

Speakers in the preceding sessions of this conference strongly reinforced the idea that activation of the NMDA receptor is necessary for the induction of long-term potentiation, developmental plasticity and associative learning. In this session, four speakers presented evidence that environmental or developmental stimuli can also trigger changes in the abundance, regulation or physiological role of the NMDA receptor itself.

Kindling is one form of long-lasting plasticity that is associated with changes in the NMDA receptor. Heinemann found that, during the kindling process, the lateral perforant path EPSP in dentate granule cells developed a large NMDA receptor-mediated component. This change had the effect of enlarging and lengthening the EPSP. Coupled with a possible loss of potassium channels in these cells, the enhanced EPSP was suggested to reduce the ability of granule cells to prevent seizure activity from propagating through the hippocampus. NMDA receptors may play a greater role in synaptic transmission, at least in part, because kindling increases the density of these receptors in hippocampal membranes (McNamara). Surprisingly, it appears to be only the antagonist-preferring form of the NMDA receptor, the form labeled by [3H]CPP, that becomes more abundant after kindling. There was some discussion of whether the NMDA receptors labeled by [3H]glutamate and [3H]CPP are indeed different molecules or simply different conformational states of a single receptor protein. Although a few findings favor their being separate molecular entities, a definitive answer will probably require isolation and sequencing of the receptor subunits to which these radioligands bind. A greater abundance of NMDA receptors on CA3 pyramidal cells could also account for the observation that kindling increases the potency of NMDA in depolarizing these cells (Nadler).

Considerable discussion centered on whether or not the observed enhancement of NMDA receptor function can account for the increased excitability associated with the kindled state. A major argument against this idea was that, although NMDA receptor antagonists effectively prevent development of the kindled state, high doses are required to block a fully kindled seizure. Even then, NMDA receptor antagonists only inhibit the behavioral expression of the seizure; they do not modify the abnormally low seizure threshold. Another negative argument was that the NMDA receptor changes might not be large enough to have much functional significance. In rebuttal, Heinemann pointed out that, since kindling enhances NMDA receptor density and possibly also the ease with which the receptor can be activated, one would expect NMDA receptor antagonists to be less potent against fully kindled seizures than against seizures in non-kindled animals. It may also be argued that kindling must produce only subtle changes in brain function, because the animals seize only in response to a strong tetanic stimulus, not spontaneously. Clearly, the role of NMDA receptors in kindling will be a matter of intense investigation in the years ahead.

A major theme of this conference was the importance of NMDA receptor activation for triggering developmental events. Nadler presented evidence that magnesium less effectively regulates NMDA channel opening in CAl pyramidal cells during development. Consistent with this idea, Brady and Swann and also Ben-Ari found that responses to NMDA in developing CA3 pyramidal cells are less voltage-dependent than in CA3 pyramidal cells from adult animals. Furthermore, Brady reported in a poster session that responses to NMDA during early development are depressed more by calcium than by magnesium. Mattson added that in his studies magnesium poorly blocked NMDA-induced cytotoxicity when hippocampal neurons were first put into culture, but became more effective as the culture matured. Thus NMDA receptors may change during cellular development from an immature, relatively magnesium-insensitive form to a mature form that is highly sensitive to magnesium.

Published evidence suggests that NMDA receptors on CAl pyramidal cells normally play little or no role in low frequency synaptic transmission. However, Wheal reported that EPSPs in these cells have a large NMDA receptor-mediated component one week after the destruction of CA3-derived afferent fibers. The surviving subpopulation of excitatory synapses may therefore utilize a different postsynaptic mechanism from the others, a mechanism that more prominently involves NMDA receptors. At least some of these residual synapses may be the local recurrent connections among CAl pyramidal cells reported on earlier in the conference by Thomson.

MARKERS OF AXON SPROUTING

Geddes and his colleagues have identified two genes that appear to be activated by lesions of hippocampal pathways. One protein, named by them SNAP-25, has a structure suggestive of a zinc-binding protein and its anatomical pattern of immunoreactivity parallels the Timm's stain. Denervation increases transcription of the mRNA for SNAP-25, as demonstrated by in situ hybridization. However, the mRNA for alpha-1 tubulin proved to be a better marker of denervation, because its level of expression in adult hippocampal neurons is normally lower than that of the SNAP-25 mRNA. Thus increases are more readily demonstrated. The consensus was that in situ hybridization of cDNA probes for protein markers such as these will aid in the analysis of excitotoxic lesions, especially in human disease.

Represa advanced the view that kainate receptor binding can serve as a marker for mossy fiber boutons. In particular, increased binding was suggested to occur in regions of the hippocampus that are invaded by mossy fiber sprouts after kindling or in human temporal lobe epilepsy. With respect to kindling, however, Nadler and McNamara noted that they had observed transient reductions in kainate binding rather than increases. This discrepancy may have resulted from labeling kainate receptors of different affinity. Kindling possibly alters kainate receptor conformation such that some receptors shift from a low affinity (30 nM KD) to a high affinity (12 nM KD) state. If this is so, however, then changes in kainate binding would not bear any simple relation to changes in the number of mossy fiber boutons.

REPAIRING EXCITOTOXIN LESIONS

The pioneering work of Bjorklund, Stenevi, Olson and their colleagues has stimulated hope that excitotoxic brain damage in humans, such as results from cerebral ischemia or temporal lobe epilepsy, can be repaired with transplants of fetal brain tissue. Zimmer's studies showed that hippocampal afferent fibers remain intact and in place for months after their target cells have been destroyed by an excitotoxin or an ischemic insult, that these fibers can innervate transplanted fetal neurons of the appropriate type and that at least some of the grafted fetal neurons can grow axons of a length sufficient to innervate distant target neurons in the host brain. Thus transplanted fetal hippocampal neurons appear to substitute in the host circuitry for homologous neurons that had been destroyed by the excitotoxic lesion. These results provide the clearest evidence to date that excitotoxic brain damage can be at least partially reversed with grafts of fetal material.

ANOXIA AND NMDA RECEPTORS

K. Krnjević

Anaesthesia Research Department, McGill University
Montréal, Québec

INTRODUCTION

Brief periods of anoxia cause a marked, but apparently fully reversible interruption of integrated brain function, whose cellular mechanism is not yet fully understood. For some 50 years it has been known that the hippocampus is one of the first brain regions to be affected by anoxia (Sugar and Gerard 1937). This disruption results from at least two major changes: one is a marked enhancement of K^+ conductance in pyramidal neurons, and a corresponding fall in excitability (Hansen et al. 1982; Misgeld and Frotscher 1984); another is sharp depression of Ca^{2+} currents (Krnjević and Leblond 1987; 1989). The latter may be of importance in explaining the block of synaptic transmission, as well as perhaps the early loss of cognitive function, in certain aspects of which hippocampal Ca^{2+} signals appears to play an essential role (eg. Teyler and DiScenna, 1987; Smith, 1987). A major component of the Ca^{2+} fluxes involved in long-term plastic changes follows activation of hippocampal NMDA receptors (Collingridge et al., 1983; McDermott et al., 1986); so it was of interest to see how anoxia and NMDA receptors might interact. This question could also be important in the light of much evidence that NMDA receptor-mediated Ca^{2+} influx is probably responsible for the selective necrosis of hippocampal pyramids (especially in CA1) induced by prolonged anoxia/ischemia (Choi, 1988; Siesjö and Bengtsson, 1989). This article briefly reviews some relevant experiments performed on hippocampal slices.

EXPERIMENTS ON CA1/CA3 PYRAMIDAL CELLS IN SUBMERGED SLICES FROM WISTAR RATS.

In these experiments (Krnjević et al. 1989), voltage-dependent Na^+ and K^+ currents and the inward rectifier (I_Q) were largely eliminated by exposing the slices to 0.5 M tetrodotoxin (TTX), 2 mM Cs^+ and 10 mM tetraethylamonium (TEA). In addition, in some cases the recording microelectrodes were filled with 3M CsCl. Under such conditions, large inward currents could be evoked by depolarizing membrane commands, as illustrated in Figs. 1B,C. As in previous experiments (Krnjević and Leblond 1987, 1989), these inward currents - especially those that were evoked from a holding potential (V_H) of -50 mV and inactivated relatively slowly - were sharply but reversibly depressed by short periods of anoxia (superfusion with saline saturated with 95% N_a and 5% CO_2).

NMDA receptors are not involved in the anoxic suppression of $Ca2+$ currents

Among the possible effects of anoxia is a release of glutamate (McIlwain, 1973; Lobner and Lipton, 1988). This might cause a sufficient influx of Ca^{2+} through NMDA channels (MacDermott et al., 1986; Ogura et al., 1988; Connor et al., 1988) to cause inactivation of the voltage-dependent Ca^{2+} currents. As shown by Fig. 1C, a prolonged application of the specific NMDA antagonist APV- at a dose that blocks the NMDA-evoked inward current (I_{NMDA}) - failed to prevent the anoxic suppression of the voltage-dependent inward currents.

Excitatory Amino Acids and Neuronal Plasticity
Edited by Y. Ben-Ari
Plenum Press, New York, 1990

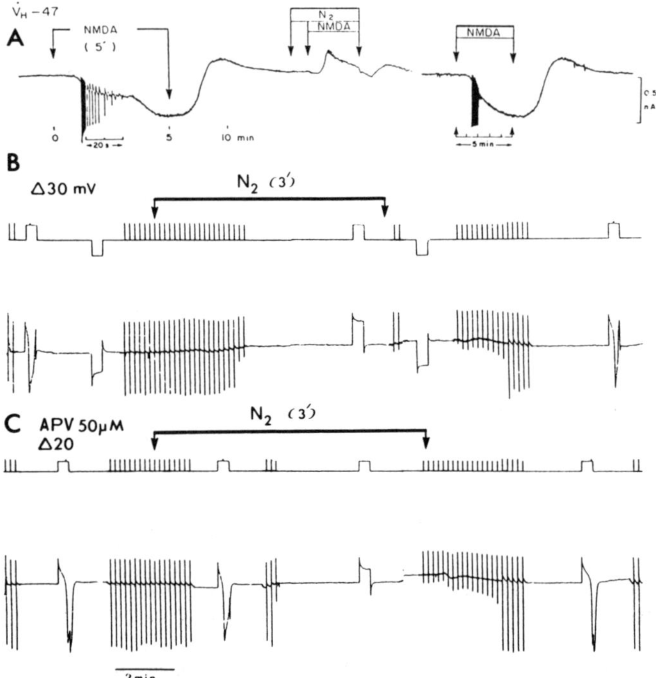

Fig. 1.   In a CA3 hippocampal neuron recorded with KCl-micro-electrode in **submerged** slice, anoxia abolished both NMDA-evoked and voltage-dependent TTX-insensitive inward currents. A: chart recording of currents evoked by 5 M NMDA; there is a slow inward shift, as well as transient inward currents near onset (shown at higher speed during lst NMDA application). When NMDA was reapplied during anoxia (middle), there was only a small outward shift. After a 3.5-min break, trace at right shows return of inward current during final NMDA application. B: as usual, anoxia suppressed inward currents evoked in same cell by 30 mV depolarizing pulses (also from V_H -47 mV). C: this effect of anoxia was not prevented by superfusion with 50 M APV. In **B** and **C**, top traces record 200-ms voltage steps (20 or 30 mV) and bottom traces membrane current changes; note accelerated traces at intervals (Fig. 3 from Krnjević et al., 1989).

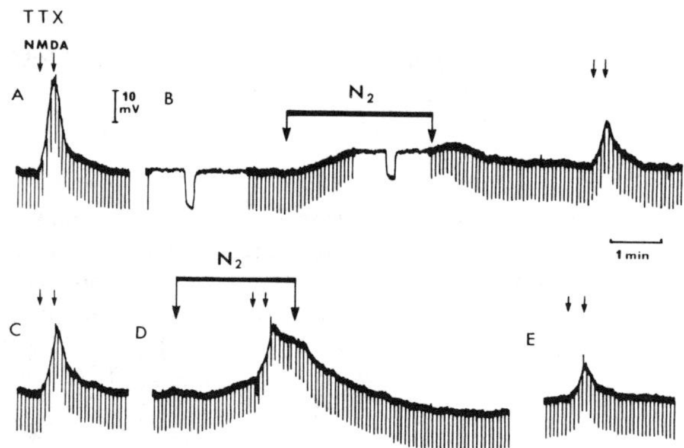

Fig. 2.   Anoxia can prolong the depolarizing effect of NMDA applications. Intracellular recording was in a CA1 neuron, in slice (from Sprague-Dawley rat) kept at 33.6°, in an "interface" chamber. Superfusing medium contained 1 M TTX throughout but no other blocking agent. Constant 0.5 nA hyperpolarizing pulses (200 ms) were applied at regular intervals to measure input resistance (R_N). NMDA was applied by iontophoresis (20 nA for 15 s, between small arrows). Initial trace (**A**) shows a characteristic large depolarization and apparent increase in R_N evoked by NMDA. In **B**, anoxia (substituting 95% N_2 for 95% O_2, while keeping CO_2 constant at 5%) produced the usual fall in R_N, as well as some depolarization - note resistance-measuring pulses displayed on accelerated paper traces before and during anoxia. In **D**, during anoxia there was a marked slowing of repolarization after the depolarizing effect of NMDA (now somewhat reduced in magnitude, cf. **C** and **A**). Note also the consistently weaker actions of NMDA during the post-anoxic phase (**B,E**).

Anoxia can block INMDA

In the presence of K channel blockers and at a V_H near -50 mV, NMDA induces a clear, TTX-insensitive inward current (Fig. 1 A). When the application of NMDA was repeated 90 sec after the onset of anoxia, NMDA-failed to evoke more than a trace of inward current. This effect was fully reversible after the return of "normal" oxygenation.

In submerged slices, anoxia produced only depressions of I_{NMDA}. This suggested that NMDA receptors, like voltage-dependent Ca^{2+} channels, may be inactivated either by an elevation of intracellular free Ca^{2+} (Eckert and Chad, 1984) or by inadequate phosphorylation (Chad and Eckert, 1986; MacDonald et al., 1989).

EXPERIMENTS ON NON-SUBMERGED SLICES FROM SPRAGUE-DAWLEY RATS

(unpublished observations by K. Krnjević and Y. Z. Xu)

The findings in interface-type chambers (Haas et al., 1979) were somewhat different. Even in slices that were not treated with K^+-channel blockers, NMDA-evoked depolarizations or inward currents were not necessarily diminished during anoxia. This can be seen in Fig. 2. In this example anoxia by itself had a mild depolarizing effect (B). When combined with an application of NMDA, the result was a prolongation of the latter's depolarizing action (D). Comparable or greater prolongations of I_{NMDA} were seen during voltage-clamping by a single electrode, even when NMDA by itself generated no detectable inward current. Moreover, if K^+ channels were blocked by intracellular Cs^+ (or extracellular TEA), anoxia could produce a dramatic potentiation of the magnitude or duration of the NMDA-evoked depolarizations (or inward currents), sometimes culminating in a spreading depression-like paroxysmal depolarization.

CONCLUSIONS

Although anoxia can produce both depression and facilitation of NMDA-induced depolarizations (or inward currents) - with the former predominant in submerged slices - the enhancement seen in non-submerged slices may throw some light on the role played by NMDA receptors in ischemic cell injury.

Because of the rapid and more efficient superfusion of submerged slices, agents such as K^+ and excitatory transmitters that are released during anoxia cannot accumulate as much or so rapidly as in non-submerged slices. In this respect, the latter more closely resemble ischemic tissue, which by definition has lost its normal circulation. One can therefore expect that in the ischemic brain, as in the non-submerged slices, anoxia may magnify Ca^{2+} influx triggered by the activation of NMDA receptors, and thus potentiate even at an early stage deleterious effects produced by the release of glutamate.

ACKNOWLEDGEMENTS

The experiments of K. Krnjević and Y. Z. Xu were financially supported by the Medical Research Council of Canada.

REFERENCES

Eckert, R., and Chad, J.E., 1984, Inactivation of Ca channels, **Prog. Biophys. Mol. Biol.**, 44: 215.

Chad, J.E., and Eckert, R., 1986, An enzymatic mechanism for calcium current inactivation in dialysed Helix neurones, **J. Physiol.**, 378: 31.

Choi, D.W., 1988, Calcium-mediated neuro-toxicity: relationship to specific channel types and role in ischemic damage, **TINS**, 11: 465.

Collingridge, G.L., Kehl, S.J., and McLennan, H., 1983, Excitatory amino-acids in synaptic transmission in the Schaffer collateral-commissural pathway of the rat hippocampus, **J. Physiol.**, 334: 33.

Connor, J.A., Wadman, W.J., Hockberger, P.E., and Wong, R.K.S., 1988, Sustained dendritc gradients of Ca^{2+} induced by excitatory amino acids in CA1 hippocampal neurons, **Science**, 240: 649.

Haas, H.L., Schaerer, B. and Vosmansky, M., 1979, A simple perfusion chamber. **J. Neurosci. Methods.** 1: 323.

Hansen, A.J., Hounsgaard, J., and Jahnsen, H., 1982, Anoxia increases potassium conductance in hippocampal nerve cells. **Acta. Physiol. Scand.**, 115: 301.

Krnjević , K., Cherubini, E., and Ben-Ari, Y., 1989, Anoxia on slow inward currents of immature hippocampal neurons, **J. Neurophysiol.**, 62, In press.

Krnjević , K., and Leblond, J., 1987, Anoxia reversibly suppresses neuronal calcium currents in rat hippocampal slices, **Can. J. Physiol. Pharmacol.**, 65: 2157.

Krnjević , K., and Leblond, J., 1989, Changes in membrane currents of hippocampal neurons evoked by brief anoxia, **J. Neurophysiol.**, 62: 15.

Lobner, D., and Lipton, P., 1988, Glutamate receptors and irreversible anoxic damage in hippocampal slices: mechanisms of interaction, **Soc. Neurosci. Abstr.**, 13: 647.

MacDermott, A.B., Mayer, M.L., Westbrook, G.L., Smith, S.J., and Barker, J.L., 1986, NMDA-receptor activation increases cytoplasmic calcium concentration in cultured spinal cord neurones, **Nature**, 321: 519.

MacDonald, J.F., Mody, I., and Salter, M.W., 1989, Regulation of N-methyl-D-aspartate receptors revealed by intracellular dialysis of murine neurones in culture, **J. Physiol.**, 414: 17.

McIlwain, H., 1973, Consequences of cerebral hypoxia examined at tissue-metabolic level. In, Biochemistry of Cerebral Anoxia, Hypoxia an Ischemia, Ed. M.M. Cohen, **Monogr. in Neural Sciences** (Karger, Basel), 1: 122-129.

Misgeld, U., and Frotscher, M., 1982, Dependence of the viability of neurons in hippocampal slices on oxygen supply, **Brain Res. Bull.**, 8: 95.

Ogura, A., Miyamoto, M., and Kudo, Y., 1988, Neuronal death in vitro: parallelism between survivability of hippocampal neurones and sustained elevation of cytosolic Ca^{2+} after exposure to glutamate receptor agonist, **Exp. Brain Res.**, 73: 447.

Siesjö B., and Bengtsson, F., 1989, Calcium fluxes, calcium antagonists, and calcium-related pathology in brain ischemia, hypoglycemia, and spreading depression: A unifying hypothesis, **J. Cereb. Blood Flow Metabol.,** 9: 127.

Smith, S.J., 1987, Progress on LTP at hippocampal synapses: a post-synaptic Ca^{2+} trigger for memory storage? **TINS,** 10: 142.

Sugar, O., and Gerard, R.W., 1938, Anoxia and brain potentials, **J. Neurophysiology,** 1: 558.

Teyler, T.J., and DiScenna, P., 1987, Long-term potentiation, **Ann. Rev. Neurosci.,** 10: 131.

MODULATION OF ATP SENSITIVE K+ CHANNELS: A NOVEL STRATEGY TO REDUCE THE
DELETERIOUS EFFECTS OF ANOXIA

Y. Ben-Ari

INSERM Unité 29, 123 Boulevard de Port-Royal, 75014 Paris

Present strategies to reduce the deleterious effects of anoxia are essentially centered on the use of agents that block the effect of excitotoxic release of glutamate . Since the anoxie lesion is thought to result from excessive Ca++ influx at least partly through the NMDA receptor channel complex - (see Choi 1988) most present attempts have been made on the use of NMDA antagonists. These strategies include 1) blockers of NMDA receptor channel complex, and 2) blockers of the glycine allosteric site of NMDA receptors in order to reduce the action of glutamate on these receptors.

These approaches which have clearly been shown to protect in experimental animals the hippocampus from anoxic dammage (references in chapters by Choi and Urban) are not without posing problems. Since NMDA receptors and Ca++ chanels are implied in a variety of important neuronal functions it is unlikely that potent NMDA or Ca++ channel blockers will be given on a prophylactic basis in patients susceptible to suffer from anoxic episodes. I would like to propose an alternative aproach which may provide this possibility, i.e. the use of agents acting on ATP sensitive K+ channels.

ATP sensitive K+ channels (K+ ATP) are inhibited by intracellular ATP (Noma 1983 ; Stanfield 1987), in fact they are sensitive to the ATP/ADP ratio. They constitute therefore a useful device to modulate the deleterious effects of anoxia. They are present in a variety of peripheral systems notably in the pancreas where they control insuline release (Ashcroft et al., 1984 ; Cook and Hales 1984) ; thus the hypoglycemic sulphonglureas which are used in the treatment of diabetes release insuline from B cells by clocking ATP-K+ channels (Sturgess et al., 1985 ; Schmid-Automarchi et al., 1987) ; this generates a Ca++ influx voltage dependent Ca++ channels and insuline release. Hyperglycemic hormones such as Galanine (De Weille et al., 1988a) but also somotostaline (De Weille et al., 1988b) or the hypotensive agent diazoxide (Aschroft 1988) open these channels. Since specific high affinity binding sites to Glibenclamide (Sturgess et al., 1985 ; Schmidt-Automarchi et al., 1987) a potent sulphonglurea have been found in the CNS (Mourre et al., 1988) as well as galanine inmunoreactivity (Melander et al., 1986), I have tested their effects on CA3 hippocampal neurons in slices using conventional intracellular physiological technique.

MATERIAL AND METHODS

Experiments were performed on CA3 hippocampal neurons recorded in slices obtained from adult male Wistar rats. Rats were anaesthetized with ether and decapitated. The brain was removed and submerged in artificial cerebrospinal fluid (ACSF) of the following composition (mM) : NaCl 126, KCl 3.5, CaCl 2.0, MgCl2 1:3, NaH2PO4 1.2, NaHCO3 25, glucose 11 (pH 7.3), and gassed with 95% O_2 and 5% CO_2. Slices, approximately 500 μm thick, were cut using a McIlwain tissue chopper and incubated at room temperature in ACSF for at least 1 h before use. Individual slices were transferred to a submerged type chamber and superfused with ACSF at 2.5-3 ml/min at 34°C (see Ben-Ari and Gho, 1988 for further details).

Intracellular recordings were made with 3 M KCl-containing microelectrodes (resistance of 60-150 MΩ). Current was passed through the recording electrode by means of an Axoclamp 2 amplifier. Bridge balance was checked repeatedly during the experiment and capacitative transcients with the electrode tip outside the neuron were reduced to a minimum by negative capacity compensation. Membrane potential was estimated from the potential observed upon withdrawal of the

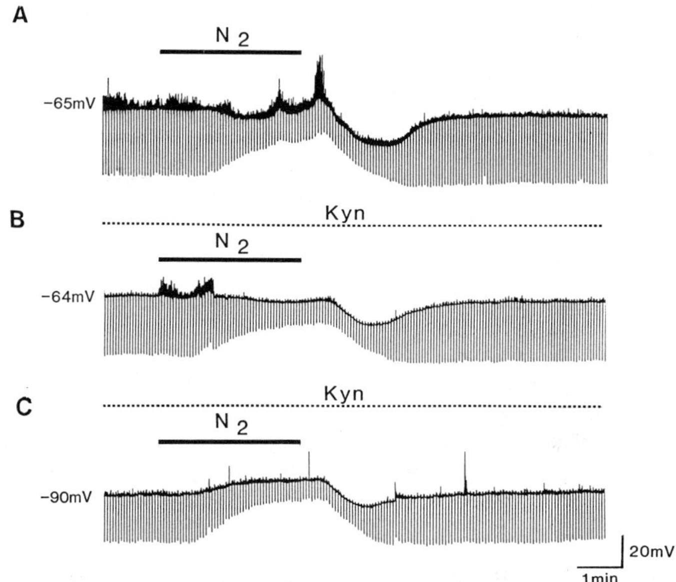

Fig. 1   Kynurenate reduces the anoxic depolarization. ABC data obtained from the same CA3 neuron; in this and following figures, the slices were exposed to anoxic episodes as indicated by the dark lines and downward deflections are hyperpolarizing potentials elicited by anodal current pulses (0.5 nA - 200 msec, 0.2Hz). A control ACSF, B, in presence of kynurenate (1mM); C the anoxic hyperpolarization shifted to the depolarizing direction at - 90 mV. KCl electrodes in this and following figures.

electrode from the cell. Stimulating electrodes (twisted bipolar NiCr insulated wires, 50 m o.d. or bipolar etched tungsten electrodes) were positioned in the hilar zone. Stimulation parameters were 50-100 μs duration. 5-50 V intensity, and 0.05 Hz frequency. Signals were digitized and displayed on a digital oscilloscope and on a computer driven chart recorder. Anoxia was produced by perfusion with an ACSF saturated with N_2 95%, CO_2 5% ; this produces a full blck of synaptic transmission in 2 min in CA1 and 3-6 min in CA3 (Cherubini et al., 1989). Anoxic tests were repeated at 10-15 min intervals. Drugs were dissolved in ACSF and superfused via a three-way tap system. Drugs used were tetrodotoxin (TTX, 1 μM, Sigma), kynurenic acid (1 mM, Sigma), glibenclamide (GLIB, 0.5-5 μM, gift of Dr. Lazdunski), and galanin (GAL, 1 μM, gift of Dr. Lazdunski).

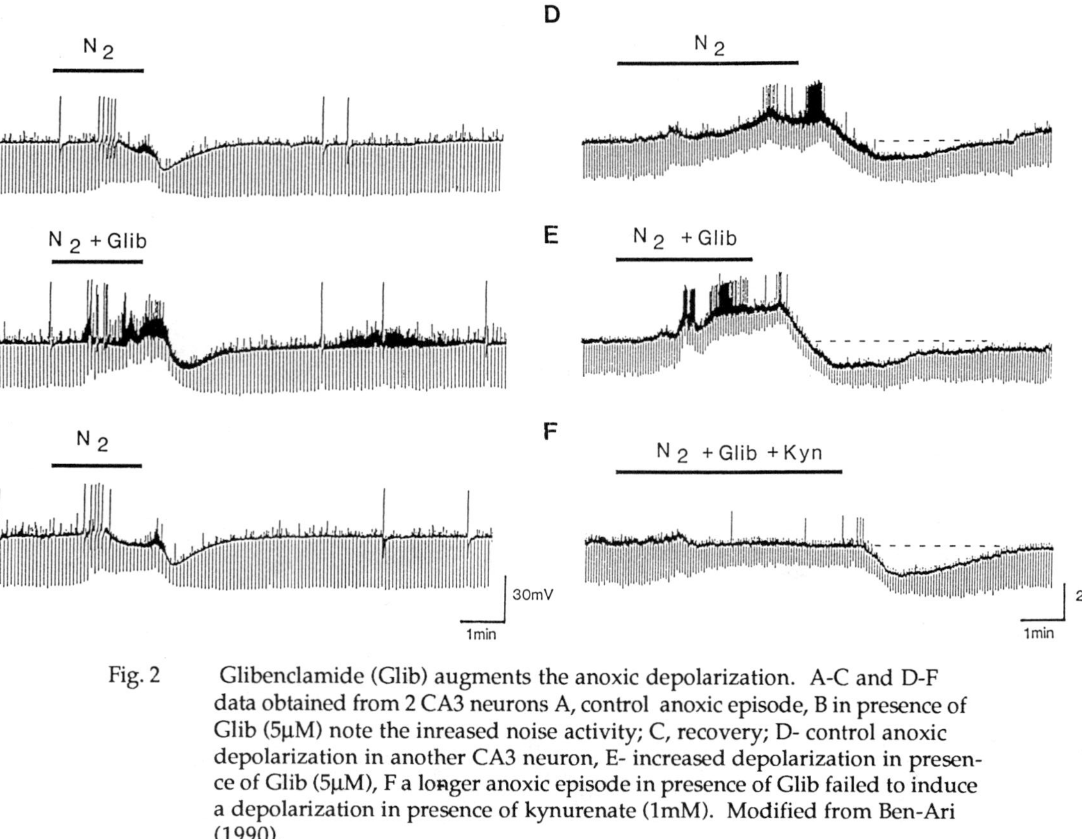

Fig. 2   Glibenclamide (Glib) augments the anoxic depolarization. A-C and D-F data obtained from 2 CA3 neurons A, control anoxic episode, B in presence of Glib (5μM) note the inreased noise activity; C, recovery; D- control anoxic depolarization in another CA3 neuron, E- increased depolarization in presence of Glib (5μM), F a longer anoxic episode in presence of Glib failed to induce a depolarization in presence of kynurenate (1mM). Modified from Ben-Ari (1990).

RESULTS

A brief anoxic episode induced in CA3 neurons in over 90% of the cases a depolarization and increase in synoptic (B1A, B2A, B3A, B4A). This was often preceeded by a small hyperpolarization and associated with a fall in input resistance (over 50% usually). Return to oxygenated solution an additional post-anoxic hyperpolarisation (ibid).

The anoxic depolarization is due to a release of glutamate since it was blocked by TTX (1μM) or the broad spectrum excitatory amino acid antagonist kynurenate (1mM). Anoxic episodes in presence of these agents induced only the hyperpolarization and the post anoxic hyperpolarization (Fig. 1A-C).

The anoxic hyperpolarization is due to the activation postsynaptycally of a K^+ conductance since it reversed polarity at approximately -75 -80 mV. A K^+ ATP conductance is not involved in this response since the amplitude or reversed potential was not affected by galanine (1μM) or glibenclamide (0.5-5 μM) which respectively open or close K^+ ATP channels (Fig. 5 see Ben-Ari 1990). The hyperpolarization is therefore most likely due to a Ca^{++} dependent K^+ inductance as suggested previously (see Hansen et al. 1982, Krnjevic 1975).

The *post anoxic hyperpolarization* was not investigated. However, as in CA1 neurons this is likely due to the activation of Na^+-K^+ ATPase electrogenic pump (see Hansen et al. 1982, Krnjevic et Leblond 1987, Fujiwara et al. 1987).

Gliberclamide augments anoxic depolarization

As shown in Fig. 2, bath application of GLIB '(O.5-5 μM) augmented consistently the synaptic noise and the anoxic depolarization. Thus Fig.2 A-C, the anoxic hyperpolarization which in this case predominated in the control anoxic episode was masked by the depolarizing effects of Glib which further increased the anoxic depolarization. The effects of Glib were prevented by TTX (not shown, see Ben-Ari 1990) or Kynureate (1mM, Fig. 2F, and Ben-Ari 1990) confirming that its effects do not take place directly at the postsynaptic level.

Galanine reduces anoxic depolarization

Bath applications of Galanine (Gal 1 μM) produced most frequently no change in membrane potential and input resistance; occasionally a small hyperpolarization was seen. In contrast Gal consistently reduced or blocked the anoxic depolarization, thus as shown in Figs. 3-B, 4-B, Gal reduced the depolarization when directly applied during the anoxic episode and fully blocked the depolarization when applied first in oxygenated then in anoxic ACSF.

Somatostatin and Diazoxide reduce anoxic depolarization

Somatostatin in contrast to galanine often had a hyperpolarizing effect in oxygenated ACSF associated with a fall in input resistance. In oxygen deficient ACSF, Somatostatin reduced or blocked like Galanine the anoxic depolarization even when it had no effect in oxygenated ACSF (not shown, see Ben-Ari et al. 1990).

Diazoxide - a relatively selective activator of K^+ ATP channels (Aschcroft 1988) - had little or no effects in oxygenated ACSF. In contrast of anoxic ACSF, diazoxide blocked the anoxic depolarization (see Fig. 6) thus revealed the anoxic hyperpolarization. Interestingly recovery from the effects of anoxia was slow (often more than 30 min.) Similar effects were seen in single electrode voltage clamp experiments (see Ben-Ari et al. 1990).

DISCUSSION

The sequence of events produced by a brief anoxic episode in CA3 neurons is reminiscent of that seen previously in CA1 hippocampal neurons (c.f. Hansen et al. 1987). It includes the activation of a K^+ conductance likely a Ca^{++} sensitive K^+ conductance (Hansen et al. 1982, Krnjevic 1975).

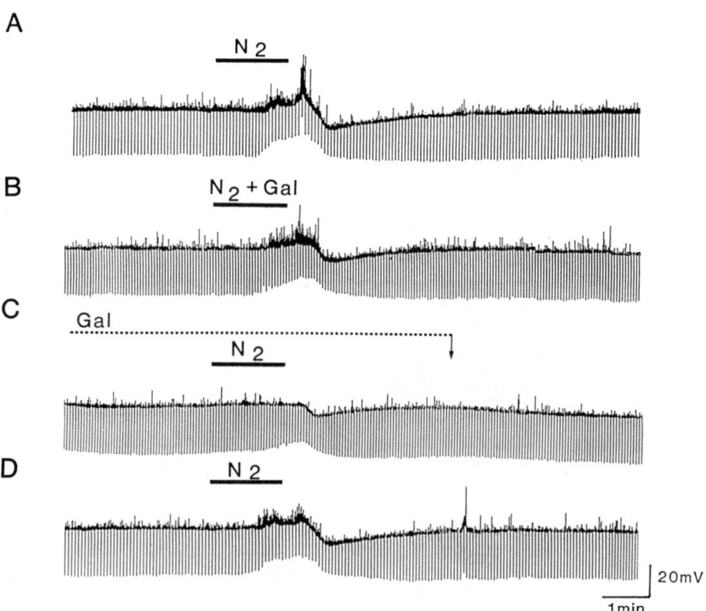

Fig. 3   Galanine (Gal) blocks the depolarizing effects of anoxia. A- anoxia (90s)
produced a depolarization and a post anoxic hyperpolarization B- galanine
(1mM) applied during the anoxic episode reduced the depolarization (and
the post anoxic hyperpolarization); C, continuous bath application of Gal in
oxygenated ACSF had no effect on membrane potential and input resistance,
the anoxic depolarization was fully blocked ; D recovery.- Adapted from Ben-
Ari (1990).

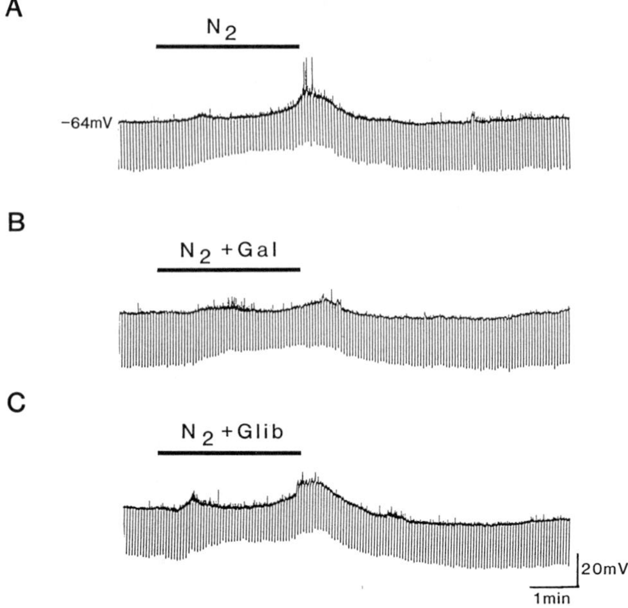

Fig. 4   Opposite effects of Gal and Glib on the anoxic depolarization in the same CA3 neuron. A : anoxia in control ACSF, B : in presence of Gal (1mM) C : in presence of Glib (5μM).

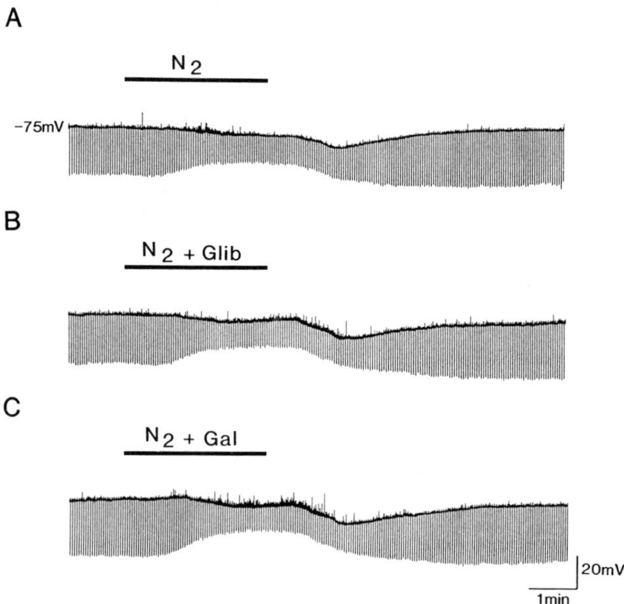

Fig. 5   Lack of effect of Glib and Gal on the anoxic induced in presence of TTX (1mM). Adapted from Ben-Ari (1990).

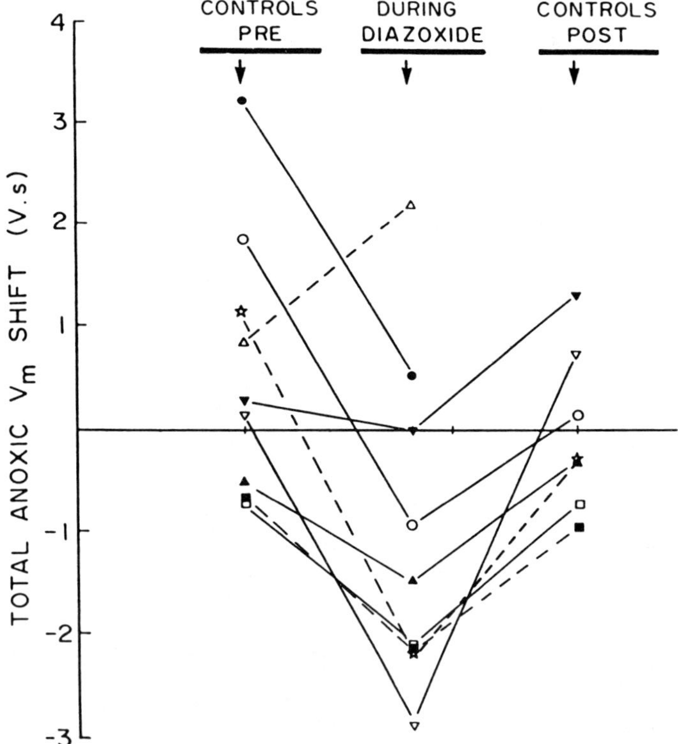

Fig. 6   Effects of Diazoxide (0.87mM) on the overall changes in Vm produced by brief anoxic episodes. Values were obtained by graphical integration of depolarizing and hyperpolaryzing shifts from the base line for 9 anoxic tests. Outward and inward current shifts were converted to voltage changes.- Adapted from Ben-Ari et al. (1990).

and not a K$^+$ATP one as suggested by Fujiwara et al. (1987). It also induces a post anoxic hyperpolarization likely due to the activation of an electrogenic pump (ibid).

The anoxic depolarization seen consistently in CA3 neurons is due to a release of glutamate since it is blocked by kynurenate; in fact a preliminary observation using a newly developped in vitro local recuparation technique (M.P. Roisin, J.L. Brassart, G. Charton, V. Crepel and Y. Ben-Ari in preparation) confirm that brief anoxic apisodes release endogenous glutamate. This depolarization was not consistently found in CA1 slices (Hansen et al. 1982, Fujiwara et al. 1982, Cherubini et al. 1989 and Krnjevic et al. 1989), at least with brief anoxic episodes suggestions important regional differences in the contribution of K$^+$ ATP channels in the control of transmitter release.

The present observations sugest that agents that open K$^+$ATP channels may be useful in reducing the enhanced transmitter - notably glutamate - release which occurs during anoxia. In spite of small differences Galanin, Somatostatin and Diazoxide all reduced the anoxic depolarization - the effects of Diazoxide are particularly interesting since this agent which has prominent hypotensive effect in the periphery is a selective activator of K+ ATP channels in contrast to the other agents which have been shown to also act on other conductances notably a reduction by galanine of voltage gated CA++ channels (see Tamura et al. 1988, Konop Ka et al. 1989) and the activation by somatostatin of an M type K+ channel (Pittman and Higgins 1981, Moore et al. 1988). However, the present observations indicate that in contrast to these actions, the effects of galanin, somatostatin and diazoxide on the anoxic response do not take place at the post synaptic level i.e. on the pyramidal neuron; this would explain why earlier studies in isolated cortical neurons on culture have failed to find a significant proportion of ATP-K+ channels (Ashford et al. 1988); also in CA1 hippocampal neurons (in slices) ATP-K+ channels modulators do not significantly alter the post-synaptic effects of brief anoxic episodes (Krnjevic and Leblond 1988).

I suggest that K+ ATP channel activation open K+ ATP channels located on the terminals, notably of the mossy fibers where a high density of glibenclamide binding sites have been found (Mourre et al. 1989). During anoxia the reduction in ATP (Kass and Lipton, 1982) and increase in ADP will enable the opening of K+ ATP channels and facilitate the action of K+ ATP channels activations. It is possible, in fact, that one of the roles of galanine and other endogenous peptides which act on K+ ATP channels is to prevent excitotoxic activation of synapses during deleterious conditions, the release of galanine acting as a protective mechanism. Whatever the exact mechanism, modulation of ATP-K+ channels may provide a novel strategy to modulate the anoxic response did reduce excessive transmitter release.

This alternative strategy may turn out to offer several advantages to the glut-NMDA blocker approach notably the fact that ATP-K+ modulators will act primarily during the anoxic episode and not in (control) oxygenated conditions; this may allow the development of prophylactic drugs to be given to patients who are susceptible to develop anoxic insults.

REFERENCES

Ashcroft, F.M. (1988) Adenosine 5'-triphosphate sensitive potassium channels. **Annu. Rev. Neurosci.** 11:97-188.

Ashford, M.L.J., Sturgess, N.C., Trout, N.J. Gardner, N.J., and Hales, C.N. (1988) Adenosine-5'-triphosphate-sensitive ion channels in neonatal rat cultured central neurones. **Pflügers Arch.** 412: 297-304.

Ben-Ari, Y. (1989) Galanin and glibenclamide modulate the anoxic release of glutamate in rat CA3 hippocampal neurons. **Eur. J. Neurosci.** In press.

Ben -Ari Y. and Gho, M. (1988) Long lasting modification of the synaptic properties of rat CA3 hippocampal neurones induced by kainic acid. **J. Physiol. (Lond.)** 404: 365-384.

Ben-Ari, Y. and Cherubini, E. (1988) Brief anoxic episodes induce long lasting changes in synaptic properties of rat CA3 hippocampal neurons. **Neurosci. Lett.** 90: 273-278.

Ben-Ari, Y. and Lazdunski, M. (1989) Galanin protects hippocampal neurons from the functional effects of anoxia. **Eur. J. Pharmacol.** 1989, 165, p. 331-332.

Ben-Ari, Y., Krnjevic, K. and Crepel, V. (1990) Activators of ATP sensitive K+ channels reduce anoxic depolarization in CA3 hippocampal neurons. **Neurosci.** in press.

Benveniste, H., Drejer, J., Schousboe, N., and Diemer, N.H. (1984) Elevation of the extracellular concentration of glutamate and aspartate in rat hippocampus during transient cerebral ischemia monitored by intracerebral microdial. **J. Neurochem.** 43: 1369-1374.

Cherubini, E., Ben-Ari, Y. and Krnjevic, K. (1989) Anoxia produces small changes in synaptic transmission membrane potential and input resistance in immature rat hippocampus. **J.N.P.**

Choi, D.W. (1989) Calcium-mediated neurotoxicity: relationship to specific channel types and role in ischemic damage. **TINS** 11: 465-469.

Cook, D.L. and Hales, C.N. (1984) Intracellular ATP directly blocks K+ channels in pancreatic B-cells. **Nature** 311: 271-273.

De Weille, J., Schmid-Antomarchi, H., Fosset, M., and Lazdunski, M. (1988a) ATP sensitive K+ channels that are blocked by hypoglycemia-inducing sulfonylureas in insuline-secreting cells are activated by galanin, a hyperglycemia-inducing hormone. **Proc. Natl. Acad. Sci. USA 85:** 1312-1316.

De Weille, J., Schmidt-Antomarchi, H., Fosset, M. and Lazdunski, M. (1988b) Regulation of ATP-sensitive K+ channels in insulinama role of cAMP. **Proc. Natl. Acad. Sci.** 86, 2971-2976.

Fujiwara, N., Higashi, H. Shimoji, K., and Yoshimura, M. (1987) Effects of hypoxia on rat hippocampal neurones in vitro. **J. Physiol. (Lond.)** 384: 131-151.

Hansen, A.J., Hounsgaard, J., and Jahnsen, H. (1982) Anoxia increases potassium conductance in hippocampal nerve cells: **Acta Physiol. Scand.** 115: 301-310.

Kass, I. and Lipton, P. (1982) Mechanisms involved in irreversible anoxic damage to the in vitro rat hippocampal slice. **J. Physiol. (Lond.)** 332: 459-472.

Kirino, T. (1982) delayed neuronal death in the gerbil hippocampus following ischemia. **Brain Res.** 239: 57-69.

Konopka, L.M., McKeon, T.W., and Parsons, R.L. (1989) Galanin-induced hyperpolarization and decreased membrane excitability of neurones in mudpuppy cardiac ganglia. **J. Physiol.** 410: 107-122.

Krnjevic, K. (1975) Coupling of neuronal metabolism and electrical activity. In: Ingvar, D. H. and Lassen, N.A. (cds) Brain Work pp 65-78. Munksgaard, Copenhagen.

Krnjevic, K. and LeBlond, K. (1987) Anoxia reversibly suppresses neuronal calcium currents in rat hippocampal slices. **Can. J. Physiol. Pharmacol.** 65: 2157-2161.

Krnjevic, K. and LeBlond, J. (1988) Are there hippocampal ATP-sensitive K channels that are activated by anoxia? **Eur. J. Physiol.** 411: R145.

Krnjevic, K., E.Cherubini, Y. Ben-Ari (1989) Anoxia in slow inward currents of immature hippocampal neurons. **J. Neurophysiol.** 1989. 62: 896-906.

Melander, T., Hökfelt, T., and Rökacus, A. (1986) Distribution of galanin-like immunoreactivity in the rat central nervous system. **J. Comp. Neurol.** 248: 475-517.

Mourre, C., Ben-Ari, Y., Bernardi, H., Fosset, M., and Lazdunski, M. (1989) Antidiabetic sulfonylureas: localization of binding sites in the brain and effects on the hyperpolarization induced by anoxia in hippocampal slices. **Brain Res.** 486: 159-164.

Noma, A. (1983) ATP regulated K+ channels in cardiac muscle. **Nature** 305: 147-148.

Pulsinelli, W. A., Brierley, J. B. and Plum, F. (1982) Temporal profile of neuronal damage in a model of transient forebrain ischemia. **Ann. Neurol.** 11: 491-498.

Rökacus, A. (1987) Galanin: a newly isolated biologically active neuropeptide. **TINS** 10: 158-164.

Schmid-Antomarchi, H., De Wzille, J., Fosset, M., and Lazdunski, M. (1987) The receptor for antidiabetic sulfonylureas controls the activity of the ATP-modulated K2 channel in insulin-secreting cells. **J. Biol. Chem.** 262: 15840-15844.

Smith, M.L., Auer, R. N., and Siesjö, B. K. (1984) The density and distribution of ischemic injury in the rat following two or ten minutes of forebrain ischemia. **Acta Neuropathol. (Berlin)** 64: 319-332.

Stanfield, P. R. (1987) Nucleotides such as ATP may control the activity of ion channels. **TINS** 10: 335-339.

Sturgess, N. C., Ashford, M. L. J., Cook, D. L., and Hales C.N. (1985) The sulfonylurea receptor may be an ATP-sensitive potassium channel. **Lancet** ii: 474-475.

Tamura, K., Palmer, J. M., Winkelmann, C. K., and Wood, J. D. (1988) Mechanismof action of galanin on myenteric neurons. **J. Neurochem.** 60: 966-979.

Trube, G. and Hescheler, J. (1984) Inward-rectifying channels in isolated patches of the heart cell and membrane: ATP-dependence and comparison with cell-attached patches. **Pflügers Arch.** 401: 178-184.

EFFECTS OF TRANSIENT FOREBRAIN ISCHEMIA IN AREA CA1 OF THE GERBIL

HIPPOCAMPUS: AN IN VITRO STUDY

L. Urban, K. H. Neill, B. J. Crain, J. V. Nadler and G. G. Somjen

Departments of Cell Biology, Pathology
Neurobiology and Pharmacology
Duke Medical Center, Durham, NC 27710

INTRODUCTION

Lasting loss of function in the CNS after episodes of cerebral ischemia is caused by neuronal cell death (Brierly and Graham, 1984). Among neurons the most vulnerable to transient ischemia are the CA1b pyramidal cells in the hippocampus. The response of these neurons to oxygen deprivation is rather unusual. Following a short period of temporary recovery the signs of degeneration appear gradually, 2-4 days after the ischemic episode (Kirino, 1982, Crain et al., 1988). This phenomen is known as "delayed neuronal death" (Kirino, 1982, Pulsinelli et al., 1982). It is hoped that experimental study of its mechanism may help to devise therapeutical interventions that could arrest the degenerative process. The delayed cell degeneration has been attributed to secondary failure of the circulation by some investigators (Ames et al., 1968) and to a slow, intrinsic neuronal process by others (Simon et al., 1984; Rothman and Olney, 1987). Cells vary greatly in their susceptibility to such damage, and this variability could also be explained either by vascular or by intrinsic cellular differences.

There is substantial evidence that elevated extracellular concentrations of excitatory amino acids damage neurons that possess the appropriate receptors (Olney, 1983). Similarities in cell damage induced by excitatory amino acids and by ischemia have led to the "excitotoxicity" hypothesis of ischemic degeneration (Rothman and Olney, 1986; Choi, 1988).

CA1 hippocampal cells receive their major excitatory inputs from pathways that release either glutamate or aspartate (Nadler et al., 1976; Corradetti et al., 1983; Burke and Nadler, 1988). In support of the excitotoxic hypothesis it has been reported that extracellular concentrations of glutamate and aspartate increase in the hippocampus during hypoxia and ischemia (Benveniste et al., 1984; Hagberg et al., 1985) and that toxic concentrations of excitatory amino acids initiate the delayed death of CA1b pyramidal cells. The damage caused either by excess excitatory amino acids or by hypoxia of cortical cells in culture was shown to be calcium dependent (Goldberg et al. 1986; Choi, 1987).

Both glutamate and aspartate are believed to exert their toxic action by activation of postsynaptic N-methyl-D-aspartate (NMDA) receptors (Nadler et al.,1981; Mayer et al., 1987; Murphy et al., 1987). Ischemic damage in area CA1 can be reduced by prior ablation of excitatory pathways to pyramidal cells (Wieloch et al., 1985; Onodera et al, 1986; Johansen et al., 1987; Jorgensen et al., 1987, Kaplan et al., 1989) or by the administration of NMDA antagonists (Boast et al., 1987, 1988; Gill et al., 1987). More remarkably, Schaffer collateral lesions or administration of NMDA antagonists provide protection even when performed after reperfusion (Johansen et al., 1987;Boast et al., 1988; Gill et al., 1988). These results imply that postischemic degeneration involves the acti-

Excitatory Amino Acids and Neuronal Plasticity
Edited by Y. Ben-Ari
Plenum Press, New York, 1990

vation of NMDA receptors during the period following reperfusion, and that by preventing this activation the degenerative process can be interrupted.

Contrary to what may be expected from the excitotoxic hypothesis of delayed neuron damage, no sign of either enhanced spontaneous neuronal activity or of seizures has been observed in area CA1 (Armstrong et al., 1989) during the period of reperfusion. Neuronal hyperactivity without seizures appears insufficient to destroy CA1 pyramidal cells in otherwise normal hippocampal formation (Vicedomini and Nadler, 1987).

The experiments we report now were designed to resolve some of the uncertainty generated by these apparently contradictory observations. We combined transient global cerebral ischemia with testing of neuronal function in the post-ischemic state by an *in vitro* technique, to test the hypothesis that there is enhanced excitatory synaptic transmission during the period after reperfusion. Experimentation *in vitro* eliminates the effects of systemic and vascular changes.

All experiments were performed on Mongolian gerbils. Transverse hippocampal slices were prepared either from control animals (intact or sham operated), or from gerbils which had undergone 5 minute bilateral carotid occlusion. To investigate early changes after reperfusion, hippocampi were obtained either immediately or within 1-4 hours after carotid occlusion, and recordings were then made at intervals for 11 to 16 hours. A second group of experiments was designed to assess course of the degeneration by both physiological and histological means. For this purpose hippocampal slices were prepared 1, 2, 3 and 10 days after carotid occlusion. All slices were maintained in an interface chamber at 36.5°C, as described in detail elsewhere (Urban et al., in press).

Evoked responses during the first 11-14 hours after ischemia

The first question is whether there are any physiologically detectable changes in area CA1 during the period between reperfusion and the appearance of cell degeneration. Repeated simultaneous recordings in the stratum radiatum for fEPSPs and in the pyramidal cell layer for population spikes evoked by Schaffer collateral stimulation revealed a transient enhancement of the fEPSP several hours after reperfusion (Fig.1). Compared to preparations taken from either intact or sham operated control animals, in slices prepared after carotid occlusion both the duration and the initial slope of the fEPSP increased at all stimulus intensities between 5 and 10 hours after ischemia.

 Although initially the amplitude of the orthodromic population spike increased also, this increase was short-lived compared to that of the fEPSP (Fig.1, 2). Moreover, even during the initially enhanced transmission, a given fEPSP always elicited a smaller population spike than in the control state. The initial increase in the population spike was followed by a rapid decline while the initial slope of the fEPSP was still double or greater compared to control, indicating that much larger EPSPs brought much fewer pyramidal cells to the firing threshold. The difference in the rates of deterioration between fEPSPs and population spikes is emphasized in Fig. 3.

In control animals the duration of the fEPSP was 16.6 ±0.4 msec (mean ±S.E.M.), and stayed constant for at least 7 hours. In contrast, the fEPSP duration measured in slices taken from experimental animals exceded the value of the control by 3 hr after the transient ischemia (20.1 ±0.9 msec) and increased progressively with time (Fig. 4). Linear regression revealed an average increase of 1.1-1.2ms/hr in fEPSP duration.

During the first 14 hours after reperfusion the amplitude of the antidromic population spike remained constant (Fig. 2). In the dentate gyrus synaptic transmission did not change in this period.

The data suggest that transient ischemia alters the synaptic efficacy of area CA1b in two ways: by enhancing fEPSP while at the same time depressing the excitability of the pyramidal cells. The enhanced fEPSP may be due to pre- and/or postsynaptic changes. The rapid down regulation of adenosine A1 receptors after transient forebrain ischemia (Lee et al., 1986) suggests an elevated transmitter release. The transient ischemia might enhance release by reducing the ability of adenosine to control the release process. The postsynaptic responsiveness could have been enhanced by activation of NMDA receptors. Normally, activation of this receptor adds a slow component to the fEPSP (Collingridge et al., 1988a,b). This effect could therefore be responsible for the increased duration of the fEPSP, but probably not its accelerated initial rate of rise after ischemia.

The reported ability of NMDA receptor antagonists to attenuate cell damage in area CA1 even when administered hours after ischemia suggests the possible involvement of NMDA receptors (Boast et al.,1988; Gill et al., 1988). To test their involvement, we administered the new specific NMDA receptor antagonist 3-((+/-)2-carboxypiperazin-4-yl)propyl-1-phosphonate (CPP) to hippocampal slices *in vitro*. CPP prevented both the increase in the duration of the fEPSP and the loss of the population spike of post-ischemic slices (Fig. 5).

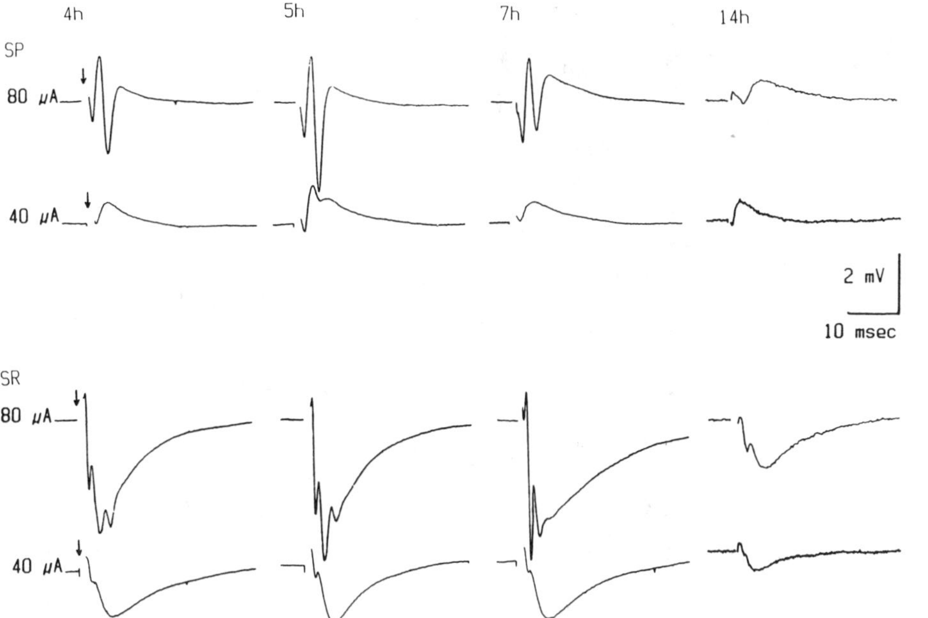

Fig. 1.   Orthodromic population spikes (SP) and fEPSPs (SR) recorded simultaneously at various times after a 5-minute carotid occlusion. Traces represent responses of CA1 pyramidal cells to Schaffer collateral stimulation at the indicated stimulus strengths. Note the transient enhancement of both responses and their consequent decline 14 hours after reperfusion. (From Urban et al.)

Unexpectedly, however, CPP enhanced the initial slope of the fEPSP and consequently the population spike. This CPP-induced increase was seen in slices taken from unoperated gerbils, and also in the earliest post-ischemic phase, before the expected onset of enhanced transmission seen in untreated post-ischemic slices. This effect cannot be explained by the known action of CPP, blockade of NMDA receptors. It could be suggested that the apparent enhancement by CPP of the "control" response is in fact due to its protection against post-hypoxic depression, because all hippocampal slices undergo a hypoxic period during their preparation and must therefore be considered "post-Hypoxic". It could also be that CPP has as yet undiscovered actions besides blocking NMDA receptors.

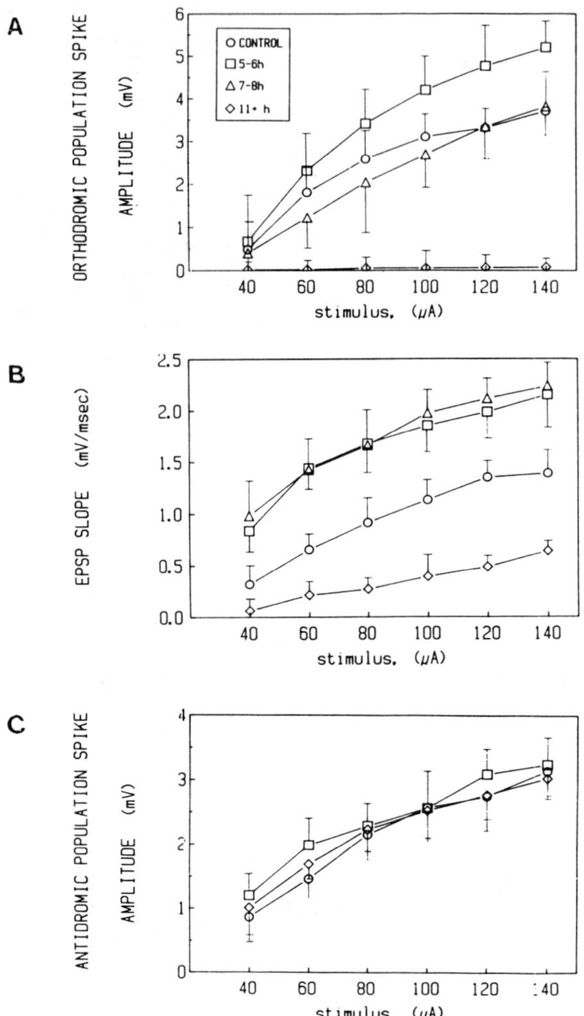

Fig. 2.   Effect of transient forebrain ischemia on the orthodromic and antidromic responses of CA1b pyramidal cells. Mean values of orthodromic population spike amplitude (A), fEPSP slopes (B) and antidromic population spike amplitude (C) are plotted as functions of stimulus intensity at various postischemic intervals. (From Urban et al.)

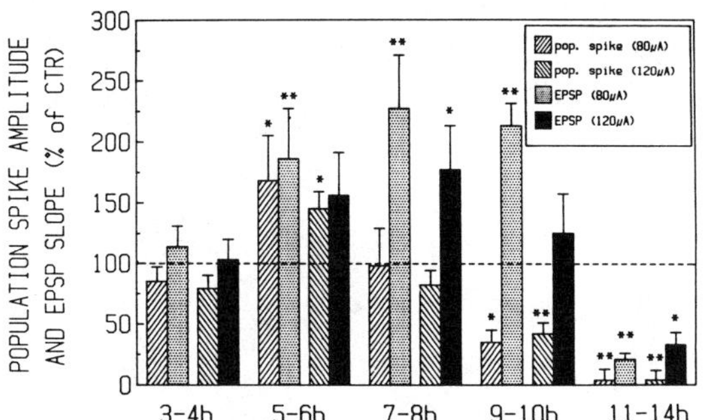

Fig. 3.   Statistical analysis of changes in evoked responses to Schaffer collateral-commissural stimulation in area CA1 after transient forebrain ischemia. Data were obtained with two representative submaximal stimulus currents. Each bar represents an averaged population spike amplitude or fEPSP slope expressed as a percentage (S.E.M.) of the control value recorded at an equivalent time. Two-way analysis of variance showed significant differences between experimental and control data ($p<0.01$ for both treatment and stimulus current). * $p<0.05$; ** $p<0.01$ (Dunnett test, df:29) (From Urban et al.)

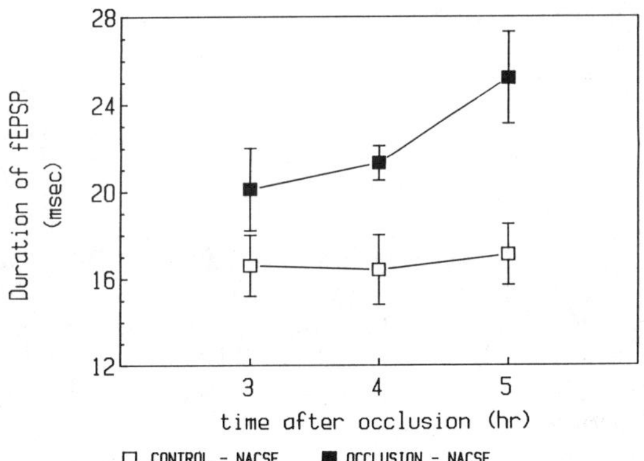

Fig. 4.   Bilateral carotid occlusion increases the duration of the fEPSP in area CA1b. Mean values of duration of the fEPSP (S.E.M.) were plotted as functions of time after reperfusion. The experimental slices were prepared immediately after the occlusion.

Loss of electrophysiological function and degeneration beyond 24 hours after transient ischemia.

Electrophysiological responses to both orthodromic and antidromic stimulation were markedly depressed in area CA1 1 day after transient carotid occlusion. All evoked responses disappeared in CA1 by the third day (Fig. 6). This finding is consistent with the histological observation of cell degeneration (Crain et al., 1988). The orthodromic population spike was affected much earlier than the antidromic (11-14 hr), which could reflect different safety factors for ortho- and antidro-

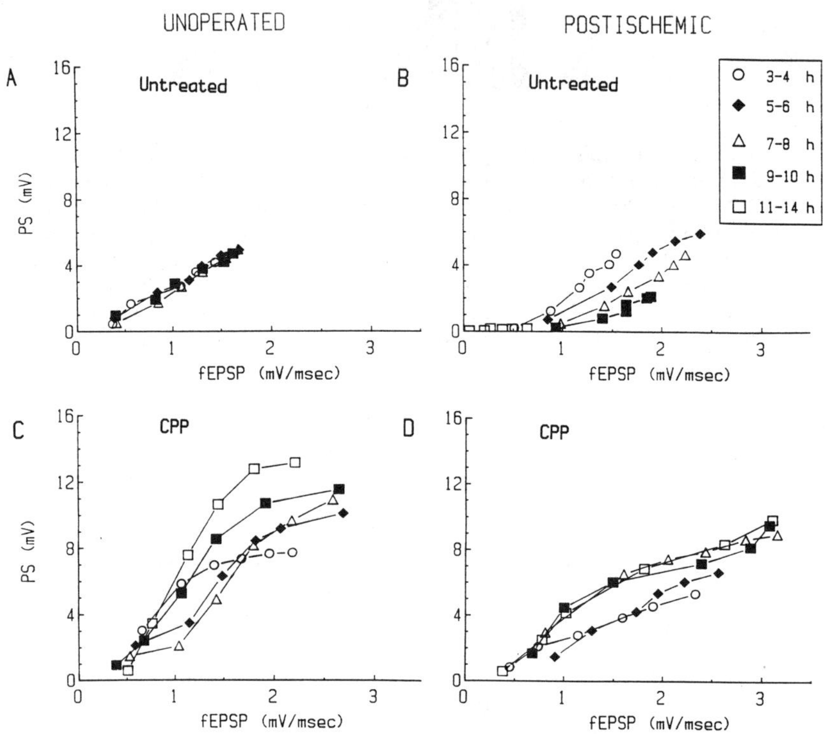

Fig. 5.   The progressive depression of pyramidal cell excitability (B) is prevented by 10 μM CPP perfused in the ACSF through the period of the in vitro recording (D). In both the control (C) and experimental (D) group the amplitude of population spike and the initial slope of fEPSP are enhanced by CPP. Input-output curves were constructed at various intervals after slice preparation (A, C) or carotid occlusion (B, D).

mic activation. The fact that pyramidal cells are still capable of generating impulses at a time when synaptic activation fails, justifies the hope that they could, in this stage, still be saved and their function recovered (Johansen et al., 1987;Boast et al., 1988; Gill et al., 1988). Indeed, the loss in pyramidal cell excitability could be a natural protective device against the excessive synaptic activation. It is noteworthy that in glucose deficient medium the ortho- and antidromic responses are similarly affected (Fan et al., 1988).

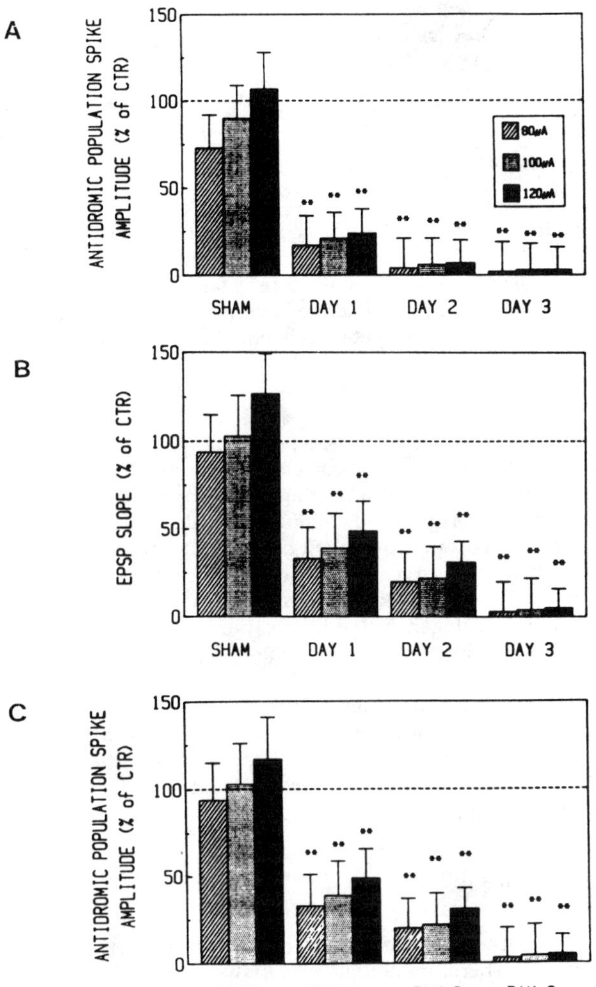

Fig. 6.   Quantitative analysis of evoked responses in area CA1b 1-3 days after transient ischemia. Each bar represents an averaged value of population spike amplitude (A, C) or initial slope of the fEPSP (B) expressed as in Fig. 3. Note the marked depression of evoked responses one day after the transient ischemia and the loss of virtually all evoked responses on day 3. **: p<0.01 (Dunnett test; df(A):119;df(B):87; df(C):71; after two way analysis of variance yielded p<0.01 for both treatment and stimulus current) (From Urban et al.)

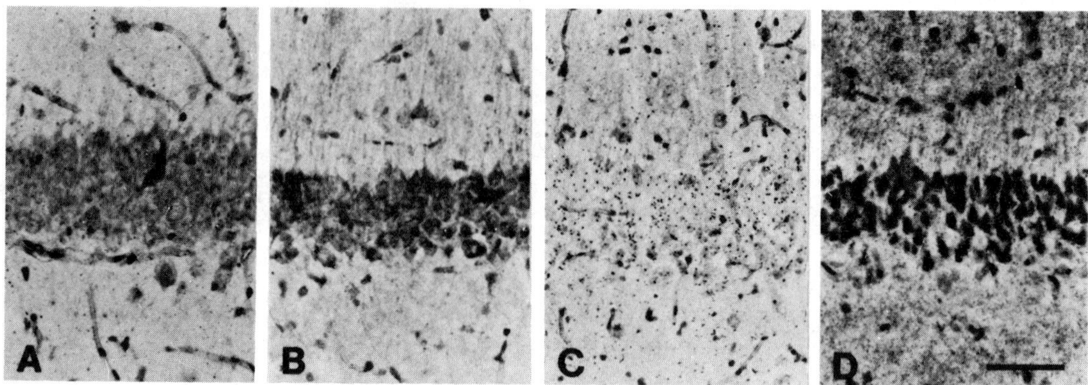

Fig. 7.   Silver impregnation of area CA1 from a control slice (A) and from slices cut 10(B), 24 (C) and 72hr (D) after transient forebrain ischemia. Cell bodies of the pyramidal cells are normally stained in A and B. In C the argyrophilic nucleoli stand against the poorly stained cytoplasm. In D massive argyrophilia marks the degenerating somata. Scale bar=100 um. (Urban et al.)

While orthodromic transmission is depressed before morphologic signs of degeneration set in, the loss of the antidromic population spike is correlated with the development of degenerative changes. 24 hours after ischemia, when the antidromic population spike declines, the nucleoli of the pyramidal cells show intense argyrophilia against the pale nuclei and cytoplasm (Fig. 7). The pallor may be the light microscopic equivalent of the disaggregation of polyribosomes and swelling of endoplasmic reticulum observed under electron microscope (Kirino and Sano, 1984; Petito and Pulsinelli, 1984). Recent studies demonstrated that CA1b pyramidal cells which were protected from ischemia-induced degeneration by ablation of excitatory pathways (Kaplan at al., 1989) show the same histological appearance for a prolonged period.

Complete loss of synaptic and antidromic activity was observed 2 days after occlusion, with consistant argyrophilia in CA1 pyramidal cell layer. This is in contrast to the relatively intact electrophysiological function and normal histological appearance of the dentate granule cell layer. In area CA1 only the presynaptic fiber volley could still be elicited, reflecting the resistance of the Schaffer collaterals to hypoxia.

SUMMARY AND CONCLUSIONS

Selective delayed post-ischemic degeneration of CA1b neurons takes place in tissue slices *in vitro* as it does in brain *in situ*. Therefore neither selectivity nor the delay of the process can be explained by vascular factors. Changes of orthodromic evoked potentials precede morphologic signs of degeneration, but antidromic activation of neurons fails *pari passu* with histopathologic degeneration. The marked, transient, enhancement of excitatory synaptic potentials is compatible with the idea that increased release of excitatory amino acids contributes to neuron damage. The fact that degeneration proceeds in the absence of spontaneous activity or overt electrographic seizures indicates, however, that increased excitation cannot be the sole cause of the damage. Postsynaptic excitability of neurons decreases even while synaptic potentials are enhanced. The mechanism of decreased excitability is not clear, but its development could be interpreted as a compensatory change, counteracting enhanced excitatory transmission. We confirmed that it is possible to save neurons by drug treatment administered after the ischemic insult, and demonstrated that such protection is not due to an effect on blood vessels.

These findings are relevant to the proposed clinical use of NMDA receptor antagonists to prevent ischemic brain damage (Meldrum, 1985; Rothman and Olney, 1986; Choi, 1988).

REFERENCES

Ames, A., Wright, R. L., Kowada, M., Thurston, J. M., and Majno, G., 1968, Cerebral ischemia: The no-reflow phenomenon, **Am. J. Pathol.**, 52:437-453.

Armstrong, D. R., Neill, K. H., Crain, B. J., and Nadler, J. V, 1989, Absence of electrographic seizures after transient forebrain ischemia in the Mongolian gerbil, **Brain Res.**, 476: 174-178.

Benveniste, H., Drejer, J., Schousböe, A, and Diemer, N.-H., 1984, Elevation of the extracellular concentrations of glutamate and aspartate in the rat hippocampus during transient cerebral ischemia monitored by intracerebral microdialysis, **J. Neurochem.**, 43: 1369-1374.

Boast, C. A., Gerhardt, S. C., and Janak, P., 1987, Systemic AP7 reduces ischemic brain damage in gerbils, in: "Excitatory Amino Acid Transmission," T. P. Hicks, D. Lodge and H. McLennan, eds., pp. 249-252, Alan R. Liss, New York.

Boast, C. A., Gerhardt, S. C., Pastor, G., Lehmann, J., Etienne P. E., and Liebman, J. M., 1988, The N-methyl-D-aspartate antagonists CGS 19755 and CPP reduce ischemic brain damage in gerbils, **Brain Res.**, 442:345-348.

Brierley, J. B., and Graham, D. I., 1984, Hypoxia and vascular disorders of the central nervous system, in: "Greenfield's Neuropathology," 4th edn, J. H. Adams, J.A.N. Corsellis and L. W. Duchen, eds., pp. 125-207, Edward Arnold, London.

Burke, S. P., and Nadler, J. V., 1988, Regulation of glutamate and aspartate release from slices of the hippocampal CA1 area: effects of adenosine and baclofen, **J. Neurochem.**, 51: 1541-1551.

Choi, D. W., 1987, Ionic dependence of glutamate neurotoxicity in cortical cell culture, **J. Neurosci.**, 7: 369-379.

Choi, D. W., 1988, Glutamate neurotoxicity and diseases of the nervous system, **Neuron,** 1: 623-634.

Collingridge, G. L., Herron, C. E., and Lester, R.A.J., 1988a, Synaptic activation of N-methyl-D - aspartate receptors in the Schaffer collateral-commissural pathway of rat hippocampus, **J. Physiol. (Lond.)**, 399:283-300.

Collingridge, G. L., Herron, C. E., and Lester, R.A.J., 1988b, Frequency-dependent N-methyl-D - aspartate receptor-mediated synaptic transmission in rat hippocampus. **J. Physiol. (Lond.)**, 399:301-312.

Corradetti, R., Moneti,G., Moroni, F., Pepeu G., and Wieraszko, A., 1983, Electrical stimulation of the stratum radiatum increases the release and neosynthesis of aspartate, glutamate and -aminobyturic acid in rat hippocampal slices, **J. Neurochem.**, 41:1518-1525.

Crain, B. J., Westerkam, W. D., Harrison, A. H., and Nadler, J. V., 1988, Selective neuronal death after transient forebrain ischemia in the Mongolian gerbil: a silver impregnation study. **Neuroscience**, 27:387-402.

Fan, P., O'Regan, P. A., and Szerb, J. C., 1988, Effect of low glucose concentration on synaptic transmission in the rat hippocampal slice. **Brain Res. Bull.**, 21: 741-747.

Gill, R., Foster, A. C., and Woodruff, G. N., 1987, Systemic administration of MK-801 protects against ischemia-induced hippocampal neurodegeneration in the gerbil, **J. Neurosci.**, 7:3343-3349.

Gill, R., Foster, A. C., and Woodruff, G. N., 1988, MK-801 is neuroprotective in gerbils when administered during the postischemic period, **Neuroscience.** 25: 847-855.

Goldberg, W. J., Kadingo, R. M., and Barrett, J. N., 1986, Effect of ischemia-like conditions on cultured neurons: protection by low Na^+, low Ca^{2+} solutions, **J. Neurosci.**, 6:3144-3151.

Hagberg, H., Lehmann, A., Sandberg, M., Nystrom, B., Jacobson, I., and Hamberger, A, 1985, Ischemia-induced shift of inhibitory and excitatory amino acids from intra-to extracellular compartments, **J. Cerebr. Blood Flow Metab.**, 5: 413-419.

Johansen, F. F., Jørgensen, M. B., and Diemer, N.-H., 1987, Ischemia-induced delayed neuronal death in the CA-1 hippocampus is dependent on intact glutamatergic innervation, in: "Excitatory Amino Acid Transmission," T. P. Hicks, D. Lodge and H. McLennan, eds., pp.245-248, Alan R. Liss, New York.

Kaplan, T. M., Lasner, T. M., Nadler, J. V., and Crain, B. J., 1989, Lesions of excitatory pathways reduce hippocampal cell death after transient forebrain ischemia in the gerbil, (Submitted for publication).

Kirino, T., 1982, Delayed neuronal death in the gerbil hippocampus following ischemia, **Brain Res.**, 239: 57-69.

Kirino, T. and Sano, K., 1984, Fine structural nature of delayed neuronal death following ischemia in the gerbil hippocampus, **Acta Neuropath. (Berl.)**, 62: 209-218.

Lee, K., Tetzlaff, W., and Kreutzberg, G., 1986, Rapid down regulation of hippocampal adenosine receptors following brief anoxia, **Brain Res.**, 380:155-158.

Mayer, M. L., MacDermott, A. B., Westbrook, G. L., Smith, S. J., and Barker, J. L., 1987, Agonist- and voltage-gated calcium entry in cultured mouse spinal cord neurons under voltage clamp measured using arsenazo III., **J. Neurosci.**, 7: 3230-3244.

Meldrum, B., 1985, Excitatory amino acids and anoxic/ischemic brain damage, **Trends. Neurosci.**, 8: 47-48.

Murphy, S. N., Thayer, S. A., and Miller, R. J., 1987, The effect of excitatory amino acids on intracellular calcium in single mouse striatal neurons in vitro, **J. Neurosci.**, 7: 4145-4158.

Nadler, J. V., Vaca, K. W.,White, W. F., Lynch, G. S., and Cotman, C. W., 1976, Aspartate and glutamate as possible transmitters of excitatory hippocampal afferents, **Nature**, 260: 538-540.

Nadler, J. V., Evenson, D. A., and Cuthbertson, G. J., 1981, Comparative toxicity of kainic acid and other acidic amino acids toward rat hippocampal neurons, **Neuroscience**, 6:2505-2517.

Olney, J. W., 1983, Excitotoxins: an overview, **in:** "Excitotoxins," K. Fuxe, P. Roberts and R. Schwarcz, eds., pp. 82-96, MacMillan Press, London.

Onodera, H., Sato, G., and Kogure, K., 1986, Lesions to Schaffer collaterals prevent ischemic death of pyramidal cells, **Neurosci. Lett.**, 68:169-174.

Petito, C. K. and Pulsinelli, W. A., 1984, Delayed neuronal recovery and neuronal death in rat hippocampus following severe cerebral ischemia: possible relationship to abnormalities in neuronal processes, **J. Cerebr. Blood Flow Metab.**, 4:194-205.

Pulsinelli, W. A., Brierley, J. B., and Plum, F., 1982, Temporal profile of neuronal damage in a model of transient forebrain ischemia, **Ann. Neurol.**, 11: 491-498.

Rothman, S. M. and Olney, J. W., 1986, Glutamate and the pathophysiology of hypoxic-ischemic brain damage, **Ann. Neurol.**, 19: 105-111.

Rothman, S. M., and Olney, J. W., 1987, Excitotoxicity and the NMDA receptor, **TINS**,10:299-302.

Simon, R. P., Swan, J. H., Griffiths, T., and Meldrum, B. S., 1984, Blockade of N-methyl-D-aspartate receptors may protect against ischemic damage in brain, **Science**, 226:850-852.

Urban, L., Neill, K., Crain B. J., Nadler J. V., and Somjen, G. G., in press, Postischemic synaptic physiology in area area CA1 of the gerbil hippocampus studied in vitro, **J. Neurosci.**

Vicedomini, J. P. and Nadler, J. V., 1987, A model of status epilepticus based on electrical stimulation of hippocampal afferent pathways, **Exp. Neurol.**, 96: 681-691.

Wieloch, T., Lindvall, O., Blomqvist, P., and Gage, F. H., 1985, Evidence for amelioration of ischemic neuronal damage in the hippocampal formation by lesions of the perforant path, **Neurol. Res.**,7:24-26.

ACUTE BRAIN INJURY, NMDA RECEPTORS, AND HYDROGEN IONS : OBSERVATIONS

IN CORTICAL CELL CULTURES

Dennis W. Choi[1], Hannelore Monyer[1], Rona G. Giffard[2], Mark P. Goldberg[1],
and Chadwick W. Christine[1]

Departments of Neurology[1] and Anesthesia[2] Stanford University Medical
Center Stanford, CA 94305, U.S.A.

INTRODUCTION

Excess stimulation of NMDA receptors by endogenous glutamate likely contributes to the
neuronal cell loss associated with several types of acute brain injury *in vivo* (Meldrum, 1985;
Rothman and Olney, 1987, Choi, 1988), including ischemia (Simon et al., 1984), hypoglycemia
(Wieloch, 1985), epilepsy (Labuyere et al., 1986) and trauma (Faden and Simon, 1988). Among the
experiments supporting this statement are those studying the controlled delivery of insults to
dispersed neuronal and glial cells in primary culture. Demonstration that a given pharmacological
manipulation is neuroprotective in such cultures establishes that a beneficial effect can be
produced directly on brain parenchyma, without involvement of systemic metabolism or
alterations in blood flow. While organizational features of the intact nervous system are not
expressed in cell culture, many intrinsic aspects of neuronal and glial cell behavior do appear to be
qualitatively preserved. In particular, basic mechanisms relevant to glutamate transmission and
glutamate neurotoxicity are present in cultured brain cells.

Hypoxia

A critical study implicating glutamate neurotoxicity in the hypoxic injury of cultured neurons
was provided by Rothman (1984), who showed that the broad spectrum glutamate antagonist D-
glutamylglycine could attenuate the hypoxia-induced injury of rat hippocampal neurons. We
found that the selective NMDA antagonist ketamine reduced hypoxic neuronal injury in
neocortical cultures derived from fetal mice (Weiss et al., 1986), and subsequently characterized
the neuroprotective effects of several NMDA antagonists against hypoxic injury (Goldberg et al.,
1987a; 1987b; 1988d).

In our experiments, cultures were transferred into a defined solution (Eagle's minimal
essential media with Earle's salts) and then simply placed in a low oxygen environment for
several hours at 37°. At the end of the desired period of oxygen deprivation, cultures were returned
to the normal culture incubator overnight prior to assessment of neuronal injury by phase-contrast
morphology, trypan blue exclusion, and efflux of lactate dehydrogenase (LDH) to the bathing
medium.

Typically, 6-8 h of exposure to hypoxia was needed to destroy most of the neurons in the dish;
glia were not damaged. Addition of any of several selective NMDA antagonists, including D-2-
amino-5phosphonovalerate, or MK-801, produced concentration-dependent reduction in resultant
neuronal injury. At high antagonist concentrations, injury was typically reduced to less than 20% of

untreated levels. This neuroprotection was robust but not absolute, as it was lost if the hypoxic exposure time was increased beyond 16 - 20 h.

The hypothesis that hypoxic induction of NMDA receptor-mediated injury reflects, at least in part, excess synaptic release of neurotransmitter glutamate, predicts that the injury might also be reduced by presynaptic approaches designed to reduce glutamate release from synaptic terminals. Two sets of experiments in cortical cultures support this prediction. First, adenosine agonists, which have a widespread depressant effect on synaptic release (Dunwiddie, 1985), can attenuate hypoxic neuronal injury *in vitro* (Goldberg et al., 1988c) as *in vivo* (Evans et al., 1988; von Lubitz et al., 1988). Second, restricting the availability of extracellular glutamine to neurons, a maneuver likely to reduce the synthesis of transmitter glutamate via the "glutamateglutamine cycle" (Bradford et al., 1978; Hamberger et al., 1979), protects cultured cortical neurons from hypoxic insult (Goldberg et al., 1988a). However, as both approaches are expected to affect cellular processes other than glutamate release, additional experiments will be required before the benefits observed in these experiments can be definitively attributed to alterations in presynaptic function.

Glucose deprivation

Just as *in vivo* (Wieloch, 1985), neurons in culture can be destroyed by glucose deprivation (Monyer et al., 1989). Characterization of this *in vitro* injury, carried out in the absence of altered oxygen supply or pH, suggests similarities to the injury induced by pure hypoxia. About 6 - 8 h exposure to glucose deprivation produces eventual neuronal cell degeneration without gross glial damage. This neuronal degeneration can be strongly, but not absolutely, prevented by NMDA antagonists. Some differences may exist between the neuronal injuries induced by glucose deprivation and by oxygen deprivation in culture, a not unexpected situation given the different implications of the two insults for cell metabolism, and the different patterns of brain injury induced by the two insults *in vivo* (Auer and Siesjo, 1988). For example, while restricting the supply of extracellular glutamine reduces NMDA receptor-mediated injury in both glucose deprivation and hypoxia, only in the former condition does this maneuver also increase overall neuronal injury (Monyer and Choi, 1989).

"Ischemia"

Combination of both oxygen and glucose deprivation ("ischemia" *in vitro*) triggers eventual neuronal degeneration about 10 times faster than either insult alone (Goldberg et al., 1988b). Experiments to date indicate that the neuronal injury resulting from 30 - 60 min exposure to combined oxygen and glucose deprivation can be largely prevent by NMDA antagonists. The speed of this paradigm was sufficient to allow direct testing of the hypothesis that resulting neuronal injury is dependent on extracellular Ca$^+$.

Extracellular ionic manipulations of the sort used to examine direct glutamate neurotoxicity (Choi, 1987) are not well tolerated by cortical cells for the several hours required to injure neurons by either oxygen or glucose deprivation alone, but are not toxic over the 30 - 60 min exposure times used with the combined insult. Under physiologic ionic conditions, this brief ischemia produced rapid neuronal cell swelling, and widespread neuronal degeneration by the next day. Removal of extracellular Ca^{++} accentuated the acute swelling, and perhaps as a result, also increased neuronal loss. If acute neuronal swelling was prevented by replacing Na$^+$ with equimolar choline, Ca^{++} removal then reduced late neuronal loss.

Trauma

Several recent studies have suggested that NMDA antagonists may reduce the brain or spinal cord damage induced by mechanical trauma (Faden and Simon, 1988; McIntosh et al., 1988; Faden et al., 1989; Hayes et al., 1989). While a portion of this protective action could be accounted for by a reduction of concurrent ischemic brain injury, a study in cortical cultures suggests that NMDA antagonist may also produce a beneficial effect against a purely mechanical insult (Tecoma et al., 1989). In that study, a plastic stylet was used to produce linear scratches in the cell layer, resulting in disruption down to the bare plastic dish surface. Immediately afterwards, neurons within 1-2 mm away from the scratch edge developed marked somatic swelling; over the next day, many of these neurons went on to degenerate. Addition of 10 -100 M D-APV or dextrorphan produced a reduction both in acute neuronal swelling, and subsequent neuronal degeneration.

Acidosis - Firebreak on the excitotoxic storm ?

Brain ischemia *in vivo* is accompanied acutely by extracellular acidosis, a derangement widely considered to contribute to resultant neuronal injury. In experimental animals, administration of sufficient glucose to worsen ischemic lactic acidosis is associated with potentiation of brain damage. However, arguing against a direct deleterious action of acidosis on ischemia neurons, recovery of neuronal function in hippocampal slices subjected to hypoxia was actually improved by addition of 10 mM lactic acid to the perfusate, sufficient to reduce pH to 6.8 - 6.9 (Schurr et al., 1988). A possible basis for a beneficial effect of acidosis on ischemic neuronal survival was provided by Morad et al. (1988), who found that moderate acidity (pH 6.6) could markedly reduce N-methyl-D-aspartate (NMDA) receptor-activated currents in cultured hippocampal neurons.

We found that moderate extracellular acidosis (pH 6.5) markedly reduced the inward whole cell current induced by NMDA on cultured cortical neurons; at pH 6.1 kainate-induced current was additionally reduced (Giffard et al., 1989). Furthermore, such acidosis reduced the cortical neuronal injury caused by toxic glutamate exposure, as well as the neuronal degeneration induced by combined oxygen and glucose deprivation (ibid.). While it remains possible that acidosis may have, additionally, deleterious effects in ischemic brain, these findings raise the possibility that moderate acidosis may act to decrease the intrinsic vulnerability of cortical neurons to acute injury by ischemia, and possibly by other insults as well.

Excitotoxicity is by nature a self-propagating event: damaged neurons release glutamate and induce further damage. Once triggered by acute insult, the innate ability of excitotoxicity to amplify and extend neuronal injury seems potentially unbounded. Endogenous protective mechanisms may thus play a critical role in damping out excitotoxicity and limiting resultant brain damage. It is intriguing to speculate that extracellular acidosis may be one such built-in "firebrake", appearing in the setting of relatively severe injury to stop further damage mediated by NMDA receptors. Thus the therapeutic utility of pharmacological NMDA antagonists may be bracketed between the initiation of excitotoxicity, and the onset of tissue acidosis. Rapid brain acidosis could conceivably contribute to a low utility of NMDA antagonists in certain settings, such as in global ischemia.

REFERENCES

Auer, R.N., and Siesjo, B.K., 1988, Biological differences between ischemia, hypoglycemia, and epilepsy, **Ann. Neurol.**, 24:699.
Bradford, H.F., Ward, H.K., and Thomas, A.J., 1978, Glutamine - A major substrate for nerve endings, **J. Neurochem.**, 30:1453.
Choi, D.W., 1987, Ionic dependence of glutamate neurotoxicity in cortical cell culture, **J. Neurosci.**, 7:369.
Choi, D.W., 1988, Glutamate neurotoxicity and diseases of the nervous system, **Neuron**, 1:623.
Dunwiddie, T.V., 1985, The physiological role of adenosine in the central nervous system, **Int. Rev. Neurobiol.**, 27:63.
Evans, M.C., Swan, J.H., and Meldrum, B.S., 1988, An adenosine analogue, 2-chloroadenosine, protects against long term development of ischaemic cell loss in the rat hippocampus, **Neurosci. Lett.**, 83:287.
Faden, A.I., Demediuk, P., Panter, S.S., Vink, R., 1989, The role of excitatory amino acids and NMDA receptors in traumatic brain injury, **Science**, 244:798.
Faden, A.I., and Simon, R.P., 1988, A potential role for excitotoxins in the pathophysiology of spinal cord injury, **Ann. Neurol.**, 23: 623.
Giffard, R.G., Monyer, H., Christine, C.W., and Choi, D.W., 1989, Acidosis reduces NMDA receptor activation, glutamate neurotoxicity, and oxygen-glucose deprivation neuronal injury in cortical cultures, **Brain Res.**, in press.

Goldberg, M.P., Monyer, H., and Choi, D.W., 1988a, Hypoxic neuronal injury in vitro depends on extracellular glutamine, **Neurosci. Lett.**, 94:52.

Goldberg, M.P., Monyer, H., and Choi, D.W., 1988b, Cortical neuronal injury in vitro following combined glucose and oxygen deprivation: ionic dependence and delayed protection by NMDA antagonists, **Soc. Neurosci. Abstr.**, 14:745.

Goldberg, M.P., Monyer, H., Weiss, J.W., and Choi, D.W., 1988c, Adenosine reduces cortical neuronal injury induced by oxygen or glucose deprivation in vitro, **Neurosci. Lett.**, 89:323.

Goldberg, M.P., Pham, P.C., and Choi, D.W., 1987a, Dextrorphan and dextromethorphan attenuate hypoxic injury in neuronal culture, **Neurosci. Lett.**, 80:11.

Goldberg, M.P., Viseskul, V., and Choi, D.W., 1988d, Phencyclidine receptor ligands attenuate cortical neuronal injury following N-methyl-D-aspartate exposure or hypoxia, **J. Pharmacol. Exp. Therap.**, 245:1081.

Goldberg, M.P., Weiss, J.W., Pham, P.C., and Choi, D.W., 1987b, N-methyl-D-aspartate receptors mediate hypoxic neuronal injury in cortical culture, **J. Pharmacol. Exp. Therap.**, 243:784.

Hamberger, A.C., Chiang, G.H., Nylen, E.S., Scheff, S.W., and Cotman, C.W., 1979, Glutamate as a CNS transmitter. I.Evaluation of glucose and glutamine as precursors for the synthesis of preferentially released glutamate, **Brain Res.**, 168:513.

Hayes, R.L., Jenkins, L.W., Lyeth, B.G., Balster, R.L., Robinson, S.E., Clifton, G.L., Stubbins, J.F., Young, H.F., 1988, Pretreatment with phencyclidine, an N-methyl-D-aspartate antagonist, attenuates long-term behavioral deficits in the rat produced by traumatic brain injury, **J. Neurotrauma.**, 5:259.

Labuyere, J., Fuller, T.A., Olney, J.W., Price, M.T., Zorumski, C., Clifford, D., 1986, Phencyclidine and ketamine protect against kainic acid-induced seizures and seizure-related brain damage, **Soc. Neurosci. Abstr.**, 12:344.

McIntosh, T., Soares, H., Hayes, R., Simon, R., 1988, The NMDA receptor antagonist MK-801 prevents edema and restores magnesium homeostasis after traumatic brain injury in rats, in: "Frontiers in Excitatory Amino Acid Research," E.A. Calalheiro, J. Lehmann, and L. Turski, eds., Alan R. Liss, New York.

Meldrum, B., 1985, Possible therapeutic applications of antagonists of excitatory amino acid neurotransmitters, **Clin. Sci.**, 68:113.

Monyer, H., Goldberg, M.P., Choi, D.W., 1989, Glucose deprivation neuronal injury in cortical culture, **Brain Res.**, 483:347.

Monyer, H., and Choi, D.W., 1989, Glucose deprivation neuronal injury in vitro is modified by withdrawal of extracellular glutamine, **J. Cereb. Blood Flow Metab.**, accepted.

Morad, M., Dichter, M., and Tang, C.M., 1988, The NMDA activated current in hippocampal neurons is highly sensitive to [H+] **Soc. Neurosci. Abstr.** 14:791.

Rothman, S., 1984, Synaptic release of excitatory amino acid neurotransmitter mediates anoxic neuronal death, **J. Neurosci.**, 4: 1884.

Rothman, S.M., and Olney, J.W. 1987, Excitotoxicity and the NMDA receptor, **Trends Neurosci.**, 10:299.

Schurr, A., Dong, W.Q., Reid, K.H., West, C.A., and Rigor, B.M., 1988, Lactic acidosis and recovery of neuronal function following cerebral hypoxia in vitro, **Brain Res.**, 438:311.

Simon, R.P., Swan, J.H., Griffiths, T., and Meldrum, B.S., 1984, Blockade of N-methyl-D-aspartate receptors may protect against ischemic damage in the brain, **Science**, 226:850.

Tecoma, E.S., Monyer, H., Goldberg, M.P., and Choi, D.W., 1989, Traumatic neuronal injury in vitro is attenuated by NMDA antagonists, **Neuron**, 2:1541.

Von Lubitz, D.K., Dambrosia, J.M., Kempski, 0., and Redmond, D.J., 1988, Cyclohexyl adenosine protects against neuronal death following ischemia in the CAl region of gerbil hippocampus, **Stroke**, 19:1133.

Weiss, J., Goldberg, M.P., and Choi, D.W., 1986, Ketamine protects culturedneocortical neurons from hypoxic injury, **Brain Res.**, 380: 186.

Wieloch, T., 1985, Hypoglycemia-induced neuronal damage prevented by an N-methyl-D-aspartate antagonist, **Science**, 230:681.

MECHANISMS OF EXCITATORY AMINO ACID NEUROTOXICITY

IN RAT BRAIN SLICES

John Garthwaite and Giti Garthwaite

Department of Physiology
University of Liverpool
Brownlow Hill, Liverpool L69 3BX, U.K.

INTRODUCTION

Excitatory amino acids (EAA), in addition to their roles in
neurotransmission and synaptic plasticity, have also been implicated in a
variety of neurodegenerative disorders of the central nervous system. The
work of Olney and colleagues who showed that glutamate and related
excitants are toxic to central neurones in vivo through acting on EAA
receptors (see Olney, 1984) coupled with the realization that EAAs
(glutamate in particular) are likely to be the principal excitatory
transmitters in the brain, provided the key theoretical link between brain
physiology and brain pathology. The most convincing experimental evidence
to date that the neurotoxic action of endogenous EAAs contributes to
neuronal loss in vivo is in the acute conditions of ischaemia (Simon et
al., 1984a; Gill et al., 1988), hypoglycaemia (Weiloch, 1985) and trauma
(Faden et al., 1989) where EAA antagonists have been shown to provide
significant protection. Their involvement in more chronic conditions such
as Alzheimer's disease, Huntington's chorea and Parkinsonism, which are
more difficult to model experimentally, is more speculative (Choi, 1988).

The focus of this article is on those mechanisms set in motion by
activation of EAA receptors which lead to neurodegeneration. As with
questions concerning EAA receptor pharmacology and physiology, in vitro
preparations have proved invaluable in addressing this problem. We have
used rat brain slices, mainly of two areas, the cerebellum and hippocampus.

METHODOLOGY

Although slices from rats of any age can be used, those from young
animals are preferred for experiments where prolonged alterations in the
ionic composition of the bathing medium are necessary because they tolerate
such conditions very well whereas adult slices are less hardy. A further
advantage of young slices is the relative paucity of excitatory nerve
terminals, the release of EAA from which could add unwanted complications
to studies of direct agonist toxicity.

When cut and incubated appropriately, the slices remain viable for up
to 24 h and, at both light and electron microscopic levels, the

Excitatory Amino Acids and Neuronal Plasticity
Edited by Y. Ben-Ari
Plenum Press, New York, 1990

preservation of the cells resembles closely that of their counterparts
fixed in situ. Thus, even subtle experimentally-induced changes can be
detected reliably using histological methods. The use of intact pieces of
functioning brain tissue, does have its drawbacks, one of which is that -
as in vivo - uptake mechanisms very effectively limit the diffusion of
glutamate and other transported amino acids through the interstitial space
(Garthwaite, 1985). This access limitation can be mitigated by exposing the
slices to high bath concentrations of the amino acids but this creates
additional problems associated with the establishment of large steady-
state concentration gradients across the tissue thickness (Garthwaite,
1985). Consequently, although some experiments have been done using
glutamate (see below) we have chosen a receptor-based approach in which the
toxicities of selective EAA receptor agonists are studied in preference to
those of putative endogenous toxins; no assumptions about the identities of
the latter are made.

The protocol used is straightforward: slices are cut and allowed to
rest for about 2 h. Excitants are added for 2-90 min and the slices are
then incubated in agonist-free medium for (usually) a further 90 min. This
recovery period was originally adopted in order to give reversibly-affected
cells time to recover and irreversibly-damaged cells time to become
unambiguously necrotic, as judged by conventional histological criteria
(Hajos et al., 1986). But, unexpectedly, with one class of agonist,
reversibly-affected cells can become necrotic during the recovery period
(see below).

DIFFERENTIAL VULNERABILITY

A distinctive feature of EAA neurotoxicity in vivo (Herndon & Coyle,
1977; Nadler et al., 1981; Kohler & Schwarcz, 1983), as well of
neurodegenerative disease (Siesjo, 1981), is that some neuronal types are
killed and some are not. This is also a obvious feature of the toxicity of
EAA receptor agonists in brain slices. In the young (8 day old) cerebellum,
for example, dividing, premigratory and migrating granule cells are
insensitive to all agonists tested; differentiating granule cells in the
internal layer are vulnerable to NMDA but not to kainate; Golgi cells are
vulnerable to kainate but not to NMDA; neurones in the intracerebellar
nuclei can be killed by both agonists whereas Purkinje cells at this age
are not permanently damaged by either (Garthwaite & Garthwaite, 1986a;
Hajos et al., 1986).

Clear differences in the toxic profiles of different agonists are
also seen in young hippocampal slices: NMDA kills all differentiating
neuronal types, quisqualate is more selective for CA3 pyramidal cells and
kainate is toxic only to a scattered population, probably of inhibitory
interneurones (Garthwaite & Garthwaite, 1989a). These results are
consistent with the effects of the agonists in vivo (MacDonald et al.,
1988; Silverstein et al., 1986; Franck & Schwartzkroin, 1984; Wolf &
Keilhoff, 1984).

The explanations for differential neuronal sensitivity are, in many
cases, not clearly known; some reasons are offered later in this article.
Nevertheless, the ability to identify specific target neurones of agonists
provides an important criterion for classifying EAA neurotoxins,
independently of any assumptions about their receptor selectivity. On this
basis, three different groups emerge: NMDA-like (NMDA, quinolinate,
glutamate), quisqualate-like (quisqualate, AMPA) and kainate-like (shown,
so far, only by kainate itself). [AMPA = α-amino-3-hydroxy-5-methyl-4-
isoxazolepropionic acid.]

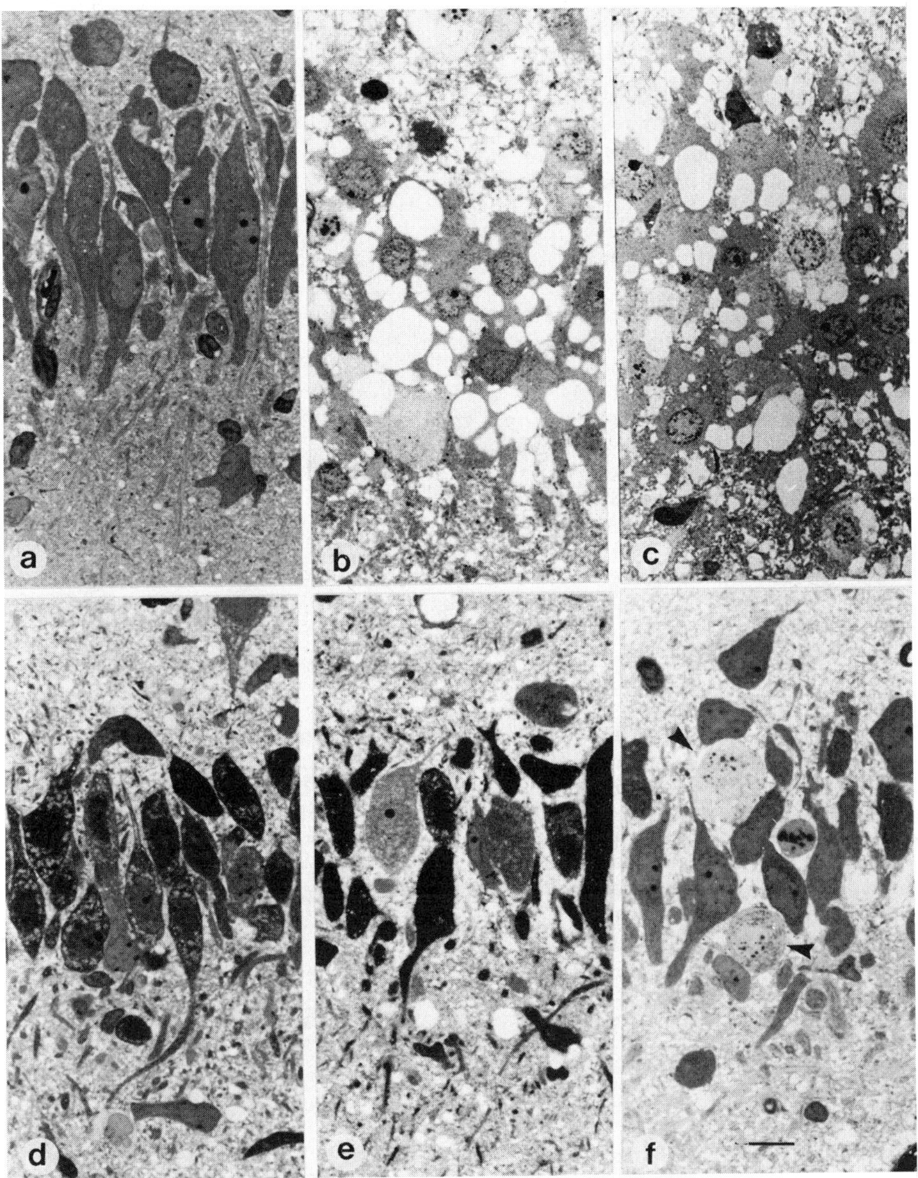

Fig. 1. Neurotoxicity of EAA receptor agonists in the CA3 region of young rat hippocampal slices: (a), control, (b) glutamate (3 mM), (c) NMDA (100 μM), (d) quisqualate (30 μM), (e) AMPA (100 μM) and (f), kainate (100 μM). The exposure periods were 30 min; scale = 10 μm.

Pathological distinctions

A further differentiation between agonist groups can be made on the basis of the pathological manifestations of their toxicity. As illustrated for CA3 cells in hippocampal slices in Fig. 1, NMDA and glutamate induce a striking oedematous degeneration of the neurones: the cell bodies and dendrites display massive swelling and vacuolation and, in the nucleus, the

chromatin is focally clumped (Fig. 1b,c). In the same neurones, quisqualate
and AMPA induce a prominent "dark cell degeneration" characterized by
darkly-stained, often condensed, cell bodies and primary dendrites and with
the neuronal cytoplasm encrusted with microvacuolations (Fig. 1d,e).
Another hallmark of quisqualate-like toxicity is an intense swelling of the
distal dendrites (Garthwaite & Garthwaite, 1986a, 1989a; Siman & Card,
1988). With kainate, the pathology is usually (but not always; Garthwaite &
Garthwaite, 1984, 1986a) of the oedematous type (Fig. 1f).

CONCENTRATION-EXPOSURE TIME INTERRELATIONSHIPS

With each agonist examined so far, the period of exposure needed to
induce irreversible damage is closely related to the concentration of
agonist used. This is illustrated in the case of NMDA toxicity in
cerebellar and hippocampal slices in Fig. 2, which plots the t_{50} (the
exposure time needed to kill 50% of the neurones) against the NMDA
concentration. The first point to be made is that NMDA receptor activation
can cause cell death quickly - within a few minutes - or slowly (the
underlying mechanism appears to be the same, see below). The speed with
which irreversible damage is inflicted depends on the NMDA concentration.
Secondly, over the low concentration range, small changes in concentration
can have a very large effect on the duration of the exposure needed to
effect cell death to an extent that, conceivably, several hours or more
could be needed (although this has not been explicitly tested). At the
higher concentrations the t_{50} becomes less dependent on concentration,
presumably as the toxic mechanism saturates.

Also apparent from this plot is that, whichever way the data are
viewed, hippocampal neurones (CA1, CA3 and dentate granule cells) are more
vulnerable than cerebellar granule cells, i.e. at a given concentration a
shorter exposure is needed, or for a given exposure period, a lower
concentration is needed. (The same holds for CA1 vs the other neurones in
hippocampal slices). These findings are consistent with the relative
densities of NMDA receptors on the different neurones (Cotman et al.,
1987). However, the magnitude of the difference in vulnerability appears to
vary with exposure time. Thus, with a 30 min exposure, a 2.5-fold higher
concentration is needed to kill 50% of cerebellar granule cells compared

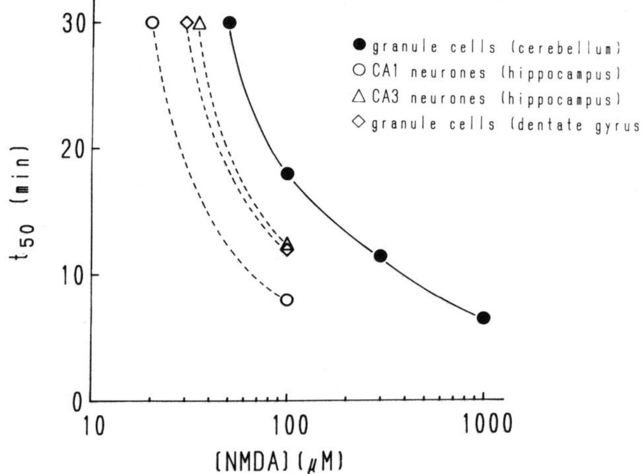

Fig. 2. Concentration-exposure time relationships for NMDA toxicity
in cerebellar and hippocampal slices; t_{50} = exposure time (min)
required to induce necrosis of 50% of the neurones. The data are
taken from Garthwaite (1989) and Garthwaite & Garthwaite (1989).

with CA1 neurones whereas with an 8 min exposure, the corresponding
concentration difference is nearly 10-fold. Although the reasons for this
are unclear, the implication is that, with brief exposures, the differences
in vulnerability between neurone types can become exaggerated.

DEVELOPMENTAL CHANGES IN VULNERABILITY

A certain degree of maturation is required before neurones become
vulnerable to EAA neurotoxicity. Cerebellar granule cells, for instance,
are resistant to NMDA toxicity until they have migrated from the germinal
layer to their destination in the internal granule cell layer (Hajos et
al., 1986). Unlike their somewhat more mature neighbours, the relatively
undifferentiated granule cells of the dentate gyrus present in the 8 day
old hippocampal formation are also unharmed by NMDA and quisqualate
(Garthwaite & Garthwaite, 1989a). In addition, glutamate neurotoxicity
gradually develops with time in fetal cerebral cortical neurones maintained
in tissue culture (Choi et al., 1987).

As the neurones mature, a number of changes can take place. In the
cerebellum, granule cells are only transiently susceptible to NMDA toxicity
(Garthwaite & Garthwaite, 1986); other neurones may also be particularly
vulnerable to NMDA during development, e.g. hippocampal pyramidal cells and
striatal neurones (MacDonald et al., 1988). Electrophysiological and
binding experiments have also suggested that the NMDA system in several
brain areas may be specially effective during the developmental period
(Tsumoto et al., 1987; Dupont et al., 1987; Garthwaite et al., 1987; Hamon
& Heinemann, 1988; Tremblay et al., 1988; Represa et al., 1989). Non-NMDA
receptor agonists, in contrast, generally become more effective toxins as
development proceeds. In vivo, this has been shown most clearly in the case
of kainate (Campochiaro & Coyle, 1978; Gaddy et al., 1979; Wolf & Keilhoff,
1984) and, with this agonist, the afferent innervation appears to play an
important part in the toxic mechanism (see Ben-Ari, 1985). Kainate also
becomes a more potent and a more generally toxic agonist in slices of both
cerebellum (Garthwaite & Garthwaite, 1986) and hippocampus (Garthwaite &
Garthwaite, 1989; Siman & Card, 1988) with maturation. Analogous results
have been obtained with quisqualate in both brain areas (Garthwaite &
Garthwaite, 1986 and unpublished observations).

MECHANISMS OF EAA NEUROTOXICITY

It was originally proposed that neurones would succumb when available
energy resources became exhausted and/or when lethal alterations in the
composition of the intracellular fluid had occurred, as a result of a
sustained increase in plasma membrane permeability (see Olney, 1978). In
other words, a ionic mechanism was suggested to be primarily responsible.
This idea was given greater specificity by Meldrum and colleagues
(Griffiths et al., 1984) who, drawing on pathological findings from a
number of peripheral tissues suggested that Ca^{2+} was the prime culprit and
that the neurones were destroyed because they became overloaded with this
ion. That Ca^{2+} is a required ion for NMDA toxicity has now been
convincingly supported by results using brain slices and some tissue
culture models. More limited data suggest the same is true for kainate
toxicity. But recent evidence implicates a different and more complex
mechanism for quisqualate and AMPA.

NMDA Neurotoxicity

In both cerebellar and hippocampal slices, NMDA toxicity is abolished
when Ca^{2+} is omitted from the bathing solution (G. Garthwaite et al., 1986;

Garthwaite & Garthwaite, 1986b, 1989b). The notion, derived from studying
dispersed cultures of fetal neurones, that there are two distinct toxic
mechanisms - a rapid one caused by Cl^--dependent osmotic lysis and a
slower, Ca^{2+}-dependent one (Rothman et al., 1987; Choi, 1987) - has not
been supported: in intact brain slices, NMDA (and other agonists) can
elicit irreversible neuronal damage quickly or slowly (depending on agonist
concentration, see Fig. 1) but the mechanism is Ca^{2+}-dependent and Cl^--
independent, irrespective of the exposure time used. The Cl^-- (and Na^+-)
dependent osmotic mechanism may thus be an artifact stemming from the lack
of physical constraints on swelling in the abnormal 2-dimensional tissue
culture environment.

Intracellular Ca^{2+} overloading. A technique which renders visible by
electron microscopy the subcellular sites which accumulate large amounts of
Ca^{2+} was used to test the hypothesis that the neurones become overloaded
with this ion (Garthwaite & Garthwaite, 1986c). The results showed a
remarkable correlation between the sites of Ca^{2+} accumulation and the loci
of the cytopathological changes taking place during the progress towards
the stage of irreversible damage (Fig. 3). The first obvious event is the
amassing of Ca^{2+} in the Golgi apparatus, and this coincides with the
swelling of these organelles. Then, Ca^{2+} starts to collect in the nucleus
and, at the same time, the chromatin starts to aggregate. At this stage,
Ca^{2+} does not appear in the cytoplasm in quantities high enough to be
detected with this technique and if the agonist is removed, the neurones
are able to recover. The onset of irreversible damage is signalled by the
additional accumulation of Ca^{2+} in populations of mitochondria and these
start to swell. At about the same time, large quantities of Ca^{2+} - probably
reaching a concentration of 100 μM or more - appear free in the cytoplasm.
This stage corresponds to the neurones being irreversibly damaged: if the
agonist is removed, the neurones become necrotic. Exposure of the slices to
NMDA in the absence of added Ca^{2+} in the medium not only prevents the
neurones becoming irreversibly damaged but also prevents the pathological
changes in the Golgi apparatus, nucleus and mitochondria, implying - since
other ionic and metabolic changes associated with the depolarizing effect
of NMDA persist in these conditions (J. Garthwaite et al., 1986) - that
these structural changes are caused specifically by Ca^{2+}.

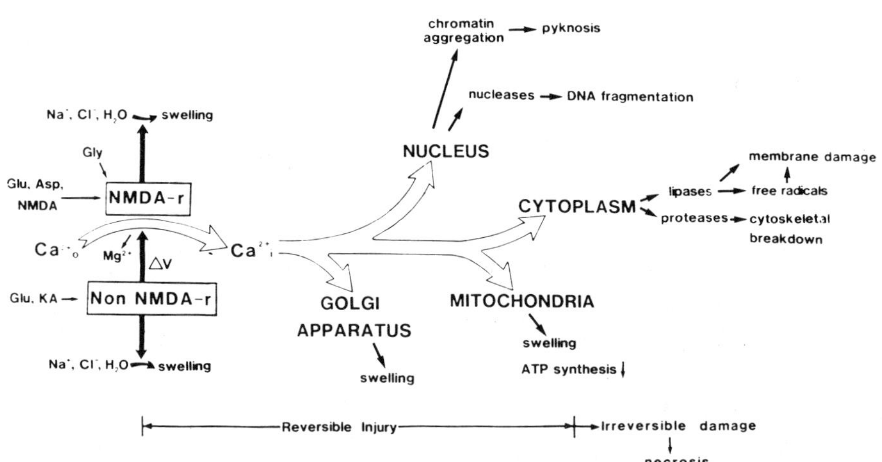

Fig. 3. Progress to the stage of irreversible neuronal damage
following sustained influx of Ca^{2+} through NMDA receptor (NMDA-r)
channels and its potentiation by depolarizing the cells through
activation of non-NMDA receptors (non-NMDA-r). Adapted from
Garthwaite (1989).

As to the route of lethal Ca^{2+} entry, the available evidence points to the NMDA receptor channels being chiefly responsible: the Ca^{2+} conductance of these channels is high (Mayer et al., 1987); depolarizing the neurones for an equivalent time with high K^+ (so inducing Ca^{2+} influx through voltage-sensitive Ca^{2+} channels) does not cause toxicity (J. Garthwaite et al., 1986) and the characteristics of toxic mechanism are as predicted from the properties of the NMDA channel: i.e. NMDA toxicity is potentiated in nominally Mg^{2+}-free medium or by simultaneously depolarizing the neurones with K^+ or kainate (Garthwaite & Garthwaite, 1987).

The precise mechanisms set in motion by high intracellular Ca^{2+} levels are still unclear. High Ca^{2+} in the nucleus can activate nucleases, leading to DNA fragmentation; uptake of Ca^{2+}, together with phosphate and water, into mitochondria leads to swelling and loss of ATP synthesis; high cytoplasmic Ca^{2+} activates proteases, which can destroy the cytoskeleton and affect the activity of several enzymes, and lipases, which can cause membrane damage and the production of damaging free radicals (Campbell, 1983; Siesjo, 1981; Siesjo & Bengtsson, 1989). The impression gained from examination of the neuronal ultrastructure is that a wide range of interrelated destructive events which affect all the cellular and subcellular compartments are simultaneously activated (Hajos et al., 1986); it remains to be determined which, if any, of these is of particular importance to the transition to irreversible damage and, hence, to determine if this transition is preventable by means other than removal of Ca^{2+}.

Recently, another facet of NMDA toxicity has been found, namely that brief, non-lethal exposure of neurones to NMDA can render them hypervulnerable to subsequent receptor activation for a long time (at least 1 h) after the initial exposure (Garthwaite, 1989). The reason for this is not yet clear. One possibility is that it takes some time for the cells to expel the Ca^{2+} accumulated during the initial exposure and so their capacity to tolerate a second Ca^{2+} load is automatically impaired. Activation of second messenger systems which increase Ca^{2+} influx (Connor et al., 1988) could also be involved.

<u>Kainate Neurotoxicity</u>

In young cerebellar slices, kainate (up to 100 μM) does not kill the cells that are vulnerable to NMDA, the granule cells, but it can very rapidly (t_{50} = 5 min at 100 μM) kill Golgi neurones, which resist NMDA toxicity (Hajos et al., 1986). As with NMDA, however, kainate toxicity towards Golgi cells is Ca^{2+}-dependent. Rather surprisingly, the Ca^{2+} concentration needed to be reduced only to 300 μM to prevent the damage caused by supramaximal exposures to kainate (Garthwaite & Garthwaite, 1986b). Kainate appears to induce Ca^{2+}-influx through two main routes: the receptor channels and the voltage-sensitive Ca^{2+} channels (Murphy & Miller, 1989). The relative importance of these routes regarding kainate toxicity is not entirely clear but since the toxicity is not replicated by depolarization with K^+ (J. Garthwaite et al., 1986) it is probable that Ca^{2+} influx through the receptor channels is a necessary event.

A puzzling question is, why is the toxicity of kainate restricted to this small population of neurones? Granule cells, for instance, are certainly capable of being killed (by NMDA) and they also possess abundant kainate receptors to judge from the high depolarizing efficacy of this agonist on these neurones (J. Garthwaite et al., 1986). Accordingly, morphological studies of the changes occurring after addition of kainate show that during the first few minutes, granule cells become highly swollen: their cell diameters increase by 50%, which corresponds to a 3-fold increase in volume. With continued exposure, however, this swelling subsides to the extent that, by 30 min (a time when Golgi neurones are

disintegrating), the granule cells almost regain their normal size (Hajos
et al., 1986). This behaviour is suggestive of some form of desensitization
taking place; but kainate receptors, reputedly, do not desensitize (Kiskin
et al., 1986).

An alternative explanation is that a kainate inhibitor is liberated.
We would like to raise the possibility that the inhibitor is glutamate,
released from the damaged Golgi cells and, perhaps, elsewhere. Three pieces
of circumstantial evidence support this proposal. Firstly, glutamate is
known to inhibit neuronal kainate responses (Ishida & Neyton, 1985; O'Brien
& Fishbach, 1986; O'Dell & Christensen, 1989). It does so because, though
active at the same receptor type, it has a much lower efficacy than
kainate. In young cerebellar slices, glutamate is a potent inhibitor of the
kainate-induced increase in cyclic GMP levels (IC_{50} = 5 μM; Garthwaite et
al., 1989). This cGMP response is predominantly a reflection of the
activation of granule cells (J. Garthwaite & Garthwaite, 1987). Secondly,
if glutamate is being released and was gaining access to granule cells in
μM concentrations, it would be predicted that a stimulation of cGMP
synthesis by the action of glutamate on granule cell NMDA receptors
(Garthwaite, 1985) would be detectable. Indeed, the cGMP response to
kainate displays a rapid, transient phase and a slow, more sustained, phase
and this slow component (but not the fast one) can be blocked by NMDA
antagonists (Garthwaite, 1982). Finally, if Golgi cells – a putative source
of the glutamate – are prevented from being killed, then glutamate should
not leak out and inhibit the action of kainate on granule cells. As
mentioned above, the toxic effect of kainate on Golgi cells can be
prevented by using media deficient in Ca^{2+}. When slices are exposed to
kainate in these conditions, the swelling of granule cells does not subside
with time but becomes greatly exaggerated (J. Garthwaite et al., 1986).

Thus, a tentative model for the differential effects of kainate in
young cerebellar slices (Fig. 4) is that the killing of one neuronal
population (Golgi cells) and the subsequent leakage of glutamate from these
cells, prevents other neurones from being killed; the inhibitory effect of
kainate on glutamate uptake (Johnston et al., 1979) should aid this
process. It should be noted, however, that damaged Golgi cells may not be

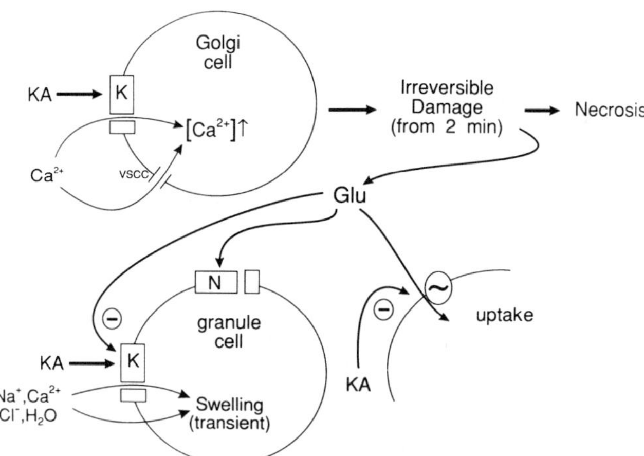

Fig. 4. Model for differential sensitivity of cerebellar neurones to
the toxic effects of kainate (KA): release of glutamate from the
rapidly necrosed Golgi neurones (and possibly other sources) inhibits
the action of kainate on granule cells. K, N: KA and NMDA receptors.

the only source of glutamate: kainate may also release glutamate from
astrocytes (Gallo et al., 1989) and nerve terminals (Ferkany & Coyle, 1983)
in a partially Ca^{2+}-dependent manner and so these structures could also
participate. Temporally, however, the inhibition of granule swelling
coincides more with the time course of irreversible damage being inflicted
on Golgi neurones than with the release of glutamate from astrocytes and
nerve terminals, which is more immediate.

Quisqualate/AMPA Neurotoxicity

The distinctive pathology of hippocampal neurones exposed to
quisqualate and AMPA (Fig. 1; Garthwaite & Garthwaite 1989a) compared with
NMDA and kainate provided the first clue that a different mechanism was at
work. Indeed, the pathology was interesting in itself because it seemed to
mimic the dark type of degeneration commonly observed in vivo after periods
of cerebral ischaemia, hypoglycaemia and prolonged seizures (Simon et al.,
1984b; Agardh et al., 1980; Sloviter & Dempster, 1985). It has been
suggested that dark cell degeneration is observed at the level of the
neuronal somata when the primary lesion is dendritic (Sloviter & Dempster,
1985) and intense dendritic swelling is certainly an obvious feature of
quisqualate-induced pathology (Garthwaite & Garthwaite, 1986a, 1989a; Siman
& Card, 1988).

In further contrast to NMDA and kainate, quisqualate toxicity was
unaffected by omitting Ca^{2+} from the exposing solution (Garthwaite &
Garthwaite, 1989b). Tests with other ions (Na^+, Cl^-, Mg^{2+}) also proved
negative. Moreover, while NMDA and kainate toxicities can be prevented by
including, during the exposure, the appropriate antagonist (2-amino-5-
phosphonovalerate, APV, with NMDA or 6-cyano-7-nitroquinoxaline-2,3-dione,
CNQX, with kainate), quisqualate toxicity was not, even through the
antagonist (CNQX) was able to markedly reduce the depolarizing action of
quisqualate. Even more surprisingly, CNQX was effective if it was included
during the 90 min recovery period after exposure to quisqualate (Garthwaite
& Garthwaite, 1989c). Another antagonist, kynurenic acid, was similarly
effective if applied during the recovery whereas tetrodotoxin or the NMDA
antagonist, APV, were not. Recent experiments showed that the ionic and
pharmacological characteristics of AMPA toxicity in hippocampal slices are
the same as those of quisqualate (unpublished observations).

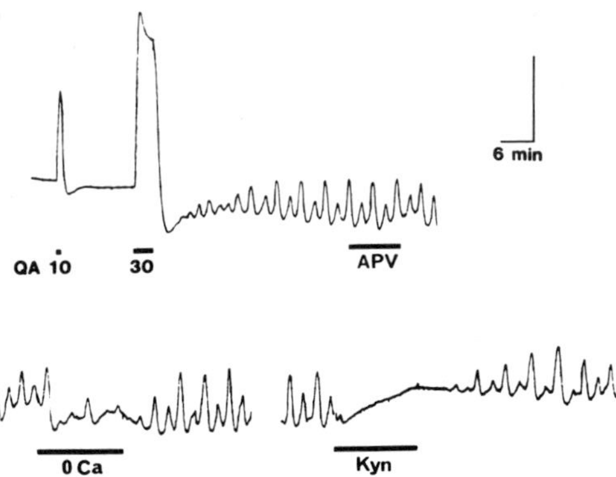

Fig. 5. Recurring oscillations following exposure of an adult rat
cerebellar surface slice to quisqualate (QA; concentrations in μM).
The recordings were made using a grease-gap method (J. Garthwaite et
al., 1986) and all solutions contained tetrodotoxin (0.5 μM). APV (D-
isomer) and kynurenate (Kyn) were perfused at 50 μM and 1 mM
respectively. Calibration: 1 mV, 6 min.

These results implied that the exposure itself is not killing the cells but that the agonists are triggering - through a CNQX-insensitive mechanism - something that manifests itself after the exposure as a delayed degeneration blockable by non-NMDA antagonists. In the case of quisqualate, release of stored quisqualate (Harris et al., 1987) or an effect of prolonged activation of the CNQX-insensitive "metabotropic" receptors (Sugiyama et al., 1987; Palmer et al., 1988) were possibilities, but since AMPA is neither transported into cells (Honoré et al., 1982) nor an agonist for metabotropic receptors (Palmer et al., 1988), other mechanisms need to be (and are currently being) sought.

If non-NMDA receptor activation is continuing after washout of the agonists to an extent that can lead to cell degeneration, it should be possible to record electrophysiological correlates of this occurrence. In adult cerebellar slices, a phenomenon having the appropriate properties has been observed using extracellular (gap) recordings of population responses (Fig. 5). Brief (1 min) applications of quisqualate lead to the usual depolarization followed by recovery; but with longer applications (4 min) spontaneous oscillations develop after washout of the agonist. These oscillations have a regular (and highly temperature-dependent) period and often last for the remainder of the experiment (up to 6 hours). Moreover, they are unaffected by APV or by tetrodotoxin but are reduced by perfusion of a Ca^{2+}-free solution and are abolished by kynurenic acid or CNQX. These results provide independent evidence for continued non-NMDA receptor activation following exposure to quisqualate and it is tempting to suggest that - in a more robust form - that this phenomenon is a reflection of events which are set in motion by quisqualate and AMPA to cause the delayed toxicity of these agonists. The presumed mediation of the toxicity by an endogenous agonist may explain why the toxic effects of quisqualate/AMPA are so pathomimetic.

CONCLUSIONS

On the basis of pharmacological and pathological properties, and of target cell identity, the toxicities of EAA receptor agonists can, at present, be classified into three main groups: (1) NMDA-like toxins, whose effects can be inhibited by including NMDA antagonists during the exposure. Their toxicities are Ca^{2+}-dependent and result from an overloading of the mechanisms serving to keep intracellular Ca^{2+} at physiological levels, probably chiefly through excessive influx of this ion through NMDA receptor channels. The mechanism is thus an extension of a physiological event, one which underlies the function of the NMDA system in synaptic plasticity. The potency of this group appears to be higher in the developing CNS than in adults, probably reflecting the physiological importance of the NMDA system during synaptogenesis. (2) Kainate-like toxins whose effects can be blocked by application of CNQX or removal of Ca^{2+} during the exposure. A reason why this very powerful excitant is not more generally toxic may be because a release of glutamate from rapidly-necrosed neurones and other sources is inhibitory. (3) Quisqualate-like toxins, whose effects are inhibitable by CNQX after, but not during, exposure and whose toxicity is likely to be mediated by an endogenous toxin active at non-NMDA receptors. The triggering mechanism is CNQX- and Ca^{2+}-insensitive.

ACKNOWLEDGEMENTS

The Research in our laboratory is supported by the Medical Research Council (U.K.). Many thanks to Geoffrey Williams, Denice Bradley and Gerard Gilligan for their much-valued assistance.

REFERENCES

Agardh, C.D., Kalimo, H., Olsson, Y. & Siesjo, B.K. (1980) Hypoglycemic
    brain injury. I. Metabolic and light microscopic findings in rat
    cerebral cortex during profound insulin-induced hypoglycemia and in
    the recovery period following glucose administration. Acta
    Neuropathol. (Berl.) 50: 31-41.
Ben-Ari, Y. (1985) Limbic seizure and brain damage produced by kainic
    acid: mechanisms and relevance to human temporal lobe epilepsy.
    Neuroscience 14: 375-403.
Campbell, A.K. (1983) The pathology and pharmacology of intracellular Ca2+.
    In A.K. Campbell (ed.), Intracellular calcium, Wiley, Chichester, pp.
    393-454.
Campochiaro, P. & Coyle, J.T. (1978) Ontogenetic development of kainate
    neurotoxicity: correlates with glutamatergic innervation. Proc.
    Natl. Acad. Sci. U.S.A. 75: 2025-2029.
Choi, D.W., Maulucci-Gedde, M. & Kriegstein, A.R. (1987) Glutamate
    Neurotoxicity in cortical cell culture. J. Neurosci. 7: 357-368.
Choi, D.W. (1987) Ionic dependence of glutamate neurotoxicity. J.
    Neurosci. 7: 369-379.
Choi, D.W. (1988) Glutamate neurotoxicity and diseases of the nervous
    system. Neuron 1: 623-634.
Connor, J.A., Wadman, W.J., Hockberger, P.E. & Wong, R.K.S. (1988)
    Sustained dendritic gradients of Ca2+ induced by excitatory amino
    acids in CA1 hippocampal neurons. Science 240: 649-653.
Cotman, C.W., Monaghan, D.T., Ottersen, O.P. & Storm-Mathisen, J. (1987)
    Anatomical organization of excitatory amino acid receptors and their
    pathways. Trends Neurosci. 10: 273-280.
Dupont, J-L., Gardette, R. & Crepel, F. (1987) Postnatal development of
    the chemosensitivity of rat cerebellar Purkinje cells to excitatory
    amino acids: an in vitro study. Devl. Brain Research 34: 59-68.
Faden, A.I., Demediuk, P., Panter, S.S. & Vink, R. (1989) The role of
    excitatory amino acids and NMDA receptors in traumatic brain injury.
    Science 244: 798-800.
Ferkany, J.W. & Coyle, J.T. (1983) Kainic acid selectively stimulates the
    release of endogenous excitatory acidic amino acids. J. Pharmacol.
    exp. Ther. 225: 399-406.
Franck, J.E. & Schwartzkroin, P.A. (1984) Immature rabbit hippocampus is
    damaged by systemic but not intraventricular kainic acid. Devl.
    Brain Research 13: 219-227.
Gaddy, J., Britt, M., Neill, D. & Haigler, H. (1979) Susceptibility of rat
    neostriatum to damage by kainic acid; age dependence. Brain Research
    176: 192-196.
Gallo, V., Giovannini, C., Suergiu, R. & Levi, G. (1989) Expression of
    excitatory amino acid receptors by cerebellar cells of the type-2
    astrocyte cell lineage. J. Neurochem. 52: 1-9.
Garthwaite, G. & Garthwaite, J. (1984) Differential sensitivity of rat
    cerebellar cells in vitro to the neurotoxic effects of excitatory
    amino acid analogues. Neuroscience Lett. 48: 361-367.
Garthwaite, G. & Garthwaite, J. (1986a) In vitro neurotoxicity of
    excitatory amino acid analogues during cerebellar development.
    Neuroscience 17: 755-767.
Garthwaite, G. & Garthwaite, J. (1986b) Neurotoxicity of excitatory amino
    acid receptor agonists in rat cerebellar slices: dependence on
    extracellular calcium concentration. Neuroscience Lett. 66: 193-198.
Garthwaite, G. & Garthwaite, J. (1986c) Amino acid neurotoxicity:
    intracellular sites of calcium accumulation associated with the
    onset of irreversible damage to rat cerebellar neurones in vitro.
    Neuroscience Lett. 71: 53-58.
Garthwaite, G. & Garthwaite, J. (1987) Receptor-linked ionic channels
    mediate N-methyl-D-aspartate neurotoxicity in rat cerebellar slices.
    Neuroscience Lett. 83: 241-246.

Garthwaite, G. & Garthwaite, J. (1989a) Neurotoxicity of excitatory amino acid receptor agonists in young rat hippocampal slices. J. Neurosci. Methods 29: 33–42.

Garthwaite, G. & Garthwaite, J. (1989b) Differential dependence on Ca2+ of N-methyl-D-aspartate and quisqualate neurotoxicity in young rat hippocampal slices. Neuroscience Lett. 97: 316–322.

Garthwaite, G. & Garthwaite, J. (1989c) Quisqualate neurotoxicity: a delayed, CNQX-sensitive process triggered by a CNQX-insensitive mechanism in young rat hippocampal slices. Neuroscience Lett. 99: 113–118.

Garthwaite, G., Hajos, F. & Garthwaite, J. (1986) Ionic requirements for neurotoxic effects of excitatory amino acid analogues in rat cerebellar slices. Neuroscience 18: 437–447.

Garthwaite, G., Yamini, B. Jr. & Garthwaite, J. (1987) Selective loss of Purkinje and granule cell responsiveness to N-methyl-D-aspartate in rat cerebellum during development. Devl. Brain Research 36: 288–292.

Garthwaite, J. (1982) Excitatory amino acid receptors and guanosine 3',5'-cyclic monophosphate in incubated slices of immature and adult rat cerebellum. Neuroscience 7: 2491–2497.

Garthwaite, J. (1985) Cellular uptake disguises action of L-glutamate on N-methyl-D-aspartate receptors. Br. J. Pharmacol. 85: 297–307.

Garthwaite, J. (1989) NMDA receptors, neuronal development and neuronal degeneration. In J.C. Watkins & G.L. Collingridge (eds.), The NMDA Receptor, Oxford University Press, in press.

Garthwaite, J. & Garthwaite, G. (1987) Cellular origins of cyclic GMP responses to excitatory amino acid receptor agonists in rat cerebellum in vitro. J. Neurochem. 48: 29–39.

Garthwaite, J., Garthwaite, G. & Hajos, F. (1986) Amino acid neurotoxicity: relationship to neuronal depolarization in rat cerebellar slices. Neuroscience 18: 449–460.

Garthwaite, J., Southam, E. & Anderton, M. (1989) A kainate receptor linked to nitric oxide synthesis from arginine. J. Neurochem. : (in press).

Gill, R., Foster, A.C. & Woodruff, G.N. (1988) MK-801 is neuroprotective in gerbils when administered during the post-ischaemic period. Neuroscience 25: 847–855.

Griffiths, T., Evans, M.C. & Meldrum, B.S. (1984) Temporal lobe epilepsy, excitotoxins and the mechanism of selective neuronal loss. In Excitotoxins, K. Fuxe, P.J. Roberts & R. Schwarcz (eds.), Macmillan, London. 331–342.

Hajos, F., Garthwaite, G. & Garthwaite, J. (1986) Reversible and irreversible neuronal damage caused by excitatory amino acid analogues in rat cerebellar slices. Neuroscience 18: 417–436.

Hamon, B. & Heinemann, U. (1988) Developmental changes in neuronal sensitivity to excitatory amino acids in area CA1 of the rat hippocampus. Devl. Brain Res. 38: 286–290.

Harris, E.W., Stevens, D.R. & Cotman, C.W. (1987) Hippocampal cells primed with quisqualate are depolarized by AP4 and AP6, ligands for a putative glutamate uptake site. Brain Research 418: 361–365.

Herndon, R.M. & Coyle, J.T. (1977) Selective destruction of neurons by a transmitter agonist. Science 198: 71–72.

Honore, T., Lauridsen, J. & Krogsgaard-Larsen, P. (1982) The binding of [3H]AMPA, a structural analogue of glutamic acid, to rat brain membranes. J. Neurochem. 38: 173–178.

Ishida, A.T. & Neyton, J. (1985) Quisqualate and L-glutamate inhibit retinal horizontal-cell responses to kainate. Proc. Natl. Acad. Sci. U.S.A. 82: 1837–1841.

Johnston, G.A.R., Kennedy, S.M.E. & Twitchin, B. (1979) Action of the neurotoxin kainic acid on high affinity uptake of L-glutamic acid in rat brain slices. J. Neurochem. 32: 121–127.

Kiskin, N.I., Krishtal, O.A. & Tsyndrenko, A.Ya. (1986) Excitatory amino acid receptors in hippocampal neurons: kainate fails to desensitize them. Neurosci. Lett. 63: 225–230.

Kohler, C. & Schwarcz, R. (1983) Comparison of ibotenate and kainate neurotoxicity in rat brain: a histological study. Neuroscience 8: 819-835.

McDonald, J.W., Silverstein, F.S. & Johnson, M.V. (1988) Neurotoxicity of N-methyl-D-aspartate is markedly enhanced in developing rat central nervous system. Brain Research 459: 200-203.

Mayer, M.L., MacDermott, A.B., Westbrook, G.L., Smith, S.J. & Barker, J.L. (1987) Agonist- and voltage-gated calcium entry in cultured mouse spinal cord neurons under voltage clamp measured using arsenazo III. J. Neurosci. 7: 3230-3244.

Murphy, S.N. & Miller, R.J. (1989) Regulation of Ca++ influx into striatal neurons by kainic acid. J. Pharmacol. exp. Ther. 249: 184-193.

Nadler, J.V., Evenson, D.A. & Cuthbertson, G.J. (1981) Comparative toxicity of kainic acid and other acidic amino acids toward rat hippocampal neurons. Neuroscience 6: 2505-2517.

O'Brien, R.J. & Fishbach, G.D. (1986) Characterization of excitatory amino acid receptors expressed by embryonic chick motoneurones in vitro. J. Neurosci. 6: 3275-3283.

O'Dell, T.J. & Christensen, B.N. (1989) A voltage-clamp study of isolated stingray horizontal cell non-NMDA excitatory amino acid receptors. J. Neurophysiol. 61: 162-172.

Olney, J.W. (1978) Neurotoxicity of excitatory amino acids. In Kainic acid as a tool in neurobiology, E.G. McGeer, J.W. Olney, & P.L. McGeer (eds.), Raven Press, New York, 95-121.

Olney, J.W. (1984) Excitotoxins: an overview. In K. Fuxe, P.J. Roberts, R. Schwarcz (eds.), Excitotoxins, Macmillan, London, 82-96.

Palmer, E., Monaghan, D.T. & Cotman, C.W. (1988) Glutamate receptors and phosphoinositide metabolism: stimulation via quisqualate receptors is inhibited by N-methyl-D-aspartate receptor activation. Mol. Brain Res. 4: 161-165.

Represa, A., Tremblay, E. & Ben-Ari, Y. (1989) Transient increase of NMDA-binding sites in human hippocampus during development. Neuroscience Lett. 99: 61-66.

Rothman, S.M., Thurston, J.H. & Hauhart, R.E. (1987) Delayed neurotoxicity of excitatory amino acids in vitro. Neuroscience 22: 471-480.

Siesjo, B.K. & Bengtsson, F. (1989) Calcium fluxes, calcium antagonists, and calcium-related pathology in brain ischemia, hypoglycemia, and spreading depression: a unifying hypothesis. J. Cereb. Blood Flow Metab. 9: 127-140.

Siesjo, B.K. (1981) Cell damage in the brain: a speculative synthesis. J. Cereb. Blood Flow Metab. 1: 155-185.

Silverstein, F.S., Chen, R. & Johnston, M.V. (1986) The glutamate analogue quisqualic acid is neurotoxic in striatum and hippocampus of immature rat brain. Neurosci. Lett. 71: 13-18.

Siman, R. & Card, J.P. (1988) Excitatory amino acid neurotoxicity in the hippocampal slice preparation. Neuroscience 26: 433-447.

Simon, R.P., Swan, J.H., Griffiths, T. & Meldrum, B.S. (1984a) Blockade of N-methyl-D-aspartate receptors may protect against ischemic damage in the brain. Science 226: 850-852.

Simon, R.P., Griffiths, T., Evans, M.C., Swan, J.H. & Meldrum, B.S. (1984b) Calcium overload in selectively vulnerable neurons of the hippocampus during and after ischemia: an electron microscopy study in the rat. J. Cereb. Blood Flow Metab. 4: 350-361.

Sloviter, R.S. & Dempster, D.W. (1985) "Epileptic" brain damage is replicated qualitatively in the rat hippocampus by central injection of glutamate or aspartate but not by GABA or acetylcholine. Brain Res. Bull. 15: 39-60.

Sugiyama, H., Ito, I. & Hirono, C. (1987) A new type of glutamate receptor linked to inositol phospholipid metabolism. Nature 325: 531-533.

Tremblay, E., Roisin, M.P., Represa, A., Charriaut-Marlangue, C. & Ben-Ari, Y. (1988) Transient increased density of NMDA binding sites in the developing rat hippocampus. Brain Research 461: 393-396.
Tsumoto, T., Hagihara, K., Sato, H. & Hata, Y. (1987) NMDA receptors in the visual cortex of young kittens are more effective than those of adult cats. Nature 327: 513-514.
Weiloch, T. (1985) Hypoglycemia-induced neuronal damage is prevented by a N-methyl-D-aspartate receptor antagonist. Science 230: 681-683.
Wolf, G. & Keilhoff, G. (1984) Kainate and glutamate neurotoxicity in dependence on the postnatal development with special reference to hippocampal neurons. Devl. Brain Research 14: 15-21.

COMMENTARY - EXCITATORY AMINO ACIDS AND ANOXIA

D. Choi and K. Krnjevic

There is not much historical precedent for considering the phenomenon of hypoxic neuronal injury together in the same symposium with receptor mechanisms, development, plasticity, long term potentiation, and regeneration. However, recent understandings suggest that hypoxic neuronal injury indeed shares many aspects of excitatory amino acid biology with these other themes; specifically, neurotoxic stimulation of glutamate receptors may account for some of the neuronal damage induced by hypoxia/ischemia. Hypoxic neuronal injury may be considered to be a pathological extension of long-term potentiation, and an important initial substrate for regeneration. The five papers presented in this session provided a representative sampling of current uses of brain slice or cell culture model systems to explore certain aspects of the neuronal cell injury induced by either oxygen deprivation or excitotoxin exposure. Some specific points raised during the discussion can be summarized :

Discussion of K. Krnjevic's presentation explored further the basis for the "paradoxical" loss of NMDA-induced responses (INMDA) on hippocampal pyramidal cells that he observed with 2 - 4 min of experimental anoxia in submerged slices, but not in an "interface" chamber where anoxia more often enhanced or prolonged INMDA. This could be explained by greater extracellular accumulation of K+ and perhaps other agents (including excitatory amino acids) in non-submerged slices. In this respect, such non-submerged slices are probably a better model of the ischemic brain in situ. The submerged slice might on the other hand offer advantages with regard to certain experimental questions, e.g. those requiring precise control of the extracellular space. Dingledine stated that in his own experience, raised [K+]o greatly enhanced the depolarizing effects of anoxia. Questions whether the effects of low pH were examined, and whether similar tests were made with quisqualate or AMPA received negative answers.

Dr. Ben-Ari's observation of reduced early anoxic depolarization with galanin, somatostatin, and especially diazoxide generated interest in the possible neuroprotective uses of these compounds in human brain ischemia. Discussion explored further the possible role of presynaptic ATP-sensitive K+ channels in limiting anoxic glutamate release. It was considered important in the future to examine the quantitative extent to which these compounds delayed anoxic depolarization with insults of increasing severity. Balazs commented that anoxia causes a major redistribution of glutamate, which moves from neurons into glia rather than remaining in the extracellular space.

Dr. Urban's study was discussed as an important bridge between in vivo ischemia, and in vitro studies, attempting to extend the time resolution of a slice study from the minute scale into the hour scale. A couple of interpretive hurdles were mentioned: the need to compare recordings in slices from different animals, and the possibility that the observed early reduction in population spike amplitude might include a contribution from altered slice geometry rather than from reduced neuronal firing. The author was asked whether unilateral ischemia had been tried, which would have left the other hemisphere as a particularly useful control. But he replied that this might not give reliable results because of significant possible vascular connections between the two sides. He was also asked whether the progressive loss of excitability and EPSPs might be explained by a relative enhancement of IPSPs, if inhibitory neurons are more resistant to ischemia. This was not investigated. Another more general question was why all slices in vitro do not show evidence of permanent damage since the isolated tissue is unavoidably exposed to ischemia during the decapitation and removal of the brain. But it was pointed out, that slices in fact do show significant impairment of oxygen consumption and ATP synthesis (cf. Lipton and Whittingham 1984, in Brain Slices, Ed. by Dingledine, Plenum Press, New York).

Dr. Choi's presentation was followed by questions regarding the location and magnitude of glutamate accumulation following hypoxic insult in cell culture. Dr. Choi suggested that this accumulation might be largely local to excitatory synapses, and that bath levels of glutamate might therefore not reach recognizably toxic levels. The possibility was raised that electrical recordings from neurons following hypoxia might be useful as a "bioassay" of excessive synaptic glutamate release or leakage. It was asked whether the effects of low pH were specific, rather than a general interference with cationic fluxes; and whether they might be exerted by changing the ionization of the NMDA receptors (or even its agonists). Choi answered that some specificity was apparent, at least compared to kainate induced currents; an effect directly on NMDA ionization was felt to be unlikely given known pK's. A suggestion that in vitro systems might not accurately reflect certain processes involved in brain ischemia in vivo met with no resistance.

Discussion after Dr. Garthwaite's presentation considered further the basis for his interesting observation that the toxicity of kainate on cultured granule cells was transient and reversible. Several discussants agreed with his suggestion that the release of an endogenous inhibitor, perhaps glutamate itself, might be triggered by the initial kainate exposure. How much glutamate is actually present at critical sites in the extracellular space? (not known; perhaps the cells' response provide the best assay for glutamate). What role does desensitization play? Is glutamate released from glia? (Maybe). Some kainate receptors mediate Ca++ influx: if present on Golgi cells, they might explain the observed selective action. What about Purkinje cells? They were little affected in these experiments. Conclusion Perhaps the most important lesson to be drawn from this session is that NMDA-receptor-mediated damage can be quite selective: certain cell types, such as hippocampal pyramids and cerebellar granule cells are highly susceptible to its effects, but others, such as dentate granules and cerebellar Golgi and Purkinje cells are much less so. (This may be a characteristic of GABAergic cells). But these may be damaged much more readily by activation of kainate receptors. A further important point is that intrinsic mechanisms may tend to limit cell damage, either by preventing excessive glutamate release (presynaptic ATP-sensitive K channels), or by limiting glutamate's excitotoxic action after release (acidosis, glutamate-desensitization). These may offer clues to possible therapeutic alternatives to blanket antagonism of NMDA-receptors.

EUROPEAN NEUROSCIENCE ASSOCIATION SATELLITE SYMPOSIUM
ON EXCITATORY AMINO ACIDS AND NEURONAL PLASTICITY
August 27-31, 1989, Fillerval, France

CONTRIBUTORS

ASCHER, P.
Laboratoire de Neurobiologie
Ecole Normale Supérieure
46 rue d'Ulm
75230 Paris, Cédex 05, France

BALAZS, R.
Netherlands Institute for
Brain Research
Meibergdreef 33
1105 AZ Amsterdam,
The Netherlands

BEN-ARI, Y.
INSERM Unit 29
123 Bd de Port-Royal
75014 Paris, France

BINDMAN, L.J.
University College of
London
Department of Physiology
Gower Street
WC1E 6BT London, England

BLISS, T.V.P.
National Institut for
Medical Research
Division of Neurophysio-
logy andNeuropharmacology
The Ridgeway, Mill Hill
NW71 AA London, England

BOCKAERT, J.
INSERM Unit 264
Centre de Pharmacologie-
Endocrinologie
rue de la Cardonille
B.P. 50-55
34033 Montpellier, France

CAMBRAY-DEAKIN, M.A.
The Physiological Laboratory
University of Liverpool
P.O. Box 147
Liverpool L69 3BX, England

CHERUBINI, E.
INSERM Unit 29
123 Bd de Port-Royal
75014 Paris, France

CHOI, D.W.
Stanford University
Medical Center
Department of Neurology
Stanford, CA 94305, USA

CLINE, H.T.
Stanford University Medical
Center
Department of Biology
Stanford,New Haven,
CT 06511, USA

COSTA, E.
FGIN, Georgetwon Univ.
School of Medicine
3900 Reservoir Road
NW, Washington,
DC 20007, USA

CREPEL, F.
Faculté des Sciences d'Orsay
Laboratoire de Neurobiologie du Développement
91405 Orsay, France

CUENOD, M.
University of Zurich
Institut of Brain Research
August Forel Strasse 1
CH 8029 Zurich, Suisse

DE GRAAN, P.N.E.
Division of Molecular Neuro-
biology
Rudolph Magnus Institute
Laboratory for Physiological
Chemistry and Institute of
Molecular Biology and Medical
Biotechnology,
Padualaan 8, 3584 CH Utrecht,
The Netherlands

DINGLEDINE, N.W.
University of North Carolina
School of Medicine
Department of Pharmacology
Chapel Hill, NC 27599, USA

FAZELI, M.S.
Institut of Psychiatry
Department of Pharmacology
and Clinical Pharmacology
St. George's Hospital Medical
School, Cranmer Terrace,
London, SW17 ORE, England

FOSTER, A.C.
Merck Sharp and Dohme Ltd.
Neuroscience Research Center
Terlings Park Easrwick Road
Essex, CM20 2QR, England

GARTHWAITE, J.
University of Liverpool
Department of Physiology
Brownlow Hill,
Liverpool L69 3BX, England

GEDDES, J.W.
University of California
Division of Neurosurgery
Irvine CA 92717, USA

GRANTYN, R.
Max Planck Institut of
Psychiatry
Department of Neurophysiology
A.M. Klopfersoitz 18 A,
8033 Planegg-Martinreid, FRG

HEINEMANN, U.
University of Koln Robert Koch
Institut for Normal and
Pathologic Physiology
Strasse 39
D-5000 Koln 41,FRG

HONORE, T.
Ferrosan Group
CNS division
Sydmarken 5
DK-2860 Søborg, Denmark

HORN, G.
University of Cambridge
Department of Zoology
Downing Street,
Cambridge CB2 3EJ, England

HUETTNER, J.E.
Harvard Medical School
Department of Neurobiology
220 Longwood Avenue
Boston, MA02115, USA

JONES, R.S.G.
Division of Neuroscience
John Curtin School of Medical
Research
Australian National University
Canberra
ACT2601, Australia

KAUER, J.A.
Departments of Pharmacology
and Physiology
University of California
San Francisco, CA 94143, USA

KEMP, J.A.
Merck Sharp and Dohme Ltd.
Neuroscience Research Center
Terlings Park Eastwick Road
CM20 2QR, England

KRNJEVIC, K.
MacGill University
Department of Physiology
3655 Drummond St.
Montreal, Que H3G, Canada

KUDO, Y.
Department of Neuroscience
Mitsubishi Kasei Institut of
Life Science
Minamiooya,
Machida, Tokyo 194, Japan

LODGE, D.
Royal Veterinary College
Department of Veterinary
Basic Sciences
London NW1 OTU, England

MacDERMOTT, A.B.
Columbia University
Center for Neurobiology and
Behavior
160 W. 168th
New York, NY 10032, USA

MacDONALD, J.F.
Playfair Neuroscience Unit
The Toronto Hospital
Department of Physiology and
Pharmacology
University of Toronto
Ontario, MST 258, Canada

MacNAMARA
Duke University Medical
Center
Departments of Medicine and
Pharmacology
Epilepsy Research Laboratory
Veterans
Durham, NC 27705, USA

MALINOW, R.
Department of Molecular and
Cellular Physiology
Beckman Center
Stanford University School
of Medicine
Stanford California, 94305 USA

MATTHIES, H.
Academy of Sciences of GDR
Institute of Neurobiology and
Brain Research
Leipziger Strasse 44
3090 Magdeburg, GDR

MATTSON, M.P.
Sanders-Brown Research
Center on Aging and
Department of Anatomy and
Neurobiology
University of Kentucky Medical
Center
Lexington, KY 40536-0230, USA

MAYER, M.L.
Unit of Neurophysiology
and Biophysics
NIH, Bldg 36, Room 2-A21
Bethesda, MD 20892, USA

NADLER, J.V.
Duke University Medical
Center
Department of Pharmacology
Neurobiology and Medicine
P.O. Box 3813
Durham, North Carolina
27710, USA

REPRESA, A.
INSERM Unit 29
123 Bd de Port-Royal
75014 Paris, France

ROISIN, M.P.
INSERM Unit 29
123 Bd de Port-Royal
75014 Paris, France

ROUTTENBERG, A.
Northwestern University
Department of Psychology and
Biological Sciences
2021 Sheridan Road
Evanston Illinois 60201, USA

SASTRY, B.R.
University British Columbia
Neuroscience Research
Laboratory
Department of Pharmacology
& Therapeutics
The University of British
Columbia
Vancouver, BC, V6T 1W5,
Canada

SHASHOUA, V.E.
Ralph Lowell Laboratories,
McLean Hospital
Harvard Medical School,
Belmont, MA 02178, USA

SOKOLOVSKY, M.
Faculty of Life Sciences
Tel Aviv University
Laboratory of Neurobio-
chemistry
Department of Biochemistry
Tel Aviv 69978, Israel

STELZER, A.
Surgeons of Columbia
University
College of Physicians
Department of Neurology
630 West 168th Street
NY 10032, New York, USA

SUGIYAMA, H.
National Institut for Physiolo-
gical Sciences
Department of Cellular Physio-
logy
Okazaki, 444, Japan

SWANN, J.W.
Wadsworth Center for Labo-
ratories and Research
New York State Department
of Health
Empire State Plaza,
Albany, NY 12201-0509, USA

TEICHBERG, V.I.
Weizmann Institut of Science
Department of Neurobiology
P.O. Box 26
Rehovot, 76100, Israel

TERLAU, H.
Max Planck Institut für Biophy-
sikalishe Chemie
Department of Neurobiology
3400 Goettingen, FRG

THOMSON, A.M.
Royal Free Hospital School
Medicine
Department of Physiology
Rowland Hill St.
London NW3 2PF, England

TRENKNER, E.
Department of Pathology
Columbia University
Physicians and Surgeons
630 West 168 Street
New York, NY 10032, USA

TSUMOTO, T.
Biomedical Research Center,
Osaka University Medical
School
Department of Neurophysio-
logy
Kitaku Osaka 530, Japan

URBAN, L.
Duke University Medical Center
Department of Cell Biology,
Pathology, Neurobiology and
Pharmacology
P.O. Box 3813
Durham North, USA

WATKINS, J.C.
The Medical School University
Walk
School Department of Pharma-
cology
Bristol B58 1TD, England

WHEAL, H.
University of Southampton
Department of Neurophysio-
logy
School of Biological
Bassett Crescent East Science
Building
Southampton S09 3TU, England

ZIMMER, J.
University of Aarhus
Institut of Anatomy B
Neurobiology
DK 8000 Aarhus, Danemark